Meson-Nuclear Physics—1976

(Carnegie-Mellon Conference)

AIP Conference Proceedings
Series Editor: Hugh C. Wolfe
No. 33

Meson-Nuclear Physics—1976

(Carnegie-Mellon Conference)

Editors

P.D. Barnes, R.A. Eisenstein, L.S. Kisslinger

American Institute of Physics
New York   1976

Copyright © 1976 American Institute of Physics, Inc.

This book, or parts thereof, may not be reproduced in any form without permission.

L.C. Catalog Card No. 76-16811
ISBN 0-88318-132-0
ERDA CONF-760561

Printed in the United States of America

Proceedings of the

International Topical Conference on

Meson-Nuclear Physics

Carnegie-Mellon University

Pittsburgh, Pennsylvania

May 24-28, 1976

PREFACE

    A central part of medium energy physics involves the interactions of mesons with nucleons and nuclei. There are so many new theoretical and experimental developments in this area that it is not possible to review and discuss them in depth at the broad conference on High Energy Physics and Nuclear Structure. Therefore, the International Topical Conference on Meson-Nuclear Physics was organized to provide an overview at this time when important new experimental facilities are becoming available. We believe that the papers in this volume show that the early promise that medium energy nuclear physics would form a unifying link between nuclear structure and particle physics - for the mutual benefit of both - seems well on the way to fulfillment.

    We would like to thank the members of the International Advisory Committee and the Organization Committee for the efforts on behalf of the Conference. In addition, we are grateful for the financial support of the Conference provided by the U. S. Energy research and Development Administration, the U. S. National Science Foundation, and the International Union of Pure and Applied Physics. We would like also to thank the Administration and Department of Physics of Carnegie-Mellon University for help at all stages in planning for the Conference and in assistance in the preparation of these proceedings. The unstinting efforts of Mrs. F. Megahan inall phases of the organization and execution of both the Conference and these proceedings are gratefully acknowledged.

INTERNATIONAL ADVISORY COMMITTEE

A. Arima	A. Kerman
M. Banerjee	V. Lobashov
L. Bertocchi	D. Measday
J. Blaser	A. Mitra
G. Brown	J. Pniewski
K. Crowe	L. Rosen
J. Deutsch	J. Schiffer
V. Dzhelepov	I. Shapiro
J. Eisenberg	F. Tabakin
G. Emery	G. Tibell
T. Ericson	C. Tzara
V. Gillet	R. Welsh
R. Haddock	D. Wilkinson
	A. Yavin

ORGANIZATION COMMITTEE

P. Barnes - Co-chairman
R. Burman
C. Dover
J. Hüfner
L. Kisslinger - Co-chairman
P. Radvanyi
L. Wilets
C. Wilkin

CONFERENCE PARTICIPANTS

J. ADOLPH, University of Virginia, Charlottesville, VA, USA
L. AGNEW, Los Alamos Scientific Lab., Los Alamos, NM, USA
M. ALBERG, University of Washington, Seattle, WA, USA
G. ALBERI, Instituto de Fisica Teorica, Miramare, Trieste, Italy
Y. ALEXANDER, University of Maryland, College Park, MD, USA
R. D. AMADO, University of Pennsylvania, Philadelphia, PA, USA
J. AMANN, Los Alamos Scientific Lab., Los Alamos, NM, USA
J. ARVIEUX, SIN, Villigen, Switzerland
E. ASLANIDES, Universite Louis Pasteur, Strasbourg Cedex, France
N. AUSTERN, University of Pittsburgh, Pittsburgh, PA, USA
H. W. BAER, Case Western Reserve University, Cleveland, OH, USA
M. K. BANERJEE, University of Maryland, College Park, MD, USA
E. BARANGER, University of Pittsburgh, Pittsburgh, PA, USA
P. D. BARNES, Carnegie-Mellon University, Pittsburgh, PA, USA
Y. A. BATUSOV, Joint Institute for Nuclear Research, Moscow, USSR
R. D. BENT, Indiana University, Bloomington, IN, USA
S. BERNSTEEN, Carnegie-Mellon University, Pittsburgh, PA, USA
A. BERNSTEIN, MIT, Cambridge, MA, USA
B. S. BHAKAR, University of Manitcha, Winnipeg, Canada
F. BINON, CERN, Geneva, Switzerland
J. S. BLAIR, University of Washington, Seattle, WA, USA
M. BLESZYNSKI, Institute of Nuclear Physics, Krakow, Poland
M. BRACK, S.U.N.Y., Stony Brook, NY, USA
D. D. BRAYSHAW, Stanford University, Stanford, CT, USA
W. J. BRISCOE, Catholic University, Washington, D.C., USA
G. E. BROWN, S.U.N.Y., Stony Brook, NY, USA
D. BRYMAN, TRIUMF, Vancouver, B.C., Canada
G. R. BURLESON, New Mexico State University, Los Cruces, NM, USA
P. T. CAHILL, University of Illinois, Urbana, IL, USA
J. B. CAMMARATA, Stanford University, Stanford, CA, USA
H. CATZ, CEN Saclay, Gif-sur-Yvette, France
C. CERNIGOI, Universita di Trieste, Trieste, Italy
S. CHAKRAVARTI, Argonne National Laboratory, Argonne, IL, USA
I. T. CHEON, McMaster University, Hamilton, Canada
F. COESTER, Argonne National Laboratory, Argonne, IL, USA
M. D. COOPER, Los Alamos Scientific Lab., Los Alamos, NM, USA
P. COUVERT, CEN Saclay, Gif-sur-Yvette, France
J. N. CRAIG, LAMPF, Los Alamos, NM, USA
H. CRANNELL, Catholic University, Washington, D.C., USA
K. M. CROWE, Lawrence Berkeley Lab., Berkeley, CA, USA
A. DELOFF, University of Guelph, Guelph, Canada
P. DEPOMMIER, University of Montreal, Montreal, Canada
B. DESPLANQUES, Carnegie-Mellon University, Pittsburgh, PA, USA
M. DILLIG, S.U.N.Y., Stony Brook, NY, USA
J. J. DOMINGO, SIN, Villigen, Switzerland
K. G. R. DOSS, Los Alamos Scientific Lab., Los Alamos, NM, USA
C. B. DOVER, Brookhaven National Lab., Upton, NY, USA
J. F. DUBACH, MIT, Cambridge, MA, USA

G. F. DUGAN, Columbia University, Irvington, NY, USA
S. A. DYTMAN, Los Alamos Scientific Lab., Los Alamos, NM, USA
J. M. EISENBERG, Tel-Aviv University, Tel-Aviv, Israel
R. A. EISENSTEIN, Carnegie-Mellon University, Pittsburgh, PA, USA
G. T. EMERY, Indiana University, Bloomington, IN, USA
G. N. EPSTEIN, Michigan State University, East Lansing, MI, USA
D. J. ERNST, Texas A&M University, College Station, TX, USA
H. W. FEARING, TRIUMF, Vancouver, B.C., Canada
R. D. FELDER, Rice University, Houston, TX, USA
E. FERREIRA, SLAC, Stanford, CA, USA
H. FESHBACH, MIT, Cambridge, MA, USA
V. N. FETISOV, Academy of Sciences, Moscow, USSR
A. FIGUREAU, Institut de Physique Nuclaire, Villeurbanne, France
W. A. FRIEDMAN, University of Wisconsin, Madison, WI, USA
B. L. FRIMAN, S.U.N.Y., Stony Brook, NY, USA
H. O. FUNSTEN, College of William and Mary, Williamsburg, VA, USA
A. GAL, The Hebrew University, Jerusalem, Israel
H. GARCILZO, Escuela Superior De Fisica & Math, I.P.N., Mexico
A. A. GERARD, CEN Saclay, Gif-sur-Yvette, France
W. R. GIBBS, Los Alamos Scientific Lab., Los Alamos, NM, USA
B. F. GIBSON, Los Alamos Scientific Lab., Los Alamos, NM, USA
V. GILLET, Centre d'Studio de Saclay, Gif-sur-Yvette, France
J. N. GINOCCHIO, Los Alamos Scientific Lab., Los Alamos, NM, USA
M. GOLDHABER, Brookhaven National Lab., Upton, NY, USA
P. GOODE, Rutgers University, New Brunswick, NJ, USA
B. GOULARD, University of Montreal, Montreal, Canada
C. GUARALDO, Laboratori Nazionali di Frascati, Frascati, Italy
S. A. GURVITZ, The Weizmann Institute of Science, Rehovot, Isreal
E. HADJIMICHAEL, Fairfield University, Fairfield, CT, USA
Y. HAHN, University of Connecticut, Storrs, CT, USA
W. C. HAXTON, Inst. fur Kernphysik der Univ., Mainz, West Germany
R. H. HEFFNER, Los Alamos Scientific Lab., Los Alamos, NM, USA
L. HELLER, Los Alamos Scientific Lab., Los Alamos, NM, USA
T. HENNINO, Institute de Physique Nucleaire, Orsay, France
P. HERCZEG, Los Alamos Scientific Lab., Los Alamos, NM, USA
A. T. HESS, Los Alamos Scientific Lab., Los Alamos, NM, USA
H. HIRABAYASHI, KEK National Lab., Ibaraki-Ken, Japan
M. HIRATA, Brown University, Providence, RI, USA
J. HUDOMALJ-GABITZSCH, Rice University, Houston, TX, USA
E. V. HUNGERFORD, University of Houston, Houston, TX, USA
C. H. Q. INGRAM, SIN, Villigen, Switzerland
K. JOHNSON, MIT, Cambridge, MA, USA
M. B. JOHNSON, Los Alamos Scientific Lab., Los Alamos, NM, USA
R. JOHNSON, TRIUMF, Vancouver, B.C., Canada
G. JONES, TRIUMF, Vancouver, B.C., Canada
D. JULIUS, Weizmann Institute of Science, Rehovot, Israel
T. KALOGEROPOULOS, Syracuse University, Syracuse, NY, USA
A. N. KAMAL, University of Alberta, Edmonton, Canada
A. KANOFSKY, Lehigh University, Bethlehem, PA, USA
P. J. KAROL, Carnegie-Mellon University, Pittsburgh, PA, USA
B. KEISTER, Carnegie-Mellon University, Pittsburgh, PA, USA

R. KELLY, Lawrence Berkeley Lab., Berkeley, CA, USA
F. C. KHANNA, Chalk River Nuclear Lab., Ontario, Canada
K. KILIAN, CERN, Geneva, Switzerland
Y. E. KIM, Purdue University, West Lafayette, IN, USA
J. KINALLY, University of Virginia, Charlottesville, VA, USA
L. S. KISSLINGER, Carnegie-Mellon University, Pittsburgh, PA, USA
W. M. KLOET, Los Alamos Scientific Lab., Los Alamos, NM, USA
J. H. KOCH, MIT, Cambridge, MA, USA
D. KOLTUN, University of Rochester, Rochester, NY, USA
T. KOPALEISHVILI, Tbilisi State University, Tbilisi, USSR
K. KUBODERA, University of Tokyo, Tokyo, Japan
E. P. LAMBERT, Institut de Physique, Neuchatel, Switzerland
R. H. LANDAU, Oregon State University, Corvallis, OR, USA
J. P. LAVINE, University of Rochester, Rochester, NY, USA
J. LAW, University of Guelph, Guelph, Canada
Y. LE BORNEC, I.P.N., Orsay, France
H. T. S. LEE, Argonne National Lab., Argonne, IL, USA
R. H. LEMMER, Rand Afrikaans University, Johannesburg, S. Africa
F. LENZ, SIN, Villigen, Switzerland
K. K. LI, Brookhaven National Lab., Upton, NY, USA
B. LIEB, George Mason University, Fairfax, VA, USA
G. LIND, Utah State University, Logan, UT, USA
P. LIPNIK, Institut de Physiquae, Louvain-la-Neuve, Belgium
L. C. LIU, University of New York, Brooklyn, NY, USA
J. T. LONDERGAN, Indiana University, Bloomington, IN, USA
R. MACHLEIDT, S.U.N.Y., Stony Brook, NY, USA
T. A. J. MARIS, Univ. Fed. Do Rio Grande DoSul, Porto Alegre, Brazil
G. R. MASON, TRIUMF, Vancouver, B.C., Canada
T. MASTERSON, TRIUMF, Vancouver, B.C., Canada
G. S. MAURER, Catholic University, Washington, D.C., USA
R. E. McADAMS, Utah State University, Logan, UT, USA
W. K. McFARLANE, Temple University, Philadelphia, PA, USA
H. McMANUS, Michigan State University, East Lansing, MI, USA
K. W. McVOY, University of Wisconsin, Madison, WI, USA
H. A. MEDICUS, Rensselaer Polytechnic Inst., Troy, NY, USA
H. O. MEYER, University of Basel, Basel, Switzerland
L. MEYER-SCHUTZMEISTER, Argonne National Lab., Argonne, IL, USA
G. A. MILLER, University of Washington, Seattle, WA, USA
L. D. MILLER, University of Virginia, Charlottesville, VA, USA
A. N. MITRA, University of Delhi, Delhi, India
T. MIZUTANI, SIN, Villigen, Switzerland
E. J. MONIZ, MIT, Cambridge, MA, USA
F. MYHRER, CERN, Geneva, Switzerland
A. NAGL, Catholic University, Washington, D.C., USA
D. E. NAGLE, Los Alamos Scientific Lab., Los Alamos, NM, USA
B. M. K. NEFKENS, University of California, Los Angeles, CA, USA
J. W. NEGELE, MIT, Cambridge, MA, USA
C. M. NEWSTEAD, Brookhaven National Lab., Upton, NY, USA
G. V. NGUYEN, Institut de Physique Nucleaire, Orsay, France
G. NIXON, Indiana University, Bloomingtom, IN, USA
J. V. NOBLE, University of Virginia, Charlottesville, VA, USA

Y. NOGAMI, McMaster University, Hamilton, Canada
W. T. NUTT, C.U.N.Y., Brooklyn, NY, USA
A. OLIN, University of Victoria, Victoria, B.C., Canada
M. OLSSON, University of Wisconsion, Madison, WI, USA
L. ORPHANOS, University of Virginia, Charlottesville, VA, USA
E. OSET, S.U.N.Y., Stony Brook, NY, USA
O. H. OTTESON, Utah State University, Logan, UT, USA
H. PALEVSKY, Brookhaven National Lab., Upton, NY, USA
C. PICCIOTTO, University of Victoria, Victoria, B.C., Canada
H. S. PICKER, Trinity College, Hartford, CT, USA
H. PILKUHN, Institut fur Theor. Physics, Karlsruhe, Germany
G. PIRAGINO, Instituto di Fisica, Torino, Italy
H. J. PIRNER, Vielteilchen Physik, Heidelberg, West Germany
J. PNIEWSKI, University of Warsaw, Warszawa, Poland
R. POWERS, CIT, Pasadena, CA, USA
H. PUGH, National Science Foundation, Washington, D.C., USA
M. RADOMSKI, Michigan State University, East Lansing, MI, USA
P. RADVANYI, Institut de Physique Nucleair, Orsay, France
K. S. RAO, Catholic University, Washington, D.C., USA
D. G. RAVENHALL, University of Illinois, Urbana, IL, USA
M. RAYET, Universite Libre de Bruxelle, Bruxelles, Belgium
A. REITAN, Fysisk Institutt, Trondheim, Norway
A. K. REJ, University of Trondheim, Trondheim, Norway
J. A. RETTER, University of Connecticut, Storrs, CT, USA
M. RHO, CEN Saclay, Gif-sur-Yvette, France
A. RINAT, Weizmann Institute of Science, Rehovot, Israel
D. O. RISKA, Michigan State University, East Lansing, MI, USA
E. T. RITTER, ERDA, Washington, D.C., USA
R. ROCKMORE, Rutgers University, New Brunswick, NJ, USA
L. P. ROSA, Univ. Fed. Do Rio De Janeiro, Rio De Janeiro, Brazil
E. ROST, University of Colorado, Boulder, CO, USA
J. E. RUSSELL, University of Cincinnati, Cincinnati, OH, USA
A. N. SAHARIA, Carnegie-Mellon University, Pittsburgh, PA, USA
M. SALOMON, University of British Columbia, Vancouver, Canada
L. H. SCHICK, University of Wyoming, Laramie, WY, USA
J. P. SCHIFFER, Argonne National Lab., Argonne, IL, USA
A. SCHIZ, Fermi National Accelerator Lab., Batavia, IL, USA
T. H. SCHUCAN, University of Basel, Basel, Switzerland
C. G. SCHUHL, CEN Saclay, Gif-sur-Yvette, France
E. SCHWARZ, Institut de Physique, Neuchatel, Switzerland
R. E. SEGEL, Northwestern University, Evanston, IL, USA
R. SEKI, California State University, Northridge, CA, USA
C. SHAKIN, Brooklyn College, Brooklyn, NY, USA
Y. A. SHCHERBAKOV, Joint Inst. for Nuclear Research, Moscow, USSR
G. SHEN, Fermi National Accelerator Lab., Batavia, IL, USA
E. B. SHERA, Los Alamos Scientific Lab., Los Alamos, NM, USA
J. D. SHERMAN, Los Alamos Scientific Lab., Los Alamos, NM, USA
R. R. SILBAR, Los Alamos Scientific Lab., Los Alamos, NM, USA
D. I. SOBER, Catholic University, Washington, D.C., USA
R. A. SORENSEN, Carnegie-Mellon University, Pittsburgh, PA, USA
M. SOYEUR, CEN Saclay, Gif-sur-Yvette, France

M. M. STERNHEIM, University of Massachusetts, Amherst, MA, USA
P. STOLER, Rensselaer Polytechnic Institute, Troy, NY, USA
C. E. STRONACH, Virginia State College, Petersburg, VA, USA
D. W. STORM, Nevis Lab., Irvington, NY, USA
R. B. SUTTON, Carnegie-Mellon University, Pittsburgh, PA, USA
F. TABAKIN, University of Pittsburgh, Pittsburgh, PA, USA
N. B. deTAKAESY, McGill University, Montreal, Canada
F. TAKEUTCHI, CERN, Geneva, Switzerland
M. THIES, S.U.N.Y., Stony Brook, NY, USA
H. A. THIESSEN, Los Alamos Scientific Lab., Los Alamos, NM, USA
A. W. THOMAS, CERN, Geneva, Switzerland
Z. D. THOME, Univ. Fed. Do Rio De Janeiro, Rio De Janeiro, Brazil
P. TRUOL, Universitat Zurich, Zurich, Switzerland
W. TURCHINETE, MIT, Middleton, MA, USA
C. TZARA, CEN Saclay, Gif-sur-Yvette, France
H. UBERALL, Catholic University, Washington, D.C., USA
S. L. VERBECK, Los Alamos Scientific Lab., Los Alamos, NM, USA
P. VERNIN, CEN, Saclay, Gif-sur-Yvette, France
B. J. VERWEST, S.U.N.Y., Stony Brook, NY, USA
R. VINH MAU, Universiti Pil M. Curie, Paris Cedex, France
P. WALDEN, TRIUMF, Vancouver, B.C., Canada
G. E. WALKER, Indiana University, Bloomington, IN, USA
S. J. WALLACE, University of Maryland, College Park, MD, USA
H. J. WEBER, University of Virginia, Charlottesville, VA, USA
C. WERNTZ, Catholic University, Washington, D.C., USA
W. R. WHARTON, Carnegie-Mellon University, Pittsburgh, PA, USA
L. WILETS, University of Washington, Seattle, WA, USA
E. J. WINHOLD, Rensselaer Polytechnic Institute, Troy, NY, USA
L. WOLFENSTEIN, Carnegie-Mellon University, Pittsburgh, PA, USA
R. WOLOSHYN, University of Pennsylvania, Philadelphia, PA, USA
C. S. WU, Carnegie-Mellon University, Pittsburgh, PA, USA
K. H. YANG, University of Maryland, College Park, MD, USA
A. I. YAVIN, University of Illinois, Urbana, IL, USA
M. ZAIDER, Los Alamos Scientific Lab., Los Alamos, NM, USA
D. ZIEMINSKA, Institut Fizyki Doswiackzalnej, Warszawa, Poland

TABLE OF CONTENTS

I
INTRODUCTION AND PREVIEW

I.1 - Introduction - *P. D. Barnes, R. A. Eisenstein and L. S. Kisslinger*................ 2

I.2 - Meson-Nuclear Physics: A Preview - *Daniel S. Koltun*.... 3

II
PION-NUCLEON INTERACTION AND PION NUCLEUS ELASTIC SCATTERING

II.1 - πN Partial Wave Analysis in the First and Second Resonance Regions - *R. L. Kelly*............ 12

II.2 - The Role of the Δ(3,3) in Low Energy πN Elastic and Photoproduction Models - *M. G. Olsson*.......... 21

II.3 - Measurement of the π^+ and π^- Total Cross Sections on Hydrogen and Deuterium for Pion Energies from 50 to 300 MeV - *E. Pedroni, K. Gabathuler, J. Arvieux, P. Corfu, J. Domingo, P. Gretillat, W. Hirt, Q. Ingram, J. P. Piffarett, P. Schwaller and N. Tanner*....... 25

II.4 - Pion-Proton Scattering Below 100 MeV - *P. Bertin, B. Coupat, J. Duclos, A. Gérard, A. Hivernat, D. Isabelle, J. Miller, J. Morgenstern, J. Picard, R. Powers and P. Vernin.*............ 34

II.5 - Differential Cross Sections for Elastic (π^+p) and (π^-p) Scattering Between 400 MeV/c and 590 MeV/c - *V. A. Gordeev, V. P. Koptev, S. P. Kruglov, L. A. Kuzmin, A. A. Kulbardis, Y. A. Malov, I. I. Strakovsky and G. V. Scherbakov*............ 36

II.6 - Pion-Nucleus Elastic Scattering - *Frank Tabakin*. 38

II.7 - Low Energy Pion Scattering From Nuclei - *R. A. Eisenstein*. 55

II.8 - Energy and Angular Dependence in Pion-Proton Bremsstrahlung - *D. I. Sober, H. C. Ballagh, P. F. Glodis, R. P. Haddock, K. C. Leung, B. M. K. Nefkens and D. E. A. Smith*. 65

II.9 - Model Calculations for Radiative Pion Proton Scattering - *Q. Ho-Kim and J. P. Lavine*. 68

II.10 - Improved π-^{12}C Optical Potential at 50 MeV - *M. Thies*. 71

II.11 - Covariant Theory of the Off-Shell Pion-Nucleon Scattering Amplitude - *J. Barry Cammarata and Manoj K. Banerjee*. 74

II.12 - Determination of Pion-Nucleon Form Factor From Pion-Nucleon Elastic Scattering - *D. J. Ernst and Mikkel Johnson*. 76

II.13 - Separable-Potential Model For the Pion-Nucleon Interaction - *W. T. Nutt*. 78

II.14 - The Klein-Gordon Equation with Optical Potentials of the Strong Interaction - *R. Seki*. 80

II.15 - Pauli-Blocking Effects in Low Energy Pion-Nucleus Scattering - *C. B. Dover and R. H. Lemmer*. 82

II.16 - PIPIT: A Momentum Space Optical Potential Code for Pions - *R. A. Eisenstein and F. Tabakin*. 84

II.17 - An Improved Optical Potential for Low and Inter-Mediate Energy Pions - *R. H. Landau and A. W. Thomas*.................. 86

II.18 - Elastic Scattering of Pion by He^4 and C^{12} - *T. W. Chen and D. W. Hoock*.................. 88

II.19 - Effects of the Pauli Suppression of the Born Amplitude in a Nuclear Medium - *W. T. Nutt*......... 90

III

NEW THEORETICAL APPROACHES

III.1 - Relativistic Description of Directly Interacting Pions and Nucleons - *Leon Heller*.............. 93

III.2 - Isobar Propagation in the Nuclear Medium - *E. J. Moniz*.................. 105

III.3 - Field Theoretic Aspects of Pion Nucleus Physics - *M. K. Banerjee*.................. 119

III.4 - Isobars in Nuclei - *H. J. Weber*.............. 130

III.5 - Pionic Degrees of Freedom in Nuclei II - *Mannque Rho*.................. 146

III.6 - Doorway-Isobar Theory for Meson-Nucleus Physics - *Leonard S. Kisslinger*.................. 159

III.7 - Crossed Pion Absorption Contribution to the π-d Scattering Length - *T. Mizutani and N. C. Mukhopadhyay*.................. 172

III.8 - Pion Exchange Currents and Nuclear Charge Form Factors - *Mark Radomski and D. O. Riska*.......... 175

III.9   - Field Theory Treatment of Pi-Nucleus Scattering - *Gerald A. Miller*. 178

III.10 - Higher Order Pion-Nucleus Optical Potential and the Effective Δ-Nucleus Potential in the (3,3) Resonance Region - *M. Hirata, F. Lenz and K. Yazaki*. 180

III.11 - The Effect of the Exclusion Principle on the Δ-Isobar - *B. L. Friman*. 182

III.12 - A Study of the Doorway Isobar Model for Pions - *L. S. Kisslinger and A. Saharia*. 184

III.13 - Resonance Damping in the Interaction of Pions with Nucleons in Nuclear Matter - *L. C. Liu, W. T. Nutt and C. M. Shakin*. 186

III.14 - Pion-Condensation and Short-Range Correlation in Nuclei - *W. T. Weng, T. T. S. Kuo and G. E. Brown*. . . 187

III.15 - Contribution of the Meson-Exchange Currents to Charge Density of ^{16}O - *Il-T. Cheon*. 188

III.16 - Axial Polarizability and Pionic Field in Nuclei - *J. Delorme, M. Ericson, A. Figureau and C. Thévenet*. 190

III.17 - Analyticity and Pion-Nucleus Scattering - *O. Dumbrajs*. 192

III.18 - Test of the New Threshold Expansion of the Pion-Nucleus S-Wave Amplitude by Means of Dispersion Relations - *H. Pilkuhn, H. G. Schlaile and N. Zovko*. 194

III.19 - Are Pionic Exchange-Current Contributions to
$p + p \to {}^2H + e^+ + \nu_e$ Well Determined? - *H. S. Picker*... 196

III.20 - Effect of Anti-Symmetrization on Back-Angle p-d Scattering - *Leonard S. Kisslinger and Chi-Shiang Wu*.. 198

III.21 - Reanalysis of the A=8 Mirror Asymmetry - *Kuniharu Kobodera*.. 200

IV

PION ABSORPTION AND PRODUCTION ON NUCLEI

IV.1 - Pion Absorption and Production Experiments - *Elie Aslanides*.. 204

IV.2 - Current Theories of Pion Production and Absorption - *J. V. Noble*... 221

IV.3 - Pion-Nucleus Total Cross-Section Data From LAMPF and BNL - *M. D. Cooper*.................................. 237

IV.4 - Interpretation of Pion-Nucleus Total Cross Sections in the Region of the πn (3,3) Resonance - *C. B. Dover*... 249

IV.5 - Positive and Negative Pion Production Near Threshold - *Y. Le Bornec, B. Tatischeff, L. Bimbot, I. Brissaud, H. D. Holmgren, J. Källne, F. Reide and N. Willis*.................................. 260

IV.6 - Angular Distribution of the Reaction ${}^{16}O(\pi^+,p){}^{15}O$ at 66 MeV - *D. Bachelier, J. L. Boyard, T. Hennino, J. C. Jourdain, P. Radvanyi and M. Roy-Stéphan*...... 262

IV.7 - Low Energy Pion Production By 400-500 MeV Protons - *D. Bryman, G. Beer, G. R. Mason, E. Mathie, A. Olin, L. P. Robertson and J. S. Vincent*. 264

IV.8 - Models for Proton Induced Pion Production - *M. Dillig and M. G. Huber*. 266

IV.9 - Relativistic PWIA for (p,π^+) Reactions - *L. D. Miller and H. J. Weber*. 268

IV.10 - S-Wave Pion Absorption by Nuclei - *F. Hachenberg, J. Hüfner and H. J. Pirner*. 270

IV.11 - The $(\pi^-,2n)$-Reaction on Light Nuclei - *B. Bassalleck, D. Engelhardt, W. Klotz, C. W. Lewis, F. Takeutchi, H. Ullrich and M. Furic*. 272

IV.12 - Pion Production on Li^6 Via the Reaction $^6Li(p,d\pi^+)He^5$ at E_p = 800 MeV - *J. Hudomalj-Gabitzsch, J. Clement, W. Dragoset, R. Felder, G. S. Mutchler, T. M. Williams, G. C. Phillips, E. V. Hungerford, M. Warneke, L. Pinsky and J. C. Allred*. 274

IV.13 - Improved Analysis of Coulomb-Nuclear Interference Experiment for Pions on ^{16}O - *M. B. Johnson and M. D. Cooper*. 276

IV.14 - Strong Absorption Effects in Pion-Nucleus Total Cross Sections - *W. A. Friedman, K. W. McVoy, J. E. Sedlak*. 278

V

PION-NUCLEUS CHARGE EXCHANGE AND OTHER REACTIONS

V.1 - Pion Induced Inclusive Reactions - *P. D. Barnes*. 281

V.2 - Nucleon Charge Exchange and (π,πN) Reactions -
R. R. Silbar. 297

V.3 - (π,πn) Reaction on Light Nuclei - Paul J. Karol. . . . 305

V.4 - Gamma Rays Following the Interaction of 70 MeV
Pions with S-D Shell Nuclei - M. Zaider, D. Ashery,
S. Cochavi, S. Gilad, M. Moinester, Y. Shamai and
A. I. Yavin. 307

V.5 - ^{4}He(p,d)^{3}He Reaction at E_p = 770 MeV - Pierre Couvert. . 310

V.6 - Pion Charge Exchange at Rest - M. D. Hasinoff, M.
Salomon and A. Reitan. 316

V.7 - Nuclear Correlations in the $^{13}C(\pi^+,\pi^0)^{13}N$ Reaction -
E. Oset. 318

V.8 - The Elastic Charge Exchange of a Pion on ^{13}C Near
the 3 3 Resonance Region - S. Furui. 320

V.9 - Pion Charge Exchange in the (3,3) Resonance Region -
N. Auerbach and J. Warszawski. 322

V.10 - Pion-Carbon 12 Wavefunctions and Inelastic Scattering -
W. R. Gibbs, A. T. Hess and G. J. Stephenson, Jr. . . . 324

V.11 - Cross-Section for the Double Charge Exchange Reaction
$\pi^+ + ^4He \to \pi^- + 4p$ at Pion Energies of 98, 135, 145 and
156 MeV - I. V. Falomkin, V. I. Lyashenko, G. B.
Pontecorvo, Yu. A. Shcherbakov, M. Albu, T. Angelescu,
O. Balea, A. Mihul, F. Nichitiu, A. Seraru, F.
Balestra, R. Garfagnini, G. Piragino, C. Guaraldo and
R. Scrimaglio. 326

V.12 - Pion Double Charge-Exchange on ^{4}He - *A. T. Hess, W. R. Gibbs, B. F. Gibson and G. J. Stephenson, Jr.* . . . 328

V.13 - Possible Evidence for Short-Range 4-N Correlations From (π,γX) Reactions - *C. E. Stronach, J. H. Stith, C. M. Dennis, B. J. Lieb, W. F. Lankford, H. O. Funsten, W. J. Kossler, H. S. Plendl and V. G. Lind.* . . . 330

V.14 - Negative Pion Absorption at Rest Leading to ^{11}Be Bound States - *B. Coupat, D. B. Isabelle, P. Y. Bertin, A. Gérard, J. Miller, J. Morgenstern, J. Picard, B. Saghai and P. Vernin.* 332

V.15 - Quasi-Free (π,πN) Scattering at the (3,3) Resonance - *V. E. Herscovitz, Th. A. J. Maris, P. M. Mors and C. Schneider.* 334

V.16 - Modifications of Cascade-Evaporation Calculation for Interpretation of Prompt-Gamma Type Experiments - *M. Zaider and D. Ashery.* 336

V.17 - Pion Induced Reactions in the Isobar Model - *J. N. Ginocchio.* 338

V.18 - A Remark on Pion Capture in Heavy Nuclei - *M. P. Locher and F. Myhrer.* 340

V.19 - A Proposed Medium Energy Data Library (MEDL) - *E. D. Arthur and R. J. Barrett.* 342

V.20 - A Compilation of Pion Absorption Data - *K. G. R. Doss, S. A. Dytman and R. R. Silbar.* 344

V.21 - Particle and Fragment Multiplicities for He and Ne Nuclei - *A. S. Kanofsky, R. C. Allen and G. Lazo.* . . . 346

VI

FEW-BODY SYSTEMS AND PION-NUCLEUS PHYSICS

VI.1 - Pion Induced Reactions on Light Nuclei - *F. G. Binon*. . 348

VI.2 - Scattering of Pions on ^{3}He and ^{4}He in the Δ_{33} Resonance Region - *Yu. A. Shcherbakov*. 365

VI.3 - Faddeev Calculations of πD Scattering - *A. W. Thomas*. 375

VI.4 - Multiple Scattering Calculations in πd Elastic Scattering - *Erasmo M. Ferreira*. 384

VI.5 - Graph Summation Method in πd Problem - *V. M. Kolybasov*. 394

VI.6 - The Theory of Pion-^{4}He Scattering - *F. Lenz*. 403

VI.7 - The Few-Body Problem and Pion-Nuclear Physics - *B. F. Gibson*. 418

VI.8 - Covariant N-N Dynamics and πD Scattering - *D. D. Brayshaw*. 443

VI.9 - Pion Production in P-P Interactions at 800 MeV - *R. D. Felder, J. Hudomalj-Gabitzsch, T. M. Williams, G. S. Mutchler, J. M. Clement, K. R. Hogstrom, W. H. Dragoset, G. C. Phillips, E. V. Hungerford, M. Warneke, B. W. Mayes, L. Y. Lee and J. C. Allred*. 446

VI.10 - Pion Production by 800 MeV Protons From Deuterium with a Spectator Neutron - *E. V. Hungerford, J. Lo, M. Warneke, J. C. Allred, B. W. Mayes, L. Pinsky, J. Clement, W. H. Dragoset, R. Felder, K. Hostrom,*

          J. Hudomalj-Gabitzsch, G. S. Mutchler, G. C. Phillips and T. Williams............... 448

VI.11 - Covariant Calculations of πD Scattering - *A. S. Rinat and A. W. Thomas*................ 450

VI.12 - Pion Deuteron Elastic Scattering - *E. M. Ferreira, L. P. Rosa and Z. D. Thomé*............ 452

VI.13 - Pion Deuteron Scattering in the Resonance Region - *H. Garcilazo*........................ 454

VI.14 - Multiple Scattering Analysis of π⁻He Scattering at 1.12 GeV - *Yukap Hahn*................ 456

VI.15 - π-^{3}He, π-^{4}He Elastic Scattering and π-^{3}He Charge Exchange: An Improved Calculation - *R. H. Landau*... 458

VI.16 - Is There an Isohelion? - *F. Nichitiu*.......... 460

VI.17 - A Lowest Order Optical Model Study on π-^{4}He Scattering - *T.-S. H. Lee*................ 462

VI.18 - Model for the Absorptive Component in Low-Energy Elastic Pion-Deuteron Scattering - *P. Goode, R. Rockmore and H. McManus*............... 463

VI.19 - Pion-Deuteron Absorption - *W. R. Gibbs, B. F. Gibson and G. J. Stephenson, Jr.*........... 464

VI.20 - Influence of Form Factors on Pionic Deuteron Disintegration - *M. Brack, D. O. Riska and W. Weise*............................. 466

VI.21 - Calculation of the Pion Production Reaction ^{3}He(p,π$^+$)^{4}He - *James H. Alexander and Harold W. Fearing*........................... 468

VI.22 - ^{3}He(p,π^+)^{4}He Reaction at 415 and 716 MeV - *B. Tatischeff, L. Bimbot, R. Frascaria, Y. Le Bornec, M. Morlet, N. Willis, R. Beurtey, G. Bruge, P. Couvert, D. Garreta, D. Legrand, G. A. Moss and Y. Terrien*................. 470

VI.23 - Soft-Pion Production in N-N Collisions - *A. W. Thomas and I. R. Afnan*................. 472

VI.24 - Slow π-Meson Elastic Scattering on Nuclei - *G. G. Bunatian and Yu. S. Pol*............ 474

VII

KAON-NUCLEON AND KAON-NUCLEUS SCATTERING AND REACTIONS-HYPERNUCLEI

VII.1 - Current Status of Kaon-Nucleon Analysis Below 3 GeV/c - *K. K. Li*................. 476

VII.2 - The Off-Shell $\overline{K}N$ T-Matrix - *M. Alberg*........ 486

VII.3 - New Data on the (K$^-$,Π^-) Reaction - *K. Kilian*...... 497

VII.4 - Hypernuclear Spectroscopy and Strangeness Analog States - *Nguyen Van Giai*................. 507

VII.5 - An Overview of Hypernuclear Physics - *H. Feshbach*... 521

VII.6 - Model For Low Energy Kaon-Nucleon Interaction in the I=0 State - *S. C. B. Andrade and E. M. Ferreira*................. 532

VII.7 - K$^+$-Nucleus Elastic Scattering - *R. A. Eisenstein and F. Tabakin*................. 534

VII.8 - Angular Distribution for the (K^-,π^-) Reaction on ^{12}C and ^{27}Al - G. C. Bonazzola, T. Bressani, E. Chiavassa, G. Dellacasa, M. Gallio, A. Musso and G. Rinaudo. 536

VII.9 - Strangeness Exchange Reaction on Nuclei - A. Bouyssy. 538

VII.10 - Strange Giant Resonance in (K^-,π^-) Reaction - N. Hoshi and T. Fujita. 540

VII.11 - The $\Lambda(1405)$ in Nuclear Matter and Implications for Kaonic Atoms - J. M. Eisenberg. 542

VII.12 - Kaonic Atoms Level Shifts: Potential Approach - A. Deloff and J. Law. 544

VII.13 - Kaonic Atoms Level Shifts: Multiple Scattering Approach - A. Deloff and J. Law.. 546

VII.14 - Application of Forward Dispersion Relation to Kaon ^{12}C Scattering - K. Arai, I. Endo and M. Kikugawa.. 548

VII.15 - Black Sphere Model For the Line Widths of Kaonic and Antiprotonic Atoms - W. B. Kaufmann and H. Pilkuhn. 550

VIII

ELECTROMAGNETIC PROCESSES AND MESON-NUCLEAR PHYSICS

VIII.1 - Recent Results in Pionic and Kaonic Atoms - R. J. Powers.. 552

VIII.2 - Photoproduction of Pions in Light Nuclei - C. Tzara.. 566

VIII.3 - Radiative Pion Capture in Nuclei - *Peter Truöl*. . . . 581

VIII.4 - Threshold Photoproduction of Pions - *Justus H. Koch*. 591

VIII.5 - Photoproduction of Negative Pions and the Ground State Wave Function of ^{12}C - *K. Srinivasa Rao*. . . . 601

VIII.6 - Spin-Flip Transition Strength of $^{12}C(\gamma,\pi^+)^{12}B$ - *K. Shoda, H. Ohashi and K. Nakahara*. 604

VIII.7 - The $^{12}C(\gamma,\pi^-)^{12}N$ Reaction Near Threshold - *A. M. Bernstein, N. Paras, W. Turchinetz, B. Chasan and E. C. Booth*. 606

VIII.8 - Threshold Photoproduction of π^+ Mesons in Deuterium - *E. C. Booth, B. Chasan, A. Bernstein and P. Bosted*. 608

VIII.9 - Near Threshold π^0 Photoproduction from the Deuteron - *J. H. Koch and R. M. Woloshyn*. 610

VIII.10 - π^- Photoproduction in ^{12}C - *Anton Nagl and H. Überall*. 612

VIII.11 - A PWIA Analysis of Charged Pion Photoproduction from ^{12}C - *K. Baba, I. Endo, M. Fujisaki, S. Kadota, Y. Sumi, H. Fujii, Y. Murata, S. Noguchi and A. Murakami*. 614

VIII.12 - Excitation of the Carbon-14 Ground State by the Charged Pion Photoproduction at Threshold in Nitrogen-14 - *A. Figureau and N. C. Mukhopadhyay*. . . 616

VIII.13 - Photo- and Electroproduction of Pions and Kaons Near Threshold from Nuclear Targets - *J. B. Cammarata and T. W. Donnelly*. 618

VIII.14 - The Charged Pion Photoproduction on ^{12}C Near the 3 3 Resonance Region - *S. Furui*. 620

VIII.15 - Radiative Pion Capture in ^{3}He - *W. R. Gibbs, B. F. Gibson and G. J. Stephenson, Jr.* 622

VIII.16 - The SIN-Pairspectrometer and Radiative Pion Capture in ^{12}C - *J. C. Alder, B. Gabioud, F. Hoop, C. Joseph, J. F. Loude, H. Medicus, N. Morel, A. Perrenoud, J. P. Perroud, D. Renker, H. Schmitt, G. Strassner, M. T. Tran, P. Truöl, B. Vaucher, H. v. Fellenberg, E. Winkelmann and C. Zupancic*. 624

VIII.17 - Investigation of 1p-Shell Nuclei with Radiative Pion Capture: ^{6}Li, ^{7}Li and ^{9}Be - *J. C. Alder, B. Gabioud, C. Joseph, J. F. Loude, H. Medicus, N. Morel, A. Perrenoud, J. P. Perroud, D. Renker, H. Schmitt, G. Strassner, M. T. Tran, P. Truöl, H. von Fellenberg and E. Winkelmann*. 626

VIII.18 - Radiative Pion Capture in Oxygen-Isotopes - *J. C. Alder, B. Gabioud, C. Joseph, J. F. Loude, H. Medicus, N. Morel, J. P. Perroud, A. Perrenoud, D. Renker, H. Schmitt, G. Strassner, M. T. Tran, P. Truöl, H. v. Fellenberg and E. Winkelmann*. 628

VIII.19 - Search for 2γ Emission in π Capture: A Progress Report - *J. Deutsch, D. Favart, P. Lipnik, P. Macq and R. Prieels*.................. 630

VIII.20 - The TRIUMF π⁰ Spectrometer and the Panofsky Ratio in Hydrogen and Deuterium - *M. Salomon, D. Berghofer, M. D. Hasinoff, R. MacDonald, D. F. Measday, J. Spuller, T. Suzuki, J. K. P. Lee, J. M. Poutissou, R. Poutissou and P. Depommier*.... 632

VIII.21 - Elastic η-Meson Photoproduction from Deutrons - *Yu. N. Krementzova and A. I. Levedev*......... 634

VIII.22 - Partial η-Meson Photoproduction Cross Section From Li⁶ - *A. I. Lebedev, V. A. Trjasuchev and V. N. Fetisov*........................ 636

VIII.23 - Invariant Impulse Approximation and Pionic Atoms - *R. Mach*............................. 638

IX

NUCLEON-NUCLEON INTERACTION

IX.1 - Some Mesonic Aspects of the Nucleon-Nucleon Interaction - *R. Vinh Mau*................. 642

IX.2 - An Overview of the N-N System - *G. E. Brown*...... 655

IX.3 - The Effect of the Δ(1236)-Resonance in NN-Scattering, Nuclear and Neutron Matter Including All Partial Waves - *R. Machleidt and K. Holinde*.... 663

IX.4 - Evidence for a Δ-Nucleon Virtual State - *G. Alberi and F. Baldracchini*................ 666

IX.5 - Charge Asymmetry of Nuclear Forces - *M. A. Alberg, E. M. Henley, G. A. Miller and J. F. Walker.* 668

X
PANEL DISCUSSION

X.1 - Pion Excitation of Nuclear Collective Modes and High Spin States - *G. E. Walker.* 674

X.2 - Nucleon-Nucleon Correlations and Elastic Pion Double Charge-Exchange Reactions - *Gerald A. Miller.* . . 684

X.3 - Quasi-Free Formation of and Supersymmetry in Light Hypernuclei - *A. Gal.*. 694

XI
NEW FACILITIES AND INSTRUMENTATION

XI.1 - Status of the Nevis Synchrocyclotron Facility and Experimental Program - *Derek W. Storm.* 706

XI.2 - Present Status of Beam Lines at KEK - *H. Hirabayashi, S. Kurokawa, A. Kusumegi, A. Maki, S. Mikamo, T. Sato, Y. Suzuki, M. Taino, K. Takamatsu, M. Takasaki, K. Tsuchiya and A. Yamamoto.* 713

XI.3 - The SIN Pion Spectrometer - *ETH/Zurich-Grenoble-Heidelberg-Karlsruhe-Neuchatel-SIN Collaboration.*. . . . 724

XI.4 - Design of A π^0 Spectrometer at LAMPF - *R. Heffner.* . . . 733

XI.5 - Performance of EPICS Pion Channel - *D. C. Slater, C. L. Morris, H. A. Thiessen, J. Källne, C. Fred Moore, Joe E. Bolger, S. Iversen, A. Obst, J. F. Amann, S. Verbeck, G. Burleson, J. Peterson and S. Greene.* 740

XI.6 - Performance of a Large Streamer Chamber for the Observation of Particle-Nuclei Interactions with All Charged Particles Detected - *A. S. Kanofsky, R. C. Allen and A. Hasan*. 742

XII
NEW DIRECTIONS

XII.1 - The Nuclear Many Body Problem as a Relativistic Field Theory - *L. Wilets*. 746

XII.2 - Bag Model of Hadron Structures - *K. Johnson*. 755

XII.3 - New Directions for Nuclear Physics--A Personal View - *Maurice Goldhaber*. 756

AUTHOR INDEX. 766

INTRODUCTION AND PREVIEW

INTRODUCTION

This volume is organized to try to make it most useful as a reference work. For the most part, the chapters are arranged according to the conference sessions, with the invited papers appearing in the order in which they were given. However, there has been some rearrangement due to the different demands of an oral vs. a written format.

The contributed papers are placed according to the subject matter after the invited papers. By allowing for two-page contributed papers rather than simply abstracts, the organizers hoped to enhance the usefulness of the proceedings. It seems that this has proved true, for the contributed as well as the invited papers contain a great deal of valuable new material. Although the majority of contributed papers could not be presented orally, much of the work was reviewed by the invited speakers.

The discussion which appears following each paper presented at the conference is taken mainly from the written discussion sheets. Although this does not give a completely accurate representation of the actual discussion, it allows for higher order corrections to both the questions and replies, and thereby enhances the scientific value of the proceedings. Some of the discussion does not appear, since the members asking a question or making a comment did not feel that it was worth submitting in written form. The discussion associated with the panel was prepared by the referee, J. Eisenberg and T. Londergan. The scientific secretaries responsible for the written discussion in the various sessions were A. Hess (Monday A.M.), B. Keister (Monday P.M.), W. Nutt (Tuesday A.M.), A. Saharia (Tuesday P.M., Thursday P.M.), S. Cotanch (Wednesday A.M.), M. Soyeur (Friday A.M.), W. Wharton (Friday P.M.). The comments by the participants of the White Water Raft Trip were not recorded due to technical difficulties.

June, 1976

P. D. Barnes
R. A. Eisenstein
L. S. Kisslinger

MESON-NUCLEAR PHYSICS: A PREVIEW

Daniel S. Koltun[*]
University of Rochester, Rochester, New York 14627

ABSTRACT

This preview gives a brief survey of the main topics to be discussed at this conference, in the context of the subject of meson-nuclear physics.

INTRODUCTION

The purpose of this Preview is to provide a quick guide to the main topics of this Conference, pointing out some highlights of the days to come. It differs from a Preview in the cinema or theater, which is put together from the finished production, and is meant to give you a sense of the drama (or comedy) to follow. No one here has seen or heard the finished production of the conference; it doesn't yet exist. What I have seen is the Conference Program and the contributed papers, and base my remarks on them. If my remarks do not correspond precisely to what the speakers will actually say, no apology is called for. After all, this is not a Conference Summary!

I have divided the subject into several general categories, corresponding to the main areas of this conference, but I shall not follow the order of the program. In the interests of economy I have not given references; the "reviews", "discussions", and "contributions" mentioned in what follows may be located in the Conference Program, or in the Contributed Papers. In some cases I have included the name of the relevant speaker, in brackets.

ELEMENTARY SCATTERING PROCESSES OF STABLE MESONS: π, K

There are new measurements of $\pi^{\pm}p$ elastic scattering for $E_\pi < 100$ MeV from Saclay as well as total πp cross sections from SIN, which will be presented here.(Bertin, Domingo) The analysis of π-N scattering will also be discussed. Theorists will certainly welcome more precise determination of the low energy π-N parameters, particularly the s-wave and non-resonant p-wave phase shifts, because of the role they play in the determination of low energy π-nucleus scattering. (Kelly, Olsson)

On the theoretical side, the problem has been to extend the πN amplitudes into kinematic domains not measured in πN scattering. There has been increased emphasis on more dynamical and less ad-hoc treatment of this problem. Field theoretic models, following, for example that of Chew and Low, or models in which resonances are treated as elementary objects, are now competing with separable interactions obtained from, say, the inverse scattering problem. The

---

[*]Research supported in part by the U. S. Energy Research and Development Administration.

interest is not so much in a better description of πN scattering as in developing a suitable dynamical picture which will apply to nuclei as well. (Cammarata, Ernst).

We shall also hear about the current status of KN scattering analysis, although here there is even more need for new experiments. (Li, Alberg).

I include pion-deuteron reactions among the elementary scattering processes. This is a favorite theoretical subject, since it provides the simplest case for multiple scattering theory, and is also accessible to treatment by three-body techniques. Both approaches are to be reviewed here. (Thomas, Farreira) Now the nuclear reactions induced by π^+d lead to three channels:

$$\pi^+d \to \pi^+d \qquad \text{elastic}$$

$$\pi^+d \to \pi^+np + \pi^0pp \quad \text{breakup and charge exchange}$$

$$\pi^+d \to pp \qquad \text{absorption.}$$

Multiple scattering theory generally treats each channel independently. Conventional three-body theory combines the first two channels. The absorption channel has generally been treated as a separate problem, and has been of interest as a relatively tractable if complicated case for the study of pion absorption with all the nuclear physics put in.

An interesting theoretical development in the last few years has been the application of new techniques for treating all three πd channels as a combined dynamical problem. This is a move in the direction of recombining the problems of pion scattering with the meson theory of nuclear forces, to which I shall refer again later. (Brayshaw).

On the experimental side, there have been recent measurements of πd elastic scattering at low energy at Berkeley, and of total cross sections at SIN. (Thomas, Ferreira, Domingo). There are also a new set of LAMPF measurements of cross sections for $pp \to \pi^+d, \pi^+pn, \pi^0pp$. (Felder).

There will also be a presentation of new experimental and theoretical results on πp bremsstrahlung. (Sober, Ho-Kim).

PION ELASTIC AND TOTAL CROSS SECTIONS

There are no contributed papers on new experimental results, although some will be presented in the invited talks. (Tabakin, Eisenstein, Cooper). Low energy elastic scattering experiments are planned or underway at LAMPF.

This leaves plenty of time for the theorists, who have much to say in this area. Over the last several years there has been a lot of energy devoted to the study of optical potentials or to multiple scattering expansions, with an eye to accurate calculation of cross sections based on known nuclear physics and known π-N amplitudes. The earliest results, based on Kisslinger's optical potentials, or on Glauber's eikonal theory, were relatively successful in giving the main features of the energy and angular dependence of cross sections. For low energies, including pionic atoms (zero energy) simple optical potentials do give good results, but their parameters are not what one would expect from π-N scattering. (This is also true for Kaonic atoms, for which the discrepancies are larger.) For pion energies around that of the 3-3 resonance, differential cross sections at smaller angles are not hard to produce, but they do not provide a good test of the approximations used, since the scattering is so diffractive.

Much recent theoretical work has been devoted to including effects left out of the simple approximations; many of these are the subjects of contributed papers. Within the context of ordinary multiple scattering theory, there are improved treatments of the kinematics of π-N scattering and of nuclear binding effects, which do seem to lead to better optical potentials at 50 MeV, where anomalous behavior has recently been pointed out. (Eisenstein, Thies). (Such effects may be very large in the case of Kaonic atoms.) Relativistic treatment, with and without full covariance, has been discussed in the literature. This improvement should be more important with increased pion energy.

There has been continued interest in the role of correlations, in terms of the Pauli Principle and the Lorentz-Lorentz effect. But here the focus of discussion has moved noticeably in recent years, from conventional multiple scattering theory to a more microscopic view of the π-N scattering in the nuclear system. This has allowed one to deal with questions like the modification of the 3-3 resonance in a nuclear system, or the influence of pion absorption on the optical potential. The methods have included field-theoretic models, like the Chew-Low model of resonant scattering. There are other models in which the πN resonance is treated as an elementary particle or excitation (Δ), the interaction of which with the nucleus modifies dynamically the π-N scattering. There will be a number of talks on these topics (Moniz, Banerjee, Mizutani, Kisslinger). They form part of the subject of the meson degrees of freedom of nuclei, to which I shall return below.

This is not to say that multiple scattering theory has been abandoned. Many of the recent applications have been to pion scattering from very light nuclei (e.g. ^{3}He, ^{4}He) which will be reviewed by a number of speakers. (Binon, Scherbukov, Lenz, Gibson).

INTERPRETING NUCLEAR REACTIONS OF MESONS AS DIRECT REACTIONS

Under this heading I list a number of inelastic reactions about which there will be some discussion at this conference. We do not have new results for all of these reactions:

$(\pi^\pm, \pi^0)$, $(\pi^\pm, \pi^\mp)$	charge exchange (Reitan, Miller)
$(\pi, \pi N)$	knockout (Silbar, Karol)
(π, π')	inelastic scattering (Walker)
(π, p), (p, π)	absorption, production (to bound states) (Aslanides, Noble, LeBornec)

(π,γ), (γ,π)	radiative capture (Truol)
	photoproduction (Tzara, Rao)
(K^-,π^-)	strangeness exchange (Kilian, Bressani, Giai)

There is new experimental information for a number of these reactions. For the radiative processes, there are new data contributed from BATES, and electroproduction (e,e'p) from Sendai, as well as (π,γ) from SIN. For $(\pi^\pm,\pi^0)$ and $(\pi,\pi N)$ we have had only measurements of residual state total cross sections by radiochemical or nuclear-γ techniques. There is the exciting possibility of future measurements of the π^0 in charge exchange by γ-γ coincidence; the proposed π^0 spectrometer will be described here (Hefner). There is also a proposal to look at recoil protons in $(\pi,\pi p)$, at LAMPF. Both possibilities would give us some much needed kinematic information about these reactions.

There will be some new data presented on the very interesting Strangeness Exchange reaction (K^-,π^-), and some new theoretical work has been contributed. This reaction has made accessible to study, for the first time, excited states of hypernuclei. Besides the presentation of new data and theory, there will be a review of hypernuclear physics. (Feshbach, Nguyen).

Now, for all the reactions listed, there is evidence that they may be direct reactions, but they need not be simple. In some cases, there may well be multistep processes competing with, or dominating the simplest one-step direct processes. For $(\pi,\pi N)$, a simple knockout picture reproduces neither the energy dependence nor the ratio of cross sections induced by π^+/π^-. One appears to get distinct improvement by including scattering of the outgoing nucleon. It may be equally possible to describe this reaction as inelastic scattering to broad states in the continuum (giant resonances), with subsequent emission, as has been pointed out by Robson and by Schiffer. Certainly, kinematic measurements would give us more understanding of this reaction. (Silbar, Karol).

For the charge exchange, again we have no kinematics. The energy dependence of the total cross sections to bound states, in light nuclei, has been a puzzle, and recent optical or multiple scattering calculations have not been successful. Here, incoherent processes in intermediate states, or correlations (which should be equivalent) may be important. This has been suggested by a number of people, and is the subject of several contributions. Here again, one would love to have differential cross sections, which may be available in the future with the π^0 spectrometer of LAMPF.

The question of the correct description of charge exchange may be quite important to developing a correct theory of the optical potential. As mentioned earlier, the elastic scattering is not particularly sensitive to all features of the optical potential, or equivalently, to the pion waves in the elastic channels. Different inelastic processes are sensitive to different features of these waves. The analog charge exchange (and double charge exchange) ought to be useful for settling questions of scattering mostly on the energy

shell for the pion, involving small momentum transfers for the nucleons.
The (π,p) and (p,π) reactions are highly inelastic, in a kinematic sense. It seems to be possible to interpret these as direct, one step "pionic stripping" (or pickup) reactions, with distorted waves. However, there are a number of multistep processes which are competetive, and which may render the extraction of nucleon transfer information subject to large uncertainties. There will be a review of this subject (Noble) and a number of contributed papers. In this connection, there is a contributed report from Saclay on a measurement of the absolute rate of $^{12}C(\pi^-,p)^{11}Be$ for stopped pions.

We have both experimental and theoretical contributions on the (γ,π) and (π,γ) reactions, with particular interest in accurate measurements of charged π photoproduction near threshold on light nuclei. The usual treatments are distorted-wave impulse calculations. Although a recent discrepancy in the threshold value may have been removed, the slope of the cross section with energy does not seem to be in agreement. Multistep processes, particularly meson exchange (or currents) between nucleons may be important here, as elsewhere in nuclear radiative interactions. (Tzara, Koch).

ABSORPTIVE REACTIONS OF PIONS

This very rich subject includes many aspects of nuclear reactions running all the way from direct to statistical. It is perhaps surprising that, with the exception of the (π^+,p) and (p,π^+) reactions already discussed, there are not a large number of new results reported here. (Aslanides, Barnes, Zaider).

Recent experimental work has largely concentrated on the measurement of nuclear gamma emissions following π-absorption, giving rates for populating given final states. Some new (π,γ(nuc)X) results have been contributed here. The interest has been in understanding the rates of populating nuclei many nucleons below the target in the mass table. Whether these correspond to direct removal of clusters of nucleons, or to some multistep processes, is not established. It seems clear that to answer such questions, one cannot rely on looking only at final nuclear states. One needs to know what particles are emitted in absorption, in what combinations, with what spectra, correlations, and so on. Although there are scattered examples of this kind of study in the past, there is little going on now in kinematic and correlation experiments or absorption. I am not sure that such work is scheduled at the new accelerators at present or in the near future. One exception is a CERN contribution to this conference, reporting on a complete kinematic study of $(\pi^-,2n)$ on light nuclei. (Bassalleck). There is strong evidence that this reaction is direct, and the new work, with complete kinematics and improved energy resolution does demonstrate the use of this reaction for the study of the nuclear structure of the final target (2 hole) states.

Perhaps one of the problems with pion absorption is that our thinking about the reaction has tended to concentrate on one very simple limit or the other, either (π,2N) as a direct reaction, or statistical cascade models at the other. Yet there are many things that can happen in absorption between the emission of a few fast particles and the evaporation of the last few low energy neutrons. There is a dynamical question of how a pion shares its momentum and energy before it is absorbed in the nucleus, as well as the problem of the nuclear deexcitation. I think the subject is ripe for theoretical, as well as experimental study.

There are good theoretical reasons for wanting a sound method for dealing with the entire absorption reaction, aside from the intrinsic interest in this dynamical problem. Absorption is a rather probable reaction, yet we have few measurements of the total absorption cross sections on nuclei, and no real theory for these cross sections. Yet, because absorption is a major reaction branch, it must have an effect on elastic scattering, and on all inelastic processes in general. It will be necessary to learn how to treat scattering and absorption together. I mentioned earlier that this problem has been attacked in the case of πd reactions. In this connection there is a contribution on the theory of absorption at rest, in pionic atoms. (Hackenberg).

NUCLEAR STRUCTURE AND MESON DEGREES OF FREEDOM

We shall have a panel discussion at this conference on the question: "Can one learn important new things about nuclear structure with pions?" This brings to mind the fact that there are questions of nuclear structure which have been addressed previously, to which we shall not have new answers given at this conference. One example is: "How can we measure the difference between the distribution of neutrons and that of protons in nuclei?" In part, we are waiting for new, accurate experimental data, like differential cross sections for elastic scattering, and perhaps for charge exchange, on appropriate medium and heavy mass nuclei. However, we also will have to understand better the theory of elastic scattering, with all its "small" corrections, before we can with confidence extract density distributions from scattering data. It is the uncertainty in the analysis which has limited our ability to use existing data, say from pionic atoms, to study neutron and proton density distributions. For kaonic atoms, this kind of analysis seems even more remote. Other related information from K-mesons, such as might be provided by extensions of the early studies of branching ratios of absorption reactions induced by stopped K⁻, are simply not available.

Another question we used to discuss more is "How can we demonstrate the presence of short-range correlations between nucleons in nuclei?" Here the problem has been even more intensely one of dealing with all the competing processes, which tend to mask any clear demonstration of correlation effects. This makes it almost more profitable to ask: "How much of mesic nuclear physics can we understand without invoking correlations?"

Now one may well be able to do without explicit short-range correlations, if one puts in something to replace them. This can be accomplished, for example, by going back to the theory of nuclear forces, and asking what produces repulsive effects at short range. If your meson theory tells you that this comes from exchange of vector mesons, like the ρ-meson, then you may try to include the ρ in your scattering theory, in place of the short-range nuclear force effect. Examples of this approach may be found in several contributed papers. (Thies, Brack). Related reviews of the N-N interaction will also be given. (Vinh Mau, Brown).

Now this is one example of the subject of Meson Degrees of Freedom of nuclei, which will be given considerable attention at this conference. Instead of picturing nuclei as composed only of interacting nucleons, we are now asked to consider the nucleus as containing mesons as well as nucleons, whose mutual interaction determines the nuclear dynamics. In other words, we are asked to return to the meson theory of nuclear forces. The elements may include not only mesons and nucleons, but also excitations of the nucleons by mesons, that is, nucleon resonances like the Δ (3-3 resonance), or for hypernuclei, the Y^* (hyperon resonances).

Now this approach may be applied either to the theory of the nuclear ground state, or to the theory of nuclear reactions induced by mesons. In the first category fall several examples, which will be discussed here: Nucleon resonances in the nuclear ground state (Weber, Machleidt), Meson exchange currents in electromagnetic and weak interactions (Rho, Riska), Field theories and many-body systems (Wilets).

The second category includes those studies of meson scattering in which excited states of mesons or nucleons are formed (Δ,ρ,Y^*) which propagate in and interact with the target nucleus. These have been mentioned earlier in connection with elastic scattering and optical models. (Moniz, Kisslinger).

It is not necessarily useful to keep these two parts of the subject separated, since one is talking about different excitations of the same system. Many of the models used, such as Δ propagation, or ρ - or π-exchange will appear both in scattering and in ground state theories. The same problems of coupling constants, form factors, and such will appear in both. Some aspects will differ, however. For example, a virtual Δ in a nuclear ground state has zero width, while a propagating Δ in a scattering state has a finite decay width, which will depend on the interaction with the nucleus.

The extended picture of nuclei will certainly have its own problems and ambiguities. In a sense more is demanded of it, since the nuclear sturcture and meson scattering must come out together. To make such a picture into a dynamical theory, one has to know how many mesons or resonances one must include to have a complete and consistent formalism. There are many problems which lead us back to field theory, and field theoretic models of scattering. And when we are there, we still have to know how to calculate useful numbers.

Beginnings have been made in developing the theory of the meson degrees of freedom of nuclei, and of applying it to problems of nuclear physics. It is not clear that this approach will replace all our other methods of dealing with nuclear structure or with nuclear reactions of mesons. But it can hardly fail to extend our theoretical tools, and to unify our understanding of the complicated interplay of mesons and nuclei.

Finally, a word of warning to those experimentalists who work with meson beams: this latest development in theory shows what happens to theorists when their attention is allowed to wander because they are not being overwhelmed by new experimental data! Such infidelity should probably be viewed with amused tolerance, but perhaps also with a touch of alarm. If one can really study meson nuclear physics with nuclear ground states, well, what next?

CONCLUSION

Enough of the Preview! On with the Show!

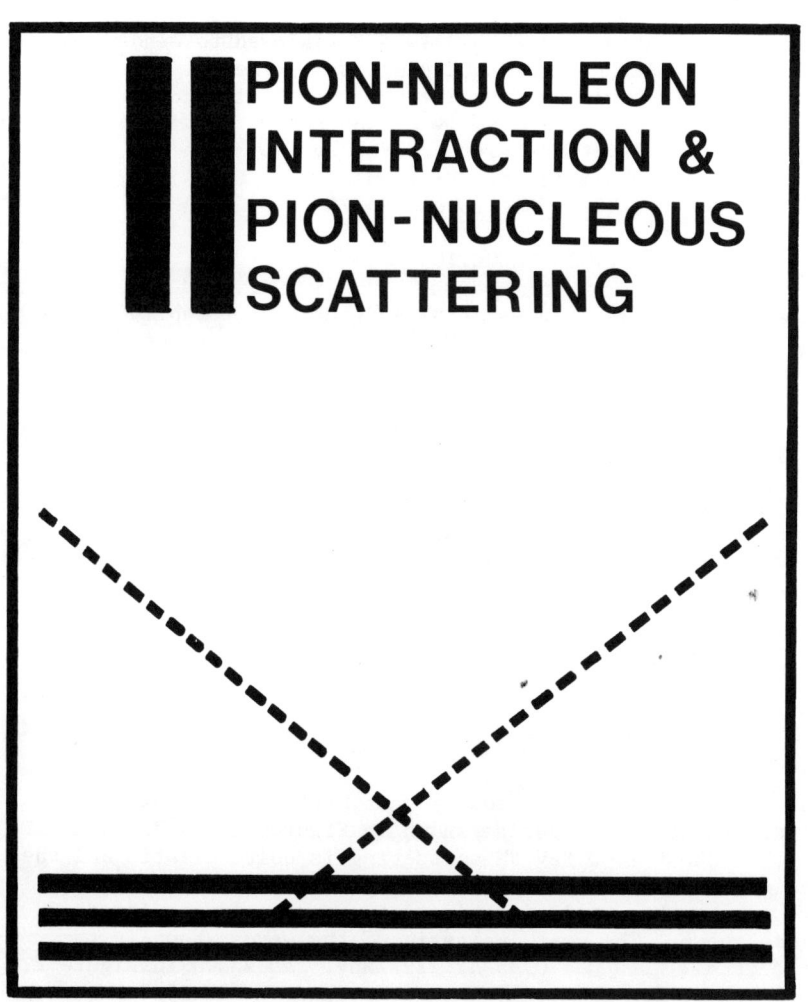

πN PARTIAL WAVE ANALYSIS IN THE FIRST AND SECOND RESONANCE REGIONS

R. L. Kelly[†]
Lawrence Berkeley Laboratory
University of California
Berkeley, California 94720

INTRODUCTION

The established resonances in πN scattering for lab momenta below about 1.0 GeV/c are shown in Figure 1. The first resonance region lies at 300 MeV/c and contains only one resonance, the well-known $\Delta(1232)$. At 700 MeV/c, or 1500 MeV center-of-mass energy, there is a second resonance region containing three $I = \frac{1}{2}$ resonances in the S11, P11, and D13 partial waves. The eta-production threshold also lies in this region. The third resonance region at 1.0 GeV/c or 1700 MeV contains seven resonances, the most elastic being the F15, D15, S11, and S31. Knowledge of the πN amplitudes in the first two resonance regions, though far advanced, is by no means complete, and we can anticipate vigorous activity in this area in the near future at the new meson factories. In this talk I will review current knowledge of the amplitudes at these energies, and point out what appear to be some promising directions for future research.

FIRST RESONANCE REGION

The results of the three most recent partial wave analyses[2-4] in the first resonance region are shown in Figures 2 and 3. At energies below about 1300 MeV πN scattering is quite elastic and the amplitudes are specified rather completely by the phase shifts. D and F waves are rather small in this region, and only the largest (the D13) is shown. The most inelastic of the waves shown is the P11 of Ayed et al. with $\eta = 0.95$ at 1292 MeV. As shown in Figure 2, the recent work of Bertin et al. has determined the I = 3/2 waves in the near threshold region, and they are presently making $\pi^- p$ measurements to fill the gap at these energies in the $I = \frac{1}{2}$ waves. Carter et al. have attempted to determine the level of isospin-breaking in the P33 wave by fitting different P33 amplitudes to the $\pi^+ p$ and $\pi^- p$ data. A small but significant difference in the energy dependence of these waves can be seen in Figure 1, corresponding to a width difference of $\Gamma^\circ - \Gamma^{++} = 6.4 \pm 1.8$ MeV.

The agreement among the different analyses for the I = 3/2 waves is generally better than for $I = \frac{1}{2}$. The worst disagreement is between the results of Refs. 2 and 3 for the P11 wave (both η and δ) and there are also significant discrepancies in the S11 wave. Several comparisons of the Carter et al. amplitudes to new data[5,6] have been made since the analysis was carried out, and though the agreement is not perfect it has been generally good. Bugg et al.[7] have used the results of Ref. 2 and other partial wave analyses at higher energies for a dispersion relation calculation of the S-wave scattering lengths. The π^+p and charge-exchange scattering lengths have also been determined directly[4,8] by measurements in the near-threshold region and these results agree with those of Ref. 7.

Most of the data available in the near-threshold and first resonance regions are elastic cross section data. Improvements in the analysis will probably require more elastic polarization data and more charge-exchange cross section data. Carter et al. comment that precise charge exchange data would be particularly useful in determining the level of isospin breaking in the S-waves.

There have been a number of recent determinations of the $\Delta(1232)$ pole parameters. Three of these determinations[9-11] have fit the phase shifts of Ref. 2 and treated the Δ^{++} and Δ^0 separately. The resulting real parts of both pole positions are 1211 ± 1 MeV, and the imaginary parts are -50 ± 1 MeV and -53 ± 1 MeV, respectively. The conventional Breit-Wigner mass of 1232 ± 2 MeV lies 21 MeV higher than the real part of the pole position. This large difference is due primarily to the proximity of the $\Delta(1232)$ to elastic threshold and the resultant energy dependence of the width. A closely related effect is the large negative phase of the pole residue, about $-46°$ to $-49°$.

SECOND RESONANCE REGION

The resonating waves in the second resonance region from the analysis of Ayed et al. are shown in Figure 4. The S- and D-wave resonances are fairly well determined, but the P11 is not very well determined by present data. Ayed et al. find two resonances in the P11 wave, at 1413 MeV and 1532 MeV, where earlier analyses have found a single resonance closer to 1470 MeV. The situation here is still controversial and further clarification will have to await better data, particularly π^-p polarization data. The non-resonant $I = 3/2$ S- and P-waves are also quite significant at 1500 MeV. All are rather elastic (though they rapidly become inelastic at higher energies) with large negative real parts due to low energy repulsive behavior in the case of the S31 and P31, and due to the high energy tail of the $\Delta(1232)$ in the case of the P33.

An important recent development is the observation of the effect of the ηn threshold (at 1488 MeV) in π^-p elastic scattering. Debenham et al.[12] have observed a striking cusp in the backward direction at this threshold. This cusp arises from a square-root threshold singularity in the backward spin-non-flip amplitude in interference with the non-singular part of the amplitude. The singularity is in the S11 partial wave, and its strength is determined, via unitarity, by the known[13] energy dependence of the $\pi^-p \to \eta n$ cross section at threshold. Fitting to the new data of Ref. 12 allows determination of the relative phase of the S11 singularity and the non-singular part of the backward non-flip amplitude. The S11 wave of Ayed et al. is consistent with the presence of an ηn cusp, but does not contain it explicitly. Chao and Bhandari[14] have made a fit to the S11 partial wave of Ayed et al.

with a parametrization which does contain the cusp explicitly, with strength and phase determined as described above, and also includes the S11(1535) resonance just above the cusp. The result is shown in Figure 5.

Besides the intrinsic interest of the ηn threshold, it is also a potentially important tool for the analysis of other effects in the second resonance region. By measuring the threshold energy dependence of the π⁻p cross section at a series of fixed angles and fitting the resulting data with an interfering threshold singularity one obtains a measure of the angular variation of the *overall phase* of the spin-non-flip amplitude at 1488 MeV. This would provide a "bench mark" which could be very useful in resolving partial wave ambiguities in the 1500 MeV region, such as the poor determination of the P11 wave. Polarization measurements of the requisite accuracy would be more difficult, but they would give additional information on the overall phase of the spin-flip amplitude.

REFERENCES

† This work was supported by the U.S. Energy Research and Development Administration.

1. Particle Data Group, Rev. Mod. Phys. 48, S1 (1976).
2. J.R. Carter et al., Nucl. Phys. B58, 378 (1973).
3. R. Ayed and P. Bareyre, private communication to the Particle Data Group, 1974.
4. P.Y. Bertin et al., preprint, 1975 and Paper I.1 submitted to this conference.
5. C. Amsler et al., Phys. Lett. 57B, 289 (1975).
6. J.C. Comiso et al., Phys. Rev. D 12, 738 (1975).
7. D.V. Bugg et al., Phys. Lett. 44B, 278 (1973).
8. J. Duclos et al., Phys. Lett. 43B, 245 (1973).
9. J.S. Ball and R.L. Goble, Phys. Rev. D 11, 1171 (1975).
10. D.B. Lichtenberg, Lett. Nuovo Cimento 12, 616 (1975).
11. S.S. Vasan, to be published in Nucl. Phys., 1976.
12. N.C. Debenham et al., Phys. Rev. D 12, 2545 (1975).
13. D.M. Binnie et al., Phys. Rev. D 8, 2789 (1973).
14. Y.-A. Chao and R. Bhandari, Carnegie-Mellon University preprint COO-3066-64, February 1976 and R. Bhandari, private communication.

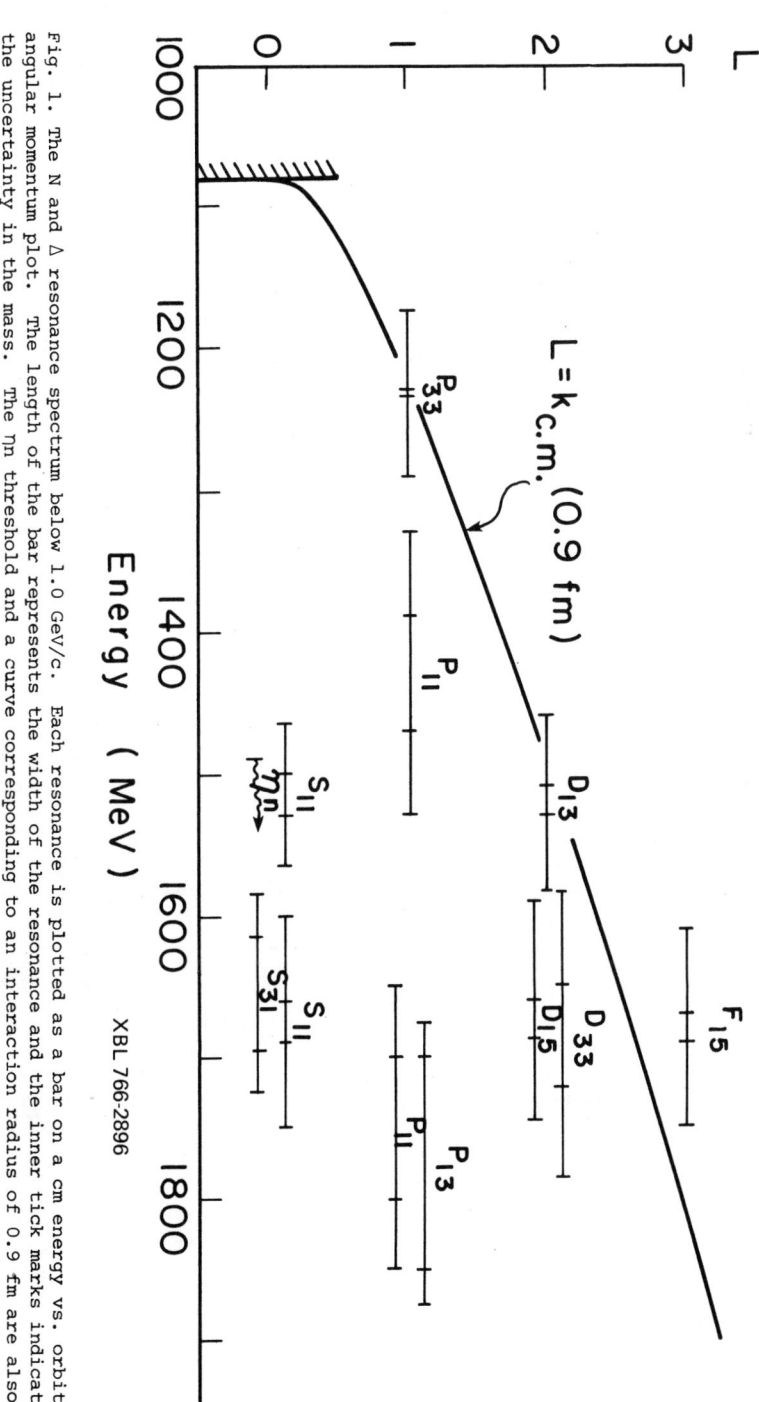

Fig. 1. The N and Δ resonance spectrum below 1.0 GeV/c. Each resonance is plotted as a bar on a cm energy vs. orbital angular momentum plot. The length of the bar represents the width of the resonance and the inner tick marks indicate the uncertainty in the mass. The ηn threshold and a curve corresponding to an interaction radius of 0.9 fm are also shown. The resonance information is taken from the 1976 Review of Particle Properties.

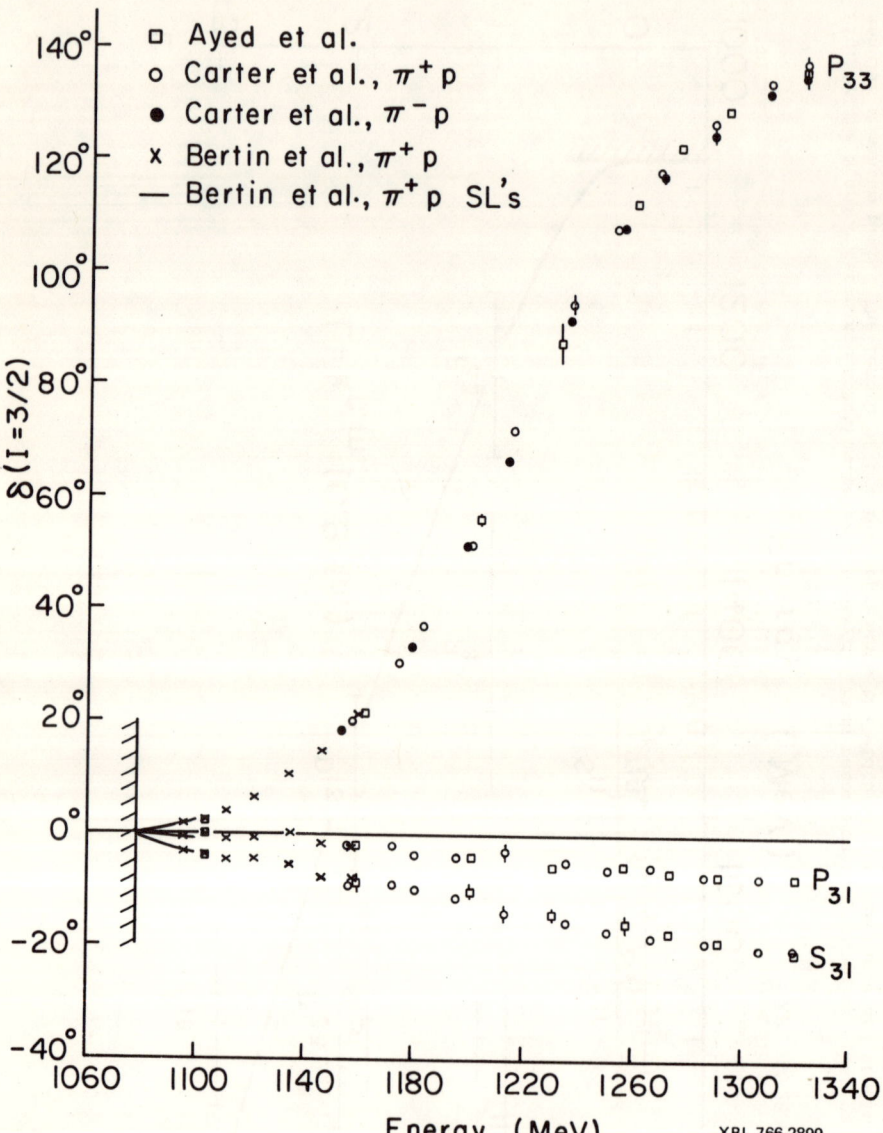

Fig. 2. I=3/2 phase shifts from the analyses of Refs. 2-4. Both sets of P33 phase shifts (corresponding to Δ^{++} and Δ^0) determined by Carter et al. are shown, and the threshold behavior corresponding to the scattering lengths of Bertin et al. is shown.

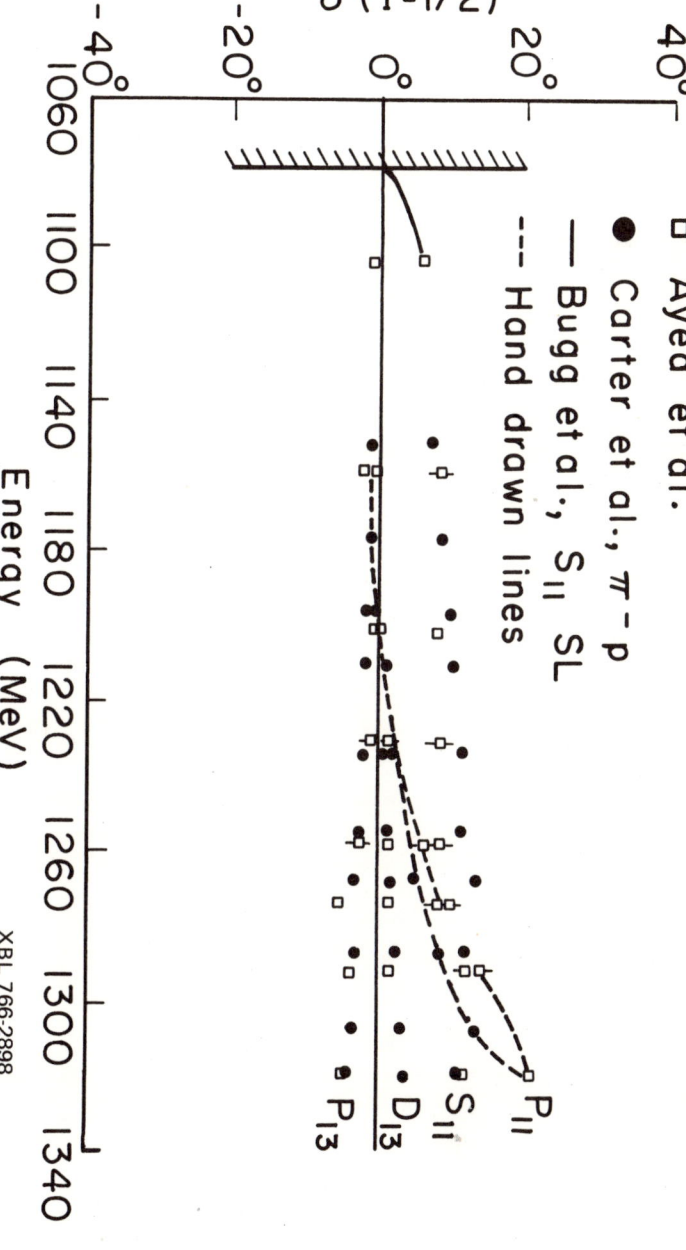

Fig. 3. I=1/2 phase shifts from the analyses of Refs. 2 and 3. The hand drawn lines trace out the P11 phases of these analyses. The threshold behavior of the S11 wave corresponding to the scattering length of Bugg et al.[7] is shown.

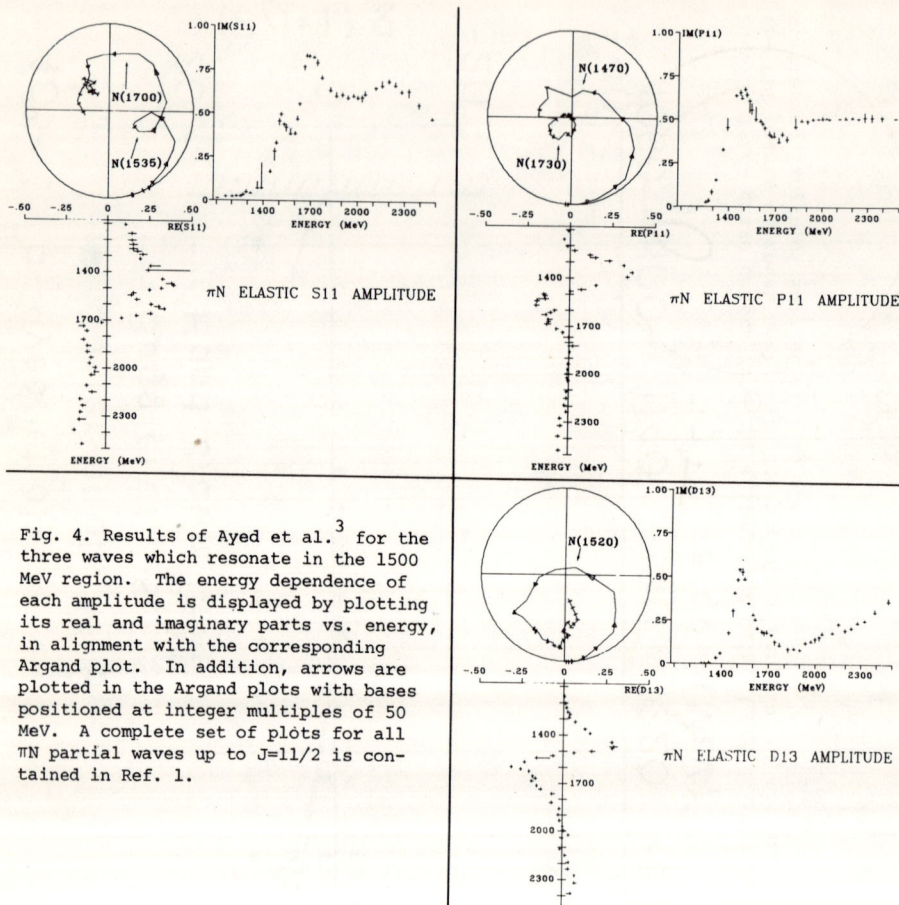

Fig. 4. Results of Ayed et al.[3] for the three waves which resonate in the 1500 MeV region. The energy dependence of each amplitude is displayed by plotting its real and imaginary parts vs. energy, in alignment with the corresponding Argand plot. In addition, arrows are plotted in the Argand plots with bases positioned at integer multiples of 50 MeV. A complete set of plots for all πN partial waves up to J=11/2 is contained in Ref. 1.

XBL 765-1955

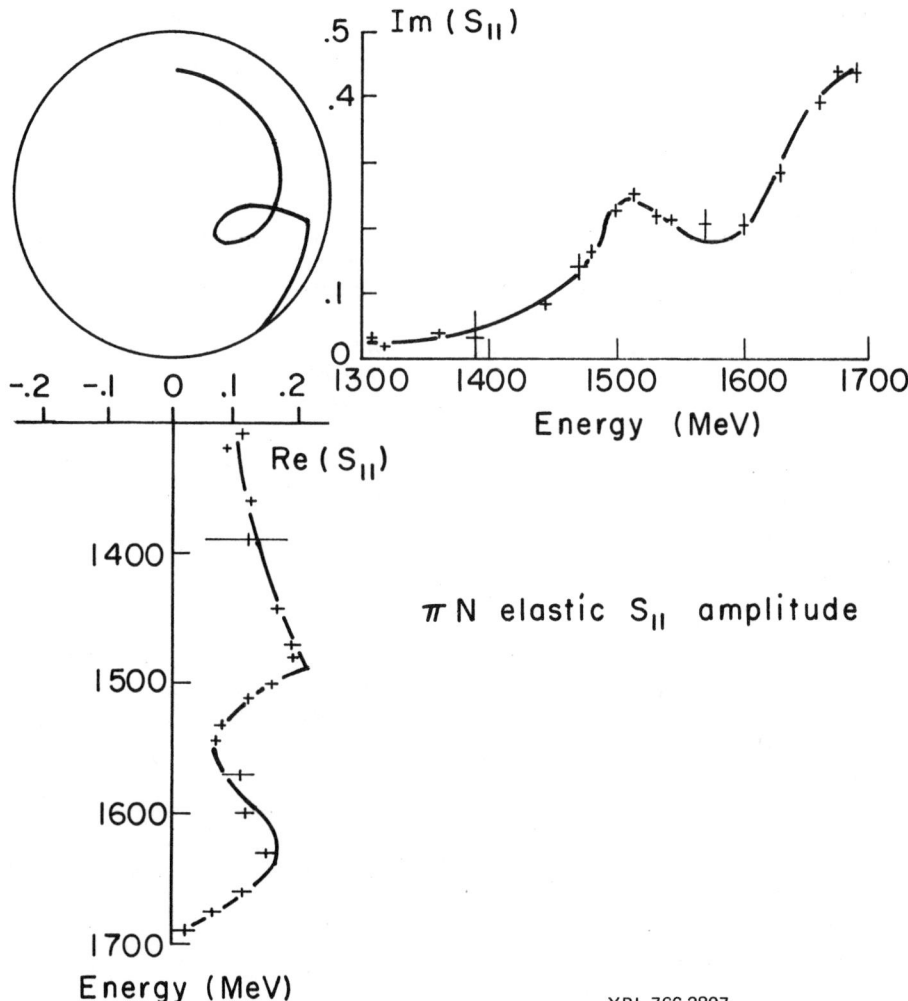

Fig. 5. Results of Chao and Bhandari[14] for the S11 wave in the ηn cusp region. The result of their fit is shown superimposed on the amplitudes of Ayed et al.[3]. The real and imaginary parts are shown in alignment with the Argand plot as in Fig. 4.

B. M. K. Nefkens (UCLA): Two comments: a) What is the status of the 180° phase degeneracy in πN scattering suggested several years ago by Ayed, Bareyere and Sonderegger?

I have calculated the $\pi^-p \to \pi^-p$ differential cross section from the phase shifts of Carter et al. and found that they disagree at T_π = 200 MeV with the experimental data by Bugg, Carter et al. but agree with the older Berkeley data at 310 MeV. Why this preference for the older experimental data?

Kelley: a) There is a new backscattering experiment (Debenham et al.) which may shed light on this question, but I don't think there has been any new analysis of the phase degeneracy since the data appeared. However, they don't find any evidence for violation of I-spin bounds. b) I would like to look at your fits before making a detailed comment, but let me say that there are difficulties with the Carter data at their highest energies, particularly in the I=1/2 wave. There are disagreements between Carter et al. and Ayed et al. at the high energies.

M. K. Banerjee (U. of Maryland): What are the latest numbers for the s wave scattering lengths?

M. Olsson: The most reliable values of the crossing even and odd s-wave πN scattering lengths found by dispersion relations are probably those of D. V. Bugg et al., Phys. Letters 44B, 278 (1973) who give $a_0^{(+)}$ = -0.005±0.002 and $a_0^{(-)}$ = 0.087±0.002. In this conference P. Bertin et al. have found the isospin 3/2 s-wave scattering length to be a_3 = -0.090±.002 by direct extrapolation to threshold which is consistent with the dispersive value of $a_3 = a_0^{(+)} - a_0^{(-)}$ = -0.092±.003.

THE ROLE OF THE Δ(3,3) IN LOW ENERGY πN ELASTIC AND PHOTOPRODUCTION MODELS

M. G. Olsson*
University of Wisconsin, Madison, Wi. 53706

ABSTRACT

A model using particle poles and current algebra can account for all the photoproduction multipoles and elastic πN phase shifts in the Δ(3,3) region. The approximate nature of the Δ(3,3) as a stable particle pole is discussed.

INTRODUCTION

There now exist accurate determinations of the low energy photoproduction (γN→πN) multipoles[1] as well as much information on elastic (πN→πN) scattering. The πN scattering phase shifts are known over a wide range of energies allowing dispersive calculation of many threshold and sub-threshold amplitude parameters.[2] This wealth of experimental information makes possible detailed comparisons with theoretical models. For πN scattering the kinematic region we would like to understand is shown in Fig. 1.

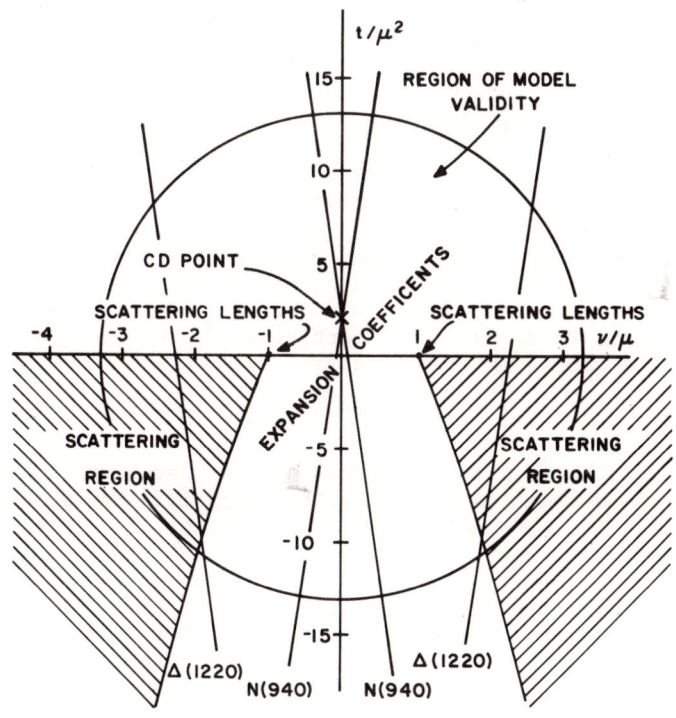

Fig. 1. Kinematic region for low energy πN scattering.

In this low energy regime the scattering amplitudes[3,4] can be considered a superposition of i) Born and current algebra terms ii) diffraction and resonance contributions. The Born and current algebra terms are well known[3,4] functions of ν and t with the exception of the σ-term which is relatively small and a function of momentum transfer t alone. The diffractive component is the analytic reflection of high energy diffraction and appears only in the $B^{(+)}$ amplitude in πN scattering. Resonance effects are dominated by the $\Delta(3,3)$ and appear in virtually every photoproduction multipole and elastic phase shift. Thus the way in which the Δ is described is extremely important. As we shall see a useful approximation for the Δ neglects its decay width into $N\pi$.

THE $\Delta(3,3)$ AS A STABLE PARTICLE

The simplest possible treatment of the Δ is to assume it is stable like the nucleon. The scattering amplitudes are then easily computed using second order perturbation theory. Because the spin 3/2 propagator is not unique and there are two types of coupling for the $\Delta N\pi$ (and $\Delta N\gamma$) vertices, the Δ contribution depends on two parameters. The Δ pole term is unique and is measured by the coupling constant g_Δ. The non-pole Δ part is due to derivative coupling of the pion and to a smaller extent it depends on a parameter Z cojointly measuring the admixture of off mass shell spin 1/2 components and the coupling properties.

As long as one avoids the immediate vicinity of the resonance pole in a resonant channel this model will provide much useful information. By projecting out the various multipoles and phase shifts we find good agreement with all of the non-resonant data in the $\Delta(3,3)$ region. The model is assumed to yield the real part of the amplitude above and below threshold.

Recently a Saclay group[5] has measured π^+p scattering very close to threshold. With the model parameters fixed by the older data[4] we compare our solution with the new S_{31} and P_{31} phase shifts in Fig. 2. It is interesting to note that the pole model also does very well for the resonant P_{33} partial wave even though this wave was not used in the original analysis.

THE $\Delta(3,3)$ PARAMETERS

There are many ways of expressing the Δ resonance mass, the most common of which is to quote the energy 1232 MeV where the (3,3) phase shift passes through 90°. This can be misleading however since a non-resonant background can shift[7] this energy changing its interpretation. A second "resonance energy" of 1212 MeV corresponds to the maximum speed point on the Argand circle. The real part of the complex pole position at 1211 MeV can also be used to characterize the resonance. Finally the low energy tail can be approximated by a stable particle pole and extrapolated to

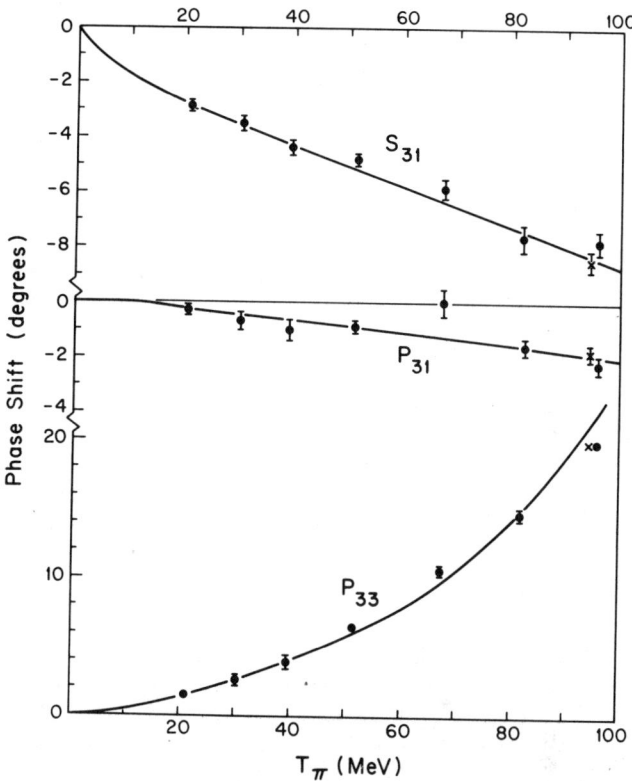

Fig. 2. Saclay isospin 3/2 πN phase shifts.

the resonance mass. Using this method Höhler et al.[6] found

$$M_\Delta = 1220 \text{ MeV} ; \quad \frac{g_\Delta^2}{4\pi} = .27 . \qquad (1)$$

Our analysis[4] is consistent with these values and the overall fit also finds that,

$$Z = -0.45 \pm 0.20 . \qquad (2)$$

The resonant photoproduction multipoles[7] M_{1+} (magnetic dipole) and E_{1+} (electric quadrupole) can be expressed in terms of the experimental (3,3) phase shift and the theoretical backgrounds by a form of Watson's theorem. The agreement between the predicted and observed resonant multipoles is excellent.

The problem remaining is to calculate the (3,3) phase shift in the vicinity of the Δ resonance using the same parameters as for the other partial waves. Since the model consists of a pole on the real energy axis plus background a width must be calculated as well as ensure that the resonance plus background is unitary. A standard reaction formalism yields,

$$\delta = \delta_r + \delta_b$$

$$\cot \delta_r = \frac{-E + M_\Delta(E)}{\Gamma(E)/2} \tag{3}$$

where the background phase shift δ_b is small and positive coming principally from nucleon exchange. The resonance mass is energy dependent[8] and given by,

$$M_\Delta(E) = M_\Delta - \frac{\pi}{2} \int_0^\infty \frac{dk'\Gamma(E')}{E'-E} \tag{4}$$

In lowest order $\Gamma(E)$ is a known monotonically increasing function causing an infinite mass shift. Higher order diagrams cannot be evaluated from field theory because a spin 3/2 field especially with a derivative coupling interaction is non-renormalizable. A dispersive estimate however indicates how all Δ mass estimates can be viewed consistently. At low energies the integral will be positive and a small effective resonant mass (say 1220 MeV) will be obtained. At 1232 MeV $M_\Delta(E)$ will have increased to about 1245 MeV so that with the addition of the positive background phase shift, the total phase shift will pass through 90° at 1232 MeV. If the experimental phase shift points are satisfactorily fitted the complex pole position will be at 1212 MeV.

REFERENCES

*Supported in part by the University of Wisconsin Research Committee with funds granted by the Wisconsin Alumni Research Foundation, and in part by the Energy Research and Development Administration under contract E(11-1)-881, COO-553.

1. W. Pfeil and D. Schwela, Nucl. Phys. B45, 379 (1972); F. A. Berends and A. Donnachie, Nucl. Phys. B84, 342 (1974).
2. H. Nielsen and G. C. Oades, Nucl. Phys. B72, 310 (1974).
3. M. G. Olsson and E. T. Osypowski, Nucl. Phys. B87, 399 (1975).
4. M. G. Olsson and E. T. Osypowski, Nucl. Phys. B101, 136 (1975).
5. P. Y. Bertin et al., Nucl. Phys. (to be published).
6. G. Höhler, H. P. Jakob and R. Straus, Nucl. Phys. B39, 237 (1972).
7. M. G. Olsson, Phys. Rev. D (1976) (to be published).
8. L. Stodolsky, Phys. Rev. D1, 2683 (1970).

MEASUREMENT OF THE π^+ AND π^- TOTAL CROSS SECTIONS ON HYDROGEN AND DEUTERIUM FOR PION ENERGIES FROM 50 TO 300 MeV.

E. Pedroni, K. Gabathuler, J. Arvieux, P. Corfu, J. Domingo
P. Gretillat, W. Hirt, Q. Ingram, J.P. Piffaretti, P. Schwaller, N. Tanner

E.T.H., S.I.N., Neuchâtel, Oxford, Karlsruhe, Grenoble.

ABSTRACT

A precision measurement of the $\pi^\pm$ Total Cross Sections on Hydrogen and Deuterium by the classical transmission method is described. The experimental arrangement and analysis procedures are explained and preliminary results presented.

This paper will be in the nature of a progress report on the total cross section measurements we are presently carrying out at SIN. We have had one four week running period in Nov - Dec 1975, and we will have another period in early June. We hope this second period will be sufficient to complete the data accumulation phase of the experiment. Since we are attempting to achieve a 1 % or better accuracy, the analysis phase takes much longer than the data accumulation phase, and we are only mid-way through this process with our December data. Thus I will not be able to present cross section values, but will only try to explain our experimental and analysis procedures.

The goal of the experiment is an accurate comparison of the total cross sections for π^+ and π^- on deuterium. Previous measurements[1] of the $\pi^\pm$ total cross sections on self conjugate nuclei have revealed quite large differences between the π^+ and π^- cross sections for low pion energies (ca. 100 MeV). Coulomb nuclear interference effects make it extremely difficult to extract any information on a possible violation of charge symmetry from the observed differences. Deuterium is the only self conjugate nucleus for which one has any hope of being able to extract such information. The hydrogen total cross sections are measured simultaneously as an experimental control. Since the π nuclear amplitudes are the main ingredient of any theoretical calculation of the deuterium total cross sections, it is important to confirm that the hydrogen total cross sections calculated from these amplitudes agree with our measurements. The hydrogen measurements are also used to correct for the 1 % hydrogen contamination of our deuterium. In addition we hope to further check Bugg's[2] observation of a width difference between Δ^{++} and Δ^0.

Figure 1 shows the experimental arrangement. The pion beam is produced in a 9 mm thick Be target disk oriented along our beam line. The beam transport system is a basically symmetric double QQ D QQ system, having a dispersion at the intermediate focal plane of 7 cm per 1 % $\frac{\Delta p}{p}$. The momentum acceptance of our beam is defined by the scintillation counter S_1 to be approximately 1 % F.W.H.M. The dispersed beam is recombined by the second QQ D QQ system and the last two quadrupoles are adjusted to produce a waist at the scattering target. The use of a double bend transport

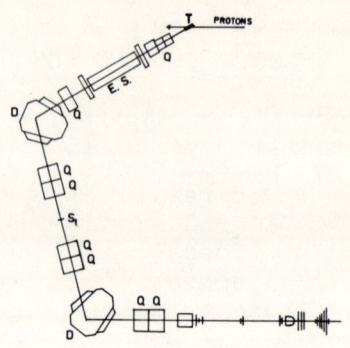

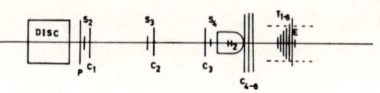

Fig. 1

system, and a 2 metre electrostatic separator (E.S.) effectively, removes the proton contamination at the target; pulse height proton rejection is also used on S_1 and S_2. The beam acceptance at the target is defined by a triple counter telescope (S_2, S_3, and S_4) consisting of circular scintillation counters having diameters of 50, 50, and 30 mm respectively and separated by approximately 1 meter. A DISC Cherenkov counter was placed approximately 2.5 meters upstream of the target and was used to electronically reject the electron, muon, and proton contamination in the beam. The DISC had a 3 cm thick radiator cell which was filled with a mixture of glycerine and water having the appropriate index of refraction for the desired pion momentum. For momenta below 190 MeV/c the DISC could not be used but here time of flight between counters S_1 and S_4 was sufficient to reject particles other than pions. Thus for momenta $\geq$ 190 MeV/c the master trigger defining an incident pion was S_1 $\bar{S}_{1p}$ S_2 $\bar{S}_{2p}$ D S_3 S_4 and for momenta $\leq$ 190 MeV/c S_1 $\bar{S}_{1p}$ S_2 $\bar{S}_{2p}$ S_3 S_4.

The total cross sections are measured by the classical transmission technique, however, we have replaced the usual scintillation transmission counters by three 40 x 40 cm multi wire proportional chambers (C4 - 6). These chambers have 2 mm wire spacing and are read out with a delay line scheme utilizing commercial delay lines having a 5 nsec delay between wire taps. We chose delay lines rather than digital read out for reasons of economy and speed of construction. The beam defining scintillation telescope is also supplemented by three 10 × 10 cm wire chambers (C1 - 3) having 1 mm wire spacing. The use of wire chambers rather than scintillation transmission counters not only allows us to obtain much better angular resolution but greatly reduces the problems caused by beam particles interacting with the detector material. The improved resolution of the beam defining wire chambers C1 - 3 allows us to make more accurate corrections for π decay and, together with C4 - 6, allows us to make an approximate determination of the interaction point in the target cell. The transmission counter assembly (T_{1-6} and E) were used in the early stages of the experiment to check the wire chamber system; we no longer use then for cross section measurements since the chambers are much superior.

There are however two problems created by the use of the wire chamber system: Firstly, the rate of data accumulation is dictated by the very long time required by the computer to read and store an event; this is essentially determined by the number of words per event and the speed of the tape unit. Secondly, because of the long readout time of our chamber delay lines (1 μsec) the rate of incident particles must be kept quite low (< 1000/sec) in order to obtain a low probability of having two events in the chamber during readout. The long time required for event storage requires us to limit readout to only those events which will be useful. In particular, since the typical target absorbtion is only 5 - 10 %, the majority of events are those where no detectable scattering has occurred. To avoid wasting time reading these events we place a small scintillation counter (usually T_1) behind the chambers at a distance such that it just covers the envelope of the "unscattered" beam. This counter is then used as a veto in the computer readout command and such "unscattered" events are simply recorded in a scalar for normalization purposes. In order to make certain that we aren't biasing our data by this procedure we also read and record a small fraction (typically 1 %) of these unscattered events. In order to avoid problems due to the long memory time of our delay lines we have placed the large scintillation counter P immediately in front of S_2. The size of P is such that any beam particle which is incident on the large chambers must pass through P. The signal from this counter is used to inhibit analysis of an event if there was another particle passing through P within a 3 μsec interval centered about the event. Since this event rejection is made on the incident beam it produces no bias in our data. By means of extra TDC channels and a routing unit on the delay line signals we are able to measure true double events in the large chambers, that is events where a single incident particle results in two charged particles passing through the large chambers.

Since the objective of the measurement is an accurate comparison of the total cross sections for the two charged pions it is crucial that the momenta of the two beams are identical. In switching beam polarities the currents in all quadrupoles were simply reversed, but the currents in the two bending magnets were adjusted to give the same magnetic field as determined by precision NMR probes. Since it was impossible to reverse the electrostatic field in the separator, we investigated possible momenta shifts due to a mistuning of the separator cross-field magnets; we found no detectable effect. As an additional control on the equality of the π^+ and π^- momenta, we recorded the total time of flight between the pion production target T and the last beam defining counter S_4 for each good event. A pick up electrode placed immediately before the production target produces a start signal when the proton burst passes through it, and the stop signal is produced by S_4 gated by the full pion identification electronics. The measured width of the pion flight time peak was approximately 1.2 nsec (F.W.H.M.). This narrow width together with a flight path of 23 meters and a TDC resolution of 100 psec per channel allows us to easily detect a momentum shift of 0.4 MeV/c at 200 MeV/c.

Figure 2.

View of Target Wagon Assembly. The targets are seen from the upstream position. The cells are from left to right: deuterium, dummy, and hydrogen. The long vertical cylinders on the deuterium and hydrogen cells contain the refrigeration expansion engines.

The complicated plumbing system is dictated by the automatic safety system. One of the 40 × 40 cm M.W.P.C. used to detect the scattered beam can be seen on the right of the hydrogen target.

The target system shown in Fig. 2 consists of two identical cryogenic vessels, each containing a large volume liquid target cell. The cells consist of a spun aluminium cylindrical body of 15 cm diameter with an integral domed exit surface and a mylar entrance window. The aluminium exit dome has a thickness of 0.5 mm and the mylar entrance window of 125 μ. The mylar entrance window is mounted on a flanged plug by means of which the effective length of the target liquid can be varied from 30 to 10 cm. Our present data was taken with a target length of 20 cm. The vacuum vessel has a 24 cm diameter 250 μ Mylar entrance window and a 40 cm diameter 375 μ Mylar exit window. In addition to the two active cryogenic targets there is also a mechanically identical dummy target used to determine the background. The cryogenic targets are equipped with independent closed cycle refrigeration systems capable of supplying 10 watts of refrigeration power at $20^{\circ}K$ simultaneously with approximately 40 watts at $70^{\circ}K$. The 20° stage is equipped with a large condensing surface in direct communication with the target cell. By admitting the appropriate gas to the condensing surface approximately 1/2 liter of liquid hydrogen or deuterium can be produced per hour. Once a system of semiconductor level monitors indicates that the target cells are full the gas input is stopped and a heating current is passed through a resistance mounted on the $20^{\circ}K$ cooling surface. A regulation system measures the liquid temperature by means of a thermistor probe and controls the current supplied to the heater in order to keep the liquid temperature constant at some preset value. The stability of this system is typically better than ± $0.02^{\circ}K$. The absolute temperature of the liquid is measured with a closed bulb vapour pressure thermometer placed in the liquid. The wagon containing the deuterium, hydrogen, and dummy target is mounted on a rail system so that by rolling the appropriate target into the beam, we can make a rapid comparison of the target transmission.

Data collection consisted of recording the xy coordinates in the 6 wire chambers and the pion time of flight for each scattered event (those not passing through the T_1 veto counter) plus 1 % of the unscattered events. After the desired number of good events had been collected (typically 100 K per momentum point and charge) several scalars containing beam and normalization information were also read onto the tape. For a given pion charge and momenta the transmission was determined in rapid succession for the hydrogen, deuterium, and dummy targets. Typical data collection time for one target was 15 to 20 minutes. After data for one pion charge had been accumulated the magnetic fields were reversed and data was taken for the other charge. Because of the long run down time required to avoid hysteresis effects in the bending magnets, the polarity switching procedure took approximately one hour. To check for possible momentum shifts caused by hysteresis, we usually repeated a deuterium run after switching back to the original polarity; these runs always agreed perfectly within the statistical error. During data acquisition, a rough on-line analysis of the events was carried out so that we could monitor the correct operation of the system. Data was taken at intervals of 15 MeV/c from 145 MeV/c to 450 MeV/c. The lower limit was dictated by the very low π^- flux at this momentum and the difficulty of correcting the data for the π decay between S_4 and C_4. The higher limit was determined by the bending magnet power supplies, but these have been modified and we will extend our measurements to at least 500 MeV/c in our next run.

In principle a single transmission wire chamber would suffice, however the use of 3 such chambers enables us to increase the overall efficiency of the detection system, improve the angular resolution, and map the efficiency of each chamber. In addition for those events in which at least two chambers before and after the scattering recorded tracks it is possible to determine the approximate interaction point in the target volume. Figure 3 shows a projection onto the beam axis of such an interaction point determination for the hydrogen and dummy targets. In the plot for the dummy one can easily recognize the peaks due to scattering from the mylar entrance window of the target cell, the aluminium dome, the

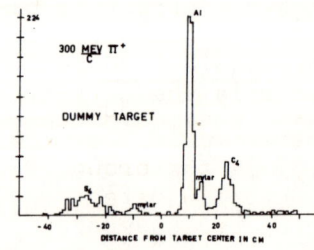

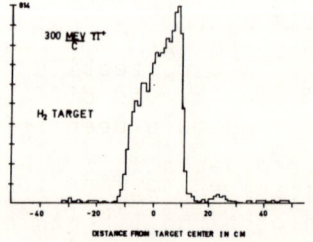

Fig. 3

mylar exit window of the vacuum can, and the wire chamber C_4. The broad bump labeled S_4 is probably a combination of C_3, S_4, and the mylar entrance window on the vacuum can since for large negative displacements the error in the position determination becomes quite large.

Such plots illustrate one of the problems of using wire chambers instead of scintillation transmission counters, namely one has much too much information and has to start worrying about how to use it. For example in the classical transmission counter experiment one essentially extrapolates the ratio of the full to empty target transmission, correcting the empty target measurements for the mean energy loss to the target center. But the dummy spectrum of Fig. 4 shows us that this mean energy correction really makes no sense. The target vessel background comes from regions either before or after the hydrogen, and thus we should apply two different energy corrections. Only in the case where the energy and angular behaviour of the material before and after the hydrogen region are identical does the mean energy correction make sense; this is clearly not the case for mylar and aluminium. Thus in our next run we shall investigate the background due to S_4 and the entrance window separately from that due to the aluminium done and the exit window. In addition it is clear that the background due to material upstream of the hydrogen will be modified due to multiple scattering in the full target.

From the raw data one forms the partial transmission ratio:
$$R(\Omega) = \frac{T_{full}(\Omega)}{T_{dummy}(\Omega)} = e^{-\sigma(\Omega)n}$$
where:
$T(\Omega)$ is the proportion of master triggers which lead to a particle detected within the solid angle Ω; $\sigma(\Omega)$ is the total cross section per target nucleus for an incoming pion to lead to no charged particle within Ω; n is the number of target nuclei per cm^2.

The corrections due to detector efficiency, π decay, and background differences between target full and empty (dummy) have been omitted here for simplicity:

Then to find the true total cross section we must extrapolate the relation:
$$\sigma(\Omega) = -\frac{\log R(\Omega)}{n} \qquad \text{to } \Omega = 0$$

Of course the cross sections we would obtain by this extrapolation would be infinite since they contain the Coulomb terms. Thus we must subtract the pure Coulomb and Coulomb-nuclear interference term before we can obtain a useful result. It is this Coulomb nuclear interference correction which is the main source of difficulty in total cross section measurements. In the case of hydrogen we can calculate the interference term from the known π proton amplitudes, and Fig. 4 shows a plot of - log R(Ω) corrected for the Coulomb effects.

In addition to make the curves relatively flat to aid the extrapolation procedure the calculated elastic scattering has been subtracted; the calculated elastic total cross section has of course been re-added to the ordinates. The ordinates are labed as percentage of absorption for conceptual simplicity, but they are actually - log R. One notes that both curves are relatively flat for a large angular range but begin to shoot up for small angles.

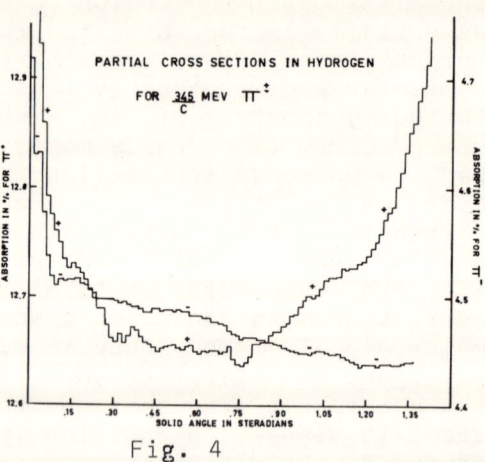

Fig. 4

This divergence almost certainly reflects inadequancies in the treatment of the Coulomb interference effects; however there are many other corrections which can become important at small angles. For example, the correction for π decay between S_4 and C_{4-6} which has not yet been applied is important for small angles; similarly the slight change in the target cell background due to energy loss and multiple scattering in the target liquid will make a contribution at small angles. The curve for π^- is very smooth reflecting the fact that approximately 2/3 of the total cross section is due to the charge exchange reaction $\pi^- + p \rightarrow \pi^0 + n$, and the final neutrals are ideal for a method of measurement essentially based on the simplistic picture that once a particle interacts it is lost from the beam. The rise at large angles for the π^+ curve is due to the falling off of the chamber acceptance for solid angles greater than .8 sterradians, we have not yet had time to calculate the correction for this effect. The fact that the curve increases rather than decreases is due to the process of subtracting the calculated elastic scattering.

Two things are apparent from Figure 4: Firstly we are presently at the point of being able to confidently extract total cross sections at a level of about 1 % accuracy. Secondly there are a lot of small bumps and shoulders especially in the π^+ curve which we do not yet understand, and we have a lot more work to do before we can confidently push the extrapolation error significantly below 1 %. Again the chambers have given us too much data, we can see many effects which would be concealed with the limited angular resolution of transmission counters. We are now in the process of trying to understand the fine details of such curves, and at present we have, perhaps too pessimistically, arrived at the point where we don't believe any one's total cross section measurements to an accuracy better than 1 %.

[1] C. Wilkin et al.; Nucl. Phys. B62 (1973) 61.
[2] J.R. Carter, D.V. Bugg, and A.A. Carter; Nucl. Phys. B58 (1973) 378.

Alan Schiz (Yale University): I would like to comment on some data from Fermilab taken by a Yale Fermilab collaboration. We have measured small angle hadron nuclei scattering at the following energies: at 175 GeV/c and 70 GeV/c $\pi^\pm, k^\pm, p, \bar{p}$ on H, Be, C, Al, Cu, Sn, Pb and at 125 GeV/c π^+, k^+, p on H, Be, Al and Pb. In all cases we measure from a minimum momentum transfer squared, t, of -.001 $(GeV/c)^2$. The maximum t varies with energy being $\sim -.08$ $(GeV/c)^2$ at 70 GeV/c, $\sim -.25$ $(GeV/c)^2$ at 125 GeV/c, and $\sim -.5$ $(GeV/c)^2$ at 175 GeV/c. We hope to extract elastic differential cross sections, total cross sections, and perhaps leading particle cross sections. We would appreciate any communications from interested parties.

PION-PROTON SCATTERING BELOW 100 MeV

P. Bertin, B. Coupat, J. Duclos, A. Gérard, A. Hivernat, D. Isabelle,
J. Miller, J. Morgenstern, J. Picard, R. Powers*, P. Vernin
(CEN Saclay and Université de Clermont-Ferrand, France)

Using the low-energy pion channel of the Saclay Linac, we made a systematic measurement of the differential cross-section of π^+ and π^- elastic scattering on hydrogen, from 20.8 to 95.9 MeV, in the angular range 60° – 145° (c.m.).

We present here the final results of our π^+ data analysis. The π^+p differential cross sections are plotted on Figs. 1 and 2. The results obtained at 94.5 MeV by P.J. Bussey et al.[1] are also plotted on Fig. 1 and show a very good agreement with ours.

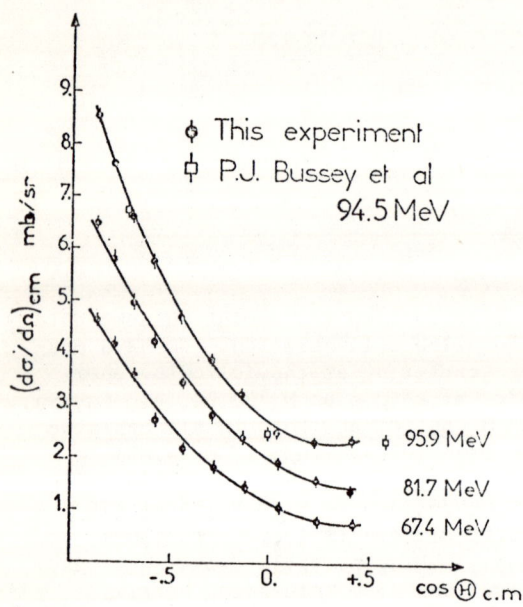

Fig. 1
π^+p differential cross-section at 67.4, 81.7, 95.9 MeV.

We extracted at each energy a set of phase shifts δ_3, δ_{31}, δ_{33}, corrected from Coulomb effects. The results of this calculation are plotted on Fig. 3 and compared to a CERN theoretical fit.[2]

The knowledge of these phase shifts allows a determination of the scattering lengths a_3, a_{31}, a_{33}, more direct than from higher energy measurements. Using our data only, we find : $a_3 = -0.090 \pm 0.002$; $a_{31} = -0.044 \pm 0.006$; $a_{33} = 0.219 \pm 0.004$. These values are consistent with the present most credible values.

* California Institute of Technology, Pasadena.

It must be emphasized that a new analysis of all the pion nucleon data below 300 MeV including ours and using energy-dependent models, has now to be performed; (for example, some interesting information may be obtained by expanding the invariant scattering amplitudes about the unphysical point ($\nu = 0$, $t = 0$) [3,4]).

Such an analysis will surely provide improved values of the fundamental constants of the pion nucleon interaction.

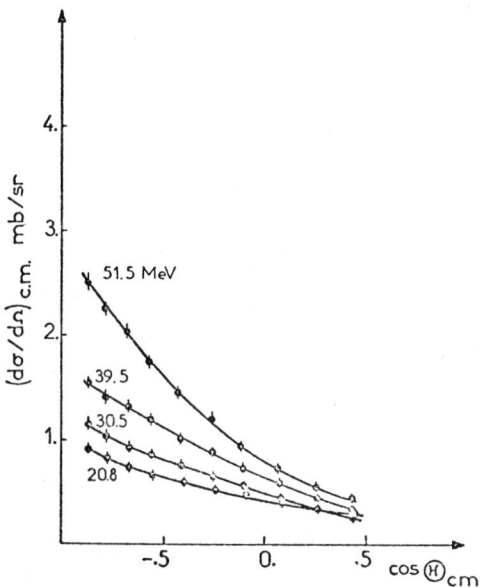

Fig. 2 : $\pi^+ p$ differential cross-section at 20.8, 30.5, 39.5, 51.5 MeV.

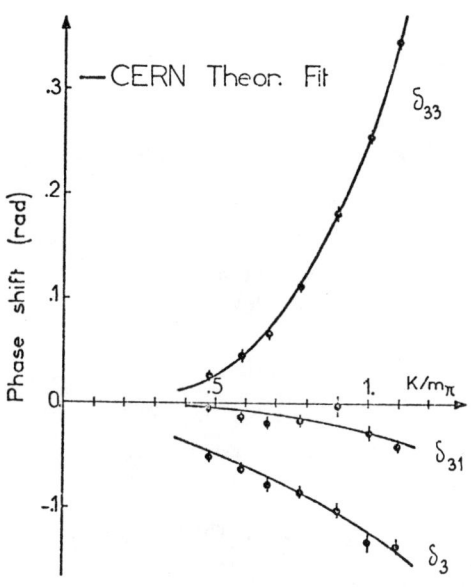

Fig. 3 : T = 3/2 phase shifts.

REFERENCES

1. P.J. Bussey *et al.*, Nucl. Phys. B58, 363 (1973)
2. Particle Data Group, UCRL 20030 πN (Feb.1970)
3. H. Nielsen and G.C. Oades, Nucl. Phys. B72, 310 (1974)
4. M.G. Olsson and E.T. Osypowski, Nucl. Phys. B101, 136 (1975).

DIFFERENTIAL CROSS SECTIONS FOR ELASTIC ($\pi^+ p$) AND ($\pi^- p$) SCATTERING BETWEEN 400 MeV/c AND 590 MeV/c

V.A.Gordeev, V.P.Koptev, S.P.Kruglov,
L.A.Kuzmin, A.A.Kulbardis, Y.A.Malov,
I.I.Strakovsky, G.V.Scherbakov

Leningrad Nuclear Physics Institute,
Gatchina, Leningrad District, 188350

Differential cross sections for ($\pi^+ p$) and ($\pi^- p$) elastic scattering have been measured at 8 momenta between 400 MeV/c and 590 MeV/c (Fig.1). The data were obtained in a wide angular range with the accuracy of (2÷5)%. The results for 404 MeV/c and 409.3 MeV/c are in good agreement with the data of Bussey et. al.[1]. It is shown that to describe (πp) scattering up to 600 MeV/c it is sufficient to take into account the (πp)-interaction in S, P and D-states. The comparison with the predictions of different phase-shift analyses allowed to favour that of CERN-EXP.

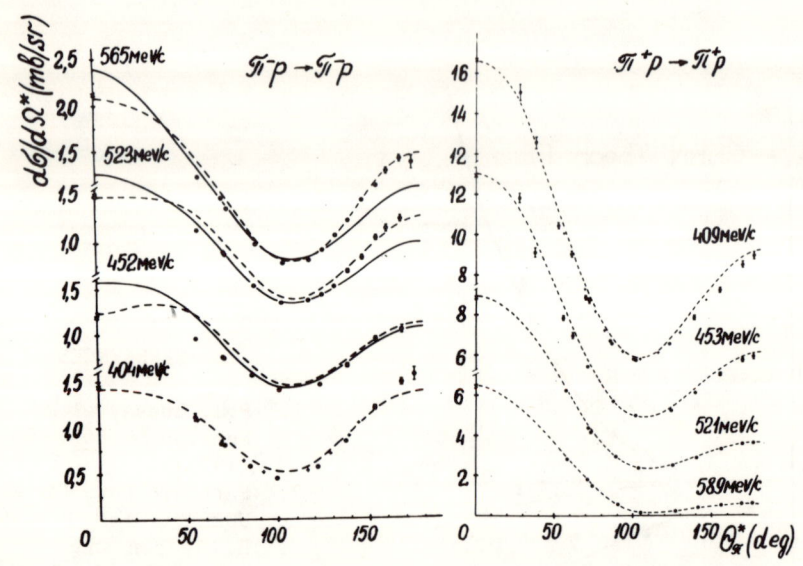

Fig.1. Differential cross sections for πp elastic scattering: ● our data; + data of Bussey et. al.[1]; dashed lines - from CERN-EXP; solid lines - from GLASGOW A.

The total cross sections for (π^+p) elastic scattering determined in this experiment are in good agreement with the data published earlier, whereas the total cross sections for (π^-p) elastic scattering are 10% less than previously published data.

Differential cross sections were extrapolated from our data to the angle 180° and are in good agreement with previously published (π^+p) data (Fig.2). What for (π^-p) backward elastic scattering our data disagree strongly with the measurements of Rothschild et. al.[2].

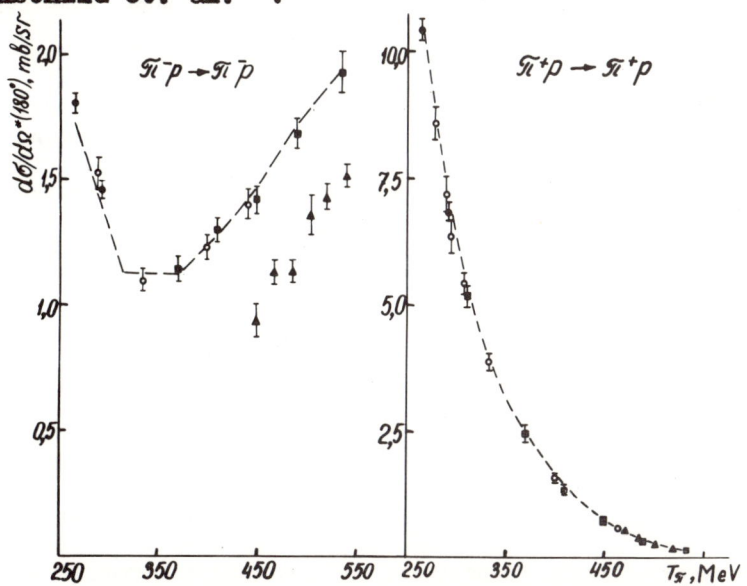

Fig.2. Differential cross sections for backward elastic scattering: ○ - extrapolated from our data, ● - from data of Bussey et. al.[1], ■ - from data of Ogden et. al.[3]; ▲ - data of Rothschild et. al.[2]; dashed lines - from CERN-EXP.

REFERENCES

1. P.J.Bussey, J.R.Carter, D.R.Dance, B.V.Bugg and A.A.Carter, Nucl. Phys. B58, 363 (1973).
2. R.E.Rothschild, T.Bowen, P.D.Caldwell, D.Davidson, E.W.Jenkins, R.M.Kalbach, D.V.Peterson and A.E.Pifer, Phys. Rev. D5, 499 (1972).
3. P.M.Ogden, D.E.Hagge, J.A.Helland, M.Banner, J.-F.Detoeuf and J.Teiger, Phys. Rev. B137, 1115 (1965).

PION-NUCLEUS ELASTIC SCATTERING

Frank Tabakin
University of Pittsburgh, Pittsburgh, Pa. 15260

INTRODUCTION

Pion-nucleus (πA) elastic scattering has been the subject of many recent studies.[1] In this talk, I would like to discuss some of those efforts and highlight some of the basic theoretical issues that need to be dealt with in preparation for the expected flood of precision data. I will focus on elastic scattering above 100 MeV; R. Eisenstein will discuss the lower energy region where some new data has become available.

The main issue is of course: what will we learn from pion-nucleus scattering? If the data is precise, I believe we will learn how to explain not only total, elastic, reaction and differential cross-sections, but also the real part of the foward scattering amplitude and various characteristic energies. For example, the characteristic energies for the total cross section peak value and width (which are not resonances since many πA partial waves contribute), the energy at which Re $f_{\pi A}(\Theta=0)=0$, and the energy where maximum diffraction occurs can serve as sensitive tests of theory--if accurately known.

The final theory might be a phenomenological one, as for many other reactions, but the present efforts are attempts to be fundamental. We seek suitable descriptions for analysis of data, from which physically meaningful quantities can be extracted. Can the theory of multiple scattering (MST) provide the insights needed for such physical description? Can optical potentials be deduced from πN (pion-nucleon) and nuclear input by simply applying a formulation developed for high energy protons? Will the many approximations needed in that case be applicable to a projectile that resonates with the constituent nucleons and can even be absorbed? The initial theoretical studies indicate that standard multiple scattering theory does encouragingly well in representing the known rough data. However, I believe the recent theoretical studies by many authors demonstrate that although a good starting point the conventional "static" multiple scattering formalism will not suffice. Refinements are demanded and are the key to learning from pions. Better dynamics is needed and is being developed.

Information about the πA dynamics is of interest not only in itself, but because it serves to describe the behavior of a pion near the nucleus, i.e., wave functions are obtained in addition to amplitudes. These wave functions (distorted waves) play an essential role in various reactions and inelastic processes. The distorted waves incorporate much information about the momentum distribution and motion of nucleons in nuclei, and perhaps also about special modes of nuclear excitation. The ability to elastically scatter both π^+ and π^- mesons from a variety of nuclei is a very special hold on nuclear matter. The shell model view of the nucleus, wherein meson degrees of freedom are suppressed, will be under close scrutiny.

In addition to providing information needed for analysis of reactions, the elastic process itself requires knowledge of such events as absorption, knockout, etc. We should not separate πA elastic scattering from the full (coupled) picture, but let us do it for now recognizing the need and some meritorious efforts to incorporate explicit absorption and inelastic effects into elastic optical potentials.[2]

MULTIPLE SCATTERING THEORY

Why are we encouraged to use multiple scattering theory? First the pion-nucleon range is small compared to the average internucleon separation, so a scattering from one nucleon can be essentially finished before the next collision begins (i.e., scattering centers exist). Therefore, the relative ranges are suggestive of multiple scattering. The pion-nucleon interaction is well determined, except at lower energies (<50 MeV). Indeed below the resonance it is weaker (and better known) that the two nucleon case. Recoil effects should be small and manageable since the pion has ∼1/7th the nucleon mass. These qualitative remarks provide motivation for trying multiple scattering theory. (I will not discuss Glauber theory approaches.)

Another reason for using MST is the success of simple optical potentials which were extracted from πN input, in incorporating the main features required by the known data. For example, the Kisslinger[3] optical potential $U = t\rho + (b_0+b_1\vec{k}'\cdot\vec{k})\rho(\vec{k}'-\vec{k}) \rightarrow (b_0\rho(r)+b_1\vec{\nabla}\rho\vec{\nabla})$ already incorporates: the requisite energy dependence from the 3-3 resonance, a surface localized absorption due to elastic pion-nucleon collisions and sizeable back angle scattering. Refinements, beyond the Sternheim-Auerbach[3] adjustments of b_0 and b_1 to fit data, are needed now only because we desire a full microscopic theory of precision comparable to the expected data, from which we can extract dependable distorted waves.

I shall now present a brief critique of conventional MST with emphasis on establishing its range of validity in the 3-3 resonance region. Several standard assumptions are called into question, namely the impulse, factorization and associated static approximations. Let us begin with a picture of the πA system. It is really a relativistic many-body problem. Virtual mesons (about 6) dance around each nucleon, virtual isobars now and then appear, and real isobars are produced by the πN collision. The incident pion is moving toward the nucleus with a speed of $v_\pi \sim .9c$, and a momentum $k_\pi \sim$ 300 MeV/c corresponding to an incident kinetic energy of $T_\pi = $ 190 MeV (near resonance). The bound nucleons are slower $v_N \sim .3c$, but their Fermi momentum is comparable to k_π, $P_N \sim P_F \sim$ 270 MeV/c. The πN width also gives a significant distance $c\hbar/\Gamma_{33} \simeq 1F$, which we will see later gives an energy-dependent nonlocal effect for the πA scattering.

In this picture the incident pion is seen to proceed from one nucleon to the next. If no nucleons overlap, either by being at rest or by hard core correlations, and the first πN collision is finished before the second is initiated, then only the pion-nucleon wave in the far zone is needed. Then the Bèg[4] theorem applies and only on-shell or amplitude (phase shift) information is required to specify the constituent dynamics. If there is some overlap albeit small, either from nucleon motion or πN range effects, then more πN dynamics is required. One way to construct such dynamics and the πN relative wave function is by introducing a potential. Inserting a pion-nucleon potential into a Schrodinger equation is one way of assuring reasonable πN unitarity, analyticity and range properties. However, the reality of pion absorption and the demands of crossing symmetry makes this step questionable. It is therefore a bold assumption to assume a potential and further to assume equality of the free and in-the-medium potential operators $v_{medium} \approx v_{free}$. Several authors have studied πN dynamics in a nuclear medium using index of refraction, Chew-Low or field theory approaches which account for the reality of the emission and absorption of mesons.[1] We will hear more about such approaches at this meeting. Indications are that for uniform nuclear matter, only small Pauli effects occur for <u>internal</u> nucleon lines. It is doubtful however that conclusions based on uniform infinite matter apply to a finite nucleus, surface-dominated reaction. [There are also some studies of the Pauli effect on external lines by Landau and McMillan, which is not the same question as modification of v by the medium.[1]] We also assume V_{NN} is unaffected by the intervention of the incident pion.

Accepting the potential concept with reservation permits us to carry out the standard multiple scattering derivations. The elastic cross section is obtained from a pion-nucleus T-matrix: $\sigma(\theta) \sim |<\vec{k}'o|T(E)|\vec{k}o>|^2$ where $|\vec{k}o>$ denotes a pion of momentum $\vec{k}$ and a nuclear ground state. The states $|\vec{k}n>$ are eigenstates of $H_0 = K_\pi + H_A$; K_π is the pion kinetic energy operator and H_A, the nuclear hamiltonian, includes nucleus recoil energy. The interaction between the pion and A- nucleons

$$V = \sum_1^A v_{\pi i}$$

can be eliminated in favor of a pion-nucleon t-matrix. The goal is to replace the many-body scattering problem by a two-body one. An optical potential is introduced and hopefully can be constructed from knowledge of the constituent πN dynamics. It is possible that <u>the resulting series in t-matrices might have a validity beyond its derivation from potentials</u>. For example, perhaps field theory results can be cast in the convential multiple scattering form to make helpful contact.

Before defining an optical potential, we can express T = V + VGT as:

$$T(E) = \sum_{i=1}^A T_i(E) \tag{1}$$

$$T_i(E) = \tau_i(E) + \tau_i(E)G(E) \sum_{i \neq j} T_i(E) \tag{2}$$

which yields a Watson multiple scattering series upon iteration. The bound-state collision operator $\tau_i(E) = v_i(E) + v_i G(E) \tau_i(E)$ represents a πN collision in the nuclear medium since the Greens function involves the nuclear Hamiltonian $G(E) = 1/(E-H_0 + i\epsilon)$. A description in terms of wave functions provides some additional insight.

$$\psi = |\vec{k}_0\rangle + G_0 \sum_{1}^{A} \tau_i \psi_i \qquad (3)$$

$$\psi_i = |\vec{k}_0\rangle + G_0 \sum_{j \neq i} \tau_j \psi_j \qquad (4)$$

Here we see that ψ_i acts as incident wave for the ith scattering and is generated from all other scattering events. Also τ_i yields the final (antisymmetrized) wave function by having one last typical collision $\psi = |\psi_i\rangle + G_0 \tau_i |\psi_i\rangle$.

Intermediate states of unphysical symmetry can appear in these steps, but in the final sum cancellations occur to yield proper indistinguishability of nucleons. Thus the convergence rate of the MS series depends on when antisymmetrization is treated. Kerman, McManus and Thaler[5] in their version of multiple scattering introduce antisymmetrization earlier. Such decisions change the organization and possibly the convergence of the associated series. The first task is to introduce the two-body (πN) operator into a multiple scattering series that hopefully is organized to converge.

The free πN collision operator $t_i(\omega) = v_i(\omega)[1+g_i(\omega)t_i(\omega)]$ for a pion striking a typical nucleon i, depends on a "starting energy" variable ω which appears in the two-body propagator $g_i(\omega) = (\omega - K_\pi - K_i)^{-1}$. Here K_π, K_i are pion and nucleon kinetic energy operators. The bound state τ-matrix can be related to the above free t-matrix using the important relation

$$\tau_i(E) = t_i(\omega) + t_i(\omega) \{G(E) - g(\omega)\} \tau_i(E) \qquad (5)$$

$$\xrightarrow[EIA]{} t_i(\omega) \xrightarrow[IA]{} t_i(E).$$

where ω can be chosen to enhance convergence, as in the reference spectrum method. Note that we do not interpret ω as the physical πN collision energy, which could be a basis for fixing ω; instead, we release ω from that interpretation and use it as a free variable. Using ω as a free variable and dropping the second term of (5), I call the extended impulse approximation (EIA). The standard impulse approximation takes ω=E, whereas in the EIA one should estimate ω by minimizing the correction terms $t_i(\omega)[G(E)-g(\omega)]t_i(\omega)$. We are essentially expanding in powers of $t(\omega)\rho$, with ω giving us extra control. Variations of 20-30 MeV due to binding and other effects can make considerable difference since $t(\omega)$ resonates. Indeed in a series of important, excellent papers Schmidt, Dedonder and Révai[6] have emphasized that ω depends on the πN momentum variables in the medium. Recently, Lenz and Moniz[7] have also shown how the ω variable relates MST to the isobar description of πA dynamics.[8] These developments indicate, as I will discuss later, the need to go beyond the usual static MST and point the way toward improvements.

The idea of an optical potential can now be introduced by separating ground and excited intermediate states, thereby generating an expansion in correlations. Omitting the details, that process using projection operators P + Q = 1 yields the Watson optical potential

$$(o|U_W(E)|o) = (o|\{\sum_1^A \hat{t}_i(E) + \sum_{i,j \neq i} \hat{t}_i(E) \frac{Q}{E-H_o+i\epsilon} \hat{t}_j(E) + \cdots\}|o) \quad (6)$$

where $\hat{t} = v(1+QG\hat{t})$ includes projection to excited intermediate states via $Q = \sum_{n>0} |n><n|$. The above potential (which is an operator in pion-space) yields T by solving a two-body Lippmann-Schwinger equation $T = U_W + U_W PG\ T$ for elastic scattering. All excitations are included in $\hat{t}$ and U_W. Of course, $\hat{t}$ is a difficult many-body operator and one still needs to invoke (5), with the replacements $\tau \to \hat{t}$, $G \to QG$ to introduce the free t-matrix. We can use (5) to introduce the free collision operator as the leading term

$$(o|U_W(E)|o) = A(o|t_1(\omega)|o) + \Delta U_W \quad (7)$$

where ΔU_W includes corrections to the impulse approximation $[t_i(G-g_i)t_i]$ and corrections for binding and excitation effects in the $t(Q/E-H_o)t$ term of (6).

KMT[5] introduce an antisymmetrization operator $\mathcal{A}$ explicitly and work with an auxiliary optical potential U' from which an auxiliary πA T' operator is constructed

$$T'(E) = U'(E)[1 + \mathcal{A}\frac{P}{E-H_o+i\epsilon} T'(E)],$$

here P projects onto the nuclear ground state. The full πA amplitude is found by scaling up T' i.e., T = (A/A-1)T'. Since it is T' that is unitarized (S' = 1+iT' $\leq$ 1) by the above Lippman-Schwinger equation, one can be sure that T is correctly unitarized (i.e., S-matrix $\leq$ 1) <u>only</u> if all 1/A effects are included - a difficult task. This point has been discussed recently by Ernst, Shakin and Thaler.[9] The KMT A/(A-1) factor can therefore yield inaccuracies, especially at low energy where S' ∿ 1 and for light targets. It is a disadvantage of KMT compared to Watson treatments. The KMT (auxiliary) optical potential for elastic scattering involves the bound collision $\tau(E)$ operator with $G \to \mathcal{A}G$

$$<o|U'(E)|o> = (A-1)<o|\tau(E)|o> + (A-1)^2 \sum_{n>0} \frac{<o|\tau|n><n|\tau|o>}{E-\epsilon_n-U_n^1+i\epsilon} + \cdots \quad (9)$$

where the appearance of $U_n = (A-1)<n|U'|n>$ accounts for distorted pion intermediate states. Using (5) to introduce the free t-matrix, we have a leading term in the KMT version

$$(o|U'_{KMT}(E)|o) = (A-1)<o|t(\omega)|o> + \Delta U_{KMT}(\omega) \quad (10)$$

where $\Delta U_{KMT}(\omega)$ arises from the second terms of (5). Corrections to this "extended impulse approximation" and from nuclear excitations are included in the $\tau(Q/e)\tau$ term of (9). Both $\Delta U_w(\omega)$ and $\Delta U_{KMT}(\omega)$ depend on the starting energy ω; also since the series differ in detail their convergence, if any, can be quite different. Which organization is best? The relative merits of the two approaches have received some attention lately,[6,9] but the conclusion waits on evaluation of the respective correction terms. Many theoretical questions remain open, not only in determining the ΔU corrections, but even concerning the first order $(A-1)<o|t(\omega)|o>$ term, which Boridy and Feshbach[10] have dubbed the Rayleigh-Lax potential for historical reasons.

RAYLEIGH-LAX POTENTIAL

The RL potential can be used to determine T and hence cross-sections provided ΔU is negligible. The hope is that the optical potential expansion is in nucleon-nucleon correlations, which are small enough to provide convergence. The T-matrix given by (1) is an expansion in probabilities and converges slower, which partly motivates the introduction of U. Postponing the discussion of ΔU, let us consider the RL potential in momentum space for the pion $\vec{k}$ $\vec{k}'$ matrix elements (the p_i momentum variable refers to nucleons):

$$<\vec{k}'o|U'|\vec{k}o> \simeq (A-1)<\vec{k}'o|t(\omega)|\vec{k}o>$$

$$= (A-1)\int \psi_o^*(p'p_2\cdots p_A)<k'p'|t(\omega)|kp>\psi_o(pp_2\cdots p_A)dpdp'\prod_2^A dp_i$$

$$\simeq (A-1)\int F_{oo}(p'p)<k'p'|t(\omega)|kp>dp\,dp' \qquad (11)$$

This expression can be represented by the following diagram, which has been discussed by Liu et al. in their covariant rendition.[11]

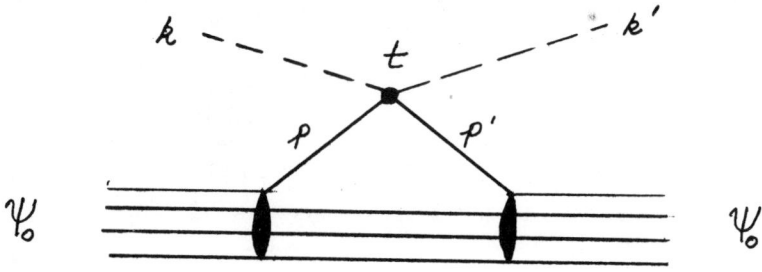

There are several aspects to this graph. First we note that the problem is formulated in the πA C.M. system (approximately equal to the πA lab system) and consequently the πN information in that frame is to be inserted for the (off-shell) matrix elements $<k'p'|t(\omega)|kp>$. We need to know the off shell ($k \neq k'$) elements because U will be inserted into an integral equation to sum the multiple scatterings. If ΔU is neglected that process is called the <u>coherent approximation</u> (no $\tau(Q/e)\tau$ terms). Also to calculate ΔU we need the off-shell ($k \neq k'$) elements of U and hence of t. The r-space version of these remarks is that we need the πN scattering wave function at short distances. From the idea that the nucleus is large compared to the πN range, one usually argues to the factorization approximation (i.e., a one-point integration of (11)).

$$<\vec{k}'o|U|\vec{k}o> \simeq (A-1)<\vec{k}'\vec{p}_f|t(\omega)|\vec{k}\vec{p}_i>\rho(q) \qquad (12)$$

where $\vec{q} \equiv \vec{k}'-\vec{k}$ and $\rho(q) \equiv \int F(p-q,p)dp$. To the extent that detailed nucleon motion matters and the πN interaction tails overlap, the above factorization is questionable. The magnitude of this effect is shown in Fig. (1), where Mach's[12] estimate of nucleon motion correction shows the need to go beyond the above factorization or static assumption. The values of $\vec{p}_i$ and $\vec{p}_f$ are often chosen to freeze the nucleons with respect to the initial and final nucleus, which is similar to the fixed scattering assumption. Although one can optimize the choice of $\vec{p}_i$, $\vec{p}_f$ to bring (11) close to (12), there is difficulty in obtaining a definitive result unless we deal with the nonstatic nucleon motion and use nuclear wave functions.

The need for an off-shell($\vec{k}' \neq \vec{k}$) model of $t_{\pi N}$ is apparent from (9)-(12). Such models have been designed using potentials and recently extended by applying inverse scattering approaches to the Chew-Low description.[13] The main features introduced by these πN models is finite, nonzero range and two-body unitarity. Cut-off form factors are introduced along with a meson propagator which, in the separable case in the πN c. m., is for each partial wave

$$<\vec{\kappa}'|\tilde{t}(\tilde{\omega})|\vec{\kappa}> = (\kappa\kappa')^\ell \frac{g_\ell(\kappa')g_\ell(\kappa)}{D_\ell(\tilde{\omega})} P_\ell(\hat{\kappa}'\cdot\hat{\kappa}). \qquad (13)$$

The g_ℓ's, which provide boundedness or unitarity, can be constructed from phase shifts provided their high energy behavior is also specified. Various types of off-shell behaviour affect the net pion-nucleus cross-sections, mainly at back angles. The form factors $g_\ell(k)$ introduce nonlocalities which yields a generalization of the Kisslinger potential to $b_0 \Omega(r/r') + b_1 \vec{\nabla}_r \Omega(r/r') \vec{\nabla}_{r'}$, where $\Omega \to \rho(r)$ in the Kisslinger limit of $g(k) \to 1$. However, the nonlocal effect is important for precision and needs to be included. The propagator $D_\ell(\tilde{\omega})$ yields the 3-3 resonance as described by the graphical version of (13)

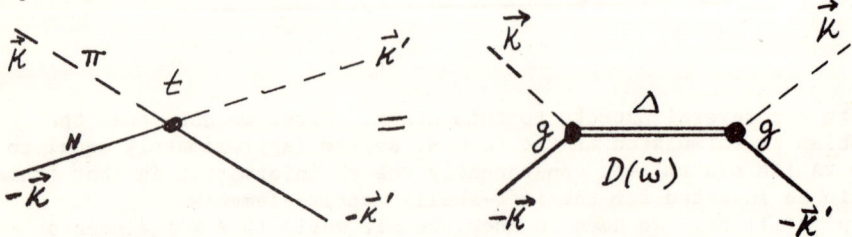

Note there is no crossing symmetry here, which is a serious omission as we shall hear from Banerjee.

The task of relating the πA c. m. t-matrix to the πN c. m. t-matrix has been the subject of much discussion. It is not a new problem, but is especially significant for pions because of the P-wave resonance. It is well known that a P-wave c. m. angular distribution maps into a different angle dependence in a laboratory frame. Thus the 3-3 strength feeds into the S and P wave parts of t. This effect needs to be included, especially for back angle scattering. It is not a relativistic effect. Among others, Landau et al.[14] emphasized the need for including this effect and claimed for their initial cases that the results were insensitive to details, provided the effect was somehow incorporated. Later work by Phatak, Miller, Landau and others[14] showed that the ambiguity in mapping $t \to \tilde{t}$ needs to be resolved. This is a technically difficult, and physically significant effect. The insights of Schmidt[6] are, I believe, a key to including this "angle transformation" effect unambiguously. Namely, one must relate $t \to \tilde{t}$ at an earlier stage, before making the factorization and frozen nucleon assumptions. Also if one wishes to conserve both energy and momentum at each vertex then off-mass shell ($E^2 \neq \vec{k}^2 + m_\pi^2$) pions must be considered. Thus, one is led to consider a covariant description. Several efforts to develop consistent dynamic schemes are underway. Reductions of the covariant Bethe-Salpeter equation is one approach,[11] relativistic potential scattering theory is another.[15] Also field theory approaches including crossing have been advocated by Cammarata and Banerjee,[16] and also by Miller.[17] In all of these approaches, if carried out completely, one should have no ambiguity in relating the pion-nucleon πA c. m. to πN c. m. dynamics. Hopefully we will learn about this problem at this meeting. My point now is that this effect leads us to go beyond the usual static MST.

Further indication of the need to develop nonstatic MST is shown in Fig. (2), where Schmidt's results for various choices of ω in the extended impulse approximation are compared. One reasonable choice for ω is $\omega = E-M-(p+k)^2/2M + \Delta E$, where a dependence on the pion plus struck nucleon momentum and on binding ΔE appears. Since ω depends on nucleon momentum it must be included in the integration (11) over nuclear motion. (Such effects were also included by Landau et al., but after factorization and frozen nucleon assumptions were invoked.[14]) This sensitivity to the variable ω is also shown in Fig. (3), where Bajaj and Nogami's results based on a pion striking a nucleon bound in a harmonic oscillator confirm Schmidt's, except for a significant difference in the real forward scattering amplitudes.[18]

The effect of nucleon binding and also of second order corrections as estimated by Schmidt are shown in Fig. (4). I do not have time to discuss various efforts to calculate the ΔU corrections.[2] I would only like to remark that the breakdown of the static MST for the RL potential indicates a need to reexamine these correction terms including nucleon motion effects.

ISOBARS AND MULTIPLE SCATTERING THEORY

The dependence of the starting energy ω on pion-nucleon variables has been emphasized here. Near a resonance this dependence is particularly significant and perhaps suggest ω be used as a fitting parameter. An alternate scheme is the isobar model of Kisslinger, Wang, and Saharia who use the isobar energy as a fitting parameter.[8] I would now like to show how the work of Lenz and Moniz relate these views.[7] That is, the choice of ω and its dependence on pion plus struck nucleon momenta $k + p \equiv P_\Delta$ is related to the isobar dynamics as illustrated by the graph:

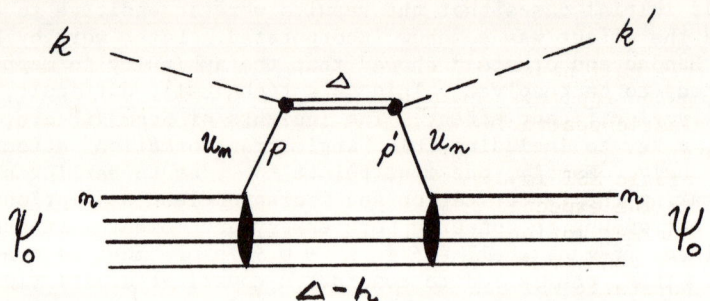

Let me present a schematic version of Lenz's demonstration of how MST yields the isobar picture. It is schematic in that the πA c. m. variables (k k'-pions, pp'-nucleons) and πN c. m. variables (κ κ'- relative pion-struck nucleon) are treated in a vague manner (see ref. 7 for the detailed relations).

We can introduce the separable form (13) to represent the Δ(1232) resonance into the RL potential (11). Also the introduction of single nucleon orbitals is made using a simple nuclear model picture

$$F(p'p) \to \sum_n U_n(p')U_n(p)$$

Thus the RL potential can be split

$$U = \int \sum_n U_n(p')\kappa'^\ell g_\ell(\kappa')\hat{\kappa}' \cdot \frac{\hat{\kappa}\kappa^\ell g_\ell(\kappa)U_n(p)}{D(\omega)}$$
$$\leftarrow \psi_{n\Delta} \rightarrow$$

where the second factor is called a delta-hole (nΔ) wave function $\psi_{n\Delta}$. Returning to the Schrödinger equation for the πA wave function ψ, we can obtain the above result from the coupled equations

$$(E-H_0)\psi = \sum_n |U_n g\rangle \psi_{n\Delta}$$

$$D(\omega)\psi_{n\Delta} = \langle U_n g|\psi\rangle \simeq (E_\Delta - \frac{P_\Delta^2}{2M_\Delta})D'(\omega)\psi_{n\Delta}$$

where Schmidt's choice of $\omega = E-(k+p)^2/2M = E-P_\Delta^2/2M_\Delta$ has been used along with the idea that $D(\omega)$ is the resonant Δ propagator. This last step and the introduction of a πN total momentum $P_\Delta = p + k$, suggest the replacement $D(\omega) \to [E_\Delta - (P_\Delta^2/2M_\Delta)]D'$, which introduces the complex isobar energy E_Δ. The label n serves to identify the nucleon hole state, so that we can speak of Δ-hole intermediate states. This schematic discussion is presented only to illustrate how the variable ω and the 3-3 resonance is included in nonstatic MST and can be related to a coupled channels isobar dynamical model, at least in first order. Hence, instead of using ω one can perhaps use E_Δ as a fitting parameter as suggested by Kisslinger and Wang.

CONCLUSION

In this half hour I have stressed a few issues and questions concerning the application of multiple scattering theory to pion-nucleus elastic scattering in the resonance region. The main points are:
1. Static MST is apparently not sufficiently precise relative to the expected precision of πA elastic scattering. Nuclear motion, recoil, and extended impulse corrections are seen to matter.
2. Nonstatic MST can be developed wherein off-shell t-matrix models, a choice of the variable ω, and the πA c. m. to πN c. m. relations need to be considered.
3. Several consistent dynamical schemes are being developed which should ultimately be unambiguous, and should include relativistic, crossing and absorption effects.
4. In a 3-3 dominance description the choice of ω can be related to the isobar energies in first order, which makes contact with the isobar models.
5. The correction terms ΔU need to be evaluated, including nucleon motion, as functions of ω to yield a hopefully convergent series. The Watson versus KMT organizations needs to be tested.
6. Fixed scattering approaches and the associated use of closure in evaluating ΔU terms needs to be examined closely.

These issues will probably be discussed here. Hopefully we will make progress toward understanding the πA dynamics with sufficient reliability to extract or confirm nuclear structure information.

REFERENCES

1. Recent review articles: J. Hüfner, Physics Reports 21C (1975); G. E. Brown and W. Weise, Physics Reports 22C (1975) 281 and references therein.
2. G. A. Miller NP A223 (1974) 477. E. Kujawski and M. Aitkin, Nucl. Phys. A221 (1974) 60.
3. L. S. Kisslinger, Phys. Rev. 98 (1955) 761. M. M. Sternheim and E. H. Auerbach, Phys. Rev. C4 (1971) 1805.
4. M. A. B. Bég, Ann. Phys. (N.Y.) 13 (1961) 110. D. Agassi and A. Gal, Ann. Phys. (N.Y.) 75 (1973) 56.
5. A. K. Kerman, H. McManus, and R. M. Thaler, Ann. Phys. (N.Y.) 8 (1959) 551.
6. C. Schmidt, Nucl. Phys. A197 (1972) 449. J. Révai, Nucl. Phys. A208 (1973) 20. J. P. Dedonder, Nucl. Phys. A180 (1972) 472.
7. F. Lenz, Ann. Phys. 95 (1975) 348. F. Lenz and E. J. Moniz, Phys. Rev. C12 (1975) 909.
8. L. S. Kisslinger and W. L. Wang, Phys. Rev. Lett. 30 (1973) 1071 and to be published; also see A. Saharia and L. S. Kisslinger (this conference).
9. D. J. Ernst, C. M. Shakin and R. M. Thaler, Phys. Rev. C9 (1974) 1374.
10. E. Boridy and H. Feshbach, Phys. Letters 50B (1974) 487.
11. L. C. Liu and C. M. Shakin (preprint April, 1976) and L. Celenza et al., Phys. Rev. C12 (1975) 1983.
12. R. Mach, Nucl. Phys. A205 (1973) 56. Also see K. Nakano and Chi-shiang Wu, Phys. Rev. C11 (1975) 1505.
13. R. H. Landau and F. Tabakin, Phys. Rev. D5 (1972) 2746. J. T. Londergan, K. W. McVoy, and E. J. Moniz, Ann. Phys. (N.Y.) 86 (1974) 147. Few Body Law cases R. A. Friedenberg (to be published).
14. E. R. Siciliano and G. E. Walker, Phys. Rev. C13 (1976) 257. R. H. Landau et al., Ann. Phys. (N.Y.) 78 (1973) 299.
15. L. Heller et al., Phys. Rev. C13 (1976) 742 and H. S. T. Lee (this meeting).
16. J. B. Cammarata and M. K. Banerjee, Phys. Rev. C13 (1976) 299.
17. G. A. Miller (preprint 1976) "Field Theory Treatment of Pi-Nucleus Scattering".
18. K. K. Bajaj and Y. Nogami, "Pion Scattering From a Bound Nucleon at Medium Energies" preprint 1976.

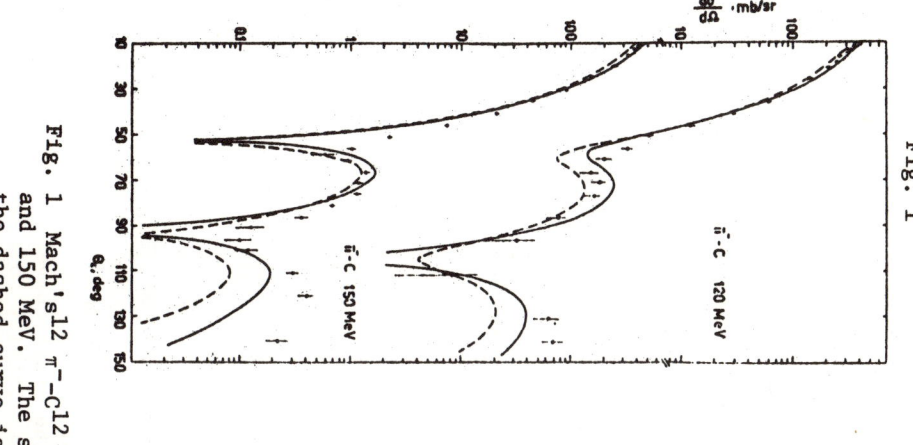

Fig. 1

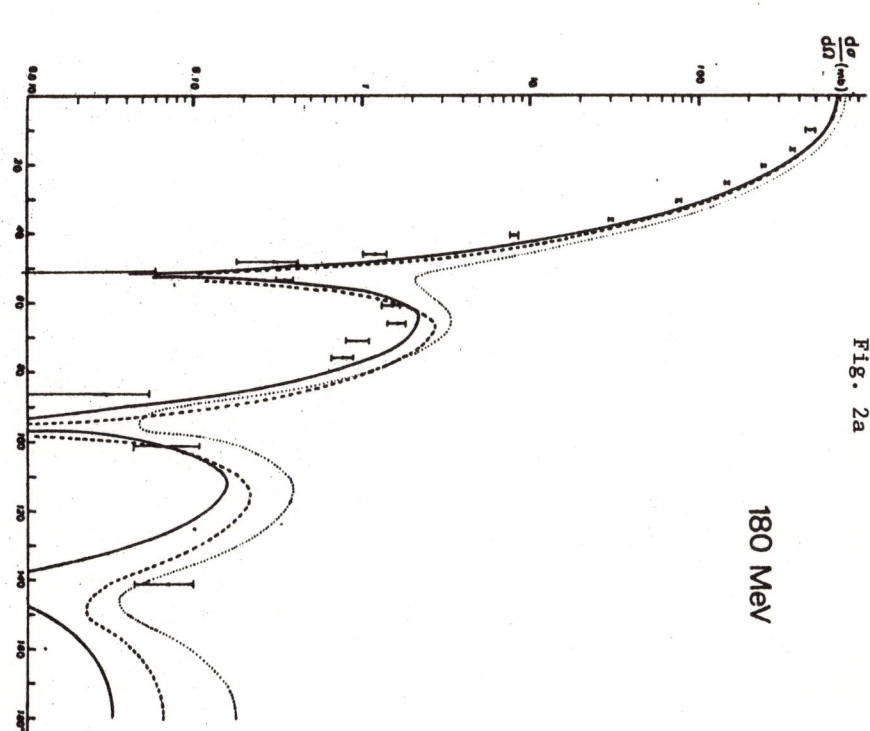

Fig. 2a   180 MeV

Fig. 1 Mach's[12] $\pi^- - C^{12}$ elastic scattering cross sections at 120 and 150 MeV. The solid curve includes nucleon motion effects, the dashed curve is from Auerbach.[3] Since these are on semi-log paper, the effects are considerable.

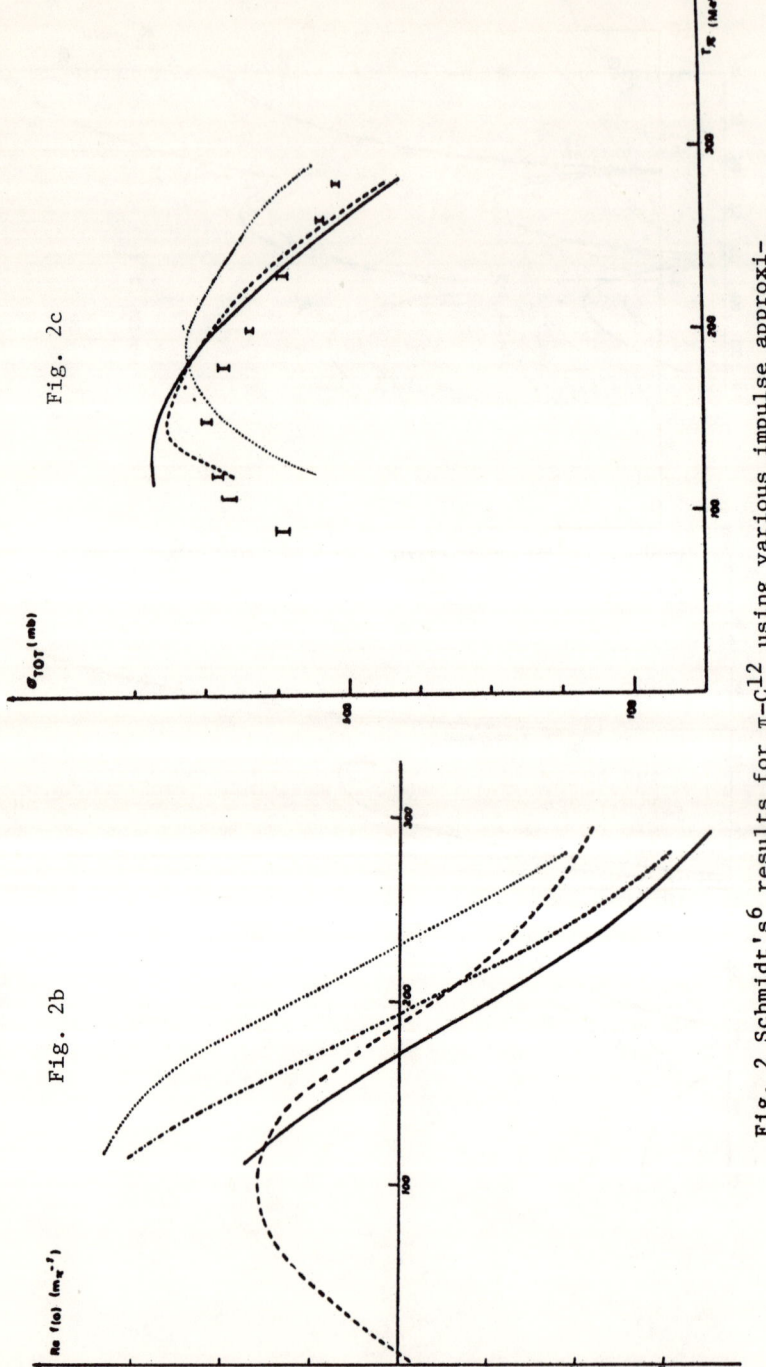

Fig. 2 Schmidt's[6] results for $\pi-C^{12}$ using various impulse approximations, i.e., various choices for ω in equation 12. The dotted curve is for the $\omega = E$ (IA) choice, the other curves are for two alternate choices of ω (EIA). Sensitivity to ω is thus demonstrated; (2a) shows the differential cross section, (2b) and (2c) show the associated Re $f_{\pi A}(0)$ and σ_{TOT}

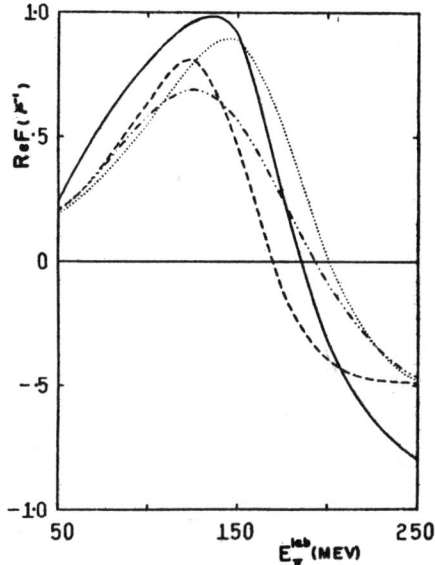

Fig. 3 Bajaj and Nogami's[18] results for various impulse approximations, i.e., again various choices for ω. The curves show the Re $f_{\pi O}16$ (Θ = 0) versus pion kinetic energy for: --- from Landau et al.[14]; -··- integration over nucleon motion; ····· use of the Chew approximation;[18] and ⎯⎯ Bajaj and Nogami's result.[18]

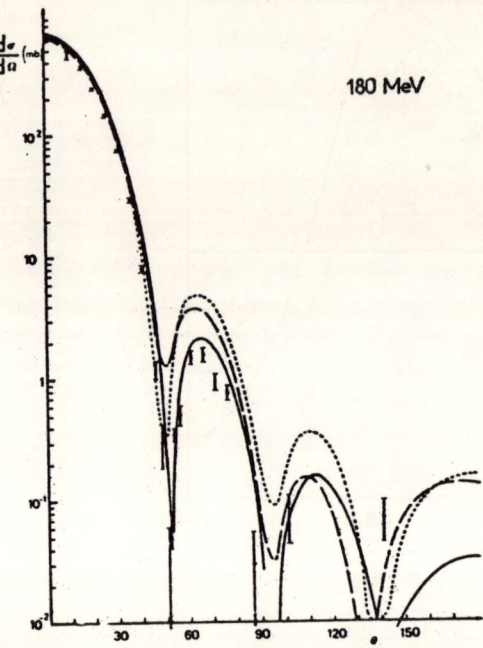

Fig. 4 Schmidt's[6] π-C^{12} differential cross section including first and second order corrections and using three versions of the impulse approximation. The cases are: ——— $\omega = E_B$ plus first order corrections; ····· $\omega = E_B$ plus first and second order corrections; ---- $\omega = E_A$ plus first and second order corrections. See ref. 6 for his choices of E_A and E_B (which are not equal to E).

Yu. A. Shcherbakov (JINR-Dubny): Is there some hope to introduce absorption processes in the optical model calculations in the near future?

F. Tabakin: It is important to learn how to incorporate real absorption of pions. A recent work by G. Miller indicates it is particularly important at lower energies, but becomes small as one goes to the 3-3 region. In the 3-3 region the absorption is apparently mostly due to elastic πN collisions that remove flux from the ground state. Concerning the use of separable potentials, they provide a convenient parametrization, but certainly are not a full dynamical description. They do not automatically incorporate crossing or absorption. (For absorption there is the possibility of going to coupled channels.) However, separable potentials in themselves are not the key to including absorption, but I do agree we should include this process especially at lower energies as one goes toward the bound state. I would also like to advocate not depending too heavily on separable potentials to sum the multiple scattering series as is done in some approaches. To make contact with field theory, covariant and crossing symmetric approaches, we will have to leave the separable potentials, so it would be good not to be too dependent on using them.

Carl Shakin (Brooklyn College): In a covariant analysis most of the problems you have raised have been solved. It is possible to find a definite prescription for ω. If you use the covariant approach there is no problem in transforming amplitudes from one frame to another. Also, we have been able to perform an integral over the Fermi motion so that the factorization assumption is not necessary.

F. Tabakin: Of course in a complete dynamical theory there is no ambiguity. I wished to emphasize in my talk that the ambiguities arise from using the static limit and then introducing various πA c.m. to πN c.m. relations. What we have all learned, and you have emphasized, is that such relations must be introduced within a full dynamical context without making a static assumption, i.e., including nucleon motion. Let me also take your comment as an opportunity to mention the various other dynamical schemes that probably will be discussed at this meeting; namely the MST with relativistic potentials (Heller, Lee and Coester), Chew-Low (Dover, Lemmer, Friedenberg), covariant reduction of Bethe-Saltpeter to Blanchenbecker-Sugar forms (Liu, Shaken, Celenza, Nutt), field theory approaches with crossing (Banerjee and Cammarata). We'll be hearing about these developments. Other approaches using the fixed scattering approach will be discussed by Gibson, I suppose.

Unfortunately, we now have several "complete dynamical formulations" - each proponent disclaims any ambiguities. I wonder if they agree. That question should be discussed here. Let me also show one of Liu's results (slide) on which Piepho-Walker, LPT cross sections are compared to your work. The differences arise from various t→t steps. Later work by Phatak closes the gap by using different steps. So perhaps the final results are starting to agree.

Kopaleishvili (Tbilisi State University): Is the problem of choosing of ω in the free pion-nucleon collision matrix connected with the relativistic (kinematical point of view) treatment of the problem? This question arises because in the non-relativistic case it seems ω is chosen to be $\omega = \frac{k^2}{2m_\pi} + \frac{p^2}{2m_N}$, where k and p are the momentum of the incident pion and of the nucleon inside the nucleus, if we want the pion-nucleus collision matrix to be Galilian-invarient.
F. Tabakin: The variable ω does not have to be related to the energy E. If you insist that ω yield Galilean invariance then it is somewhat restricted, but one could add complex Galilean invariant terms to it. My point is one can release ω from any particular model dependent choice and make it available for generalizing the impulse approximation and improving convergence in analogy to the "starting energy" in nuclear matter studies.

L. S. Kisslinger (CMU): It is not adequate to go part way with relativity. In practice one must use non-relativistic wave function. Then all of the ambiguities in the variables which have been mentioned reappear.
F. Tabakin: That's right. Even in the covariant theories one comes back to non-relativistic wave functions at which stage the benefits of covariance are endangered. There is a work by Dudly Miller on relativistic many-body wave function which might be relevant to keeping the theory covariant all the way.

Herman Feshbach (MIT): The use of multiple scattering theory at low energy, say below the (3,3) resonance is hard to justify. The corrections to which the speaker referred is simply in the direction of returning the problem to a many body problem for finite nuclei and the connection to the on shell scattering amplitudes is completely lost. Perhaps one should use the theory above the resonance as Rinat has suggested.

LOW ENERGY PION SCATTERING FROM NUCLEI

R. A. Eisenstein[†]
Carnegie-Mellon University
Pittsburgh, Pennsylvania 15213

ABSTRACT

A review of the current experimental situation in low energy π elastic scattering is given. Emphasis is placed on the very large discrepancies between the data and the predictions of optical model theories which work well at energies near the (3,3) resonance. Guidance as to the type and size of corrections to the multiple scattering series for the low energy domain is sought from an examination of theories pertaining to pionic atoms. An examination is made of several recent attempts to improve the theoretical situation. Predictions for forthcoming low energy inelastic scattering experiments are reviewed briefly.

INTRODUCTION

One of the major hopes of medium energy physics is that the relatively simple nature of the two-body πN interaction will allow us to make a viable theory of the A-body π-nucleus problem via a multiple scattering theory or some other construct. This hope seems particularly well founded since the πN interaction[1] has a very weak repulsive s-wave isoscalar part at low energies and is dominated by the (3,3) resonance at π kinetic energies up to ~300 MeV. Ideally, the π-nucleus system could be used as a laboratory for testing our ideas about multiple scattering and nuclear structure.

It has been known for some time that elastic scattering data obtained in the resonance region can be described equally well by theoretical models which obtain from very different physical origins. On the one hand, the high energy theory of Glauber[2] has been applied; on the other, the low energy, low density model of Kisslinger[3] has been used. That both simple models achieve success is thought to be due to the dominance of the πN resonance which renders the nucleus black[4]. Because of the p-wave nature of the resonance, the absorption occurs mostly in the surface region of the nucleus.

However, at lower energies, below 100 MeV say, the resonance is much less dominant and so we expect the π mean free path in nuclear matter to increase. As the pion samples the nuclear interior, nuclear structure effects that are obscured at higher energies may come into play. These include, among others, nucleon short- and long-range correlations, binding effects, and Fermi motion. Using a reasonable πN interaction of finite range, these effects could be included in a multiple scattering series to

[†] Work supported by the USERDA.

generate an optical potential for π elastic scattering. An examination of low energy scattering therefore is warranted.

An examination of the data from both of these regions should help us answer the following important question: In what energy region is the multiple scattering theory most reliable? At low energies, the πN interaction is weak, but the corrections are likely to be large, while at high energies the interaction is strong and the corrections expected to be small.

Recently, Amann et al.[5] scattered 50 MeV π^+ from ^{12}C using the LEP channel at LAMPF. The angular distribution obtained is shown in Fig. 1 compared to predictions (using free πN information) for the Kisslinger (K), local Laplacian (LL) and separable potential of Londergan, McVoy and Moniz[6] (LMM). None of these models is at all adequate in reproducing the data, the chief difficulty being the position and width of the minimum. The dip is at ~60°, some 15-20° further forward than any theory, and broader than predicted. The calculated magnitudes are also much larger than the data at forward angles. However, good phenomenological fits can be obtained by allowing free variation of the complex optical model parameters b_0 and b_1 in the K and LL models. No variation of the LMM theory was attempted. Although these fits are qualitatively excellent, they are physically deficient, because the resulting potentials either violate unitarity in one or more partial wave or else produce pions in some regions of the nucleus.

A subsequent analysis by Cooper and Eisenstein[7] of this data as well as $\pi^\pm$-^{12}C data[8] at 30 MeV and $\pi^\pm$-^{4}He data[9] at 51 MeV found that the difficulties encountered by Amann et al. are common to all data sets on light nuclei at these energies. The minimum is consistently much further forward and shallower in the data than the simple theories using free πN information will allow. In all cases, phenomenological fits can be obtained, but all of them are at best pion producing and at worst in violation of unitarity. The chief feature of the analysis is the apparent need for a much stronger and more repulsive π-nucleus s-wave interaction.

Said another way, we see that in order to move the minimum in elastic scattering to more forward angles, it is necessary to make the parameter b_0 much larger and more repulsive than is predicted by the free πN information. This is similar to the situation[10] in pionic atoms, where it is also found that the part of the optical potential generated by the s-wave πN interaction must be made more strongly repulsive.

In neither case is this too surprising, since the πN s-wave isoscalar scattering length $1/3(a_{1/2}+2a_{3/2})$ is fortuitously nearly zero. Therefore in any expansion of the amplitude, we would expect that higher order terms in the multiple scattering series would have appreciable effects. These terms of necessity force us to face up to questions neglected until now: nucleon correlations, binding Fermi motion, and πNN absorption. With a stronger (repulsive) s-wave contribution to the potential, we may also expect interference between the s-wave and p-wave pieces, especially in this low energy region where the resonance is not dominant.

In the case of π elastic scattering, it has been known for some time that in order to generate a correct π-nucleus optical potential, the π-N amplitude must be kinematically transformed to the π-nucleus frame. This procedure leads to an admixture of the p-wave amplitude into the s-wave piece and causes an enhancement of the s-wave repulsion. Although this procedure has fundamental ambiguities when transforming the off-shell π-N t-matrix, several prescriptions[12] for doing so exist in the literature.

I would like now to look in more detail at the corrections to multiple scattering that apply at zero π kinetic energy - that is, in the case of pionic atoms. Hopefully this will be a useful indicator of the kind and size of correction to be used at higher energy.

It was first pointed out by Ericson and Ericson[13] that short range correlations between nucleons play an important role in determining the way in which a low energy pion propagates through a nucleus of low density. The essential effect is that the local p-wave pion field is a modification of the external field due to the presence of dipole emitters (nucleons) in the same way that a free electromagnetic field would be modified by the presence of dielectric material. The local fields thus obviously depend on the (correlated) positions of the nucleons or molecules. In the electromagnetic case, this effect was first calculated by Lorentz and Lorenz and so the name has carried over to the pion case. Ericson and Ericson were led to a pion optical potential with a non-linear density dependence; neglecting differences between neutron and proton densities and kinematic factors, the potential has the form (Krell and Ericson[10])

$$V_N(r) = -2\pi \frac{\hbar^2}{2\mu} \{(b_0\rho(r) + i\text{Im}B_0\rho^2(r)) \quad (1)$$
$$+ \nabla \frac{\alpha(r)}{1+\frac{4\pi}{3}\xi\alpha(r)} \nabla\} \text{ where}$$

$$\alpha(r) = c_0\rho(r) + i\text{Im}C_0\rho^2(r)$$

The parameters $\text{Im}B_0$ and $\text{Im}C_0$ represent πNN absorption and are obtained from the widths of the s- and p-states. The parameter ξ is related to the two-nucleon correlation function: $\xi=0$ means no correlations; $\xi=1$ means full short range anticorrelations. Such a potential form has given good fits to π-atom data with fit parameters agreeing well with theoretically predicted values. $\xi=1$ is favored.

The potential given above has been shown by Eisenberg, Hüfner and Moniz[14] to be, under certain restrictions, an exact summation of the Watson multiple scattering series. The most important restrictions in the derivation are the zero range nature of the πN amplitude and the use of Jastrow type correlations. When the πN interaction is allowed a finite range, as a realistic calculation must do, the effects of short range correlations are largely quenched. This result has been obtained by Eisenberg et al.[14],

Iachello and Lande[15], and Hüfner and Iachello[16]. Eisenberg et al. find that a pion amplitude of reasonable range leads to $\xi \approx 0.1$ and so the preference of π atom data for $\xi=1$ may be fortuitous.

This may indicate that the non-linear density dependence is required, but on other physical grounds. For example, Delorme and Ericson[17] have recently shown that inclusion of the Pauli principle leads to corrections in the optical potential of exactly the same form as the Lorentz-Lorenz effect above. In addition, it is much less sensitive to the range of π-N forces. Another example of non-linear density dependence is two nucleon absorption, whose treatment may require a more sophisticated theory than the ansatz given above.

Koltun and Myhrer[18] have recently made a theory of π-nucleus scattering lengths including all effects through second order in the pion-nucleon scattering amplitude in a simple single particle model. Single scattering, double scattering from different nucleons, and double scattering on the same bound nucleon are included. The Pauli principle is put into the calculation by using properly antisymmetric nuclear ground state wavefunctions. The results for T=0 nuclei indicate that: (1) the single scattering contributions from s- and p-wave nearly cancel; (2) the total binding effect, including the unitarity correction of Myhrer and Silbar[19], is very small; (3) the principal remaining term is the double scattering in s-waves, the other double scatterings being small. Although the results for the scattering lengths are smaller than the experimental values, these authors indicate that the difference can probably be made up from absorptive contributions. I think this calculation offers convincing proof of the need for careful calculations of higher order effects in multiple scattering.

This brief examination of calculations related to the pionic atom case has shown that the calculation of π elastic scattering at larger values of the energy will require a careful exposition of the ideas mentioned above. An attempt to make contact between pionic atom work and low energy π scattering was made by Ericson and Krell[20], who used the best fit values from the pionic atom data in a potential of form (1) to calculate the scattering at finite k. Since no energy variation in the parameters was attempted, the calculation was not intended to be predictive in a quantitative sense. In fact their predictions for the behavior of the s-wave amplitude as a function of energy disagree with existing data.

More recently, Barmo and Pilkuhn[21] have calculated low energy π scattering in a model which explicitly accounts for the energy variation of the πN isoscalar amplitude b_0. Absorption is included by fixing ImB_0 and ImC_0 at the π-atom values. The effect of nucleon binding is included crudely by shifting the momentum at which the imaginary term is evaluated. Figure (2) shows the predicted behavior of $\exp(2i\delta_0)$ for π-^{12}C compared to data of Refs. (5,8,22). The fit is only qualitatively correct and must await further data for a better test.

In a contribution to this conference, Landau and Thomas discuss the fact that substantial improvement to the fit for the data of

Ref. (5) may be had if one makes simple adjustments in the kinematics to account for the angle transformation and binding energy. (See Fig. 3) In addition, the πN phase shifts used in the calculation have large effects on the fit, especially at forward angles. The "smooth" phase shifts of Salomon[23] improve the fit in this calculation by a factor of 2 for the forward angle data over the results using CERN phases. New low energy phase shift information from Saclay and LAMPF will help settle this matter.

Recently Gibson et al.[24] have calculated the low energy elastic π scattering from ^{4}He using a fixed nucleon approximation and a Monte Carlo technique in the evaluation of the full multiple scattering series. Using a separable form for the πN t-matrix with an off-shell cutoff form factor, and taking into account initial and final nucleon momenta recoil (i.e., the angle transformation), reasonably good fits to the data are obtained. (See Fig. 4) However, the fits worsen as the energy is lowered, culminating in a poor fit to the 24 MeV data. The principal difficulty is the depth of the minimum. The problem may require a better treatment of absorption or a relaxation of the fixed nucleon approximation, which is the major shortcoming of this calculation.

In the talk following this one, M. Thies will describe calculation he has done which includes to second order the effects of short- and long-range correlations, finite πN range, ρ-meson exchange, and absorption. The angle transform is also included. When applied to the 50 MeV ^{12}C data, the fit is really quite good.

In an approach fundamentally different from the multiple scattering theories given above, Kisslinger and Saharia[25] have used a doorway isobar model to calculate the elastic scattering over a wide energy region. While at higher energies the closure approximation gives reasonable agreement with the data, it is found that at low energies (~50 MeV on ^{12}C) closure must be violated substantially in order to fit the data at forward angles. This presumably reflects the fact that fewer doorways are available at lower energies. The angle transformation is included. More will be said about this calculation later in the conference.

Turning briefly to inelastic scattering, a contribution to this conference by Gibbs et al. notes the substantial effect on inelastic scattering to be expected when the elastic channel is drastically modified as in Ref. (5). These authors note, as has G. Miller[26], that one result of the phenomenological adjustments to the optical model is a smoothing of the π-nucleus s-wave radial wavefunction. Since the inelastic form factor is peaked in the surface region, such wavefunction smoothing can have a drastic effect on the cross section. A proposal to LAMPF[27] to study inelastic scattering to the 2$^+$ state at 4.4 MeV in ^{12}C has been approved for beam time this summer. Figure (5) shows the predicted inelastic cross sections[27] using a phenomenological form factor with[28] β_2=0.56. While the discrepancies between the free πN and adjusted optical model are large, they are not as substantial as those predicted by Gibbs et al., especially with regard to shifts in cross section magnitude. The differences in calculation could be due to the relative position of the wavefunction kink with regard to form factor position alluded to above.

From what has been said, it seems clear that low energy π scattering will provide a sensitive test of π-nucleus interaction theory. In order to obtain agreement with experiment, these theories will require quantitative and detailed nuclear structure input - a desirable state of affairs. Experiments at LAMPF, SIN and TRIUMF within the next year will hopefully bear out this prediction.

REFERENCES

1. D. Koltun, Advances in Nuclear Physics, Vol. 3, Plenum Press.
2. V. Franco, Phys. Rev. C9 (1974) 1690.
3. M. Sternheim and E. Auerbach, Phys. Rev. Letts. 25 (1970) 1500.
4. J. Hüfner, "Pions Scatter by Nuclei" in High Energy Physics and Nuclear Structure - 1975, American Institute of Physics Conference Proceedings No. 26.
5. J. Amann, P. Barnes, M. Doss, S. Dytman, A. Thompson, Phys. Rev. Letts. 35 (1975) 426.
6. J. Londergan, K. McVoy and E. Moniz, Ann. Phys. 86 (1974) 147.
7. M. Cooper and R. Eisenstein, Phys. Rev. C13 (1976) 1334.
8. M. Nordberg, J. Marshall and R. Burman, Phys. Rev. C1 (1970) 1685.
9. K. Crowe, A. Fainberg, J. Miller and A. Parsons, Phys. Rev. 180 (1969) 1349.
10. M. Krell and T. Ericson, Nuc. Phys. B11 (1969) 521.
11. D. Bugg, A. Carter, J. Carter, Phys. Lett. 44B (1973) 278.
12. R. Mach, Nucl. Phys. A205 (1973) 56.
    L. Kisslinger and F. Tabakin, Phys. Rev. C9 (1974) 188.
    E. Kujawski and G. Miller, Phys. Rev. C9 (1974) 1205.
    S. Phatak, PhD. Thesis, Univ. Pittsburgh (1974).
    G. A. Miller, Phys. Rev. C10 (1974) 1242.
13. M. Ericson and T. Ericson, Ann. Phys. 36 (1966) 323.
14. J. Eisenberg, J. Hüfner and E. Moniz, Phys. Letts. 47B (1973) 381.
15. F. Iachello and A. Lande, Phys. Letts. 50B (1974) 313.
16. J. Hüfner and F. Iachello, Nuc. Phys. A247 (1975) 441.
17. J. Delorme and M. Ericson, Phys. Lett. 60B (1976) 451.
18. D. Koltun and F. Myhrer (Preprint).
19. F. Myhrer and R. Silbar, Phys. Letts. 50B (1974) 299.
20. M. Ericson and M. Krell, Phys. Letts. 38B (1972) 359.
21. S. Barmo and H. Pilkuhn, Phys. Letts. 60B (1976) 324.
22. R. Edelstein, et al., Phys. Rev. 112 (1961) 252.
23. M. Saloman, TRIUMF report TRI-74-2.
24. B. Gibson, W. Gibbs, A. Hess, G. Stephenson and W. Kaufmann, Los Alamos Preprint LA-VR-75-2159.
25. L. Kisslinger and A. Saharia, Contribution to this conference.
26. G. Miller, Nuc. Phys. A224 (1974) 269.
27. "Studies of π^+ Scattering at 50 MeV from Light Nuclei" LAMPF Proposal #246, R. Eisenstein, spokesman.

28. E. Rost, Proceedings of the LAMPF Summer School, LASL Report LA-5443-C.

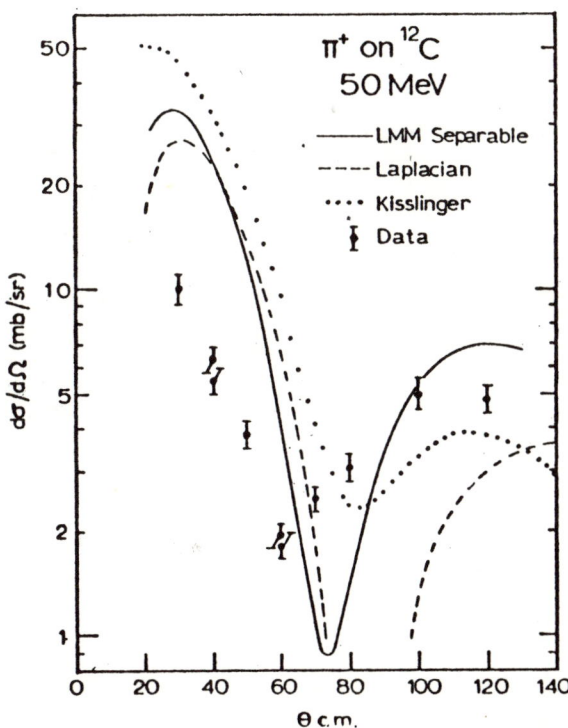

Figure 1. The angular distribution obtained by Amann et al.(ref. 5) compared to the predictions of the Kisslinger, Local Laplacian, and separable (Londergan, McVoy, Moniz) models, all constructed using free πN information.

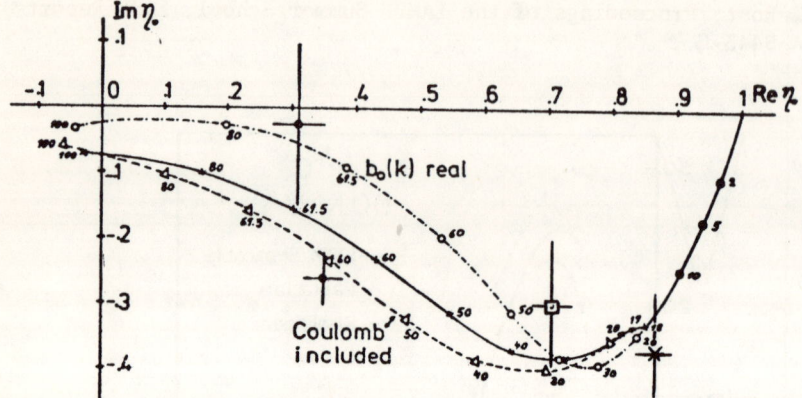

Figure 2. The Argand diagram of the S-matrix element $\eta_0 = \exp(2i\delta_0)$ for s-wave π-^{12}C scattering obtained by Barmo and Pilkuhn (ref. 21). compared to data of refs. (5,8,22). Data of ref. (5) (*) and ref. (8) (□) is the output of a best fit optical model code and not a direct phase shift fit. Comparison should be made with the dashed curve.

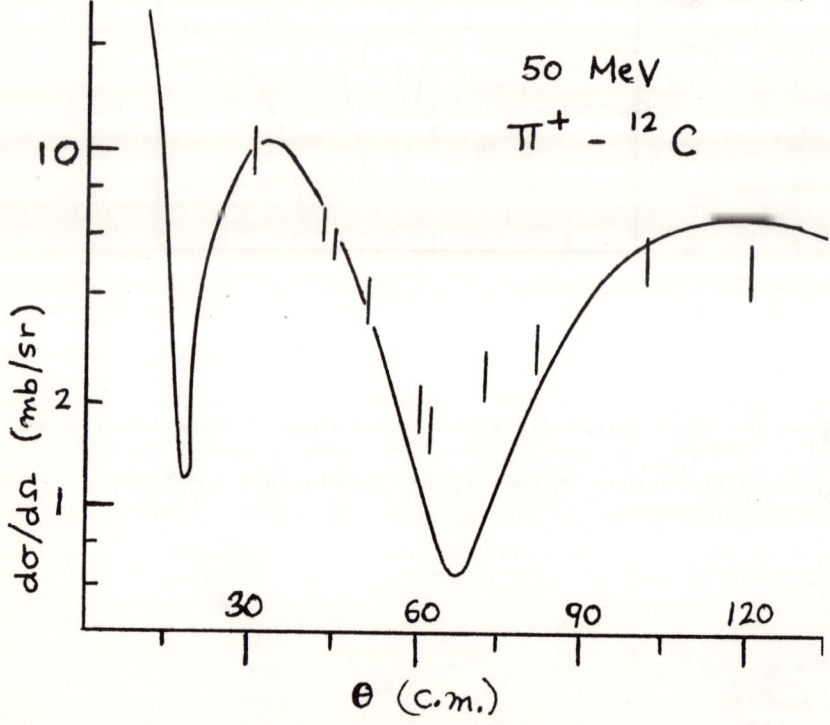

Figure 3. Fit obtained by Landau and Thomas to the data of ref. (5). Both the angle transformation and a crude binding energy correction have been made. See the text and the contribution of Landau and Thomas to the Conference.

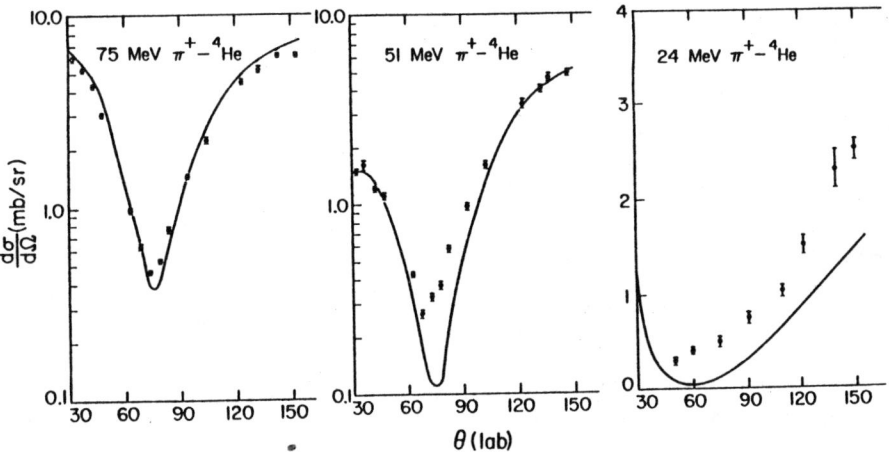

Figure 4. The fits to the π^+-^{4}He data of Crowe et al. (ref. 9) obtained by Gibson et al. (ref. 24). See text.

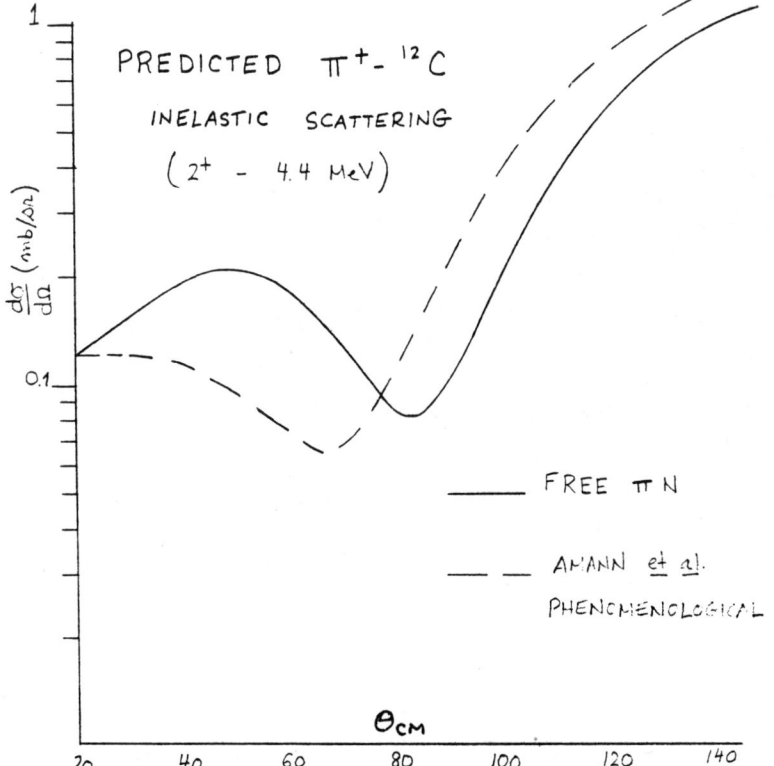

Figure 5. Predicted (ref. 27) inelastic cross sections for excitation of the 2+ state at 4.4 MeV in ^{12}C. The solid curve is the prediction using free πN information; the dashed curve is the prediction of the phenomenological model of Amann et al. Both curves are DWBA calculations.

M. Rho (SACLAY): Why is it that your list does not contain the theory of the Lorentz-Lorenz effect proposed by Baym, Brown, and Weise which in my opinion might be the only correct paper regarding this matter? The fact that the LL term is related to the Landau-Migdal g' parameter may be of some relevance in weeding out many wrong papers published on this matter.

R. Eisenstein: The list was not meant to be exhaustive but rather illustrative. The point was to outline the types of correction to be expected. Many corrections are involved and a definitive treatment has not yet emerged. Also, M. Thies will discuss the LL effect next.

M. K. Banerjee (U. of Maryland): I would like to make two comments.

1. In 1972, Huang, Levinson, and I explained the π-nucleus scattering lengths. The explanation is based on the soft pion theory. The soft pion limit (the σ commentator term) for the charge symmetric π-N scattering length is 0.08 m_2^{-1}. The hard pion correction which is attractive practically cancels it out. This hard pion correction involves rescattering. The continuation from the low energy part of the intermediate state spectrum is strongly quenched when the π-N system is imbedded in the nucleus. This effectively enchances the s-wave repulsion.

2. I wish to be on record that in my talk yesterday I made the point that it is not enough to make the angle transformation One must add on the canonical s-wave piece dictated by the requirement of charge conjugation symmetry.

Yu. A. Shcherbakov (Jinr Dubna): You have shown the Avgand diagram for $\pi^{12}C$ scattering. What can you say about the present situation for the so-called "size resonance" predicted by Ericson and Krell?

R. Eisenstein: The calculation of Ericson and Krell was really an attempt to understand the qualitative features only. The optical potential used was the one obtained from a best fit to the pionic atom data and so is not really suitable for the finite energy scattering problem. To the best of my knowledge, there is no existing evidence for the size resonances you mention.

ENERGY AND ANGULAR DEPENDENCE IN PION-PROTON BREMSSTRAHLUNG[†]

D. I. Sober
UCLA and Catholic University, Washington, DC 20064[*]

H. C. Ballagh, P. F. Glodis, R. P. Haddock, K. C. Leung[**],
B. M. K. Nefkens and D. E. A. Smith
University of California, Los Angeles, CA 90024

ABSTRACT

New measurements of $\pi^{\pm}p \rightarrow \pi^{\pm}p\gamma$ at three incident beam energies and for a wide range of photon angles are presented and discussed.

INTRODUCTION

Pion-proton bremsstrahlung ($\pi^{\pm}p \rightarrow \pi^{\pm}p\gamma$) has been proposed as a laboratory in which to test a variety of phenomena, including off-mass-shell effects and the magnetic moment of the $\Delta(1232)$ resonance[1]. The first published experimental data[2,3], for 298 MeV incident pions and backward photon angles (where cancellation between external charge radiation from the π^+ and proton should give the greatest sensitivity to internal structure effects), showed a very simple behavior. The cross sections for both reactions decrease smoothly with photon energy and fail to show the large peak at $E_\gamma \approx 70$ MeV predicted by several models.[1] This result has led to considerable theoretical activity[4-9].

We have made new measurements at additional photon angles[10] and at other incident pion energies. The experimental arrangement, in which all 3 final-state particles were detected with spark chambers and lead-glass photon counters, is described in Ref. 3.

RESULTS

1. Photon energy spectrum

At all photon angles, the differential cross section falls monotonically with photon energy, and up to $E_\gamma \approx 50$ MeV agrees with the soft-photon theorem of Low[11]. At higher E_γ, the cross section is substantially below the predictions of extended soft-photon models and isobar calculations, especially in the backward photon region.[1] Several recent static model calculations[4-6] now agree with the backward photon data, but have not yet been extended to other photon angles. We have found[12] that an approximate description of <u>all</u> the data is provided by an "external emission dominance" calculation (EED), derived from the leading

---

[†] Work supported by U. S. Energy Research and Development Admin.
[*] Present address.
[**] Present address: Computer Sciences Corp., Silver Spring, Md.

Figure 1.
a) Photon angular distribution for $E_\gamma = (50-70)$ MeV
b) Cross section vs. incident energy, backward photon counters.
(Curves show EED calculation.)

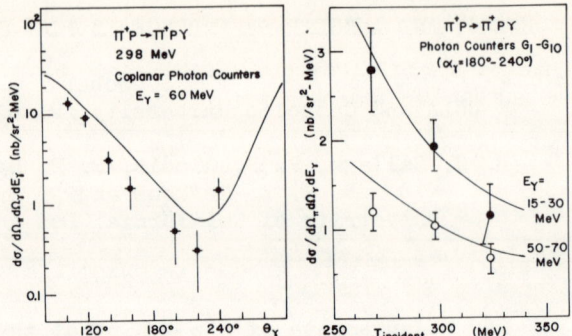

term of the Low expansion.

2. Photon angular distribution

The angular distribution of photons has now been explored at 298 MeV[10] and other incident energies. The interference between pion and proton radiation leads to a rapid variation of cross section with photon angle (Fig. 1a), which is predicted by the soft-photon expansion and EED.

3. Dependence on incident pion energy

Measurements have been made at incident beam energies of 269, 298 and 324 MeV. For a given geometry and photon energy, the differential cross sections for π^+ tend to decrease smoothly with increasing T_π (Fig. 1b), while the π^- cross sections are nearly constant. This variation is essentially that of the elastic scattering cross sections.

One motivation for repeating the experiment at several incident energies was to determine whether the simple photon spectrum at 298 MeV resulted from a fortuitous cancellation of effects. The new data imply that the 298 MeV results are consistent with the general behavior in the energy region above the $\Delta(1232)$.

REFERENCES

1. See references in Ref. 3.
2. M. Arman et al., Phys. Rev. Letters 29, 962 (1972).
3. D. Sober, M. Arman, D. Blasberg, R. Haddock, K. Leung, B. Nefkens, B. Schrock, J. Sperinde, Phys. Rev. D 11, 1017 (1975).
4. R. H. Thompson, Nuovo Cimento 16A, 290 (1973).
5. B. Bosco et al., Physics Letters 60B, 47 (1975).
6. Q. Ho-Kim and J. P. Lavine, Physics Letters 60B, 269 (1976).
7. M. M. Musakhanov, Sov. J. Nucl. Phys. 19, 319 (1974).
8. D. S. Beder, Nuclear Phys. B84, 362 (1975).
9. C. Picciotto, Nuclear Phys. B89, 357 (1975), and TH-1997-CERN, unpublished (1975).
10. K. C. Leung et al., to be published. Also, B. Nefkens et al., Proc. 5th Conf. High Energy Phys. & Nucl. Structure, G. Tibell, ed. (North Holland, 1973).
11. F. E. Low, Phys. Rev. 110, 974 (1958).
12. B. M. K. Nefkens and D. I. Sober, to be published.

J. Negele (MIT): Would you please elaborate in a little more detail precisely what physics goes into the EED approximation?

Sober: The EED ("external emission dominance") calculation is what we get by keeping only the <u>first</u> term in the Low soft-photon amplitude. Squaring and summing over photon polarizations, this gives

$$|\mathcal{M}_{\pi p \gamma}|^2 = -e^2 \left| -\frac{q_1^\mu}{q_1 \cdot k} - \frac{p_1^\mu}{p_1 \cdot k} + \frac{q_2^\mu}{q_2 \cdot k} + \frac{p_2^\mu}{p_2 \cdot k} \right|^2 |\mathcal{M}_{el}(\bar{s},\bar{t})|^2$$

where the q's, p's and k are the pion, proton and photon 4-momenta; $|\mathcal{M}_{el}(\bar{s},\bar{t})|^2$ is the square of the elastic amplitude and is proportional to the elastic scattering cross section evaluated at suitable kinematics. It is the simplest calculation one could think of doing, and there is no reason to expect that it should work as well as it does. For more details, see the UCLA preprint by Nefkens & Sober.

MODEL CALCULATIONS FOR RADIATIVE PION PROTON SCATTERING

Q. Ho-Kim[†]
Université Laval, Québec, Canada G1K7P4

J. P. Lavine[*]
University of Rochester, Rochester, N.Y. 14627

ABSTRACT

We present calculations of $\pi^{\pm}p$ bremsstrahlung cross-sections at an incident pion laboratory kinetic energy of 298 MeV and at backward photon angles. Our formalism incorporates off-the-energy-shell πp scattering amplitudes and on-the-mass-shell electromagnetic vertices. We compare our theoretical cross-sections with recent experimental data.

INTRODUCTION

Recently $\pi^{\pm}p$ radiative scattering cross-sections have been measured for pion beams of 298 MeV kinetic energy and for the backward photon region.[1] This kinematic situation was investigated since it was predicted[2] to be favorable for extracting the magnetic dipole moment of the $\Delta^{++}(1236)$ resonance from the observed bremsstrahlung cross-sections. The lack of the predicted peak for photon energies of 70 to 90 MeV has led to further theoretical work (see 3,4 and the articles cited therein). We have performed a series of bremsstrahlung calculations that include off-the-energy-shell hadronic amplitudes and on-the-mass-shell electromagnetic vertices. The πp scattering is described by models which have evolved from approximations to the Chew-Low equations[5], but which take account of proton recoil and inelasticity. These models give good predictions for $\pi^{\pm}p$ elastic scattering cross-sections. Our bremsstrahlung formalism is non-relativistic except for the kinematics. We consider external emission radiation from the pion and proton charges and from the proton magnetic moment. We choose an interaction current term that is based on considerations of current conservation[6], but which does not contribute numerically to the bremsstrahlung amplitude due to regauging. Our radiative scattering model is improved with respect to our first efforts.[4]

RESULTS

We present bremsstrahlung cross-sections for pions of 298 MeV laboratory kinetic energy. The πp interaction model involves a

---

[†]Work supported in part by the National Research Council of Canada and the Ministeré de l'Education du Québec.
[*]Work supported in part by the U.S. Energy Research and Development Administration.

cut-off function

$$v(q) = \exp(-q^2/2q_o^2), \qquad (1)$$

with q_o = 32.725 μ, where μ is the pion mass in fm^{-1}. Our calculated π^+ and π^- cross-sections for selected geometries are shown in Figs. 1 and 2. They compare rather well with the experimental data[1], which are averaged over 10 photon counters. In the π^+p reaction the P_{33} wave is dominant, which explains the appearance of a resonance in some geometries, and the charge terms interfere destructively, which explains the relatively small cross-section. In the π^-p reaction other P-waves also contribute, and the charge terms interfere constructively. We need to evaluate the role of internal radiation terms and of off-the-mass-shell proton electromagnetic form factors. Preliminary calculations with crude models for the latter show that our cross-sections are less affected by uncertainties in the form factors than are those of Picciotto.[3] We are continuing our systematic study of πp radiative scattering.

REFERENCES

1. D. I. Sober et al., Phys Rev. D11, 1017 (1975).
2. L. A. Kondratyuk and L. A. Ponomarev, Sov. J. Nucl. Phys. 7, 82 (1968).
3. B. Bosco et al., Phys. Lett. 60B, 47 (1975); D. S. Beder, Nucl. Phys. B84, 362 (1975); C. Picciotto, Nuovo Cimento 29A, 41 (1975).
4. Q. Ho-Kim and J. P. Lavine, Phys. Lett. 60B, 269 (1976).
5. G. F. Chew and F. E. Low, Phys. Rev. 101, 1570 (1956).
6. F. E. Low, Phys. Rev. 110, 974 (1958).

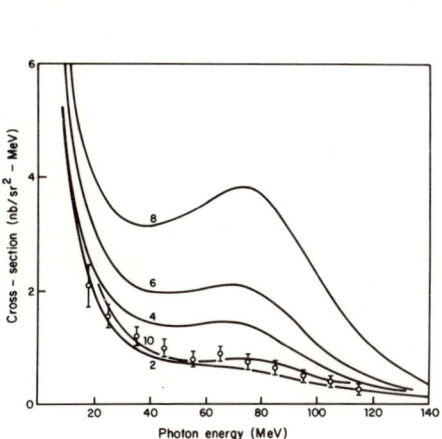

Fig.1. Differential cross-sections for π^+p radiative scattering.

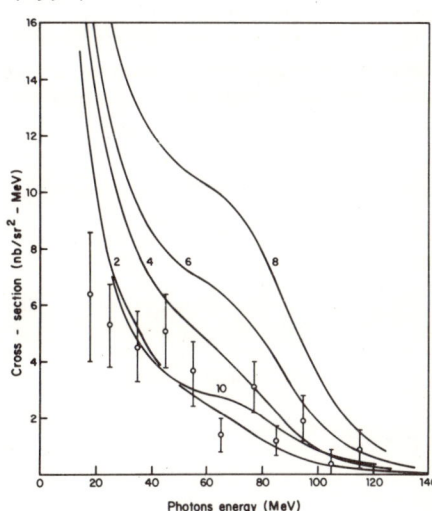

Fig.2. Differential cross-sections for π^-p radiative scattering.

B. M. K. Nefkens (UCLA): All the predictions for $\pi^+p \to \pi^+p\gamma$ that you just have shown indicate a small but unmistakenly evident bump in the photon spectrum related to the $P_{33}(1232)$ resonance. There is no evidence in any of our experimental data for such a bump. Does this not worry you?

Lavine: No. Our theory is still incomplete with respect to the double-scattering terms. There are indications in the calculations of B. Bosco et al. (Phys. Lett. 60B, 47 (1975)) and of A. H. Huffman (Bull. Amer. Phys. Soc. 20, 88 (1975)) that double-scattering will lower the cross section at large photon energies through destructive interference. Also the data for some of the individual photon detectors show some slight indications of a bump. So perhaps further experiments will clarify this.

B. F. Gibson (LASL): Since you have not included double scattering and internal emission diagrams, can you relate your calculation to the low energy theorem:
$$\sigma \sim \frac{A}{k} + B + ck + \ldots$$

Lavine: Our choice for the interaction terms can arise from the double-scattering terms. This is discussed in J. H. McGuire, A. H. Cromer and M. I. Sobel (Phys. Rev. 179, 948 (1969)). They also discuss the connection of such terms with Low's theorem. I believe our interaction term gives a non-analytic term in the photon energy k, as k goes to zero.

A. W. Thomas (CERN): You have described a calculation in which there is no elementary Δ. It seems that the calculations which agree with experiment are those in which there is no coupling to an elementary Δ. Is this a fair summary?

Lavine: A qualified yes. In our calculations we only use pion-proton off-the-energy-shell scattering amplitudes. I do not know how our results would change if we were to include explicitly radiation from the magnetic dipole moment of the Δ.

IMPROVED π-^{12}C OPTICAL POTENTIAL AT 50 MeV

M. Thies

State University of New York, Stony Brook, LI, New York 11794

ABSTRACT

The order-of-magnitude discrepancy between experiment and a Kisslinger potential calculation of π-^{12}C elastic scattering at 50 MeV disappears if kinematical corrections and higher order effects (Pauli- and short-range correlations, absorption) are taken into account.

The aim of this work is to demonstrate that corrections of higher order in the density to the pion optical potential (OP) are important for an understanding of low energy π-nucleus scattering. The situation is much more favorable than in the region of the 33-resonance, where the strong absorption of the 1.order OP drastically reduces the sensitivity of $d\sigma/d\Omega$ with respect to effects of higher order.

A calculation of π-^{12}C elastic scattering at 50 MeV was done and compared to the experimental data (ref.1). The Kisslinger potential misses the small-angle cross-section by a factor 5-7, the position of the minimum by 20°. The following corrections have been applied:

1) Lab/CM-transformation. An approximate Lorentz-transformation was used to construct the π-N scattering amplitude entering the 1.order OP in the right frame. The angular transformation in the p-wave part is crucial for the observed shift of the minimum.

2) S-wave parameters. In analogy to the OP used for pionic atoms, we have evaluated the correction arising from Pauli-correlations and found it to be less important than at threshold. The πNN-absorption and dispersion term (quadratic in ρ) was taken into account simply by using the best-fit parameter B_o from pionic atoms.

3) P-wave parameters. The 'Lorentz-Lorenz correction' (ref.2) to 2nd order in ρ was estimated, allowing for finite-range effects of the interaction, rho-meson exchange, short-range and Pauli correlations. We end up with a 'strength-parameter' $\xi \simeq 1.0$-1.5, which is much less cut-off dependent than the part resulting from pions and short-range correlations alone. Pauli correlations tend to increase the pionic contribution, but reduce the effect of the rho-meson by more than 50%.

The resulting optical potential reads

$$-\frac{2\omega}{4\pi} U_{opt}(\underline{r}) = \left\{ \frac{M+\omega}{M} b_o - \left\langle \frac{1}{r} \right\rangle (b_o^2+b_1^2 2) + B_o \rho(r) \right\} \rho(r) - c_o \nabla \rho(r) \left\{ 1 - \frac{1}{3} 4\pi\xi c_o \rho(r) \right\} \underline{\nabla} + \frac{\omega}{M+\omega} c_o \Delta\rho(r) \quad (1)$$

In fig.1,curve 1,we display the $\pi-^{12}$C differential cross-section as calculated with the improved optical potential,for $\xi=1.2$.Curve 2 does not include the angular transformation,whereas curve 3 corresponds to the case $\xi=0$ (no correlations).Without the πNN absorption term,the minimum would be much deeper.

Due to uncertainties in the input (πN phase shifts at low energies),it would be premature to draw definitive quantitative conclusions from this calculation.Nevertheless,the result is encouraging and lends support to the expectation that elastic π-nucleus scattering at low energies may become a useful testing ground for theories of the π-nucleus interaction.

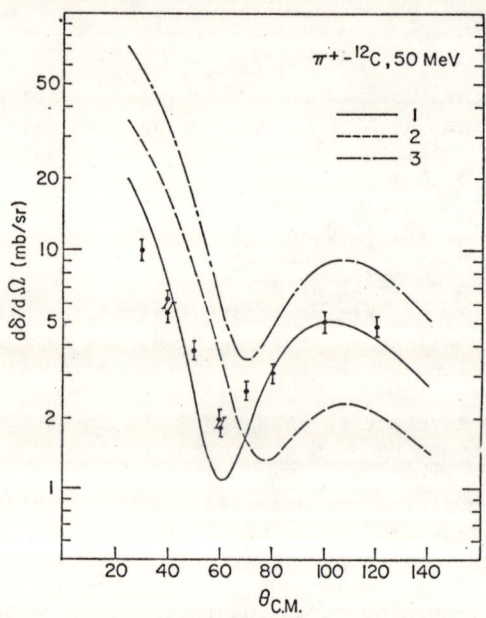

Fig. 1. Differential cross-section for elastic $\pi-^{12}$C scattering at 50 MeV.Curve 1: Full calculation.Curve 2: Same as 1,without angular transformation. Curve 3: Same as 1,without the Lorentz-Lorenz correction.The experimental points are taken from ref. 1.

REFERENCES

1. J. F. Amann, P. D. Barnes, M. Doss, S. A. Dytman, R. A. Eisenstein, and A. C. Thompson, Phys. Rev. Lett. <u>35</u> ,426 (1975).
2. M. Ericson and T. E. O. Ericson, Ann. Phys. (N.Y.) <u>36</u> ,323 (1966)

Mikkel Johnson (Los Alamos): Your calculation seems to be quite sensitive to the addition of your small corrections. Doesn't this mean that small additional improvements in the theory can destroy your nice fit?

M. Thies: No. I prefer this sensitivity to the complete lack of it in the region of the (3,3) resonance.

Rubin Landau (Oregon State U.): This comment is as relevant to this talk as it is to the previous one. Probably the main conclusion of the calculation Thomas and I did was that the uncertainty in the basic π-N phase shifts is large enough to reduce discrepancy between theory and experiment by a factor of ~2. What this means is that the comparison between theory (N1 theories) and π-nucleus data are somewhat premature of their point. We really must have much better π-N data before we can do that. Aside from this warning, Thomas and I have found that the remaining discrepancies can be accounted for by very simple bindings and recoil corrections which are not very important at higher energy.

L. Heller (NASL): Since the phrase "angle transformation" has been used many times, I want to make some comments about it, in the way of advertising my talk on Thursday. 1. There is no meaning to Lorentz-transforming an off-energy-shell T matrix element because it is not a covariant object. 2. Nevertheless, there is a definite rule for evaluating such a matrix element where the total three-momentum of the two-particles is not zeros. 3. This rule is not just a matter of using a correct angle for the scattering; it is considerably more complicated. 4. There is an approximation which may be good for pion-nucleus scattering at medium energy, which consists of two parts: a) use a properly defined relativistic relative momentum in both the initial and final states. (Both the magnitudes and directions are different from the nonrelativistic expressions); b) there is an important multiplicative factor.

COVARIANT THEORY OF THE OFF-SHELL PION-NUCLEON SCATTERING AMPLITUDE*

J. Barry Cammarata**
Institute of Theoretical Physics, Department of Physics
Stanford University, Stanford, California 94305

Manoj K. Banerjee

Department of Physics and Astronomy
University of Maryland, College Park, Md. 20742

In analyzing pion-nucleus scattering and reactions in the impulse approximation it is necessary to specify an off-shell pion-nucleon scattering amplitude. Without a dynamical theory of the πN interaction the off-shell behavior is unknown and has to be hypothesized. The off-shell amplitudes which have been most widely used (e.g. the Kisslinger form) are merely extrapolations of the on-shell transition matrix, and generally the off-shell behavior in the momentum variables of these amplitudes does not reflect the physical occurrence of resonances or other known properties of the πN interaction. Since it is now well established[1] that the off-shell behavior will significantly affect pion-nucleus calculations it is desirable to have a more reasonable model for this behavior. To accomplish this we have extended the classic work of Chew and Low[2] on the off-shell pion-nucleon scattering amplitude at low energy to a covariant theory which includes the effects of nucleon recoil and the seagull terms. The latter were evaluated with the help of the soft-pion limit of the off-shell amplitude, obtained by letting the four momentum k_f of the final pion vanish. This soft-pion amplitude contains the σ-commutator and nucleon pole terms (in the $k_f = 0$ limit). The soft-pion amplitude is subtracted from the off-shell amplitude so that the latter now is determined from what may be called a once-subtracted Low equation. The Born or driving term of this equation is composed of a vector current term, the σ-commutator term, plus the sum of all nucleon and antinucleon pole terms. The latter are included in the Low expansion and the subtraction terms in the soft-pion amplitude and, when summed, have the simple and manifestly covariant form

$$\bar{u}(p_f) \left[\frac{\not{k}_f \not{k}_i}{(p_i+k_i)^2 - M^2} \tau_f \tau_i + \frac{\not{k}_i \not{k}_f}{(p_i-k_f)^2 - M^2} \tau_i \tau_f \right] u(p_i) ,$$

---

*Work supported by the U.S. Energy Research and Development Administration and the National Science Foundation.
**Address after 1 September 1976: Department of Physics, Virginia Polytechnic Institute, Blacksburg, Virginia 24061

which can be shown to be a covariant generalization of the driving term of the Chew-Low static theory. The validity of this approach rests on the assumption that the πN form factor has approximately the same magnitude for invariant momentum transfers of 0 and $4M^2$, where M is the nucleon mass. A crude dispersion relation analysis of this form factor[3] is consistent with this assumption.

REFERENCES

1. R. H. Landau, S. C. Phatak, and F. Tabakin, Ann. Phys. (N.Y.) 78, 299 (1973).
2. G. F. Chew and F. E. Low, Phys. Rev. 101, 1570 (1956).
3. J. B. Cammarata and M. K. Banerjee (unpublished).

DETERMINATION OF PION-NUCLEON FORM FACTOR FROM PION-NUCLEON ELASTIC SCATTERING

D. J. Ernst
Texas A&M University, College Station, TX 77843

Mikkel Johnson
Los Alamos Scientific Laboratory, Los Alamos, NM 87545

ABSTRACT

The pion-nucleon form factor in the static Chew-Low model of pion-nucleon scattering is determined from the elastic pion nucleon data. Crossing symmetry and the effect of the inelastic pion production channels on the elastic channel are included. The form factor is determined up to a momentum of $k = 10m_\pi$ by adjusting it to fit the pion-nucleon data.

THEORY AND RESULTS

The interaction hamiltonian for a pion interacting with a nucleon is assumed to be of the form

$$H_1 = \sum (V_k^{(0)} a_k + V_k^{(0)\dagger} a_k^\dagger)$$

$$V_k^{(0)} = if_r^{(0)} \vec{\sigma} \cdot \vec{k} \, \tau_k v(k)/(2\omega_k)^{1/2} \qquad (1)$$

where $a_k^\dagger$ creates a pion of momentum $\vec{k}$, energy ω_k, $f_r^{(0)}$ is a pion-nucleon coupling constant and $\vec{\sigma}$ and $\vec{\tau}$ are nucleon spin and isospin operators. Following Chew and Low, we may relate $v(k)$ to the scattering amplitude $t_{pq}(\omega)$. If we introduce $h_\alpha(\omega)$ via

$$t_{pq}(\omega) = -4\pi \frac{v(p)v(q)}{\sqrt{4\omega_p \omega_q}} \sum_{\alpha=1}^{4} P_\alpha(q,p) h_\alpha(\omega), \qquad (2)$$

where P_α is the projection operator onto a state of total angular momentum and isospin, then $h_\alpha(\omega)$ satisfies the disperion relation

$$h_\alpha(\omega) = \frac{\lambda\alpha}{\omega} + \frac{1}{\pi} \int_\mu^\infty d\omega' \frac{Imh_\alpha(\omega')}{\omega'-\omega-i\eta} + \frac{1}{\pi} \int_\mu^\infty d\omega' \frac{Imh_\alpha(-\omega')}{\omega'+\omega} \qquad (3)$$

Chew and Low then keep only one-nucleon, one-pion intermediate states in Eq. (3) and derive a non-linear equation for $h_\alpha(\omega)$. We solve the Chew-Low equation with crossing symmetry and without making this "one-meson approximation."

From Eq. (2) we may parameterize $h_\alpha(\omega)$ in terms of the elastic scattering data by

$$h_\alpha(\omega_p) = \hat{\eta}_\alpha(p) e^{i\hat{\delta}_\alpha} \sin\hat{\delta}_\alpha / p^3 v^2(p), \quad (4)$$

with $\hat{\delta}_\alpha$ and $\hat{\eta}_\alpha$ real. Below inelastic threshold, $\hat{\eta}_\alpha$ equals one and $\hat{\delta}_\alpha$ is the usual phase shift. If we use Eq. (4) and otherwise follow Chew and Low we may convert the dispersion theory into an equation which determines $h_\alpha(\omega)$. The resulting equations are essentially the same as the equations obtained in the one-meson approximation except that they explicitly contain the function $\hat{\eta}_\alpha(p)$. Using $\hat{\eta}_\alpha$ as determined by experiment, this equation can be solved to yield $\hat{\delta}_\alpha$ for a given $v(p)$. We adjust $v(p)$ so that $\hat{\delta}_\alpha$ (theory) = $\hat{\delta}_\alpha$ (experiment) in the dominant (3,3) channel. The numerical procedure of Salzman and Salzman[1] is followed. The form factor $v(k)$ is pictured in Fig. (1).

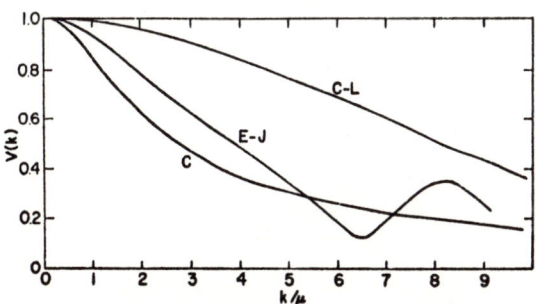

Fig. 1. Our form factor (E-J) compared to the Salzman-Chew-Low (C-L) and the Carter, et al[2] (c) determinations.

REFERENCES

1. G. Salzman and F. Salzman, Phys. Rev. <u>108</u> (1957) 1619.
2. A. A. Carter, et al, Nucl Phys. <u>B26</u> (1971) 445.

SEPARABLE-POTENTIAL MODEL FOR THE PION-NUCLEON INTERACTION

W. T. Nutt
Department of Physics, Brooklyn College, Brooklyn, N. Y. 11210

ABSTRACT

A separable potential which fits the low and intermediate π-N scattering is proposed which is more convenient for application than those separable models which use Regge parameterizations of the very high energy phase shifts. The form factors for this model are equal to zero for momenta q>1 GeV/c, and are expected to provide more reasonable off-shell behavior than the form factors obtained from those models based on the Regge extrapolation.

Recently a potential model has been proposed which provides an exact description of the on-shell scattering data for energies less than 2 GeV.[1] This model is characterized by form factors which have extremely large extensions in momentum space due to the phase-shift extrapolation used at high energy. In this contribution we propose an alternate treatment of the high-energy behavior which leads to a potential model with form factors whose limited extension in momentum space make them more convenient for various applications of the separable model.

Following ref. 1 we introduce a quasi phase $\hat{\delta}$ and a parameter $\hat{\eta}$, such that the scattering amplitude is given by

$$f \equiv \frac{\hat{\eta} e^{2i\hat{\delta}} - 1}{2iq} = \frac{\hat{\eta} e^{i\hat{\delta}} \sin\hat{\delta}}{q}$$

Then one has for $D(E)$, and the form factors, $v(q)$[1],

$$\ln D(E) = \frac{1}{\pi} \int \frac{\hat{\delta}(E') dE'}{E - E' + i\epsilon}$$

$$v^2(q) = -\frac{\lambda (E_q + \omega_q)}{2\pi q (E_q \omega_q)} \hat{\eta} \sin\hat{\delta} \exp \frac{P}{\pi} \int_{E_{th}} \frac{\hat{\delta}(E') dE'}{E - E'}$$

The extrapolation used for $\hat{\eta}$ and $\hat{\delta}$ in ref. 1 makes the form of $v(q)$ quite sensitive to the scattering in the multi-GeV region, where the separable model is of questionable value. The calculation of the function $D(E)$ then becomes very difficult for $E<E_{th}$ since integrations must be carried out to energies greater than 30 GeV to obtain convergence. (It is worth noting that in a complete dynamical analysis based upon a careful treatment of off-shell effects, an accurate evaluation of D at energies below the physical threshold is particularly important for understanding the low-energy pion-nucleus scattering data.)

We propose to avoid some of these problems by using an extrapolation such that $\hat{\delta}=0$ and $\eta=1$ for energies $\geq$ 2.4 GeV. (This implies $\delta=0$ or $\delta=\pi$ and $\eta=1$ for E> 2.4 GeV). This procedure yields form factors which are zero for q>1 GeV/c and allows for a more accurate calculation of D(E) for $E<E_{th}$. We also believe that our model provides a more reasonable specification of the T matrix for highly off-shell values of the momentum and energy parameters.

1. J. T. Londergan, K. W. McVoy and E. J. Moniz, Ann. Phys. (N.Y.) 86, 147 (1974).

THE KLEIN–GORDON EQUATION WITH OPTICAL POTENTIALS OF THE STRONG INTERACTION

R. Seki
California State University, Northridge, California 91324

ABSTRACT

The Klein-Gordon equation with the external electromagnetic field is unambiguously formulated by the two component theory as shown by a lucid presentation of Feshbach and Villars[1]. Following Hamiltonian formalism, they wrote down a definite form of the Klein-Gordon equation with a real, non-momentum dependent potential for the strongly interacting pion system nearly two decades ago. In this pedagogical note we describe the Klein-Gordon equation with less restricted optical potentials, relying on our guidance in field theoretical Lagrangian.

We start with a Lagrangian for a scalar boson field ϕ,

$$\mathcal{L} = -(\partial_\mu \phi^\dagger \partial_\mu \phi + \mu^2 \phi^\dagger \phi) - S(x)\phi^\dagger \phi + T(x)\partial_\mu \phi^\dagger \partial_\mu \phi + i U_\mu(x)[(\partial_\mu \phi^\dagger)\phi - \phi^\dagger(\partial_\mu \phi)] - U_\mu(x)U_\mu(x)\phi^\dagger \phi, \quad (1)$$

which gives

$$(\partial_\mu - i U_\mu(x))^2 \phi = \mu^2 \phi + S(x)\phi + \partial_\mu(T(x)\partial_\mu \phi). \quad (2)$$

In analogy to the field equation above, we write one particle equation for a scalar boson (artificially charged +e) of mass m as

$$\left(i\frac{\partial}{\partial t} - e\phi - u(\underline{r})\right)^2 \psi = \left[(\underline{p} - e\underline{A}(\underline{r}))^2 + m^2 + m s(\underline{r}) + m \underline{\nabla} t(\underline{r}) \cdot \underline{\nabla}\right]\psi, \quad (3)$$

where we made the replacements and assumptions

$$S(x) \text{ and } T(x) \longrightarrow m s(\underline{r}) \text{ and } m t(\underline{r})$$

$$U_\mu(x) \longrightarrow i u(\underline{r})\delta_{4\mu} + A_\mu(\underline{r}),$$

and the external potentials $s(\underline{r}), t(\underline{r})$ and $u(\underline{r})$ are allowed to be complex functions.

We interpret the relativistic equation Eq. 3, via the two component theory: It yields $\psi = \varphi + \chi$, where $\varphi(\chi)$ is the positive (negative) energy wave function. Equation 3 is rewritten as

$$i\frac{\partial}{\partial t}\underline{\Psi} = H\underline{\Psi} = (\sigma_3 + i\sigma_2)\left[\frac{\underline{\pi}^2}{2m} + s(\underline{r}) + \underline{\nabla} t(\underline{r}) \cdot \underline{\nabla}\right]\underline{\Psi} + \left[e\phi + u(\underline{r})\right]\underline{\Psi} + m\sigma_3\underline{\Psi}, \quad (4)$$

where $\Psi = \begin{pmatrix} \varphi \\ \chi \end{pmatrix}$ and $\vec{\Pi} = \vec{p} - e\vec{A}$. Following Ref. 1, we see that the Hamiltonian above satisfies the condition of Hermiticity in a sense of $\sigma_3 \tilde{H}^* \sigma_3 = H$ when s,t,u are real. Furthermore, in this case, the interaction does not remove the particle-antiparticle degeneracy for a neutral boson when $u(\underline{r}) = 0$: $u(\underline{r})$ is definitely excluded for the pion system. For the kaon system, the neutral kaons do not have this degeneracy and a further care has to be taken in order to include the strangeness in Eq. 4.

We note that the equation for eigenvalue problems in mesonic atoms is Eq. 3 with the replacement $i\partial/\partial t \to E$ and that it is not an equation just for the positive energy wave function as has been commonly interpreted. The equation is obtained strictly for the positive energy wave function via Foldy-Wouthuysen transformation: It yields one component Hamiltonian for φ,

$$H' = \left(m + \frac{\Omega^2}{2m} - \frac{\Omega^4}{8m^3} + \cdots \right) + \xi + \frac{1}{32m^4} [\Omega^2, [\Omega^2, \xi]] + \cdots,$$

where $\Omega^2/2m = \Pi^2/2m + S(\underline{r}) + \underline{\nabla} t(\underline{r}) \cdot \underline{\nabla}$, $\xi = e\phi + u(\underline{r})$, and the last term explicitly shown is the Zitterbewegung term.

The imaginary part of the potentials can be justified by writing the continuity equation

$$\frac{\partial}{\partial t} \bar{\psi} \psi + \underline{\nabla} \cdot \underline{j} = \mathcal{S}(\underline{r}),$$

where $\bar{\Psi} = \psi^* \sigma_3$. The current $\underline{j}$ and the volume and surface absorption $\mathcal{S}$ are given by

$$\underline{j} = i \left[\underline{\nabla} \bar{\psi} \Sigma (1/2m + \text{Re } t) \psi - \bar{\psi} \Sigma (1/2m + \text{Re } t) \underline{\nabla} \psi \right] + (e/m) \bar{\psi} \Sigma \underline{A} \psi,$$

$$\mathcal{S} = 2 \left\{ \bar{\psi} \Sigma \text{Im}(s+u) \psi + \underline{\nabla} \cdot [\bar{\psi} \Sigma \text{Im} t \underline{\nabla} \psi - \underline{\nabla} \bar{\psi} \Sigma \text{Im} t \psi] \right\},$$

where $\Sigma = \sigma_3 + i\sigma_2$ and s = Re s + i Im s and similarly for t and u.

Further sophistications such as an introduction of the isospin with the charge exchange terms and of the center-of-mass motion correction in the lowest order can be done along the methods shown in Ref. 1 and Ref. 2, respectively.

1. H. Feshbach and F. Villars, Rev. Mod. Phys. 30, 24 (1958). See also J. D. Bjorken and S. D. Drell, Relativistic Quantum Mechanics (McGraw-Hill, N.Y. 1964), Chapter 9.
2. R. Seki, Phys. Lett. 58B, 49 (1975).

PAULI-BLOCKING EFFECTS IN LOW ENERGY PION-NUCLEUS SCATTERING

C. B. Dover*
Brookhaven National Laboratory, Upton, NY 11973

R. H. Lemmer
Rand Afrikaans University, Johannesburg, S. Africa

ABSTRACT

It is shown that a hole-line expansion of the pion self-energy overestimates the quenching of the πN scattering amplitude in nuclear matter, and that the inclusion of RPA-type correlations is essential for a correct description.

INTRODUCTION

In a qualitative discussion of pion-nucleus scattering near the (3,3) resonance, it has been suggested[1] that the main effect of the Pauli principle is to "block" (quench) the free πN scattering amplitude. We indicate in this note that (i) this suggestion is equivalent to a single hole-line expansion of the pion self-energy π in nuclear matter, and (ii) how this expansion has to be augmented by RPA-type correlations that are necessarily present in a field-theoretic description and which appreciably modify the effective πN amplitude at low energies.

PION SELF-ENERGY

We obtain a hole-line expansion for $\pi(q)$ at pion momentum and energy $q = (\vec{q},\omega)$ by restricting the π-diagrams to be single bubble diagrams, containing <u>one</u> independent hole-line, but any number of internal pion lines. Then $\pi \simeq 4\pi\rho_0 <\tilde{f}>$, ρ_0 = density of nuclear matter, $<\tilde{f}>$ = Fermi average of the πN forward-scattering amplitude $\tilde{f}$ in the medium that is given by diagrams like 1(a) to 1(e). One can show[2] that $<\tilde{f}>$ satisfies a Chew-Low equation[3] of the form

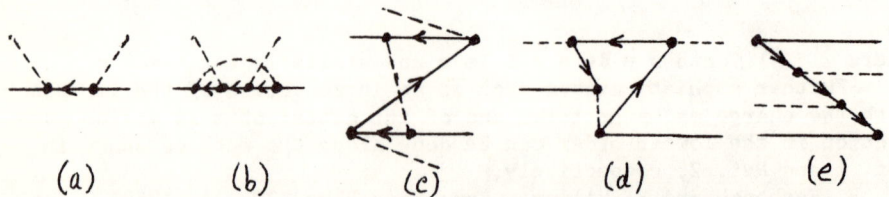

Fig. 1 Typical diagrams for the effective πN amplitude. Arrows pointing left (right) denote particles (holes). Each diagram has a counterpart with crossed external meson lines.

* Supported, in part, by Energy Research and Development Administration.

$$\tilde{h}_\alpha(\omega) = \frac{\lambda_\alpha}{\omega} F(q) + \frac{1}{\pi} \int_\mu^\infty d\omega_p \, p^3 \, F(p) \left\{ \frac{|\tilde{h}_\alpha(\omega_p)|^2}{\omega_p - \omega - i\delta} + \sum_\beta A_{\alpha\beta} \frac{|\tilde{h}_\beta(\omega_p)|^2}{\omega_p + \omega} \right\} \quad (1)$$

that contains the blocking factor $F(q)$ explicitly.[2] $\tilde{h}_\alpha(\omega)$ refers to that part of $<f>$ in spin-isospin channel α. The other notation[3] is standard. Equation (1), which would suggest a strong quenching of the $\tilde{h}_\alpha$ near threshold $q \sim 0$, is incomplete in two respects. (i) Simple Fermi-averaging is too crude to properly account for the fact that the "crossed" Born diagram of 1(a) (not shown) makes available as many states as are blocked out in 1(a), and thereby removes the $F(q)$-factor in the Born term. (ii) Backward-going graphs involving pion exchange between nucleons in the medium are absent in the first term of the hole-line expansion, which includes only 1(b). By allowing additional hole-lines, we introduce processes such as shown in Fig. 1(c-e). Including both (i) and (ii) leads to an equation of the form

$$h_\alpha(\omega) = \frac{\lambda_\alpha}{\omega} + \frac{1}{\pi} \int_\mu^\infty d\omega_p \, p^3 \left\{ \frac{F(p)}{\omega_p - \omega - i\delta} - \frac{1-F(p)}{\omega_p + \omega} \right\} |h_\alpha(\omega_p)|^2 + \text{crossing term} \quad (2)$$

in which the Born term is not blocked, and the dispersion term is modified. Near threshold ($\omega \approx \mu$), one can obtain the fourth order correction to the free space amplitude $h_\alpha^{free}(\omega)$ in the form ($\alpha = \{T,J\} = \{11, 13, 31, 33\}$)

$$h_\alpha(\mu) - h_\alpha^{free}(\mu) = -4\pi N[\rho_0] \left(\hat{\lambda}_\alpha^2 + \sum_\beta A_{\alpha\beta} \hat{\lambda}_\beta^2 \right)$$

$$N[\rho_0] = \sum_\ell (1-F(p))/\omega_p^2 \quad ; \quad \hat{\lambda}_\alpha = \frac{f^2}{3}(1,-2,-2,4) \quad (3)$$

Using Eq. (3), we have studied the influence of Pauli effects on the low energy pion-nucleus optical potential.[2] Even for nuclear matter densities ($\rho_0 \approx 0.48\mu^{-3}$), the net effect is small.

REFERENCES

1. H.A. Bethe, Phys. Rev. Lett. 30, 105 (1973); J. Eisenberg and H.J. Weber, Phys. Lett. 45B, 110 (1973).
2. C.B. Dover and R.H. Lemmer, to be published.
3. G.F. Chew and F.E. Low, Phys. Rev. 101, 1570 (1956).

PIPIT: A MOMENTUM SPACE OPTICAL POTENTIAL CODE FOR PIONS.[†]

R. A. Eisenstein
Carnegie-Mellon University, Pittsburgh, PA 15213

F. Tabakin
University of Pittsburgh, Pittsburgh, PA 15260

Momentum space methods[1,2] for solving scattering problems offer a useful alternative to the standard[3,4] coordinate space approach. This is so because finite range, non-local potentials, which would be cumbersome in $\vec{r}$ space often can be expressed simply in $\vec{p}$ space. Thus, a momentum space optical potential code is in general a more fertile theoretical testing ground for diverse reaction mechanisms. This is especially true now that it is possible to include the Coulomb-interaction in an essentially exact manner using the method of Vincent and Phatak.[5]

The optical model program PIPIT[6] is an outgrowth and extention of earlier versions (unpublished) by Landau, Lee, Phatak and Tabakin. The code solves the Lippmann-Schwinger equation with relativistic kinematics, incorporating an optical potential specialized to π-nucleus scattering. However, the structure of the program is quite general and is in fact a numerical realization of Fredholm theory. It can be converted to other applications[7] in a straightforward way.

The present configuration of the program is fully described in ref. 6. It solves a Fredholm equation (generalized to a complex potential) for the K-matrix by means of a matrix inversion technique. The optical potential used is taken from first order multiple scattering theory:[8]

$$U(\vec{k}',\vec{k}) = \frac{A-1}{A}\{\rho_p(\vec{q})t_{\pi p}(\vec{k}',\vec{k},\vec{k}_o) + \rho_n(\vec{q})t_{\pi n}(\vec{k}',\vec{k},\vec{k}_o)\}$$

The ground state neutron and proton densities can be of p-shell Gaussian, Woods-Saxon or "Wine-bottle" form. Their size parameters are independently variable. The on-shell t-matrix is constructed from available π-N phase shifts; the off-shell extrapolation is made using: (1) the separable model of Londergan, McVoy and Moniz (2) separable form factors of Gaussian type incorporating proper threshold behavior, and (3) a Kisslinger model with Gaussian off-shell damping to control divergence. The kinematic transformation from π-N to π-nucleus frame is accomplished using the method of Phatak.[9] The Coulomb interaction is included via the method of ref. 5. In addition to cross sections and π-nucleus T-matrix elements for each partial wave, the program also calculates $\vec{r}$ space wavefunctions for any partial wave of interest.

The generality of PIPIT allows its use as a theoretical tool for the study of π-nucleus[1] and K-nucleus[7] scattering and is a useful adjunct to standard $\vec{r}$-space codes such as PIRK.[3] Although PIRK is much faster and more compact than PIPIT, it is also much more restricted in the class of problems it can solve. The generality of PIPIT suggests future application of momentum space methods to bound state problems such as π^-, K^- and $\bar{p}$ atomic systems. Work along these lines is currently in progress.

REFERENCES

1. R. Landau, S. Phatak, F. Tabakin, Ann. Phys. <u>78</u>, 299 (1973).
2. T.S.H. Lee and F. Tabakin, Nuc. Phys. <u>A226</u>, 253 (1974).
3. R.A. Eisenstein, G.A. Miller, Comp. Phys. Comm. <u>8</u>, 130 (1974).
4. R.A. Eisenstein, G.A. Miller, Comp. Phys. Comm. (to be published).
5. C.M. Vincent and S. Phatak, Phys. Rev. <u>C10</u>, 391 (1974).
6. R.A. Eisenstein and F. Tabakin (submitted to Comp. Phys. Comm.) Copies of the write up and deck may be obtained by writing to R. Eisenstein.
7. F. Tabakin and R. Eisenstein, "K^+-Nucleus Elastic Scattering" Contribution to this conference.
8. A. Kerman, H. McManus, R. Thaler, Ann. Phys. <u>8</u>, 551 (1959).
9. S. Phatak, "Pion-Nucleus Elastic Scattering" Ph.D. Thesis, University of Pittsburgh (1974).

† Work supported by the ERDA and the NSF.

AN IMPROVED OPTICAL POTENTIAL FOR LOW AND INTERMEDIATE ENERGY PIONS

R. H. Landau[*]
Dept. of Physics, Oregon State University, Corvallis, OR 97331

A. W. Thomas
Theory Division, Cern, Geneve 23, Switzerland

TEXT

In a recent letter[1] we reported that a good part, but not all, of the failure of the theoretical optical potential to describe low energy pion-nucleus elastic scattering may be due to uncertain πN input. The residual large discrepancy in the $\pi\pm$ - 4He and the $\pi+$ - 12C data near $30°$, however, could all easily be connected by a simple and reasonable kinematic and binding correction. We present here a further study of this three-body approximation to the optical potential at both low and intermediate energies, a study which not only demonstrates the usefulness of the approach, but also provides empircal information on the interaction of a pion-activated nucleon with the nuclear core.

All calculations are based on the first order optical potential, given in terms of the free πN t-matrix by

$$\langle \vec{k}'|U|\vec{k}\rangle = (A-1)\int d\vec{p}\,\phi^*(\vec{p}-\frac{(A-1)}{A}\vec{q})\phi(p)\langle \vec{k}',\vec{p}-\vec{q}-\vec{k}/A|t^{\pi N}(\omega)|\vec{k},\vec{p}-\vec{k}/A\rangle \quad (1)$$

$t^{\pi N}(\omega)$ is the pion-nucleon t-matrix, $\phi(p)$ is the nuclear wave function, $\vec{q} = \vec{k}'-\vec{k}$, and $\vec{k}$ and $\vec{k}'$ are the initial and final pion momenta in the pi-nucleus center of mass frame. If the nuclear wave function is of gaussian form, a simple change of variable permits (1) to be rewritten as

$$\langle \vec{k}'|U|\vec{k}\rangle = (A-1)\,F(q)\int d\vec{p}|\phi(p)|^2\,\langle \vec{k}',\,\vec{p}+\vec{p}_o-\vec{q}|t^{\pi N}(\omega)|\vec{k},\,\vec{p}+\vec{p}_o\rangle \quad (2)$$

where $F(q)$ is the (average) nuclear form factor, and $\vec{p}_o = -\frac{\vec{k}}{A} + (\frac{A-1}{2A})\vec{q}$. In the "factored approximation" (and for non Gaussian Form Factors)

$$\langle \vec{k}'|U|\vec{k}\rangle \simeq (A-1)\,F(q)\langle \vec{k}',\,\vec{p}_o-\vec{q}|t^{\pi N}(\omega_o)|\vec{k},\,\vec{p}_o\rangle. \quad (3)$$

We have investigated two possibilities for the important choice of πN collision energy ω_o. The first is the "optimal" choice[1,2] $\omega_o^2 = [E_\pi(k) + E_N(p_o)]^2 - (\vec{k} + \vec{p}_o)^2$ which follows from (2) and conserves πN energy when there is pi-nucleus energy conservation. Because of this fact, our "angle transformation" is quite reliable since it always relates physical or near-physical quantities. The second choice of sub-energy is based on a three-body picture in which the pion ($\vec{k}$) interacts with an active nucleon ($\vec{p} + \vec{p}_o$) outside of a passive core ($\vec{P} = -(\vec{k}+\vec{p}+\vec{p}_o)$), $\tilde{\omega} = E_\pi(k) + k^2/2Am_N + m_N - E_B - P^2/2(A-1)m_N - P^2/2(m_N + m_\pi)$. Note that the q-dependence of $\vec{p}_o$ and $\vec{P}$ makes both choices ω_o and $\tilde{\omega}$ raise the value of collision energy for larger scattering angles in addition to changing the energy for different values of $\vec{k}$ and $\vec{k}'$. The essential difference between these energy choices, lies in a "$-E_B$" term and in a "$-p^2[(m_\pi + m_N)^{-1} + ((A-1)m_N)^{-1}]$" term, obtained after substituting $\vec{P}$. If shell model values for these parameters were to be used, the πN sub-energy would be lowered by about 22 MeV and 16 MeV, respectively. Since we employ a separable πN t-matrix which exactly reproduces the πN phase shifts and solve a relativistic Lippmann-Schwinger equation in momentum space, it is straightforward to include the above modifications.

We have investigated $\pi^\mp$ 12 C at 50 MeV, and $\pi^\pm$ 4He at energies from 24 to 260 MeV, in all cases comparing our results with data.[3] Since we base our investigation of the recoil and binding effects upon a calculational approach which has shown itself to be (somewhat surprisingly) successful in its agreement with experiment[2], we believe the results and conclusions obtained may provide a valid insight into the required corrections to (2). We find that at lower energies the "smooth" πN phaseshifts of M. Salomon (TRIUMF Report-74-2) lower the calculated cross sections (as compared to those calculated using the CERN theoretical phases) and thus bring the theory closer to the data. There is not yet, however, good agreement. If a typical shell model value of nucleon binding energy, $E_B \simeq 22$ MeV, is used with the three body definition of energy, the cross sections are generally reduced by too great an amount. However, we have found that <u>using the three-body choice of energy with a smaller value of E_B provides a distinct improvement in the fit to data over the entire energy range</u>! The value of E_B tends to monotonically decrease with increasing energy, from a value $\sim$ 5 MeV at T_π = 24 MeV to a value of - 5 MeV at 200 MeV. In addition there is a progressively decreasing sensitivity to E_B, and to the energy variable prescription with increasing energy. This latter behavior is of course reasonable for what amounts to be a binding connection to the impulse approximation. The former behavior is somewhat puzzling since it indicates that the interaction of an active nucleon with the core is significantly less attractive than that of a passive nucleon in its shell model state. Perhaps higher order connections to the impulse approximation due to this nucleon-core interaction can considerably reduce the downward shift suggested by a naive interpretation of ω.[4] This has been shown to occur in a somewhat different problem and is presently under investigation for the energies and nuclei considered here.

REFERENCES

1. R. H. Landau and A. W. Thomas, submitted to Phys. Letters, Nov. 1975.
2. R. H. Landau, S. C. Phatak and F. Tabakin, Ann. of Phys. <u>78</u> (1973) 229; R. H. Landau, Ann. of Phys. <u>92</u> (1975) 205; Phys. Letters <u>57B</u> (1975) 13; and references cited within.
3. J. F. Amann et al, Phys. Rev. Letters <u>35</u> (1975); Yu. A. Shcherbakov et al., Dubna-Torino pre-print (to Nuovo Cimento); F. Binon et al., CERN Preprint, (to Phys. Rev. Letters); K. Crowe et al., Phys. Rev. <u>180</u> (1969) B49.
4. R. D. Amado; C. Schmit; G. Fäldt; F. Myhrer.

*Supported in part by the O.S.U. Research Council (NSF Gu 3662) and Computer Center.

ELASTIC SCATTERING OF PION BY He^4 AND C^{12}

T.W. Chen and D.W. Hoock
Department of Physics
New Mexico State University
Las Cruces, New Mexico 88003

The eikonal approximation is known to work well for small angle scattering at high-energy. Recently the single-backscattering approximation method has been shown to work well for high-energy scatterings in the backward region.[1] In this method a large-angle scattering is accounted for by considering just those multiple scattering processes in which backscatterings occur only once. In a pion-nucleus large-angle scattering, the pion is therefore allowed to be rescattered at the same nucleon, however, only once. The elastic scattering amplitude from the single-backscattering method is very much similar to the one from the eikonal method. In the pi-nucleus C.M. system we can write:

$$f = -\frac{1}{4\pi} \sum_{j=1}^{A} \int e^{i\vec{q}\cdot\vec{r}_j} u_j \prod_{i \neq j}^{A} (1 + \chi_i)^\gamma \qquad (1)$$

where u_j is the Fourier transform of the two-body pi-nucleon scattering amplitude,

$$\chi_i = \frac{1}{2ki} \int_{-\infty}^{z} u_i(x,y,z')dz' \qquad (2)$$

with the z-direction taken to be along the incoming momentum $\vec{k}$ and $\vec{q}$ is the momentum transfer. If $\gamma = 1$, Eq. (1) is the eikonal multiple scattering formula from a nucleus of A nucleons and, if $\gamma = 2$, it becomes the large angle scattering amplitude obtained in Ref. 1. In general, γ is, therefore, expected to be a function of the scattering angle θ and lies between 1 and 2. We have chosen

$$\gamma = 1 + \sin^4 \frac{\theta}{2} \qquad (3)$$

and applied Eq. (1) to π-He^4 and π-C^{12} elastic scatterings.[2] The differential cross section is obtained by averaging Eq. (1) with nuclear ground-state wave function, a harmonic oscillator wave function, and computing $|<\psi|f|\psi>|^2$.

The results are in good agreement with the data. Fig. 1 shows the results for π-He^4 at 180 MeV and π-C^{12} at 180 MeV. The π-nucleon amplitude is determined from phase shifts in accordance with the sudden passage approximation (solid curve) or the fixed scatterer approximation (dotted curve).

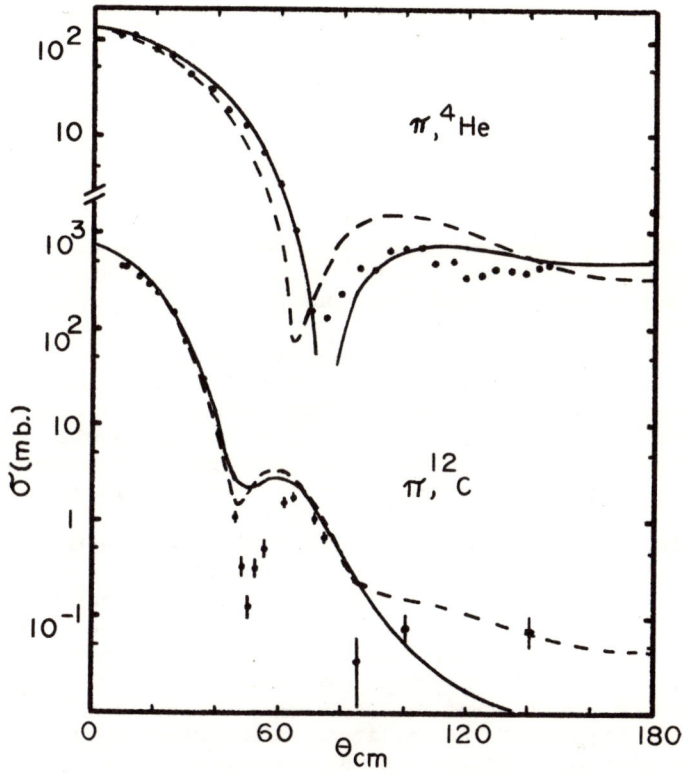

Fig. 1.

REFERENCES

1. T.W. CHEN, Phys. Rev. C. (May, 1976).
2. For more details, see D.W. Hoock, Ph.D. Thesis, New Mexico State University (1976).

EFFECTS OF THE PAULI SUPPRESSION OF THE BORN AMPLITUDE IN A NUCLEAR MEDIUM

W. T. Nutt
Department of Physics, Brooklyn College, Brooklyn, N. Y. 11210

ABSTRACT

It is noted that the suppression of the Born term in the pion-nucleon interaction which is expected due to the action of the Pauli Exclusion Principle in a nuclear medium gives rise to a downward shift of the (3,3) resonance.

The role of the Pauli Principle in suppressing the Born term of the pion-nucleon amplitude in nuclear matter has been discussed previously.[1] The main interest has been in estimating the importance of this suppression in quenching the pion-nucleon scattering amplitudes. In this work we note that the suppression of the Born term leads to a downward shift of the (3,3) resonance in a nuclear medium.

The pion-nucleon scattering amplitude may be thought of as composed of the Born amplitude and a series of higher-order graphs which impose unitarity and also a resonant structure in some channels. It has recently been suggested that the amplitude in the (3,3) channel may well be described by a resonance at about $\sqrt{s}$ = 1220 MeV and a background which interferes with the resonance to produce the observed resonant amplitude.[2] In this work we identify the background amplitude with that of Brown, Puff and Wilets (BPW).[3] (The BPW Born amplitudes are similar to those of pseudovector coupling theory for the s-waves and have a rather standard form in the p-wave channels.)

If one modifies the free π·N amplitudes in a nuclear medium by quenching the BPW Born amplitudes, one finds important effects in the P_{11} and P_{33} channels. There is no longer a 1570 MeV resonance in the P_{11} channel since that resonance appeared in the BPW Born amplitude. In the case of the P_{33} channel, one finds that the resonance in the modified T matrix is shifted downward to 1212 MeV. This value is consistent with a dispersion theory analysis of Hohler et al.[4]

In summary, we conclude that if the Born amplitude is indeed suppressed by Pauli blocking, and the rest of the amplitude containing higher order graphs is hardly modified, one can predict a downward shift of the (3,3) resonance in nuclear matter of about 20 MeV.

1. C. B. Dover, D. J. Ernst and R. M. Thaler, Phys. Rev. Letters 32, 557 (1974) and references given therein.
2. M. G. Olsson, Univ. of Wisconsin, preprint.
3. W. D. Brown, R. D. Puff and L. Wilets, Phys. Rev. C2, 331 (1970).
4. G. Höhler, H. P. Jakob and R. Strauss, Nucl. Phys. B39, 232 (1972).

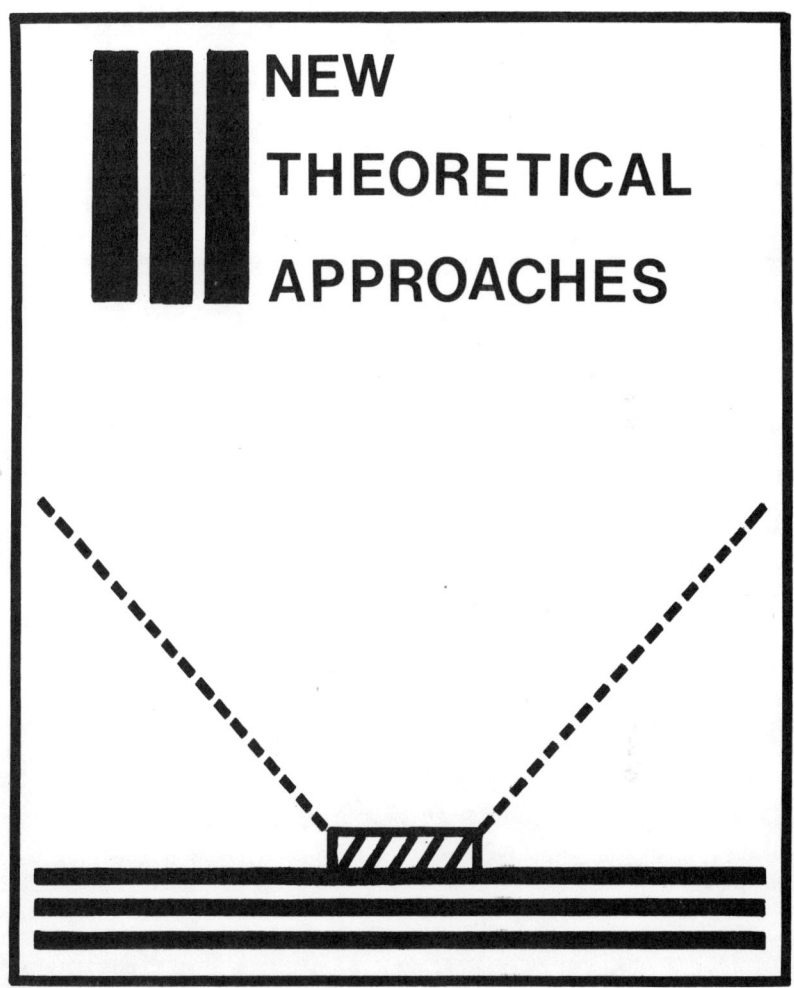

III NEW THEORETICAL APPROACHES

RELATIVISTIC DESCRIPTION OF DIRECTLY INTERACTING PIONS AND NUCLEONS*

Leon Heller
Theoretical Division, Los Alamos Scientific Laboratory
Los Alamos, New Mexico 87545

ABSTRACT

The expected magnitudes of the leading relativistic effects on an off-energy-shell T matrix element are estimated using the Bakamjian-Thomas formulation of relativistic potential theory. For pion-nucleon scattering at medium energy, the two largest corrections are expected to result from the use of relativistic relative momenta rather than nonrelativistic values. The importance of additional terms depends upon the detailed behavior of the T matrix.

INTRODUCTION

In the Introduction to his 1961 paper entitled "Relativistic Particle Systems with Interaction," Foldy[1] pointed out some common misconceptions which exist concerning the requirements of Lorentz invariance on physical theories. One of these is the notion that they must be constructed out of four-dimensional tensors and spinors in such a way that it is obvious that all physical quantities are Lorentz invariant, i.e., that they must be "manifestly invariant." This misconception has already led to considerable confusion in the literature of pion-nucleus scattering at relativistic energies. On one side there appears the statement that the conventional multiple scattering theories or the use of an optical potential, formalisms which were developed using potential theory, cannot be carried over into the relativistic domain. On another side are a host of attempts to take the fundamental structure of these formalisms, the collection of <u>off-energy-shell T matrix elements</u>, and treat it as though it had a covariant meaning. It is here that one finds the phrase "transformation of the T matrix from the pion-nucleon center of mass system to the pion-nucleus center of mass system."

When manifestly invariant apparatus is used--e.g., relativistic quantum field theory treated by covariant perturbation methods (Feynman diagrams), or the Bethe-Salpeter equation--every subunit of a calculation is Lorentz invariant. In pion-nucleus scattering, for example, one might consider the diagram shown in Fig. 1. The upper vertex is called an <u>off-mass-shell amplitude</u> for pion-nucleon scattering, and (neglecting spin) is a function of six scalar variables which can be constructed from three independent four-momenta. Similarly the lower

---

*Work performed under the auspices of the U. S. Energy Research and Development Administration.

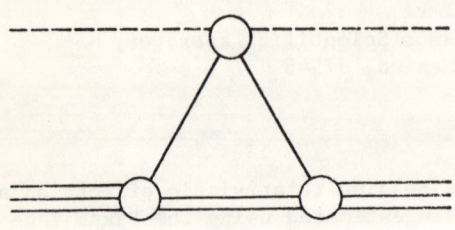

Fig. 1. A Feynman diagram for pion-nucleus scattering.

vertices are Lorentz invariant functions of three scalar variables. Although this property of subdividing into Lorentz invariant pieces is very neat, it is not essential.

In 1949 Dirac[2] clearly stated the conditions for a Lorentz invariant classical theory to be expressible in Hamiltonian form. Going over to quantum mechanics they became the requirement that the set of ten generators H, $\vec{P}$, $\vec{J}$ and $\vec{K}$ of infinitesimal time and space translations, three-dimensional rotations, and Lorentz boosts satisfy a particular set of commutation relations .

At this point there occurs a fundamental division of theories into two classes[2], according to the choice of dynamical variables out of which the generators are constructed. These are (i) the position, momentum, and spin operators for a set of particles, and (ii) field operators at points in space-time, or their Fourier components. Theories in the first class are said to be 'directly interacting',[1,3] as opposed to those in the second category where the interactions are mediated by fields. We shall be concerned with class (i) only.

THE TWO-BODY PROBLEM

Bakamjian and Thomas[4] were the first to write down the generators for two directly interacting particles. They made a <u>canonical</u> transformation from the operators $\vec{r}_1$, $\vec{p}_1$, $\vec{r}_2$ and $\vec{p}_2$ (we neglect spin for simplicity) to a set consisting of $\vec{R}$ and $\vec{P}$, which describe (relativistically) the center of mass of the system, and $\vec{\rho}$ and $\vec{k}$, which refer to the internal motion. They chose $\vec{P} = \vec{p}_1 + \vec{p}_2$, and their choice for $\vec{k}$ is of the identical <u>form</u> as the Lorentz transformation which would take two particles with momenta $\vec{p}_i$ and energies $E_i(\vec{p}_i) = (m_i^2 + \vec{p}_i^2)^{\frac{1}{2}}$ to their center of mass system. But note that this is a canonical transformation and <u>not</u> a Lorentz transformation on operators or state vectors. One expression for $\vec{k}$ is

$$\vec{k} \equiv \vec{k}(\vec{p}_1, \vec{p}_2) = \vec{k}_E + \frac{\vec{k}_E \cdot \vec{P}}{\omega(\omega + E)} \vec{P}$$

where (1)

$$\vec{k}_E \equiv \vec{k}_E(\vec{p}_1, \vec{p}_2) = \frac{E_2 \vec{p}_1 - E_1 \vec{p}_2}{E}$$

$E = H_0 = E_1 + E_2$ and

$$\omega = (E^2 - \vec{P}^2)^{1/2} = (m_1^2 + \vec{k}^2)^{1/2} + (m_2^2 + \vec{k}^2)^{1/2} \equiv \omega(k) \tag{2}$$
$$= \omega_1(k) + \omega_2(k) \ .$$

In the instant form of dynamics which they used[4], both $\vec{P}$ and $\vec{J}$ have their free particle forms; interaction appears only in the generators H and $\vec{K}$. The interaction in H is produced via the sequence

$$h = \omega + v(\vec{\rho},\vec{k}) \ , \quad H = (h^2 + \vec{P}^2)^{1/2} \tag{3}$$

and the potential is defined by $V = H - H_0$. v is an arbitrary (rotationally invariant) function of the internal operators.

Due to the fact that the boost generator $\vec{K}$ contains interaction, neither H_0 nor V separately satisfy commutation relations appropriate to the fourth component of a 4-vector. [Only the <u>sum</u> $H_0 + V$ does.] The same remark applies <u>a fortiori</u> to the T matrix obtained from the Lippmann-Schwinger equation

$$T(W) = V + V \frac{1}{W + i\varepsilon - H_0} T(W) \tag{4}$$

Consequently the concept of "Lorentz transforming" an off-energy-shell T matrix element is a useless one.[5]

If the impulse approximation is made, then the objects which are needed in multiple scattering theory are the off-energy-shell matrix elements $\langle \vec{q}_1',\vec{q}_2'|T(E)|\vec{q}_1,\vec{q}_2 \rangle$ where the states $|\vec{q}_1,\vec{q}_2\rangle$ are eigenstates of the individual particle momentum operators and also of the unperturbed Hamiltonian H_0. In any particular frame of reference, such as the pion-nucleus center of mass system, one needs these matrix elements for arbitrary values of $\vec{Q} = \vec{q}_1 + \vec{q}_2$. The dependence on Q can be found because the dependence of V upon $\vec{P}$ is specified by Eq. (3). [In a Galilean invariant theory V commutes with $\vec{K}$ as well as with $\vec{P}$.]

Since the Bakamjian-Thomas formulation has no interaction term in $\vec{P}$, the T matrix elements can be written

$$\langle \vec{q}_1',\vec{q}_2'|T(W)|\vec{q}_1,\vec{q}_2 \rangle = (2\pi)^3 \delta^{(3)}(\vec{Q}' - \vec{Q}) T(W;\vec{q}_1',\vec{q}_2',\vec{q}_1,\vec{q}_2) \tag{5}$$

with the normalization $\langle \vec{q}_1',\vec{q}_2'|\vec{q}_1,\vec{q}_2 \rangle = (2\pi)^6 \delta^{(3)}(\vec{q}_1' - \vec{q}_1) \delta^{(3)}(\vec{q}_2' - \vec{q}_2)$. Since only three of the four vector arguments of T are independent, they can be replaced by $\vec{Q}$ and two relative momenta. In reference 5 this function is written as

$$T(W;\vec{q}_1',\vec{q}_2',\vec{q}_1,\vec{q}_2) = NT(W;\vec{Q};\vec{q}',\vec{q}) \tag{6}$$

where $\vec{q} = \vec{k}(\vec{q}_1,\vec{q}_2), \vec{q}' = \vec{k}(\vec{q}_1',\vec{q}_2')$, and $\vec{k}$ is the function defined in and above Eq. (1). The main result of reference 5 is an explicit expression for $T(W;\vec{Q};\vec{q}',\vec{q})$ as a <u>functional</u> of $t(\omega;\vec{p}',\vec{p}) \equiv T(\omega,0;\vec{p}',\vec{p})$. That is to say, if all the T-matrix elements with zero total momentum are known, then all the matrix elements with arbitrary values of $\vec{Q}$ can be calculated. The expression for N is[5] where

$$N = [J(\vec{q}_1,\vec{q}_2)J(\vec{q}_1{}',\vec{q}_2{}')]^{-\frac{1}{2}} \quad (7)$$

$$J(\vec{q}_1,\vec{q}_2) = \left|\frac{\partial(\vec{q}_1,\vec{q}_2)}{\partial(\vec{Q},\vec{q})}\right| = \frac{E_1(q_1)E_2(q_2)}{E_1(q_1)+E_2(q_2)} \cdot \frac{\omega_1(q)+\omega_2(q)}{\omega_1(q)\omega_2(q)} .$$

The formula for T is[5]

$$T(W,Q;\vec{q}',\vec{q}) = F(Q;q',q)t(\omega(q);\vec{q}',\vec{q})$$
$$+ \int \frac{d^3p}{(2\pi)^3} F(Q;q',p)F(Q;q,p)t(\omega(p);\vec{q}',\vec{p})t^*(\omega(p);\vec{q},\vec{p})$$
$$\times \left[\frac{F^{-1}(Q;s,p)}{\omega(s)+i\varepsilon-\omega(p)} - \frac{F^{-1}(Q;q,p)}{\omega(q)+i\varepsilon-\omega(p)}\right] \quad (8)$$

where $\omega(s) \equiv (W^2-\vec{Q}^2)^{\frac{1}{2}}$ and

$$F(Q;q',q) \equiv \frac{\omega(q')+\omega(q)}{E(Q,q')+E(Q,q)} \quad (9)$$

with $E(Q,q) \equiv (\omega^2(q)+Q^2)^{\frac{1}{2}} = E_1(q_1)+E_2(q_2)$. [For the half-off-energy-shell case, $W = E(Q,q)$, the first term on the right hand side of Eq. (8) provides the entire answer.[6]]

We now want to examine the differences between the structure of the relativistic and nonrelativistic expressions for T, and also the leading relativistic corrections. For the latter we will only assume that $\vec{q}_1$ and $\vec{q}_2$ (also $\vec{q}$ and $\vec{Q}$) are small compared to the sum of the energies $E_1 + E_2$ (and similarly for the primed variables). For pion-nucleon scattering we do not assume that q_π is small compared to E_π.

In a Galilean invariant theory, the function T on the right side of Eq. (5) would have the form

$$T(K;\vec{q}_1',\vec{q}_2',\vec{q}_1,\vec{q}_2) = T(K - \frac{Q^2}{2(m_1+m_2)}; \vec{q}_{NR}',\vec{q}_{NR}) \quad (10)$$

where K includes just kinetic energy, and $\vec{q}_{NR} = (m_2\vec{q}_1 - m_1\vec{q}_2)/(m_1+m_2)$. From Eqs. (6)-(9) it is seen that there are four specific places where relativistic effects enter: (i) the use of the relativistic relative momentum variable $\vec{q}$ rather than $\vec{q}_{NR}$; (ii) the factor N; (iii) the relativistic energy variable $\omega(s)$ in Eq. (8), and (iv) the presence of F in that same equation.

Before examining the expected magnitude of these effects, it is worth pointing out that Eq. (8) has a moderately complicated structure. One obvious point, but worth mentioning, is that the right hand side of Eq. (8) does not contain just one set of arguments for the function t. It is necessary to do an integration over a whole range of those arguments. [This has as one important consequence the fact that even if $t(\omega,\vec{q}',\vec{q})$ is a separable function, $T(W,Q;\vec{q}',\vec{q})$ is not.] It will be shown below, however, that for pion-nucleon scattering at medium energy, there may be a good approximation which does involve just one set of arguments for t.

The magnitude of the four effects mentioned above in any particular problem depends upon the details of the t matrix, including how far off-shell the needed matrix elements are.[7] It will now be shown, however, that the fractional correction due to (i) is of order (kinetic energy/total energy); those due to (iii) and (iv) are of order (total momentum/total energy)^2; and that due to (ii) is enhanced (for the pion-nucleon system) compared to (iii) and (iv) by the ratio of the target nucleon's mass to the projectile's energy, thereby making it more similar in size to (i). <u>For the nucleon-nucleon system, on the other hand, all four corrections are comparable.</u>

The claim about point (iv) is obvious since $F(Q;q',q) - 1 = -Q^2/2\omega(q')\omega(q) + \ldots$; and putting this into Eq. (8) yields[5]

$$T(W,Q;\vec{q}',\vec{q}) = t(\omega(s);\vec{q}',\vec{q}) + \frac{Q^2}{2}\left[-\frac{t(\omega(q);\vec{q}',\vec{q})}{\omega(q')\omega(q)}\right.$$

$$+ \left(\frac{1}{\omega(s)} - \frac{1}{\omega(q')} - \frac{1}{\omega(q)}\right) g(\omega(s);\vec{q}',\vec{q})$$

$$\left. + \frac{1}{\omega(q')} g(\omega(q);\vec{q}',\vec{q})\right] + \mathcal{O}(Q^4) \tag{11}$$

where

$$g(\omega;\vec{q}',\vec{q}) \equiv \int \frac{d^3p}{(2\pi)^3} \frac{1}{\omega(p)} \frac{t(\omega(p);\vec{q}',\vec{p}) t^*(\omega(p);\vec{q},\vec{p})}{\omega + i\varepsilon - \omega(p)} . \tag{12}$$

The significance of point (iii) can be seen by identifying W with $K + m_1 + m_2$. Expanding $\omega(s)$ then leads to the following difference from the nonrelativistic energy given in Eq. (10)

$$\omega(s) - (m_1 + m_2) - \left[K - \frac{Q^2}{2(m_1+m_2)}\right]$$

$$= \frac{Q^2}{2(m_1+m_2)}\left[\frac{K}{m_1+m_2} - \frac{Q^2}{4(m_1+m_2)^2}\right] + \ldots . \tag{13}$$

Irrespective of whether K or $Q^2/2(m_1+m_2)$ is larger, the fractional difference is indeed of order $Q^2/(m_1+m_2)^2$.

To demonstrate the assertion about the size of (i) it is sufficient to compare $\vec{q}_E = \vec{k}_E(\vec{q}_1,\vec{q}_2)$ with q_{NR} since, according to Eq. (1), $\vec{q}$ and $\vec{q}_E$ only differ by the fractional amount $(\vec{Q}/E)^2$. Since

$$\vec{q}_{NR} - \vec{q}_E = \frac{m_2 E_1 - m_1 E_2}{(E_1 + E_2)(m_1 + m_2)} \vec{Q} , \tag{14}$$

examination of the special case in which the target nucleon is at rest shows that the fractional difference is indeed proportional to the kinetic energy of the projectile over the total energy. Figs. 2 and 3 show the differences between q, q_E, and q_{NR} for a pion or nucleon (particle 1) incident on a nucleon at rest (particle 2). For a pion of 200 MeV kinetic energy, for example, $1 - q/q_{NR} = 0.13$, whereas $(q/q_E) - 1$ is only 0.031.

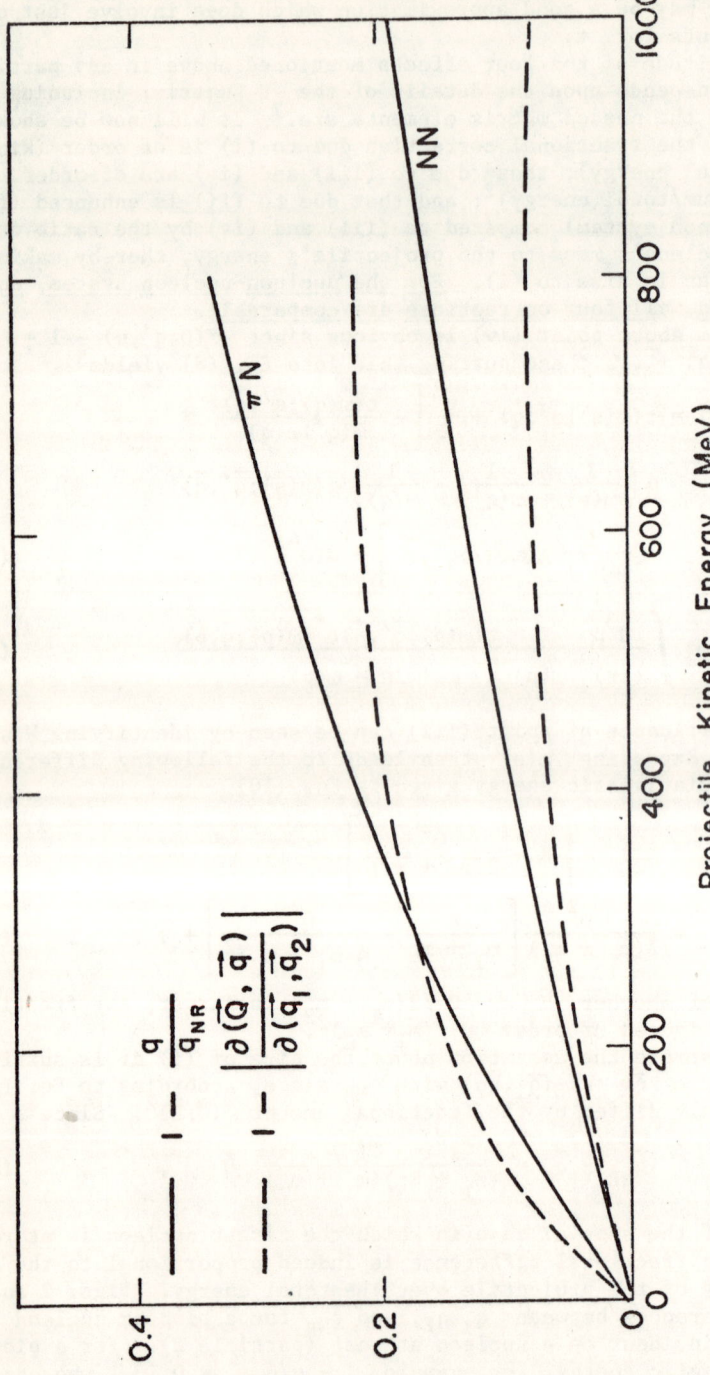

Fig. 2. Solid curves: the fractional difference between the nonrelativistic and relativistic values of the relative momentum for a pion or nucleon incident on a nucleon at rest. Dashed curves: the Jacobian of the transformation from total and (relativistic) relative momenta to the individual particle momenta, also for the target nucleon at rest. The upper curves are for an incident pion, and the lower curves for an incident nucleon.

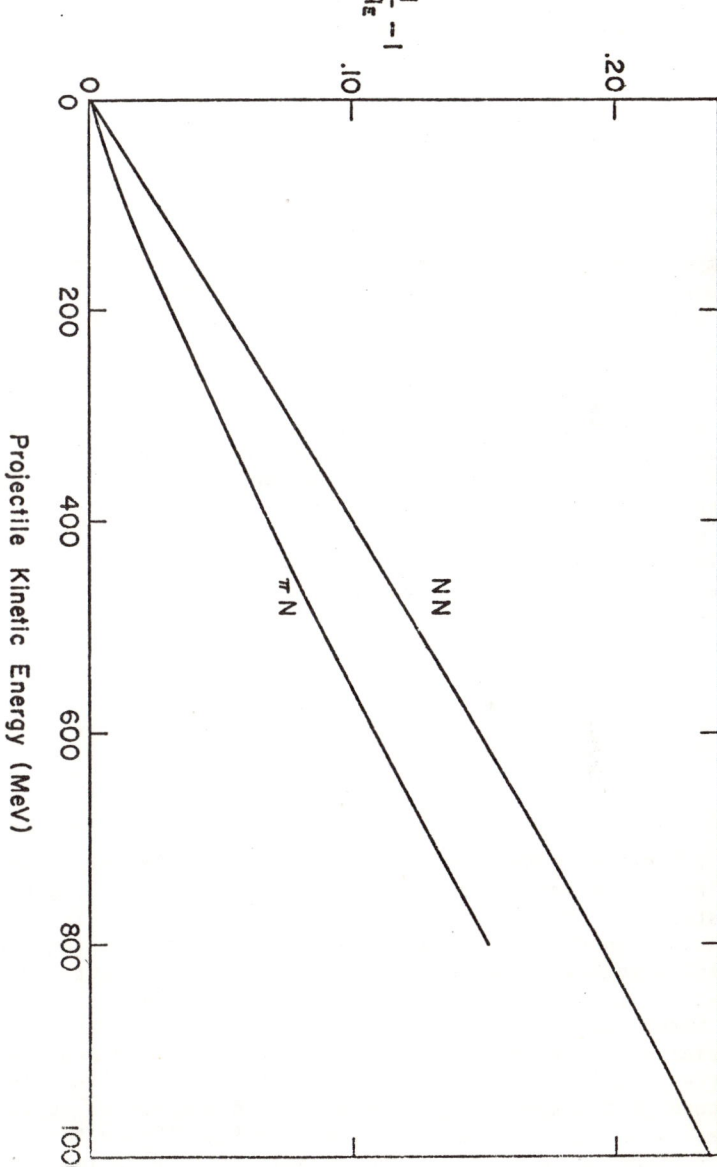

Fig. 3. The fractional difference between the relativistic relative momentum and the auxiliary quantity q_E defined in Eq. (1), for a pion or nucleon incident on a nucleon at rest. For this arrangement the vertical scale is also equal to $E(E^2 - \vec{Q}^2)^{-\frac{1}{2}} - 1$ where E is the total energy and $\vec{Q}$ the total momentum.

The factor N, which involves the Jacobian of the transformation from the variables $\vec{q}_1, \vec{q}_2$ to $\vec{Q}, \vec{q}$, was defined in Eq. (7), and the leading relativistic correction can be written

$$J^{-1}(\vec{q}_1, \vec{q}_2) - 1 = -\frac{1}{2}\frac{Q^2}{E^2} + \frac{E_1 - E_2}{E_1 E_2 E}\vec{q}\cdot\vec{Q} + \ldots \quad . \tag{15}$$

The $\vec{q}$ in the second term on the right of Eq. (15) can be replaced by q_E to that order. It is that second term which enhances this effect, compared to Q^2/E^2, by the ratio of the nucleon's mass to the pion's energy. As seen on Fig. 2, the departure of this Jacobian from unity is comparable in size to the fractional difference between q and q_{NR}.

In summary, the two largest relativistic effects on the pion-nucleon off-energy-shell T matrix elements at medium energies appear to be items (i) and (ii) above, namely, using relativistic relative momenta, and the multiplicative factor N involving the Jacobian of the transformation of the momentum variables. Since it is just as simple to use $\omega(s)$ in Eq. (11) as any other energy variable, one may as well do so. The only part of Eq. (11) which requires work to calculate is the term in brackets multiplying $Q^2/2$. Its importance clearly depends on the detailed behavior (and size) of the matrix $t(\omega, \vec{q}', \vec{q})$. If that term could be neglected, then one would have the very simple approximation[8]

$$T(W; \vec{q}_1', \vec{q}_2', \vec{q}_1, \vec{q}_2) \cong Nt(\omega(s); \vec{q}', \vec{q}) \quad . \tag{16}$$

MORE THAN TWO PARTICLES

The reason for studying the two-body off-energy-shell T matrix is to use it in some process such as bremsstrahlung or projectile-nucleus scattering. But the requirements of Lorentz invariance must be re-examined in the new setting. At the present time there does not exist a fully relativistic description for a directly interacting many-body system. To order $1/c^2$, however, such a framework has been provided.[9] In reference 9 it is shown that in order to satisfy both Lorentz invariance (to order $1/c^2$) and separability, there must be three-body forces, in general, and the form of these forces is dictated by the nature of the two-body potentials.[10] But there is still more freedom whereby one can do away with the Bakamjian-Thomas form of the interaction given in Eq. (3), _at the expense of introducing still more three-body forces._[9] The effect of three-body forces of relativistic origin has not yet been studied.

There is an extension of the Bakamjian-Thomas approach to several coupled channels. In the most straightforward version there are two elementary particles in each channel,[11,5] so that reactions such as A + B → C + D can be included. Presumably additional particles can be permitted in these channels following the work of reference 9.

In another extension[12] which seems especially attractive for a pion interacting with nucleons, the potential couples the noninteracting two particle state $|\pi N\rangle$ to a state of a single bare particle $|N_o\rangle$. This is similar to the Lee model,[13] but has an arbitrary vertex function (which depends on the relative momentum $\vec{k}$ of the πN state), and is Lorentz invariant. Consequently it is similar in spirit to the work of reference 14, but has a Hamiltonian foundation.

A discussion of pion-nucleon scattering has been given, using this isobar model,[15] and a study is underway[16] which incorporates it into pion-nucleus scattering. The Hamiltonian is written[16]

$$H = H_{NUCLEUS} + H_{o,\pi} + \sum_{i=1}^{A} V_{i\pi} \qquad (17)$$

where each $V_{i\pi}$ depends upon the coordinates of the pion and the ith nucleon according to Eq. (3),

$$V_{i\pi} = (h_{i\pi}^2 + \vec{P}_{i\pi}^{\,2})^{\frac{1}{2}} - (\omega_{i\pi}^2 + \vec{P}_{i\pi}^{\,2})^{\frac{1}{2}} \qquad (18)$$

where $\vec{P}_{i\pi} = \vec{P}_i + \vec{P}_\pi$. Based on the work of reference 9, one would guess that pion-two-nucleon forces are needed for Lorentz invariance, but Eq. (17) may include the leading relativistic effects for pion-nucleus scattering at medium energies.

DISCUSSION

One approach to studying projectile-nucleus scattering at relativistic energies expresses the generators of the Lie algebra of the Poincaré group directly in terms of the dynamical variables of the particles. The Lippmann-Schwinger equation and the off-energy-shell two-body T matrix elements play an important role, just as in nonrelativistic quantum mechanics. The Bakamjian-Thomas[4] formulation of the two-body problem leads to a definite dependence of these matrix elements on the total momentum of the pair of particles,[5] and we have discussed the sizes of the leading relativistic corrections in a semi-quantitative fashion. In the work of Lee and Coester[15,16] the importance of these effects is being studied in an actual pion-nucleus scattering calculation. The effect of the three-body forces[9] is still to be determined.

Approaches to scattering at relativistic energies which start from a four-dimensional equation deal with an entirely different object, an off-mass-shell amplitude. Even after some three-dimensional reduction is made, it is <u>still</u> an off-mass-shell amplitude, but with a definite restriction on the fourth components of some vectors which reduces the number of independent scalars from six to four. It is possible to make this reduction in such a way that (after redefining the amplitude and Born term by removing some kinematic factors) the equation has a similar appearance to the matrix elements of Eq. (4), <u>but only in the center of mass system</u>. For example, the propagator in an arbitrary frame would not be equal to $[W + i\varepsilon - (m_1^2 + \vec{p}_1^{\,2})^{\frac{1}{2}} - (m_2^2 + \vec{p}_2^{\,2})^{\frac{1}{2}}]^{-1}$, but would instead equal $[W + i\varepsilon - (m_1^2 + \vec{k}^2)^{\frac{1}{2}} - (m_2^2 + \vec{k}^2)^{\frac{1}{2}}]^{-1}$. That is to say, the quantities which appear are invariant, but the three-dimensional equation is not covariant, i.e., it does not have the same form in all frames of reference.

We do not claim that one of these approaches is intrinsically better than the other, but urge that they not be confused with each other, as has sometimes been done.[17]

ACKNOWLEDGMENTS

I want to thank G. E. Bohannon for many clarifying conversations about this subject, and T.-S. H. Lee for discussions about his work with F. Coester, and for providing me with some of the results of their work in advance of publication. Stimulating discussions with W. M. Kloet and A. Rinat on the relationship to the work in reference 14 are gratefully acknowledged.

REFERENCES AND FOOTNOTES

1. L. L. Foldy, Phys. Rev. 122, 275 (1961).
2. P.A.M. Dirac, Revs. Mod. Phys. 21, 392 (1949).
3. See the introduction to the reprint collection "The Theory of Action-at-a-Distance in Relativistic Particle Dynamics", edited by E. H. Kerner, Gordon and Breach, Science Publishers, Inc. (1972).
4. B. Bakamjian and L. H. Thomas, Phys. Rev. 92, 1300 (1953). Also see H. Osborn, ibid. 176, 1514 (1968).
5. L. Heller, G. E. Bohannon, and F. Tabakin, Phys. Rev. C 13, 742 (1976).
6. In reference 5, the half-off-shell answer was gotten first, and Eq. (8) obtained from it assuming no bound states.
7. The on-energy-shell T matrix element is Lorentz invariant, of course, if the states are normalized invariantly.
8. See the discussion in reference 5 of some *ad hoc* prescriptions which have appeared in the literature having the form of Eq. (16).
9. L. L. Foldy and R. A. Krajcik, Phys. Rev. Letts. 32, 1025 (1974), and Phys. Rev. D 12, 1700 (1975).
10. This 3-body force is not unique. In Ref. 9 it is shown that any function of the internal coordinates which satisfies the separability condition can be added on.
11. F. Coester, Helv. Phys. Acta 38, 7 (1965).
12. F. Coester, "Meson Theory of Nuclear Forces", in Quanta, edited by P.G.O. Freund, C. J. Goebel, and Y. Nambu, The University of Chicago Press (1970), p. 147.
13. T. D. Lee, Phys. Rev. 95, 1329 (1954).
14. R. Aaron, R. D. Amado, and J. E. Young, Phys. Rev. 174, 2022 (1968).
15. T.-S. H. Lee and F. Coester, Bull. Am. Phys. Soc. 21, 619 (1976), and to be published.
16. Private communication from T.-S. H. Lee.

17. See the discussion in Section V of reference 5. In L. Celenza, L. C. Liu, and C. M. Shakin, Phys. Rev. C $\underline{12}$, 1983 (1975), comparisons are made between the kinematics which applies to the off-mass-shell pion-nucleon amplitude needed at the upper vertex of Fig. 1, and the kinematics which enters an off-energy-shell T matrix element. We believe such comparisons of formulas from completely different kinds of theories to be meaningless, and reject the implication that the former procedure is correct and the latter incorrect.

B. S. Bhakar (U. of Manitoba): Are the three-body forces you have mentioned required only for a 3-body problem or is it a result valid for the N-body problem? As I understand, this result is a consequence of the requirement to close the Lorentz algebra to order $1/C^2$ if one goes from two-body to three-body.
L. Heller: Foldy and Krajcik have shown that the N-body problem to order $1/C^2$ needs 3-body forces, in general, to satisfy Lorentz invariance and separability; but 4- or more-body forces are not required to this order.

Carl Shakin (Brooklyn College): How do you generalize your theory when you consider more than two particles?
L. Heller: If the 3-body forces which are required by relativity to order $1/C^2$ are neglected, then the Hamiltonian for pion-nucleus scattering can be written $H = H_A + H_{o\pi} + \sum_{i=1}^{A} H'_{i\pi}$ where $H_{o\pi}$ and $H'_{i\pi}$ have the form discussed in the two-body portion of the talk. This structure for H is all that is needed to derive Watson's multiple scattering theory.

Manoj K. Banerjee (U. of Maryland): Problems arise when you allow for the creation and annihilation of pions. Now the eigenstates of the free Hamiltonians are the bare nucleons. Your off the energy shell T matrix elements refer to these bare particle states. When we use the shell model or extract the nuclear density from electron scattering we are talking about nearly physical nucleons and not about bare nucleons. So those who believe that creation and annihilation of pions are important parts of the dynamics should use a formalism where one can avoid referring to bare nucleons. Field theory provides such a method. Those who believe that creation and annihilation of pions are not important parts of the dynamics should use your method with all the cautions you so lucidly described.

L. Heller: I do not know how to solve the field theory problem with its infinitely many degrees of freedom. What I have discussed is an attempt to use a description with a finite number of degrees of freedom, which hopefully includes the most important physical effects and satisfies Lorentz invariance. In this approach there is no such thing as a bare nucleon. But one can allow for pion absorption by introducing a bare isobar eigenstate of the free Hamiltonian. All scatterings involve physical particles, i.e., eigenstates of the full Hamiltonian.

Rubin Landau (Oregon State Univ.): I wonder if these relativistic effects don't increase with energy, more so than exists with the 20% effects found at 50 MeV. In particular, would this have any effect on the high energy total cross sections?

L. Heller: The relativistic effects undoubtedly increase with increasing pion energy above the 3-3 resonance. In the written version of this talk, curves are presented which show that for 300 MeV pions the relativistic effects on the relative momentum variable, and also on the Jacobian of the transformation (which is a multiplicative factor in the amplitude), is approximately 20%. I would expect 40% effects in the cross section, but actual calculations have not been performed.

ISOBAR PROPAGATION IN THE NUCLEAR MEDIUM*

E. J. Moniz
Theoretical Division, Los Alamos Scientific Laboratory
Los Alamos, New Mexico 87544
and
Laboratory for Nuclear Science and Department of Physics
Massachusetts Institute of Technology, Cambridge, Massachusetts 02139

ABSTRACT

It is argued that introduction of the isobar degree of freedom in describing pion-nucleus interactions provides a convenient, unified framework within which to discuss both many-body corrections to the standard multiple scattering approach and the properties of the $\Delta(1232)$ in nuclear matter. Important aspects of isobar-nucleus dynamics, namely, isobar-hole interactions and Δ self-energy modifications, are discussed in the context of pion elastic scattering and incoherent pion production.

I. INTRODUCTION

Meson-nucleon interactions at low to intermediate energies are dominated by resonances or, effectively, by the formation of short-lived "particles" with well-defined quantum numbers. For example, the pion-nucleon system exhibits resonances in all spin-isospin channels for invariant energies less than 2 GeV, while the $\bar{K}$-p interaction near threshold is controlled by the (subthreshold) $\Lambda(1405)$ resonance. Consequently, it is clear that any attempt to relate meson-nucleus and meson-nucleon interactions implicitly contains the isobar (i.e., in the energy variation of the elementary amplitudes), and furthermore it might seem natural to introduce explicitly the isobar degree of freedom in theoretical approaches to meson-nucleus interactions. However, the second step certainly is not necessary, as is evident from the fact that most treatments of pion-nucleus scattering are based upon fixed-scatterer multiple scattering theory, in which the isobar plays no explicit role. More precisely, since we never directly observe the resonance in our detectors, we can in principle project out these channels without altering the physics in any way. Nevertheless, it will be argued here that introduction of the isobar degree of freedom is very <u>convenient</u> in a many-body approach to the meson-nucleus problem, providing a unified framework within which both kinematical corrections to the standard multiple scattering approach and dynamical modifications of the elementary amplitudes can be discussed. We shall confine our attention to pion-nucleus interactions in the energy regime dominated by the $\Delta(1232)$ resonance and will discuss some specific reactions involving the important aspects of isobar-nucleus dynamics, namely, Δ-hole interactions and Δ self-energy interactions. A more detailed outline of the paper follows.

After a brief description of the free isobar propagator, we shall present some simple examples in which the dynamics of isobar propagation can lead to significant quantitative effects in the description of pion-nucleus scattering. First, the importance of nucleon recoil as a kinematical correction to the fixed-scatterer formalism will be emphasized, and then modifications due to the Pauli principle of the Δ self-energy in nuclear matter (and therefore of the πN amplitude in the medium) will be discussed. This is intended to motivate the ensuing discussion, in which these physical considerations are incorporated into somewhat more sophisticated approaches to pion elastic scattering and to the isobar propagator in nuclear matter. Specifically, we shall discuss a Tamm-Dancoff approach to π-^{16}O elastic scattering, in which the isobar-hole states are diagonalized with the result that only one collective Δ-hole (doorway) state dominates the scattering in each partial wave. This is encouraging for a unified approach to several coherent reactions on the same nucleus. Finally, we calculate the Δ self-energy in nuclear matter and show how this enters directly in incoherent reactions such as total photoabsorption.

The discussions will, of necessity, be brief and unfortunately much work in the field cannot be presented. A number of relevant references can be found in the proceedings of this conference and in the recent review article of Brown and Weise.[1]

II. THE FREE ISOBAR PROPAGATOR

Before discussing the properties of the isobar in nuclei, it is useful to start with a dynamical model for the free Δ. Our approach is to introduce an elementary or bare isobar, with bare mass M_Δ, coupled to the πN channel with a vertex function $v(k)$, as depicted below. The assumption is that, since the πN channel is the only open

channel in the energy region of interest, the important physics will be included if we explicitly handle only this channel. Then, the "dressed" Δ-propagator is given simply in terms of the πN "bubble" self-energy:

$$\Delta^{-1}(s) = \Delta_0^{-1}(s) - \Sigma(s)$$

$$= (s - M_\Delta^2) - \int_0^\infty \frac{dq\, q^2}{6\pi^2} \frac{\omega_q + E_q}{2\omega_q E_q} \frac{q^2 v^2(q^2)}{s^+ - (\omega_q + E_q)^2} \quad (1)$$

where s is the invariant energy squared and where a Blankenbecler-Sugar prescription for the propagator has been assumed[2] in writing Eq. (1). Our task now is to choose the available parameters so that the observed Δ-propagator is reproduced.[3] This is easily done by noting that πN elastic scattering in the 3-3 channel proceeds through the Δ-propagator:

$$\Rightarrow \Delta(s)^{-1} = |\Delta(s)^{-1}| e^{i\delta_{33}(s)}$$

For example, a Yukawa form factor $v(k) = g/(\alpha^2 + k^2)$ with the parameters $g = 3.14\ M_\Delta^2$, $\alpha = 1.8\ \text{fm}^{-1}$, and $M_\Delta = 6.83\ \text{fm}^{-1}$ yields an extremely accurate representation of the 3-3 phase shift. The important point is that Eq. (1) leads to an explicit form for the real and imaginary parts of the Δ self-energy; we shall return to this point later.

III. SOME SIMPLE EXAMPLES

A. Kinematical corrections: the effect of nucleon recoil.

In the conventional multiple scattering approach to pion scattering, the nucleon coordinates are "frozen" at fixed positions during the scattering process, thereby eliminating propagation of the isobar. However, precisely because the two-body interaction is resonant, the interaction time is long and we can expect the resonating πN system (i.e., the isobar) to propagate, on the average, about 2/3 fm. This is not negligible compared to internucleon separations and, in fact, simple exercises demonstrate the importance of this effect.

First consider pion-deuteron elastic scattering in the single-scattering approximation. The amplitude is given by

$$T(\vec{k}, \vec{k}'; s) = \int \frac{d\vec{q}}{(2\pi)^3} \psi_d^* \left[\frac{v(m^2)\ \vec{m} \cdot \vec{m}'\ v(m'^2)}{\Delta^{-1}(\sigma_q)} \right] \psi_d \qquad (2)$$

where $\vec{q}$ is the spectator nucleon momentum, $\vec{m}$ and $\vec{m}'$ are relative momenta at the πNΔ vertices, ψ_d is the deuteron wavefunction, and $\Delta(\sigma_q)$ is the isobar propagator evaluated at the invariant sub-energy

$$\sigma_q = (\sqrt{s} - E_q)^2 - \vec{q}^{\,2} \qquad (3)$$

The details can be found in the paper of Woloshyn, Moniz, and Aaron,[2] but we wish to emphasize here only the fact that the momentum dependence in σ_q corresponds to spatial propagation of the Δ. In the fixed nucleon approximation, σ_q would be fixed at some average value, and the result is shown in Figure 1: at large momentum transfers (i.e., back angles), the approximation is rather poor. In Reference 2, it is seen that this remains true when the full three-body problem is solved and indeed that the multiple scattering series converges much more rapidly when the isobar kinematics is correctly accounted for. In the language of the optical potential, the factorization approximation is not adequate for precise quantitative predictions.

Another simple example of the importance of properly treating isobar propagation is provided by studying pion propagation in infinite nuclear matter. The wave equation is just

$$G^{-1}(p)\psi(p) = [p^2 - k^2 - \pi(p;k)]\psi(p) = 0 \qquad (4)$$

where k is the on-shell pion momentum (fixed by the energy), p is the (complex) wavenumber in the medium, and π is the pion self-energy or optical potential. Any nonlocality will be reflected by a p-dependence in π and will result in multiple solutions p_i of Eq. (4). Therefore, we can write a multiple eigenmode representation of the Green's function[4]

$$G(r) = \sum_i g_i \frac{e^{ip_i r}}{r} \qquad (5)$$

with g_i the residues ($\Sigma_i g_i = 1$), and expect a strong mixing between the eigenmodes whenever the pion mean free path becomes comparable to the nonlocality.[4] For example, isobar propagation corresponds to a nonlocality in the pion coordinate and, following the paper by Lenz,[5] we can study this by writing[6]

$$\pi(p;k) = \frac{k \rho \sigma_{\pi N} \Gamma/2}{E - E_R + i\Gamma/2 - p^2/2M^*} \qquad (6)$$

This corresponds to the lowest order optical potential being generated by resonant scattering, and the isobar kinetic energy term in the Breit-Wigner denominator to isobar propagation. Inserting this into Eq. (4), it is clear that two eigenmodes emerge, which, in the low density limit, can be readily identified as "pion" and "Δ-hole" modes. At resonance, the solutions are

$$p_i^2 = k^2 + \frac{iM^*\Gamma}{2}\left[1 \pm \sqrt{1 - \ell_\Delta/\ell_\pi}\right], \quad i = 1,2$$

$$\ell_\pi = (\rho \sigma_{\pi N})^{-1}, \quad \ell_\Delta = 2\frac{k/M^*}{\Gamma/2} \qquad (7)$$

where ℓ_π and ℓ_Δ are the pion mean free path and the isobar propagation distance, respectively. Clearly, when $\ell_\pi \gg \ell_\Delta$ (i.e., $\rho \to 0$), the conventional pion mode dominates, whereas when ℓ_Δ and ℓ_π are comparable, the eigenmodes are mixed. The net effect of all this is that, at large densities, the pion is much less damped in the medium when the Δ-hole channel is offered as a means of propagation.

It should be clear from these examples that isobar propagation is an important ingredient in the description of pion-nucleus interactions. We repeat that this could be incorporated without explicit introduction of the Δ but that recoil is most easily conceptualized in terms of isobar propagation.

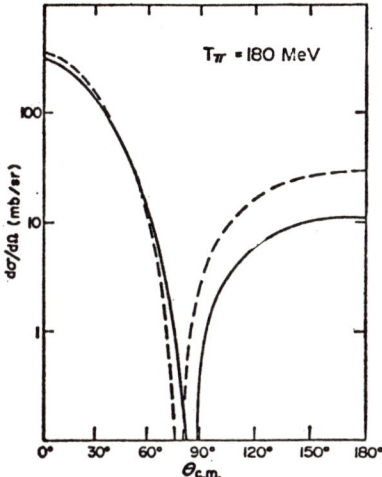

Figure 1. Impulse approximation for πd scattering: dashed and solid curves are without and with isobar propagation, respectively. Curves taken from Reference 2.

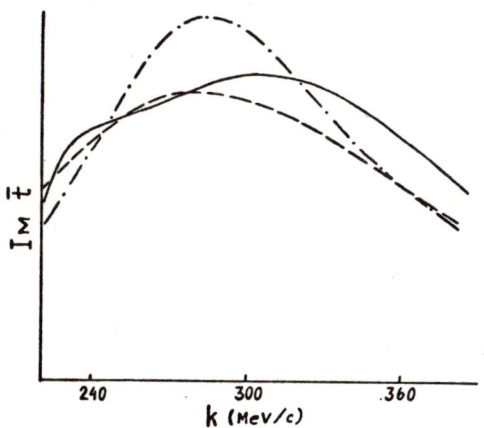

Figure 2. Imaginary part of the forward on-shell πN amplitude in the Fermi sea, as a function of the pion momentum k. Dot-dash, dashed, and solid curves are the free space, Fermi averaged, and full results, respectively.

B. Dynamical modification of the Δ-propagator: effect of the Pauli principle.

In considering propagation of the isobar in nuclei, we must expect dynamical modifications due to interaction of the constituents (i.e., the pion and nucleon) with the nuclear medium. In the simplest approximation, the pion and nucleon propagate in optical potentials and the nucleon is subject to restrictions imposed by the Pauli principle. We consider here only the Pauli blocking effect; that is, we consider the problem of computing the isobar propagator in a free Fermi gas:

$$\Delta^{-1} = \Delta_0^{-1} - v G_0 Q v \qquad (8)$$

where v represents the πNΔ vertex functions, G_0 is the free πN propagator, and Q prevents the intermediate nucleon from occupying states in the filled Fermi sea, thereby modifying the self-energy. In fact, this calculation can be carried out in closed form for simple vertex functions even with the inclusion of spin;[7] here, we just present in Figure 2 the results for the imaginary part of the forward πN t-matrix in the medium (of course, this is directly proportional to Im Δ). The dot-dash curve is the free space value for a stationary nucleon, while the dashed curve includes Fermi averaging over the target nucleon (k_F = 270 MeV/c, corresponding to nuclear matter density). The solid curve represents the full calculation including Pauli blocking. We see that there is a substantial modification of the amplitude in the medium, including a "shift of the resonance" towards higher energy. As emphasized in Reference 1, pion and nucleon interactions must be included before definite conclusions can be drawn, and this will be discussed in section V.

In ending this section, we repeat that both kinematical and dynamical effects arising from isobar propagation can be quantitatively significant and must be incorporated into a reliable theoretical approach to the pion-nucleus problem. We now go on to outline some preliminary calculations of coherent and incoherent reactions in which this has been attempted.

IV. π-^{16}O ELASTIC SCATTERING IN THE Δ-h FORMALISM

We present here results of a many-body approach to pion elastic scattering which is very similar in spirit to Tamm-Dancoff calculations of nuclear excited states; this work, still in an early stage, is being carried out in collaboration with M. Hirata, J. Koch, and F. Lenz. Basically, Δ-hole states are treated as doorway states[8] and a diagonalization is performed to find the eigenstates. This approach will be useful if, after diagonalization, only a few collective Δ-h states carry most of the strength in each partial wave. The calculation is carried out for ^{16}O, treated as a harmonic oscillator closed shell nucleus, with the intermediate Δ state also expanded in a harmonic oscillator basis.[9] The approach is summarized diagrammatically below:

Therefore, the Δ-h states act as entrance and exit channels for the pion and the problem is to compute the intermediate Δ-h propagator. The first three terms inside the brackets correspond to the free isobar propagator with the Pauli blocking effect; the fourth term corresponds to pion propagation between Δ-h states (this should be distinguished from pion "exchange" in the sense that the pion here is "real"); the next term accounts for a background potential acting upon the isobar (e.g., creation and destruction of nuclear p-h pairs would contribute here); the sixth term represents some additional effective interaction between Δ-h states, such as exchange of a ρ-meson; and the last term shown is one example of a multi-hole contribution. To this point, we have included the first five diagrams, and the connection to standard multiple scattering theory can be seen by expanding the Δ-h propagator:

where the dressed propagator now contains both the Pauli effect and the effect of the background Δ potential; this potential is taken to have the shape of the nuclear density and the strength is treated as the only free parameter. It is clear that we have the coherent approximation to the multiple scattering series, but including effects beyond those usually treated in the lowest order optical potential, such as isobar propagation (Fermi motion), binding effects, and Pauli restrictions; in addition, other contributions to the Δ-h interaction are trivially added.

The next step is to diagonalize the Δ-h propagator:

$$\langle D_i | G_{\Delta h} | D_j \rangle = \frac{\delta_{ij}}{E - E_R + i\Gamma/2 - \varepsilon_i} \quad , \quad |D_i\rangle = \sum_j C_j^i |(\Delta\text{-}h)j\rangle$$

$$T_{\pi-{}^{16}O} = \sum_i \frac{\langle 0|\vec{k}'\cdot S^\dagger|D_i\rangle\langle D_i|\vec{k}\cdot\vec{S}|0\rangle}{E - E_R + i\Gamma/2 - \varepsilon_i} \qquad (9)$$

where $\vec{S}$ is the operator creating a Δ from a nucleon.[1] The results of the diagonalization are the complex eigenvalues ε_i and the coefficients C_j^i relating the eigenstates to the original Δ-h basis states. With a Δ background potential of strength roughly $(-70 - 70i)$ MeV,[10] we obtain the total cross section results shown in Figure 3; furthermore, the fit to the differential cross section data is also quite good and we note that the partial wave decomposition of the nuclear amplitude is significantly different from that obtained with a standard optical potential code producing comparable results for the total cross section.

We would like to discuss the results in some more detail. As an example, consider the $\pi - {}^{16}O$, $J^P = 0^-$ state. Truncating the Δ-space at $8\hbar\omega$, we have a 17-dimensional space, and the relative contributions of the eigenstates to the nuclear amplitudes are listed in Table 1 for pion kinetic energy 140 MeV. While the pion coupling to Δ-h states is distributed over many states, we see from Table 1 that virtually all the strength is carried by one collective eigenstate (number 9 in the table), which we can characterize as a 0^- collective state[11] in ${}^{16}_\Delta O$! In Table 2, we list the energy eigenvalues for the important eigenstates in each partial wave, again (the imaginary parts of the eigenvalues are fairly energy dependent, while the real parts increase with pion energy). In the case of the 4^- and 5^+ states, more than one eigenstate contributes appreciably and all the relevant eigenvalues are given; i.e., the Λ-h strength is fragmented over a few states in the higher partial waves. Note that the imaginary parts of the eigenvalues are strongly L-dependent, corresponding to a large broadening in the 0^- case, and a slight narrowing in the 5^+; this of course points out the importance of having available partial wave decomposition of the pion-nucleus elastic amplitude. Further theoretical study will focus upon understanding the dynamical origin of the Δ single-particle potential, but the results are already quite encouraging. In particular, several dynamical effects have been included in the calculation and the dominance of one collective eigenstate in each partial wave implies both that a simple, theoretically motivated parameterization of the pion-nucleus amplitude may be possible[8] and that we can consequently learn about Δ-nucleus dynamics. Furthermore, having obtained the collective Δ-h states, several coherent processes which go via the Δ can be calculated in a unified way with comparatively little additional effort; for example, a calculation of the high energy $\gamma({}^{16}O, {}^{15}N_{gs})p$ reaction is being carried out.[12]

V. THE Δ SELF-ENERGY IN NUCLEAR MATTER

Finally, we turn to the questions of the Δ self-energy in nuclear matter and of the manner in which we can expect to observe the modifications from the free space value. We start by addressing the second question.

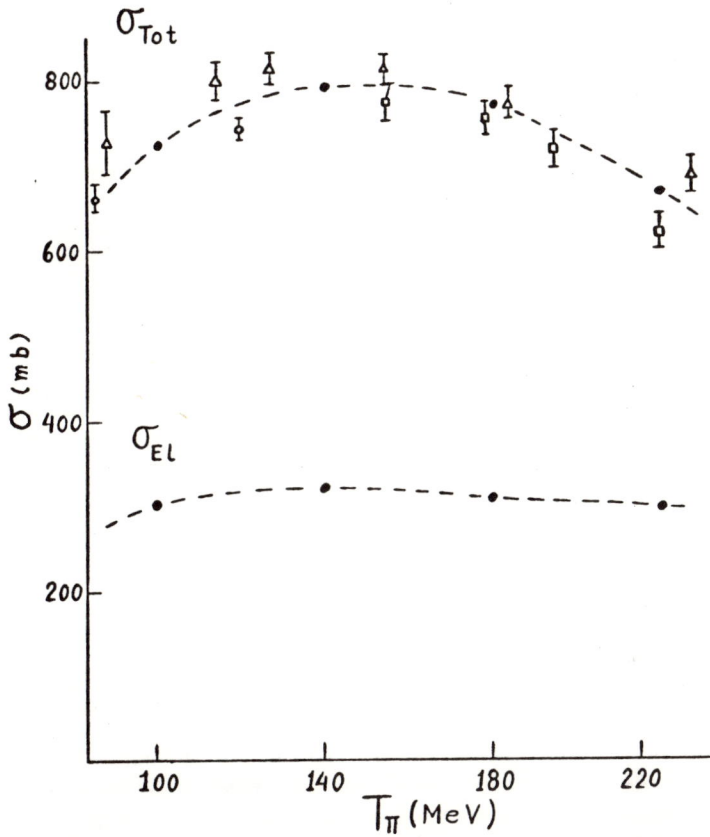

Figure 3. Total and total elastic cross sections for π-^{16}O scattering. Circles connected by the dashed lines represent the theoretical predictions in the Δ-h model.

Table 1. Relative contribution of the eigenstates to the $0^- \pi-{}^{16}O$ amplitude at T_π = 140 MeV. The N_i label the eigenstates.

N_i	T_i	N_i	T_i	N_i	T_i
1	0.02 - i 0.03	7	0.13 - i 0.05	13	0.17 - i 0.12
2	0.00 + i 0.01	8	0.00 + i 0.00	14	0.00 - i 0.01
3	0.01 - i 0.00	9	-0.52 + i 2.81	15	0.00 + i 0.00
4	-0.02 - i 0.04	10	0.16 - i 0.27	16	0.00 - i 0.01
5	0.00 + i 0.00	11	0.01 + i 0.00	17	0.00 - i 0.00
6	0.00 + i 0.01	12	-0.03 + i 0.01		

Table 2. Energy eigenvalues in MeV of the dominant eigenmodes in the contributing partial waves at T_π = 140 MeV.

J^P	0^-	1^+	2^-	3^+	4^-	5^+
ε_i	-39 - 176i	-12 - 118i	-13 - 100i	-9 - 64i	19 - 27i	42 + 11i
					46 - 22i	45 - 14i
						27 + 11i

In studying the nuclear response to electromagnetic probes, it is known that incoherent processes are the best way to learn about single particle properties in the nucleus. For example, the electron quasi-elastic cross section yields direct information on the nuclear Fermi momentum[13] and on the effective mass of nucleons in nuclear matter;[14] i.e., this data provides information on the nucleon propagator in the medium. Following this line of reasoning, we propose that total photoabsorption measurements in the energy regime characterized by Δ excitation may be the best avenue to information on the Δ propagator in nuclear matter.[15] Furthermore, simple kinematics reveals that the low energy side of the quasi-free Δ excitation peak will probe only those situations in which the Δ is nearly at rest (i.e., head-on collisions of photons with energy $E\gamma \approx$ 250 MeV with nucleons at the Fermi surface). This is a particularly interesting kinematical condition since we can expect the self-energy modifications to be greatest for Δ's at rest; for example, the Pauli blocking effects are clearly maximized in this situation. In summary, therefore, the total photoabsorption cross section in the region of the Δ, the calculation of which is denoted by the diagram below,

$$\sigma_T(\gamma A) \propto \text{Im}\left[(\gamma) \!+\!+\!+\!+\!\!\bigcirc\!\!+\!+\!+\!+\! (\gamma)\right]$$

offers a fairly direct measurement of the Δ-propagator in nuclear matter, and in particular the low energy side singles out isobars of low momentum. We now present a simple calculation of the modified self-energy for a Δ at rest; this work has been performed with F. Lenz and K. Yazaki, and is still in progress.

Calculation of the Δ-propagator in the nuclear medium involves a problem of self- consistency, since the Δ self-energy involves the pion propagator in the medium and the pion self-energy in turn involves the Δ propagator in the medium:

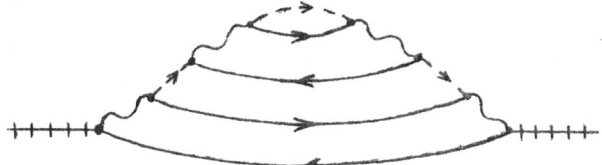

Here, the double-dashed line represents the pion propagator in the medium and the slash on the nuclear line denotes Pauli blocking. We present here only results for a noninteracting Fermi sea of nucleons (as is done in References 13 and 14 for quasi-elastic nucleon knockout). Then solution of the coupled equations includes terms in the photon polarization propagator such as that shown below:

Note, however, that interference diagrams obtained, for example, by crossing the hole lines are not included. If we adopt the multiple eigenmode representation for the pion propagator (see Equation 5) and a Yukawa-like πNΔ vertex function, then the self-energy in the medium for a Δ at rest can be written as

$$\sum(E) = -i\frac{g^2}{4\pi}\sum_j g_j \left\{ \frac{p_j^3}{(\alpha^2+p_j^2)^2}\left[1 + \frac{i}{\pi}\ln\left(\frac{p_j - k_F}{p_j + k_F}\right)\right] \right.$$
$$\left. -i\frac{\alpha}{2}\frac{(\alpha^2+3p_j^2)}{(\alpha^2+p_j^2)^2}\left[1 - \frac{2}{\pi}\tan^{-1}\frac{k_F}{\alpha}\right] - \frac{i}{\pi}\frac{\alpha^2 k_F}{(\alpha^2+p_j^2)(\alpha^2+k_F^2)} \right\} \quad (10)$$

where the eigenmodes p_j are functions of energy.[16] The element of self-consistency resides in the fact that the p_j in turn depend upon $\Sigma(E)$ in a non-trivial way. However, we can start by considering the simpler case in which the eigenmodes are calculated with the first order optical potential including isobar propagation. We then find that there are two important eigenmodes (i.e., those with small imaginary parts) and their trajectories as a function of energy are shown in Figure 4. As in Section II.A, the rather weak damping is caused by a mixing between the pion and Δ-h eigenmodes. Inserting these results into Eq. (10), we obtain the imaginary part of the Δ self-energy shown in Figure 5. The significant features here are the

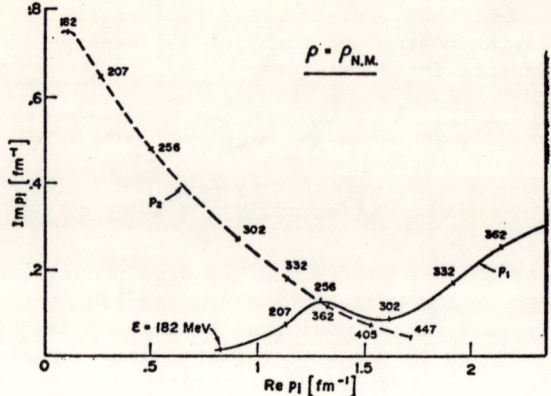

Figure 4. Trajectories of the two most important eigenmodes. Numbers along the trajectories represent the total pion energy.

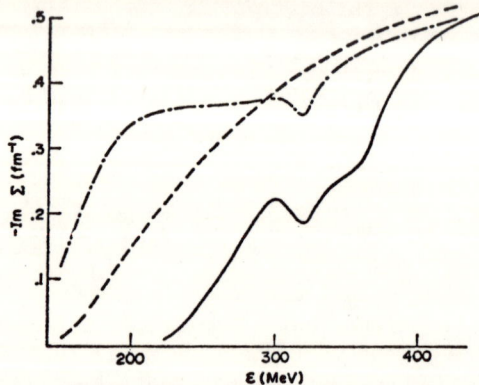

Figure 5. Imaginary part of the Δ self energy as a function of pion total energy. Dashed curve is the free space value, while the dot-dash and solid curves are the values in the medium both without and with the Pauli restriction, respectively. These have been evaluated at nuclear matter density.

important role played by the Pauli principle at the lower energies, the structure introduced into the self-energy through the eigenmode mixing, and the fact that Im $\Sigma(E)$ is reduced considerably in the medium (corresponding crudely to a narrowing of the resonance). It must be stressed that these results pertain only to nuclear matter density and to isobars at rest and that much more is to be done before any comparison to data would be sensible: nucleon-nucleon correlations, self- consistency, and background terms must be introduced and the integral over allowed hole momentum must be performed in evaluating $\sigma_T(\gamma A)$. Nevertheless, these results point to the importance of a detailed theoretical treatment of Δ self-energy interactions in nuclei.

VI. CONCLUDING REMARKS

We have seen that introduction of the isobar degree of freedom in a many-body approach to the problem of pion-nucleus interactions provides a unified framework within which both kinematical corrections to the standard multiple scattering approach and dynamical modifications of the elementary amplitudes can be discussed. Further, these effects are quantitatively important and more refined calculations of the isobar-nuclear dynamics are called for.

*Work supported by the U.S. Energy Research and Development Administration.

REFERENCES

1. G. E. Brown and W. Weise, Phys. Reports $\underline{22}$, (1975), p. 909.
2. R. Woloshyn, E. J. Moniz, and R. Aaron, Phys Rev. $\underline{C13}$ (1976), p. 286.
3. Even with the introduction of inelastic channels, the inverse scattering problem can be solved to give the vertex function and M_Δ in terms of the πN (complex) phase shifts (J. T. Londergan, and E. J. Moniz, to be published). The momentum dependence of the vertex function turns out to be insensitive to the high energy phase shifts but, not surprisingly, the opposite is true of M_Δ.
4. F. Lenz and E. J. Moniz, Phys Rev. $\underline{C12}$ (1975), p. 909.
5. F. Lenz, Ann. Phys. $\underline{95}$ (1975), p. 348.
6. Here, we are using M^*= 1232 MeV.
7. E. J. Moniz and A. Sevgen, to be published.
8. L. Kisslinger and W. Wang, Phys Rev. Lett. $\underline{30}$ (1973), p. 1071, and preprint.
9. A similar calculation has been carried out in ^{4}He by Hirata, Lenz, and Yazaki and is discussed in the talk by Lenz at this conference.
10. In contrast to the results in Reference 9, the real part of the Δ potential is weakly energy-dependent here.
11. It must be remembered that the eigenvalues are energy-dependent, since the Δ-h interaction is itself energy-dependent (in contrast to the situation in nuclear structure calculations[1]).
12. M. Hirata, J. H. Koch, F. Lenz, and E. J. Moniz, to be published.
13. E. J. Moniz, et al., Phys Rev. Lett. $\underline{26}$ (1971). p. 445.

14. E. J. Moniz, Phys. Rev. 184 (1969) p. 1154.
15. A more detailed discussion of this paper is given in an invited talk by the author at the Saclay Meeting on Mesonic Effects in Nuclei (May 1975).
16. Several authors, in considering the so-called Kisslinger catastrophe in the pion optical potential, have advocated working with a short range vertex function (α large) and with Im $\Sigma(E)$ expressed in terms of the pion wavenumber in the medium. However, we can see from Eq. (10) that this is inconsistent since the modification of the real part of $\Sigma(E)$ is then greater. In this situation, the results will depend sensitively upon the treatment of nuclear-nucleon correlations.

Frank Tabakin (U. of Pittsburgh): Why do you use infinite nuclear matter to study the eigenmodes when we know that elastic scattering of pions is a surface dominated process where the density is not uniform but varying rapidly?

Moniz: It is certainly true that elastic scattering of pions takes place mainly in the nuclear surface, where the density is both small and rapidly varying and therefore where we would have a difficult time in unravelling the eigenmode structure. However, reactions which probe the nuclear interior and which involve large momentum transfer should be quite sensitive to the eigenmode structure. For example, Equation (7) reveals that at high density one obtains two weakly damped modes, one with large and one with small wavenumber; i.e., the pion wavefunction should have a structure in momentum space quite different from that predicted without isobar propagation and large momentum transfer processes should reflect this.

FIELD THEORETIC ASPECTS OF PION NUCLEUS PHYSICS

M. K. Banerjee*
University of Maryland, College Park, Md. 20742

ABSTRACT

Field theoretic aspects of pion nucleus physics are illustrated with a discussion of a phenomenological form of the pion-nucleon off-shell scattering amplitude and with comments on multiple scattering.

A pion and a nucleon interact through two kinds of mechanism-- (i) exchange of bosons with even G parity, scalar-isoscalar and vector-isovector, (I will refer to this as the t-channel mechanism) and (ii) absorption and emission of the pion by the nucleon and the crossed process (s- and u-channel mechanisms). This second mechanism necessarily introduces many bare pions in the intermediate stages of the interaction and converts the problem of interaction of two physical particles into an essentially many-body problem. Field theory is a very natural way of dealing with such a problem.

An orthogonal viewpoint is to believe that the phenomenological pion-nucleon and pion-nucleus interactions can be understood in terms of a fixed number of pions. The work of Afnan and Thomas[1] is an example of this approach. The title of the talk suggests I should compare the two approaches. But I am afraid that then this talk will appear to have a negative view. So, instead of doing that I will mainly discuss how field theoretic considerations can help us in developing a phenomenological approach to the problem.

The ideas discussed here were developed in collaboration with J. B. Cammarata and S. J. Wallace.

Let us first discuss the off-mass-shell pion nucleon scattering amplitude, defined as

$$F(k',p',\beta;k,p,\alpha) = -i(k^2-m_\pi^2)(k'^2-m_\pi^2)\int d^4x\, e^{ik'\cdot x} <p'|T(\phi_\beta(x),\phi_\alpha(0))|p> \quad (1)$$

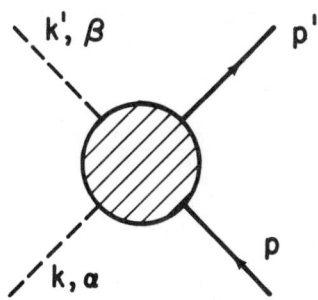

associated with the scattering event shown in Fig. 1. ϕ is the pion field. The nucleons are on the mass shell. In pion nucleus scattering the nucleons are not quite on the mass shell. But so far as the pionic degree of freedom is concerned it is not a bad idea to consider the nucleons to be on the mass shell with a spectrum appropriate for a nucleon moving in an average potential. The lowest order optical potential

Fig. 1. Pion-Nucleus Scattering Amplitude.
*Supported by U. S. Energy Research and Development Administration.

for use in a Klein-Gordon equation is obtained by adding such amplitudes.[2] Thus

$$\langle \vec{k}'\beta|U(\omega)|\vec{k}\alpha\rangle = \sum_\lambda F(k',\lambda,\beta;k,\lambda,\alpha) \qquad (2)$$

where $U(\omega)$ is the optical potential for pion lab energy $\omega = k_o = k_o'$ (neglect recoil energy of the nucleon). $F(k',\lambda\ \beta;k,\lambda,\alpha)$ is the off shell amplitude for elastic scattering of a pion off a nucleon in the orbit λ.

Clearly we need to know what $F(k',p',\beta;k,p,\alpha)$ is to develop a theory of pion nucleus scattering. Let us try to make a model.[3] To simplify our discussion we consider only the isoscalar, spin independent part of the amplitude.

We know that the (3,3) resonance is of central importance in the low energy ($\omega < 300$ MeV) region. The resonant character and the successful static model of Chew and Low[4] suggest that we use a factorable form

$$-4\pi\, A^{(+)}(\omega)\, \vec{Q}'\cdot\vec{Q} \qquad (3)$$

$$\text{where } \vec{Q} = \frac{M\vec{k} - \omega\vec{p}}{M+\omega} \text{ and } \vec{Q}' = \frac{M\vec{k}' - \omega\vec{p}'}{M-\omega} \qquad (4)$$

are the relative momenta. For simplicity I am omitting the pion-nucleon form factors. The nucleon motion is being assumed to be non-relativistic.

Attractive as this form is it has a serious flaw. It violates a symmetry property which the scattering amplitude possesses, namely crossing symmetry. It is directly a consequence of the fact that π^+ and π^- are charge conjugates of each other and that π^o is self-charge conjugate. The crossing symmetry states that

$$F(k',\beta,p';k,\alpha,p) \equiv F(-k,\alpha,p';-k',\beta,p) \qquad (5)$$

It arises purely due to the property of the pion. The symmetry is true for any pion in-pion out process regardless of the nature of the initial and final states of the other particles.

Since we are neglecting the recoil energy and discussing only the isoscalar term the crossing symmetry says that the amplitude must remain unchanged under the interchange $\vec{k} \leftrightarrow -\vec{k}'$ and $\omega \to -\omega$. But note that under this change

$$\vec{Q} \to \vec{Q}_c = -\frac{M\vec{k}' - \omega\vec{p}}{M-\omega} \text{ and } \vec{Q}' \to \vec{Q}'_c = -\frac{M\vec{k} - \omega\vec{p}'}{M-\omega} \qquad (6)$$

so

$$A^{(+)}(\omega)\, \vec{Q}\cdot\vec{Q}' \longrightarrow A^{(+)}(-\omega)\, \vec{Q}_c\cdot\vec{Q}'_c \neq A^{(+)}(\omega)\, \vec{Q}\cdot\vec{Q}' \qquad (7)$$

Let us conjecture that the remedy is to symmetrize our original guess, So, we take

$$-2\pi\{A^{(+)}(\omega)\,\vec{Q}\cdot\vec{Q}' + A^{(+)}(-\omega)\,\vec{Q}_c\cdot\vec{Q}'_c\}$$

$$= -4\pi\,\frac{A^{(+)}(\omega) + A^{(+)}(-\omega)}{2}\,\frac{\{\vec{Q}\cdot\vec{Q}' + \vec{Q}_c\cdot\vec{Q}'_c\}}{2}$$

$$+\; -4\pi\,\frac{A^{(+)}(\omega) - A^{(+)}(-\omega)}{2}\,\frac{\{\vec{Q}\cdot\vec{Q}' - \vec{Q}_c\cdot\vec{Q}'_c\}}{2} \tag{8}$$

In the static limit Chew-Low theory tells us that $A^{(+)}_{static}(\omega)$ is an even function of ω. This suggests that $\frac{A^{(+)}(\omega) - A^{(+)}(-\omega)}{2}$ is $\sim \frac{\omega}{M}$. So is $\frac{\vec{Q}\cdot\vec{Q}' - \vec{Q}_c\cdot\vec{Q}'_c}{2}$. So the second term is $\sim \frac{\omega^2}{M^2}$. We can develop a simple phenomenology if we go only one order beyond the static limit by keeping terms $\sim \frac{\omega}{M}$, but throwing out terms $\sim \left(\frac{\omega}{M}\right)^2$ and higher. Then our form for the s- and u-channel mechanism part of the amplitude is

$$-4\pi\, H^{(+)}(\omega)\,\frac{\vec{Q}\cdot\vec{Q}' + \vec{Q}_c\cdot\vec{Q}'_c}{2} \tag{9}$$

where

$$H^{(+)}(\omega) = \tfrac{1}{2}\{A^{(+)}(\omega) + A^{(+)}(-\omega)\} \tag{10}$$

Let us examine the amplitude in the c.m. frame where $\vec{p} = -\vec{k}_{cm}$ and $\vec{p}' = \vec{k}'_{cm}$.

$$-4\pi\, H^{(+)}(\omega)\left\{(1+\tfrac{\omega}{M})\,\vec{k}'_{cm}\,\vec{k}_{cm} + \tfrac{\omega}{M}(\vec{k}^2_{cm} + \vec{k}'^2_{cm})\right\} \tag{11}$$

Notice that we have acquired an s wave term which is $\sim \frac{\omega}{M}$ relative to the p wave term. It is easy to understand how it arises. In the c.m. frame the nucleon pole term of the Low expansion where the pion is absorbed first and then emitted (Fig. 2a) is a purely p wave term. But the crossed term (Fig. 2b) does have an s wave part of the order of $\frac{\omega}{M}$. It has higher partial waves also in higher order of $\frac{\omega}{M}$. Thus we find that the crossing symmetry requires a certain relationship between the s and p wave parts of the scattering amplitude.

The discussion also shows why we should not consider

$$-2\pi\, A^{(+)}(\omega)(\vec{Q}'-\vec{Q})^2 \tag{12}$$

with

$$A^{(+)}(\omega) = A^{(+)}(-\omega) \tag{13}$$

as a crossing symmetrized form of the s-u-channel mechanism. We will be then adding an s wave term which is of the same order as the p wave term.

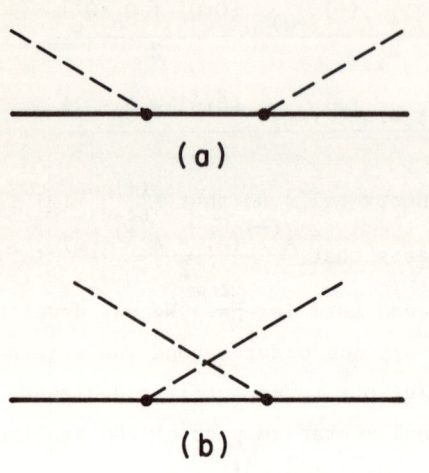

Fig. 2. Nucleon Pole Terms of the Chew-Low Theory.

Let us also note that the s wave term we have generated by crossing symmetrizing is attractive in nature. Since at low energy the isoscalar part of the s wave amplitude is repulsive we will require a relatively larger repulsion from the t-channel mechanism.

Before discussing the t-channel mechanism let us examine the form of our crossing symmetrized amplitude due to the s- and u-channel mechanism in a frame where the nuclear velocities are small. It is

$$-4\pi\, H^{(+)}(\omega)\{\vec{k}'\cdot\vec{k} - \frac{\omega}{2M}(\vec{k}+\vec{k}')\,(\vec{p}+\vec{p}')\} \qquad (14)$$

The second term has vanishing contribution if the target nucleus is unchanged under time reversal, e.g., J = 0 state. Thus, for ^{12}C or ^{16}O we have only the Kisslinger term.

Now let us turn to the part of the amplitude which arises because of the t-channel mechanisms. The sources of these terms are the sea gull terms

$$-i\int d^4x\, e^{-ik\cdot x}\langle p'|\delta(x_0)[\dot\phi_\alpha(x), j_\beta(0)] + ik_0\delta(x_0)[\phi_\alpha(x), j_\beta(0)]|p\rangle$$

which we do not know. But if we assume PCAC, by which I mean that

$$\frac{\sqrt{2}}{f_\pi}\frac{\partial A^\alpha_\mu(x)}{\partial x_\mu} = \phi_\alpha \qquad (15)$$

is the canonical pion field, we can evaluate the soft pion limit $F(k'=0,\beta,p';k=p'-p,\alpha,p)$ and then eliminate the unknown sea gull terms with the help of the known soft pion limit. Cammarata and I have developed such an approach.[5] Cammarata will talk about it in the contributed paper session. So I will be brief. Our idea was to extend the Chew-Low theory so that the driving terms are approximately covariant. To do this the first step is to include the process shown in Fig. 3 (and its crossed term) where the physical intermediate state is an anti-nucleon. The usual nucleon pole terms plus these anti-nucleon pole terms of the Low expansion give us a covariant driving term. In fact we get the Feynman diagrams which look like Figs. 2a and 2b. We get covariance at an apparent price.

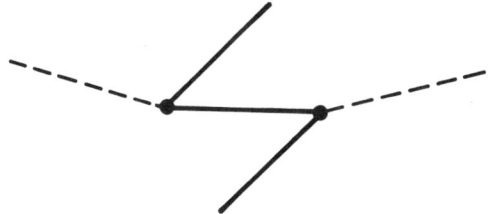

Fig. 3. Anti-nucleon Pole Terms.

The price is the large s wave part that the Feynman graph has. But this is largely cancelled by the sea gull terms. As stated before we eliminate the unknown sea gull term by using the soft pion limit
$F(k'-0,\beta,p';k-p'-p,\alpha,p)$. Note that this is not the same soft pion limit one uses to predict the charge exchange scattering length. The latter is obtained by evaluating F in the limit $k + k' \to 0$. Our soft pion limit contains the σ commutator term and nucleon pole terms. Our integral equation has the form

$$\langle p',k'\beta|j_\alpha(0)|p\rangle = \frac{g_\pi^2(m_\pi^2-t)}{M^2 m_\pi^2} \langle p'|\frac{\sqrt{2}}{f_\pi}\sigma_{\text{commutator}}(0)\delta_{\alpha\beta}|p\rangle$$

$$+ \frac{g_\pi^2}{M^2}\frac{1-g_A^2}{g_A^2} i\varepsilon_{\beta\alpha\lambda}\langle p'|k\cdot v^\lambda|p\rangle$$

$$+ \frac{g_\pi^2}{2M}\bar{u}(\vec{p}')\{\frac{\not{k}'\not{k}}{s-M}\tau_\beta\tau_\alpha + \frac{\not{k}\not{k}'}{u-M^2}\tau_\beta\tau_\alpha\}u(\vec{p})$$

+ {Chew-Low type rescattering terms
for πN intermediate states minus
soft pion terms} (16)

The main isoscalar driving term is the first one. Notice the factor m_π^2-t. Its origin is easy to understand. F has a factor $(k^2-m_\pi^2)(k'^2-m_\pi^2)$. When $k' \to 0$ conservation of momentum fixes $k \to p'-p$. So the factor becomes $m_\pi^2(m_\pi^2-t)$. $\frac{\sqrt{2}}{f_\pi}\sigma_{\text{commutator}}$ is a local field which is scalar, isoscalar. So one expects the t dependence of $\langle p'|\frac{\sqrt{2}}{f_\pi}\sigma_{\text{commutator}}(0)|p\rangle$ to be of the form $\frac{1}{(t-m_1^2)(t-m_2^2)}$ with $m_2^2 > m_1^2 > (4m_\pi)^2$. Thus the term goes to zero as $|t| \to \infty$. But for low t the numerator factor is significant. Wallace and I feel that this suggests that we use the form $-4\pi(a_o + a_\sigma \vec{q}^2)$, $(\vec{q} = \vec{k}'-\vec{k})$, for the t channel mechanism term. Thus the full phenomenological form for the isoscalar, spin independent amplitude becomes

$$-4\pi[a_o + a_\sigma\vec{q}^2 + H^{(+)}(\omega)\{\vec{Q}\cdot\vec{Q}' + \vec{Q}_c\cdot\vec{Q}'_c\}/2] \qquad (17)$$

Note that now both the t-channel mechanism and the s-u-channel mechanism terms are combinations of s and p wave terms. Unfortunately π-N scattering data give us only two complex quantities

$$f_s^{(+)} = \frac{f_{11}(s) + 2f_{31}(s)}{3}$$

and (18)

$$f_p^{(+)} = \frac{f_{11}(p) + 2f_{31}(p) + 2f_{13}(p) + 4f_{33}(p)}{3}$$

while we have three complex parameters. The extra constraint, which is required, must be obtained from the theory of π-N scattering. Since the problem has not been solved we can only indulge in speculation.

To illustrate what may be happening let us consider two possibilities, viz, (i) $a_\sigma = 0$ and (ii) $a_o = a_\sigma$. Undoubtedly in reality a_o and a_σ are both non-zero and unequal. To evaluate the parameters we go to the c.m. frame, multiply our invariant amplitude by $-M/(4\pi\sqrt{s})$ and equate the s wave part to $f_s^{(+)}$ and the p wave part to $f_p^{(+)}$. Keep in mind that we throw out terms $\sim (\frac{\omega}{M})^2$. Thus we have

$$\frac{M}{\sqrt{s}}[a_o + 2a_\sigma(\omega^2-m_\pi^2) + \frac{\omega}{M} H^{(+)}(\omega^2-m_\pi^2)] = f_s^{(+)}$$

(19)

$$\frac{M}{\sqrt{s}}[-2a_\sigma(\omega^2-m_\pi^2) + (1+\frac{\omega}{M}) H^{(+)}(\omega^2-m_\pi^2)] = f_p^{(+)}$$

The imaginary part of $f_s^{(+)}$ is usually small but always positive. As we approach the (3,3) resonance the imaginary part of $\frac{\omega}{M} H^{(+)}(\omega^2-m_\pi^2)$ increases and can easily be larger than that of $f_s^{(+)}$. So we should not be surprised if the imaginary parts of the a_o and a_σ terms become negative. This is illustrated in Fig. 4.

The form (17) gives the following optical potential for scattering of pion by a J = 0, I = 0 target,

$$-4\pi[a_o\rho(r) + a_\sigma(\nabla^2\rho(\vec{r})) - H^{(+)}\vec{\nabla}\cdot\rho(\vec{r})\vec{\nabla}] \ldots, \quad (20)$$

containing a volume term, a Laplacian term and a Kisslinger term. There are, of course, enormous number of calculations with the volume and the Kisslinger terms which corresponds to $a_\sigma = 0$ in our scheme. Once in a while in an attempt to fit data it has been found necessary to use a_o with negative imaginary part. I am certain there are many among you who, like me, had thought the wrong sign of the imaginary part must have implied violation of unitarity for pion-nucleon scattering amplitude. Now we see that we cannot jump to that conclusion from the sign alone. It is a quantitative question. From Fig. 4 it appears that below 80 MeV a_o and a_σ should have positive imaginary parts.

Inevitably the question of unitarity in the π-nucleus scattering comes up--particulary in the s wave. Before commenting on this let us remind ourselves of the approximate nature of our phenomenology. We have thrown out $(\frac{\omega}{M})^2$ terms. It is very likely that s

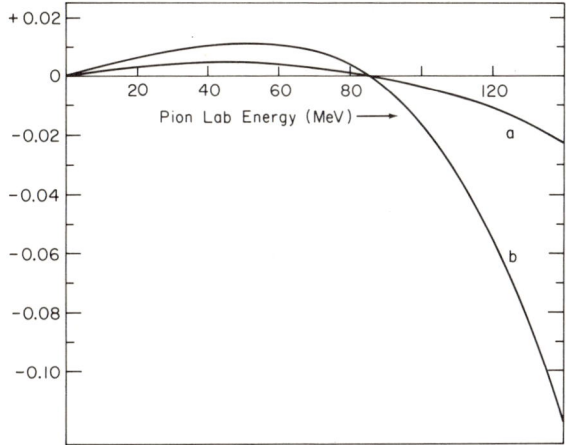

Fig. 4. Imaginary part of the t-channel terms. Curve (a) is for the $a_o = a_\sigma$ case and curve (b) is for the $a_\sigma = 0$ case. The amplitudes are in $c = h - m_\pi = 1$ units.

wave terms which are of the order of $(\frac{\omega}{M})^2 H^{(+)}$ are not negligible compared to the t-channel mechanism terms. We should keep this caveat in mind. If our simple procedure leads to violation of unitarity in pion nucleus scattering we should abandon our approximate theory and try to develop a more accurate method of evaluation of the parameters. Fortunately, we have not found any violation of unitarity up to ω = 200 MeV for either the case $a_\sigma = 0$ or the case $a_o = a_\sigma$. This does not mean that our approximation of neglecting $(\frac{\omega}{M})^2$ terms is necessarily good up to ω = 200 MeV.

Let us now apply these ideas to understand pion nucleus elastic scattering data. The low energy data are likely to be more sensitive to the new term, $4\pi a_\sigma (\nabla^2 \rho)$, we have introduced. At higher energy the Kisslinger term is dominant and more partial waves come into play. So I will discuss the elastic angular distribution of 50 MeV π^+ on ^{12}C. The experiment was performed by Amann, et al.[6] Dr. Eisenstein will also talk about it tomorrow.

Using the CERN experimental phase shifts[7] we find values of a_o and a_σ given in Table I. The first row gives the value for

Curve	a_o	a_σ	$H^{(+)}$
Dashed	-0.043+i0.01	0	0.174+i0.016
Solid a	-0.018+i0.0044	-0.018+i0.0044	0.145+i0.023
Dotted	-0.086+i0.01	0	0.174+i0.016
Solid b	-0.032+i0.004	-0.032+i0.004	0.142+i0.031

Table I. All numbers are in mesic units.

$a_\sigma = 0$ case and the second for the $a_o = a_\sigma$ case. In Fig. 5 we show the resulting differential cross sections. For the density of ^{12}C we used

$$\rho(r) = \frac{4}{b^3 \pi^{3/2}} (1 + \frac{4}{3} \frac{r^2}{b^2}) e^{-r^2/b^2} \ldots \qquad (21)$$

with $b = 1.18\ m_\pi^{-1}$. No form factor was used.

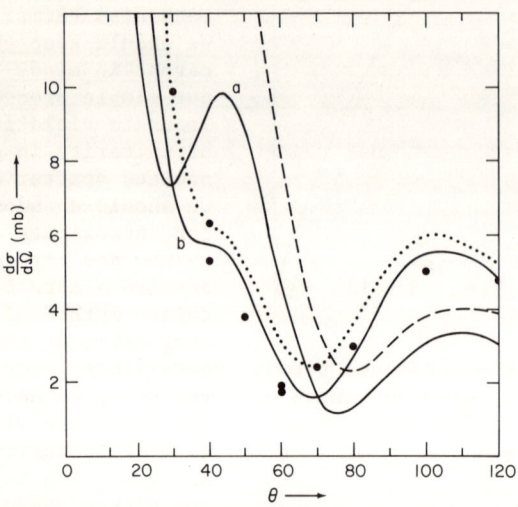

Fig. 5. 50 MeV $\pi^+ - {}^{12}C$ differential cross section.

As long as the form factor is hard it should not play a significant role in the scattering problem. The black circles represent data. The dashed curve is the first row of Table I and the solid curve (a) due to those in the second row. Notice the very marked difference in the angular distribution. So we must know what a_o and a_σ are before we can claim to have explained the $\pi^+ - {}^{12}C$ data in terms of π-N data.

None of the two curves fit the data. More repulsion in the s wave is required. For the $a_\sigma = 0$ case if we double the value of a_o we get the dotted curve. For the $a_o = a_\sigma$ case if we increase the real part from -0.018 to -0.032 the solid curve marked b is obtained. It is possible that an equally good fit can be obtained for the $a_\sigma = 0$ case by searching for a better value of a_o. Moreover, we have not included the contribution to the optical potential from the mechanism of absorption of the pion by a correlated pair. This is likely to be mostly imaginary. The real part of the isoscalar term coming from this mechanism is small becuase of large cancellation between the direct and the crossed processes.

Because of these considerations we cannot claim to have justified the case for the a_σ term on the basis of improved fit to the 50 MeV $\pi^+ - {}^{12}C$ data. The justification must come from the theory of π-N scattering. The presence of the σ commutator term as a driving term in the integral equation (16) is not in much doubt. In fact, one has a reasonably good idea of its size. It is believed to be $\sim 63 \pm 9$ MeV.[8] This gives a value of $(-0.08 \pm 0.01) m_\pi^{-1}$ for the driving term. But the rescattering terms will change this number. The contribution of the rescattering terms can be divided into two parts in terms of the energy spectrum of the physical intermediate states. The high energy part of the spectrum, where the s wave resonances are, will give a contribution which is very weakly dependent on the energy ω of the pion for which we want to determine the scattering amplitude. The contribution from the low energy intermediate states will be more strongly dependent on ω. Huang, Levinson and I[9] had used these ideas to explain π-nucleus scattering lengths. Our calculations indicate that the high energy intermediate state contribution reduces the value of the σ term from $-0.08\ m_\pi^{-1}$ to $-0.035\ m_\pi^{-1}$. But we had studied only the scattering length problem we really cannot tell how the $\vec{q}^2$ part is changed. All that I can say at this stage is that the parameters which give the best fit to the $\pi^+ - {}^{12}C$ data at 50 MeV are not unreasonable.

Now let me make a few comments on the many body aspects of pion nucleus interaction. Wick[10] and Chew and Low[4] taught us how to deal with problem in pion-nucleon scattering. So the main question in pion-nucleus scattering is whether we are plagued by the many body aspect when a pion is in flight between two nucleons. A nucleon is a source of many virtual pions and carries a pionic cloud around it. The spatial extension of the pionic clous is m^{-1} where m is the mass which characterizes the damping of the pion nucleon form factor. Dispersion theory suggests that this mass $m \gtrsim 7m_\pi$. There are two length parameters associated with the nucleus which we have to keep in mind. One is the mean internucleon distance ~ 1.2 fm. The other is the distance within which, because of the short range repulsion, the correlated wave function of a pair of nucleons is very small compared to the uncorrelated wave function. This distance is ~ 0.3 fm. If the spatial extension of the pionic cloud is smaller than both these distances we have very little to worry about. If on the other hand, the extension of the pionic cloud is larger than 0.3 fm we have to re-examine the method of evaluating the contribution of multiple scattering to the otpical potential. Mere inclusion of the form factors is not enough.

Even when the pionic cloud is very short ranged it has a role to play in the contribution of multiple scattering to the optical potential. We all know about the Lorentz-Lorenz Ericson-Ericson effect.[11] If the a_σ term, discussed before, is absent the optical potential with the LLEE effect is

$$4\pi\ H^{(+)}(\omega)\vec{\nabla}\ \frac{\rho(\vec{r})}{1 + \frac{4\pi}{3} H^{(+)}(\omega)\rho(r)}\ \vec{\nabla} \tag{22}$$

Roughtly half of it arises from Pauli correlation. An alternative way is to incorporate the Pauli correlation effect into the single nucleon term and exhibit separately the terms arising out multiple scattering from correlated clusters of nucleons. Let $\hat{t}^{(+)}(\vec{k}',\vec{k})$ represent the Pauli corrected π-N scattering amplitude. Then the first order optical potential is

$$\langle\vec{k}'|U^{(1)}|\vec{k}\rangle = \sum_\lambda (\lambda|\hat{t}^{(+)}(\vec{k}',\vec{k})|\lambda) \tag{23}$$

and the second order multiple scattering term is

$$-\left(\frac{3}{4}\right)\left(\frac{7}{12}\right)\frac{1}{3}(4\pi H^{(+)}(\omega))^2 \vec{\nabla} \rho^2(\vec{r}) \vec{\nabla} \tag{24}$$

The factor of $\frac{3}{4}$ reminds us that only $\frac{3}{4}$ of the rest of the nucleons can be brought in contact with a nucleon with given spin and isospin. The factor $\frac{7}{12}$ reflects the spin and isospin dependence of the π-N scattering amplitude. Here I have used a purely (3,3) amplitude. If we could describe the dynamics of π-N scattering in terms of a single pion in all intermediate states then it is easy to see that the Pauli corrections will be proportional to $\rho^2(r)$[12] and that when added to the second order multiple scattering term will give

$$-\frac{1}{3}(4\pi H^{(+)}(\omega))^2 \vec{\nabla} \rho^2(\vec{r}) \vec{\nabla} \tag{25}$$

which is the $(H^{(+)})^2$ term of Ericson and Ericson.[11] However, when we recognize that there are many pion intermediate states we have a different result. For our purpose it is fair to consider the nucleus to be an assembly of physical nucleons and not bare nucleons and pions. Thus the electron scattering data, after correcting for the nucleon electromagnetic form factors, measures the density of these physical nucleons. When we use shell model we assign orbits to these physical nucleons. So the Pauli correction must be introduced in a Chew-Low type theory.[13] A perturbative calculation yields the following non-local form for the correction to the optical potential due to Pauli correction

$$\delta U = -\pi \vec{\nabla}_r \rho^2(\vec{r},\vec{r}') F(\vec{r}-\vec{r}') \vec{\nabla}_{r'} \tag{26}$$

where $\rho(\vec{r},\vec{r}')$ is the single nucleon density matrix and

$$F(\vec{r}-\vec{r}') = \frac{1}{\pi}\int \frac{d\omega_q^2 \, \text{Im}\, H^{(+)}(\omega_q)}{\omega_q^2 - \omega^2 - i\eta} j_0(q|\vec{r}-\vec{r}'|) \ldots \tag{27}$$

In deriving this form I have used the static Chew-Low theory and have assumed $\rho(\vec{r},\vec{r}')$ to be spin independent. In the usual potential theory approach with fixed number of pions the integral for $F(\vec{r}-\vec{r}')$ has $\frac{3}{4}q^3(H^{(+)}(\omega))^2$ in place of $\text{Im}\, H^{(+)}(\omega_q)$. Note the difference in the argument of the amplitude function. The former gives the local form of Ericson and Ericson. With the latter we get a non-local form with a fairly large range of non-locality.

I hope that these illustrative comments will show the importance of the field theoretic aspects of pion-nucleus physics.

REFERENCES

1. I. R. Afnan and A. W. Thomas, Phys. Rev. C10, 109 (1974).
2. J. B. Cammarata and M. K. Banerjee, Phys. Rev. Lett. 31, 610 (1973).
3. J. B. Cammarata and M. K. Banerjee, Phys. Rev. C13, 299 (1976).
4. G. F. Chew and F. E. Low, Phys. Rev. 101, 1570 and 1579 (1956).
5. J. B. Cammarata and M. K. Banerjee, Bull. Am. Phys. Soc. II, 20, 1192 (1975).

6. J. F. Amann, et al., Phys. Rev. Lett. <u>35</u>, 426 (1975).
7. D. J. Herndon, et al., LRL Report, UCRL-20030πN, Feb. 1970.
8. R. L. Burman, High Energy Physics and Nuclear Structure, 1975 (Santa Fe and Los Alamos) AIP Conference Proceedings, no. 26, p. 41.
9. W. T. Huang, C. A. Levinson and M. K. Banerjee, Phys. Rev. <u>C5</u>, 651 (1972).
10. G. C. Wick, Rev. Mod. Phys. <u>27</u>, 339 (1955).
11. M. Ericson and T. E. O. Ericson, Ann. Phys. <u>36</u>, 323 (1966).
12. J. Delorme and M. Ericson, Phys. Lett. <u>60B</u>, 451 (1976). There is a difference between their result and ours.
13. This result was arrived at by C. B. Dover and R. H. Lemmer, Phys. Rev. <u>C7</u>, 2312 (1973). However, we disagree with their logic.

M. Rho (SACLAY): I understand from your discussion that the σ-term is very important. Is the connection between the π-N σ-term and the π-nuclear known? In other words, is a simple scaling in mass number A assumed?
Banerjee: Yes. To the σ term one adds the rescattering terms to get the hard pion result. The low energy intermediate state rescattering terms are quenched when the πN system is in the nucleus. The high energy contribution remains unaffected. The latter added to the σ commutator term gives 0.035 m_π^{-1}. In our explanation of the scattering lengths we assumed that the σ term is linear in A.

Kisslinger (CMU): 1. Are your conclusions with the new term consistent with the higher energy cross section? 2. Might not there be corrections to order $\frac{\omega}{M}$ arising from the nucleons being off the mass shell?
Banerjee: At higher energy the a_σ term is not of great importance. The off mass shell effects will alter $\frac{\omega}{M}$ to $\frac{\omega}{M-B.E.}$ where B.E. is the binding energy. But it will not give rise to new terms.

Il-T. Cheon (McMaster Univ.): In your optical potential you do not have the term associated with the real absorption of the pion which is usually described with $\rho^2(r)$. Could you give a comment of it?
M. Banerjee: Dr. Mizutani will speak on this question.

Martin Cooper (LASL): In view of the mixing of s and p wave π-nucleon terms by the crossing symmetry requirements, does that mean that a phenomenological potential fit to data may violate unitarity in one π-nucleus partial wave?
M. Banerjee: No. Unitarity must be maintained at the π-N level and at the π-nucleus level. What I was pointing out was that our approximate method of evaluating the parameters could lead to bad numbers which violate unitarity in π-nucleus scattering. Fortunately it does not happen up to ω = 200 MeV.

ISOBARS IN NUCLEI

H.J. Weber*
University of Virginia, Charlottesville, Va. 22901

ABSTRACT

Recent theoretical studies of (NΔ), (NN*) and ($\Delta\Delta$) isobar admixtures to the two-nucleon bound and continuum states are reviewed, including a critical survey of spectator experiments designed to search for the $\Delta(1236)$ in the deuteron ground state.

INTRODUCTION

Two decades of high energy experiments have revealed the rich and complicated excitation spectrum of the nucleon[1] which, except for much larger mass differences and strong N* decay widths, looks like that of a nucleus. The first excited level is the (3,3) resonance $\Delta(1236)$, which has become a pillar of medium energy and pion physics, and lies 0.3 GeV/c^2 above the nucleon ground state N(938). The Roper resonance N*(1470) plays the role of a breathing mode; the N*(1520) has $J^\pi = (3/2)^-$ and, as a 1^- photo-excitation, is the analog of the giant resonance, while the Regge recurrence N*(1688) of the nucleon is the first rotational 2^+ excitation. Yet nuclear theory treats nucleons in nuclei as point-like particles, structureless except for a few static properties such as their mass, charge and magnetic moment. As nuclei are being probed at higher energies and momentum transfer, nucleon collisions inside nuclei are recorded that are increasingly inelastic, via NN $\rightarrow$ NΔ, NN* and NN $\rightarrow$ $\Delta\Delta$. Thus a compromise between such radical views as using hydrogen targets exclusively or ignoring N*'s and mesons, except to the extent they may be included in nonrelativistic phenomenological potentials, becomes desirable.

ISOBAR CONFIGURATIONS IN THE NN SYSTEM

In the past few years nuclear models have been developed which incorporate explicitly nucleon resonances, usually called N*'s or isobars, as nuclear constituents besides protons and neutrons.[2] This idea was first raised in the context of proton deuteron backward elastic scattering at 1 GeV [3] in an attempt to supplement the neutron stripping mechanism by N* exchange at large momentum transfer. Isospin invariance prevents the exchange of $\Delta(1236)$ but several low lying T = 1/2 isobars may actually contribute.[4]

The basic assumptions of the nonrelativistic isobar model are (i) in the medium range part of nuclear forces, N's and N*'s in conjunction with their (transition) potentials saturate mesonic degrees of freedom, except for meson exchange currents which transfer an incoming high momentum through at least two mesons to two nucleons.

*Supported in part by the National Science Foundation.

(ii) The velocity of N^*'s is estimated to be $c/3$ to $c/2$ so that non-relativistic calculations retain at least qualitative validity.
(iii) Short isobar life times are generally ignored. Thus the isobar model generalizes the wave function of any bound two-nucleon system $(NN)_b$ to include the lowest lying N^*'s, in particular $\Delta(1236)$,

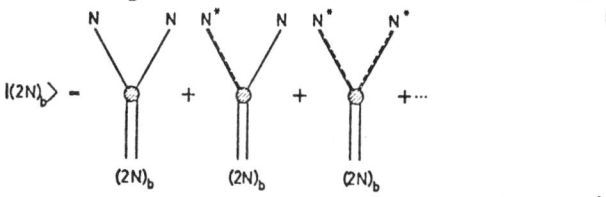 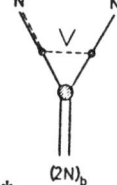

Fig. 1a. Generalized bound two nucleon $(NN)_b$ wave function.

Fig. 1b. (NN^*d) vertex in the impulse approximation.

$N^*(1520)$ and $N^*(1688)$. Each diagram in Fig.1 represents the relative wave function of two point-like particles moving in a potential. If N^*'s actually exist in nuclei and are nuclear constituents, then we expect them to play an important role already in NN scattering and in the bound two-nucleon system $(NN)_b$, i.e. the deuteron ground state. Yet the scattering state, $(NN)_c$, and $(NN)_b$ differ substantially from each other with respect to their Δ content. The lowest isobar configuration (IC) is the $(N\Delta)$ with isospin $T = 1,2$. It occurs only in $(NN)_c$ because the deuteron is isoscalar. The lowest IC in $d = (NN)_b$ is the intrinsically more difficult $(\Delta\Delta)$ whose (free) mass is degenerate with $(NN^*(1520))$. This, and possible $\Delta\Delta \to NN^*$ mixing via pion and ρ meson exchange, suggests that the deuteron ground state has a different and more complicated isobar structure than the two-nucleon continuum. The peripheral or impulse approximation (IA) in Fig.1b provides crude estimates for N^* wave functions in terms of the (NN) wave function and an $NN \to NN^*$ transition potential V, which corresponds to using the nonrelativistic Schrödinger equation. Since isobars are virtually produced on N's which are relatively far apart, admixture of isobars becomes a perturbation problem except for the detailed shape of radial wave functions. The results for the four $(\Delta\Delta)$ LS-states are typical.[2] The $NN \to \Delta\Delta$ potential V has a central (V_c) and tensor part (V_t). From the $(np)^3S, ^3D$ states, V_c generates $(\Delta\Delta)^3S, ^3D$ states while V_t generates $(\Delta\Delta)^7D, ^7G$ states. Except for the $(\Delta\Delta)^3S$ state which is quite sensitive to the less well known high momentum components of the (np) wave function, the short range structure of the $(\Delta\Delta)$ wave functions originates from the transition potential (see Fig.7). They peak at ~ 2 fm^{-1} so that the Δ Fermi momentum is about three times that of a nucleon in the deuteron. The $(\Delta\Delta)$ amplitudes $\tilde{\psi}_{\Delta\Delta}$ are only $\sim 10\%$ of the 6.5% $(np)^3D$ state. The $(\Delta\Delta)$ probability $P_{\Delta\Delta}$, defined as $\langle\tilde{\psi}_{\Delta\Delta}|\tilde{\psi}_{\Delta\Delta}\rangle$, lies between 0.4 and 2%. Inclusion of ρ (and ω) meson exchange provides a natural short range cut off, considerably weakens the OPE tensor force and thus removes much of the unphysical short range cut-off dependence of V in one-pion exchange (OPE).

Recently nonrelativistic coupled channel calculations[6] (CC) have been carried out for the (np), (ΔΔ) and several (NN*) IC including π,ρ,ω exchange. With a weak direct ΔΔ potential which is consistent with the deuteron binding energy B, the IA is found to be within 10-25% of the CC results. Coupling a single LS channel of one IC to the (np) states at a time turns out not to be a valid approximation in conjunction with a strong direct ΔΔ interaction. From the Table 1 we see that both ρ-exchange and the diagonal ΔΔ interaction, which is repulsive in all ΔΔ channels, decrease the ΔΔ probability. The (np)^{3}D state probability is ~6.4%, and the short range cut-off parameter is 5 fm^{-1}.

NN→ΔΔ	ΔΔ→ΔΔ		method	ΔΔ-prob. (%)				
ρ	π	ρ+ω		^3S	^3D	^7D	^7G	total
-	-	-	IA	.038	.012	.878	.073	1.00
-	-	-	CC	.036	.013	.856	.062	.97
-	√	-	CC	.023	.014	.649	.060	.74
√	-	-	IA	.075	.014	.476	.040	.60
√	-	-	CC	.089	.014	.458	.036	.60
√	√	√	CC	.059	.017	.344	.034	.45

Table 1. ΔΔ probabilities from various NN→ΔΔ and ΔΔ→ΔΔ potentials via π,ρ,ω exchange (marked by √).

In this context a very interesting feature was discovered. To fit the deuteron binding energy, one has to weaken the intermediate range attraction of the central potential for NN scattering to account for the additional attraction due to the two-pion-exchange (TPE) dispersion contribution with intermediate Δ's (Fig 7). This strongly suggests re-examining NN scattering which has been started recently.

There is indeed a long-standing technical and conceptual difficulty associated with the generally successful, semi-phenomenological one-boson exchange (OBE) model for NN scattering. Two fictitious scalar mesons σ, σ' (~400) are required to describe the medium range attraction. The only scalar meson actually found, ε(1250)[1], lies too high in mass and does not seem to couple strongly enough to have much of an impact in NN scattering. Furthermore, OBE with σ's misrepresents the (πNΔ) tensor coupling which is quite important in neutron star matter at higher than nuclear matter densities. Thus it was rather satisfying, and consistent with the isobar idea, when it was found in coupled channel calculations that explicit inclusion of intermediate (NΔ) and (ΔΔ) channels in the TPE contribution of NN scattering[7] does provide most if not all of the required attraction in the crucial δ($^1S_0$) phase shift in pp scattering (see Figs.2,3). The fits for the P-wave phase shifts are of similar quality. Although it is true that scalar-isocalar meson exchange between two slow nucleons can be represented by an iterated OBE potential this is no longer true for exchange of isovector mesons which lead also to intermediate NΔ and ΔΔ states. Thus the role of the σ'mesons' in NN scattering is substantially reduced; this pushes the merely phenomenological treatment of NN scattering deeper into the hard

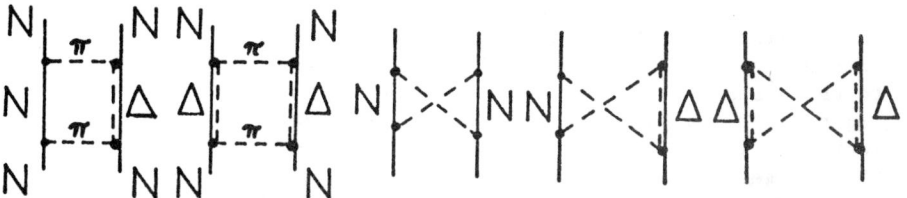

Fig.2a. Two-pion exchange contributions with intermediate $N\Delta$ and $\Delta\Delta$ channels in NN scattering.

Fig.2b. Crossed TPE contributions in NN scattering.

core region. Although the connection of the TPE with intermediate Δ's and the medium range attraction in the NN force is an old concept,[8] two new developments have brought about the more recent renewal of interest. First it was discovered that the crossed TPE diagrams in Fig. 2b almost cancel each other over a wide domain, and in particular for momentum transfer $q \sim 3-5$ fm^{-1} in the medium range regime. This provides the justification for CC calculations which ignore these terms and supports the basic idea of the isobar model. Secondly, in contrast with a dispersion treatment, the Δ's in TPE are off their mass shell. (The box diagrams in Fig. 2a may be obtained as the 2nd Born iteration of the NN $\to$ NΔ,$\Delta\Delta$ potentials and are, of course, included in coupled NN, NΔ, $\Delta\Delta$ channel calculations.) Thus, the NN $\to$ NΔ,$\Delta\Delta$ interactions involve the static propagator of the exchanged pion so that the large Δ - N mass difference does not entail drastic retardation effects. This crucial aspect is confirmed in the following relativistic 3-body model for (NN*) IC's with N, N, π as basic constituents.

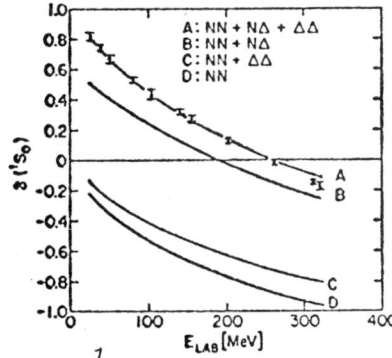

Fig.3. $^1S_0$ phase shifts calculated with various intermediate channels (NN), (NΔ) and ($\Delta\Delta$).

There are, however, difficulties with the physical interpretation of this isobar model. First, the nonrelativistic isobar probability is only meaningful in the weak binding limit. Yet, if the two Δ's are actually bound, then a strong $\Delta\Delta$ attraction is required to make the $\Delta\Delta$-NN mass difference small. Secondly, in lowest order perturbation theory, the $\Delta\Delta$ admixture is $\sim V_{NN\to\Delta\Delta}/2(m_\Delta-m)$, with the time uncertainty of $1/2 \,(m_\Delta-m)^{-1} \simeq 1/3$ fm/c. Since the size of the ($\Delta\Delta$) configuration is ~ 1 fm, this indicates that the ($\Delta\Delta$) state lives only 1/3 of the time it takes to cross it at the speed of light. That is to say, Δ's are not actually bound but are "continuously" created from nucleons via NN $\to$ NΔ,$\Delta\Delta$ transitions. Therefore, if the instability of N*'s is neglected in the isobar model, $\langle \tilde\psi_{\Delta\Delta} | \tilde\psi_{\Delta\Delta} \rangle$ measures not only the probability that two Δ's appear simultaneously in the deuteron but includes also those cases where, e.g., one nucleon virtually dissociates into Δ + π ; but this Δ decays already into N + π before the first exchanged π reaches the next nucleon and excites it to the Δ state, etc. In this sense, then, the isobar model may be regarded as a description of the off-mass-shell behavior of bound nucleons and is an attempt to shed light on the structure of nuclei in the regime of moderately small inter-nucleon distances.

Besides NN scattering there is other experimental evidence supporting such an inclusive interpretation of the presence of Δ(1236) in the deuteron. A long standing 10% discrepancy between σ_{exp} = 334.2 ± 0.5 mb for the radiative capture of thermal neutrons, np → dγ, and the best theoretical σ_{th} = 302.5 ± 4.0 mb, based on the single-particle magnetic moment operator and a Hamada-Johnston (np) wave function, was removed[9] upon including pair and π-exchange current contributions and the (NΔ)configuration, with each contributing about equally. The appearance of the Δ(1236) at low momentum transfer strongly motivates searching for it in the inverse reaction, the photo disintegration of the deuteron γd → np, at higher momentum transfer up to the pion production threshold. In fact, only when the Δ is included in the deuteron bound state and continuum[10] does the total cross section agree with the data at E_γ ~ 140 MeV (Fig.4). It is consistent that for E_γ ~ 100 - 140 MeV the angular distributions markedly improve upon including the Δ.

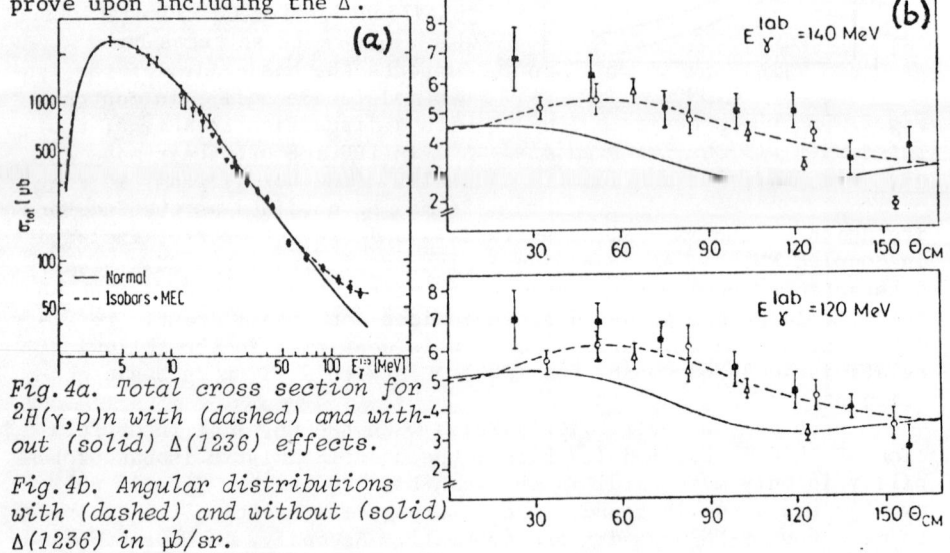

Fig.4a. Total cross section for $^2H(\gamma,p)n$ with (dashed) and without (solid) Δ(1236) effects.

Fig.4b. Angular distributions with (dashed) and without (solid) Δ(1236) in μb/sr.

RELATIVISTIC ISOBAR MODELS

Kinematic relativistic corrections, retardation and off-shell effects may be studied using a covariant Lippmann-Schwinger (LS) equation

$$\langle T \rangle = \langle V \rangle + (2\pi)^{-4} \int d^4Q_1 \langle V \rangle G \langle T \rangle \quad . \quad (1)$$

In one model[11] the deuteron is taken as a bound state of a 3-particle system of two N's and one pion such that the π always forms a quasi-

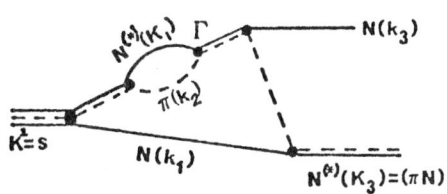

Fig.5 (NN*d) vertex in a relativistic π,N,N model.

particle (N or N*) with one nucleon. The 3-particle propagator G and the driving potential V are determined from dispersion relations whose input, the discontinuities of V and G, follow from imposing elastic and 3-body unitarity. V turns out to be generated by π exchange with essentially the static pion propagator, while the solution for $G \propto g \delta(Q_1^0)$ involves the renormalized quasi-particle propagator g; the δ-function reduces Eq.(1) in its homogeneous bound-state form to a three-dimensional formalism. The relative (Wightman-Garding) nucleon-isobar momentum variable Q_1 is essential to get rid of the relative energy in a covariant way and enforce the cluster properties consistently. The results of the coupled channel calculations for the (NN*(1520)) IC shows that the IA is still reasonable (Fig.6); off-shell effects are quite large but do not change wave functions qualitatively as non-relativistic results are an order smaller in magnitude. The restriction to three particles is too narrow, though, to study the (ΔΔ)IC as well and configuration mixing effects.

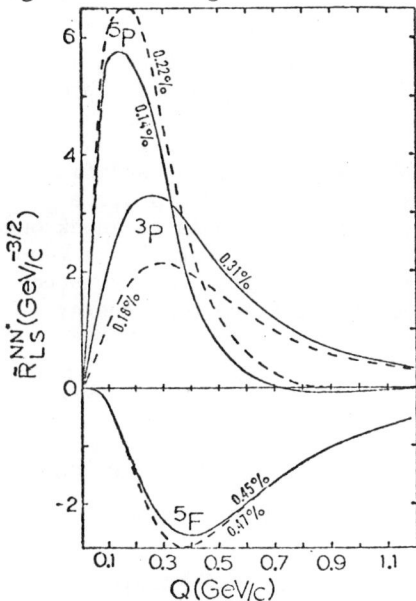

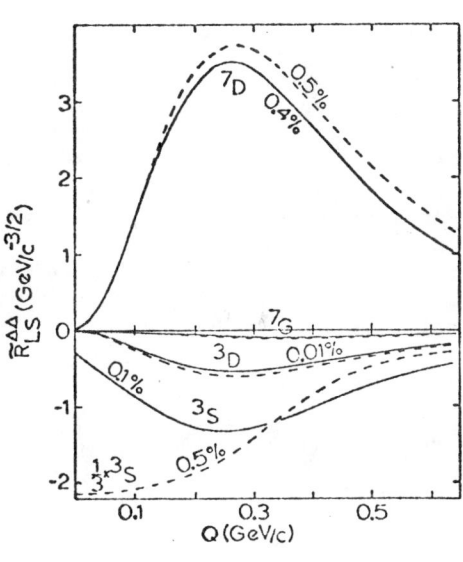

Fig.6. Radial NN*(1520) wave functions of relativistic (πNN) model in IA (dashed) and CC (solid).

Fig.7. Radial ΔΔ wave functions of relativistic 2-body model in IA using Gourdin (dashed) and McGee (solid) (np) wave function.

Furthermore, the reduction of the Bethe-Salpeter equation to a three-dimensional, linear formalism is not unique upon enforcing elastic unitarity at all energies only. Recently an 'optimal' 2-body LS-type equation[12] has been discussed. In its homogeneous form

$$(-\frac{1}{2\mu} q^2 + V)\psi = E\psi , \quad q \cdot P = 0, \quad (2)$$

it involves the Wightman-Gårding relative momentum $q = 1/2 (k_1-k_2) - (m_1^2-m_2^2)P/2s$, where $P = k_1 + k_2$ and $s = P^2$. Again, the constraint $q \cdot P = 0$ eliminates the relative energy. The eigenvalue $E = -q^2/2\mu$ when both particles are on their mass shell, $k_i^2 = m_i^2$; μ is their reduced mass. Eq.(2) has the following attractive properties. (i) If $s = (m_1 + m_2 - B)^2$ and $B << m_i$, then $E \simeq -B$. (ii) Eq.(2) reduces to the Schrödinger equation in the nonrelativistic limit. (iii) When $m_2 \to \infty$, $q \cdot P = 0$ becomes equivalent to $k_1^2 = m_1^2$ (i.e. the Klein-Gordon equation for particle 1) and Eq.(2) reduces to that of Fronsdal, Gross, Todorov.[13] (iv) The constraint $(q_f - q_i) \cdot P = 0$ for the quasi-potential $V(q_f-q_i)$ implies that both the direct and crossed ladders are included. (v) The two-body propagator $(\frac{1}{2\mu} q^2/_{q \cdot P=0} + E)^{-1}$ is identical to Weinberg's eikonal propagator[14] and that of Coester et al[15] in relativistic many-body theory. Results based on Eq.(2) for the ($\Delta\Delta$) configuration in OPE[11] are shown in Fig.7. Since $q^2 < 0$ when $q \cdot P = 0$, V involves again the static pion propagator. $\Delta\Delta - NN^*$ (1520) mixing in OPE amounts to a 5% correction.

Δ-SPECTATOR EXPERIMENTS

Earlier it was emphasized that the proper interpretation and corresponding measurement of isobar probabilities is a basic problem. Pictorially, one wants to make a series of snapshots of the nucleus in rapid succession and count the fraction of photos that show a Δ. Recent nuclear matter calculations[16] do essentially that and find 2.8 to 3.7% for the probability of a nucleon to be in a Δ state. For a deuteron target, the closest analog of a photo is to knock out one Δ, while the spectator Δ and its momentum distribution are measured assuming or attempting to verify that the latter pre-existed in the target and retained its memory throughout the high energy collision.

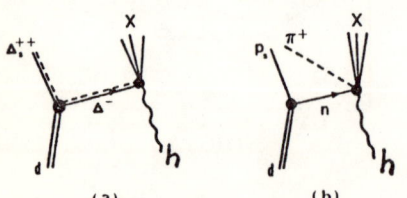
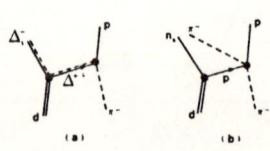

Fig.8(a) Δ^{++} spectator signal, (b) nonresonant (π^+p) background of a spectator experiment.

Fig.9(a) Δ^{++} exchange for $\pi^-d \to p\Delta^-$, (b) (π^-n) background.

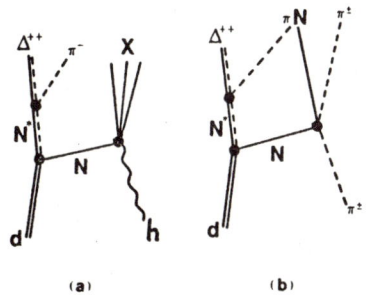

Fig.10. Contribution of (NN^*) IC's to inclusive (a) and quasi-elastic (b) Δ^{++} spectator production.

The signature of most spectator experiments is a Δ^{++} recoiling backward in the laboratory system because the p, π^+ decay products of the Δ^{++} can both be readily identified and Δ^{++} production is favored by a Clebsch-Gordan. More importantly, the most dangerous background of nonresonant $p\pi^+$ systems involves a slow π^+ generated from the bound neutron in the deuteron target by the projectile h (see Fig.8b) together with a spectator proton Ps, whereas slow π^-'s are much more copiously produced than π^+'s in the fragmentation of a neutron. Finally a backward Δ in the lab is rather unlikely to be produced in a high energy reaction so that such Δ's might have pre-existed in the deuteron.

Several such spectator experiments have recently been carried out to probe the $(\Delta\Delta)$ configuration. M. Goldhaber et al.[17] analyzed quasi-elastic scattering of $\pi^\pm$ and K^+ mesons off the virtual Δ^- constituent of the $(\Delta^{++}\Delta^-)$IC at 15 and 12 GeV/c beam momentum, respectively. Their report triggered similar studies for $\pi^+ + (d=\Delta\Delta) \to \pi^+\Delta\Delta$ at 4 GeV/c[18], $\bar{p} + (d=\Delta\Delta) \to \bar{p}\Delta\Delta$ at 5.5 GeV/c[19] and $p + (d=\Delta\Delta) \to \Delta NN$ at d_L = 3.33 GeV/c.[20] Each experiment recorded a significant number of (πN) events in the $\Delta(1236)$ mass region recoiling backward in the lab (see Fig.11). Fig.12a shows the coincident Δ^- events of

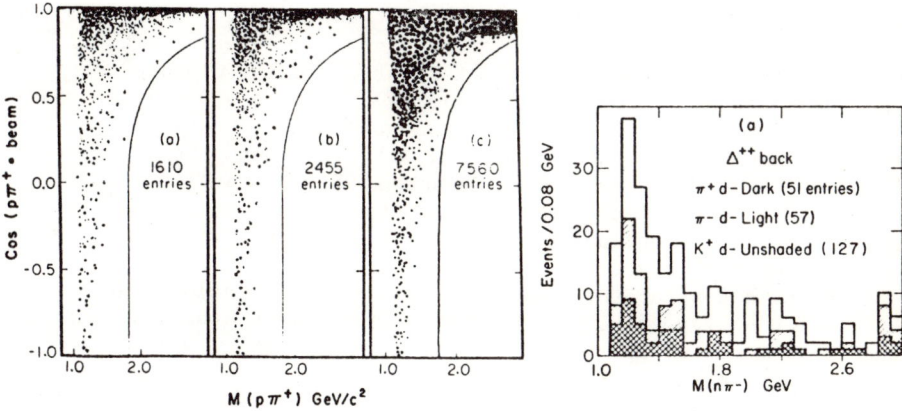

Fig.11. $p\pi^+$ mass distribution versus $\cos(p\pi^+$ lab angle) for $hd \to h(\pi^+p)(\pi^-h)$ from ref.17, where h is (a) π^+, (b) π^-, (c) K^+. Curves show kinematic boundaries.

Fig.12(a). $n\pi^-$ mass distribution for events with backward-lab $p\pi^+$ in $\Delta(1236)$ region of quasi-elastic spectator experiment of ref.17.

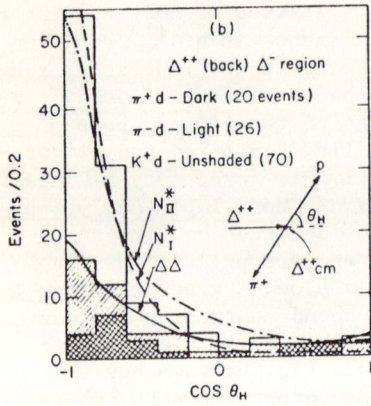

Fig.12(b). Decay angular distribution of the backward (pπ^+) sample in the pπ^+ rest frame. θ_H is the angle of p relative to the pπ^+ lab direction. The fits $N^*_{I,II}$ are Monte Carlo generated background using a pure (I) and modified (II)[21] Hulthén spectrum for protons.

ref. 17 in the forward hemisphere which also concentrate in the $\Delta(1236)$ mass region. Yet the angular distributions of the πN decay products of the Δ-spectator in its rest frame (Fig.12b) consistently reveal considerable forward-backward asymmetry for these experiments, which suggests strong background contamination to the genuine Δ-spectator signal. It appears that backward-lab production of a Δ by an incident high energy projectile does not suffice to separate off the non-resonant (πN) background. As a consequence, only an upper limit of 0.7 to 0.9% total ($\Delta\Delta$) probability could be extracted from the data.

Subsequently doubts have been raised as to the sensitivity of this group of spectator experiments to the ($\Delta\Delta$) configuration in the deuteron.[22,23] We refer to ref.22 for a more detailed discussion. On the basis of the peripheral isobar model in IA the measured cross section for deuteron breakup, with a Δ-spectator observed at momentum p in the lab, is $\sigma_{br} \propto 2|f_{h\Delta}|^2 P_{\Delta\Delta}(p)$ involving the $\Delta\Delta$ probability density and the elementary projectile-Δ amplitude. There are a number of different effects which tend to reduce the IA estimate of the quasi-elastic σ_{br} at small and medium spectator momenta. For example, for the (np) configuration coherent elastic scattering with $\sigma_{coh} = 4\, P_{np}|f_{hN}|^2 F_N^2(q/2)$ competes with N-breakup. Using closure one finds $\sigma_{br} = 2\, P_{np}|f_{hN}|^2 (1 + F_N(q) - 2F_N^2(q/2)) \to 0$ as $q \to 0$ and the elastic (np) form factor $F_N \to 1$. An analogous reduction factor may apply to Δ-breakup. Since the Δ's are not actually bound but are generated from nucleons, N-breakup with subsequent NN $\to \Delta\Delta$ final-state interaction (FSI) must also be included, however, and may substantially modify the reduction, particularly so if h-Δ scattering is large and coherent with hN scattering.

At high Δ-spectator momentum p there is kinematic suppression. For a high energy beam, the maximum possible laboratory backward Δ-spectator momentum $p_c \sim (m_d^2 - m_\Delta^2)/2m_d = 531$ MeV/c, which seems to play a role in the high momentum tail of Fig.15d.

Another reduction of σ_{br} by about 10-20% is supplied by shadowing which is $\propto 1/\langle r^2_{\Delta\Delta}\rangle$ and thus more effective for the short range ($\Delta\Delta$)IC than for the loosely bound (np) configuration. A kinematic reduction of $\sim 25\%$ is related to the flux factor $(m_d - E_s)/E_s$ in the breakup cross section, where E_s is the spectator energy. For a Δ-spectator, E_s is generally higher than for a nucleon spectator, where this correction is only $\sim 8\%$.

Probably the most important reduction factor is related to the fact that the struck Δ constituent is far off its mass shell. As a consequence, σ_{br} must be weighted by the probability for the virtual Δ to become real in the collision, which is a squared form factor f. Choosing f as the proton's elastic form factor $\sim \exp(5t)$, where $\sqrt{|t|}$ measures the difference between the real and virtual Δ masses, this easily leads to a reduction of σ_{br} by a factor $\sim$ 10.

There is another final-state scattering effect which could lead to a large asymmetry of the Δ-spectator decay in its rest frame which may be mistaken for background. If a nearly on-mass-shell pion between slow particles is exchanged, this could knock nucleons from Δ spectator decay forward in the Δ-rest frame. Recent experiments on deuteron targets struck by protons and pions lead to a depletion of backward protons of 10-15%. For the short range ($\Delta\Delta$) wave function this asymmetry induction is expected to be even larger, whereas knocking Δ's off a backward angle is less effective in view of the generally high Δ-spectator momentum. Finally suppose the Δ's were actually bound in the deuteron. Then the attractive $\Delta\Delta$ potential of 0.6 to 1 GeV is expected to be supported by intense meson exchange. As a result, knockout of the Δ's will in general be accompanied by ejection of several mesons. Several of these reduction factors, in particular the overwhelming virtual-Δ form factor, can be avoided, however, by summing over the breakup channels of the knocked-out Δ in an inclusive experiment, while quasi-elastic experiments are restricted to the less likely ($\Delta\Delta$) channel alone. In conclusion, then, the $\Delta\Delta$ probability extracted from an inclusive spectator experiment is expected to come closer to that calculated in the isobar model when corrected for the isobar decay width. This seems to be borne out by the data in Figs.14,15 with one caveat, though. An inclusive experiment (Fig.10a) also counts the decay Δ^{++}'s originating from (NN*)IC's, whose N* has a decay branch to the $\Delta(1236)$ like N*(1520), N*(1688). Therefore it measures not only the $\Delta\Delta$ probability density but a weighted sum of all IC's in the deuteron ground state. On the other hand, a quasi-elastic spectator experiment (Fig.10b) would reject most of these Δ^{++} events because the associated πN systems in the forward hemisphere are mostly not in the $\Delta(1236)$ mass region.

As to the correction for the short life time of isobars, an interesting suggestion has recently been reported[24] how to remedy this deficiency. Using the OPE and IA approximations, the ($\Delta\Delta$) wave function is directly related to the $\Gamma_{\pi N\Delta}$ vertex which depends both on the momentum p and the Δ energy E, $\tilde{\psi}_{\Delta\Delta}(p,E) \propto \Gamma_{\pi N\Delta}\Gamma_{\pi N\Delta}/(m_\pi^2 - q^2 - (E-m_N)^2 - i\epsilon)$. An energy value $E = m_\Delta - i\Gamma/2$ is proposed as appropriate for a ($\Delta\Delta$) wave function in non-spectator type processes for which the ($\Delta\Delta$) probability is defined as $P_{\Delta\Delta}^{LS} = \langle\tilde{\psi}_{\Delta\Delta}^{LS}(E)|\tilde{\psi}_{\Delta\Delta}^{LS}(E)\rangle$. For spectator experiments, however, the energy spectrum of the knocked-out Δ suggests replacing the $\delta(E-m_\Delta + i\Gamma/2)$ by a Breit-Wigner,

$$P_{\Delta\Delta}^{LS} = \int_{m+m_\pi}^{\infty} \frac{\Gamma/2\pi}{(E-m_N)^2 + \Gamma^2/4} \langle\tilde{\psi}_{\Delta\Delta}^{LS}(E)|\tilde{\psi}_{\Delta\Delta}^{LS}(E)\rangle \, dE \quad . \qquad (3)$$

The results in Fig.13 show that the Δ probabilities corresponding to the knock-out situation could be much larger than for the bound case, $E = m_\Delta - i\Gamma/2$, should retardation be important. The actual energy

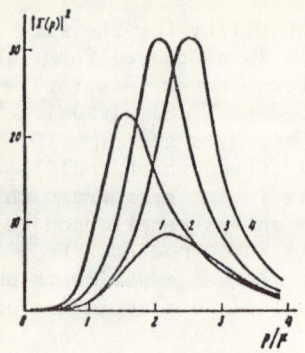

Fig.13. $(\Delta\Delta d)$ vertex in IA for $^7D_1$ state. Curves 1,2, 3,4 correspond to energies $E = m_\Delta - i\Gamma/2$, $m_\Delta - \Gamma/2$, m_Δ, $m_\Delta + \Gamma/2$ in OPE with a short range cut-off 2.1 fm^{-1}.

spectra of Δ spectators from the deuteron differ somewhat from a Breit-Wigner shape and are shifted relative to m_Δ.

Finally, it has been suggested[25] to suppress the background of nonresonant (πN) systems by capitalizing on the narrow momentum distribution of spectator protons (in Fig.8b) at 0 of the (np) wave function. The decay protons from the Δ^{++} spectator have the broader momentum distribution of the $(\Delta\Delta)$IC which is centered at ~0.35 GeV/c. Hence eliminating all protons with momentum $\leq$ 0.2 GeV/c will not cause a significant loss of Δ-spectators but reduces the (πN) noise drastically. In contrast, some of the quasi-elastic spectator experiments actually have a bias against fast protons. This technique was first successfully used by Benz and Söding[26] who studied the only inclusive spectator reaction so far, $\gamma d \to \Delta^{++} +$ anything, with the DESY bremsstrahlung beam at 5.5 GeV maximum energy. Again, the signature is a Δ^{++} recoiling backward in the laboratory. In fact, the whole ridge of concentrated background events in their mass-versus-momentum plot of $(p\pi^+)$ spectators in Fig.14a disappears when all slow spectator protons with momentum $\leq$ 0.2 GeV/c are eliminated (Fig. 14b). The remaining events concentrate in the $\Delta(1236)$ mass region (Fig.15a). The integrated background (dashed line in Fig.15a with measured neutron data) is smaller than the Δ^{++} signal by a factor of 3. The azimuthal and polar decay angular distributions of the Δ^{++}

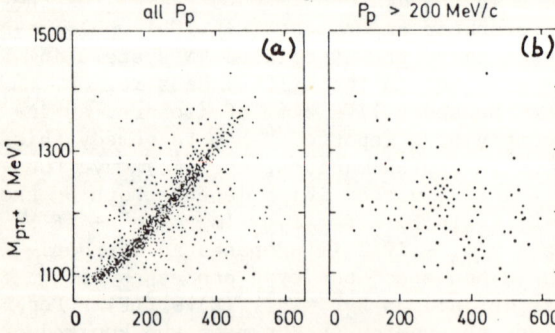

Fig.14. Momentum/mass correlation plot of backward-lab $(p\pi^+)$ systems in $\gamma d \to p\pi^+ +$ anything (from ref.26).

spectator in its rest frame now are nearly isotropic rather than asymmetric (Fig. 15b,c). The shape of the measured momentum distribution in Fig.15d is in reasonable agreement with the $(\Delta\Delta)$ model.[27] The high momentum tail is probably kinematically suppressed, as we mentioned earlier. An isobar probability of 3.1% is extracted from the data which is somewhat higher than theoretical estimates for the $(\Delta\Delta)$IC[27] because the inclusive reaction measures a weighted sum of all IC's (see Fig.10) with qualitatively similar probability densities.

An alternative method to remove the background was used for forward proton production in π^- deuteron collisions, $\pi^- + (d = \Delta^{++}\Delta^-)$

→ p∆⁻. Its ∆⁺⁺ exchange mechanism[28] is shown in Fig.9a. The background is mostly caused by quasi elastic backscattering of the incident π⁻ from the bound proton in the deuteron (Fig.9b). Therefore

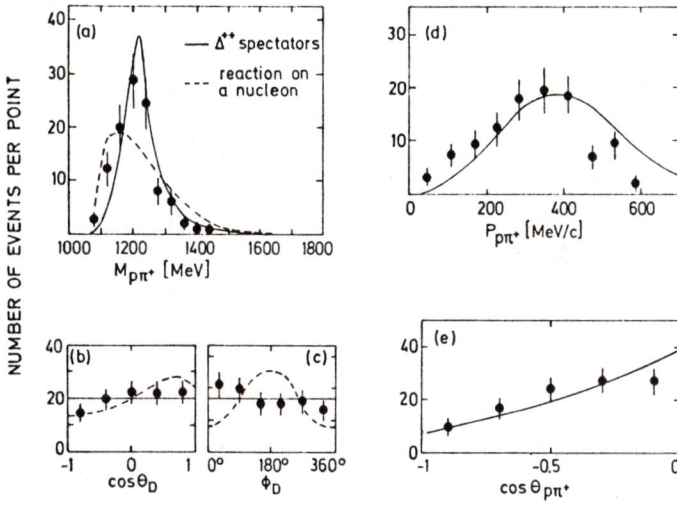

Fig.15a. ∆⁺⁺ spectator mass distribution; (b,c) polar, azimuthal π⁺ distribution in the pπ⁺ rest frame relative to beam direction; (d) ∆⁺⁺ lab momentum plot; (e) Angular distribution in the lab relative to the beam direction, after removal of slow spectator protons. Solid curves are (∆∆) isobar model predictions,[27] dashed curves correspond to the background process of Fig.8b.

a π⁻ scattered at large angle in coincidence with an incident π⁻ and forward proton event is counted as background, whereas the absence of such a backscattered π⁻ is taken as the signature of a ∆⁻ spectator event. To minimize the background, the experiment was done[29] with the Saclay π⁻ beam at ~1.1GeV/c, where π⁻p backward elastic scattering has a minimum (dashed arrow in Fig.16), which remains when the

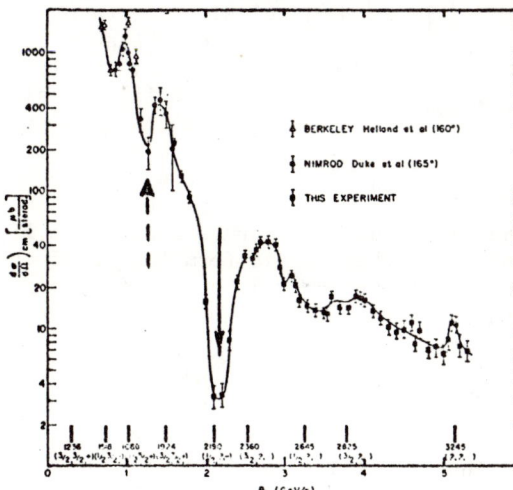

Fig.16. π⁻p elastic scattering at 180° versus π lab momentum (ref.30).

Fermi motion of the bound neutron is included according to Fig.9b. 1% (∆∆) state would fill the dip. Predictions for the angular distributions are also qualitatively different. The free π⁻p distribution, whose shape closely resembles the quasi-elastic cross section from the deuteron has a minimum for 0° protons. The (∆∆)IC generates a peak at $\theta_p = 0°$ in the lab instead (see Fig.17). The observed differential cross section for the ∆⁻ signal in Fig.18 has the negative slope predicted by the isobar model. The measured background is either flat or has positive slope. So, to some

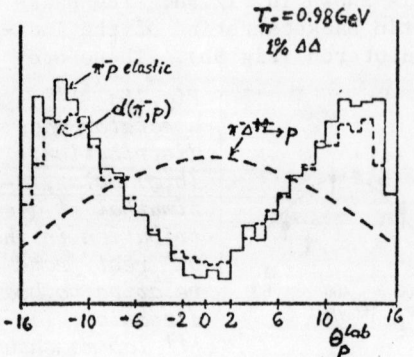

Fig.17. Schematic angular distributions for $\pi^-p \to \pi^-p$ (solid), $\pi^-d \to p\Delta^-$ via Δ exchange (dashed) and quasi-elastic (dotted).

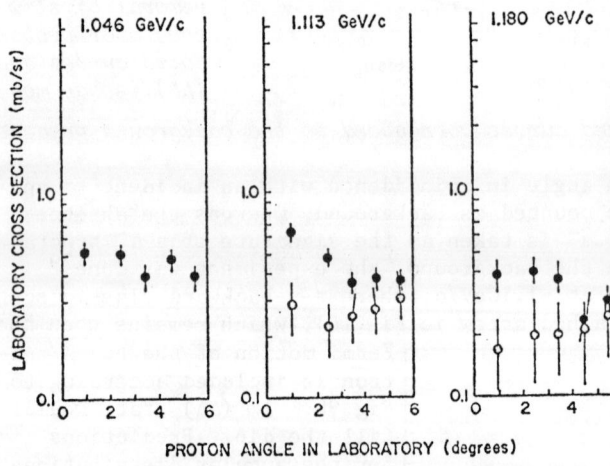

Fig.18. Lab angular distributions for deuterium (full circles) and hydrogen (open circles).

extent at least, background and Δ^- spectator signal seem to have been separated. An upper limit of 0.4% ($\Delta\Delta$) probability has been extracted from these data. This estimate is subject to some of the foregoing reservations pertinent to the quasi-elastic spectator experiments. The background is expected to be minimal at ~2.1 GeV/c, where the π^-p backward elastic cross section (solid arrow in Fig.16) has its deepest minimum. It could be further suppressed by eliminating all events with a slow spectator neutron. The recent work[31] on $\pi^-d \to p\Delta^-$ at 1.68 GeV/c is inconclusive because neither the slope of the proton angular distribution at 0° nor the energy dependence were measured.

In conclusion, it appears that a number of different experiments qualitatively support the basic concept of the $\Delta(1236)$ in the deuteron. There does not seem to be disagreement between several quasi-elastic and one inclusive spectator experiment because the former very likely are much less sensitive to the ($\Delta\Delta$) state than was anticipated. Finally, we emphasize that, contrary to an erroneus implication in the literature,[32] isobars contribute little, even at high momentum transfer, to the elastic form factor of the deuteron (see Fig.24 of the N^* review in Springer Tracts, ref.2); certainly they do not flatten out the form factor at high q as meson exchange currents have the potentiality of doing.

REFERENCES

1. Particle Data Group, Rev.Mod.Phys. 48, (1976) S1.
2. For reviews, see H. Arenhövel and H.J. Weber, Springer Tracts 65 (1972) 58; H. Arenhövel, in Proc.Symp.Interaction Studies in Nuclei, Mainz (1975) 727, H. Jochim and B. Ziegler, eds. (North Hol-

land); H.J. Weber, "Spectator Experiments as Probes of N^* Admixtures," ibid.p.749, and in Proc.VII Internat.Conf. on Few Body Problems in Nuclear and Particle Physics, ed. A.N. Mitra, Delhi North-Holland, 1976); R.Beurtey, in Proc.VI'th Int.Conf. on High Energy Physics and Nuclear Structure, Santa Fe (1975) p.653, D. Nagle et al. eds. (AIP, NY); L.S. Kisslinger, in Proc.Topical Meeting on High Energy Collisions, Trieste (1974).
3. A. Kerman and L.S. Kisslinger, Phys.Rev. 180, (1969) 1483.
4. J.S. Sharma, V.S. Bhasin and A.N. Mitra, Nucl.Phys. B35 (1971)466.
5. P. Haapakoski and M. Saarela, Phys.Lett. 53B (1974) 333; H. Arenhövel, ibid.p.224.
6. S. Jena and L.S. Kisslinger, Ann.Phys.(NY) 85 (1974) 251; E. Rost, Nucl.Phys. A249 (1975) 510; H. Arenhövel, Z.Phys.(1976)[Mainz preprint KPH 14/75].
7. A.M. Green and P.Haapakoski, Nucl.Phys. A221 (1974)429; R. Smith and V.Pandharipande, Nucl.Phys. A256(1976)327; H. Sugawara and F. von Hippel, Phys.Rev. 172(1968) 1764; D. Riska and G.Brown, Nucl. Phys. A153 (1970) 8.
8. M. Konuma, A. Miyazawa and S. Otsuki, Progr.Theor.Phys. 19 (1958) 17; A. Klein and B. McCormick, ibid. 20, 870; W. Cottingham and R. Vinh Mau, Phys.Rev. 130(1963)735; M.Partovi and E.Lomon, Phys. Rev. D2 (1970) 1999; Y. Nogami and E. Satoh, Nucl.Phys.B19(1970)91.
9. D.O. Riska and G.E. Brown, Phys.Lett.38B(1972)193; see also ref.10.
10. H. Arenhövel, W. Fabian and H.G. Miller, Phys.Lett.52B(1974)303, Z.Phys. 271 (1974)91.
11. H.J. Weber, U.Virginia preprint, Nucl.Phys.A(1976), and unpublished work.
12. J. Namyslowski, priv.comm. and Warsaw preprint IFT/9/75.
13. I. Todorov, Phys.Rev.D3 (1971) 2351.
14. S.Weinberg, Phys.Rev. 150 (1966) 1313.
15. F. Coester, S. Pieper and F. Serduke, Phys.Rev. C11 (1975)1.
16. A.M. Green and J.A. Niskanen, Nucl.Phys.A249(1975)493; B.D. Day and F.Coester, Phys.Rev. C13 (1976) 1720.
17. C.P. Horne et al., Phys.Rev.Lett. 33 (1974) 380.
18. M.J. Emms et al., Phys.Lett. 52B (1974) 372.
19. H. Braun et al., Phys.Rev.Lett. 33 (1974) 312.
20. B.S. Aladashvili et al., Nucl.Phys. B89 (1975) 405.
21. For nucleon spectator events a recent analysis by G. Alberi et al, CERN preprint TH.2113 of binding, multiple scattering and FSI effects concludes that the strong channel dependence can best be explained by inelastic FSI via intermediate isobars. Consequently any ad hoc modification of high momentum components of the (np) deuteron wave function, as in the N_{II}^* fit, is not legitimate and does not explain the asymmetry in terms of background.
22. A.S. Goldhaber and L.S. Kisslinger, Los Alamos preprint, May(1976).
23. P. Söding, priv. comm.
24. V.E. Markushin, Zh ETF 21(1975)463(JETP Lett. 21(1975)211).
25. N.R. Nath, P.K. Kabir and H.J. Weber, Phys.Rev.D10(1974) 811.
26. P. Benz and P. Söding, Phys.Lett. 52B(1974) 367.
27. N.R. Nath and H.J. Weber, Phys.Rev. D6 (1972) 1975.
28. N.R. Nath, H.J. Weber and P.K. Kabir, Phys.Rev.Lett.26(1971)1404.
29. C.F. Perdrisat et al., Phys.Lett. B(1976), in press.

30. S.W. Kormanyos et al., Phys.Rev.Lett. 16 (1966) 709.
31. B. Abramov et al., Moscow preprint ITEP-38 (1974).
32. R.G. Arnold et al., Phys.Rev.Lett. 35 (1976) 776, and in Proc. VI'th Internat.Conf. on High Energy Physics and Nucl. Structure, Santa Fe (1975), D. Nagle et al.eds. (AIP, NY) p.385.

D. Koltun (U. of Rochester): Would you clarify your remark on the unimportance of Δ in the elastic electric form-factor of the deuteron, relative say, to the importance (3%) in $n+p \to d+\gamma$?

H. Weber: The electric form factor of the deuteron involves only the Δ current $<\Delta|J_\mu|\Delta>$ and not the transition current $<\Delta|J_\mu|N>$ because the deuteron ground state is isoscalar. That is to say, the (γdd) vertex contains the ($\Delta\Delta$d) vertex in 2nd order. So, the triangle in the (γdd) vertex consists of three Δ's and is correspondingly small.

Pierre Radvanyi (Institut de Physique Nucleaire, Orsay): When you take any specific reaction where a Δ should appear, you find in the detailed analysis of each case that there are many difficulties and ambiguities which lead to very uncertain conclusions. You presented in your talk calculations of Δ percentages in the ground states of nuclei: what are the uncertainties on such calculations? I would also be very happy if L. Kisslinger could tell us his comments on this problem.

H. Weber: For Δ-knockout experiments, detailed and reliable estimates for pion (and ρ-meson) rescattering and other final state interaction effects have not yet been systematically calculated. For quasi-elastic Δ-breakup, some numerical estimates have been given by A. S. Goldhaber and L. S. Kisslinger, mostly in the range of 10-25%. The virtual $\Delta \to$ real Δ form factor is the worst uncertainty which, however, may be avoided in inclusive spectator experiments. Estimates for π^- rescattering for the Virginia Δ^- spectator reaction $\pi^- d \to p \Delta^-$ in the minima of the $p\pi^-$ backward elastic cross section indicate that the effect is not large at these particular energies. Certainly, the opposite predictions of the ($\Delta\Delta$) model and the background for the angular distributions remain qualitatively unchanged.

For processes where the Δ is not knocked out, such as NN scattering and $\gamma d \to np$, the short life time of isobar configations and the proper interpretation of calculated isobar probabilities are among the dynamical uncertainties. One such estimate I have discussed in Fig.13 of my talk. Another relating to the uncertainty of high momentum components of the ($\Delta\Delta$) state induced by that of the (np) state is shown in Fig. 7.

Thus, your global condemnation of the isobar model on the basis of uncertainties and ambiguities appears to me to be unjustified in this generality. I agree that much hard theoretical work remains to be done.

L. Kisslinger (CMU): In reply to Dr. Radvanyi's question, I believe that the study of virtual N^*'s in nuclei is in about as good shape as the study of the two pion exchange contribution to the N-N interaction. The question of experimental detection, especially in going from virtual to real N^*'s is complicated and deserves a long discussion.

A. W. Thomas (CERN/TRIUMF): A follow up from your explanation of the small effect of isobars in the deuteron form-factor: Although the Δ-Δ must come in second order, the N-N component can be in first order. Since you've suggested that $P_{\Delta\Delta}$ is 1%, and the total isobar probability is ~3%, the contribution from $I=1/2$ resonance must be significant.

H. Weber: There are two reasons. (i) Isobar configurations generally contribute little at high momentum transfer to the elastic form factor of the deuteron because, in the (γdd) vertex, the bound (NN^*) wave function is folded with the (np) bound wave function. The (NN^*) wave functions typically peak at 300 to 400 MeV/c. Thus at high q(≥ 2.5 fm^{-1}), the increasing momentum mismatch between the two wave functions prevents a sizeable contribution to the elastic form factor. Such a constraint does not apply to breakup reactions like γd$\rightarrow$np, where the Δ is not knocked out. (ii) (NN^*) probabilities with isospin $T=1/2$ are estimated to be less important than the ($\Delta\Delta$) state in the deuteron ground state.

Your conclusion from the DESY data that the combined (NN^*) probability is significant relative to the ($\Delta\Delta$) appears to me to be premature at this stage.

A. N. Kamal (Alberta): In n+p$\rightarrow$d+γ does ω exchange where ω decays radiatively contribute?

H. Weber: In principle, the π-ω exchange current involving an isovector photon contributes to the photodiintegration of the deuteron. If I remember correctly, R. Adler has estimated this effect to be small near threshold, though.

Dan O. Riska (Michigan State University): One $\pi\omega$ exchange current does not contribute to the electromagnetic form factor of the deuteron because it is an isovector operator.

H. Weber: I agree.

Editors Comment: In answer to A. W. Thomas's question, it should be noted that the most important N-N components at high momentum transfer involve N*'s of high spin. This reduces the contribution to the electric transition (first order) moment. For the three-body form factor the N*(1688) contribution is larger than the D contribution (unpublished result of Ho-Kim and Kisslinger). With regard to D. Koltun's question, the effects on elastic form factors is far greater than 3%, but one must learn how to make more accurate predictions at large momentum transfer.

PIONIC DEGREES OF FREEDOM IN NUCLEI II

Mannque Rho
Service de Physique Théorique, CEN.Saclay,
B.P.n°2, 9190 Gif-sur-Yvette, France

I

This is a continuation of my talk given a year ago at Santa Fe, which explains the title. My talk at Santa Fe [1] was incomplete for two reasons. One, the SLAC experiment on the deuteron electromagnetic form factor at large momentum transfers [2] raised a great deal of questions on the structure of the two-nucleon system, in particular the role of mesonic cloud in responses to electromagnetic probes. It seemed at that time that the whole foundation of meson-exchange currents which otherwise seemed to be fairly sound was crumbling under the beautiful data. I could not explain the embarrassingly big disagreement between the data and the meson-exchange currents at the time. We now know how to go about reconciling those conflicting observations, the meson-current description on one hand and the quark scaling description on the other. The other subject which I could not discuss at the time was an intimate connection between weak axial currents in nuclear matter and pion-nucleus interaction [3]. I shall briefly describe the recent development on the matter, although the qualitative feature has not changed since a year ago. Although there is an extremely interesting novel development on the second-class axial current in connection with mesonic currents, I shall not talk about this since here one is more interested in the fundamental structure of weak currents rather than in nuclear structure and furthermore a detailed discussion will be given at the forthcoming Sussex meeting on weak interaction [4]. Due to lack of time, I shall not discuss mesonic effects in complex nuclei [5].

II. THE DEUTERON

There are roughly three ways of describing the deuteron electromagnetic structure : a) Heitler-London-Cutkosky picture ; b) meson-exchange currents ; c) quarks.

a) Heitler-London-Cutkosky picture [6]

When the proton and the neutron are far apart, say, beyond the distance of m_π^{-1}, each of which carrying its meson cloud, fluctuation in meson density is negligible, as probably is the case with the deuteron most of the time. However as they come together at a distance (r) smaller than m_π^{-1}, the meson clouds start overlapping while exchanging mesons between the two, as a consequence of which fluctuation in meson density will increase as r decreases. One would then expect that at large momentum transfers (or equivalently at short distances), a virtual photon will see increasing pion density fluctuation, so that the electromagnetic form factors would look more different from the impulse approximation at high momentum transfers than at small momentum transfers. There is no quantitative calculation corraborating or refuting this expectation, so it is not possible to say whether or not the observed form factor is in accord with this description. Qualitatively, though, Cutkosky observed many years ago [6] that the overlapping meson clouds (and the meson density fluctuation) are expected to be less effective in the T=0 state than in the T=1 state of the two-nucleon systems. Thus while the deuteron form factor may be uncomplicated as the experimental data indicate, the electron scattering on ^{3}H and ^{3}He which involve both T=0 and T=1 channels might show a deviation. It would be interesting to see whether or not this qualitative prediction is borne out in the forthcoming experiments with three-body nuclei. Even if it is, it would be extremely important to understand how the increasing pion density conspires to give a negligible effect in the deuteron.

b) Meson currents

Blankenbecler and Gunion (BG) [7] argued that in analogy with pion-deuteron scattering at high energy (at GeV energies) where double scattering makes an important contribution, the electromagnetic form factor of the deuteron should also show a flattening-out at about $q^2 \sim (10-12) fm^{-2}$. This prediction based on the vector meson dominance (VMD) and multiple scattering and which is closely related to the shadowing effect expected from the hadronic component of the photon would imply that at large momentum transfers, the electromagnetic form factors (both electric and magnetic) fall less rapidly than the impulse approximation results. The Blankenbecler-Gunion prediction is given, and compared with the Stanford data, in fig. 1. The experiment clearly shows that some or all of the Blankenbecler-Gunion assumptions are wrong : there is no flattening-out. Note in this connection that no nuclear shadowing is observed in electroproduction of hadrons [8].

The meson-current description of Chemtob, Moniz and Rho (CMR) [9] was motivated by the BG idea of momentum-sharing (resulting in a flattening-out) and makes use of the VMD hypothesis. In terms of the Feynman graphs representing the exchange currents, fig. 2, this amounts to ignoring possible form factors at each meson-nucleon vertices. Suitably rescaling the original CMR result using the new experimental value for $G_{\rho\pi\gamma}^2$, one finds that the CMR and BG results are roughly equivalent. Again we now know that the CMR description is incomplete !

It was pointed out by Woloshyn and others [10] that ignoring the meson-nucleon form factors gives an incorrect fall-off behavior. Woloshyn obtains a correct asymptotic behavior by counting the number of monopole form factors at meson-nucleon (MNN) vertices and γ_{MM} vertex, and the number of meson propagators (these are indicated by crosses in fig. 3) ; thus fig. 3 tells us that for $q^2 \to \infty$, the

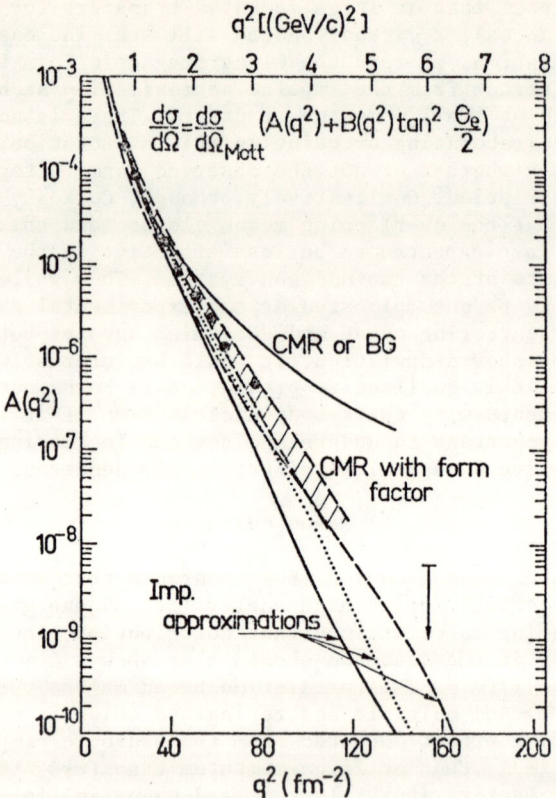

Figure 1 : Mesonic current calculations compared with experiment.

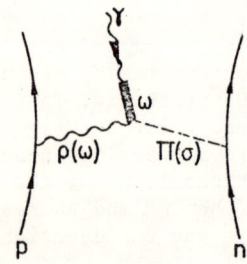

Figure 2
Mesonic current calculated by CMR

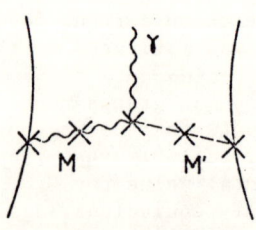

Figure 2
Scaling laws from mesonic currents. Count monopoles for the crosses

form factor should scale $\sim (q^2)^{-5}$ in accord with the data and with the quark counting rule (described below). Putting in suitable MNN vertices, one finds that the meson-current description is in quantitative agreement with the experimental data : fig. 2 shows the results of Desplanques and Kisslinger [10] in which the theoretical values are bracketed by different values of M_A in the πNN form factor $(1+q^2/M_A^2)^{-1}$.

c) Quark picture [11]

In renormalizable field theories with quarks and gluons and in exact "color" symmetry, the quark (constituent) exchange process (fig. 4) dictates the scaling behavior of the deuteron electromagnetic form factor [11]. The quark counting rule tells us that the elastic form factors of hadrons should scale as $(q^2)^{-p}$ where p is the minimum number of off-shell quark propagators. In the case of the deuteron p=5. Equivalently one can express p in terms of the number of constituents n(=number of valence quarks and antiquarks) as

$$p = n - 1$$

Thus p is 1 for pion, 2 for proton, and 3A-1 for nucleus of A nucleons. That this scaling rule works well for mesons as well as baryons can be seen in fig. 5. The best fit [12] of the deuteron data leads to p = 5.0±0.6.

It is both surprising and intriguing that the meson current and the quark description give essentially equivalent results and that the quark scaling matches smoothly into the low momentum transfer region. How good is it in general to consider the deuteron as a state of six quarks in a bag [13], namely the M.I.T. bag [14] ? We do not know the answer yet. Considered as a classical 6-quark bag, the deuteron would have a binding energy of $(2-\sqrt{2})m_N \approx 500$ MeV, much too large to be acceptable ! Quantum effects may be responsible for the discrepancy [13], indeed a rough inclusion of quantum effects lowers this to ~ 130 MeV, not too bad on hadronic mass scale. One is then led to a model for the deuteron in which the nucleus spends most of its time as a proton and a neutron, and a fraction of time as a classical 6-quark bag. The fact that the deuteron deviates from a classical 6-quark bag is interpreted as a source of nuclear force [13,14].

III. NEUTRON STAR

At very short distances, the deuteron would be more like six free quarks in asymptotically free theories. The same effect is obtained by squeezing nucleons into dense matter [15]. In the current gauge theories, asymptotic freedom is also obtained as the nucleon density ρ becomes large. This has a consequence on neutron stars where the core density can become very large ; viz, it raises the possibility that for asymptotically large densities, there might

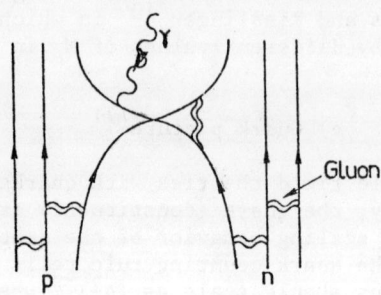

Figure 4 : Constituent (quark) - exchange graph contributing to the deuteron form factor. Solid lines are quarks.

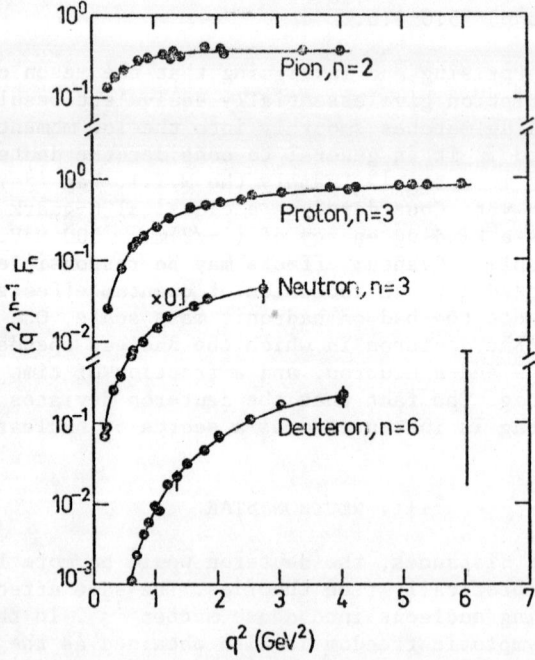

Figure 5 : Application of quark counting rule to other hadrons (from ref. 11)

be formed a star made of quarks, rather than of neutrons (perhaps a giant M.I.T. bag ? [16]). Whether or not such a quark star is stable against gravitational collapse is a controversial issue[17]. And so is whether the phase transition from neutron to quark phase occurs at densities commensurate with stable stars [16]. In any event, it is a fascinating issue which will surely find its place in the future description of nuclear structure.

IV. AXIAL CURRENT IN NUCLEAR MEDIUM

The PCAC relates the divergence of the axial current ($\partial_\mu A_\mu$) to pion field (ϕ_π). Thus one can study the off-shell pion-nuclear interaction through weak processes in nuclei. There is an interesting development in associating the renormalization of weak currents in nuclear matter with the manifestation of certain aspects of nuclear force [3]. I shall describe only briefly the present status of our knowledge on this matter.

The problem may be defined in the following way. Suppose that one wants to describe the weak interaction in nuclei solely in terms of nucleon degrees of freedom, others such as mesons, N^*'s etc being incorporated into "effective" form factors. More precisely, if one writes the space component of the axial current (the time component is small) for the processes

$$n \rightarrow p + e^- + \bar{\nu}_e$$
$$\mu^- + p \rightarrow n + \nu_\mu$$

as

$$\vec{A}_\pm = - g_A(q^2) \tau_\pm \vec{\sigma} + (g_p(q^2)/2m_N) \tau_\pm \vec{q} \sigma \cdot \vec{q}$$

where $\vec{q}$ is the momentum transfer, and $g_A(q^2)$ and $g_p(q^2)$ the axial and pseudoscalar form factors respectively, then one would like to calculate $g_A^*(q^2)$ and $g_p^*(q^2)$ which are defined by

$$\vec{A}(\vec{r}) = - g_A^*(q^2) \sum_{i=1}^{A} \tau_\pm(i)\, \vec{\sigma}(i)\, \delta(\vec{r}-\vec{r}_i)$$
$$+ \frac{g_p^*(q^2)}{2m_N} \sum_{i=1}^{A} \tau_\pm(i)\, \vec{q}\, \vec{\sigma}_i \cdot \vec{q}\, \delta(\vec{r}-\vec{r}_i)$$

for nuclear processes, say,

$$^{16}N \rightarrow {}^{16}O + e^+ + \nu_e$$
$$\bar{\mu}^- + {}^{16}O \rightarrow {}^{16}N + \nu_\mu.$$

This is a well-defined problem only for infinite nuclear matter. For finite system, one probably has to resort to a systematic calculation of the meson-exchange currents [18].

An intriguing – and simple – result is obtained for nuclear matter. $g_A^*(q^2)$ is short-ranged as is $g_A(q^2)$. Therefore what modifies g_A to g_A^* is the short-range part of nuclear force. This is equivalent to the Lorenz-Lorenz correction[19] in pion-nuclear optical potential and is related to the Landau-Migdal parameter g' (occuring in Γ^ω as $g'\tau_1\cdot\tau_2\ \sigma_1\cdot\sigma_2$)[20]. The net effect is to modify $g_A = 1.25$ to $g_A^* \sim 1$, the precise value depending on the degree of sophistication in calculation.

A lot more drastic effect is expected in $g_p^*(q^2)$. The standard lore tells us that the pseudoscalar form factor $g_p(q^2)$ for free nucleons can be described by the graph

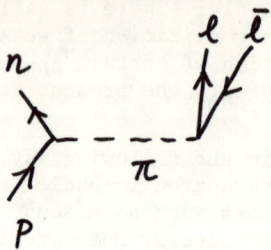

where $\ell\bar{\ell}$ are lepton pairs. This leads to the Goldberger-Treiman relation

$$m_\mu g_p(q^2) = \frac{2m_\mu m_N g_A(q^2)}{q^2+m_\pi^2}$$

which is about 7 $g_A(q^2)$ for μ-capture ($q^2 \approx \frac{1}{2} m_\mu^2$). Now consider what can happen to this mechanism when weak interaction occurs in nuclear medium. It is now understood that the main thing that happens is to change the pion-decay vertex by Δ-hole excitations,

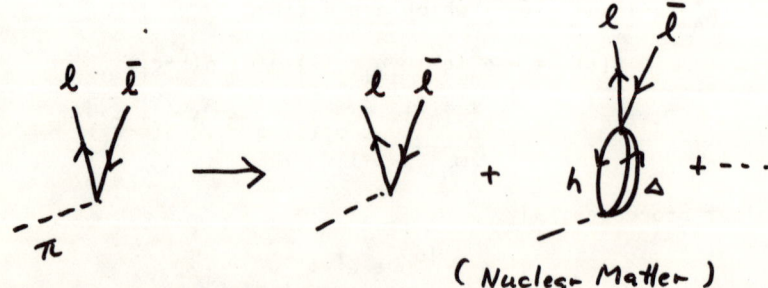

(Nuclear Matter)

By comparing with the pion-nuclear optical potential
$\pi(q,\omega) = 2\omega V_{opt}(q,\omega)$, one can easily see that the net effect is to multiply $g_p(q^2)$ by

$$\left[1+ q^{-2}\pi(q,\omega)\right]$$

Because of the p-wave attraction, this is in general less than 1. Furthermore for the nuclear matter density $\rho \approx 0.17$ f_m^{-3}, there is almost an exact cancellation, again the degree to which this occurs depending upon how one goes about calculating $\pi(q,\omega)$.

The present situation vis-à-vis experiments is not quite clear. The Lorentz-Lorenz effect, properly scaled in density, appears to be consistent with the observed queuching of g_A for light nuclei, which is about 10 % [21]. Whether or not all the other effects not included in the Lorentz-Lorenz (such as tensor force configuration mixing etc [22]) conspire to cancel as conjectured previously is not yet clear, though quite plausible. On the other hand, there is no clear signal that g_p^* is as dratically quenched as predicted. In view of the main mechanism for the suppression being the pion-decay vertex renormalization, the experiment of the type $\pi^- + \pi^+ \to 2\gamma$ in nuclei [23] can tell us whether or not our idea of the quenching of the pseudoscalar form factor is a viable one. (Note that the $\pi^+ + \pi^- \to 2\gamma$ vertex is <u>a priori</u> renormalized quite differently from the $\pi \to \ell\bar{\ell}$ vertex).

The fact that in nuclear matter $g_A^* \sim 1$, $g_p^* \sim 0$ suggests an interesting possibility that the nuclear matter somehow screens strong interaction. We do not yet know whether this is profound or just an accident. One can also conjecture that such a renormalization phenomenon would be crucial in the formation of pion condensation in nuclear and neutron matters. In the condensed phase, pion field appears explicitly ; therefore the axial current which finds its place via the PCAC will be subject to the kind of renormalization we have been talking about. In as much as the critical density ρ_c at which condensation first sets in is extremely sensitive to an effective $g_A(q^2)$ [24] the renormalization must be examined with a great care.

V. CONCLUDING REMARKS

To understand the mesonic degrees of freedom in nuclei is in my opinion the ultimate goal of meson-nuclear physics. Thus it is perhaps not premature to think of nuclei in terms of quarks and other constituents (if any). There is a consistency between the meson currents and the quark descriptions, but at the moment we have very poor understanding on how it comes about. We clearly need more data on the electromagnetic structure of nuclei, in particular on that of the tri-nucleon systems. It should surprise no one that in order to understand weak processes in nuclei, we need to understand better pion-nuclear interactions and vice versa. Pion is, after all, an essential degree of freedom for nuclear force. The crucial question that theorists must answer is : which experiments can exhibit those predicted effects in a clear and convincing manner ?

REFERENCES

1. M. Rho, in proceedings of the VI International Conference on High Energy Physics and Nuclear Structure (American Institute of Physics, New York, 1975).
2. R. G. Arnold et al, ibid.
3. M. Ericson, A. Figureau and C. Thévenet, Phys. Letters $\underline{45}$ B, 19 (1973) ;
K. Ohta and M. Wakamatsu, Phys. Letters $\underline{51}$ B, 325 (1974) ;
Nucl. Phys. A $\underline{234}$, 445 (1974) ;
M. Rho, Nucl. Phys. A $\underline{231}$, 493 (1974) and unpublished ;
J. Delorme, M. Ericson, A. Figureau and C. Thévenet, April 1976, to be published.
4. For a complete discussion of the second-class current problem treated in terms of meson-exchange currents, see J. Delorme, K. Kubodera and M. Rho, Physics Reports (in preparation).
5. M. Radomski and D.O. Riska, Contribution to this conference.
6. R.E. Cutkosky, Phys. Rev. $\underline{112}$, 1027 (1958) ; $\underline{116}$, 1272 (1959).
7. R. Blankenbecler and J. Gunion, Phys. Rev. D $\underline{4}$, 718 (1971).
8. J. Eickmeyer, S. Michalowski, N. Mistry, R. Talman and K. Ueno, Phys. Rev. Letters $\underline{36}$, 289 (1976).
9. M. Chemtob, E. Moniz and M. Rho, Phys. Rev. C $\underline{10}$, 344 (1974) ; also R. Adler and S. Drell, Phys. Rev. Letters $\underline{13}$, 349 (1964).
10. R.M. Woloshyn, Phys. Rev. Letters $\underline{36}$, 220 (1976) ;
M. Gari and H. Hyuga, Phys. Rev. Letters $\underline{36}$, 345 (1976) ;
B. Desplanques and L.S. Kisslinger, Private Communication.
11. S. Brodsky and G. Farrar, Phys. Rev. D $\underline{11}$, 1309 (1975) ;
S. Brodsky, SLAC-PUB-1699, December 1975.
12. Quoted by S. Brodsky, Ref. 11.
13. G.T. Fairley and E.J. Squires, Nucl. Phys. B $\underline{93}$, 56 (1975).
14. K. Johnson, Talk given at this Conference.
15. J.C. Collins and M.J. Perry, Phys. Rev. Lett. $\underline{34}$, 1353 (1975) ;
J.R. Ipser, M.B. Kislinger and P.D. Morley, EFI preprint 75-38 (University of Chicago) ;
B.D. Keister and L.S. Kisslinger, Carnegie-Mellon Preprint, 1976.
G. Baym and S.A. Chin, University of Illinois preprint 1976.
16. G. Baym and S.A. Chin, Ref. 15.
17. See Ipser et al, Keister and Kisslinger, Ref. 15.
18. M. Chemtob and M. Rho, Nucl. Phys. A $\underline{163}$, 1 (1971) ;
A. Barroso and R.J. Blin-Stoyle, Nucl. Phys. A $\underline{251}$, 446 (1975).
19. M. Ericson and T.E.O. Ericson, Ann. of Phys. (N.Y.) $\underline{36}$, 323 (1966) ;
G. Baym and G.E. Brown, Nucl. Phys. A $\underline{247}$, 395 (1975).
20. See Baym and Brown, Ref. 19.
21. D.H. Wilkinson, Phys. Rev. C $\underline{7}$, 930 (1973) ; Nucl. Phys. A $\underline{209}$, 470 (1973) ; Nucl. Phys. A $\underline{225}$, 365 (1974).
22. See M. Rho, Ref. 3 for details.
23. T.E.O. Ericson and C. Wilkin, Phys. Lett. $\underline{57}$ B, 345 (1975).
24. D.K. Campbell, R.F. Dashen and J.T. Manassah, Phys. Rev. D $\underline{12}$, 979, 1010 (1975).

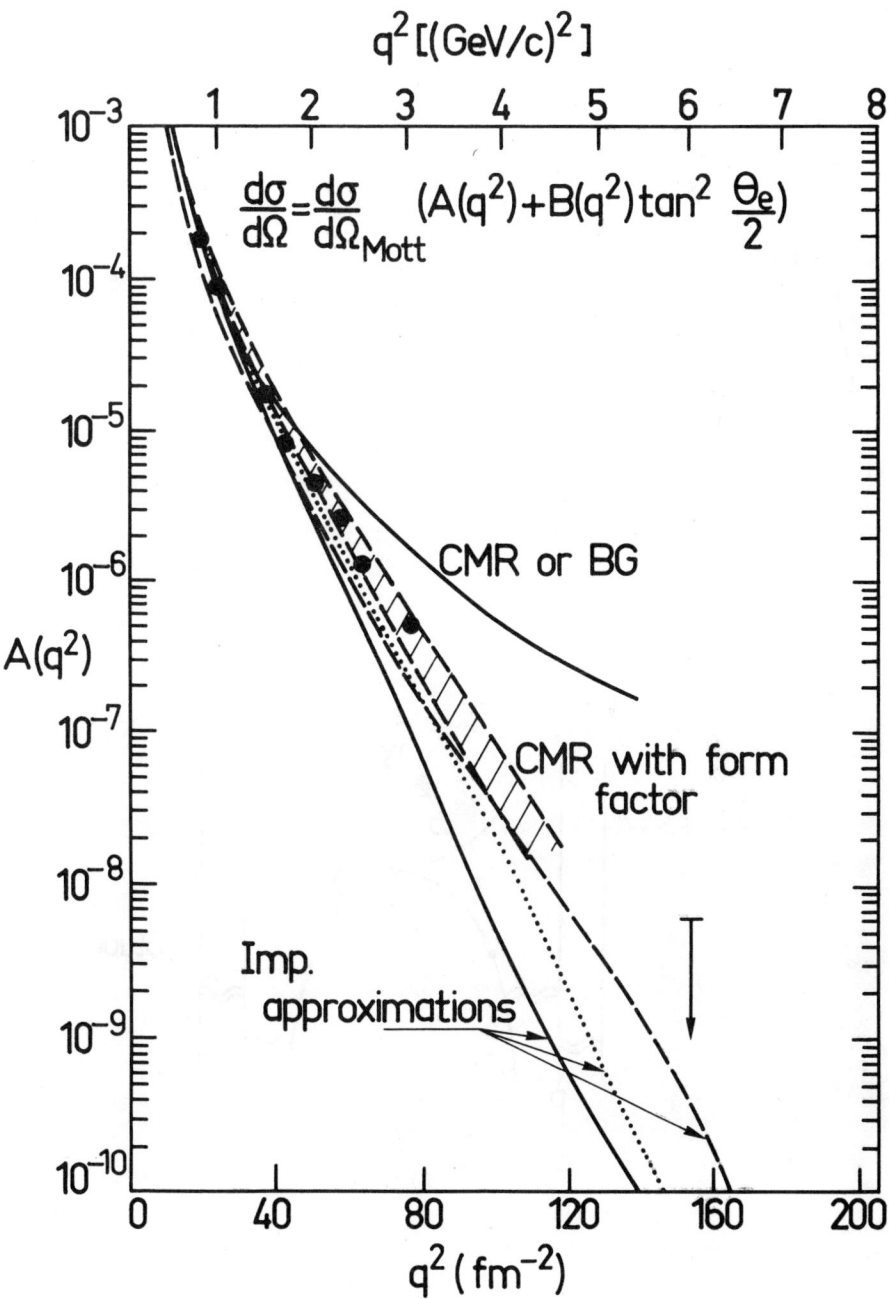

Fig. 1

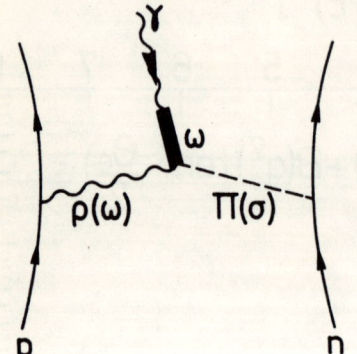

Fig. 2

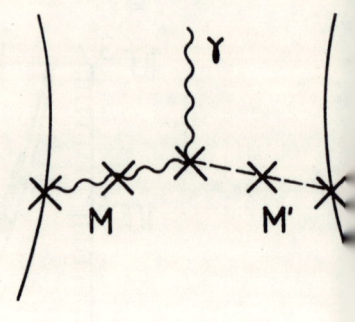

Fig. 3

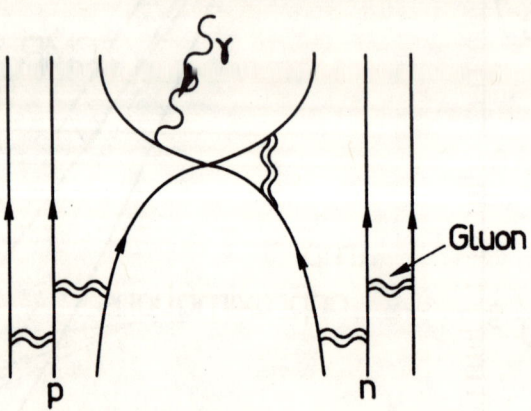

Fig. 4

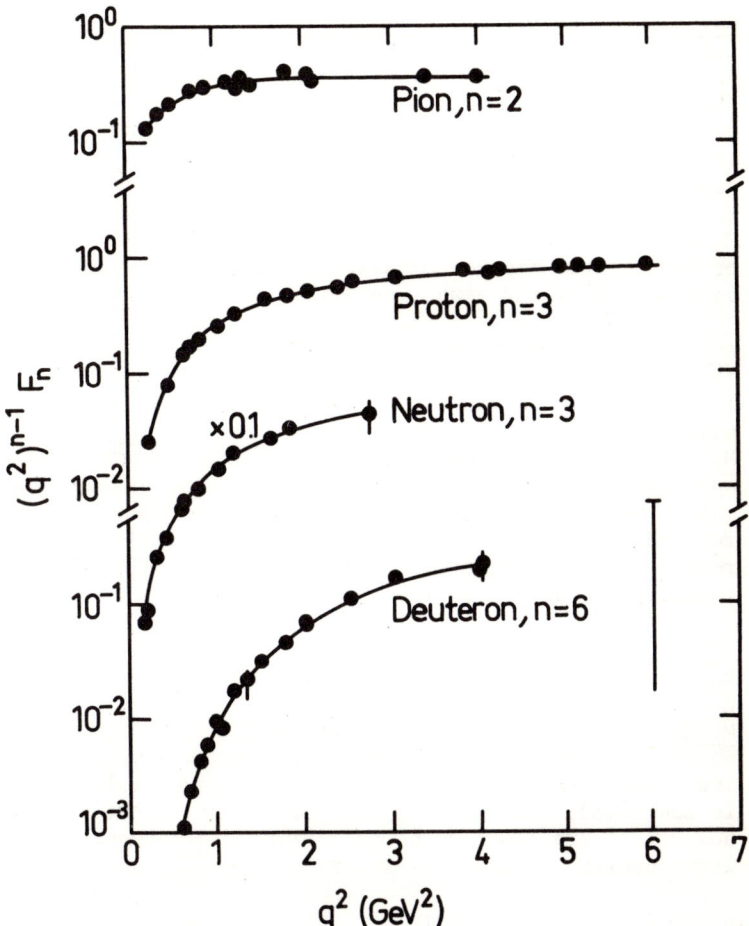

Fig. 5

Ralph Amado (U. of Pennsylvania): It is reassuring that the q^{-10} fall of $A(q)$ comes out of softening the verteces in the exchange picture or from the quark picture but it is still a puzzle as to why the q^{-10} fall fits so naturally on to the low q $A(q)$. In principle there could be a shoulder in $A(q)$ representing the transition from the "classical" n-p picture of the deuteron to the quark bag picture. Do you have any insight on why such a shoulder is not seen?

M. Rho: It's a good question. I wish I knew the answer. The present status may be summarized by saying that the more agreement between theory and experiment, the less we understand.

Wolfenstein (CMU): Does the vanishing of g_p in nuclear matter hold as the pion pole is approached? I ask this since the best way to look for g_p is in radiative muon capture where one is closer to the pole.

M. Rho: The quenching of g_p as I described here occurs only for space-like pions, so it applies only to the normal μ-capture. The quenching is not effective for a time-like pion which is relevant for radiative μ capture. Thus the Ohta's argument given in a recent Phys. Rev. Letters paper which gives a similar effect in radiative μ-capture is probably incorrect.

Dan O. Riska (Michigan State Univ.): Could you comment on the experimental evidence for the quenching of the form factors q_A and g_p?

M. Rho: Evidence exists in light nuclei for a quenching of g_A by about 10%. Whether or not this is explained by a modern version of the Lorentz-Lorenz effect is still controversial. As far as g_p is concerned, the second-class current complicates the matter; however, there seems to be no glaring discrepancy in μ-capture by light nuclei.

H. J. Weber (U. of Virginia): Would you comment on non-local contributions to the pair current which are usually ignored? We (D. Drechsel & H. J. Weber, Nucl. Phys. A256 (1976)) have found, for the isoscalar current, that at low momentum transfer they could be larger than the local term which is usually the only term retained in non-relativistic calculations.

M. Rho: I was not aware of this. As far as the isovector magnetic exchange current is concerned, the non-local term is not so important at least for small momentum transfer. I don't understand how the non-local terms become important for isoscalar currents, unless one is disturbing something sacred like gauge invariance.

DOORWAY-ISOBAR THEORY FOR MESON-NUCLEUS PHYSICS

Leonard S. Kisslinger*
Carnegie-Mellon University, Pittsburgh, Pa. 15213

I. THE ISOBAR-DOORWAY PICTURE

In the medium energy region pion-nucleon dynamics is dominated by resonances. In fact, a useful definition of medium energy is the region of baryon resonances. The spirit of meson-nucleus physics is to treat meson-nucleus scattering and reactions in terms of the fundamental interactions. One reason that this has been fairly successful is that this resonance dominance enables one to understand qualitatively some of the characteristic energy dependences, strong surface interactions, and other features which have been observed. On the other hand, one must not be naïve. For example, the large bumps in the total π-nucleus cross sections are not true nuclear resonances, but giant phenomena echoing the π-N resonance. Careful theoretical treatment is needed to unravel this.

The main two objectives of Isobar-Doorway Theory are to take full advantage of the resonance dominance to improve our treatment of meson-nucleus phenomena and to extend our understanding of the baryon resonances and their interactions. Both of these objectives are accomplished by recognizing that to the extent that a resonance (N*) dominates, all reactions proceed through doorway states in which a nucleon has been replaced by an N*. The simplest examples are Δ-nucleon hole states depicted in Fig. 1.

The formalism has been described in great detail in a recent paper with Wayne Wang[1] which will soon appear, so I will only sketch the results of the theory. The only detailed results I give are those obtained in the Isobar-Doorway Model with W. Wang and more recently with A. Saharia. Very closely related work including a microscopic treatment of Δ-hole states for a multiple scattering approach will be discussed by E. Moniz and by F. Lenz[2] at this conference.[3] Detailed references to other related theoretical work is given in Ref. 1, and will not be repeated here.

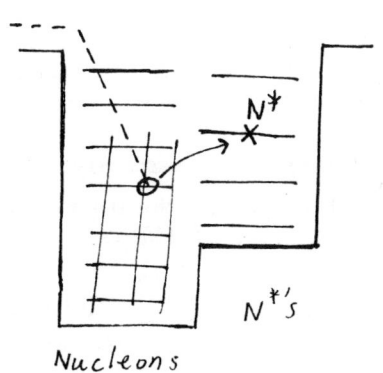

Fig. 1

II. BARYON RESONANCES

The first purpose of this section is to carefully define the parameters which are employed to depict the baryon resonances. Especially for the K-N(Y*) resonances there is an extraordinary lack of clarity in the definitions of these parameters in the published papers, and I hope that this section will make the numerical results given later more useful. Also, the operator relationship is given for the two-body system which is needed for the nuclear I-D treatment.

The resonance amplitude coupling channels α and β is given by

$$t^R_{\alpha\beta} = \frac{\sqrt{\Gamma_\alpha} \sqrt{\Gamma_\beta}/2}{M_R - E - i\,\Gamma(E)/2} \quad, \tag{1}$$

where Γ_α and Γ_β are taken as constants and the energy-dependent width $\Gamma(E)$ is defined as

$$\Gamma(E) = \Gamma_o \left(\frac{k\,\gamma_o}{k_o\gamma_o}\right)^{2\ell+1} \frac{D_\ell(k_o\gamma_o)}{D_\ell(k\,\gamma_o)} \tag{2}$$

in terms of the barrier penetration factors

$$D_o = 1$$
$$D_1(x) = 1 + x^2$$
$$D_2(x) = 9 + 3x^2 + x^4$$

The quantity Γ_o is the reduced width (generally taken as constant), and k, k_o, and γ_o are the momentum, momentum at resonance ($M_R = \sqrt{k_o^2 + M_N^2} + \sqrt{k_o^2 + m^2}$), and interaction radius, respectively.

If one introduces interactions H_{N*Nm} which couple the meson and nucleon to form the N*, then an alternate form of (1) is

$$t^R = \frac{\langle MN|H|N^*\rangle\langle N^*|H|MN\rangle}{E - M_{N*} + i\,\Gamma(E)/2} \tag{3}$$

The matrix elements $\langle N^*|H|MN\rangle$ between the meson nucleon states and the $|N^*\rangle$ states thus give the vertex quantities Γ_i.

III. ISOBAR DOORWAY FORMALISM FOR PION-NUCLEUS SYSTEMS

A. General Formalism

The theory is most conveniently represented in the projection operator formalism[4] which has been widely used for nuclear reactions. We introduce the projection operators P, which project onto the states of the meson and the nuclear ground state (for elastic scattering) or other particular states of interest, D, which projects onto the doorway states described in Section I, and Q, which projects onto all other [inelastic] states. One can represent the meson-nuclear T-matrix by

$$T = T^{NR} + \sum_{jk} \frac{\langle \psi_o^{(-)}|H_{PD}|D_j\rangle\langle D_k|H_{DP}|\psi_o^{(+)}\rangle}{(E-E_{D_j})\delta_{jk} - (\Delta_{jk}^{e\ell} + \Delta_{jk}^{in}) + i(\Gamma_{jk}^{e} + \Gamma_{jk}^{in})/2} \quad (4)$$

where $|\psi_o\rangle$ are solutions in P-space, $(E-H_{PP})/\psi_o^{(\pm)}\rangle = 0$, $|D_j\rangle$ are the doorway states which are solutions in D-space, $(E_{D_J} - H_{DD})/D_j\rangle = 0$, and the width and shifts are defined by

$$\Delta_{jk}^{e} + i\Gamma_{jk}^{e}/2 = \langle D_j|H_{DP}\frac{1}{E-H_{PP}+i\varepsilon}H_{PD}|D_k\rangle$$

$$\Delta_{jk}^{in} + i\Gamma_{jk}^{in}/2 = \langle D_j|H_{DQ}\frac{1}{E-H_{QQ}}H_{QD}|D_k\rangle , \quad (5)$$

with the H representing effective Hamiltonians which include the effect of Q-space. T^{NR} is a nonresonant term obtained by neglecting the resonances. We restrict our discussion to energies $\lesssim 300$ MeV, so only the $\Delta(1232)$ is under consideration. The only approximation used in obtaining Eq. (4) is the "doorway assumption", that all inelastic reaction take place through the doorway states, i.e., $H_{PQ} = 0$.

The optical potential corresponding to the T-matrix (4) is

$$V = V^{NR} + \sum_{jk} \frac{\langle o|H_{PD}|D_j\rangle\langle D_k|H_{DP}|o\rangle}{(E-E_{D_j})\delta_{jk} - \Delta_{jk}^{in} + i\Gamma_{jk}^{in}/2} \quad (6)$$

The nonresonant potential V^{NR} is obtained using the usual multiple scattering techniques with the nonresonant π-N t-matrix. Since the interaction is weak, and the potential is better approximated as an "ordinary" one proportional to the matter distribution, this can be done with better accuracy than for the complete potential.

At this stage one can reintroduce a multiple scattering expansion by choosing a set of doorway states labeled by filled orbits.[1] The new element provided by the theory is a form for the bound state t-matrix which might be convenient for introducing dynamics and correlations. This aspect of the theory is discussed by F. Tabakin at the conference.[5] There has also been some work done in this area by W. Friedman.[6] However, I will restrict myself hereafter to the Isobar-Doorway model, which is not a multiple scattering expansion but a representation which includes all orders of multiple scattering within the approximations of the theory.

B. The Pion-Nucleus Isobar-Doorway Model

In this model one chooses the doorway states to diagonalize the propagator in the doorway space; i.e., one diagonalizes the Hamiltonian matrix $<D_i|H_{DD}|D_j>$, so that

$$<D_i|E-H_{DD}|D_j> = \delta_{ij}(E-E_{D_i} + i\, \Gamma_i(E)/2) \;. \tag{7}$$

This gives a simplified form for the T-matrix

$$<k'|T|K> \cong <K|T^{NR}|K> + \sum_i \frac{<K'|t|K>[E-M_\Delta + i\,\Gamma_\Delta/2]}{E-E_{D_i} + i\,\Gamma_i(E)/2} F_i(\underset{\sim}{k}',\underset{\sim}{k}) \tag{8}$$

and for the optical potential

$$<k'|V|k> = <k'|V^{NR}|k> + \sum_i \frac{<k'|t|k>[E-M_\Delta + i\,\Gamma_\Delta/2]}{E-E'_{D_i} + i\,\Gamma_i^{in}(E)/2} F_i(\underset{\sim}{k}',\underset{\sim}{k}) \;, \tag{9}$$

with $F_i(\underset{\sim}{k}',\underset{\sim}{k})$ a mixed nuclear-Δ density function modified by Δ-propagation

$$F_i(\underset{\sim}{k}',\underset{\sim}{k}) = \int d^3r_1 e^{i\underset{\sim}{k}'\cdot\underset{\sim}{r}_1} \phi_\eta^*(r_1) \int d^3r_2 \; \rho_{\Delta_i}(r_1,r_2) e^{-i\underset{\sim}{k}\cdot\underset{\sim}{r}_2} \phi_N(r_2) \tag{10}$$

In expression (10) $\phi_N(r_1)$ is a nucleon single-particle wave function and $\rho_{\Delta_i}(r_1,r_2)$ is a mixed density function for the Δ, given by $\phi_{\Delta_i}^*(r_1)\phi_{\Delta_i}(r_2)$, where $\phi_{D_i}(r)$ is the wave function of the Δ in the ith doorway state.

The diagonilization process can be carried out in a model. If one restricts oneself to Δ-hole states then one can easily diagonalize potentials given by meson exchange processes such as those represented by Fig. 2. The exchange interaction is fairly well known, since the vertices are given on-shell by the $\Delta \to N\pi$ decay. However, the direct term is not well known, as the $\pi\Delta\Delta$ vertex is not at all well understood. One can compare this, e.g. to the analogous situation for inner bremsstrahlung from the delta.[7] Such diagonalizations have been discussed for Δ-h states,[8] but with the uncertainty in the interactions there is general difficulty in carrying out realistic dynamic calculations. There is additional information from the two-body calculations including virtual resonances, but it is still difficult to carry out accurate dynamic calculations.

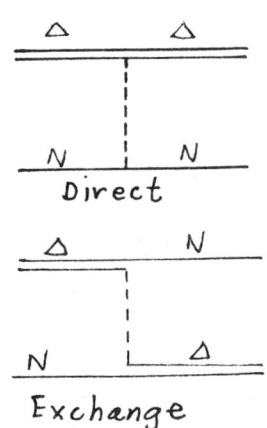

Fig. 2

We therefore proceed by treating the expressions (8) and (9) as phenomenological, with certain parameters to be determined by systematic fits to experiment. We assume that the doorway states which give significant contribution are spread over an energy range small compared to the width and replace the sum by a single term which represents this group of states. This is accomplished by introducing an average doorway energy $\bar{E}_D$ and average widths and a phenomenological mixed density function taken as a Gaussian distribution

$$\rho_D(r_1, r_2) = \rho_\Delta(r_1-r_2) \cong e^{-|r_1-r_2|^2/\lambda^2}. \qquad (11)$$

With these forms one can determine the parameters by comparison with experiment.

i. Closure

If one takes an average energy and width in Eqs. (9) and (10) and assumes that the doorway state sum goes over the complete set of states, then $\rho_\Delta(r_1-r_2) \to \delta(r_1-r_2)$ and the function $F(\underline{k}',\underline{k})$ goes to the Fourier transform of the nuclear density. In this approximation the resonant part of the potential is related to the resonant part of the ordinary first order potential in a simple way.

$$V^{Res} = V^{Res}_{First\ order} \frac{E-M_\Delta + i\,\Gamma_\Delta/2}{E-M_\Delta-\Delta E + i\,\Gamma^{in}/2} \qquad (12)$$

This closure form, which corresponds to the limit of $\lambda \to 0$ in the expression for the Δ-mixed density (11), was the form of the Isobar Doorway model originally used in comparison with experiment.[9]

ii. Recent Results

In the study of Ref. 1 it became obvious that the closure approximation is not adequate. During the past year fits to the ^{12}C data have been carried out including the nonlocality arising from the effective Δ-propagation represented by (11). This has been done using the momentum space computer program developed by Phatak and Tabakin. There are three parameters of the theory, which we now discuss.

a) The Energy Shift ΔE. The quantity ΔE is the difference in energy between the average doorway state and the ground state after subtracting mass difference. Thus, it gives the difference between the binding energy of the nucleons active in the scattering and the average Δ binding energy. If one uses the pion exchange interaction of Fig. 2 then ΔE < 0, since the Δ-N direct term is considerably stronger than N-N interaction in SU(6). However, it was observed in the earliest calculations[9] that ΔE>0 in the resonance region. The interpretation given is that the nucleon interaction "bonds" are broken as the nucleon becomes a short-lived Δ.

It is important to note that there are characteristic aspects of the data which enable one to extract ΔE, since they are far more sensitive to this parameter than the other parameters. It is the vanishing of the real part of the resonant amplitude which best singles out ΔE. This can show up in the Coulomb-nuclear interference, the sharpening of the diffraction minima near resonance, and other aspects of the angular distribution.

b) <u>The Width Parameter, β(E)</u>. It is expected[1,9] that the inelastic width will be similar to the free width. We introduce the parameter β(E) by

$$\Gamma^{in}(E) = \beta(E) \, \Gamma_\Delta(E). \qquad (13)$$

In the previous work this parameter was determined by fitting total widths.

c) <u>The Density or Nonlocality Parameter</u>. The parameter λ (Eq.11) is a measure of the nonlocality arising from Δ-propagation. It is not to be thought of as simply the distance traveled by a Δ before it decays, but the result of the multiple interaction. It is determined from the angular distribution.

We can summarize these results by the picture given in Fig. 3. The three parameters λ, β, and ΔE represent interaction effects to all orders within the model.

Typical results are shown in the contributed paper of Kisslinger and Saharia.[10] Referring to the figures in that paper one observes first that over a broad energy region

$$\Delta E = +10 \text{ MeV}.$$

The simplest interpretation of this result is that the Δ is free after

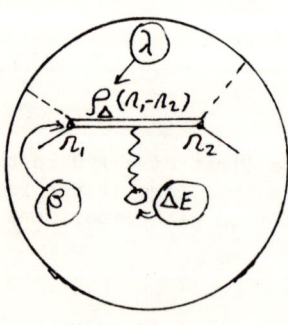

Fig. 3

it is formed in the nucleus. Next we note that

$$\beta \simeq 1.1 \quad ,$$

so that the width of the Δ is only slightly modified in the medium. The λ parameter, however, has an important energy dependence. For the higher energies in the resonance region (E $\gtrsim$ 200 MeV) the closure approximation is adequate and λ is quite small. At 120 MeV there are important corrections to closure and $\lambda \approx 0.4$ Fm. As one goes further down in energy to 50 MeV there are large effects of ths nonlocality, and the value $\lambda \approx 1.1$ gives a much better fit to the data than the closure result. This will be discussed further by R. Eisenstein.[11]

The interpretation of these results is that at the higher energies there are many doorway states contributing and thus the closure approximation is good. As one goes to lower and lower energies, farther down on the low wing of the resonance, fewer doorway states are effective, and the nonlocality becomes increasingly more important. One needs a number of states to form a wave pocket and localize the Δ.

The results particularly for the lower energies are far from quantitative fits. What the model discussed offers is a framework which avoids the impulse approximation and provides a method for including interaction effects to all orders. There are important corrections, especially the kinematic corrections associated with a correct Lorentz transformation and proper inclusion of the momenta of the bound nucleons. For the optical potential accurate off-shell extrapolations of the T-matrix are important, although this is not critical for the elastic scattering, where using the T-matrix form one only goes off shell by the nonresonant scattering (see Eq.(4)). There are other relativistic considerations such as pure absorption and crossing properties which could be important.

IV. ISOBAR-DOORWAY MODEL FOR KAON-NUCLEUS SYSTEMS

A. Resonances

The K^--N system exhibits a number of resonances. The analysis is complex and not yet entirely satisfactory in large part because of the lack of complete data. We use the analysis of R. Armenteros et al.[12] The parameters are shown in the Table, with the notation of Sec. II, for $M_R \lesssim 1800$ MeV and L < 3. The labels for the quantum numbers are I and 2J, the isospin and 2x the spin; e.g., D_{13} is an L=2, Isospin=1, spin=3/2 resonance. The parameters differ a bit from the latest parametrization but give satisfactory fits to the data.

TABLE I
K⁻N Resonance Parameter

Resonance	M_r	Γ_o (MeV)	Γ_i (MeV)
D_{03}	1519	16	7.2
S_{01}	1662	38	5.3
D_{13}	1662	49	3.9
D_{03}	1691	31	5.6
D_{15}	1768	128	58.0
D_{05}	1807	123	11.0

One observes that this rich spectrum contains a number of low-lying resonances with widths considerably smaller than the $\Delta(1232)$. Thus the Doorway-Isobar game seems even more promising than for pions. Note also the occurrence of low-lying D-state resonances. These will give characteristic forms to the optical potentials which are derived for the K⁻-nucleus system. We make the following off-shell extrapolation for the D-wave amplitudes.

$$t^{L=2} = A_2 P_2(\cos\theta) \to 1 - \frac{3}{2}\frac{|k-k'|^2}{P_c^2} + \frac{3|k-k'|^4}{8 P_c^4} \quad . \quad (14)$$

This will give terms in the optical potential of the type ρ, $\nabla^2 \rho$, and $\nabla^4 \rho$, respectively. There are also background amplitudes which we take from Ref. 12, which contains S and P waves. The resulting optical potential is thus

$$2EV = b_o\, p_L^2\, \rho(r) + b_1 \nabla\cdot\rho\nabla + b_2 \nabla^2\rho + b_3 \nabla^4\rho \quad , \quad (15)$$

with the parameters obtained from the K⁻-N amplitudes at each energy. The parameters b_o, b_1, b_2, and b_3 contain both nonresonant and resonant parts.

From the form (14) and the Table I one can anticipate a number of interesting aspects of K-nuclear scattering in general and the Isobar-Doorway picture in particular. The high derivatives tell us that there are strong surface interactions. Because of the resonances contributing to the b_i parameters, there will be strong energy dependences. Thus the ability of the Kaon to penetrate the nucleus will be strongly energy dependent. One can exploit this not only for the elastic scattering, but for reactions as well. This can be quite significant for the (K⁻,π⁻) excitation of Unitary Analog states as will be discussed later in the conference.[13]

Typical results using the parameters of Ref. 12 and the of Eq.(13) for elastic K⁻-nucleus elastic scattering are shown in Fig. 4 and 5. These results were obtained by using a modified version of the computer code PIRK, which contains a $\nabla^4\rho$ term. In

Fig. 4 one notes the strong energy variation at the $D_{03}(1519)$ resonance. The lab energy of 135 MeV corresponds to a K^--N center of mass energy of 1.519 beV, the position of this narrow (16 MeV) resonance. The tendency for sharpening of the minima at resonance which was observed in pion-nucleus scattering is seen here as one changes the energy by only 5 MeV. The third curve shows the effect of choosing $\Delta E=10$, corresponding to free Y* resonances. There is quite a marked effect, which might make it possible to explore such dynamic questions for the Y*'s. Fig. 5 shows similar effects near other resonances.

V. SUMMARY

The Doorway Isobar theory gives closed expressions for the T-matrices and optical potentials, rather than a multiple scattering series. It enables one to incorporate binding effects and Δ-propagation effects to all orders. The phenomenological form for the T-matrix will be useful in that the off-shell ambiguities are minimized and one can determine the parameters from elastic scattering.

The optical potential should be quite useful as it gives a convenient form which incorporates effects difficult to calculate directly and which result in very complicated forms in the multiple scattering approach. A number of effects must still be included to obtain quantitative fits. It will be essential to make systematic comparisons with experiment to see if consistent parameters result. It would then be most interesting to compare these parameters with theoretical models for the energy shifts, widths, and propagation properties. It is here that one begins to genuinely extend our understanding of strong interactions. The present results indicating that the Δ is unbound with a width broadened by say 10%, and with a nonlocality which is strongly energy dependent already offers some intriguing possibilities.

Thus the Isobar-Doorway Model can provide a convenient framework for including effects which are quite difficult to represent in a multiple scattering formulation. It is the resonance dominated nature of the meson-nucleon interactions which make this feasible. The theory has the nice feature of avoiding the question "but what about the higher order terms?" It replaces it by new questions: "How many doorway states are important and what is their nature?" It is the answer to these questions which will give us our deepest understanding.

REFERENCES

* Supported in part by the National Science Foundation.
1. L. S. Kisslinger and W. L. Wang, Ann. Phys. (N.Y.) (1976).
2. E. Moniz, invited paper at Conference; F. Lenz, ibid.
3. See also G. E. Brown and W. Weise, Phys. Reports C22(1975)279.
4. H. Feshbach, A. K. Kerman, and R. Lemmer, Ann. Phys.(N.Y.) 41, 230 (1967).
5. F. Tabakin, invited paper at Conference.

6. W. Friedman, Phys. Rev. C12, 1294 (1975).
7. Discussed at the conference by D. Sober and by Q. Hokim, and J. Lavine.
8. F. Lenz, Ann. Phys. (N.Y.) (1976). See also Ref. 2 and Ref. 3.
9. L. S. Kisslinger and W. L. Wang, Phys. Rev. Letters 30, 1071 (1973).
10. L. S. Kisslinger and A. Saharia, contributed paper at conference.
11. R. Eisenstein, invited paper at conference.
12. R. Armenteros et al., Nucl. Phys. B14, 91 (1969).
13. Nguyen Van Giai, invited paper at conference.
14. R. Eisenstein and G. Miller, Computer Physics Comm. 8 (1974) 130.

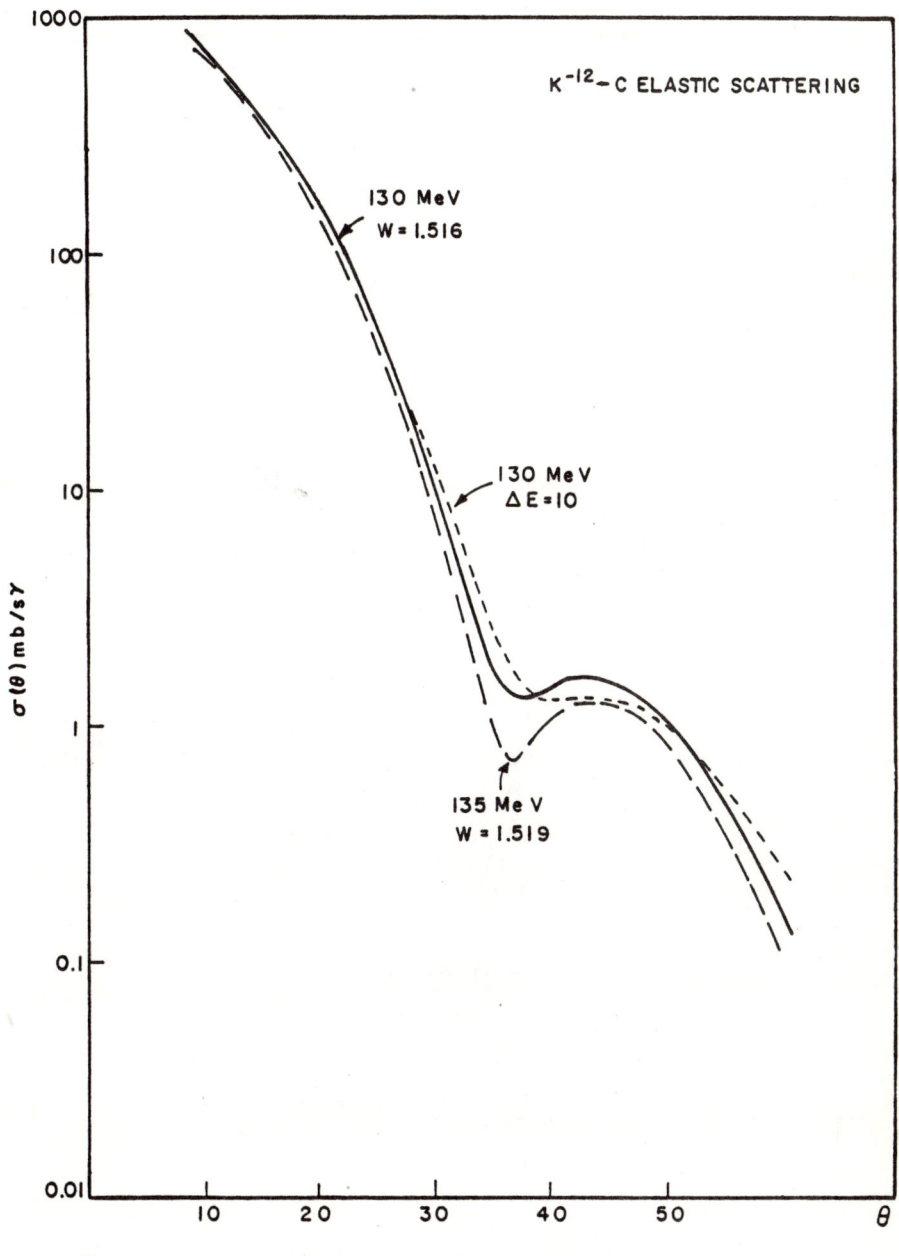

Fig. 4

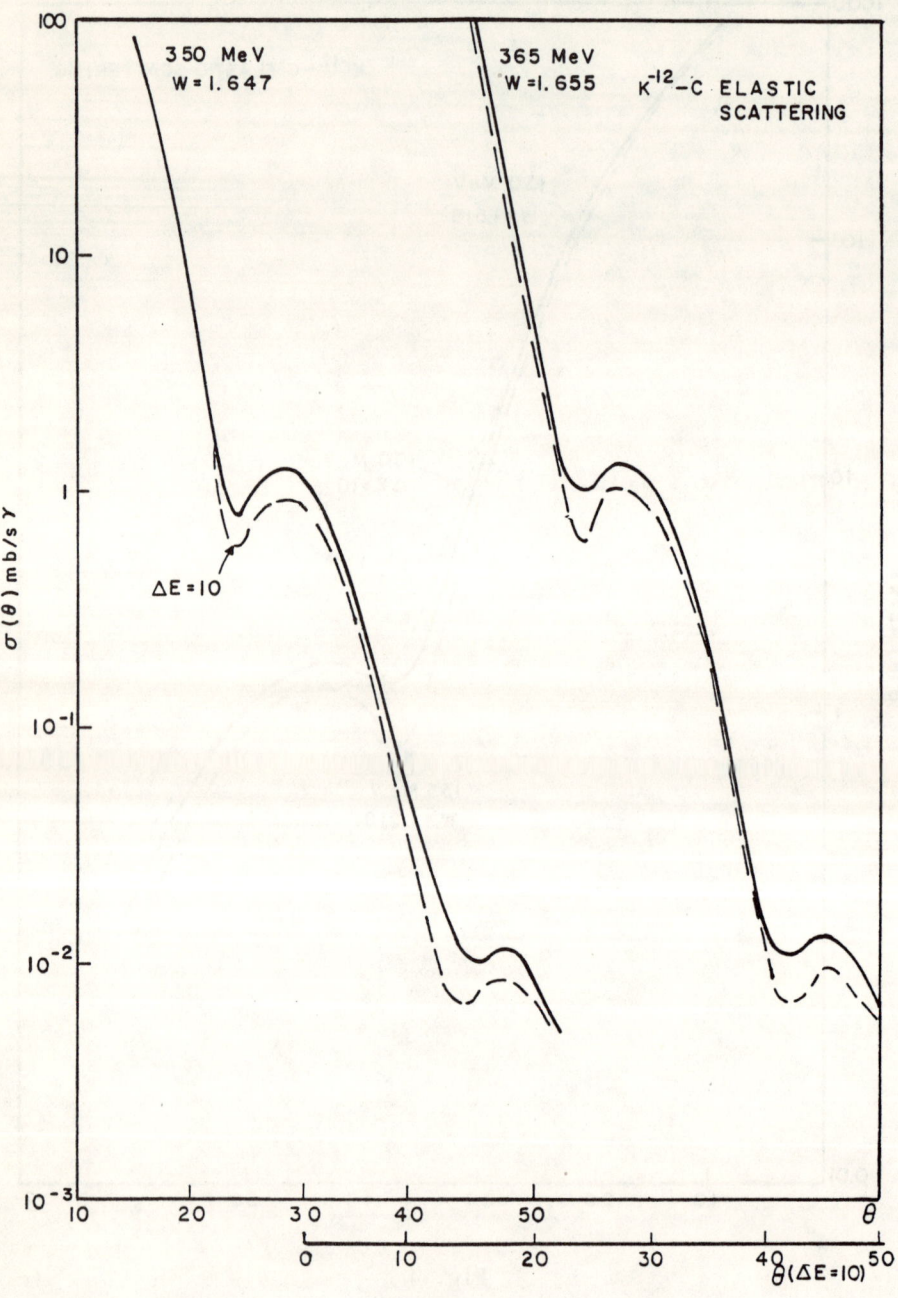

Fig. 5

Carl Shakin (Brooklyn College): We have made a calculation at Brooklyn College of the collision broadening of the Δ through studying the π-N T matrix in nuclear matter. We find that in your notation $\beta=2$ is reasonable at full nuclear density, however this damping is strongly density dependent and a value of $\beta=1.1$ or $\beta=1.2$ is reasonable for an interaction that takes place in the nuclear surface.

L. Kisslinger: That is an interesting result. However, as I pointed out in my talk, I believe that there is considerable uncertainty in the interaction which should be used.

Rubin Landau (Oregon State Univ.): You have commented that the closure approx. is not good below resonance. If this is so then is it also true that its use in calculating the second order optical potential for π's is suspect, and in just what way would it breakdown? In particular, the present calculations of $U^{(2)}$ are local and even eikonalized.

L. Kisslinger: Yes, the non-locality is an essential feature in the results which I have just shown. Your question has brought up an important point: that regardless of the interaction, the second order term in a Watson expansion of the optical potential is always non-local. The non-locality represented by our parameter λ is a many-body effect, including nuclear correlations. Although it is represented as a Δ-propagation property, I believe that it is quite different from the non-locality pointed out by Frieder Lenz, mentioned in Frank Tabakin's talk this morning. This must be studied further.

Yu. A. Shcherbakov (JINR-DUBNA): The $\pi + C^{12}$-data is reproduced badly at the minimum by the isobar doorway model. Have you tried a comparison of the theory with data on lighter nuclei such as ^{3}He or ^{4}He?

L. Kisslinger: We have not yet studied He. We shall. We shall also incorporate a number of corrections discussed at this conference in our more nearly quantitative calculations in the future.

CROSSED PION ABSORPTION CONTRIBUTION TO THE π-d SCATTERING LENGTH

T. Mizutani and N.C. Mukhopadhyay
Theory Group, SIN, CH-5234 Villigen, Switzerland

There have been a number of investigations [1] on the possible importance of the crossing in the low-energy pion-nucleus scattering. Here we are interested in the π-d scattering at threshold, and the possible effect of crossing on the scattering length, $a_{\pi d}$, via the pion absorption in the intermediate states (figs. 1a,b) [in the figures, <u>blobs</u> indicate $\pi d \rightleftarrows NN$ half shell t matrices without N-N interactions; hatched squares denote the NN propagators, with the inclusion of N-N scattering effects].

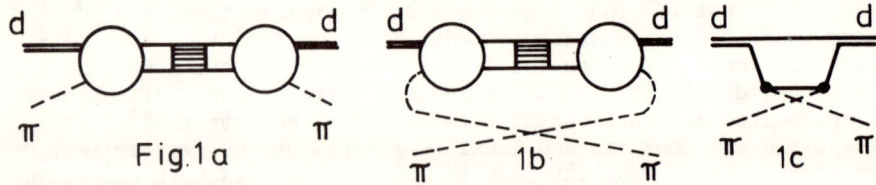

Fig.1a   1b   1c

The π-d scattering length can be written as

$$a_{\pi d} = a^S_{\pi d} + a^A_{\pi d}, \qquad (1)$$

where $a^S_{\pi d}$ is the contribution of the process with no intermediate pion absorption, $a^A_{\pi d}$ is that of the process with pion absorption, and consequently with intermediate states having no pions. Double scattering and explicit Faddeev calculations [2] yield $a^S_{\pi d} \simeq -0.035 \, m_\pi^{-1}$. On the other hand, $a^A_{\pi d}$ is obtained to be $(-0.007+0.0043i) \, m_\pi^{-1}$ in a recent calculation [3]. These results include no crossing effects.

We evaluate the crossed absorption contribution (fig. 1b) in the following way: we take, for the πd ⇄ NN t-matrices, the direct pion absorption (emission) <u>plus</u> one π-N scattering before (after) the pion absorption (emission). This is because of the apparent dominance of the pion absorption (emission) by the two-nucleon mechanism. For deuteron, we use the Reid soft-core wave function and, for the intermediate N-N interaction, we use the Mongan (types 1 and 2) potentials. The πNN vertices are Galilei-invariant.

Computed results are shown in table I. The contributions of various components of the deuteron wave functions in the initial and final states are labelled SS, SD, and DD. The role of the deuteron D-state is clearly very small.

The effect of the intermediate N-N interaction turns out to be only a few percent of the total crossing contribution; its magnitude appears insensitive to the N-N interaction, not a surprising result for such a highly off-energy-shell two-nucleon state.

To Re $a_{\pi d}^A$, the "crossed effect" contributes ~ 0.0022 m_π^{-1}, in comparison to -0.007 m_π^{-1}, for the uncrossed contribution. Since the impulse contribution to $a_{\pi d}^S$ already includes the crossed Born graph of the type fig. 1c, this should be subtracted from the gross crossing effect calculated by us. This gives the pure crossed pion absorption contribution to $a_{\pi d}$ to be ~ 0.0015 m_π^{-1}, about 21% of the uncrossed contribution.

Our calculation suggests the smallness of crossed absorption effect on the scattering length in other nuclei. It is of interest to compare the expectation of a static Chew-Low type model, for which the crossed and uncrossed absorption contributions cancel for the T=S=0 nuclei [1] even at finite energy, with our result.

We note that there is some confusion in the literature on the pion energy dependence of the t-matrix [4], which can be avoided.

TABLE I

Crossed contribution to the π-d scattering lengths in units of 10^{-3} m_π^{-1}

SS	SD	DD	Total
2.484	-0.298	0.035	2.221

REFERENCES

1) J. Cammarata and M.K. Banerjee, Phys. Rev. C13, 299 (1976), and references therein
2) I. Afnan and A. Thomas, ibid. C10, 109 (1974)
3) T. Mizutani and D.S. Koltun, to be published
4) K. Nishimoto et al., Prog. Theoret. Phys. 46, 135 (1971)

I1. T. Cheon (McMaster Univ.): In the process of pion absorption by a deuteron, the pion rescattering effect is taken into account. Can one distinguish this rescattered pion from the virtual pion exchanged between two nucleons. If this is not clearly considered, you may take a double count.

Mizutani: We have studied systematically how to avoid the pion overcounting in π-nucleus scattering (Mitzutani & Koltun to be published) and the conclusion is that the overcounting is associated with the π-N P_{11} Born term. The idea is that if one subtracts out this P_{11} Born term from the total P_{11} π-N t-matrix contributing to the rescattering contribution, there is no danger of pion overcounting. We first tried this and later found that the residual P_{11} t-matrix is very small around the energy of our interest. So we don't have P_{11} contribution in our present calculation. At any rate there's no overcounting here in the calculation.

PION EXCHANGE CURRENTS AND NUCLEAR CHARGE FORM FACTORS[*]

Mark Radomski and D. O. Riska
Michigan State University, East Lansing, MI 48824

ABSTRACT

The effect of the pion exchange current due to the Born term in the photoproduction amplitude on the charge form factors and distributions of ^{4}He, ^{16}O and ^{40}Ca is calculated in the simple harmonic oscillator shell model. The exchange current contribution is most notable in the case of the α-particle while it is rather unimportant in ^{16}O and ^{40}Ca. The contribution to the nuclear mean square radius is small (≈0.03 fm^2) and roughly independent of the mass number.

INTRODUCTION

We consider the effect of the simplest (and probably dominant) one-pion exchange current, which has been called the "pair current," on the charge form factors of ^{4}He, ^{16}O and ^{40}Ca. We use the independent particle harmonic oscillator shell model, in which these nuclei are doubly closed shells. Such a model cannot be quantitative, but should show the qualitative effect of exchange currents in complex nuclei.

We also consider the same exchange current in a finite Fermi gas model in order to investigate the variation with nuclear mass of the contribution to the mean-square charge radius.

EXCHANGE CURRENTS AND CHARGE FORM FACTORS

The charge component of the pion exchange current in question is as given in reference[1], if the isoscalar magnetic form factor is normalized to $G_M^S(0)=0.44$. The expectation value of the two-body operator is taken using a nuclear wave function being a Slater determinant of harmonic oscillator functions. Center of mass motion of the final nucleus may be accounted for by the method of Tassie and Barker[2]. Only the exchange term in the two-body density contributes in the even-even nucleus, and for closed shells this is found not to depend on the angle between the two-nucleon center-of-mass and relative coordinates. The numerical evaluation of the pion exchange form factor is then easy, and details will be published elsewhere[3]. The contribution to the mean square charge radius $<r^2>$ is determined by the slope of the form factor at the origin. An asymptotic expression for the form factor agrees with the numerical results at large q^2.

A finite Fermi gas model is constructed by multiplying the two-body density of the infinite Fermi gas by θ-functions of the nucleon coordinates. A value of the contribution to $<r^2>$ is taken from the resulting pion exchange form factor.

[*]Research supported by the National Science Foundation

RESULTS AND CONCLUSIONS

Figs. 1 and 2 display the form factors for ^{4}He and ^{40}Ca in the impulse approximation (IMP) and with this pion exchange effect included (IMP+EXCH). The oscillator parameters have been readjusted so as always to give the observed total $<r^2>$. The points and dashed curve represent various reductions of the experimental data[4,5]. In helium one notes a substantial effect producing a new diffraction minimum near where one is observed. In calcium the effect is smaller at observed values of q^2 but large in the range above 10 fm^{-2} and produced a new diffraction minimum around 16 fm^{-2}. The oxygen case is similar to calcium, but no new minimum is produced. The pion exchange contributions to $<r^2>$ are +0.032, +0.031, and +0.032 for He, O, and Ca, respectively. The pion exchange charge densities, obtained by numerical Fourier-transform of the form factors, for all three nuclei are substantial only in the central region. The effects at the center are -7%, +2%, and -2% in He, O, and Ca.

In the Fermi gas model contributions to $<r^2>$ are nearly independent of nuclear mass and about 0.05 fm^2. This effect is much smaller than that of the finite nucleon size (~ 0.6 fm^2).

Our main conclusion is that the pion exchange effect on the charge form factor is smaller in heavy nuclei than few-nucleon systems, but at large $q^2 \gtrsim 15$ fm^{-2} dominates the impulse approximation.

REFERENCES

1. A.D. Jackson, A. Lande and D.O. Riska, Phys. Lett. 55B,29(1975).
2. L.J. Tassie and F.C. Barker, Phys. Rev. 111,940(1958).
3. M. Radomski and D.O. Riska, submitted to Nuclear Physics A.
4. R.F. Frosch, et al., Phys. Rev. 160,874(1967).
5. R.F. Frosch, et al., Phys. Rev. 174,631(1968).

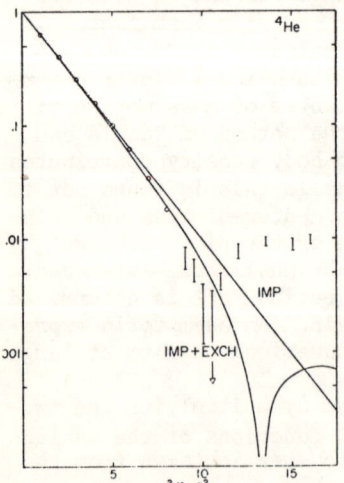

Fig.1. Form factor of ^{4}He

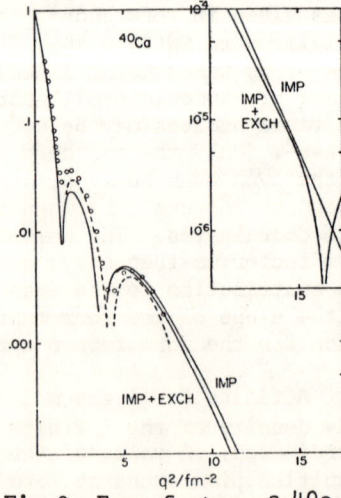

Fig.2. Form factor of ^{40}Ca.

C. Shakin (Brooklyn College): How is your result to be used in understanding something further concerning the structure of ^{4}He? The wave function you have used is quite unrealistic so that it is difficult to know whether your correction will improve agreement with experiment. It is possible that a better wave function will already give form factors similar to your "shell model plus exchange current" result.

Radomski (Michigan State U.): Our intention with this calculation was not necessarily to improve agreement with experiment of our unrealistic wave functions. The impact of our ^{4}He result is to verify the finding of Borysowicz and Riska of a large pion exchange current effect. This means that when trying to fit data with realistic wave functions, you make a mistake if you use only impulse approximation. Calculations of this effect with realistic wave functions are very hard and would be interesting; but there is no reason to think they will make the exchange charge form factors much smaller.

David Brayshaw (SLAC): One should be cautious in using overly simplified wave functions in light nuclei in drawing conclusions regarding the importance of exchange terms. In the ^{3}He-^{3}H system, the incorporation of the NN d-wave interaction (tensor term in triplet state) makes the experimental form factors much more difficult to account for theoretically.

Radomski (Michigan State U.): Such effects more likely increase than decrease the pion exchange effect.

Negele (MIT): Since many-body wave functions are under discussion, I should call to your attention a recent calculation by Gari, Zabolitzky and others from the Bochum group, in which exchange corrections are calculated for ^{4}He, ^{16}O and ^{40}Ca using the density matrix obtained from solving Brueckner and Faddeev equations. The two-body correlations in all relevant partial waves should be included quite carefully, and in this work one also sees very large ^{4}He corrections and rather smaller ^{16}O and ^{40}Ca effects.

Radomski (Michigan State U.): No reply to this comment.

A. Rinat (Weizmann Institute): To the caution regarding the calculation of exchange current contributions to form factors already uttered one might add the neglect of strong form factors. In particular, one may expect that the position of the generated minimum in ^{4}He might be influenced by the strong form factors.

Radomski (Michigan State U.): I don't think hadronic form factors would be very important at these momentum transfers.

Dan O. Riska (Michigan State Univ.): The two body matrix elements are not very sensitive to short range correlations (or hadronic vertex functions). Correlations do strongly affect the one-body matrix elements and introduction of such would cut down the impulse approximation results. In the α-particle the situation can only improve - the minimum moves towards the experimental one - when correlations are included.

FIELD THEORY TREATMENT OF PI-NUCLEUS SCATTERING[*]

Gerald A. Miller
Physics Department, University of Washington, Seattle, WA 98195

Most theoretical treatments of pion-nucleus scattering use a multiple scattering theory which gives the T-matrix as an expansion in elementary off-shell pi-nucleon amplitudes. Derivations of such theories assume the existence of a projectile-nucleon potential. However, mesons interact by being singly absorbed by or emitted from nucleons, and the assumption of a potential may not be justified. Furthermore many-body effects (such as meson annihilation on two or more nucleons) not contained in standard theories may occur.

In order to include the dynamics presented by the above assumptions, one is led to using the non-linear Low equations for both pi-nucleon and pi-nucleus scattering. The procedure for developing a multiple-scattering theory by directly inserting the two-body wave equation into the many-body problem breaks down for non-linear wave equations. In this contribution we propose an approach which is based on finding equivalent linear equations for both pi-nucleon and pi-nucleus scattering. The derivation of the resulting field-theoretic pi-nucleus T-matrix then proceeds by straightforward techniques. (In this contribution we neglect the non-linear cross term in pi-nucleon and pi-nuclear scattering.)

The first step is to examine pi-nucleon scattering. The Low equation for this process is schematically given as

$$t_z = \frac{v}{z} + t^\dagger D_z t \quad . \tag{1}$$

We claim that an equivalent linear-wave equation is

$$t_z = \frac{v}{z}(1 + \frac{z}{H_0} \frac{1}{z-H_0} \frac{z}{H_0} t_z) \quad . \tag{2}$$

By making the transformation $H_0 t'_z = z\, t_z$, one can show that

$$t'_z = \bar{v}(1 + \frac{1}{z-H_0} t'_z) = \bar{v} + \bar{v} \frac{1}{z-H_0-\bar{v}} \bar{v} \tag{3}$$

where $\bar{v}$ is an energy-<u>in</u>dependent but <u>non</u>-Hermitian potential. Upon constructing the necessary bi-orthogonal basis, using the properties of $\bar{v}$, and the definition of t'_z of Eq. (3) may be shown to be equivalent to Eq. (1). Note that the <u>only</u> assumption about the driving term is in the form of its energy dependence.

The many-body problem is described by the following Low equation

$$T_z = V_z + T^\dagger D_z T_0 \quad . \tag{4}$$

The second term of (4) includes T-matrices for pi-nucleus inelastic

[*]Supported by the Energy Research and Development Administration.

scattering and appropriate pi-nucleus energy denominators are used. The driving term, V_z, consists of a sum over single nucleons of medium-modified, pi-nucleon potentials and a sum of terms in which a pion is absorbed in one nucleon and emitted by another. We proceed by first solving Eq. (6) without this latter term and then revising the resulting multiple scattering expansion to include this term.

We find that the remaining part of the driving term is well approximated by $V_z \approx A(v+\Delta v)/z$ where Δv includes the effects of inelastic channels (π,N) in which a pion is annihilated by depositing its entire energy on a single nucleon. By following the reasoning of Eqs. (2) and (3) we have

$$T_z = AV_z^{(0)} \left(1 + \frac{z}{H_0} \frac{1}{z-H_0} \frac{z}{H_0} T_z\right) . \qquad (5)$$

The free Hamiltonian of Eq. (5) is not the same as that of Eq. (2). Ignoring this difference is equivalent to neglecting the binding and recoil effects of standard theories. Equation (5) is now in a form ameable to standard treatments and one may derive expressions for the optical potential in terms of t. In particular, the first-order (pseudo)optical potential is simply $(A-1)t$. The difference between this and standard theories is the presence of the factors $(z/H_0)^2$ in the propagators of Eqs. (2) and (5).

It is also easy to show that the change in the first-order optical potential given by the effects of the (π,N) reaction is given by

$$\Delta U' = \Omega^{(-)\dagger} \Delta v\, \Omega^{(+)} \qquad (6)$$

where the wave matrices are calculated from Eq. (5).
Explicit calculations of Eq. (6) have been performed and significant effects are obtained for pion energies less than 75 MeV.

Generalizations of the present discussion to include recoil, binding, pi-nucleon inelasticities, and the effects of crossing will also be discussed. We believe that this approach provides a formal basis for a consistent study of these effects which arise from the unique aspects of the assumed pi-nucleon interaction.

HIGHER ORDER PION-NUCLEUS OPTICAL POTENTIAL AND THE EFFECTIVE Δ-NUCLEUS POTENTIAL IN THE (3,3) RESONANCE REGION

M. Hirata*
Department of Physics, Brown University, Providence, R. I. 02912

F. Lenz
Laboratorium fur Hochenergiephysik, SIN

K. Yazaki
Department of Physics, University of Tokyo

Recently the description of an improved first-order optical potential for pion-nucleus scattering was presented in the (3,3) resonance region relaxing the fixed-scatterer approximation, and taking into account the intermediate propagation of the interacting π-N system (Δ) and the Pauli-quenching effect for the decay of Δ to π and N. This was applied to the case of the pion -^{4}He scattering.[1] In this description the Δ-residual interaction appeared in the Δ-hole propagator. This interaction was treated as an averaged one-body potential $V_{\Delta A-1} = (V_o + iW_o)e^{-\beta\gamma^2}$. V_o and W_o were fitted to reproduce data of the total cross-section and the real part of the amplitude for each energy. β was mainly determined from data of tha angualr distribution and was $0.25 f_m^{-2}$. The energy dependence of V_o and W_o are strong as shown in the set (1) of Fig. 1. Contributions from various processes may be incorporated into this effective one body potential. A typical example is one from the inelastic scattering in which a target nucleus is excited into comparatively low-lying excited states. Here, we take into account the contribution from this process in the higher order pion-nucleus optical potential and investigate whether this contribution can be practically replaced by one from a one-body potential, and the resulting modification of V_o and W_o. The Argand plots of the resonance amplitude for each partial wave in which the distortion effect due to the background potential is not included are shown in Fig. 2. The excitation energy of the target nucleus ^{4}He is limited to 30 MeV. (a) and (b) in Fig. 2 are results obtained respectively from the first order optical potential only, and from the first plus higher order. The same parameters set (1) in Fig. 1 are used in both cases. (c) is the result obtained from the first plus higher order optical potential with the modified parameters of the set (2) in Fig. 1.

This kind of higher order potential term may be possible to absorb approximately into the effective one-body potential $V_{\Delta A-1}$, although this effect is quite large. The character of the strong dependence of V_o and W_o on the energy cannot however be removed by this kind of higher order correction.

---

*Work supported in part by the U.S. Energy Research and Development Contract E(11-1)-3235.

FIGURES

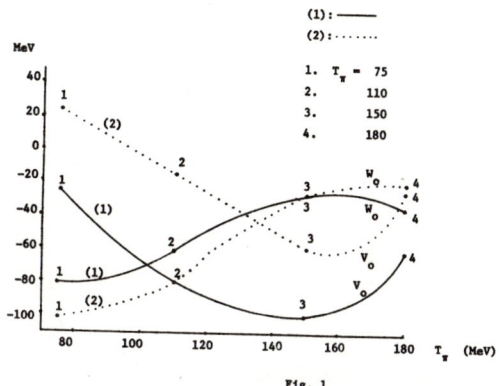

Fig. 1. Energy dependent of effective
Δ-residual nucleus potential parameters.

$$V_{\Delta A-1} = (V_o + iW_o) e^{-\beta \gamma^2}$$

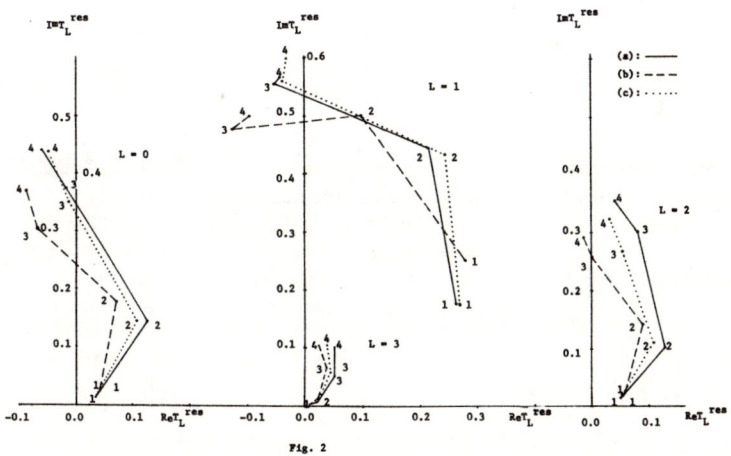

Fig. 2. Argand plots for the resonance amplitude of pion $-^4$He scattering.

REFERENCES

1. F. Lenz, invited talk to this Conference.

THE EFFECT OF THE EXCLUSION PRINCIPLE ON THE Δ-ISOBAR

B. L. Friman*
State University of New York at Stony Brook, New York 11794

ABSTRACT

We study the effect of a surrounding Fermi sea of non-interacting nucleons on the Δ-isobar, in a simple model, derived from Chew-Low theory.

If one thinks of the isobar as a composite system of a pion and a nucleon, one expects Pauli corrections because some processes are forbidden by the exclusion principle. In addition some many-nucleon processes are allowed in a Fermi sea[1]. To evaluate the sum of these corrections, which we call the effect of the exclusion principle, we study a simple model for the isobar, where it is easy to calculate the corrections.

The Chew-Low equation[2] sums up an infinite set of πN-scattering graphs to make up the amplitude. It has been shown[3] that one should identify the isobar with the sum of all these terms, except the Born-term. Our model for the isobar consists of keeping just the lowest order terms in this expansion (see Fig. 1).

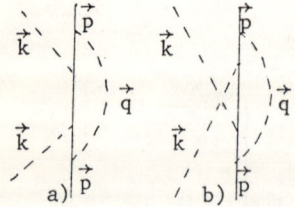

Fig. 1. The diagrams included in our model for the isobar.

In the nuclear medium we get analogous terms, with the free propagators replaced by

$$G(\vec{p},\varepsilon) = \frac{1-n(\vec{p})}{\varepsilon - \varepsilon_{\vec{p}} + i\delta} + \frac{n(\vec{p})}{\varepsilon - \varepsilon_{\vec{p}} - i\delta} \quad (1)$$

where $\varepsilon_p = \frac{\vec{p}^2}{2m}$. Thus we get a density dependent scattering amplitude $f(\rho)$. In the limit $\rho \to 0$ it reduces to the free amplitude. In Fig. 2 we show some of the graphs contributing to the isobar in a medium of nucleons.

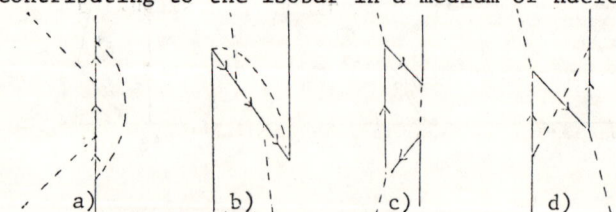

Fig. 2. Some of the graphs contributing to the density dependent scattering amplitude. Backward-going lines are holes in the Fermi sea.

---

*Supported by USERDA under Contract No. E(11-1)-3001; on leave from Åbo Akademi, 20500 Åbo 50, Finland.

The effect of the exclusion principle is repulsive, the relative size of it $\frac{f(\rho)}{f(0)} - 1$ is given in Table I for nuclear matter density $\rho_0 = 0.5 m_\pi^3$, and for $\tfrac{1}{4}\rho_0$, at threshold and at the resonance, $\omega = 2m_\pi$.

Table I. The effect of the exclusion principle for $\rho=\rho_0$ and $\rho=\tfrac{1}{4}\rho_0$, at $\omega=m_\pi$ and $\omega=2m_\pi$.

	$\omega=m_\pi$	$\omega=2m_\pi$
$\rho=\rho_0$	0.34	0.13
$\rho=\tfrac{1}{4}\rho_0$	0.13	0.04

At threshold only graphs a and b of Fig. 2 (and their respective crossed graphs) contribute to the amplitude, because all intermediate states have the same momentum. In the static limit the contribution from Fig. 2b is equal to the difference between Fig. 2a and 1a, i.e. the Pauli-blocking of intermediate particle states. Thus they both give 17% repulsion. This is consistent with the results of ref.[4]. These authors found 15% repulsion from the exclusion principle at ρ_0 but they did not evaluate the many-nucleon corrections.

The effect would be diminished by the inclusion of nucleon-nucleon short range correlations[1]. Also inclusion of the ρ-meson might cut down the effect, but one has to solve the Chew-Low equation taking the ρ-meson into account first to see how this affects the value of the cut-off mass. A calculation along these lines is presently in progress.

REFERENCES

1. G.E. Brown and W. Weise, Phys. Report 22, 279 (1975).
2. G.F. Chew and F.E. Low, Phys. Rev. 101, 1570 (1956).
3. S. Barshay, G.E. Brown and M. Rho, Phys. Rev. Lett. 32, 787 (1974).
4. C.B. Dover, D.J. Ernst and R.M. Thaler, Phys. Rev. Lett. 32, 557 (1974).

A STUDY OF THE DOORWAY ISOBAR MODEL FOR PIONS*

L. S. Kisslinger and A. Saharia
Carnegie-Mellon University, Pittsburgh, PA 15213

In the Doorway-Isobar Model[1,2] for pion-nucleus interactions, one explicitly introduces the pion-nucleon isobars (Δ's here) as nuclear constituents. States consisting of such isobars and nuclear hole configurations form doorways through which all inelastic reactions involving the Δ-resonances are assumed to take place. Although one is not able to carry out accurate calculations of the properties of such states, it has been possible to express the pion-nucleus T-matrix and the optical potential in terms of quantities which are well defined, given a nuclear model and a model for the pion-nuclear dynamics.

The form of the pion-nucleus optical potential in momentum space averaged over doorways is[2]

$$\langle k'|V|k\rangle = \langle k'|V^{N.R.}|k\rangle + \frac{\langle k'|t|k\rangle [E - M_\Delta + i\Gamma_\Delta/2]}{E - M_\Delta + \Delta E + i\Gamma^{in}(E)/2} F(\lambda, \underline{k}', \underline{k}), \quad (1)$$

with M_Δ and Γ_Δ the free Δ mass and width. The quantities ΔE and $\Gamma^{in}(E)^\Delta \equiv \beta\Gamma_\Delta(E)$ are the average energy shift of the doorway due to the difference in the nuclear vs. Δ binding and the inelastic width, respectively. The function $F(\lambda, \underline{k}', \underline{k})$ is a nuclear form factor modified by the Δ, with the parameter λ representing the nonlocality arising from the effective Δ-propagation. In the closure approximation $\lambda \to 0$ and $F(\lambda, \underline{k}', \underline{k})$ just becomes the nuclear form factor (Fourier transform of the density).

In previous work the closure approximation was used because of the numerical difficulty in evaluating the expression (1). In the present work we have analyzed the π-^{12}C elastic scattering from 50 MeV over the resonance region including a Gaussian nonlocality parameterized by λ. In all cases the parameter $\Delta E = 10$ MeV is satisfactory. This indicates that the Δ is essentially unbound in nuclear matter as was pointed out in the original work using closure[1]. The values of β approximately equal to unity are consistent with this, with $\beta = 1.1$, indicating that the Δ-width is increased by about 10% in nuclear matter. The striking new results are the large departure from closure at lower energy. The value of $\lambda = 0.9$ Fm indicates a strong nonlocality. As the energy increases, presumably more doorway states enter and the closure results are once more obtained.

REFERENCES

1. L. S. Kisslinger and W. L. Wang, Phys. Rev. Lett. 30, 1071 (1973).
2. L. S. Kisslinger and W. L. Wang, Ann. Phys. (N. Y.) (1976).

*Supported in part by the NSF

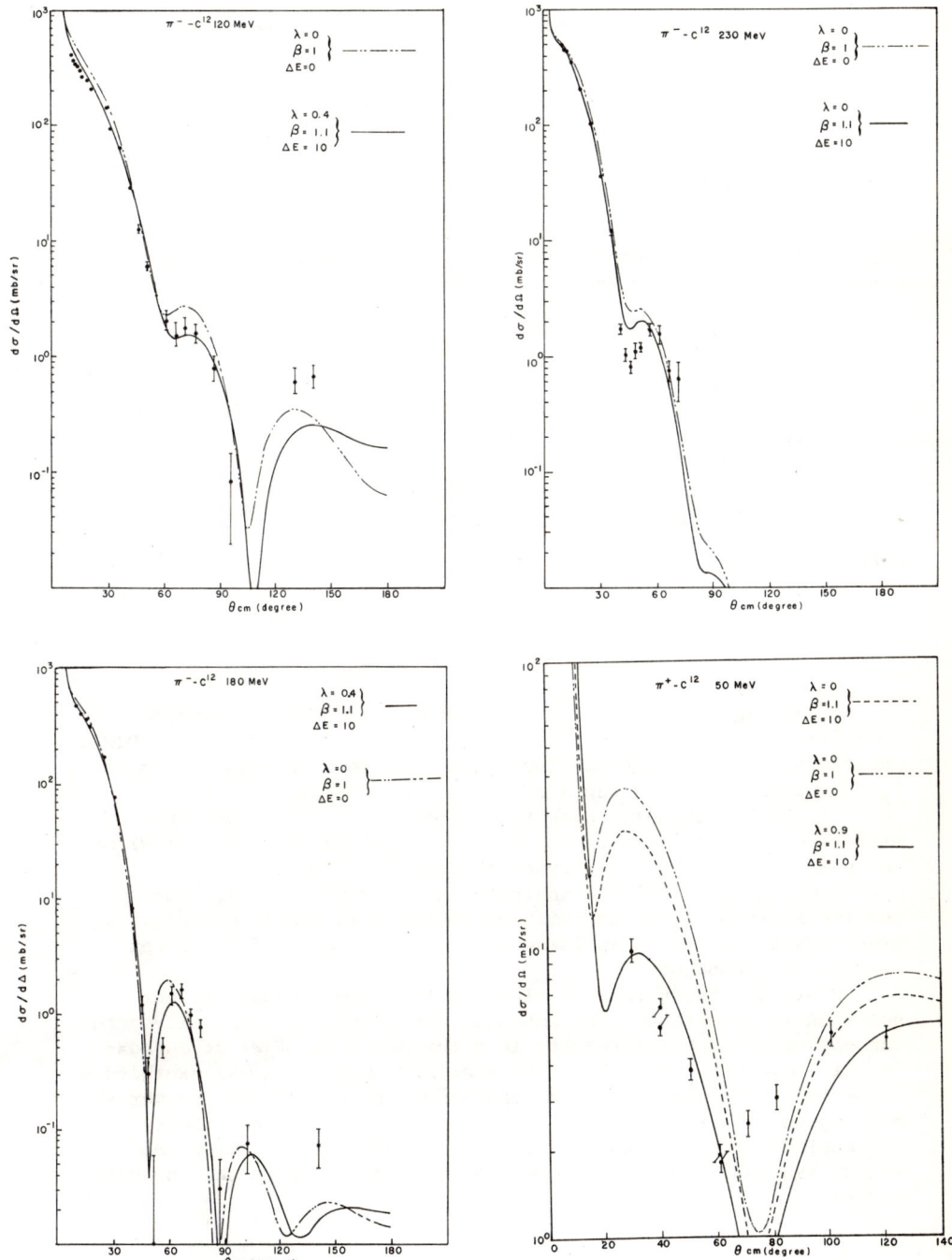

RESONANCE DAMPING IN THE INTERACTION OF PIONS WITH NUCLEONS IN NUCLEAR MATTER

L. C. Liu, W. T. Nutt and C. M. Shakin
Department of Physics, Brooklyn College, Brooklyn, N.Y. 11210

ABSTRACT

A calculation of the pion-nucleon interaction in nuclear matter including effects of propagator modification results in a significant damping width for the (3,3) resonance.

When a pion interacts with a nucleon in nuclear matter, one expects to see effects due to Pauli blocking and collision broadening. In this work we report on calculations in which we have included these features in determining an effective π-N interaction. In addition to using the Pauli exclusion principle to limit the states available to the nucleon, we also introduce self-energy terms for the pion and nucleon propagators in the T matrix equation which describes the π-N interaction in the medium. These terms represent effects due to collisions of the pion and nucleon with other nucleons in the medium. We can express the T matrix in the medium, T, in terms of the free T matrix, T_f, and the difference between the π-N propagator in the medium and the free propagator,

$$T_N = T_f + T_f (G_N - G_F) T_N$$

If the free space T matrix is approximated by a separable form specified by a form factor, $v(q)$, and a denominator function, $D(W)$, one finds that the solution for T_N has the same separable form as T_f except that $D(W)$ is replaced by another function, $\mathcal{D}(W)$.

In the P_{33} channel, near the resonance, $D(W)$ can be parameterized by the resonance position and a width, Γ. Then $\mathcal{D}(W)$ may be related to $D(W)$ by specifying the shift in the resonance position, $E_R \to E_R + \Delta$, and the increase in the width $\Gamma \to \Gamma + \Gamma\downarrow$. Our results indicate that the shift parameter Δ is small, but $\Gamma\downarrow$ is significant and may be as large as $\Gamma/2$. The value of $\Gamma\downarrow$ is not very model dependent.

Near the threshold ($\sqrt{s} \sim 1078$ MeV) a second aspect of our solution is important. The introduction of the nucleon self-energy (an optical potential) results in a threshold for $\mathcal{D}(W)$ at approximately 1010 MeV. Therefore the imaginary part of $\mathcal{D}(W)$ exhibits a significant increase over the imaginary part of $D(W)$ in the threshold region. This feature is necessary for a consistent theory if one includes the effects of nucleon binding in calculating the energy available for the fundamental π-N collision in the nuclear medium.

PION-CONDENSATION AND SHORT-RANGE CORRELATION IN NUCLEI

W. T. Weng*
University of Arizona, Tucson, AZ 85721

T. T. S. Kuo and G. E. Brown
State University of New York, Stony Brook, NY 11794

Should there be pion-condensation at normal nuclear matter density, the T=1, 0^- states of ^{16}O would be shifted away from the conventional RPA predictions. It has been pointed out that the closeness of the energy of the first T=1, 0^- state of ^{16}O, 12.78 MeV, to that of unperturbed energy, 12.32 MeV, is an indication of the absence of pion-condensation in finite nuclei.[1] The shifts can be evaluated by the approximation that the particle-hole pairs interact with real pion through a static one-pion-exchange potential. Then the confusion arises from the incorrect treatment of two-body wavefunctions in nuclei. Due to the strong short-range repulsion between nucleon-nucleon interaction the contact term in OPEP plays no role at all in the coupling of pion to nuclear particle-hole pairs.[2]

We performed three calculations to the energy shifts of two T=1, 0^- states of ^{16}O obtained from RPA: (1) with full OPEP and uncorrelated two-body wavefunctions, (2) with OPEP without delta term and uncorrelated wavefunctions, and (3) with full OPEP and correlated two-body wavefunctions. The results of the calculations are given in table I. It is clear that the consistent calculation (3) gives little change to the energies of T=1, 0^- states of ^{16}O. This result rules out the possibility of pion-condensation at normal nuclear matter density.

Table I OPEP matrix elements between two PPA states of ^{16}O

Case	Interaction	Wavefunction	Matrix Element		
			11	12	22
1	V_{OPEP}	uncorrelated	-3.98	1.32	-10.26
2	$V_{OPEP} - \delta$	uncorrelated	.25	-.16	.63
3	V_{OPEP}	correlated	.47	-.66	1.02

*Work supported by the National Science Foundation (Grant No. MPS75-07320).

References

1. S. Barshay and G. E. Brown, Phys. Lett. 47B (1973) 107.
2. W. T. Weng, T. T. S. Kuo and G. E. Brown, Phys. Lett. 46B (1973) 329.

CONTRIBUTION OF THE MESON-EXCHANGE CURRENTS TO CHARGE DENSITY OF ^{16}O

Il-T. Cheon
Department of Physics, McMaster University
Hamilton, Ontario, Canada

ABSTRACT

It was recently reported that contributions of meson-exchange currents, i.e., one-pion exchange and $\rho(\omega)$-π exchange, were significant to the electromagnetic form factors of the two- and three-nucleon systems [1]. In the framework of the independent pair model [2], we have investigated the contributions of the meson-exchange currents to the charge density and rms radius of ^{16}O. The distribution of the center-of-mass of the nucleon pair was assumed to be the same as the proton distribution obtained by the Hartree-Fock calculation [3]. Our numerical results show that the one-pion exchange current is very important in calculation of the nuclear charge density. In calculation of $\Delta\rho_n(r)$, we used the dipole type distribution for the neutron charge [4]. The rms radius of ^{16}O is suppressed by 8.93%, 0.81% and 0.25%, respectively by the contributions of one-pion exchange current, neutron charge and ρ-π exchange currents. Hence, it turns out that the meson exchange currents should be included in the Hartree-Fock calculation of the nuclear charge density.

[1] M. Chemtob et al., Phys. Rev. $\underline{C10}$('74)344.
W. H. Kloet et al., Phys. Lett. $\underline{49B}$('74)419.
A. D. Jackson et al., Phys. Lett. $\underline{55B}$('75)23.
[2] Il-T. Cheon, to be published in Phys. Lett. (1976).
[3] X. Campi et al., Nucl. Phys. $\underline{A194}$('72)401.
[4] W. Bertozzi et al., Phys. Lett. $\underline{41B}$('72)408.

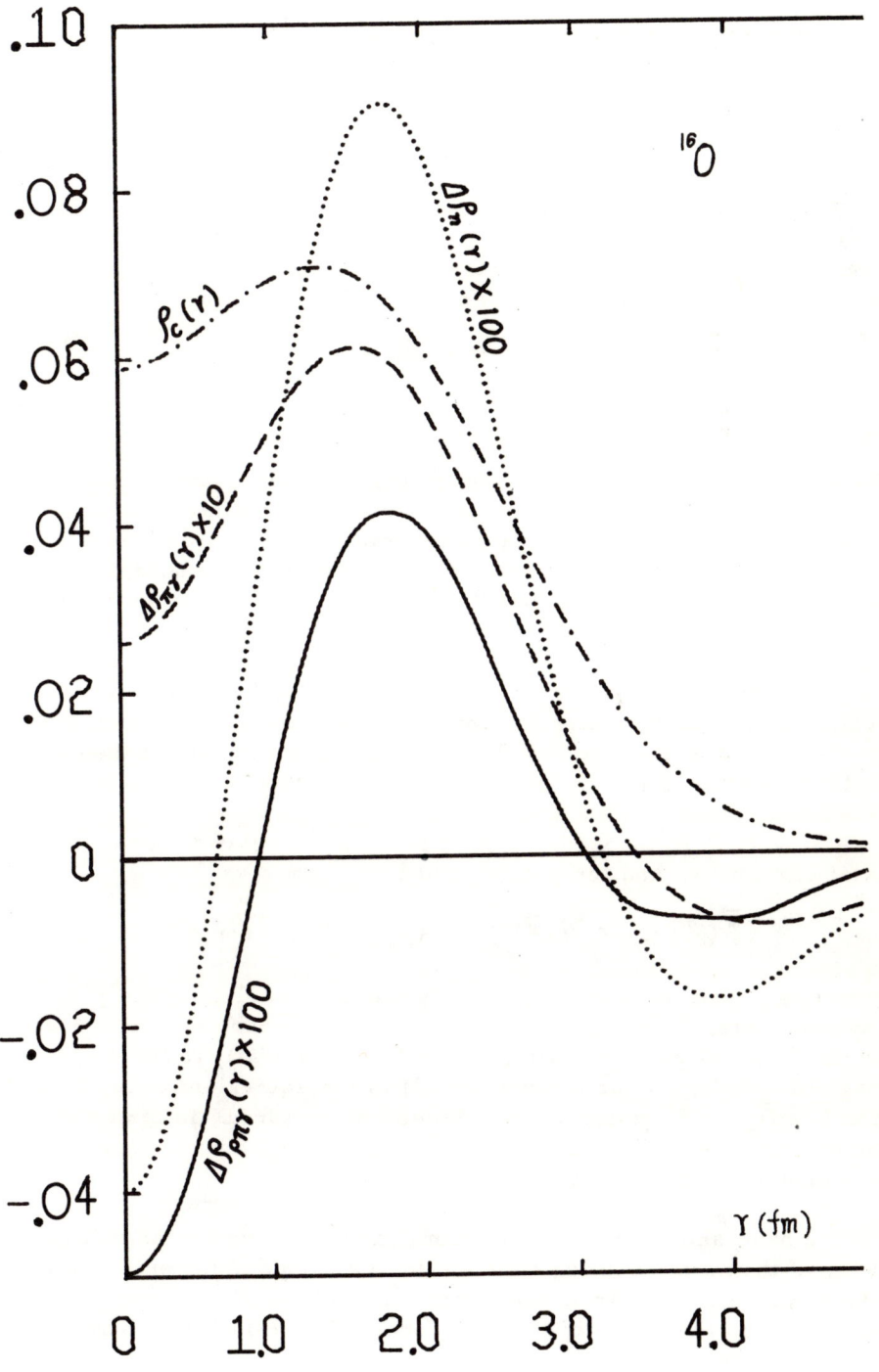

AXIAL POLARIZABILITY AND PIONIC FIELD IN NUCLEI

J. Delorme, M. Ericson, A. Figureau and C. Thévenet
Institut de Physique Nucléaire de Lyon, Université Claude Bernard
69621 Villeurbanne, France.

The effect of the isobaric excitations on the weak axial coupling constants in nuclei has been studied through P. C. A. C. We have first established the Klein-Gordon equation for the virtual pion field in the nucleus ; it takes into account pion rescattering :

$$[-\vec{\nabla}(1+\alpha(\vec{z}))\cdot\vec{\nabla} + m_\pi^2]\varphi(\vec{z}) = -\frac{g_r}{\sqrt{2}M_N}\vec{\nabla}\cdot[\eta(\vec{z})\vec{\sigma}(\vec{z})] \quad (1)$$

where g_r is the pion-nucleon coupling constant, m_π and M_N the pion and nucleon masses, and $\vec{\sigma}(\vec{x})$ the spin-isospin transition density of the nucleus. The influence of isobar excitation is contained in the axial polarizability coefficient α which is linked to the p-wave π-N scattering volume ; the vertex renormalization factor η is a function of α but it depends also on the nucleon-nucleon correlation function. This equation has been derived in a way which stresses its analogies with electromagnetism : indeed, in the limit $m_\pi = 0$, it corresponds to the equation obeyed by the potential created by a distribution of dipoles (here $\vec{\sigma}(\vec{x})$) in a dielectric medium (the quantity $1 + \alpha$ plays here the role of the dielectric constant ε).

Using P. C. A. C. we have obtained a basic relation between the axial current $\dot{\vec{A}}$ and the pionic field :

$$\vec{A}(\vec{z}) = -g_A \eta(\vec{z})\vec{\sigma}(\vec{z}) + f_\pi[1+\alpha(\vec{z})]\vec{\nabla}\varphi(\vec{z}) \quad (2)$$

where g_A and f_π are respectively the Gamow-Teller and pion decay constants. Here again the effects of the isobars are incorporated in the axial polarizability, which leads naturally to an electromagnetic analogy : our expression (2) is the equivalent of the relation $\vec{D} = \vec{D}_o + \varepsilon\vec{E}$ giving the displacement vector $\vec{D}$ in terms of the electric field $\vec{E}$ in a dielectric medium which possesses a spontaneous polarization $\vec{D}_o$.

We have shown that this relation leads in heavy nuclei to a quenching of the axial coupling constant by the Lorentz-Lorenz factor η which may originate from the short range of the Pauli correlations, depending on the range of the π-N forces. Therefore this quenching may have a different origin than the existence of short-range correlations and may arise from a Pauli blocking effect. On the other hand, the pseudoscalar coupling constant is found to be strongly suppressed.

In finite nuclei, these basic quenchings can be masked by surface effects, the general features of which have been studied with the help of a solvable model. The same model has been further used to obtain the asymptotic pion field which is linked to the effective pion-nucleus coupling constant and can be determined experimentally through π-nucleus dispersion relation. We have found that this quantity is quenched, in agreement with recent experimental data.

ANALYTICITY AND PION-NUCLEUS SCATTERING

Institute of Theoretical Physics, Warsaw University

O. Dumbrajs[+]

ABSTRACT

Some more refined methods than simple forward dispersion relations are suggested for studying the pion-nucleus scattering.

The hypothesis of analyticity of the scattering amplitude automatically provides a relativistic framework for studying correlations and interpretations of experimental data. In the case of the pion-nucleus scattering this hypothesis has been exploited so far only in its simplest form: by evaluating the forward dispersion relations for the scattering amplitude (see reviews [1,2] and papers [3,4]). In this way the real parts of the forward $\pi^4 He$, $\pi^6 Li$ and $\pi^{12} C$ scattering amplitudes have been calculated and values of the effective coupling constants $\pi^7 Li$ and $\pi^9 Be$ have been estimated using experimental data on the corresponding total cross sections and on the real parts from relevant phase shift analyses and from mesic atoms as the input.

We would like to call attention to some other, more refined, ways of exploiting the hypothesis of analyticity which are well known in particle physics and which could be of use also in pion-nucleus scattering.

(i) Derivative sum rule[5] as a consistency test of

[+] On leave of absence from the Institute of Nuclear Physics, Moscow State University,

the low-energy pion-nucleus scattering models. It provides a non-trivial consistency test[6] even when evaluated at a single energy, since the value of coupling constant is not involved in calculations.

(ii) Dispersion relations for the logarithm of the amplitude[7]. These dispersion relations correlate data on the phase and on the modulus of the scattering amplitude and bring into a play the unjustifiably ignored important characteristics of the low-energy scattering: zeros of the amplitudes[8]. In pion-nucleus scattering these dispersion relations turn out to be useful[9] not only in those cases when no data on total cross sections exist (e.g. π^3He scattering), but also in cases already examined by ordinary dispersion relations because they make use of experimental information on differential cross sections ignored in the usual approach.

REFERENCES

1. T.E.O.Ericson, M.P.Locher, Nucl. Phys. $\underline{A148}$, 1 (1970).
2. S.Dubnička, O.Dumbrajs, Phys. Reports $\underline{19C}$, 141(1975)
3. H.Pilkuhn, N.Zovko, H.G.Schlaile, Karlsruhe University preprint TKP 27/75.
4. S.Dubnička, V.A.Meshcheryakov, JINR preprint E2-9399
5. N.M.Queen, S.Leeman, F.E.Yeomans, Nucl. Phys. $\underline{B11}$, 115 (1969).
6. O.Dumbrajs, (in preparation).
7. R.Odorico, Nuovo Cimento $\underline{54A}$, 96 (1968).
8. O.Dumbrajs, M.Staszel, Journal of Phys. $\underline{G1}$, 172 (1975).
9. O.Dumbrajs, (in print).

TEST OF THE NEW THRESHOLD EXPANSION OF THE PION-NUCLEUS S-WAVE AMPLITUDE BY MEANS OF DISPERSION RELATIONS

H. Pilkuhn and H.G. Schlaile
University of Karlsruhe, Federal Republic of Germany

N. Zovko
Institute Ruder Boskovic and University of Zagreb, Jugoslavia

ABSTRACT

The new threshold expansion of the crossing-symmetric s-wave scattering amplitude is qualitatively confirmed by means of dispersion relations for the forward scattering amplitude.

Recently it has been shown[1] that the threshold expansion of the s-wave amplitude of pion scattering on a nucleus of zero isospin should be of the form

$$k^{-1} \tan \delta_o = A + k^2 B, \quad \text{Im } B = 0 \qquad (1)$$

rather than the conventional effective range expansion, with an effective range $- 2B/A^2$. The argument was based on the empirical validity of the corresponding formulas in pion-nucleus scattering, and the optical potential of low-energy pion-nucleus scattering was modified accordingly, by making the parameter b_o in the charge-symmetric local part of the potential linear in k^2:

$$b_o(k^2) = b_o - \frac{1}{3} 0.1 \, k^2 m_\pi^{-3} \qquad (2)$$

For $\pi-{}^{12}C$ scattering, a value of $B = -0.57 \text{ fm}^3$ was predicted.

We have examined the forward elastic scattering data of pions on ${}^4\text{He}$, ${}^6\text{Li}$, ${}^7\text{Li}$, ${}^9\text{Be}$ and ${}^{12}\text{C}$ by means of dispersion relations, with the aim of extracting the parameters of the threshold expansion[2]. Below the threshold ω_n for neutron emission the charge-symmetric part of the forward scattering amplitude is taken as a sum of s-wave and p-wave parts

$$F^{(+)} = \left[1 - ik(A + k^2 B)\right]^{-1} (A + k^2 B) + 3k^2 (A_1^{-1} - ik^3) \qquad (3)$$

and the complex scattering lengths A as well as the imaginary parts of the p-wave scattering volumes A_1 are taken from the complex energy shifts of pionic atoms[3] if necessary by interpolation. The dispersion integral over Im $F^{(+)}$ is carried down to $\omega^2 = m_\pi^2 - k^2 = (2\omega_n)^2$ with an extra factor $(\omega - 2\omega_n)(m_\pi - 2\omega_n)^{-1}$ to account for the fact that pion absorption in mesic atoms leads to the emission of at least two nucleons. A possible one-nucleon emission is included in a phenomenological pole far below threshold.

The resulting values of A, B and A_1 are collected in Table 1. The only fitted parameters are B and Re A_1. The negative sign of B confirms the model of[1], but the expected systematic rise of B with baryon number is not seen.

Table 1: Values of the parameters A, B and A_1 as deduced from pionic atoms[3] and dispersion relations.

target	A[fm]	B[fm]	A_1 [fm^3]
^{4}He	-.138 + .04i	-.34	.37 + .06i
^{6}Li	-.183 + .055i	-.37	.69 + .12i
^{7}Li	-.235 + .065i	-.58	1.08 + .17i
^{9}Be	-.393 + .076i	-.38	1.14 + .24i
^{12}C	-.451 + .132i	-.34	1.52 + ..34i

REFERENCES

1. S. Barmo and H. Pilkuhn, Physics Letters 6oB (1976), 324
2. H. Pilkuhn, N. Zovko, H.G. Schlaile, Karlsruhe preprint (1975)
3. J. Hüfner, L. Tauscher, C. Wilkin. Nucl.Phys. A 231 (1974),455
   G. Backenstoss, Ann.Rev.Nucl.Sci. 2o (197o), 467.

ARE PIONIC EXCHANGE-CURRENT CONTRIBUTIONS TO $p + p \rightarrow {}^2H + e^+ + \nu_e$ WELL DETERMINED?

H. S. Picker*

Physics Department, Trinity College, Hartford, Connecticut 06106

ABSTRACT

In connection with the puzzle of the missing solar neutrinos, it is important to know whether current theoretical estimates of the rate of the unmeasurably slow proton-proton reaction, $p + p \rightarrow {}^2H + e^+ + \nu_e$, are reliable to within 10%, as assumed in construction of solar models. I find that a unitary transform of range 1 fm applied to typical phenomenological ${}^1S_0$ p-p and ${}^3D_1$ deuteron radial functions can increase the S-D non-Born pionic exchange-current term, which dominates the interaction-current correction to the impulse approximation, by a factor of five. The resulting rate is 40% higher than the presently accepted value.

THE PROTON-PROTON REACTION RATE

Aside from the Gamow-Teller coupling constant and the positron Fermi function, the rate of the proton-proton reaction is determined by the magnitude-squared of a dimensionless radial matrix element. In impulse approximation, this is

$$\Lambda_{imp.}(E) = F(k) \int_0^\infty dr \; u_d(r) \; w_{pp}(k,r), \qquad (1)$$

where $E = k^2$ is the laboratory energy in fm^{-2}, $F(k)$ summarizes constants including the Coulomb Gamow factor and the deuteron range, and u_d and w_{pp} are deuteron S-wave and p-p radial functions. As Haftel and I have recently shown[1], at the kilovolt energies appropriate to the center of the sun, the value of the magnitude-squared of the RHS of (1) can be enhanced by at most 6% by short-range modifications of typical u_d and w_{pp}, such as those generated by the Reid soft-core potential[2]. However, Gari and Huffman[3] find that exchange-current corrections to (1) increase the rate by nearly 10%. By far the most important correction arises from the S-D "non-Born" pionic current term; it is

$$\Lambda_{SDNB}(E) = F(k)(2\sqrt{2}/3) \, \xi \, [\alpha(0) + \tfrac{1}{2}\gamma(0)] \int_0^\infty dr \; w_d(r) \; h_2(i\mu r) \; w_{pp}, \qquad (2)$$

where $h_2(ix) = (1 + 3/x + 3/x^2) x^{-1} \exp(-x)$, w_d is the deuteron D-wave radial function, and the interaction constants $\xi\alpha(0) = (.0614) \times (0.7)$ and $\xi\gamma(0) = (.0614) \times (3.1)$ are those used in Ref. 3. (Note two misprints in Ref. 3: in Table 1, the third-column, second-row entry

---

*Work supported by a Research Corporation Cottrell College Science Grant and by a Trinity College Junior Faculty Summer Research Grant.

should contain $Y_2(\mu r)$, not Y_0; and on the RHS of eq. (8), a minus sign should precede the factor $\sqrt{2}/3^{4)}$.) Since $h_2(i\mu r)$, with $\mu = 0.7$ fm^{-1}, emphasizes the small-r features of w_d and w_{pp}, I wondered how much the RHS of (2) would change when the standard phenomenological radial functions used in Ref. 3 are replaced by their images under short-range unitary transforms which parametrize our ignorance of the details of the two-nucleon interaction inside about 1 fm. To find out, I applied the transform

$$\tilde{w}(r) = w(r) - 2 g(r) \int_0^R dr' \, g(r') \, w(r') , \quad (3)$$

where $R = 1$ fm, to both w_d and w_{pp}. For $g(r)$ I took

$$g(r) = g_6 \, r^3 \, (r - R)^3 \, (1 - \theta(r-R)), \quad (4)$$

where g_6 is determined by the normalization $\int_0^\infty dr \, g^2(r) = 1$, as required if (3) is to be a unitary transform. This form ensures smooth matching at $r = R$, since $g(R) = g'(R) = g''(R) = 0$. To avoid possible large and meaningless contributions from the short-range part of the pionic non-Born current, I set this current to zero for $r < 0.4$ fm. For $E = 12$ KeV (lab energy), corresponding to the Gamow peak at the center of the sun, I obtain the results shown in Table I.

Table I. Proton-proton reaction rates

	Impulse only	Exchange current, no transform	Exchange current, transform
Λ^2(12 KeV)	7.16	7.63	9.84

By comparison, the values of $\Lambda^2(E=0)$ (essentially the same energy) obtained in Ref. 3 are 7.23 for the impulse term using Reid soft-core radial functions, as above, and 7.79 with all exchange corrections.

Clearly, the result in the third column of Table I should not be taken too seriously. It does indicate that work remains to be done on exchange-current corrections to the proton-proton reaction rate.

I thank J. P. Lavine for many helpful letters and conversations. Support from Research Corporation and from Trinity College is gratefully acknowledged, as is computing time provided by Trinity College.

REFERENCES

1. H. S. Picker and M. I. Haftel, to be published.
2. R. V. Reid, Ann. Phys. (N.Y.) 50, 411 (1968).
3. M. Gari and A. H. Huffman, Ap. J. 178, 543 (1972).
4. J. P. Lavine, private communication.

EFFECT OF ANTI-SYMMETRIZATION ON BACK-ANGLE p-d SCATTERING*

Leonard S. Kisslinger and Chi-Shiang Wu
Carnegie-Mellon University, Pittsburgh, Pennsylvania 15213

ABSTRACT

We have examined the effect of anti-symmetrized p-p amplitude on the p-d scattering and found that the back angle scattering must be treated more carefully than having been done in the conventional approach.

For a free p-p scattering, assuming that a non-symmetrized amplitude is given by $T_{pp}(s,t)$, the properly anti-symmetrized amplitude would be

$$\hat{T}_{pp}(s,t) = T_{pp}(s,t) \pm T_{pp}(s,u), \qquad (1)$$

s, t, and u are the Mandelstam variables and the sign between the two terms depends upon the spin state. At high energies, the forward scattering is dominated by the first (direct) term while the backward scattering is dominated by the second (exchange) term. In the conventional impulse approximation one writes the pp contribution to the p-d single scattering amplitude as

$$T_p(s,t) = \hat{T}_{pp}(s',t) \, F(\tfrac{t}{4}) \qquad (2)$$

where F is the deuteron form factor and $\sqrt{s'}$ is an effective scattering energy to be determined by an optional choice of kinematics. For example, the initial proton in the deuteron is considered as at rest[1], or the proton momentum is taken at the center of overlap of the initial and final deuteron wave function.[2] The purpose of the present paper is to test if such uniform choice of s' can produce a reasonable approximation. The more accurate form of the impulse approximation for the p-p contribution to the p-d scattering can be shown to be

$$T_1(s,t) = \int d^3q \; \phi_d^*(\vec{q}-\tfrac{\vec{\Delta}}{2}) \; T_{pp}(s_1,t_1) \; \phi(\vec{q}) \qquad (3)$$

for the direct term, and

$$T_2(s,t) = \int d^3q \; \phi_d^*(\vec{q}-\tfrac{\vec{\Delta}}{2}) \; T_{pp}(s_1,u_1) \; \phi(\vec{q}) \qquad (4)$$

for the exchange term. s_1, t_1 and u_1 are invariant variables for the two protons with nucleon motion and recoil fully taken into account, and $\vec{\Delta}$ is the momentum transfer. For an unambiguous determination of s_1 and u_1, we take the spectator neutron on the mass shell. Note that $t_1=t=-\vec{\Delta}^2$ and that T_{pp} is relatively insensitive to s_1, and therefore the small-angle scattering dominated by T_1 may be well approximated by (2). However, for the back angle scattering that is dominated by T_2, u_1 is a function of $\vec{q}$ and T_{pp} varies considerably in the region ϕ is significant. (Note that arguments of the two ϕ's are wide apart for large $\vec{\Delta}$.) Fig. 1 and 2 show T_2 of (4) along with its factorized value when s' is determined at $\vec{q}=\Delta/4$ for incident proton beam momentum of 1.7 and

3.3 GeV/c, respectively. The S-wave Hulthen wave function was assumed. The simple factorization over-estimates the backward exchange amplitude by a factor of about 2. Thus the cross-section is over-estimated by a factor of abour 4. This difference may increase with the energy. Considering such a large discrepancy between the two, one should be very careful in treating the high energy backward scattering, in particular, when one tries to draw a conclusion on the possibility of estimating N* contributions in comparison with experimental data[2].

REFERENCES

1) Bertocchi and Capella, Nuovo Cim. 51A , 369 (1967)
2) S. A. Gurvitz and A. S. Rinat, Phys. Lett. 60B, 405 (1976)

*Supported in part by the NSF

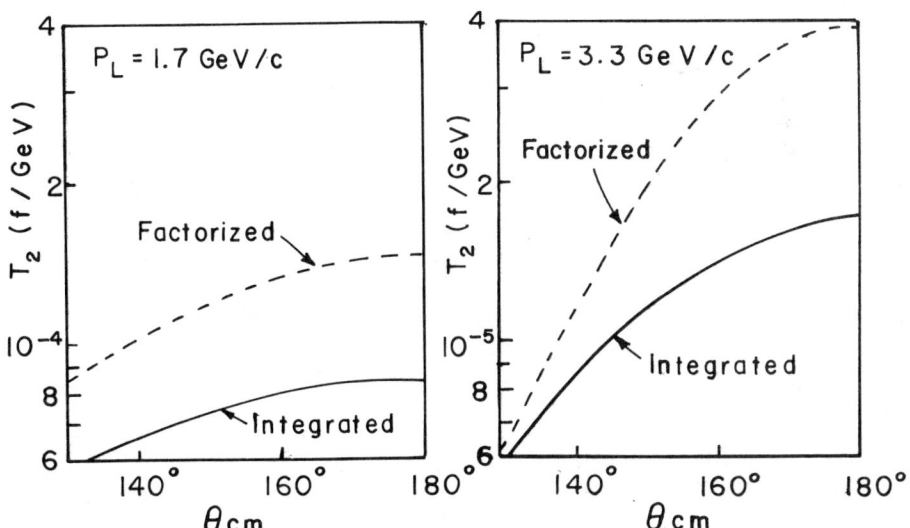

Reanalysis of the A=8 mirror asymmetry

Kuniharu Kubodera

Department of Physics, University of Tokyo, Tokyo, JAPAN

Recent β-decay correlation experiments for various nuclear systems[1-4] have intensified the controversy on the existence of second-class currents (SCC) in weak currents. It was emphasized in Ref.5 that in the discussion of SCC in nuclear β-decays it is extremely important to take account of nuclear many-body effects such as off-energy-shell and meson-exchange effects. These effects can also affect the asymmetry δ of the ft-values of mirror β-decays. In the impulse approximation which ignores them, the smallness of the energy-dependence of δ experimentally found by Wilkinson and Alburger (WA) sets a stringent limit on the strength of the possible SCC.[6] An improved analysis[5] allows for larger SCC but still the result does not seem concerting with those correlation experiments that speak for the SCC of appreciable magnitude.

A recent experiment[7] has revealed that the Gamow-Teller matrix element in the A=8 system may be energy-dependent. This can bring in an additional energy-dependence of δ which influences the above analysis. We have presented[8] a simple argument to explain the origin of the non-constancy of the G-T matrix element. Furthermore, a reanalysis of the WA data taking into account this non-constancy has been attempted[9]. A typical result is shown in Fig.1. It turns out that the WA data do not necessarily imply small SCC.

REFERENCES

1. K.Sugimoto et al.,Phys.Rev.Lett.__34__,1533(1975).
2. F.Calaprice et al.,Phys.Rev.Lett.__35__,1566(1975).
3. R.Tribble and G.Garvey,Phys.Rev.__C12__,967(1975).
4. F.Calaprice,Phys.Rev.__C12__,2016(1975).
5. K.Kubodera, J.Delorme and M.Rho, Nucl.Phys.__B66__,253 (1973).
6. D.Wilkinson and D.Alburger,Phys.Rev.Lett.__26__,1127(1971).
7. A.Nathan et al.,Phys.Rev.Lett.__35__,1137(1975).
8. K.Kubodera and A.Arima, to be published.
9. K.Kubodera, to be published.

Figure 1.

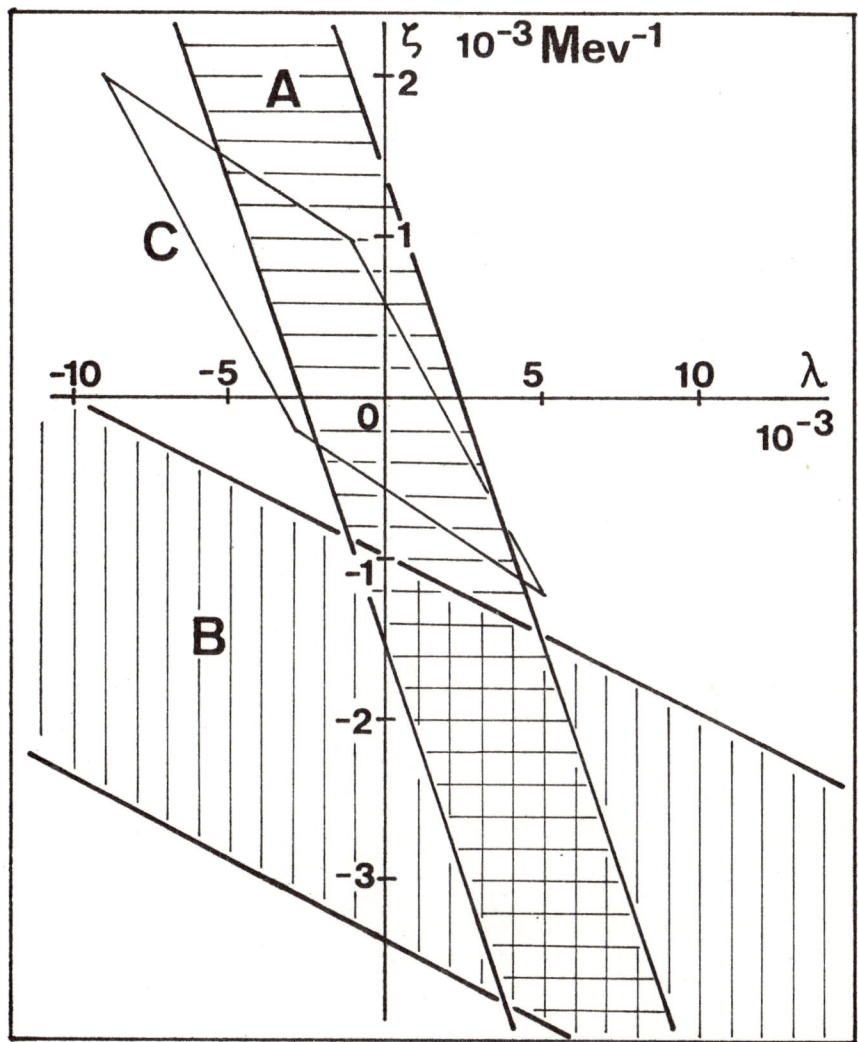

Limits on the SCC parameters ζ and λ (see Ref.5 for definition) determined from the WA experiment; region A corresponds to the result with the energy-dependence of the Gamow-Teller matrix element taken into account while region C to the result without energy-dependence. Region B represents the restriction Sugimoto et al.'s [1] experiment puts on ζ and λ.

IV. PION PRODUCTION AND ABSORPTION ON NUCLEI

PION ABSORPTION AND PRODUCTION EXPERIMENTS

Elie Aslanides

Centre de Recherches Nucléaires, 67037 Strasbourg, France
and
Université Louis Pasteur, Strasbourg, France.

ABSTRACT

The experimental situation on pion absorption and production, as established by recent $(\pi, \pi N)$, (π, NN), $(\pi, \text{charged } X)$, $(\pi, \gamma X)$, (π, N) and (N, π) reaction experiments, is reviewed.

The study of pion induced reactions on nuclei constitutes with the pion nucleus scattering studies, all the ingredients for the knowledge of the interaction of pions with nuclei. Conversely, pion physics is thought to bring new, nuclear structure information once the reaction mechanisms in the pion interactions with nuclei are understood.

In this review of the experimental situation on several π absorption and production processes, the $(\pi, \pi N)$ reaction has been included in the pion absorption family. As a boson, the pion can also be absorbed by nuclei in a "true" absorption process, in which the nucleus receives at least the pion mass energy of 140 MeV. Simple absorption processes like (π, γ) or (π, N) are weak, the first because of the electromagnetic coupling constant and the second because of the high momentum to be supplied to the nucleon. More complex processes in which two or more nucleons or aggregates are emitted are thought to be dominant absorption modes.

The $(\pi, \pi N)$, (π, NN), $(\pi, \text{charged } X)$, $(\pi, \gamma X)$ and the (π, N) reactions as well as the (N, π) reaction will be considered. Radiative pion capture (π, γ) and pion photoproduction (γ, π) will be discussed elsewhere.

We shall concentrate on the experimental work reported after the Santa Fe Conference on High Energy Physics and Nuclear Structure (June 1975).

The proceedings of the conference series on High Energy Physics and Nuclear Structure and the review articles by D. S. Koltun[1] and J. Hüfner[2] give a complete set of references and reviews of the field of pion nucleus interactions.

THE (π, πN) REACTION

It was originally thought that the (π, πN) reaction is of simple knock-out nature. It turned out that the experiments didn't verify this picture and indicated the important role initial and final state interactions might play. The understanding of the (π, πN) reaction is closely related to the problems involved in all interactions of pions with nuclei.

In most experiments, the method is basically the same and so are the problems to be solved. The pion beam intensity must be known and the composition of the beam must be determined by time of-flight, dE/dx and Čerenkov spectroscopy. Lepton contamination (μ, e) is a problem in low energy π⁻ beams, the proton contamination being a worry at higher-energy π⁺ beams. Generally, the measured quantity is the β^+ radioactivity of the residual nucleus. This activity is measured by detecting the two annihilation γ's with two NaI(Tl) crystals in a 180° geometry. The γ-γ detection efficiency, the neutron background contribution to the produced activity and target thickness effects have to be known.

Experimental data exist on ^{12}C[3-5] and the heavier targets ^{14}N, ^{16}O, ^{19}F and ^{31}P[6-10]. The most recent experiments on this reaction[6,10] used the stacked target technique which minimizes the normalization problems in activating simultaneously the targets under study and a carbon target. The $^{12}C(\pi, \pi N)$ reaction on the latter is then used for calibration.

Fig. 1 shows the results of Jacob and Markowitz[6] scaled up or down to the ^{12}C data[3] for comparison. All π⁺ and π⁻ data have a common energy dependence, reflecting the π-N resonances, although the latter are narrower (∼140 MeV) than the resonances of Fig. 1 (∼250 MeV).

Fig. 2 summarizes all the experimental values of the ratio $R = \sigma(\pi^-, \pi^-n)/\sigma(\pi^+, \pi^+N)$ in the 33 resonance region, for which the quasi free nucleon knock-out value at the resonance is $R \simeq 3$. At the present level of precision, there is no detectable A-dependence for R.

From the 250 MeV width of the (π, πN) resonances, one deduces the value of ∼180 MeV/c for the average momentum of the struck nucleon. This is consistent with the value 160-170 MeV/c for 1p shell nucleons in light nuclei[13] and in favour of a quasi free knock-out mechanism. Such a mechanism is also suggested by the similarity of the (π, πN) and the (p, pn) reaction between 200 and 460 MeV[6].

Finally, the earlier experiment of Lieb et al.[9] on ^{16}O must be mentioned. Performed on ^{16}O it looked at specific final nuclear states in the mirror nuclei ^{15}O and ^{15}N. A high resolution Ge(Li) detector was used for the detection of prompt γ rays produced by 180 MeV π^+ and 215 π^-. The data showed strong excitation of the $3/2^-$ states in ^{15}N (6.323 MeV) and in ^{15}O (6.177 MeV). Knowing that these states are predominantly $(p_{3/2})^{-1}$ states, this result suggests a direct quasi free knock-out process. Quantitatively, the Lieb et al. results[9], indicate that the $3/2^-$ nucleon knock-out represents approximately 1/4 of the total ^{16}O $(\pi^{\pm},\pi N)$ strength to bound ^{15}O levels.

A quantitative discussion of the $(\pi,\pi N)$ reaction mechanism cannot be made before new experiments, looking at well defined final states are successful. In lieu of such results,

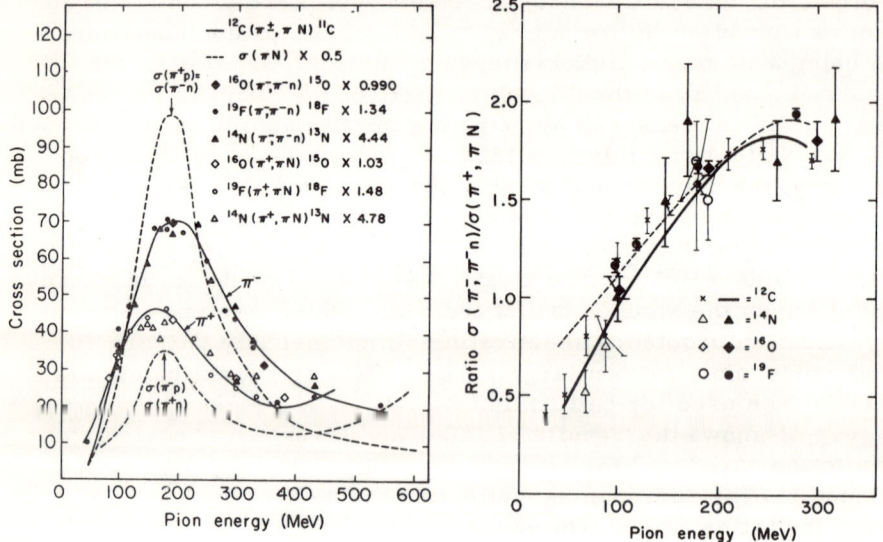

Fig. 1. Energy dependence of the reactions ^{14}N $(\pi^{\pm},x)$ ^{13}N (β^+), ^{16}O $(\pi^{\pm},x)$ ^{15}O (β^+) and ^{19}F $(\pi^{\pm},x)$ ^{18}F from Jacob and Markowitz[6]. The solid lines are ^{12}C $(\pi^{\pm},x)$ ^{11}C(β^+) data from Dropesky et al.[3].

Fig. 2. The ratio R for ^{14}N, ^{16}O and ^{18}F from Jacob and Markowitz[6]. Other data are taken from references 3-5 and 7-10. The dashed curve is a theoretical calculation by Sternheim and Silbar[12] (see text).

one must consider many possible reaction mechanisms, most of them leading to important final state interactions. When nucleon charge exchange (NCE) is taken into account[12] good agreement is obtained with the experiment (a parameter accounting for Pauli principle inhibition on the NCE has been fitted at 180 MeV) (Fig. 2). However, besides NCE, coherent target excitation, giant resonances, compound nucleus effects can be important. The validity of such hypotheses has been shown in a recent, no free parameter estimate of the ^{12}C$(\pi,\pi N)$ reaction, by Hüfner et al.[11]

THE (π, NN) REACTION

In this process a pion with small kinetic energy is captured by the nucleus, so that high energy ~ 140 MeV but very little momentum is transfered to the nucleus. The two nucleons sharing this energy are emitted in approximately opposite directions, each with a momentum ~ 360 MeV/c. As a consequence high momentum components (or short range) properties of the NN system can be probed. In fact, the distortion of the incident pion, the interactions of the two nucleons with the residual nucleus and the final state NN interaction have to be taken into account before one gets access to short range correlations.

From previous work, the ratio of the (π^+, pp) reaction cross sections on ^{4}He and ^{16}O, compared to the $\pi^+ d \rightarrow pp$ reaction cross section, was measured to be constant and independent of the pion energy[14]. This fact and the energy dependence of the latter reaction[15], reflecting the πN resonance, has shown the importance of the socalled rescattering model in the (π, NN) reaction.

Arthur et al.[16] studied the (π^+, pp) reaction on ^{6}Li, ^{14}N and ^{16}O with 70 MeV pions at LBL. Position and angle of the two protons were measured with spark chambers, dE/dx and E measurements were made with two sets of scintillation counters and two 130 mm x 100 mm NaI(Tl) crystals respectively. The apparatus could detect protons in the 35-183 MeV range with a missing mass resolution of 4.5 MeV (fwhm).

The important and new information of this experiment was the observation of ^{14}N$(\pi^+, pp)^{12}$C and ^{16}O$(\pi^+, pp)^{14}$N excitation spectra, in contradiction with simple shell model expectations, showing the influence of the nuclear structure on the (π, NN) reactions.

The general conclusion of the Arthur et al. work was that PWIA calculations of all three reactions could successfully reproduce the distributions of the data over nuclear recoil, relative proton momentum, proton-opening angles and Treiman-Yang angles, on and off the $\pi^+ d \rightarrow pp$ kinematical region.

The results of a recent experiment[17] at CERN on the ^{6}Li, ^{10}B, ^{12}C, ^{14}N and ^{16}O (π^-, nn) reactions confirm these conclusions. In particular the disagreement with fractional parentage predictions[18] of the ^{12}C and ^{14}N excitation spectra is noted.

Future experiments on the $(\pi^\pm, NN)$ reaction with variation of the pion energy and high resolution for the final state identification will help clarify the observed influence of the nuclear structure on the (π, NN) reaction mechanism.

THE $(\pi^-, \gamma X)$ REACTION

Despite the simplicity of such experiments and the high resolution (few keV) of the Ge(Li) γ-ray detectors this type of experiments suffer from the multiplicity of possible processes which can, a priori, lead to the γ-ray emitting nuclear state. Experiments of this type using stopped π's have, over the fast pion induced ones, the advantage to guarantee the pion absorption.

Recently, Engelhardt et al.[19] reported a complete study of prompt γ-rays, observed in the reactions ^{9}Be, ^{10}B, O, ^{19}F, ^{31}P, Ca, ^{92}Nb(π$^-$,γX) with stopped π$^-$ at CERN. Segel et al.[20] reported measurements of prompt γ-rays from reactions induced by 380 - MeV π$^-$ on S, Ar, Ca, V, ^{60}Ni and As at the Argonne ZGS.

The experimental setup typically includes a beam telescope of plastic scintillators and a Čerenkov counter and a high resolution Ge(Li) detector. Charged particles are eliminated using a plastic scintillator counter in front of the Ge(Li) detector. Because of the large electron contamination ($\sim$2/3) in the incident beam, the absolute normalisation of the cross sections in the Segel et al.[18] experiment is $\pm$ 50 %. The absolute uncertainty in the stopped pion experiment was $\pm$ 20 %.

The stopped pion data[19] show that the probability to remove a given number of nucleons, N, has its maximum at N = 2. The average number of removed nucleons lies between three and six with about equal number of neutrons and protons[19,20]. The removal of more than two nucleons is of comparable size to the 2N removal and can be understood in the framework of intranuclear cascade or evaporation models.

The surprisingly high probability for one nucleon removal by stopped pions[19], $\sim$ 1/4 of the 2N yield, could be due to a (π$^-$,γN) reaction or to a rescattering mechanism, where the pion interacts with two nucleons one of them being subsequently absorbed. Such a mechanism should lead to 2h - 1p states of the target instead of the simple 1h states as observed for ^{16}O and ^{40}Ca[19].

Alpha removal is observed in general agreement with conventional α-pick up reactions. The yield of γ-rays corresponding to multi α removal by stopped pions is low[19] ($\sim$1 %) and consistent with earlier slow[21,22] or stopped-pion[23,24] data. Fast pions yield more γ-rays corresponding to multi α removal[20] again in agreement with earlier data[25,26].

The preliminary analysis of a (π$^+$,γX) experiment at $T_\pi \simeq$ 36 MeV[27] using enriched ^{58}Ni, ^{60}Ni and ^{62}Ni targets, seems to indicate relatively high multi-α removal rates. For the ^{58}Ni case the cross sections $\sigma(1\alpha) : \sigma(2\alpha) : \sigma(3\alpha) : \sigma(4\alpha)$ are : 36 : 24 : 18 : 0 millibarn, and they amount to 24 % of the observed total γ yield due to particle removal. These values are to be compared to the 70 : 34 : 24 : 0 mb values for ^{60}Ni at 380 MeV[20].

THE REACTION $\pi + A \rightarrow$ HEAVY CHARGED $+ X$

Like all inclusive type reactions, experiments looking only at one charged particle, suffer from the undefined reaction final state. They should, nevertheless, provide us with the general features of the emitted particle spectra, which in turn might have an energy dependence caracterizing the reaction mechanism involved.

A year ago, Amann et al.[28] studied the energy spectra of charged particles emitted by 235 MeV π^+ bombardment of Mg, Ni and Ag targets. Protons, deuterons, tritons, ^{3}He and ^{4}He particles were measured in the energy range 15-400 MeV. The detector system was a four-element telescope of two surface barrier detectors followed by two high purity Ge detectors. It could stop 400 MeV alphas with 2 - 20 MeV resolution, depending on energy straggling in the target. Absolute cross sections were determined to $\pm$ 20 %.

The energy spectra extending up to $\sim$ 140 MeV (Fig. 3) exhibit an exponential fall off with a slope independent of the emitted particles mass. Intranuclear cascade calculations[29] reproduce neither the shape of the proton spectra, nor the magnitude of the cross sections, in contrast to their success in similar proton induced reactions.

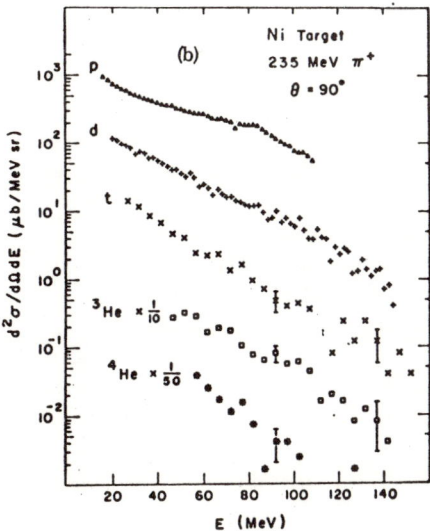
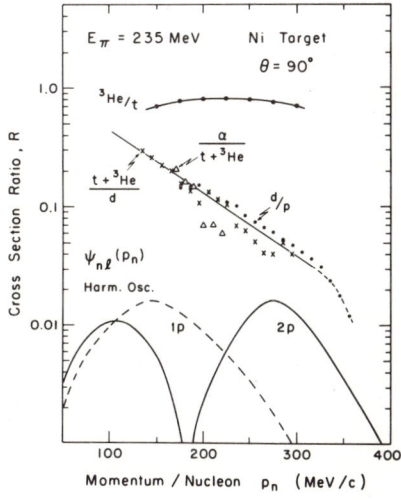

Fig. 3. Energy spectra for the different particles types, taken from Amann et al.[28].

Fig. 4. The ratios $R = \sigma^{A+1}/\sigma^A$ as a function of the momentum per nucleon p_n [28]. The harmonic oscillator wave function for the 2p orbit is shown to have a large amplitude up to 300 MeV/c.

An interesting feature of the data[28] which had already been observed in proton induced particle spectra[30] concerns the probability of removing a heavier mass particle relative to a single proton. If heavier particle emission is due to nucleon pick up induced by the primary knocked out particles, then the probability for removing a (A+1) mass particle, $\sigma^{A+1}(p_n)$ is given by

$$\sigma^{A+1}(p_n) = \sigma^A(p_n) P(p_n)$$

where the probability, $P(p_n)$, for removing a nucleon of momentum p_n is independent of A. Fig. 4 shows that such a scaling law exists for all the data. The ratio $R = \sigma^{A+1}/\sigma^A$ is independent of A and decreases with p_n. The ^{3}He/t ratio is independent of p_n.

Very recently[31] the helium spectra emitted by C, Al, Ni, Ag and Au, after $\pi^{\pm}$ bombardment at T_π = 75 MeV, have been measured at Saclay with a three surface barrier silicon detector telescope. Preliminary analysis, without ^{3}He - ^{4}He separation, indicates that the $\sigma(\pi^-)/\sigma(\pi^+)$ ratio is 1.3 - 2.0 (depending on the target) for T_{He} > 6 MeV and $\simeq$ 1 for T_{He} > 14 MeV.

A careful analysis of future experiments looking at the complete energy spectrum of the emitted particles and also detecting the eventually outgoing pion would be of great help in understanding the mechanism of multi-nucleon removal by pions with and/or without pion absorption.

THE (π, N) REACTION

The notation implies a two body reaction, i.e. the absorption of the pion and the emission of a nucleon leaving the residual (A-1) system bound. In such a process the momentum transfer to the nucleon is high ($\sim$500MeV/c) and the experiment should probe high momentum components of the nucleon wave function. In analogy to its inverse reaction, the (N, π) reaction, negative pion induced (π^-, p) reactions could be due to the absorption of the pion on a nucleon pair (p p) with subsequent emission of one of them.

Coupat et al.[32], searched for proton emission following π^- absorption at rest in ^{9}Be, ^{12}C, ^{14}N, ^{16}O and ^{27}Al. The activation method was used to identify the residual nuclei, with the exception of the ^{14}N(π^-p) reaction, where protons have been detected with a NaI(Tl) crystal.

Positive identification was established for the ^{12}C(π^-,p)^{11}Be reaction (integrated over the 1/2$^+$ g.s. and the 1/2$^-$ 0.32 MeV - state). Taking into account the contribution of the ^{13}C(π^-,pn)^{11}Be reaction in the natural carbon activation measurements, the probability (3.4 $\pm$ 0.8) 10^{-4} per stopped pion was determined for the ^{12}C(π^-,p)^{11}Be reaction. A preliminary analysis of the ^{14}N data leads to a similar (π^-,p) probability[33]. In the absence of any theoretical calculations the experimental result is comparable to Chung's estimate, 10^{-4} - 5 10^{-4}, for the ^{16}O(π^-,p) reaction, using

Jastrow correlation functions to describe the short range part of the NN force in ^{16}O, but neglecting final state interactions[34].

Bachelier et al.[35] studied the $^{16}O(\pi^+, p)^{15}O$ reaction at $T_{\pi^+} = $ 66 MeV. Protons were detected with a 13 element scintillator range telescope which could stop up to 185 MeV protons, covering 16 MeV excitation in ^{15}O with a resolution of 3.2 MeV. dE/dx measurements in the first telescope counters allowed discrimination against parasite particles. The angular distributions of protons leaving the ^{15}O nucleus in its $1/2^-$ ground state and in its

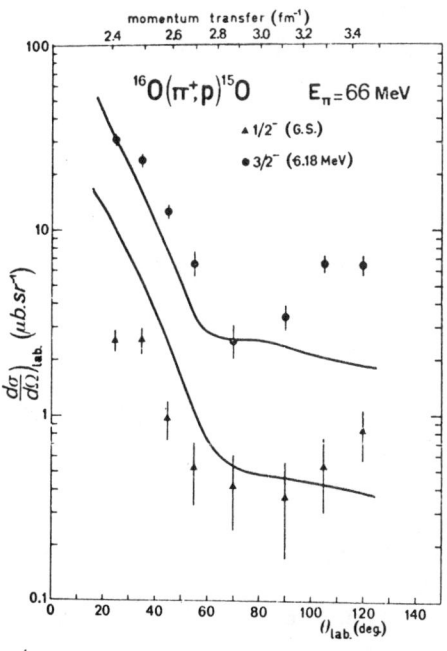

Fig. 5. Angular distributions of protons, from the $^{16}O (\pi^+, p)^{15}O$ reaction measured by Bachelier et al.[35]. The full lines are a preliminary DWBA calculation by G.A. Miller[36], using a local optical potential for the pion distortion.

$3/2^-$, 6.18 MeV state are shown in Fig. 5. The absolute cross sections are much higher than the $<17 \mu bsr^{-1}$ limit given by Amato et al.[37]. (This discrepancy seems[35] to be due to an uncertainty in the absolute proton energy scale of the Berkeley experiment[37].)

The angular distributions are similar in shape, but, the $3/2^-$ 6.18 MeV state is excited more than 8 times stronger then the $1/2^-$ g.s., in contradiction to the shell model expectation value of ~ 2, confirmed by lower energy neutron pick-up experiments. It is difficult to understand this high cross section ratio on the basis of differences in the high momentum components of $p_{3/2}$ and $p_{1/2}$ nucleons. On the other hand, the reaction mechanism for (π,p) reactions, involving possibly more than one nucleon, can be very different from a simple neutron pick up process.

In view of our ignorance about the reaction mechanism and the difficulties to treat the πNN vertex involved in such reactions, there is a great need for ($\pi^{\pm}$, N) experiments such as described above, with variation of the pion energy and the target mass.

THE (p, π) REACTION

The interest in studying the (p, π) reaction is mainly due to the high momentum transfered to the final nucleus. The observation of bound residual nuclear states should be closely related to high momentum features of the nucleus. Nevertheless, the use of this reaction to nuclear structure studies, requires the knowledge of the reaction mechanism involved. Basically, we don't know if the production involves a single nucleon or two (or many) nucleons. In the single nucleon mechanism (SNM) the pion is emitted by the projectile, which is it self captured by the target nucleus. The cross section is then strongly dependent on the high momentum components of a single nucleon in the nucleus. In the two-nucleon mechanism (TNM), a target nucleon participates in the process. In emitting the pion which leaves after all sorts of interactions. The cross section is now reflecting less of the individual and more of the high momentum components averaged over several nucleons of a nuclear shell.

A large amount of experimental work has been devoted to the study of this reaction.

The study of the ^{9}Be, ^{12}C, ^{13}C, ^{16}O, ^{28}Si and ^{40}Ca (p, π^+) reactions near threshold[38] established that the π^+ production depends on the nuclear properties of the residual nucleus, and that the differential cross sections are strongly momentum transfer, q, dependent. The (p, π^-) reaction on ^{9}Be and ^{13}C is strongly suppressed (a few nbsr^{-1}) and in contrast to the (p, π^+) reaction has very little q dependence. The π^+/π^- cross section ratios at $0°$ are ~ 30 - 45.

At 600 MeV[39], the forward angle differential cross sections on ^{9}Be, ^{12}C, ^{13}C and ^{14}N are of the order of ~0.5 μb/sr[39]. The π^+/π^- ratio in the case of the ^{10}Be - ^{10}C mirror final states is ~100.[39]

The partial success of the SNM and the TNM calculations and the fact that π^- production was low but still observable and has little q-dependence leads to the assumption that eventually both types of processes could contribute to the pion production[50,51].

New experiments were motivated from these conclusions, aiming at the energy dependence of the pion production and the measurement of complete angular distributions far from threshold energies.

Dahlgren et al.[40] measured the angular distribution of positive pions from the ^{10}B(p, π^+)^{11}B (g.s.) reaction at 185.6 MeV, 174.8 MeV and 167.3 MeV. A double focusing, $90°$, n = 0, magnetic spectrometer and a ten-channel scintillator hodoscope, located in its image plane were used for the pion detection. The apparatus allows pion detection between 17 and 48 MeV with a resolution of ~0.55 MeV. In a preliminary report of these results[40], there is a marked tendency of the angular distributions to a steeper slope, as the proton energy nears threshold in contrast with what one would naively expect for the energy dependence of such angular distributions.

Le Bornec et al. have studied the energy dependence of the $^{40}Ca(p,\pi^+)$ reaction between 149 and 154 MeV[41] and the π^+ reaction on ^{10}B, ^{13}C, ^{14}N, ^{25}Mg, ^{28}Si and ^{32}S at $T_p = 154$ MeV[42]. Their experimental set up consists of a magnetic pion spectrometer and a three scintillator hodoscope to detect the pions. Pion stopping power and total energy are measured with two thick scintillators behind the hodoscope. The dE/dx requirement, and a time-of-flight measurement was used to identify the pions. The overall energy resolution is ~1.5 MeV and the normalisation uncertainty ±30 %.

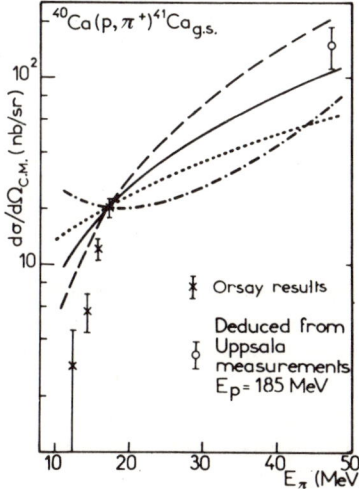

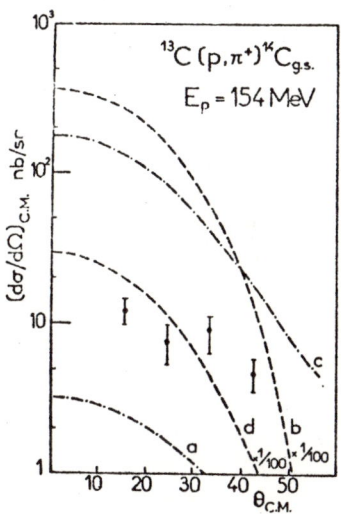

Fig. 6. Energy dependence of the $^{40}Ca(p,\pi^+)$ ^{41}Ca reaction [41] at $q_{CM} = 2.42$ f^{-1}. The dotted line corresponds to the phase space of the two-nucleon mechanism. The solid, dashed and dotted-dashed lines correspond to the phase space variation of the one nucleon mechanism.

Fig. 7. The comparison of the $^{13}C(p,\pi^+)$ ^{14}C g.s. angular distribution with different DWBA calculations within the stripping formalism [42]. The difference in the curves a, b, c, d is due to different choices of the pion nucleus optical potential a) $V_{\pi-nucl.} = 0$, b) potential I [36], c) potential II [36], d) potential I with the Kroll-Kisslinger correction.

Fig. 6 gives the energy dependence as measured by Le Bornec et al. for ^{40}Ca. It is plotted for constant momentum transfer q = 2.42 f^{-1}, as a function of the pion energy. Theoretical calculations with a SNM and the plane wave approximation, as well as with a TNM, fail to reproduce the experimental data.

The results of different DWBA calculations of the $^{13}C(p, \pi^+)^{14}C$ reaction within the stripping formalism are shown in Fig. 7. In this calculation the proton nuclear and Coulomb potentials were identical and only the pion-nucleus optical potential was changed. None of these results reproduces the experimental data, illustrating our ignorance about the choice of convenient π-nucleus potentials and the importance of the latter in SNM, DWBA calculations.

The lack of data on the (p, π) reaction far from threshold energies, motivated a series of experiments at Saclay, on D, ^{3}He, ^{6}Li, ^{7}Li and ^{9}Be.

The $pD \rightarrow T\pi^+$ reaction was studied at 410, 605 and 809 MeV[44]. The $p^3He \rightarrow ^4He\pi^+$ reaction was studied at 415 and 716 MeV[45]. The ^{6}Li, ^{7}Li(p, π^+) reaction were studied at 600 MeV[43] and the ^{9}Be $(p, \pi^+)^{10}$Be reaction at 410 and 605 MeV[44]. Only the heavier target data will be briefly discussed, since pion production on lighter targets will be reviewed elsewhere at this Conference.

All these experiments were carried out at the Saclay Synchrotron SATURNE. The pion spectra were measured with the high resolution energy-loss type spectrometer SPES I[46] with a solid angle of 3.3 10^{-3} sr and a momentum resolution of $\simeq 10^{-4}$.

The particle identification was made by three lucite Čerenkov counters, following four planes of plastic scintillators. Beam intensity monitoring was made with two six-counter telescopes at 45° and 135° and a secondary emission monitor at 0°. Overall detection efficiencies are 0.90, pion decay losses are $\simeq$50 - 60 %. Corrections for losses through interactions in the apparatus are $\simeq$20 %. The absolute normalisation of the monitors was obtained by $^{12}C(p, pn)^{11}C$ activation measurements. Depending on the experiments, the absolute normalisation uncertainty is $\pm$ 10 - 20 %.

Fig. 8 shows the π^+ angular distributions of pions feeding the $3/2^-$ ^{7}Li g.s. the $1/2^-$, 0.48 MeV level and the $7/2^-$, 4.63 MeV level and Fig. 9 shows the results on the ^{9}Be(p, π^+) reaction leading to the excitation of the ^{10}Be 0^+, g.s. and the 2^+, 3.37 MeV level. Two features are typical of these high energy data
a) high angular momentum states are preferentially excited
b) the angular distributions are slowly decreasing and their overall shape is very little dependent on the residual nuclear state. This is in contrast to the low energy data.

Assuming a single nucleon mechanism, the excitation of high angular momentum states could be due to multi-step processes, where target or residual nucleus excitation provides the transition between initial and final states of extreme angular momentum mismatch. A coupled channel DWBA calculation[47] result is in favour of such hypotheses.

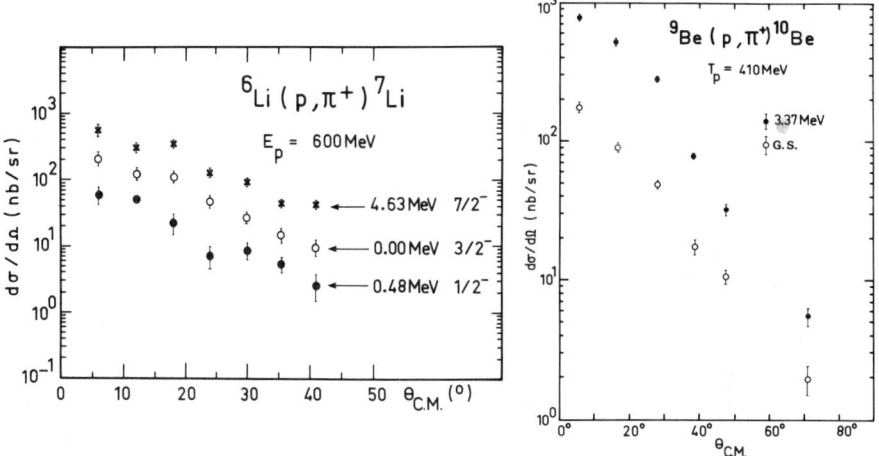

Fig. 8. Angular distribution of pions from the ^{6}Li$(p,\pi^+)^7$Li reaction at $T_p = 600$ MeV [43].

Fig. 9. Angular distributions of pions from the ^{9}Be$(p,\pi^+)^{10}$Be reaction at $T_p = 410$ MeV [44].

The slow decrease of the differential cross sections with angle, can be suggestive of a two nucleon mechanism. However, a preliminary estimate made by Dillig[48] on the basis of a two-nucleon model including proton and pion distortions cannot reproduce the experimental data beyond $20°_{C.M.}$. (A three nucleon mechanism was necessary in reproducing the data at large angles[48].)

Detailed distorted wave calculations will be necessary for the interpretation of the high energy data, the analysis of which is still in progress.

Following a suggestion by Kisslinger and Miller[49] the (p, π^-) reaction might proceed via Δ^{++} transfer, in which case the π^- production would have a much steeper q dependence than in the competing charge exchange process $(p, n)(n, \pi^-)$. The absolute π^- yield was directly related to the Δ^{++} component of the target wave function, but even small ($\sim 10^{-4}$) Δ^{++} probabilities were sufficient to make the Δ^{++} transfer process competitive. A search for copious π^- production on ^{9}Be, ^{12}C, ^{13}C and ^{25}Mg was undertaken at 613 MeV by the SPES I collaboration group[52]. A preliminary analysis of the ^{12}C data, shows that the cross section at $5°$ is a few nbsr^{-1} in agreement with the older CERN result[39]. Furthermore the ratio of the differential cross sections at $5°$ and $25°$ is $\simeq 2$. Both results are in contradiction with the theoretical predictions[49]. On the other hand, the differential cross sections for the ^{12}C(p, π^-) and ^{13}C(p,π^-) reactions at $5°_{LAB}$ are very similar, ~ 2 nb/sr, and one order of magnitude higher than the ^{9}Be(p, π^-) cross section, of 0.2 nb/sr, at $25°_{LAB}$, as qualitatively predicted[49].

CONCLUSION

The experimental information acquired in the recent years accomplished a great task in confirming qualitatively some basic ideas about the mechanism of the pion interaction with nuclei, like the pion absorption, and the influence of the nuclear structure on high momentum transfer phenomena, like the pion production. The interpretation of the experiments revealed the particular difficulties in this field, which can be summarized in our ignorance about the pion-nucleus potential and about the role πN, πNN or disguised πNN mechanisms (with one nucleon reabsorption) play in pion absorption.

From this point of view, future experiments should be less "inclusive". Complete experiments, with well defined initial and final states, studying the energy dependence of the different reactions on a large variety of targets, should help making a choice in the long list of reaction mechanisms suggested by the presently available data. Only then, will it be possible to use pion interactions for the discovery of new nuclear structure information.

I would like to thank Jörg Hüfner for his critical discussion of the present paper.

REFERENCES

1. D.S. Koltun, Adv. Nucl. Phys. 3, 71 (1969)
2. J. Hüfner, Phys. Reports 21C, 1 (1975)
3. B.J. Dropesky, G.W. Butler, C.J. Orth, R.A. Williams, G. Friedlander, M.A. Yates, S.B. Kaufman, Phys. Rev. Letters 34, 821 (1975)
4. L.H. Batist, V.D. Vitman, V.P. Koptev, M.M. Makarov, A.A. Naberezhnov, V.V. Nelyubin, G.Z. Obrant, V.V. Sarantsev, G.V. Scherbakov, Nucl. Phys. A254, 480 (1975)
5. M.A. Moinester, M. Zaider, J. Alster, D. Ashery, S. Cochavi A.I. Yavin, Phys. Rev. C8, 2039 (1973)
6. N.P. Jacob, Jr., and S.S. Markowitz, Phys. Rev. C13, 754 (1976)
7. M. Zaider, J. Alster, D. Ashery, S. Cochavi, M.A. Moinester A.I. Yavin, Phys. Rev. C10, 938 (1974)
8. G.W. Butler, B.J. Dropesky, A.E. Norris, C.J. Orth, R.A. Williams, G. Friedlander, G.D. Harp, J. Hudis, N.P. Jacob, Jr., S.S. Markowitz, S. Kaufman, M.A. Yates, High Energy Physics and Nuclear Structure, Santa Fe, 1975
9. B.J. Lieb, H.S. Plendl, H.O. Funsten, W.J. Kossler, C.E. Stronach, Phys. Rev. Letters 34, 965 (1975)

10. P. J. Karol, M. V. Yester, R. L. Klobuchar, A. A. Caretto Jr., Phys. Letters 58B, 489 (1975)
11. J. Hüfner, H. J. Pirner, M. Thies, Phys. Lett. 59B, 215 (1975).
12. M. M. Sternheim and R. R. Silbar, Phys. Rev. Letters 34, 824 (1975)
13. M. Riou, Rev. Mod. Phys. 37, 375 (1965)
14. T. Bressani, G. Charpak, J. Favier, L. Massonnet, W. F. Meyerhof, T. Zupančič, Nucl. Phys. B9, 427 (1969)
15. C. Richard-Serre, W. Hirt, D. F. Measday, E. G. Michaelis, M. J. M. Saltmarsh, P. Sharek, Nucl. Phys. B20, 413 (1970)
16. E. D. Arthur, W. C. Lam, J. Amato, D. Axen, R. L. Burman, P. Fessenden, R. Macek, J. Oostens, W. Shlaer, S. Sobottka, M. Salomon, W. Swenson, Phys. Rev. 11, 332 (1975)
17. B. Bassallek, D. Engelhardt, W. Klotz, F. Takeutchi, H. Ullrich, M. Furic, Contribution to this conference.
18. S. Cohen and D. Kurath, Nucl. Phys. A141, 145 (1970)
19. H. D. Engelhardt, C. W. Lewis, H. Ullrich, Nucl. Phys. A258, 480 (1976)
20. R. E. Segel, L. R. Greenwood, P. Debevec, H. E. Jackson, D. G. Kovar, L. Meyer-Schützmeister, J. E. Monahan, F. J. D. Serduke, T. P. Wangler, W. R. Wharton, B. Zeidman, Phys. Rev. C13, 1566 (1976)
21. D. Ashery, M. Zaider, Y. Shamar, S. Cochavi, M. A. Moinester A. I. Yavin, J. Alster, Phys. Rev. Letters 32, 943 (1974)
22. H. Ullrich, E. T. Boschitz, H. D. Engelhardt, C. W. Lewis, Phys. Rev. Letters 33, 433 (1974)
23. A. Doron, J. Julien, M. A. Moinester, A. Palmer, A. I. Yavin Phys. Rev. Letters 34, 485 (1975)
24. J. Comiso, T. Meyer, F. Schlepnetz, K. O. H. Ziock, Phys. Rev. Letters 35, 13 (1975)
25. H. E. Jackson, D. G. Kovar, L. Meyer-Schützmeister, J. P. Schiffer, R. E. Segel, S. Vigdor, T. P. Wangler, R. L. Burman P. A. M. Gram, D. M. Drake, V. G. Lind, E. N. Hatch, O. H. Otteson, R. E. McAdams, B. C. Cook, R. B. Clark, Phys. Rev. Letters 35, 1170 (1975)
H. E. Jackson, D. G. Kovar, L. Meyer-Schützmeister, R. E. Segel, J. P. Schiffer, T. P. Wangler, R. L. Burman, D. M. Drake, P. A. M. Gram, R. P. Redwine, V. G. Lind, E. N. Hatch O. H. Otteson, R. E. McAdams, B. C. Cook, R. B. Clark, 35, 641 (1975)
26. V. G. Lind, H. S. Plendl, H. O. Funsten, W. J. Kossler, B. J. Lieb, W. F. Lankford, A. J. Buffa, Phys. Rev. Letters 32, 479 (1974)
H. E. Jackson, L. Meyer-Schützmeister, T. P. Wangler, R. P. Redwine, R. E. Segel, J. Tonn, J. P. Schiffer, Phys. Rev. Letters 31, 1353 (1973)

27. Y. Cassagnou, H. Jackson, J. Julien, R. Legrain, A. Palmeri L. Roussel, private communication.
28. J.F. Amann, P.D. Barnes, M. Doss, S.A. Dytman, R.A. Eisenstein, J. Penkvot, A.C. Thomson, Phys. Rev. Letters 35, 1066 (1975)
29. G.D. Harp, Phys. Rev. C8, 581 (1973)
30. J.F. Amann et al., to be published
31. J. Julien, (Saclay-Tel-Aviv-Catane collaboration) private communication.
32. B. Coupat, P.Y. Bertin, D.B. Isabelle, G. Kawadry, P. Vernin, A. Gérard, J. Miller, J. Morgenstern, J. Picard, B.Saghai, High Energy Physics and Nuclear Structure, Santa Fe,(1975)
    B. Coupat, D.B. Isabelle, P.Y. Bertin, A. Gerard, J. Miller J. Morgenstern, J. Picard, B. Saghai, P. Vernin, contribution to this conference.
33. D.B.Isabelle (Clermont-Saclay collaboration), private communication.
34. K. Chung, M. Danos and M.G. Huber, Phys. Letters 29B, 265 (1969) and Z. Phys. 240 ,195(1970)
35. D. Bachelier, J.L. Boyard, T. Hennino, J.C. Jourdain, P. Radvanyi, M. Roy-Stéphan, contribution to this conference.
36. G.A. Miller, Nucl. Phys. A224, 269 (1974)
37. J. Amato, R.L. Burman, R. Macek, J. Oostens, W. Shlaer, E. Arthur, S. Sobottka, W.C. Lam, Phys. Rev. C9, 501(1974)
38. S. Dahlgren, B. Höistad, P. Grafström, Phys. Letters 35B, 219 (1971)
    S. Dahlgren, P. Grafström, B. Höistad, A.Asberg, Nucl. Phys. A204, 53 (1973)
    ibid A211, 243 (1973); Phys. Letters 47B, 439 (1973); High Energy Physics and Nuclear Structure, Uppsala, 1973
    Y. Le Bornec, B. Tatischeff, L. Bimbot, Î. Brissaud, H.D. Holmgren, J. Källne, F. Reide, N. Willis, Phys. Letters 49B, 434 (1974)
39. J.J. Domingo, B.W. Allardyce, C.H.Q.Ingram, S. Rohlin, N.W. Tanner, J. Rohlin, E.M. Rimmer, G. Jones, J.P. Girardeau-Montaut, Phys. Letters 32B, 309 (1970) ; J. Rohlin K. Gabathuler, N.W. Tanner, C.R. Cox, J.J. Domingo, Phys. Letters 40B, 539 (1972)
40. S. Dahlgren, T. Johansson, O. Jonsson, Uppsala, Internal report N°
41. Y. Le Bornec, B. Tatischeff, L. Bimbot, I. Brissaud, H.D. Holmgren, J. Källne, F. Reide, N. Willis, Phys. Letters 61B, 47 (1976)

42. Y. Le Bornec, B. Tatischeff, L. Bimbot, I. Brissaud, H.D. Holmgren, J. Källne, F. Reide, N. Willis, contribution to this conference.
43. T. Bauer, R. Beurtey, A. Boudard, G. Bruge, A. Chaumeaux P. Couvert, H.H. Duhm, D. Garreta, M. Matoba, Y. Terrien L. Bimbot, Y. Le Bornec, B. Tatischeff, E. Aslanides, R. Bertini, F. Brochard, Ph. Gorodetzky, F. Hibou, High Energy Physics Nuclear Structure, Santa Fe, 1975
44. E. Aslanides, T. Bauer, R. Bertini, R. Beurtey, L. Bimbot, O. Bing, A. Boudard, F. Brochard, G. Bruge, H. Catz, A. Chaumeaux, P. Couvert, J.C. Duchazeaubeneix, H.H. Duhm, D. Garreta, Ph. Gorodetzky, J. Habault, F. Hibou, G. Igo, Y. Le Bornec, M. Matoba, B. Tatischeff, Y. Terrien, Note CEA-N-1861
45. B. Tatischeff, L. Bimbot, R. Frascaria, Y. Le Bornec, M. Morlet, N. Willis, R. Beurtey, G. Bruge, P. Couvert, D. Garreta, D. Legrand, G.A. Moss, Y. Terrien, contribution to this conference.
46. J. Thirion, P. Birien, J. Saudinos, Note CEA-N-1248
47. H.H. Duhm, Symposium on Nuclear Structure, Balatonfüred, Hungary, Sept. 1975
48. M. Dillig, private communication
    M. Dillig, M.G. Huber, High Energy Physics Nuclear Structure, Santa Fe, 1975
49. L.S. Kisslinger and G.A. Miller, Nucl. Phys. $\underline{A254}$, 493(1975)
50. J. Letourneux, J.M. Eisenberg, Nucl. Phys. $\underline{87}$, 331 (1966) ; W.B. Jones, J.M. Eisenberg, ibid $\underline{A154}$, 49 (1970) ; E. Rost, P.D. Kunz, Phys. Letters $\underline{43B}$, 17 (1973) ; J.M. Eisenberg, R. Guy, J.V. Noble, H.J. Weber, ibid $\underline{43B}$, 20 (1973) ; M. Dillig, H.M. Hoffmann, M.G. Huber, ibid $\underline{44B}$, 484 (1973) ; J.M. Eisenberg, J.V. Noble, H.J. Weber, 5th International Conference High-Energy, Physics Nuclear Structure, Uppsala (June 1973)
51. A. Reitan, Nucl. Phys. $\underline{B29}$, 525 (1971) ; ibid $\underline{B50}$, 166 (1972) ; B.R. Wienke, Progr. Theor. Phys. $\underline{49}$, 1220 (1973) ; M.P. Locher, 5th International Conference High-Energy, Physics Nuclear Structure, Uppsala (June 1973) ; M. Dillig, M.G. Huber, ibid ; Z. Grossmann, F. Lenz, M.P. Locher, Ann. Phys. (NY) $\underline{84}$, 348 (1974) ; A. Reitan, Nucl. Phys. $\underline{A237}$, 465 (1975) M. Ruderman, Phys. Rev. $\underline{87}$, 383 (1952) ; S.A. Bludman, ibid $\underline{94}$, 1722 (1954) ; C.H.Q. Ingram, N.W. Tanner, J.J. Domingo, Nucl. Phys. $\underline{B31}$, 331 (1971) ; G.W. Barry, Phys. Rev. $\underline{D7}$, 1441 (1973) ; H.W. Fearing, Phys. Rev. $\underline{C11}$, 1210 (1975)
52. T. Bauer et al., private communication.

J. Ginocchio (Los Alamos): The reason the cascade calculation did not give the observed proton spectrum at 90° in pion induced reactions is that the pion absorption was treated incorrectly. I have improved the treatment of the propagation and absorption of the isobar in the nucleus. The agreement with the observed proton spectrum has improved. I describe, in a contribution to this conference, these improvements and I believe Peter Barnes describes them briefly in his talk.

T. Masterson (TRIUMF): You have mentioned single nucleon and 2-N removal, but not α removal which should also be quite probable, particularly in view of the high probability of np removal. There have been one or two measurements indicating that a large fraction of deuterons were produced.

Aslanides: To my knowledge, there hasn't been any (π,d) measurement. Also, the high probability of the 2N removal lies in the favorable kinematics, and it is not so large in deuteron removal.

Harold W. Fearing (TRIUMF): It is my understanding that the theoretical curves given for ^{40}Ca(p,π^+) which compare SNM and TNM are basically just phase space curves. If this is the case, then it is premature to make statements about the success or failure of either SNM or TNM. We need to have complete calculations in both approaches.

CURRENT THEORIES OF PION PRODUCTION AND ABSORPTION

J. V. Noble
University of Virginia, Charlottesville, VA 22901

ABSTRACT

This paper reviews recent calculations of (π^+,p) and (p,π^+) reaction cross sections leading to low-lying states of light nuclei. The experimental evidence for or against various models is discussed, and it is concluded that the bulk of the differential cross section may be described as "pionic stripping." The outlook for nuclear spectroscopy at large momentum transfer, by means of these reactions, is reviewed.

INTRODUCTION

When we speak of a "probe of nuclear structure" we generally mean a reaction which has the property that we can with reasonable accuracy divide its theoretical treatment into three distinct parts: 1) the reaction mechanism; 2) rescattering effects; and 3) the nuclear structure. Our hope is to understand enough about #1) and #2) that we can make statements about #3) which are not merely novel and interesting, but also correct. In this talk I shall review the theoretical work of the past few years, in which various investigators have tried to determine whether pion production and absorption can serve as a nuclear structure probe in the sense described above.

In order to produce or absorb a physical meson, quantum field theory tells us that we need an interaction between pions and nucleons which is linear in the meson field. The two popular candidates for the form of this interaction are the pseudoscalar coupling

$$H^{PS}_{\pi NN} = g \int d^3x \, \bar{\psi}(\vec{x}) \gamma^5 \tau_i \, \psi(\vec{x}) \, \phi_i(\vec{x}) \tag{1}$$

and the pseudovector coupling

$$H^{PV}_{\pi NN} = \frac{f}{m_\pi} \int d^3x \, \bar{\psi}(\vec{x}) \gamma^5 \gamma^\mu \tau_i \, \psi(\vec{x}) \, \partial_\mu \phi_i(\vec{x}) \,. \tag{2}$$

We see that theories based on these interactions (which have proven successful in describing the tail of the NN potential, as well as in p-wave π-N scattering and pion photoproduction) demand that, to leading order, the emission or absorption takes place on a single nucleon. Now, in order to produce or absorb a pion on a single nucleon, say one interacting with an external potential, requires a large transfer of momentum, as we see from the diagram Fig. 1.

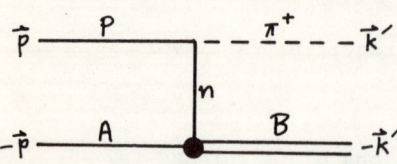

Fig. 1 Lowest-order diagram for the reaction $A(p,\pi^+)B$.

Just to give some feeling for the numbers involved, the momentum transfer in the 185 MeV $^{12}C(p,\pi^+)^{13}C$ experiment runs from 2.4 to 3.4 fm^{-1} as θ, the c.m. scattering angle, varies from 0° to 180°. In the parallel $^{12}C(d,p)^{13}C$ experiment conducted at, say, 50 MeV, the corresponding momentum transfer runs from 0.52 fm^{-1} to 3.9 fm^{-1}.

Many workers,[1] noting this large momentum transfer to a single nucleon, and cognizant also of the well-known experimental fact that the cross-section for multi-nucleon emission in π^- absorption is much larger than that for 1-nucleon emission, concluded that pion absorption and emission must involve at least two nucleons at a time, as in Fig. 2 and Fig. 3. In particular, it was felt that a

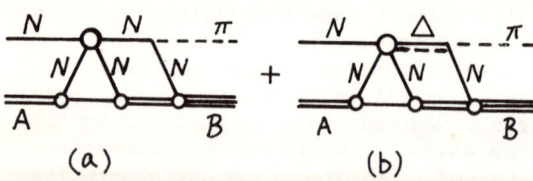

Fig. 2 Two-nucleon process for $A(p,\pi^+)B$ or $A(p,\pi^-)C$.

distinct advantage could be obtained by sharing the momentum transfer among three nucleon single-particle wave functions. It turns out, however, that momentum sharing is only a meaningful concept for harmonic oscillator wave functions. When finite-well wave functions are convoluted, the result

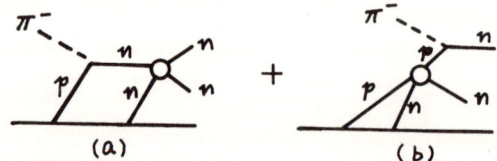

Fig. 3 Leading amplitudes for pion absorption with two-nucleon emission.

is something like the same wave function back again. Thus one important motivation for the multinucleon reaction mechanism disappears.[2]

I would like to show you precisely how a microscopic description of (π,NN) works: Thus, let us calculate the amplitudes for (π,NN) and (π,N), assuming the nucleus can be described in terms of independent-particle wave functions; for simplicity we shall suppose there are only 2 nucleons present. The (π,NN) amplitude may be represented diagrammatically as in Fig. (3a,b), whereas the (π,N) one is the inverse of Fig. 1. Then we have, at threshold

$$\frac{R_{2N}}{R_{1N}} \cong \left\{ \int d\vec{p} \int d\vec{p}' \, \delta\left(\frac{p^2}{2M} + \frac{p'^2}{2M} - m_\pi c^2\right) \left| \int d\vec{q} \, \vec{\sigma}\cdot(\vec{q}-\vec{p}) \, \psi_p(\vec{q}) \times \right. \right.$$

$$\left. \left. \psi_n(\vec{p}+\vec{p}'-\vec{q}) \left[\frac{\tilde{V}(\vec{p}-\vec{q}) + \tilde{V}(\vec{p}'-\vec{q})}{m_\pi c^2} \right] \right|^2 \right\} \left\{ \int d\vec{p} \, \delta\left(\frac{p^2}{2M} - m_\pi c^2\right) \times \right.$$

$$\left. \left| \psi_p(\vec{p}) \, \vec{\sigma}\cdot\vec{p} \right|^2 \right\}^{-1} \quad (3)$$

$$\cong \frac{\int d\vec{p}\int d\vec{p}' \, |F(\vec{p}+\vec{p}')|^2 \, (\vec{p}-\vec{p}')^2 \left[\frac{\tilde{V}(p) + \tilde{V}(p')}{m_\pi c^2}\right]^2 \, \delta\left(\frac{p^2}{2M} + \frac{p'^2}{2M} - m_\pi c^2\right)}{\int d\vec{p} \, p^2 \, |\psi(\vec{p})|^2 \, \delta\left(\frac{p^2}{2M} - m_\pi c^2\right)}$$

where F(q) is the form-factor of the single-particle density. If there were A nucleons we would have, assuming $\tilde{V}(p)$ comes from OPE,

$$\frac{R_{2N}}{R_{1N}} \cong (A-1)\frac{4}{\pi\sqrt{2}} \left[\frac{V_0}{m_\pi c^2}\right]^2 \left[\frac{\hbar^2}{Mm_\pi c^2}\right]^2 \rho_{Nuc}(r=0)/\langle|\psi(\sqrt{2Mm_\pi})|^2\rangle \quad (4)$$

where $\langle|\psi(p)|^2\rangle$ is the single-nucleon momentum distribution. Reasonable models for this distribution give a ratio of cross sections between 5(A-1) and 10(A-1). Note that even in this simple model with no nucleon-nucleon short-range correlations, the $(\pi^-, 2n)$ differential cross section is strongly peaked about $\vec{p}' \cong -\vec{p}$, i.e. when the nucleons are emitted back to back. In other words, on closer examination the dominance of 2 nucleon over 1 nucleon emission in pion absorption owes more to kinematics and phase space than to either short-range correlations or momentum sharing; and it may be understood reasonably well using the simplest microscopic model. A similar remark holds also for inclusive production, $p + A \rightarrow \pi + $ anything.

However, the amplitude implied in Fig. 2b, sometimes called the Mandelstam mechanism,[3] is not necessarily small. If we dissect it a little bit, as in Fig. 4, we see that it really involves off-shell π-N elastic scattering, which is expected to be large whenever the appropriate sub-energy is near a πN resonance. Although it turns out never to be the <u>dominant</u> piece of the amplitude, at the energy of the 3,3 for example, this term is as large as any other contribution and

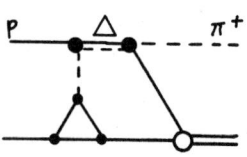

Fig. 4 Microscopic dissection of the Mandelstam mechanism.

therefore must be included as Locher and Weber[4] note. Far from the 3,3 resonance, however, the Mandelstam term is small compared with the rest of the (p,π^+) amplitude. Thus, in examining the Uppsala data at 185 MeV,[5] the Orsay data at 154 MeV,[6] the CERN[7] and SREL[8] data at 600 MeV, the Saclay data[9] at 600 MeV and above, or the PPA data at 2 and 3 GeV,[10] we can safely ignore it.

I have given some theoretical reasons why I think the various multinucleon models may be disregarded. But experimentalists generally find experimental proof more compelling. Thus I offer the $(p,\pi^\pm)$ reactions leading to mirror final states, for which the Mandelstam model, for example, would predict small π^-/π^+ ratios (~1/40), but similar angular distributions, whereas other multinucleon models predict large π^-/π^+ ratios (~1/10) but not necessarily the same angular distributions. We see in Fig. 5 that neither is satisfied. On the other hand, the single-nucleon model forbids (p,π^-) in lowest order, and only allows it through initial-state or final-state charge-exchange, Fig. 6, or by configuration mixing in the nuclear wave functions, Fig. 7. I have shown elsewhere[2] that the data are consistent with this hierarchy of effects.

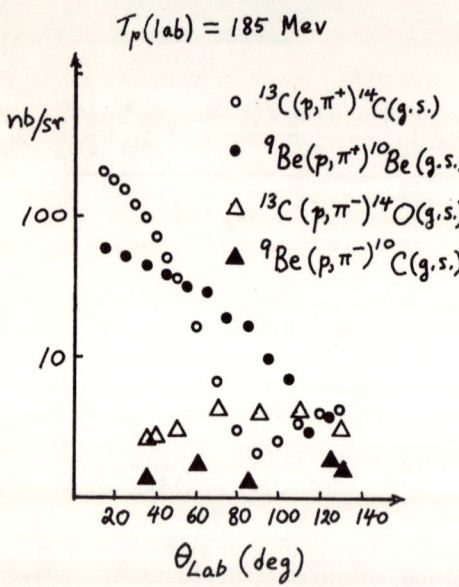

Fig. 5 Experimental differential cross sections for (p,π^+) and (p,π^-) to mirror states, at 185 MeV. (Ref.5)

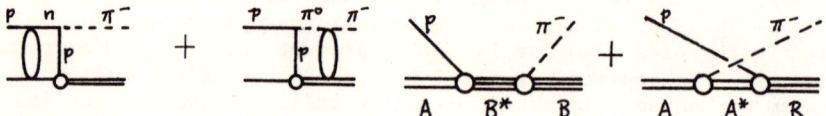

Fig. 6 Charge exchange contribution to (p,π^-).

Fig. 7 Nuclear configuration mixing contribution to (p,π^-).

And, of course, there is also the recent suggestion by Kisslinger and Miller[11] of direct Δ_{++} exchange, Fig. 8. However, recent experiments from Saclay[12] indicate that the Δ_{++} admixture in low-lying final states is so small that this amplitude is unimportant. It is important to note that the configuration mixing contribution, Fig.7, also is present in (p,π^+). Since on both theoretical and experimental

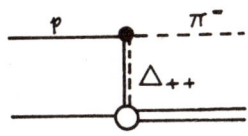

Fig. 8 The Δ_{++} exchange mechanism of (p,π^-) suggested by Kisslinger and Miller.

grounds (Fig.5) we expect it to be nondescript in angular dependence, as well as comparable to pionic stripping above 60° or so, there is no reason to trust the latter mechanism except at forward angles.[2]

By this tortuous process of elimination, we recover the DWBA for calculating the forward part of exclusive absorption or production amplitudes. In this way we have dealt with the problem of approximating the hadronic rescattering corrections. Unfortunately, the reaction mechanism question is proving a bit trickier. Even though we believe (p,π^+) goes mainly by one-nucleon exchange, there is still the question of how to calculate the πNN vertex. The problem arises because the relativistic pseudoscalar or pseudovector couplings require knowledge of the small as well as the large components of the 4-component single-nucleon Dirac spinors. The precise relation of the small to the large components is a dynamical question and in principle requires a relativistic wave function model (such as that presented at this conference by Miller and Weber[13]) for its resolution. To my knowledge, the first person recently to point out the importance of dynamics was G. A. Miller,[14] who noted it in passing. More recently, Bolsterli, et al.,[15] Friar,[16] Eisenberg, et al.[17] and Ho, et al.[18] have worried about how one should go about deriving a phenomenological non-relativistic $H_{\pi NN}$. [Of course the problem was recognized many years before this recent activity by Drell and Henley[19] as well as by Foldy.[20] Also, Barnhill[21] pointed out the ambiguity in the Foldy-Wouthuysen transformation which is the usual starting point for a non-relativistic reduction.] Since the question remains open as yet, most authors have elected to use provisionally one or another of the extreme possibilities

$$H^{eff}_{\pi NN} = \frac{f}{m_\pi} \int d^3x \, \psi^\dagger(\vec{x}) \, \vec{\sigma} \cdot \vec{\nabla}_\pi \, \tau_i \, \psi(\vec{x}) \, \phi_i(\vec{x}) \qquad (5)$$

(static model)

or

$$H^{eff}_{\pi NN} = \frac{f}{m_\pi} \int d^3x \, \psi^\dagger(\vec{x}) \, \vec{\sigma} \cdot \left[\vec{\nabla}_\pi + \frac{\omega_\pi}{2M} (\vec{\nabla}_N - \overleftarrow{\nabla}_N) \right] \tau_i \psi(\vec{x}) \phi_i(\vec{x}) \qquad (6)$$

("Galilean-invariant" model)

Because of the dynamical ambiguity mentioned above, the non-relativistic phenomenological vertex will be a linear combination of the forms (5) and (6), with a strong possibility that the mixing parameter is state-dependent. We also note that of course any Lorentz-invariant amplitude will reduce to a Galilean-invariant (GI) amplitude, i.e. one which is a function only of differences of velocities of physical particles. However, which velocities combine in this way is a strictly dynamical question and there is no need, in particular, for the πNN vertex to have an explicitly GI form. It turns out that there is a significant difference whether the "static" or "GI" form is chosen for calculations at incident energies below 200 MeV, since the "GI" model involves the difference between pion and nucleon velocities and these are nearly equal at these energies. (Thus the 185 MeV calculations which correctly employed the GI form tended to dip at $\theta = 0°$.) Some recent experiments[6] (Fig. 9) give information on the energy dependence of (p,π^+) cross-sections near threshold and, in my opinion, tend to support the static model[22] (although that can't be the whole story because it is obvious that pions at rest can be absorbed by a nucleus, followed by one- or multi-nucleon emission).

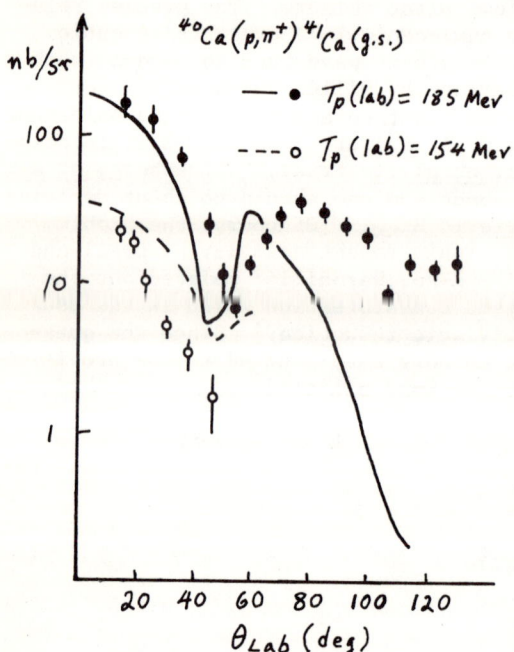

Fig.9 Data on $^{40}Ca(p,\pi^+)^{41}Ca(g.s.)$ and theoretical fits, for E_p(lab)=185 MeV and E_p(lab)=154 MeV, demonstrating the threshold behavior of the "static" model πNN coupling and DWBA.

Having sketched the ideas which have led to the presently favored phenomenology of (p,π^+) and (π^+,p) reactions, I would now like to review the results of recent calculations and discuss the outstanding problems as I see them. There are several difficulties in trying to compare the calculations of various authors. First, different optical model and bound-state wave function parameters have been used by different authors. Second, there are severe numerical difficulties associated with the coordinate-space radial numerical integrals which must be evaluated in the conventional DWBA codes; since only a few authors have discussed their methods for getting around these problems, it is far from clear that all the published results are of equal numerical accuracy. Third, several authors have explicitly used incorrect forms for the πNN vertex and therefore did not calculate what they thought they were calculating.

In Figs. 10, 11 and 12 I show some typical fits by Miller,[14] by Höistad, et al.[23] and by Keating and Wills.[24] There have been many other calculations, of course, using pionic stripping.[25] Because the earlier calculations have been done using substantially different parameters, I recently decided to calculate all of the (p,π^+) differential cross sections for which measurements were available as well as some for which no data yet exist, using a single parameter set, in order to try to learn something about the systematics. As you can see, I felt pretty confident at this point that at least for the forward (p,π^+) cross sections, the effects of rescattering and reaction mechanism could be sifted out pretty thoroughly, leaving only the nuclear structure to be learned. In fact, my level of confidence was that the disentangling could be accomplished only to within a factor of 2. Thus it seemed pointless to spend a lot of computer time to perform this task. I chose to use a fast approximation based on the first two terms of an asymptotic expansion in inverse powers of momentum transfer (the large parameter). This approximation is about 10% accurate overall, compared with direct numerical evaluation of the 3-dimensional DWBA integral, and gave results which in general agreed with those of the aforementioned authors. In the spirit of this rough survey, only deviations substantially greater than a factor of two from expectations were taken to be serious. The calculations were performed by Z. Gromadzki as part of his Ph.D. thesis research.[26]

The first surprise was the comparison of the reactions $^9Be(p,\pi^+)$ $^{10}Be(g.s.)$ with $^{10}B(p,\pi^+)$ ^{11}B at 185 MeV. (Fig.13). The measured differential cross sections differ by a factor of 10, but the angular distributions are almost identical in shape.[5] Now the nuclear spectroscopy of stripping to these states, either in the j-j coupling shell model or the intermediate coupling model of Cohen and Kurath, predicts that (all other things being equal) the cross sections should be the same. The systematics of binding energy difference and A→A+1 in the nuclear radii yields a ^{10}B cross section which is twice as large as the ^9Be cross section, and both with shapes agreeing with the data. However, although the magnitude of the ^9Be calculation is in reasonable agreement (i.e. a factor of 2) with the corresponding data, the ^{10}B one is too low by a factor of at least 5. Lest the reader suspect an error in the spectroscopic factors, in Fig. 14 I exhibit the (d,p) cross sections at 30 MeV on these two nuclei.[27] They are identical to within experimental error. The next surprise occurs in the $1p_{1/2}$ shell. Spectroscopy predicts a ratio of 2 to 1 for the cross sections of $^{12}C(p,\pi^+)$ $^{13}C(g.s.)$ and $^{13}C(p,\pi^+)$ $^{14}C(g.s.)$. Experimentally, this is found with remarkable precision (Fig. 15). However, the theory predicts the cross sections should be the same and, what is worse, much smaller than the data. [In order to fit the $1p_{1/2}$ differential cross section, it was necessary to increase greatly the spin-orbit potential in the bound state well - this was found also by Höistad, et al.] Moving on to the higher shells, we find quite reasonable fits using the common parameter set, both for the $1d_{5/2}$ (Fig. 16) and $2s_{1/2}$ (Fig. 17) shells in ^{13}C and ^{17}O, as well as for the $1f_{7/2}$ state of ^{41}Ca

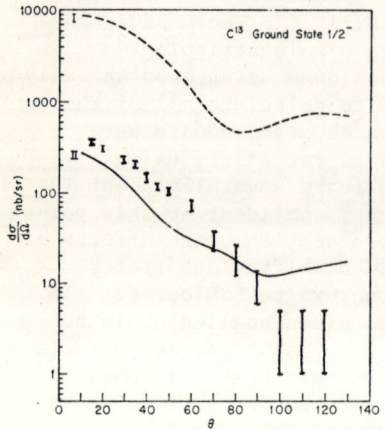

Fig. 10 Fits to $^{12}C(p,\pi^+)^{13}C(g.s.)$ at 185 MeV using DWBA with two different pion optical potentials. (Miller, Ref. 14)

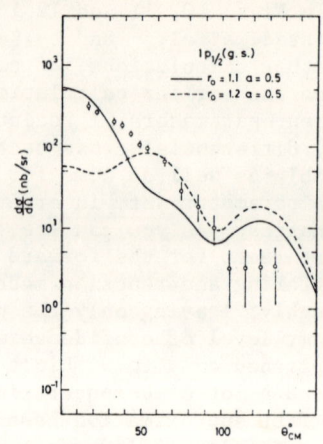

Fig. 11 Same as Fig. 10 except using two sets of bound-state parameters and a fixed pion potential (Höistad, et al., Ref. 23)

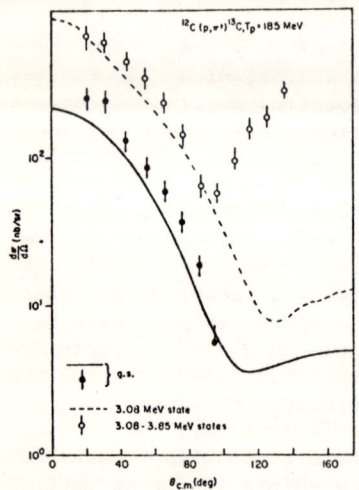

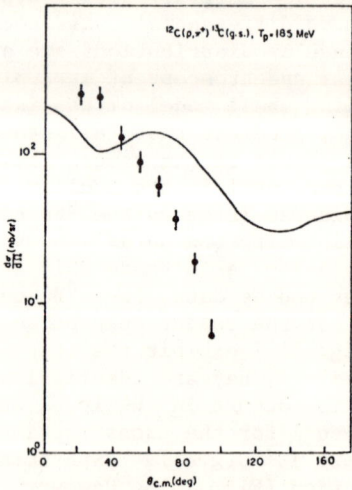

Fig. 12 Fits to $^{12}C(p,\pi^+)^{13}C$(g.s. and 3.08-3.85 MeV) at E=185 MeV with various parameters. (Keating and Wills, Ref. 24).

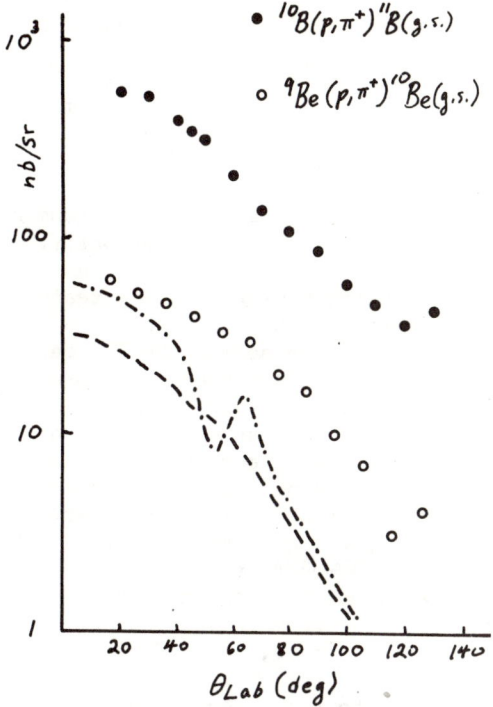

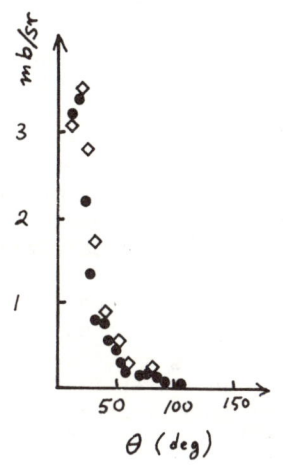

Fig. 14 Comparison of data on $^{10}B(d,p)^{11}B(g.s.)$ (◊) and $^{9}Be(d,p)^{10}Be(g.s.)$ (●) at E_d = 30 MeV (Slobodrian, Ref. 27).

Fig. 13 Comparison of data on $^{10}B(p,\pi^+)^{11}B(g.s.)$ (dots) and $^{9}Be(p,\pi^+)^{10}Be(g.s.)$ (open circles) at E_p =185 MeV, and their respective fits (^{10}B —·—, ^{9}Be ----).

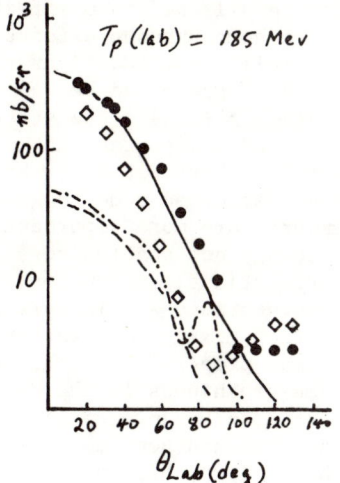

Fig. 15 Comparison between data on $^{12}C(p,\pi^+)^{13}C(g.s.)$ (●) and $^{13}C(p,\pi^+)^{14}C(g.s.)$ (◊). Note the similarity between the angular distributions as well as the factor of two in their magnitudes. The solid curve is a calculation for the ^{12}C case using Woods-Saxon bound-state parameters r_0 = 1.3fm, a = 0.5fm and $V_{s.o.}$ = -0.44V_c. The other two curves are calculations for ^{12}C (—·—·—) and ^{13}C (-----) using $V_{s.o.}$ = -0.19 V_c.

(Fig.9). In general, the 2s-1d and $1f_{7/2}$ shells are much easier to understand than the $1p_{3/2}$ and $1p_{1/2}$ shells.

Calculations have also been undertaken for higher energies. In Fig. 18 I show the recent data on ^{6}Li(p,π^+) ^{7}Li at 600 MeV from Bauer, et al.[9] and Gromadzki's calculation[26] using a pure $1p_{3/2}$ wave function. Unfortunately, the $1p_{3/2}$ spectroscopic factor reduces this cross-section by a factor of 8, so it is probably necessary to use the full intermediate-coupling wave function (which has a substantial $1p_{1/2}$ component) to do the calculation properly.

In summary, the efforts of the past few years to obtain an overall understanding of the dominant features of the (p,π^+) reaction have been successful. We now understand the hierarchy of magnitudes of various effects well enough to be able to tell when they are important and when they are not. We have seen on both experimental and theoretical grounds that the forward (p,π^+) cross sections are dominated by pionic stripping, so that we now have a handle on single-nucleon wave functions at large momentum transfer. In order of increasing importance in determining exclusive amplitudes we find rescattering effects; reaction mechanism; and overwhelmingly, the wave-function of the residual nucleus. The difference between harmonic oscillator and Woods-Saxon wavefunctions can be several orders of magnitude at these momentum transfers. Even when we restrict ourselves to more-or-less realistic wavefunctions such as Woods-Saxon ones, we find great sensitivity to the parameters, particularly the surface thickness parameter, where a 20% change can alter the cross section by a factor of 3 to 10. The spin-orbit interaction generates high-momentum components, so the results are quite sensitive to the magnitude of $V_{s.o.}$. Moreover, examination of the Pauli reduction of the Dirac equation makes clear that relativistic effects, in addition to providing the spin-orbit interaction through Thomas precession, also introduce extra momentum-dependence in the wave-function, even in s-waves. In addition to all the above effects which we know how to put in, we also have some unknown effects: Configuration mixing and short range correlations. Configuration mixing can be important because, say, a $5/2^+$ state can be $|1d_{5/2}\rangle \otimes |g.s.\rangle + |2s_{1/2}\rangle \otimes |2^+\rangle + |1f_{7/2}\rangle \otimes |1^-\rangle +...$, and the states of higher angular momentum are relatively more important at large momentum transfer. Both Miller,[14] and Eisenberg, et al.[25] have discussed this possibility and have even performed some rough calculations which indicate that such admixtures are more important at large angles, but the results were certainly not conclusive and more work needs to be done. Short range correlations, as mentioned by Miller,[14] introduce high momentum components into single particle wave functions first, through configuration mixing, as mentioned above; and second, by changing the average properties of the nuclear medium. We note that if there were some graininess in the density of finite nuclei, it would not have been seen by any experiments (e.g. electron scattering) to date, as Friar and Negele[28] have pointed out, yet would nevertheless change (probably increase) the single particle wave functions at large momentum.

I shall conclude by stating where I think more work needs to be

231

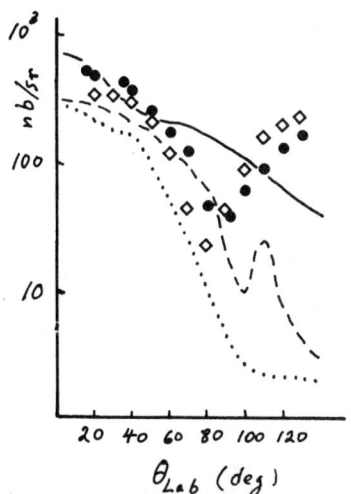

Fig. 16 Comparison of data on (p,π^+) to $1d_{5/2}$ final states in ^{13}C (●) and ^{17}O (◊) at E_p = 185 MeV. Calculations were performed for ^{13}C ($V_{s.o.}$ = -0.44 V_c ——— and $V_{s.o.}$ = -0.19 V_c -----) and ^{17}O ($V_{s.o.}$ = -0.19 V_c ······) In all cases r_0 = 1.3 and a = 0.5.

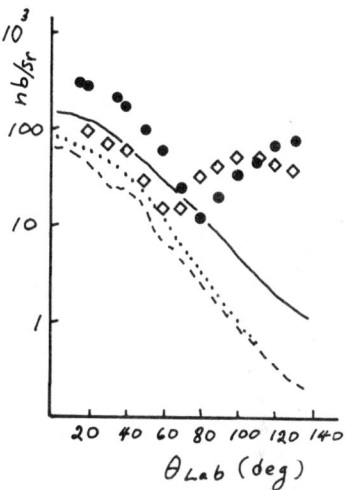

Fig. 17 Comparison of data on (p,π^+) to $2s_{1/2}$ final states in ^{13}C (●) and ^{17}O (◊) at E_p = 185 MeV. The theory was carried out with r_0 = 1.3 fm and a = 0.5 (———) and a = 0.65 (-----) for ^{13}C, and with a = 0.5 (······) for ^{17}O.

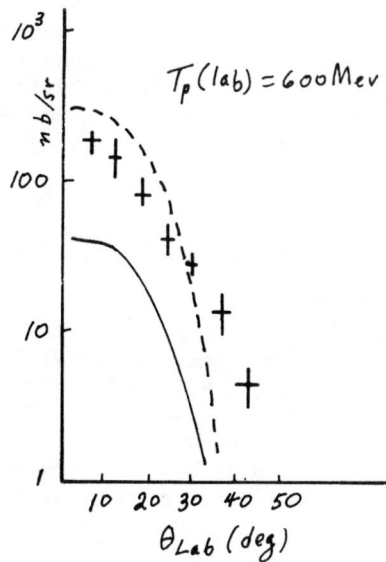

Fig. 18 Comparison of data on ^{6}Li(p,π^+) ^{7}Li(g.s.) at E_p = 600 MeV with calculations based on pure $(1p_{3/2})^3$ for ^{7}Li, with (———) and without (-----) the shell model spectroscopic factor.

done, both on the experimental and on the theoretical side. Experimentally, the systematics of spectroscopic effects, resulting from shell filling with both neutrons and protons, should be explored in a variety of neighboring nuclei. It appears as though the p-shell nuclei are very interesting and they need to be explored in more detail. As we saw with ^{9}Be and ^{10}B, we seem to be learning something here which (d,p) did not tell us. We need to have data for the same nuclei at different energies. For example, there are nice angular distributions for ^{6}Li(p,π^+) ^{7}Li at 600 MeV, but none at 185 MeV. There are good measurements for ^{9}Be, ^{10}B, ^{12}C, ^{13}C, ^{16}O, ^{28}Si and ^{40}Ca at 185 MeV, but none at 600 MeV. If the Mandelstam mechanism is ever to be important in reactions on these heavier nuclei, it will be in the 3,3 region, 300-450 MeV (we already know it is important there in $p+p \rightarrow \pi^+ d$ and $p+d \rightarrow \pi^+ + t$). Thus this looks like a promising energy range. The inverse reaction, (π^+,p), needs to be done more systematically, since it is our only way to investigate the high momentum properties of closed-shell nuclei. At present, the data are rather sparse. And finally, although I have not had time or space to discuss it here, experiments with polarized beams and/or targets look very promising, as I have emphasized elsewhere.[2] At present there exists one (p,π^+) datum, published in 1958 by Heer, et al.[29], taken with a polarized beam. And it is proving surprisingly hard to fit, so I have no doubt such measurements will provide a stringent test of theory. Finally, as we have emphasized earlier,[30] (p,γ) and (γ,p) is a useful complement to (p,π) and should be pursued.

On the theoretical side, there needs to be a lot more work on detailed wavefunctions and on calculations which are constrained by being required to fit the low momentum properties of nuclei. Also, since calculations are so sensitive to the parameters of the models it would be useful to publish all the details of one's wavefunctions when publishing one's results. Finally, there need to be calculations pertinent to polarized-beam experiments, in order to give the experimentalists something to shoot at.

REFERENCES

1. C.H.Q. Ingram, N. W. Tanner and J. J. Domingo, Nucl. Phys. B31 (1971) 331.
   A. Reitan, Nucl. Phys. B29 (1971) 525.
   A. Reitan, Nucl. Phys. B50 (1972) 166.
   Z. Grossmann, F. Lenz and M. P. Locher, Ann. Phys. 84 (1974) 348.
   I. R. Afnan and A. W. Thomas, Phys. Rev. C10 (1974) 109.
   H. W. Fearing, Phys. Rev. C11 (1975) 1210.
   M. Dillig and M. G. Huber, Phys. Lett. 48B (1974) 417.
2. J. V. Noble, "Proton-Induced Pion Production at Medium Energies - Status of Current Theories," Rev. Mod. Phys. (to be published).
3. S. Mandelstam, Proc. Roy. Soc. A244 (1958) 491.
4. M. P. Locher and H. J. Weber, Nucl. Phys. B76 (1974) 400.
   E. Borie, D. Drechsel and H. J. Weber, Z. Physik 267 (1974) 393.
   D. Drechsel and H. J. Weber, Nucl. Phys. B25 (1971) 159.

5. S. Dahlgren, P. Grafström, B. Höistad and A. Åsberg, Nucl. Phys. A204 (1973) 53.
   S. Dahlgren, P. Grafström, B. Höistad and A. Åsberg, Nucl. Phys. A211 (1973) 243.
   S. Dahlgren, P. Grafström, B. Höistad and A. Åsberg, Phys. Lett. 47B (1973) 439.
   S. Dahlgren and P. Grafström, Physica Scripta 10 (1974) 104.
   S. Dahlgren, P. Grafström, B. Höistad and A. Åsberg, Nucl. Phys. A227 (1974) 245.
6. Y. LeBornec, B. Tatischeff, L. Bimbot, I. Brissaud, H. D. Holmgren, J. Källne, F. Reide and N. Willis, in Sixth International Conference on High Energy Physics and Nuclear Structure, Los Alamos, June, 1975, contributed paper #IV. C. 23, p. 265; and to be published.
   Y. LeBornec, B. Tatischeff, L. Bimbot, I. Brissaud, J. P. Garron, H. D. Holmgren, F. Reide and N. Willis, Phys. Lett. 49B (1974) 434.
7. J. J. Domingo, B. W. Allardyce, C.H.Q. Ingram, S. Rohlin, N. W. Tanner, J. Rohlin, E. M. Rimmer, G. Jones and P. Girardeau-Montant, Phys. Lett. 32B (1970) 309.
   J. Rohlin, K. Gabathuler, N. W. Tanner, C. R. Cox and J. J. Domingo, Phys. Lett. 40B (1972) 539.
8. W. Dollhopf, C. Lunke, C. P. Perdrisat, W. K. Roberts, P. Kitching, W. C. Olsen and J. R. Priest, Nucl. Phys. A217 (1973) 381.
9. T. Bauer, R. Beurtey, A. Bondard, G. Bruge, A. Chaumeaux, P. Couvert, H. H. Duhm, D. Garreta, M. Matoba, Y. Terrien, L. Bimbot, Y. LeBornec, B. Tatischeff, E. Aslanides, R. Bertini, F. Brochard, P. Gorodetsky, F. Hibou, High Energy Physics and Nuclear Structure-1975 (AIP Conference Proceedings No. 26, New York, 1975), 437.
10. H. Brody, E. Groves, R. VanBerg, W. D. Wales, B. Maglich, J. Norem, J. Oostans, M. Silverman and G. B. Cvijanovich, Phys. Rev. D9 (1974).
11. L. S. Kisslinger and G. A. Miller, Nucl. Phys. A254 (1975) 493.
12. P. Couvert, private communication.
13. L. D. Miller and H. J. Weber, contributed paper, this conference.
14. G. A. Miller, Nucl. Phys. A224 (1974) 269.
15. M. Bolsterli, W. R. Gibbs, B. F. Gibson and G. S. Stephenson, Phys. Rev. C10 (1974) 1225.
16. J. L. Friar, Phys. Rev. C10 (1974) 955.
17. J. M. Eisenberg, J. V. Noble and H. J. Weber, Phys. Rev. C11 (1975) 1048.
18. H. W. Ho, M. Alberg and E. M. Henley, Phys. Rev. C12 (1975) 217.
19. S. D. Drell and E. M. Henley, Phys. Rev. 88 (1952) 1053.
20. L. L. Foldy, Phys. Rev. 84 (1951) 168.
21. M. V. Barnhill, III, Nucl. Phys. A131 (1969) 106.
22. J. V. Noble, "Threshold Production of Positive Pions and the Form of the πNN Vertex," Phys. Rev. Lett. (to be published).
23. B. Höistad, S. Dahlgren, P. Grafström and A. Åsberg, Physica Scripta 9 (1974) 201.
24. M. P. Keating and J. G. Wills, Phys. Rev. C7 (1973) 1336.
25. J. Letourneaux and J. M. Eisenberg, Nucl. Phys. 87 (1966) 331.

W. B. Jones and J. M. Eisenberg, Nucl. Phys. A154 (1970) 49.
J. M. Eisenberg, R. Guy, J. V. Noble and H. J. Weber, Phys. Lett. 45B (1973) 93.
E. Rost and P. D. Kunz, Phys. Lett. 43B (1973) 17.
A. Reitan, Nucl. Phys. A237 (1975) 465.
J. V. Noble, Nucl. Phys. A244 (1975) 526.
T.-S. H. Lee and S. Pittel, Nucl. Phys. A256 (1976) 509.
J. M. Eisenberg, J. V. Noble and H. J. Weber, Proc. of Fifth Int'l Conf. on High-Energy Physics and Nuclear Structure, Uppsala, Sweden, June, 1973, ed. by G. Tibell (North-Holland, Amsterdam/American Elsevier, New York, 1974), p. 270.
26. Z. Gromadzki, Ph.D. dissertation, U. of Virginia, 1976; Z. Gromadzki and J. V. Noble, in preparation.
27. R. J. Slobodrian, Phys. Rev. 126 (1962) 1059.
28. J. L. Friar and J. W. Negele, Advances in Nuclear Physics 8 (1975) 219.
29. E. Heer, A. Roberts and J. Tinlot, Phys. Rev. 111 (1958) 640.
30. J. M. Eisenberg, J. V. Noble and H. J. Weber, Proc. Int. Conf. on Photonuclear Reactions and Applications, Asilomar, March, 1973, ed. by B. L. Burman (lawrence Livermore Lab., U. of Cal., 1973) p. 957.

Gerald A. Miller (Univ. of Washington): I'd like to make a comment about the proper use of pion distorted waves. As Noble has pointed out, the restriction to single particle final state means that a large part of the two-nucleon production mechanism is contained in the use of π distorted waves. Even with the use of reasonable π distorted waves obtained from the LMM, LPT potentials the cross section is enhanced by a factor of 2 or 3 over that of the Born approximation . Not only does this effect change the magnitude of the cross section, but it reduces the sensitivity to the final state neutron wave function because the pion has a different momentum inside the nucleus than its free value.

Julian Noble: Unless one uses a pion-nucleus scattering matrix with quite singular off-shell properties, the pion momentum inside the nucleus is not very different from its free value (for pions of energy below, say, 60 MeV). The reason the optical potentials have the effect of increasing the cross section is that the momentum space single nucleon wave function typically has a zero at momentum transfer corresponding to the forward angles. Both the proton and pion distortion have the effect of mixing in the derivative of this wave function, which is large.

Kisslinger (CMU): One can do (d,p) at the same momentum tranfers as the (p,π^+). Are the results of (p,π^+) with s.p. mechanism consistent? Can the distortions account for the conflict with a simple transfer mechanism using the spectroscopic factors from (d,p). Also there are nodes in the s.p. wave functions. Could this be involved in these striking results for ratios?

Noble: I know of no case in which the 0° momentum transfer for (p,d) or (d,p) is nearly as large as that for the corresponding (π^+,p) or (p,π^+) reaction. Moreover, medium energy deuterons are much more strongly absorbed than either low energy pions or medium energy protons. Thus the (p,d) and (d,p) reactions are much more peripheral, and their forward cross sections are dominated by an entirely different region of the wave function of the transferred nucleon.

With regard to the nodes in the single-particle wave functions, I do not believe that, once distortion is included, the result is very sensitive to their precise location. When one changes the spin-orbit admixture or the surface thickness parameter, one moves these modes a trifle, but can greatly increase or decrease the overall scale of the wave function in this region.

H. J. Weber (U. of Virginia): I would like to mention that L. D. Miller and I have calculated (see contributions to this conference) a relativistic one-neutron exchange mechanism for the (p,π^+) reaction which shows surprising sensitivity to the elementary (πNN) vertex, i.e., to the off-shell exchanged neutron, in addition to the nuclear wave function.

J. Schiffer (Argonne - U. of C.): I would like to express the outrage of a nuclear structure physicist at the violence you are doing to the 1p shell. If we understand one thing in nuclear structure then it is the structure of p-shell nuclei. You are not free to introduce a 40% spin orbit force at will. If there are order of magnitude discrepancies one whould look to the reaction theory for corrections and not allow the structure calculation to change arbitrarily, unless one can make certain that the fit to the vast body of experimental data is preserved.

Noble: Let me note, first of all, that I have just finished saying that one of the things theorists must do is be less cavalier about their wave functions, and constrain them by what is known at low momentum transfer. This includes binding energies and a few low-order moments of the probability distribution. As I noted, if one replaced the usual Woods-Saxon wells by ones with small-amplitude saw-teeth on the bottom, it would not affect any of the low-momentum properties, but would greatly alter the asymptotic behavior of momentum space single particle wave functions.

Insofar as the existence of order-of-magnitude discrepancies is concerned, let me remind you that five years ago the discrepancies were two orders of magnitude and were common. Today the agreement is generally much better than a factor of 10, systematically, and I consider that a good sign. I hope that because I have chosen to display the worst, and therefore most interesting cases, I have not given you the impression that the theory is in disarray. I feel that, within its limitations, pionic stripping is a good theory.

Pierre Radvanyi (Institut de Physique Nucleaire, Orsay): In the $^{16}O(\pi^+,p)^{15}O$ experiment, the ratio of the cross sections for the excitation of the two hole states $(P3/2)^{-1}$ and $(P1/2)^{-1}$ is about 10 even at small angles. These small angles correspond in momentum transfer (about 2.5 fm^{-1}) to the largest angle measured for the classical $(^3He,\alpha)$ pick-up reaction on the same nucleus, where a ratio close to two has been found for the two hole states. This seems to indicate that the large factor observed in (π^+,p) arises from the reaction mechanism rather than from a nuclear structure affect.

J. Noble: The subject of why (p,π^+) or (π^+,p) is different from (d,p), (p,d) or $(^3He,\alpha)$ is a fascinating one and I would be glad to give, say, an hour's seminar on the subject. The forward differential cross section in either kind of reaction is determined by the forward momentum transfer, because that, together with distortion, determines which region of the wave function is important. At backward angles, many secondary mechanisms contribute. Thus it is impossible to compare back-angle $(^3He,\alpha)$ with forward (π^+,p).

Gerald A. Miller (U. of Washington): Comment in response to Dr. Radvanyi - We do understand the reaction mechanism. The discrepancies have been reduced from factors of 100 to factors of 2. The $P_{3/2}$, $P_{1/2}$ ratios in $O^{16}(\pi^+,p)N^{15}$ is almost completely understood in terms of the crudest thing you can do. The $P_{3/2}$ nucleon has more angular momentum than a $P_{1/2}$ nucleon and is better able to make up the momentum mismatch. The (π,p) reaction is much more sensitive than a (p,d) reaction even if the momentum transfer in the (π,p) reaction is only .5 fm^{-1} greater than in the (p,d) reaction. This is because you are in a region where the nucleon wave function falls very rapidly.

T. S. H. Lee (Argonne National Lab): Can one deal with the ambiguities involved in production operator? Can we live with those ambiguities and still hope to learn nuclear structure information from the (p,π^+) reaction? The ambiguity is not resolved. Galilean invariants form and static form can have 20% difference in forward angle.

J. Noble: At larger energies (say 600 MeV) the ambiguity is much less crucial. Also, the difference between "GI" and "static" models is state-dependent and distortion-dependent. Finally, I think several authors have found small differences because they evaluated the GI incorrectly. But without being able to dig into their computer codes, I cannot say for sure who did what.

PION-NUCLEUS TOTAL CROSS-SECTION DATA FROM LAMPF AND BNL[*]

M. D. Cooper
University of California, Los Alamos Scientific Laboratory
Los Alamos, New Mexico 87545

ABSTRACT

New measurements of pion-nucleus total cross sections have been made at LAMPF and BNL. The results from LAMPF include measurement of the difference of the rms neutron and proton radii of ^{48}Ca to be 0.08 ± 0.02 and that of ^{18}O to be 0.19 ± 0.02. The BNL measurements provide a new phenomenology on the downshift and spreading of the (3-3) resonance in nuclei from the first data on heavy nuclei. A new technique for handling the Coulomb effects in total cross-section measurements is discussed.

INTRODUCTION

The two new experiments to measure pion-nucleus total cross sections at LAMPF and Brookhaven National Laboratory (BNL) complement four previous measurements[1-4] of this kind by other groups. For the most part, the early measurements concentrated on the effects of the (3-3) π-nucleon resonance on the nuclear cross sections in light nuclei. Although there is significant overlap, the new measurements emphasize different features of the cross sections.

The LAMPF effort has concentrated on the comparison of isotopes and isotones to extract information about nuclear radii, the measurement of cross sections at low energies (down to 25 MeV), and on the measurement of heavy element cross sections. Of the heavy elements, special attention is being paid to ^{165}Ho, where measurements are being obtained for aligned and unaligned targets. The BNL group has precisely measured the location and width of the (3-3) resonance in nuclei, with special attention being paid to the high-Z nuclei.

Due to the limited time available, this talk will be restricted to three aspects which are new results. The first is the measurement of the neutron radii of ^{48}Ca and ^{18}O from the comparison of total cross sections for ^{48}Ca-^{40}Ca and ^{18}O-^{16}O. The second aspect is the BNL measurements and their interpretation of the location and width of the (3-3) resonance. The last subject will deal with a new technique for handling Coulomb-nuclear interference in the measurements. The proper handling of this part of the data analysis is essential in order to determine reliable cross sections at low energies or for heavy nuclei.

All of the total cross-section measurements are of the beam attenuation type.[5] The ideal measurement of total cross sections by the attenuation method would have two features: (1) a beam of N_0 stable particles incident on the target, and (2) counters of unit

---

[*]Work performed under the auspices of the U. S. Energy Research and Development Administration.

efficiency which do not disturb a pencil beam. If a small counter after the target measures N particles, the total cross section is given by

$$\sigma_{tot} = \frac{1}{nT} \ln \frac{N_0}{N} . \qquad (1)$$

Despite the simplicity of the idea, reality poses enough complexities to make the measurement a major undertaking. Five areas which require special attention are: (1) counter efficiencies, absorption and accidentals, (2) interactions in the beam detectors which make target-in and -out measurements necessary, (3) finite geometry of the beam, (4) pion decay which must be handled by elaborate Monte Carlo computer programs, and (5) the charge of the pion. The charge of the pion provides difficulties because it allows multiple Coulomb scattering and because the Coulomb amplitude has a singularity at zero degrees. In Sec. IV, special attention will be paid to the proper handling of this singularity.

NEUTRON RADII OF ^{48}Ca AND ^{18}O

The accurate determination of neutron root mean square (rms) radii from pion total cross sections is made possible by the isospin coupling of the π-nucleon system through the (3-3) resonance. The strong π⁻-neutron coupling accentuates effects due to neutrons, which gives one the sensitivity necessary for an accurate measurement. The use of accurately measured proton rms radii from electron-scattering experiments[6] separates the problem of finding neutron radii from that of just learning properties of the matter distribution. To the extent that Coulomb effects can be properly accounted for in theoretical interpretations of the cross sections, obtaining the same result with both π⁻ and π⁺ is a built-in systematic check of the pion physics.

Even with the special π⁻ neutron coupling, the absolute cross sections are not very sensitive to neutron distributions. However, the comparison of cross sections from isotopes of the same element are sensitive to differences in the neutron radii of these isotopes. Such a comparison cancels many experimental and theoretical uncertainties. For example, at kinetic energies above 100 MeV for pions on ^{40}Ca, the contribution of Coulomb-nuclear interference to the absolute cross section is large, but in the ^{48}Ca-^{40}Ca difference, it is negligible. On the theoretical side, models which predict very different values for the ^{40}Ca cross section predict nearly the same value for the difference. It is this model independence which allows the determination of relative radii.

In the cases where a good educated guess can be made for the neutron radius of one isotope, then the neutron radius of the other can be determined. Two such cases are ^{40}Ca and ^{16}O. In what follows, it is assumed that the proton and neutron rms radii

$$r_p(^{48}Ca) = r_p(^{40}Ca) = r_n(^{40}Ca) \text{ and}$$
$$r_p(^{18}O) = r_p(^{16}O) = r_n(^{16}O) . \qquad (2)$$

The equalities to the left in Eq. (2) are verified by electron scattering[6] to ±0.01 fm. The equalities to the right are believed true because of the closed-shell nature of ^{40}Ca and ^{16}O and are verified by theoretical estimates (±0.02 fm). If the theories are correct, the assumptions in Eq. (2) lead to smaller uncertainties than those from other sources in the extraction of radii for ^{48}Ca and ^{18}O.

Predictions for the difference between the total cross sections for ^{48}Ca and ^{40}Ca or for ^{18}O and ^{16}O can be obtained from the optical model. The solid curves in Fig. 1 are calculated using the Kisslinger[7] model with a Fermi density distribution. Calculations on oxygen employ a modified Gaussian density. The curves are labeled according to the difference between the rms neutron and proton radii of ^{48}Ca.

Some qualitative features of the curves may be understood from free π^- nucleon scattering. The difference in the predicted cross sections for a given $r_n - r_p$ is approximately the free π^--neutron cross section times the number of unshielded neutrons outside the absorbing core of ^{40}Ca. The point where all the curves cross is where the model predicts the nuclei are largely transparent, and eight times the π^- nucleon cross section at this energy is in fair agreement with the value of 200 mb at the crossing point. The energy of the crossing point is model-dependent. The divergence at lower energies is probably a Coulomb effect.

The thick band which is the $r_n = r_p + 0.4$ curve represents the ranges of predictions as a function of the parameters of a Fermi distribution when the rms radii are held fixed. The change which the band represents in the 10-90% skin thickness is ±0.2 fm. This band puts a limit of ±0.02 fm on the measurement of $r_n - r_p$ until measurements are made which further restrict the models.

The dashed curve for $r_n - r_p$ is calculated in the local Laplacian[8] model with a Fermi density distribution. Two features are evident: (1) above 150 MeV there is very little difference between the models, and (2) below 150 MeV there is large model dependence. These characteristics are found to be quite general for a large class of calculations. For example, they persist in the ^{13}C-^{12}C system when calculated using the phenomenological parameters of

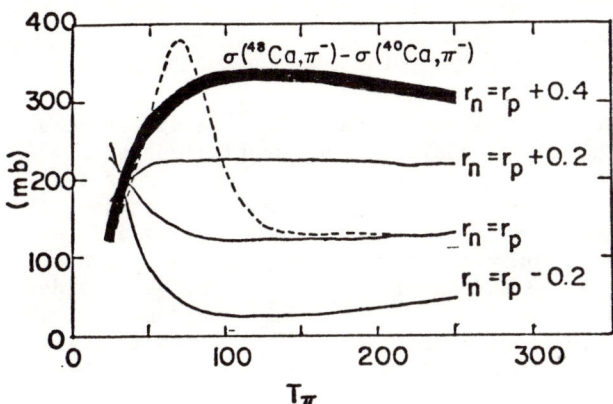

Fig. 1. Predictions for the differences in the π^- cross section of ^{48}Ca and ^{40}Ca. The solid lines are calculated with the Kisslinger model and the dashed with the local Laplacian model. The band represents the uncertainty in the rms radii due to variations of parameters in the Fermi density distribution.

Sternheim and Auerbach,[9] even though the predicted total cross sections are quite different from those calculated from free π^--nucleon parameters. It appears therefore that the difference of cross sections provides a happy separation of phenomena. At higher energies, structural information may be extracted without model dependence. At low energies, details of the dynamics can be learned by employing the structure information learned at higher energies. The reason that all reasonable models predict the same values for the differences is not quite clear, but may be related to the blackness of the nucleus in this energy region.

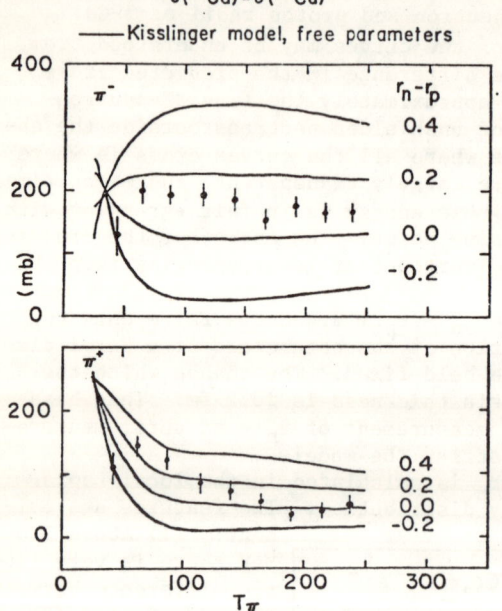

Fig. 2. <u>Preliminary</u> data from LAMPF for the ^{48}Ca-^{40}Ca cross-section differences. The solid curves are the same predictions as those of Fig. 1 for the differences for several values of $r_n - r_p$.

The <u>preliminary</u> data from LAMPF for the ^{48}Ca-^{40}Ca differences are shown in Fig. 2. The errors shown include statistics and systematic errors arising from the extrapolation of measured cross sections to zero degrees and the Coulomb-nuclear interference correction. For the π^+ statistics dominate, and for the π^- the small systematic contribution will probably be removed after a more complete analysis. The lowest energy point at 43 MeV should, at this time, be regarded with some skepticism as it requires a large model-dependent Coulomb-nuclear interference correction. Application of the technique given in Sec. IV to the data and the calculated curves should give this point equal credibility with the others.

Qualitatively, the π^- data agree quite well with the Kisslinger model, but the π^+ data diverge from a single curve at lower energies. The π^- data do not show the pronounced peak which is calculated from the local Laplacian model (shown as the dashed curve in Fig. 1).

The five highest energy points in the model-independent region have been combined to obtain a best value of $r_n - r_p$ for ^{48}Ca:

$$\pi^-: \quad r_n(^{48}\text{Ca}) - r_p(^{48}\text{Ca}) = 0.08 \pm 0.02 \text{ fm}$$
$$\pi^+: \quad r_n(^{48}\text{Ca}) - r_p(^{48}\text{Ca}) = 0.03 \pm 0.04 \text{ fm} \quad .$$
(3)

The errors are experimental and do not reflect uncertainties in the models used to extract the radii. The goal of obtaining the same value for both pion signs is barely achieved, but any discrepancy can be removed by a slight adjustment of the relative proton radii of ^{48}Ca and ^{40}Ca. This will raise the π^+ value in Eq. (3) without affecting the π^- value substantially. The necessary change is less than the uncertainty in the measured radii from electron scattering.

The values for $r_n - r_p$ in ^{48}Ca are in good agreement with the latest alpha particle[10] and proton[11] scattering measurements, as well as Coulomb energy differences.[12] Most theoretical Hartree-Fock calculations[13] [which find $r_n(^{40}\text{Ca}) - r_p(^{40}\text{Ca}) = 0.2$ in contrast to Eq. (2)] find 0.19 for $r_n(^{48}\text{Ca}) - r_p(^{48}\text{Ca})$, in disagreement with this measurement.

Whether this discrepancy between theory and experiment persists will, in the least, depend on the completion of the LAMPF group's analysis, which includes (1) refined data reduction to further reduce the error bars and remove Coulomb effects, (2) more detailed calculations on the sensitivity of the results to the value of $r_n - r_p$ in ^{40}Ca and the value of $r_p(^{48}\text{Ca}) - r_p(^{40}\text{Ca})$, (3) optical-model calculations with the Hartree-Fock matter densities,[13] (4) further tests of the pion physics and high-energy model independence by calculation with models which put in form factors to cut off pathologic off-shell behavior,[14] and (5) consistent understanding of data taken on ^{44}Ca, ^{45}Sc, and ^{51}V.

A case which has not been studied so well, either experimentally or theoretically, as the calcium isotopes, is that of ^{18}O and ^{16}O. With analogous qualifications to those made about the Ca isotopes, the data are presented in Fig. 3. Under the assumptions in Eq. (2), the highest five energy points yield values for $r_n(^{18}\text{O}) - r_p(^{18}\text{O})$ of

$$\pi^-: \quad r_n(^{18}\text{O}) - r_p(^{18}\text{O}) = 0.185 \pm 0.015 \text{ fm}$$
$$\pi^+: \quad r_n(^{18}\text{O}) - r_p(^{18}\text{O}) = 0.202 \pm 0.040 \text{ fm} \quad .$$
(4)

PHENOMENOLOGY OF π-NUCLEUS TOTAL CROSS SECTIONS

The cross sections obtained by the BNL group were analyzed in the conventional way.[5] For the heavy nuclei, the Coulomb-nuclear interference correction is very large. It was found that

$$\sigma_{AV} = [\sigma(\pi^+) + \sigma(\pi^-)]/2 \qquad (5)$$

was largely insensitive to the model used to calculate the Coulomb-nuclear interference correction even for heavy nuclei; this average

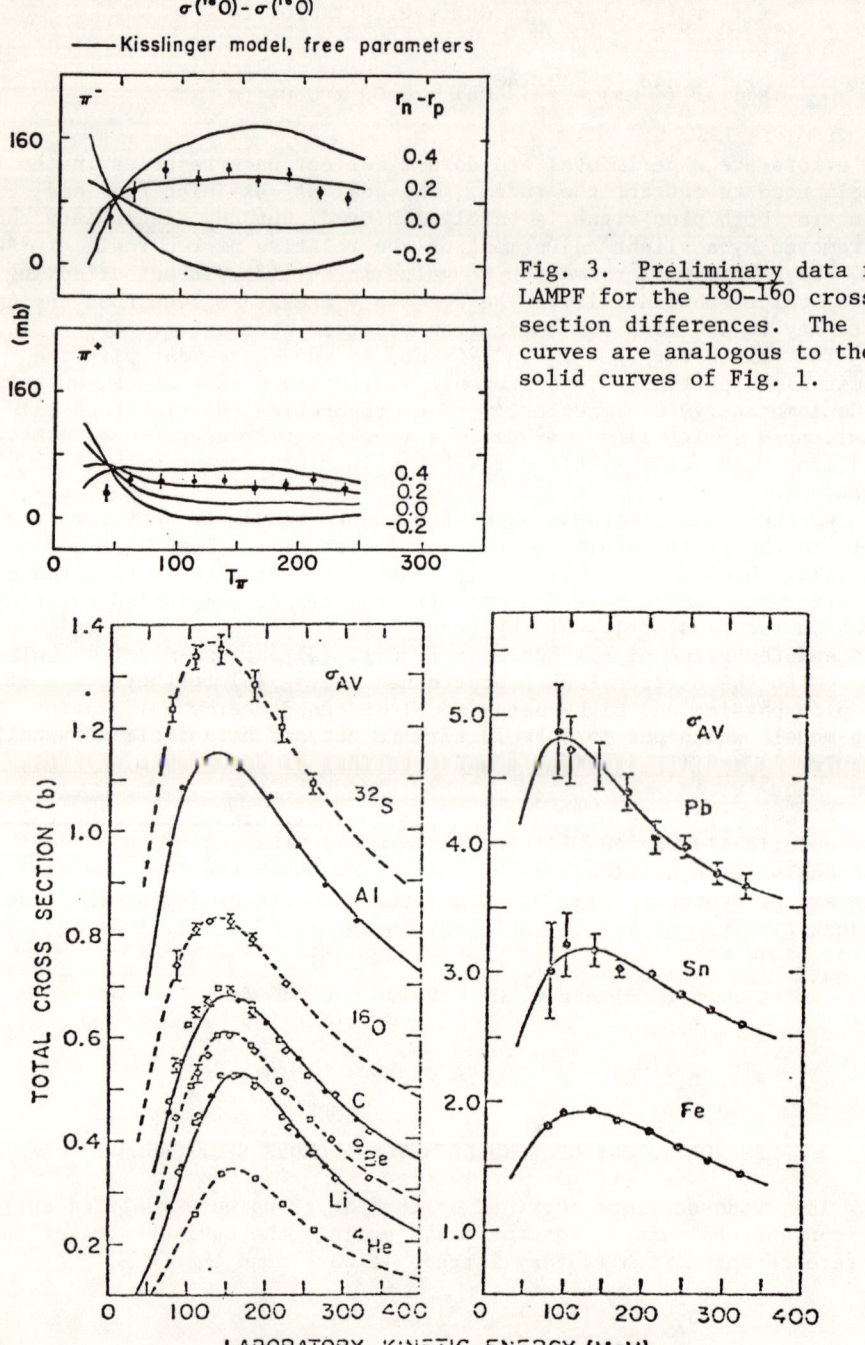

Fig. 3. <u>Preliminary</u> data from LAMPF for the $^{18}O-^{16}O$ cross-section differences. The curves are analogous to the solid curves of Fig. 1.

Fig. 4. π^+ and π^- average total cross sections from the BNL group (solid lines) and two previous experiments (dashed lines).

cross section does not necessarily satisfy an optical theorem. The individual cross sections are not reliably extracted, although the application of the techniques of Sec. IV may overcome this difficulty. The results for σ_{AV} are shown in Fig. 4 along with the results of two previous experiments.

Two qualitative features are evident from Fig. 4: (1) the peak position shifts downward with increasing mass A, and (2) the peak broadens for very heavy nuclei. The BNL group has developed a simple phenomenological model for the resonance shape in order to extract a peak maximum and width. The model is a representation of the cross section in each partial wave (up to some maximum) as a resonance. Thus, it reflects the effects of the π-nucleon resonance in each π-nucleus partial wave. Using an L-dependent resonance energy

$$E_R = E_0 + E_1 L(L + 1) \tag{6}$$

and a parameterization of the width in pion nucleus c.m. wave number k

$$\Gamma = \Gamma_1 + \Gamma_2 = ck + \Gamma_2 , \tag{7}$$

the result is

$$\sigma_{tot}(E) \approx \frac{2\pi \Gamma_1}{k^2 E_1} \left[\tan^{-1}\left(\frac{E_0 + E_1 L_{max}^2 - E}{\Gamma/2}\right) - \tan^{-1}\left(\frac{E_0 - E}{\Gamma/2}\right) \right] . \tag{8}$$

Equation (8) has the form of a modified Breit-Wigner line shape

$$\sigma_{tot}(E) \approx \frac{\pi(R + 1/k)^2 \Gamma_1 \Gamma}{[(E - E_0)^2 + (\Gamma/2)^2]} \cdot \left(1 + \frac{E_1 L_{max}^2 (E_0 - E)}{(\Gamma/2)^2}\right) . \tag{9}$$

The quantities E_0, E_1, c, and Γ_2 are parameters of the fit and $L_{max} \approx KR + 1$, where R is the equivalent spherical radius of the nucleus (1.291 times the rms radius). E is the total energy in the c.m. minus (A-1) times the nucleon mass.

The parameters which result from their fits with Eq. (8) (the curves in Fig. 4) are roughly parameterized as

$$\begin{aligned} E_0 &\approx 1227 - 16\ A^{1/3} , \\ E_1 &\approx 15\ A^{-2/3} , \\ \Gamma_1(E_0) &\approx -67 + 54\ A^{1/3} , \\ \Gamma_2 &\approx -9 + 55\ A^{1/3} . \end{aligned} \tag{10}$$

The coefficients in Eq. (10) are largely dependent on the Fe cross sections and would be improved if there were better data at the upper end of the periodic table. To obtain this data, one will need to experimentally control the amount of Coulomb multiple scattering and to handle the Coulomb-nuclear interference.

The form of Eq. (10) allows the cross section to exceed the geometric limit of $2\pi(R + 1/k)^2$ at $E = E_0$ when Γ_1 is greater than Γ_2. This is presumably due to the tails of the density distribution which deviate from a black disk. Dover[15] will discuss what can be learned from the excesses beyond the geometric limit.

Equation (9) does not describe a resonance in the usual sense since it occurs in many partial waves as indicated by the modulating factor. Therefore, one should use care in the use of the term "resonance energy." In particular, the energy E_R is not simply related to the point where the real part of the π-nucleus amplitude is zero. In any case, any successful theory of π-nucleus scattering will need to agree with the phenomenology of Eq. (10).

COULOMB-NUCLEAR INTERFERENCE

The conventional way to handle the Coulomb singularity at zero degrees has been to separate the elastic scattering amplitude

$$f = f_c + f_N \tag{11}$$

into a Coulomb amplitude f_c and a residual nuclear amplitude f_N. Then the experimental data, $\sigma_{exp}(\Omega)$, is calculated for finite angles[5] in analogy with Eq. (1) by taking N as the number of events detected by a circular counter of solid angle Ω about the beam. The effects of the singularity in f_c are removed according to

$$\sigma(\Omega) = \sigma_{exp}(\Omega) - \int_\Omega^{4\pi} d\Omega' \left\{ |f_c|^2 + 2\mathrm{Re}\, f_c^* f_N \right\} . \tag{12}$$

Then σ_{tot} is identified as the constant term in a polynomial fit in Ω to $\sigma(\Omega)$ in Eq. (12), i.e., extrapolation to zero solid angle.

Up to the present time there has been no practical alternative to using a theoretical model of the π-nucleus interaction to estimate f_N in the calculation of the last term of Eq. (12) (the "Coulomb-nuclear interference integral"). Cooper and Johnson[16] have addressed three questions regarding the above procedure. First, what optical theorem does $\sigma(0)$ satisfy? Second, is Eq. (12) well represented by a polynomial when f_N is taken from a model? And lastly, is there any way to extract model-independent quantities from $\sigma_{exp}(\Omega)$?

The results of Ref. 16 may be summarized as follows. Write the amplitude f_N in a partial wave expansion

$$f_N = \frac{1}{2ik} \sum_L (2L + 1)\, e^{2i\sigma_L} \left(e^{2i\delta_L^N} - 1 \right) P_L(\cos\theta) , \tag{13}$$

and introduce the new amplitude $\tilde{f}_N$, which does not contain the Coulomb phases σ_L,

$$\tilde{f}_N = \frac{1}{2ik} \sum_L (2L+1) \left(e^{2i\delta_L^N} - 1 \right) P_L(\cos\theta) \quad . \tag{14}$$

Both f_N and $\tilde{f}_N$ contain Coulomb distortion via the nuclear phase shifts δ_L^N. Then it can be shown that

$$\sigma(0) = \frac{4\pi}{k} \text{Im } \tilde{f}_N(0) \quad , \tag{15}$$

and that

$$\sigma(0) + \int_0^{4\pi} d\Omega' \left\{ 2\text{Re } f_c^* f_N \right\} = \frac{4\pi}{k} \text{Im } f_N(0) \quad . \tag{16}$$

Equations (13), (14), and (15) state that $\sigma(0)$, the normal definition of the total cross section, is only easily related to f_N through a model; f_N is the amplitude to which many other experiments are sensitive, e.g., elastic scattering.

Equation (16) suggests that f_N might be obtained directly by fitting the manifestly model-independent quantity

$$\sigma_{exp}(\Omega) - \int_\Omega^{4\pi} d\Omega' |f_c|^2 \quad . \tag{17}$$

The residual of Eq. (17) still contains the Coulomb-nuclear interference integral, which is not well behaved at the origin, and so a polynomial is not a good representation for Eq. (17), especially at low energies for high-Z targets. In fact, if the theoretical approximation for f_N used to evaluate Eq. (12) is not exceedingly good (it is very hard to calculate f_N accurately enough in practice), then $\underline{\sigma(\Omega)}$ <u>will not be well represented by a polynomial</u>, and this can lead to substantial error in the determination of $\tilde{f}_N(0)$. The solution to the difficulty is to fit the data with a function which explicitly contains the effects of the interference integral, i.e., to fit Eq. (17) with

$$\left(\frac{4\pi}{k} \text{Im } f_N(0) + \sum_{i=1} A_i \Omega^i \right) \cos W + \left(\frac{4\pi}{k} \text{Re } f_N(0) + \sum_{i=1} B_i \Omega^i \right) \sin W$$
$$+ \sum_{i=1} C_i \Omega^i \quad , \tag{18}$$

where

$$W = \gamma \ln\left(\frac{\Omega}{4\pi}\right) - 2\sigma_0 \quad , \tag{19}$$

and $f_N(0)$, A_i, B_i, and C_i are free parameters. The functions γ and σ_0 are the Coulomb parameter and the s-wave Coulomb phase shifts. The number of free parameters needed is determined on statistical grounds. Note that in the limits of high energy and small charge, W goes to zero and the method is just polynomial extrapolation applied to Eq. (17). However, W is significantly different from zero for pions in oxygen at T_π = 170 MeV. Note that with sufficiently accurate data, both the real and imaginary part of $f_N(0)$ can be determined. In certain energy and angle regimes, a total cross-section experiment may actually be more sensitive to Re $f_N(0)$ than to Im $f_N(0)$, i.e., where the average value of W is near 90°. The anticipated success of this method should remove many of the Coulomb difficulties alluded to in Secs. II and III.

The presence of the factor $\exp(2i\sigma_\ell)$ in Eq. (13) implies that for heavy elements and low energies, f_N may have some unusual behavior. In Pb at 50 MeV, Im $f_N(0)$ can be negative! This is different from the behavior of Im $\bar{f}_N(0)$, which is bound to be positive by unitarity. The fact that Im $f_N(0)$ can be negative is purely a Coulomb effect due to the $\exp(2i\sigma_\ell)$ factor. A more complete discussion of the physical implications of this effect can be found in Ref. 16.

The above technique has yet to be applied to data, but analogous ideas[17] have been applied to small-angle elastic scattering experiments. The nearly model-independent technique finds that the real part of the charge-independent amplitude is zero at T_π = 178 ± 4 MeV for [16]O. This value is substantially different from the model-dependent analysis[18] which found the zero at T_π = 163 ± 3. The difference is important in weighing the validity of pion-scattering theories and emphasizes the need to keep theoretical models as distant from data reduction as possible.

The author would like to express his appreciation to the members of the LAMPF group, which includes G. Burleson, J. Calarco, M. Cooper, D. Hagerman, I. Halpern, M. Jakobson, R. Jeppeson, K. Johnson, L. Knutson, R. Marrs, H. Meyer, and R. Redwine. The aligned [165]Ho measurements are being performed as a collaboration between the LAMPF group and J. Becker, T. Fisher, and B. Watson from Lockheed Palo Alto Laboratories and H. Marshak of the National Bureau of Standards. This research is supported by Lockheed Independent Research Funds and by the National Bureau of Standards. The cooperation of the scientists at the Brookhaven National Laboratory is greatfully acknowledged. Mikkel Johnson, with whom I have collaborated on the Coulomb problem, has been a constant source of good ideas on the entire subject.

REFERENCES

1. F. Binon, P. Duteil, J. P. Garron, J. Gorres, L. Hugon, J. P. Piegneux, C. Schmit, M. Spighel, and J. P. Stroot, Nucl. Phys. B17, 168 (1970).
2. B. W. Allardyce, C. J. Batty, D. J. Baugh, E. Friedman, G. Heymann, M. E. Cage, G. J. Pyle, G. T. A. Squier, A. S. Clough, D. F. Jackson, S. Murugesu, V. Rajaratnam, Nucl. Phys. A209, 1 (1973); B. W. Allardyce, C. J. Batty, D. J. Baugh, E. Friedman, G. Heymann, J. L. Weil, M. E. Cage, G. J. Pyle, G. T. A. Squier, A. S. Clough, J. Cox, D. F. Jackson, S. Murugesu, V. Rajaratnam, Phys. Lett. 41B, 577 (1972); A. S. Clough, G. K. Turner, B. W. Allardyce, C. J. Batty, D. J. Baugh, J. D. McDonald, R. A. J. Riddle, L. H. Watson, M. E. Cage, G. J. Pyle, and G. T. A. Squier, Phys. Lett. 43B, 476 (1973).

3. C. Wilkin, C. R. Cox, J. J. Domingo, K. Gabathulev, E. Pedroni, J. Rohlin, P. Schwaller, and N. W. Tanner, Nucl. Phys. $\underline{B62}$, 61 (1973).
4. N. D. Gabitzsch, G. S. Mutchler, C. R. Fletcher, E. V. Hungerford, L. Coulson, D. Mann, T. Witten, M. Furić, G. C. Phillips, B. Mayes, L. Y. Lee, J. Hudomalj, J. C. Allred, and C. Goodman, Phys. Lett. $\underline{47B}$, 234 (1973).
5. J. P. Stroot, "Experiments in Pion Nucleus Physics" in <u>Lectures from the LAMPF Summer School on the Theory of Pion Nucleus Scattering</u>, W. R. Gibbs and B. F. Gibson, V, editors (1973).
6. H. R. Collard, L. R. B. Elton, R. Hofstadter, <u>Nuclear Radii II</u> in the Landolt-Börnstein Series, Springer-Verlag (Berlin, 1967); R. C. Barrett, Rep. Prog. Phys. $\underline{37}$, 1 (1974).
7. L. S. Kisslinger, Phys. Rev. $\underline{98}$, 761 (1955).
8. H. K. Lee and H. McManus, Nucl. Phys. $\underline{A167}$, 257 (1971); G. Fäldt, Phys. Rev. $\underline{C5}$, 400 (1972); and J. H. Koch and M. Sternheim, Phys. Rev. $\underline{C6}$, 1118 (1972).
9. M. M. Sternheim and E. H. Auerbach, Phys. Rev. Lett. $\underline{25}$, 1500 (1970).
10. G. M. Lerner, J. C. Hiebert, L. L. Rutledge, Jr., C. Papanicolas, and A. M. Bernstein, Phys. Rev. $\underline{C12}$, 778 (1975).
11. G. D. Alkhazov, S. L. Belostotsky, P. A. Domchenkov, Yu. V. Dotsenko, N. P. Kuropatkin, M. A. Schuvaev, and A. A. Vorobyov, Phys. Lett. $\underline{57B}$, 47 (1975).
12. J. A. Nolan, Jr. and J. P. Schiffer, Annu. Rev. Nucl. Sci. $\underline{19}$, 1560 (1969).
13. J. W. Negele, Phys. Rev. $\underline{C1}$, 1260 (1970).
14. R. H. Landau, S. C. Phatak, and F. Tabakin, Ann. of Phys. $\underline{78}$, 299 (1973); S. C. Phatak, F. Tabakin, and R. H. Landau, Phys. Rev. $\underline{C7}$, 1803 (1973).
15. C. Dover, companion paper on total pion-nucleus cross sections at this conference.
16. M. D. Cooper and M. B. Johnson, Nucl. Phys. $\underline{A260}$, 352 (1976).
17. M. B. Johnson and M. D. Cooper, contributed papers to this conference.
18. G. S. Mutchler, C. R. Fletcher, L. V. Coulson, D. Mann, N. D. Gabitzsch, G. C. Phillips, B. W. Mayes, E. V. Hungerford, L. Y. Lee, C. Goodman, and J. C. Allred, Phys. Rev. $\underline{C11}$, 1873 (1975).

J. Schiffer (Argonne National Lab): The pion total cross sections are important and valuable data. But it seems to me a mistake to consider only effects of neutron radii. As we saw from the elastic scattering analysis, there are many questions and complications. There are likely to be similar problems here. There are reliable data on n uclear radii, not the ones you mentioned but from sub-Coulomb transfer, and they tend to be consistent with Hartree-Fock calculations.

M. Cooper: It is too early in the field of pion physics to discount pions as a useful tool for determining radii, especially based on experiments done on other nuclei. However, caution is warranted, and the results described are framed in this language merely to clarify the problems. If it turns out that the pion-nuclear theories which led to the quoted radii differences are wrong, that may be as interesting a problem to understand as any verification of Hartree-Fock calculations. I would like to re-emphasize my statement in the text that total cross section differences may be a much better handle on radii than absolute cross sections because many theoretical uncertainties present in fitting individual cross sections, e.g. elastic cross section are likely to drop out or be reduced in the differences.

M. K. Banerjee (U. of Maryland): Did you include the LLEE effect which is important in determining possible variation of radius?

M. Cooper: The LLEE effect was not included in these calculations, although it should not be as important at energies above 140 MeV as it is at very low energies. The proposition that the LLEE effect is important in determining radii differences remains to be demonstrated.

INTERPRETATION OF PION-NUCLEUS TOTAL CROSS SECTIONS IN THE REGION OF THE πn (3,3) RESONANCE*

C. B. Dover
Brookhaven National Laboratory, Upton, NY 11973

ABSTRACT

We consider some aspects of the problem of the interaction of a probe with a nucleus in the energy region where the elementary probe-nucleon interaction resonates. The case of pion-nucleus scattering is treated as an illustration.

Several promising mesonic probes of nuclei, such as $\pi^{\pm}$ and K^-, exhibit resonance structure in the elementary probe-nucleon cross sections in the low and medium energy region. An important theoretical problem is then how to extract useful nuclear information from the observed scattering of a probe in the region where the elementary amplitude varies rapidly with energy. In this talk, I will address this question as well as others pertinent to how an elementary resonance is reflected in nuclear scattering:

a) How is the energy dependence of the pion-nucleus total and reaction cross sections related to that of the elementary amplitude?

b) Can we simply characterize the pion-nucleus interaction in the resonance region as "black sphere scattering" and be done with it? (No)

c) Can pion-nucleus cross sections exceed the naive "geometric limit"? (Yes)

d) Does pion-nucleus scattering data enable us to extract the effective (density dependent) pion-nucleon interaction in the medium?

The emphasis of this talk is largely phenomenological. Most of our observations are based on simple semiclassical considerations or on the results of typical optical model calculations. We illustrate our comments with examples drawn from pion-nucleus scattering, although the type of analysis presented here may be equally well applied to K^- or $\bar{p}$ - nucleus scattering, in which narrow elementary resonances also play a role.

We conduct our discussion within the framework of the optical model, in which the pion-nucleus interaction is represented by a potential V, which in general is non-local and energy dependent. For illustrative purposes, we assume that the elementary πn resonance in the $\ell=1$, $J=3/2$, $T=3/2$ state at a lab energy of about 180 MeV dominates pion-nucleus scattering in this energy region. In the simplest approximation, this leads to a potential[1]

$$2\omega V(r,\omega) = b_1(\omega)\underset{\sim}{\nabla}\rho(r)\underset{\sim}{\nabla} \qquad (1)$$

where ω is the total pion energy, $\rho(r)$ is the total nuclear density and $b_1(\omega)$ is related to the elementary (3,3) forward scattering amplitude $\bar{f}_{\ell=1}(0)$ by

---

* Work supported by Energy Research and Development Administration.

$$b_1(\omega) = \frac{4\pi}{pk^2}(kf_{\ell=1}(0))_{c.m.} = 8\pi\Gamma(\omega)/3pk^2/(E_0-E-i\Gamma(\omega)/2)$$

$$\Gamma(\omega) = 4m_p k^3 a^2 \gamma^2/(E_0+E)/(1+k^2 a^2) \qquad (2)$$

Here we have used the parametrized form for the width $\Gamma(\omega)$ due to Carter et al.[2]; k is the πn c.m. momentum, p is the lab momentum, E is the total c.m. energy of pion and nucleon, and m_p is the proton mass. The actual (3,3) resonance has E_0 = 1231 MeV, a = 0.6277 fm and γ^2 = 0.1709. For our model studies of the effect of using a narrower width for the resonance, we have used the same values for E_0 and a, but decreased γ^2 by a factor of 10, 20, or 100.

As is well known,[3] the non-local potential of Eq. (1) is equivalent to a local potential $V_L(r)$ plus a radius dependent effective mass $\mu^*(r)$ for the pion. $V_L(r)$ has the structure of a classical dipole layer,[3] i.e., a potential which changes sign as a function of r. Since $\text{Re}b_1(\omega)$ changes sign at $E = E_0$, $V_L(r,\omega)$ also changes sign at E_0. For $E < E_0$, V_L is attractive at large r and repulsive at small r. The amplitudes $\eta_L = \exp(2i\delta_L)$ corresponding to the various π-nucleus partial waves L execute clockwise spirals in the Argand plot. A typical example is shown in Fig. 1 for $\pi^- + {}^{12}C$. Note that an elastic resonance would execute a counterclockwise circular spiral of unit radius in Fig. 1. Thus the reflection of the elementary resonance in the pion-nucleus partial wave L does not correspond to a single particle or a compound nucleus resonance. Equivalently, as we pass through the free space resonance energy E_0, the radial wave functions $u_L(k,r)$ of the pion-nucleus system do not become sharply localized within the nuclear radius. This would be a signature for the formation of a long-lived pion-nucleus state.

The Argand plot of Fig. 1 is representative of a strong absorption situation. Near 180 MeV, the spiral passes near $\eta_L = (0,0)$, indicating that $\text{Im}\delta_L$ is sizable. This of course reflects the maximum in the strength of the absorptive part of $V(r,\omega)$. This aspect of the problem is also illustrated in Fig. 2, which displays the partial reaction cross sections σ_L^r as a function of L for π^+ scattering at 134 MeV on various nuclei. Note that the total reaction cross section is given by $\sigma_R = \sum_L \sigma_L^r$, where $\sigma_L^r = \pi\lambdabar^2(2L+1)(1-|\eta_L|^2)$. For a strong absorption situation, $|\eta_L|$ is small for all partial waves L for which $L \lesssim kR$, where R characterizes the size of the system. Thus $\sigma_L^r \approx \sigma_L^e \approx \pi\lambdabar^2(2L+1)$, as shown in Fig. 2. For $L > kR$, the contributions σ_L^r drop off rapidly in a manner which is sensitive to the diffuseness of the nuclear surface. If the nuclear surface were made arbitrarily sharp, we would obtain cross sections close to the naive geometrical limit $\sigma_E \approx \sigma_R \approx \pi(R+\lambdabar)^2$. Here R would be the equivalent spherical radius of the nucleus ($R = (5/3)^{\frac{1}{2}} < r^2 >^{\frac{1}{2}}$). However, we now show that due to the presence of a non-zero nuclear density at large radii, we can in principle exceed the naive geometric limit by an arbitrary amount. One sees this most simply in the context of a simple model. Consider an optical potential $V(r,\omega) = -\frac{2\pi}{\omega}\rho(r)\, f_{\ell=0}(\omega)$ generated by the s-wave part of the elementary amplitude. In the optical limit of Glauber theory, the reaction cross section is given by

$$\sigma_R(E) = 2\pi \int_0^\infty (1-\exp(-X(b)))b\,db$$

$$X(b) = \sigma_T(E) \int_{-\infty}^{+\infty} dz\, \rho(r) \qquad (3)$$

where $\sigma_T(E)$ is the s-wave part of the elementary total cross section. If we consider a Gaussian density $\rho(r) = A(\alpha/\pi)^{3/2}\exp(-\alpha r^2)$, then we can do the integral in Eq. (3) analytically to obtain

$$\sigma_R/\sigma_G = \gamma + \ell n\, y + \text{Ei}(-y) \qquad (4)$$

where $\gamma = 0.5772$, $\text{Ei}(x) = \int_\infty^{-x} e^{-t}\, dt/t$ and $\sigma_G = \frac{2\pi}{3} <r^2>$ is a geometric cross section for this model. The quantity y, defined by[4]

$$y = A\sigma_T/\sigma_G \qquad (5)$$

is useful in determining to what extent a given process is "geometric". For $y \ll 1$, we may expand Eq. (4) to recover the single scattering limit $\sigma_R \approx A\sigma_T$, while for very large y, the $\ell n\, y$ term in Eq. (4) dominates, and σ_R can exceed σ_G. In principle, y can be made arbitrarily large, either by increasing A or σ_T (the latter subject to the unitarity limit). In practice the values of y for ^{208}Pb range from about 0.6 for K^+ to about 8-10 for $\bar{p}$ and $\pi^\pm$ (near the (3,3) resonance). As we shall see later, the observed total cross sections for $\pi^\pm$ on the heavier nuclei (Sn and Pb) do in fact exceed the naive geometrical limit $2\pi(R+\lambda)^2$.

For large y, Eq. (4) implies that σ_R grows slightly faster than $A^{2/3}$, i.e., $\sigma_R \sim A^{2/3} \ell n(A^{1/3})$. The detailed form of the A dependence depends on the form of $\rho(r)$ for large r. For instance, if we use $X(b) \sim \exp(-\eta b)$ in Eq. (3), and define $\sigma_G = 2\pi/\eta^2$, we obtain

$$\sigma_R/\sigma_G = \gamma\, \ell n\, y + (\ell n\, y)^2/2 - \Sigma$$

$$\Sigma = e^{-y}/y^2\, (1-3/c+\ldots) \qquad (6)$$

and hence $\sigma_R \sim A^{2/3}(\ell n\, A^{1/3})^2$ for large A.

The physical reason why σ_R can exceed σ_G is that the incident pion is absorbed further and further out in the surface of the nucleus as we increase the elementary cross section $\sigma_T(E)$. Alternatively, the effective absorption radius R_a for the pion is energy dependent, and for large y it becomes large compared to the r.m.s. radius of the nucleus. For example, a consideration of the range of impact parameters b which contribute to σ_R as per Eq. (3) yields $R_a^2 \approx <r^2> \ell n\, y$ for a Gaussian density.

We consider now the energy dependence of the partial cross sections σ_L^r and σ_L^e obtained by solving the Klein-Gordon equation with a schematic potential of the form (3). In Fig. 3, we show σ_L^r as a function of energy for $\pi^- + {}^{12}C$, using the (3,3) amplitude of Eq. (2) as input. Several features are evident: (a) the elementary p-wave resonance is reflected as a peak (not a resonance) in each σ_L^r, (b) the peaks in σ_L^r occur at different energies for each L (higher L corresponds to a peak at higher energy) (c) the peaks are generally broader than the free space resonance. In order to achieve some insight into these results, it is useful to investigate the systematics of the energy shifts and widths as a function of the input width Γ and the mass number A. To change Γ, we use Eq. (2) with various values of γ^2. For three values of Γ (full width at half max of the free resonance), we show in Fig. 4 the energy dependence of σ_L^e and σ_L^r for the L = 2 wave in $\pi^- + {}^{12}C$. We note that the shift in peak position is strongly dependent on Γ. A narrow input resonance at $E = E_0$ (well above threshold) is reflected as a Breit-Wigner peak in each σ_L^e or σ_L^r, also at $E = E_0$. The dependence of the downward energy shift ΔE_L (from E_0) on Γ is shown in Fig. 5. We see that lower partial waves are shifted more than higher ones and the peak in σ_L^r experiences a greater shift than σ_L^e. An interesting question is whether the total elastic and reaction cross sections could be resolved experimentally into a series of bumps corresponding to different L values. Figure 5 shows that this is unlikely, since both for a narrow elementary resonance (the Y*(1520), for instance) and for a wide resonance (the (3,3)), the splitting between adjacent L values remains small compared to Γ. Thus total cross sections will appear as a single bump, even for a very narrow input resonance.

For each L, one can extract full widths at half maximum Γ_L^e and Γ_L^r from the corresponding cross sections σ_L^e and σ_L^r. We display the systematics of these widths as a function of L in Fig. 6, for various values of Γ. For small Γ, Γ_L^r/Γ approaches unity for large L, corresponding to larger impact parameter $b \sim L/k$. A pion moving along a classical trajectory with small L will thus experience more collisions (multiple scattering) than a pion with large L. This increased multiple scattering for small L leads to the broadening of Γ_L^r observed in Fig. 6. For large Γ, the situation is more complicated since the peaks in σ_L^r occur at different energies, so the widths may not be directly comparable. However, for the physical width ($\Gamma = 125$ MeV) of the (3,3) resonance, the values of Γ_L^e and Γ_L^e are relatively independent of L for $L \lesssim kR$. We make use of this fact in parametrizing the total cross section (see Eq. (8)).

The A dependence of Γ_L^r, Γ_{tot}^e and Γ_{tot}^r is shown in Fig. 7, for a fixed input width $\Gamma = 1.35$ MeV. For a fixed L, the widths increase approximately as $A^{1/3}$. This increase is expected, since a pion on a trajectory corresponding to a given $L \sim b$ will experience more collisions in a heavy nucleus than a light nucleus, since it must plow through more nuclear matter. The $A^{1/3}$ dependence seen in Fig. 7 is consistent with a simple picture of collision broadening developed by Bugg.[5] The $A^{1/3}$ dependence of widths also characterizes the observed total cross sections, at least for $A \lesssim 56$ (see Fig. 9 and later discussion).

For the observed (3,3) resonance ($\Gamma = 125$ MeV), we show the elastic and reaction peak energies E_L^e and E_L^r and the corresponding widths Γ_L^e and Γ_L^r for three typical nuclei (^{12}C, ^{56}Fe, and ^{208}Pb). For a wide elementary resonance, the dependence of $\Gamma_L^{e,r}$ on L is not simple, since each partial cross section $\sigma_L^{e,r}$ peaks at a different energy. In fact, Fig. 8 shows that the peaks in $\sigma_L^{e,r}$ are spread out over a wide energy range (40-240 MeV for L = 0 to 11 in ^{208}Pb), both above and below the free space resonance energy. Low partial waves are pushed very close to threshold in heavy nuclei. An important point to note is that the observed broadening of the peak in the total π-nucleus cross section (with respect to the free space resonance) is only partly due to a broadening of the widths $\Gamma_L^{e,r}$. As Fig. 8 shows, the shifts $\Delta E_L^{e,r}$ are at least as important in determining the apparent width of the peak. For $\Gamma = 125$ MeV, the peak energies $E_L^{e,r}$ display an approximately <u>linear</u> dependence on L(L+1), i.e.,

$$E_L^{e,r} = E_0^{e,r} + E_1^{e,r} L(L+1) \tag{7}$$

It is clear from Fig. 8 that both $E_0^{e,r}$ and $E_1^{e,r}$ decrease with increasing A. This feature is also evident in the fits to the data (See Fig. 9).

The above considerations suggest a simple phenomenological parametrization of pion-nucleus cross sections. For simplicity, we assume that Γ_L^e and Γ_L^r are approximately independent of L for the range of L values which contribute significantly to the cross section (L $\lesssim L_{max} \approx$ kR + 1). We further assume that $E_0^e \approx E_0^r \approx E_0$ and $E_1^e \approx E_1^r \approx E_1$. Neglecting any smooth background contributions, we write the total cross section $\sigma_T(E)$ as[6]

$$\sigma_T(E) \approx \pi \lambda^2 \sum_{L=0}^{L_{max}} \frac{(2L+1)\Gamma_1\Gamma}{(E-E_L)^2 + (\Gamma/2)^2} \tag{8}$$

where $E_L = E_0 + E_1 L(L+1)$, $\Gamma = \Gamma_1 + \Gamma_2$, $\Gamma_1 = ck$ and c and Γ_2 are energy independent width parameters. For kR $\gg$ 1, we can replace the discrete sum in Eq. (8) by an integral over L and obtain

$$\sigma_T(E) \approx \frac{2\pi\lambda\Gamma_1}{E_1}\left[\tan^{-1}\left(\frac{E_0+E_1 L_{max}^2 - E}{\Gamma/2}\right) - \tan^{-1}\left(\frac{E_0-E}{\Gamma/2}\right)\right] \tag{9}$$

This four parameter form has been used[6] to fit the recent Brookhaven total cross section data as well as most of the older data. A more transparent form of Eq. (9), valid for small $E_1 L_{max}^2/E_0$, is

$$\sigma_T(E) \approx \frac{\pi(R+\lambda)^2 \Gamma_1 \Gamma}{[(E-E_0)^2 + (\Gamma/2)^2]}\left(1 + \frac{E_1 L_{max}^2 (E-E_0)}{(\Gamma/2)^2}\right) \tag{10}$$

Equation (10) displays the familiar Breit-Wigner form, modulated by an energy dependent shape correction. Taking R to be the equivalent spherical radius for each nucleus, the values of E_0, E_1, Γ_1 and Γ_2 obtained by a least squares fit of Eq. (9) to the average cross sections ($\pi^+ + \pi^-$) are shown in Fig. 9, along with the experimental data. The Brookhaven data enable one to extract four parameters with fairly small error bars, except for Sn and Pb, which do not display a clear peak.

The A dependence of E_0, E_1, Γ_1 and Γ_2 conforms roughly to our expectations. The average width parameters Γ_1 and Γ_2 vary approximately as $A^{1/3}$ (at least up to ^{56}Fe), which suggests that a simple collision broadening picture[5] may have some validity. The broadening of the peak is accompanied by a downshift in the peak position (both E_0 and E_1 decrease with A, roughly as $A^{-1/3}$ and $A^{-2/3}$, respectively). For a heavy nucleus like Pb, the peak is pushed below 60 MeV lab energy.

More extensive cross section data would warrant a less primitive analysis than we have presented here. However, even our simple approach has yielded some insights: Even though the nucleus is strongly absorptive throughout the region considered here (60-320 MeV in the BNL data), there is more information to be gleaned from the data than just the radius of a "black sphere". We are able to extract four parameters from a modest amount of data, two of which characterize the <u>asymmetry</u> of the "bump" in σ_T. We have also indicated that σ_T can (and does for heavy nuclei) exceed the naive geometric limit $2\pi(R+\lambda)^2$, due to absorption in the tail of the nuclear density. Thus the "effective radius" of the nucleus is energy dependent in the presence of an elementary resonance. The concept of collision broadening (and shifting) has some validity in the discussion of widths Γ_L^r and peak energies E_L^r. Since an optical model[1] based on Fermi-averaged free-space amplitudes predicts peak energies which are too low and widths which are too small, the present data indicate the need for using an effective πn amplitude in the nuclear medium which is characterized by an increased effective resonance energy and width.

ACKNOWLEDGMENTS

I would like to thank Phil Moffa for his invaluable assistance in the preparation of this talk, and also E. H. Auerbach for supplying a modified version of ABACUS-M.

REFERENCES

1. L.S. Kisslinger, Phys. Rev. <u>98</u>, 761 (1955).
2. A.A. Carter et al., Nucl. Phys. <u>B26</u>, 445 (1971).
3. M. Ericson and T.E.O. Ericson, Ann. Phys. (N.Y.) <u>36</u>, 323 (1966); M. Krell and T.E.O. Ericson, Nucl. Phys. <u>B11</u>, 521 (1969).
4. S. Barshay, C.B. Dover and J.P. Vary, Phys. Rev. <u>C11</u>, 360 (1975).
5. D.V. Bugg, Nucl. Phys. <u>B88</u>, 381 (1975).
6. A.S. Carroll, I-H. Chiang, C.B. Dover, T.F. Kycia, K.K. Li, P.O. Mazur, D.N. Michael, P.M. Mockett, D.C. Rahm, and R. Rubinstein, to appear in Phys. Rev. C.

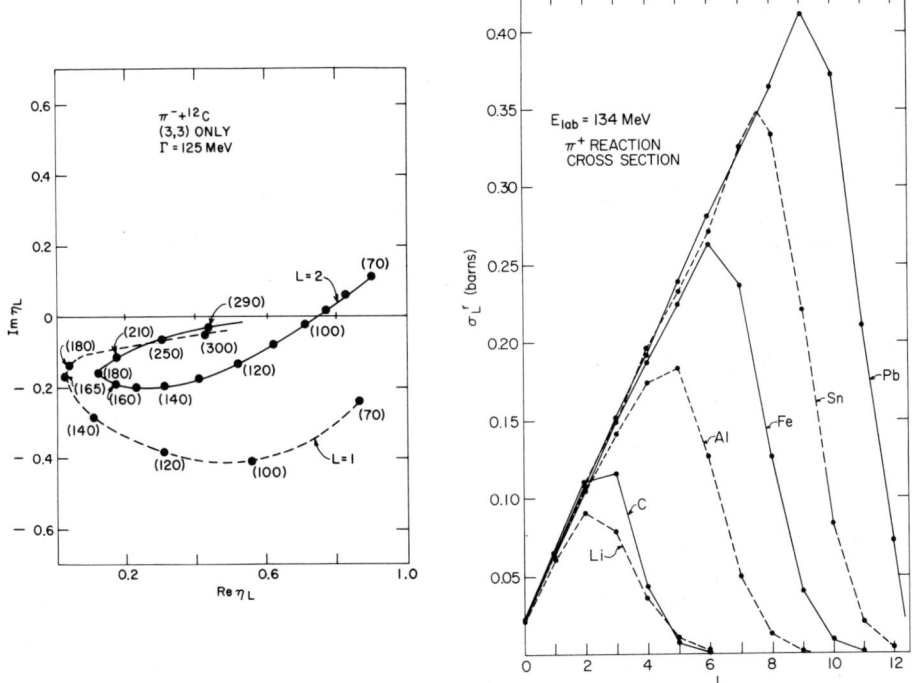

Fig. 1

Fig. 2

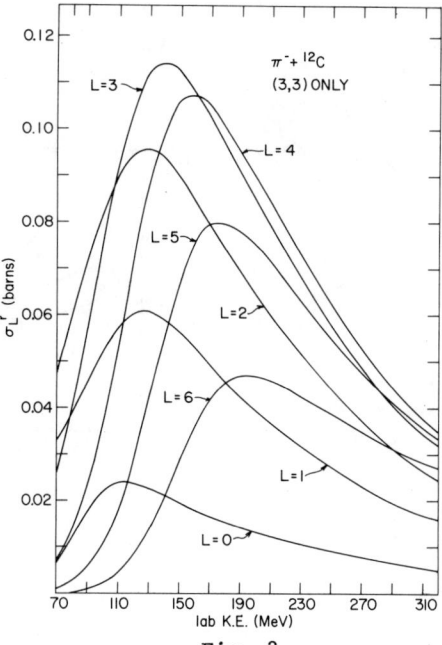

Fig. 3

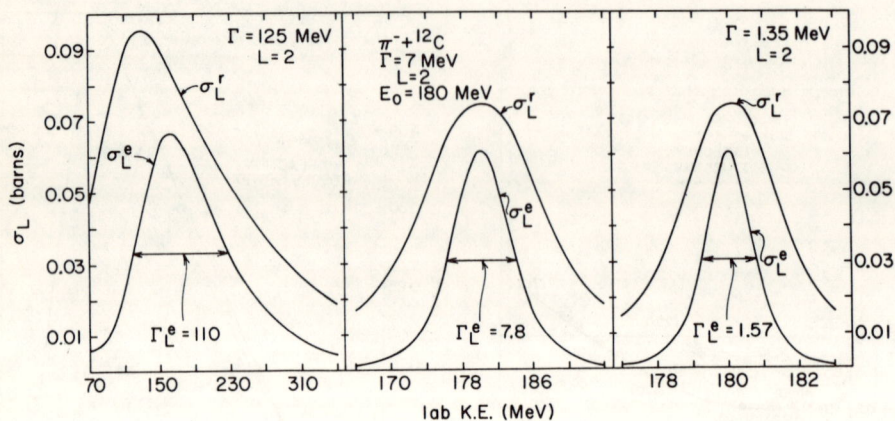

Fig. 4

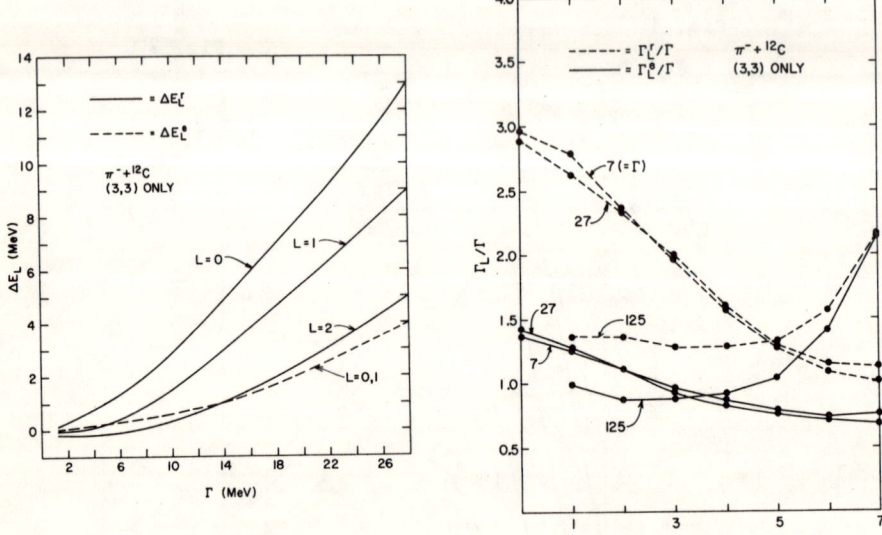

Fig. 5

Fig. 6

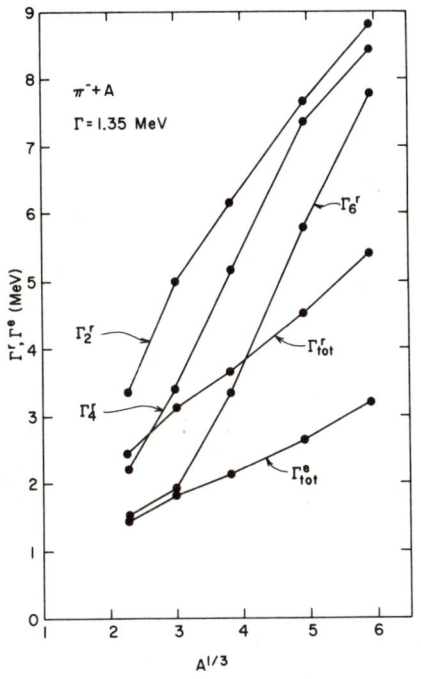

Fig. 7

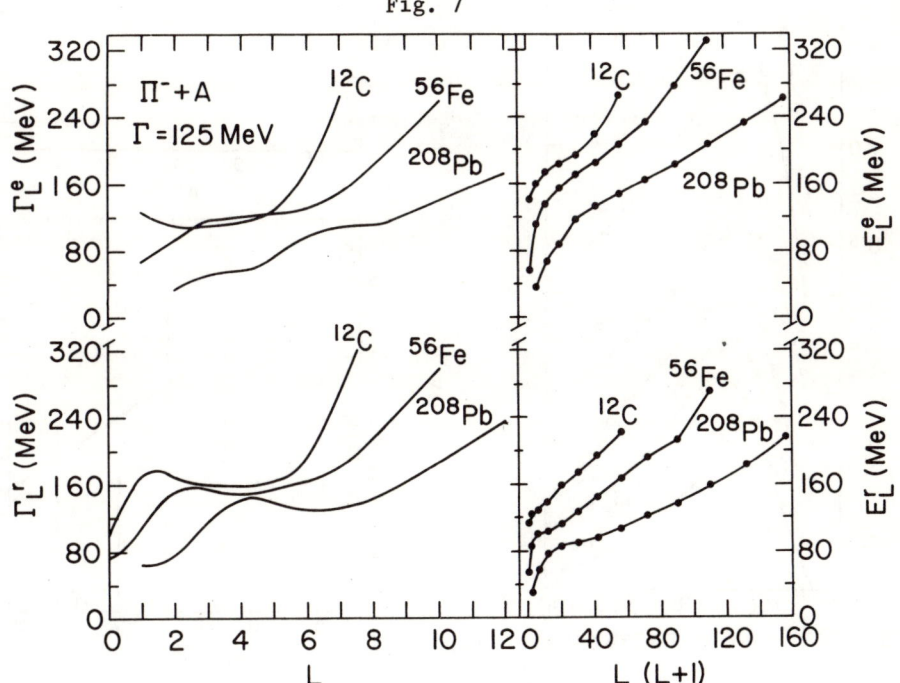

Fig. 8

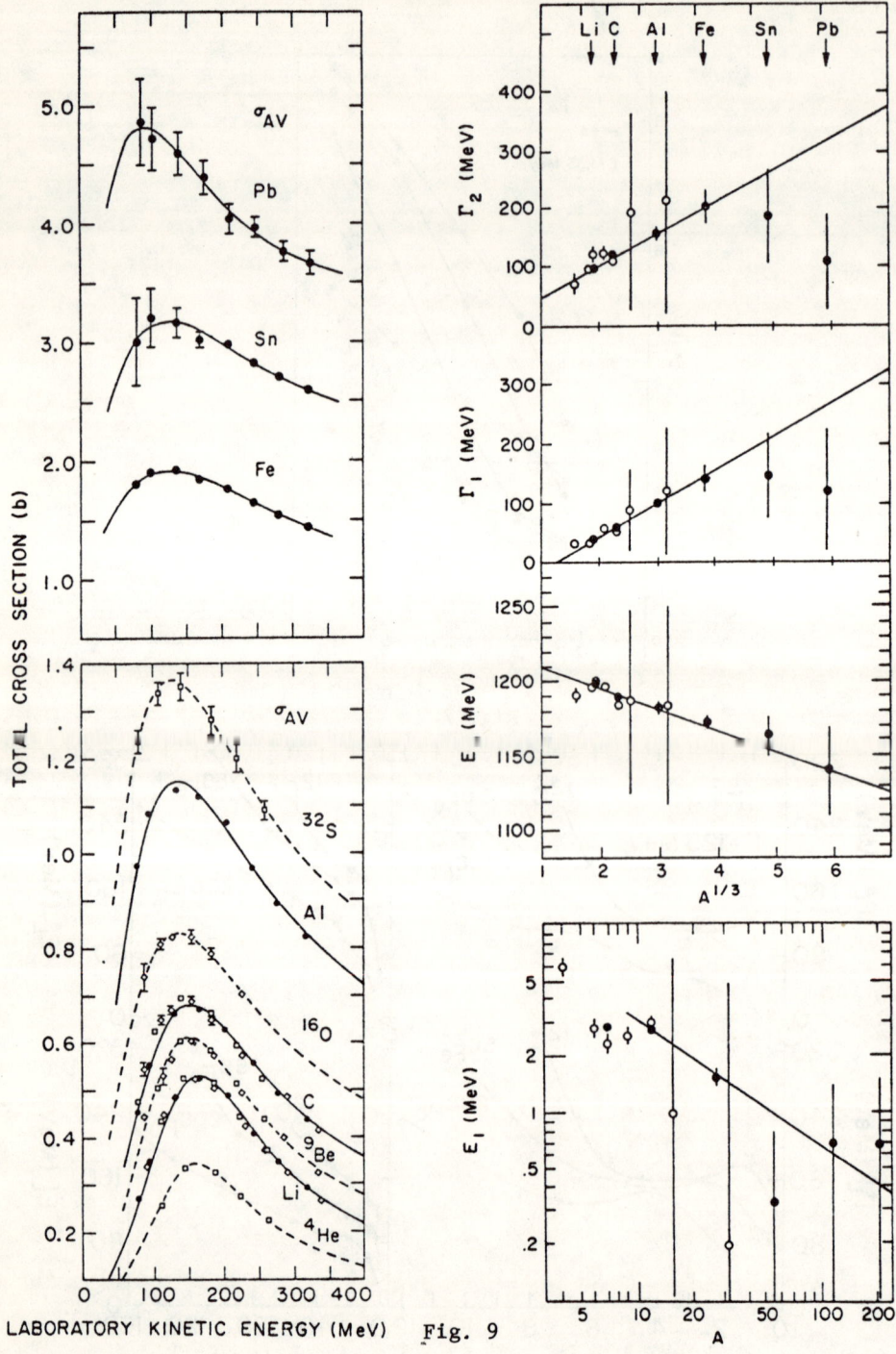

Fig. 9

Mikkel Johnson (Los Alamos): I would like to amplify on Dover's comment that the resonance may yield interesting information about pi-nucleus interactions. There also exist measurements on the real part of the forward nuclear amplitude, and these proved an entirely different type of information from that in measurements in the imaginary part.

Dover: This is certainly true. The position of the zero of ReF(0) seems to have relatively little to do with the position of the peak in the total cross section.

POSITIVE AND NEGATIVE PION PRODUCTION NEAR THRESHOLD

Y. Le Bornec, B. Tatischeff, L. Bimbot, I. Brissaud, H.D. Holmgren[*]
J. Källne[**], F. Reide and N. Willis
Institut de Physique Nucléaire, B.P. n°1, 91406, Orsay, France

Angular distributions of π^+ production induced by 154 MeV protons have been measured for several light targets (^{10}B, ^{13}C, ^{14}N, ^{25}Mg, ^{28}Si, ^{32}S and ^{40}Ca) at the IPN synchrocyclotron. An additionnal measurement has been performed at one angle on the $^{25}Mg(p,\pi^-)^{26}Si$ reaction.

For a given angle it can be observed that the differential cross section of π^+ production depends greatly on the target nucleus. Since the pion energies involved here (~ 15 MeV) are such that only S and P partial waves are acting the spin and parity conservation laws restrict to a small number the possible proton partial waves l_p included in the reaction. The corresponding quantum momentum transfer $\Delta l_q = (l_p - l_\pi)_{max}$ can be estimated and compared with the classical momentum transfer $\Delta l_{class} = (k_p - k_\pi) \cdot R_{Nucl}$. Fig. 1 emphasizes that the greatest π^+ production cross-sections correspond to the smallest differences between the two calculated momentum transfers. This effect seems to depend on both the angular momentum matching conditions and on the pion energy. This latter criterion alone, would have given the following increasing order : ^{13}C, ^{10}B, ^{14}N, ^{28}Si, ^{25}Mg, ^{32}S and ^{40}Ca. This empirical rule implies that the most favoured reactions involve high spin for the initial or final states.

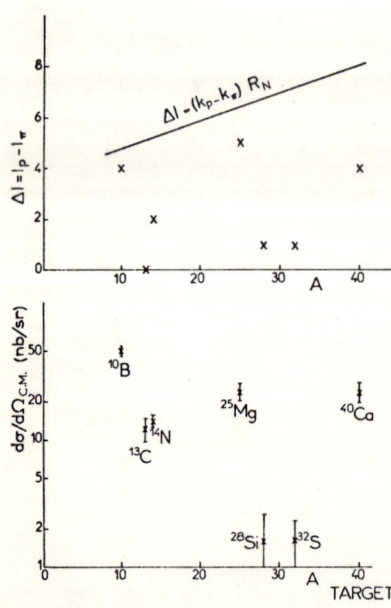

For the calculation of angular distributions, the great importance of the pion optical potential must be pointed out within the context of a stripping formalism DWBA calculations[2], the uncertainty in this potential can lead to modification of several orders of magnitude for light nuclei such as ^{10}B[1] or ^{13}C, even though for a heavier nucleus such as ^{40}Ca only small variations would arise. This great instability shows dramatically the serious consequences of our lack of knowledge of such low energy optical potentials.

This is illustrated in fig. 2. The experimental results for $^{13}C(p,\pi^+)^{14}C_{g.s.}$ are plotted together with 4 curves calculated with the same proton nuclear and Coulomb potentials, pion coulomb potential and the following pion-nucleus potentials :

[*] On leave from the University of Maryland, College Park, Maryland 20742
[**] Present address : Los Alamos Scientific Laboratory, Los Alamos, New Mexico 87544

a) $V_{\pi-Nucl.} = 0$ (no pion distortion)

b) $V_{\pi-Nucl.}$ = potential I from G. Miller [3]

c) $V_{\pi-Nucl.}$ = potential II from the same author

d) $V_{\pi-Nucl.}$ = potential I with the Kroll-Kisslinger correction.

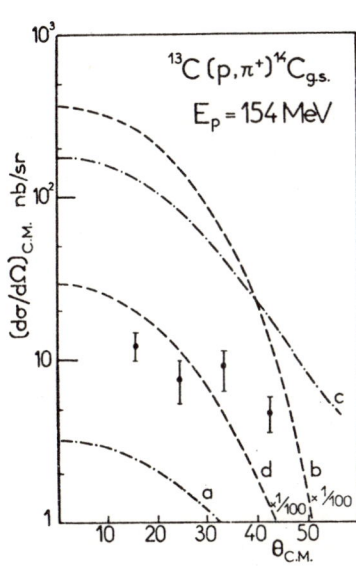

$^{13}C(p,\pi^+)^{14}C_{g.s.}$
$E_p = 154$ MeV

The study of negative pion production has been recently proposed [4] as a mean of checking the importance of the isobar Δ^{++}(1236 MeV) especially in the Mg region. We have measured the cross-section of $^{25}Mg(p,\pi^-)^{26}Si$ at a laboratory angle of 14°. This target has been chosen because it fullfilled the experimental requirements of not being too thin, having high isotope purity, emitting pions with energy >10 MeV, and leading to a level spacing in the residual nucleus not too small relatively to our energy resolution. We have found an upper limit of 1 nb/sr. This small value indicates either that the supposed probability (0.01 %) has been overestimated or that some approximations made in the calculations are not justified. Moreover it seems that this result is in better agreement with a one-nucleon mechanism than with a two-nucleon mechanism.

References :

1/ Y. Le Bornec, B. Tatischeff, L. Bimbot, I. Brissaud, H.D. Holmgren, J. Källne, F. Reide and N. Willis, Phys. Lett. 49B, 434 (1974) and 61B, 47 (1976)

2/ "PIUCK" code of P.D. Kunz and E. Rost (Univ. of Colorado)

3/ G.A. Miller, Nucl. Phys. A 224, 260 (1974)

4/ L.S. Kisslinger and G.A. Miller, Nucl. Phys. A 254, 493 (1975)

Harold W. Fearing (TRIUMF): Jim Alexander and I have done a more complete calculation in the impulse approximation model corresponding to your solid curves which we submitted as a contribution to this conference. It differs from your calculation in that distortion effects have been included, spin and antisymmetrization effects are included, and we have been very careful with the normalization, so there is no free renormalization. At 415 MeV the results agree well with the data, without the renormalization factor you found necessary. At 716 the results are ~ a factor of 5 too low, but the shape is good. Using wave functions which reproduce the electromagnetic form factors of ^{3}He and ^{4}He gives results which aren't too much different than for gaussian wave functions for angles less than 90°.

ANGULAR DISTRIBUTION OF THE REACTION $^{16}O(\pi^+, p)^{15}O$ AT 66 MeV

D. Bachelier, J. L. Boyard, T. Hennino, J. C. Jourdain,
P. Radvanyi and M. Roy-Stéphan
Institut de Physique Nucléaire, B. P. n° 1, 91406 Orsay (France)

ABSTRACT

Angular distributions of $^{16}O(\pi^+, p)^{15}O$ have been obtained for the two hole states of ^{15}O : the $(p\,3/2)^{-1}$ state at 6.18 MeV and the $(p\,1/2)^{-1}$ ground state. The ratio of the two differential cross-sections is larger than 8 at all angles. A comparison is made with preliminary calculations of G.A. Miller. A measurement has also been performed at two angles on the reaction $^{16}O(\pi^-, p)^{15}C$.

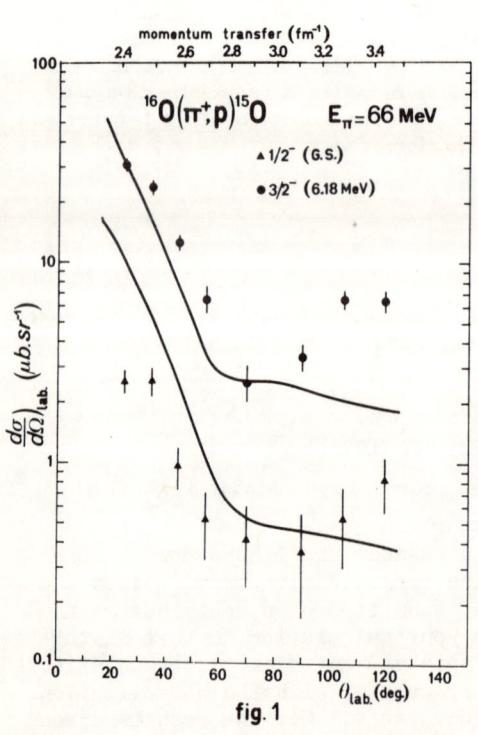

fig. 1

We have measured, at a mean pion energy of 66 MeV, the angular distributions, from 25° to 120°, for the reaction $^{16}O(\pi^+, p)^{15}O$, leading to the $(p\,3/2)^{-1}$ hole state at 6.18 MeV, and to the $(p\,1/2)^{-1}$ hole ground state of ^{15}O. The measurements have been performed with a 13 scintillators range telescope; the target was a cell filled with water. The flux of incident pions was measured with a monitor telescope, at 90° to the beam, viewing the target, previously calibrated at low intensity with in-beam detectors. We used a maximum pion intensity of 600.000 π^+/s (for a total momentum interval of 3%) corresponding to a 14 mA peak current 2% duty cycle, electron beam of the Saclay linear electron accelerator.

The two angular distributions are shown on figure 1. A minimum appears near 70°, the cross-sections rising again at backward angles. The striking fact is that the ratio of the two cross-sections stays equal or larger than 8 at all angles. This seems in contradiction with a pure single neutron pick-up mechanism which should lead to a ratio close to 2, as observed in fact in high momentum (up to 2.5 fm^{-1}) ordinary single pick-up reactions [1].

On figure 1 is also shown a preliminary calculation by G.A. Miller [2], with a crude pion optical potential ; the parameters chosen, $b_0 = 3.0 + 1.7 i$ and $b_1 = 0.0$, have not yet been checked against the appropriate pion elastic scattering data.

The fit is not yet very good, but seems able to explain the high ratio at the larger angles. This might arise from a different behaviour of p 3/2 and p 1/2 wave functions at momentum transfers larger than 3 fm^{-1}, and from the importance of the pion rescattering in the pion-nucleus optical potential.

Two differential cross-sections have also been measured for the $^{16}O(\pi^-, p)^{15}C$ reaction at the same mean energy, for the ^{15}C ground state and first excited states region ; the values obtained are : $d\sigma/d\Omega = 0.5 \pm 0.2 \mu b.sr^{-1}$ at 45° and $\leqslant 0.2 \mu b.sr^{-1}$ at 105°.

(1) D. Bachelier, M. Bernas, I. Brissaud, C. Détraz and
P. Radvanyi, Nucl. Phys. A 126, 60 (1969) ; E. Gerlic,
J. Van de Wiele, H. Langevin-Joliot, J.P. Didelez and
G. Duhamel, Phys. Lett. 52 B, 39 (1974)

(2) G.A. Miller, private communication.
G.A. Miller, Nucl. Phys. A 224, 269 (1974)
G.A. Miller and S.C. Phatak, Phys. Lett. 51 B, 129 (1974)

LOW ENERGY PION PRODUCTION BY 400-500 MeV PROTONS

D. Bryman*, G. Beer, G.R. Mason*, E. Mathie, A. Olin,
L.P. Robertson, J.S. Vincent*, TRIUMF, University of Victoria,
Victoria, B.C. Canada

ABSTRACT

Using TRIUMF proton beams of 400-500 MeV bombarding carbon and copper targets we have measured positive pion production cross sections for pion energies 20 to 100 MeV and pion angles 60 to 150°. The pions were stopped in a range telescope in which time-of-flight, energy loss, and the detection of the pion decay were used for particle identification.

EXPERIMENT

Using the variable energy TRIUMF proton beam we have measured a range of positive pion production cross sections. Fig. 1 shows the setup. Pions produced in the target were detected in a five-counter range telescope[1]. Pions stopping in the fourth counter were separated from muons by detecting the subsequent $\pi^+ \to \mu^+ \nu$ decay. The pion detection efficiency was 38%. Time-of-flight and dE/dx information was used to distinguish pions from protons and electrons. The initial energy of the pions accepted in the telescope was varied by changing the thicknesses of copper degrader in front of the stopping counter. The pion energy range of the telescope was 20 to 100 MeV.

The proton beam intensity was monitored using a three-counter telescope to detect particles scattered from an aluminum foil placed in the beam several meters downstream of the target. It was calibrated by activation of a carbon foil using $C^{12}(p, pn)C^{11}$ reaction cross sections[2]. The intensity was also measured using a He-filled ion chamber calibrated by proton-proton elastic scattering[3].

DATA ANALYSIS

The cross sections were calculated using the following equation:
$$\frac{d^2\sigma}{d\Omega dT_\pi} = \frac{N_\pi}{N_p \, \Omega \, \Delta T_\pi \, \varepsilon_d \, \varepsilon_\pi \, \varepsilon_a \, \varepsilon_m \, (N/\sin\theta)}$$

where N_π is the number of pions detected; N_p is the number of protons incident on the target as determined from the monitor calibration and the number of monitor counts recorded; $N/\sin\theta$ is the number of target atoms/μb presented to the proton beam by a target at an angle θ to it; Ω is the solid angle of the pion range telescope; ΔT_π is the energy acceptance of the telescope; ε_d is the

---

* Present Address TRIUMF, University of B.C., Vancouver, B.C. Cnd.

pion detection efficiency, the ratio of detected pions to pions stopping in the telescope; ε_a is the probability that a pion will not be lost because of nuclear absorption or scattering; ε_m is the probability that a pion will not be lost because of multiple coulomb scattering, calculated using a Monte Carlo technique. The number of pions detected, N_π, was obtained after applying time-of-flight and energy-loss limits for pions and making background corrections.

RESULTS

The differential cross sections were measured for production angles of 60°, 100° and 150° for carbon at 400, 450 and 500 MeV proton energies and for copper at 450 and 500 MeV. Measurements were also made of the pp → π^+D cross section using a CH_2 target for proton energies 400, 450 and 500 MeV at available pion energies and angles. These agreed with known values[4]. Fig. 2 shows carbon data at 60° from this experiment and earlier experiments done at Chicago[5], Berkeley[6] and SREL[1]. The 450 MeV cross sections do not agree with the large values reported in ref. 5.

Comparisons of the results of this experiment with a theoretical estimate made by Beder and Bendix[7] show general agreement at 150°, but at 100° and 60° the calculations overestimate the production of higher energy positive pions.

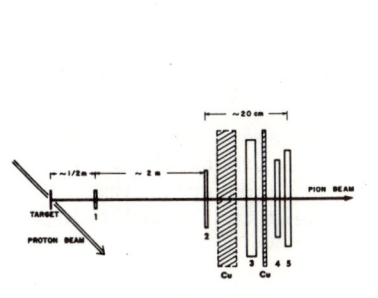

Fig. 1 Schematic of the π^+ telescope.

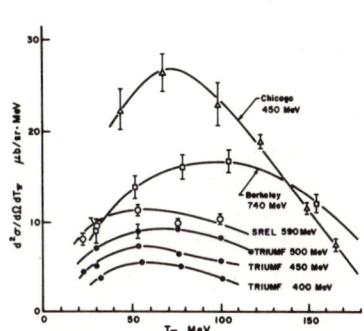

Fig. 2 π^+ production cross section from carbon at 60°.

REFERENCES

1. A similar setup was used in a previous measurement at 590 MeV by P. James et al. Univ. of Victoria preprint VPN-75-1 to be pub.
2. J.B. Cumming, Ann. Rev. Nucl. Sci. 13, 993 (1963).
3. P. Kitching, private communication.
4. J. Spuller and D.F. Measday, to be pub.
5. E. Lillethun, Phys. Rev. 125, 665 (1962).
6. D.R.F. Cochran et al. Phys. Rev. D6, 3085 (1972).
7. D.S. Beder and P. Bendix, Nucl. Phys. B26, 597 (1971).

MODELS FOR PROTON INDUCED PION PRODUCTION

M. Dillig[+)]
State University of New York, Stony Brook, New York 11794

M.G. Huber
University of Erlangen-Nuernberg, Erlangen, Germany

ABSTRACT

Different models for (p,π) production are discussed.

I. THE ONE-NUCLEON-MODEL IN THE DWBA

In this model, which is most frequently used in the literature, the π is assumed to be emitted exclusively from the projectile; all multinucleon effects are summed up as initial and final state interactions. In spite of some good fits[1,2] the model is not convincing as

- the conventional DWBA does not include proper antisymmetrization, time ordering and ρ exchange[3];
- the model mixes and disguises nuclear structure and reaction mechanism as (in an harmonic oscillator basis)[4]

$$d\sigma/d\Omega \propto \exp\{a^2_{eff}(a_{HO}, ReV_{opt}, ImV_{opt})q^2\} \qquad (1)$$

- the $\pi^\pm$ ratio at different emission angles is not well reproduced[4,5].

II. THE MULTIPLE SCATTERING APPROACH

A general idea about the microscopic nature of the (p,π) reaction is obtained from the multiple scattering expansion of the nuclear transition amplitude

$$<f|T|i> = <f|V_{N\Delta\pi}(\rho) + V_{N\Delta\pi}(\rho)\, G_\pi(\rho) G_\Delta\, V_{N\Delta\pi}(\rho) + \ldots |i> \qquad (2)$$

Closer inspection shows that far below and far above the 33 resonance this expansion is fastly converging while around the 33 region strong collective effects are expected to show up[4]. The different contributions in eq. (2) are discussed in the following.

1. The One-Nucleon-Model (in PWBA)

The model cannot predict the trends in the experimental data. However, there is some hope to test in a relativistic version with[5]

$$d\sigma/d\Omega \propto u^2_\alpha(q) + v^2_\alpha(q) + C(\alpha)u_\alpha(q)v_\alpha(q) \qquad (3)$$

the ratio of the large to small components in a wave function of a bound nucleon from (p,π) forward scattering data near threshold.

2. The Two-Nucleon-Model

The characteristic points of that model[3,4,6], which explains satisfacorily both $\pi^\pm$ data, are

- the momentum in the nucleus is shared dominantly by π and ρ exchange

- the kinematical situation allows a zero range approximation for the transition potential

$$V_{NN \to NN\pi}(r_{12}) = (V_\pi(\bar{q}) + V_\rho(\bar{q}))\delta(\underline{r}_1 - \underline{r}_2) \qquad (4)$$

- non static effects in the propagator are small[4]
- the cross section depends additionally on the transition density $\rho_{fi}(r)$

$$d\sigma/d\Omega \; \alpha \; |\int \phi(\underline{r}) \; \rho_{fi}(\underline{r}) \; e^{i\underline{qr}} \; d\underline{r}|^2 = I^2(q) \qquad (5)$$

Several modifications of the model are possible
  i) $\pi^\pm$-production in a quasifree pN collision[4]

$$d\sigma^\pm/d\Omega \; \alpha \; \overline{\sigma_{pN \to NN\pi^\pm}} \; I^2(q) \qquad (6)$$

 ii) direct isobar exchange[3,7]

$$d\sigma/d\Omega \; \alpha \; |\phi_\Delta(q)|^2 \qquad (7)$$

iii) knock out of virtual pions[8]

$$d\sigma/d\Omega \; \alpha \; \sigma_{\pi N}^+ \; |\int \phi_{j_\pi m_\pi}(\underline{r}) \; \phi_\alpha(\underline{r}) \; e^{i\underline{qr}} d\underline{r}|^2 \qquad (8)$$

for a test of the virtual pion field in a nucleus

$$A^{JM} = A^{JM}(N) + [\phi^{j\pi}A^{J'}]^{JM} + [\phi^{j\rho}A^{J'}]^{JM} + \qquad (9)$$

3. The Three-Nucleon-Model

Due to the strong πN interaction high order rescattering effects are important in the 33 region. The problems from a calculation of these effects in RPA[9] are studied presently. First results show that collective effects strongly influence the isobar self energy; especially a strong quenching of the isobar width is expected in specific nuclear states[8].

An extension of these models to a wider class of high momentum transfer processes, for example π production with heavier projectiles[8] or hypernuclear formation in (p,K) reactions[8], is in progress.

1. G.A. Miller, Nucl. Phys. A224 (1974) 269.
2. J.M. Eisenberg et al., Phys. Lett. 343 (1973) 20.
3. M. Dillig, M.G. Huber, (submitted to Phys. Lett.)
4. M. Dillig, M.G. Huber, (submitted to Nuov. Cim. Lett.).
5. R. Brockmann, M. Dillig, (submitted to Phys. Rev. C).
6. Z. Grossmann, F. Lenz, M.P. Locher, Ann. Phys. 84 (1974) 348.
7. L.S. Kisslinger, G.A. Miller, Nucl. Phys. A254 (1975) 493.
8. M. Dillig, M.G. Huber, (in preparation)
9. M. Dillig, M.G. Huber, Phys. Lett. 48B (1974) 417.

+) Work supported in part by USERDA Contract No. E(11-1)-3001

RELATIVISTIC PWIA FOR (p,π⁺) REACTIONS

L.D. Miller and H.J. Weber
University of Virginia, Charlottesville, Va. 22901

ABSTRACT

The covariant neutron-exchange mechanism is formulated for $^{12}C(p,\pi^+)$ ^{13}Cg.s. using relativistic wave functions obtained from nuclear models based on the Dirac equation for the (A+1) system. Ambiguities associated with the nonrelativistic reduction of the (πNN) vertex are eliminated. It is found that the (p,π⁺) results are surprisingly sensitive to the choice of pseudoscalar or axial-vector (πNN) coupling.

COVARIANT PION STRIPPING MODEL

The simplest model for the (p,π⁺) reaction on nuclear targets is the pionic stripping mechanism (ONE). It has attracted considerable attention because it directly involves nuclear single-particle wave functions at large momentum, viz. 2-3 fm⁻¹ for 185 MeV incident protons. Thus the recent (p,π⁺) data,[1] which yield characteristic angular distributions for resolved final states of several light to medium heavy nuclei, have usually been analyzed in terms of nonrelativistic DWIA.[2] Before one may tap the nuclear structure information, however, one must eliminate ambiguities resulting from the nonrelativistic reduction of the (πNN) vertex.[3] These involve not only "static" versus "Galilean-invariant" forms but also the relativistic transformation properties of the interaction which binds the captured nucleon.

Avoiding the nonrelativistic reduction, we present results for a covariant formulation of the ONE mechanism in PWIA. Nuclear wave functions including large and small components are generated from the Dirac equation for the (A+1) system. Rather than attempting systematic fits to the experimental data at this stage of the investigation, we study the sensitivity of the ONE to various nuclear models and off-shell parametrizations of the (πNN) coupling. Thus our results are restricted to the 185 MeV data for $^{12}C(p,\pi^+)^{13}C$g.s., where the transferred neutron is captured in a 1p 1/2 orbit. Both for γ_5 and $\gamma_5\gamma_\mu$ coupling, the (p,π⁺) amplitude consists of a static (T_S) and a non-static (T_{NS}) amplitude. For pseudoscalar coupling $T_{NS} \to 0$, if one eliminates the small component $G(Q)$ in favor of $F(Q)$ via the Dirac equation, provided one ignores the potentials, kinetic and binding energies compared to the nucleon mass. No such cancellation occurs for axial vector coupling. Nor does it occur in γ_5 or $\gamma_5\gamma_\mu$ coupling for the realistic $G(Q)$ generated from the Dirac equation and shown in Fig.1 even though the potentials generating these wavefunctions are weak Woods-Saxon wells. This is a consequence of the large momentum transfer Q involved in the (p,π⁺) reaction. Even at forward angles, $F(Q)$ is no longer "large" nor is $G(Q)$ "small". In the γ_5 case, the static T_S (∝ $F(Q)$ which happens to be near its node in Fig.1) is always small at forward angles compared to T_{NS}. At medium angles, where T_S and T_{NS} become comparable but other mechanisms beside ONE are expected to contribute, their interference is noticeable. In the $\gamma_5\gamma_\mu$ case, the dependence of T_S and T_{NS} on $F(Q)$ and $G(Q)$ is different, T_{NS} being generally smaller than for the γ_5 model. With T_S and T_{NS} roughly comparable, their interference is important at all angles for $\gamma_5\gamma_\mu$ coupling. The results are displayed in Fig.2 for the simple Woods-Saxon wells of Fig.1 as well as for a self-consistent calculation[4] where the wells are quite different. The striking sensitivity of this covariant ONE to the off-shell properties of the (πNN) vertex suggests that the (p,π^+) reaction may be a powerful tool for elucidating its structure in addition to its prospective role as an important probe of short range nuclear structure.

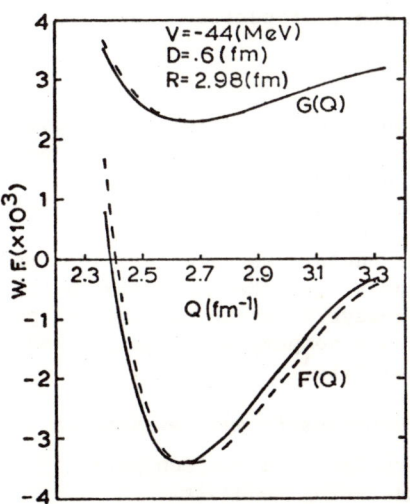 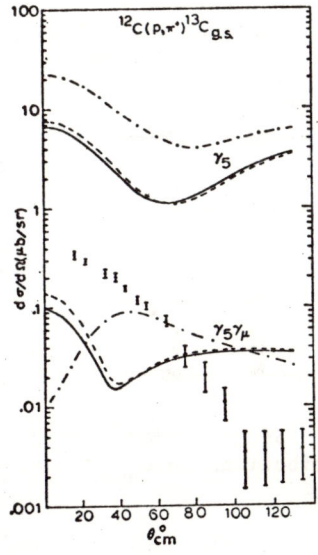

Fig 1: $1p_{1/2}$ large (F(Q)) and small (G(Q)) radial wave functions in Woods-Saxon scalar (solid) and vector (dashed) wells.

Fig 2: $d\sigma/d\Omega$ for $\gamma_5\gamma_\mu$, γ_5 models in scalar (solid), vector (dashed), and self-consistent[4] (dot-dash) wells.

REFERENCES

1. S. Dahlgren, et al., Nucl. Phys. A211, 243 (1973).
2. For a review see e.g., J.V. Noble, U. Virginia preprint. For alternative mechanisms, see Z. Grossman, F. Lenz and M.P. Locher, Ann. Phys. (N.Y.) 84, 348 (1974); L.S. Kisslinger and G.A. Miller, Carnegie-Mellon preprint.
3. M. Bolsterli, et al., Phys. Rev. C10, 1225 (1974); J.L. Friar, Phys. Rev. C10, 955 (1974); J.M. Eisenberg, J.V. Noble and H.J. Weber, Phys. Rev. C11, 1048 (1975); H.W. Ho, M. Alberg and E.M. Henley, Phys. Rev. C12, 217 (1975).
4. L.D. Miller and A.E.S. Green, Phys. Rev. C5, 241 (1972).

S-WAVE PION ABSORPTION BY NUCLEI

F. Hachenberg, J. Hüfner and H. J. Pirner

Institut für Theoretische Physik, Universität Heidelberg
and Max-Planck-Institut für Kernphysik, Heidelberg

ABSTRACT

The absorption of pions by nuclei leads to an imaginary part in the optical potential for pionic atoms. We calculate the imaginary part by assuming the rescattering mechanism to dominate (Fig. 1): The pion scatters off-shell by one nucleon and is absorbed by a second one. The πN scattering amplitude is constructed from a field theoretical model. Its off-mass shell properties prove important to reproduce the data.

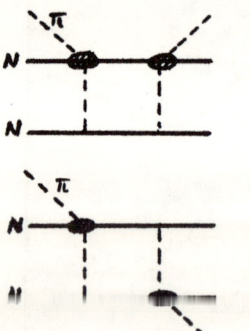

Fig. 1 Second order contributions to U_{opt}.

Pion absorption by nuclei is believed to proceed via a two-nucleon mechanism ("quasi deuteron model"). Many experimental data support this view but few quantitative calculations have been done for nuclei. We calculate the imaginary part (and the dispersive part) of the s-wave pion-nucleus optical potential and compare with the phenomenological fits.

The assumptions in our calculation are

a) absorption by a nucleon pair
b) rescattering mechanism
c) nuclear matter with nucleons at rest

The emphasis of our work is to put in a π-N scattering amplitude with unambiguous off-shell continuation since the scattered pion is far off-shell ($p_\pi^2 \simeq -(m_\pi m_N) = -(360$ MeV/c$)^2$). We use a pole model including N-pole (pseudo-vector coupling), N^*-pole, ϱ- and σ-exchange. With a σ-mass of 750 MeV we reproduce all experimental scattering lengths and volumes. As is well known, the on-shell isospin antisymmetric scattering length a^- is much larger than the symmetric one a^+. At the kinematic point of pion absorption we find the situation reversed.

The charge ratio for nucleon pairs emitted back to back is

given in terms of a^+, a^- by

$$R: = \frac{R(n,n)}{R(n,p)} = \frac{(a^+)^2 + 2(a^+ - 2a^-)^2}{(a^+)^2} \qquad (R_{exp} = 3-5)^2$$

Our result is $R = 3.1$, while using the on-shell values for a^+, a^- one gets $R \sim 10^3$. We consider this a clear indication how off-shell effects in πN-scattering become accessible through π-nucleus physics.

The coefficient B_o in the optical potential is

$\text{Im } B_o = 0.07 \; m_\pi^{-4}$, without formfactors

$\phantom{\text{Im } B_o} = 0.04 \; m_\pi^{-4}$, with formfactors

$\exp^1: \text{Im } B_o = 0.04 \; m_\pi^{-4}$

(Formfactors $F(p^2) = (m_\pi^2 - \Lambda^2)/(p^2 - \Lambda^2)$, $\Lambda = 700$ MeV, have been used at the pion-nucleon vertices)

Values for $\text{Re}B_o/\text{Im}B_o$ are presently calculated.

REFERENCES

1. L. Tauscher and W. Schneider, Z. Phys. <u>271</u>, 409 (1974)
2. M. E. Nordberg et al., Phys. Rev. <u>165</u>, 1096 (1968)

THE (π^-,2n)-REACTION ON LIGHT NUCLEI*

B.Bassalleck, D.Engelhardt, W.Klotz, C.W.Lewis, F.Takeutchi,
H.Ullrich, CERN and Institut für Exp. Kernphysik, Karlsruhe, Germany
M.Furic**, CERN, Switzerland

Negative pions from the SC II have been stopped in targets of Li-6, B-10, C-12, N-14 and O-16. In a kinematically complete experiment two neutrons were detected in coincidence by large-area position-sensitive TOF-counters with subnanosecond resolution. The data are available in the form of excitation spectra of the residual nuclei, and as function of different other variables, like: recoil momentum, relative momentum, angle θ spanned by these momenta, and opening angle. These distributions are for specific states of the residual nuclei and they are fully corrected with respect to geometrical and phase space factors. The results obtained so far indicate the existence of selection rules with respect to isospin and orbital angular momentum of the absorbing nucleon pair.

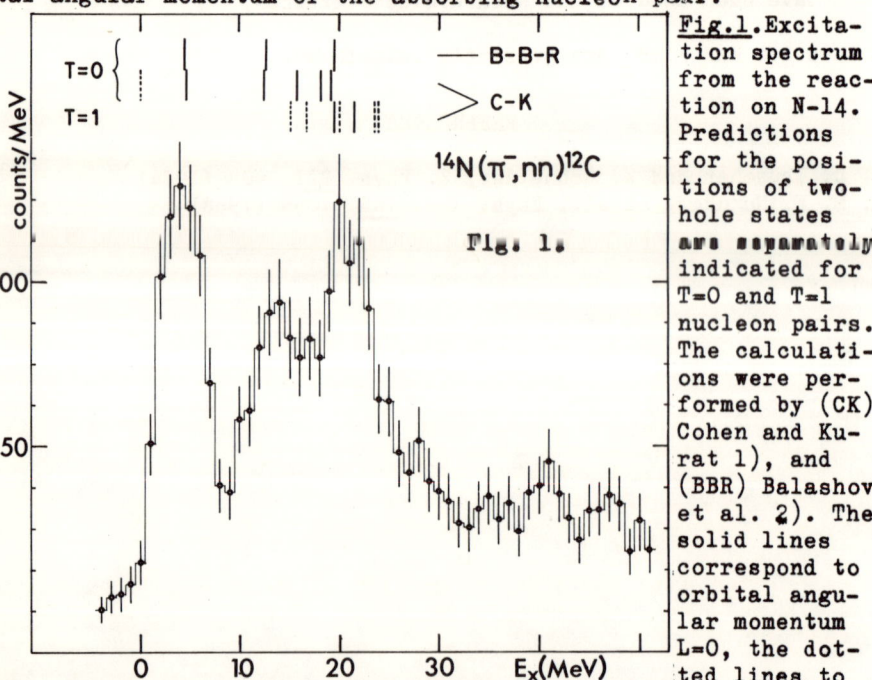

Fig.1. Excitation spectrum from the reaction on N-14. Predictions for the positions of two-hole states are separately indicated for T=0 and T=1 nucleon pairs. The calculations were performed by (CK) Cohen and Kurat 1), and (BBR) Balashov et al. 2). The solid lines correspond to orbital angular momentum L=0, the dotted lines to L=2. The spectrum shows three peaks, corresponding to excitation energies of 4 MeV, 13 MeV and 20 MeV.

*Work supported in part by the Bundesministerium für Forschung und Technologie of the Federal Republic of Germany
**On leave of absence from Institute R. Boskovic, Zagreb, Yugoslavia

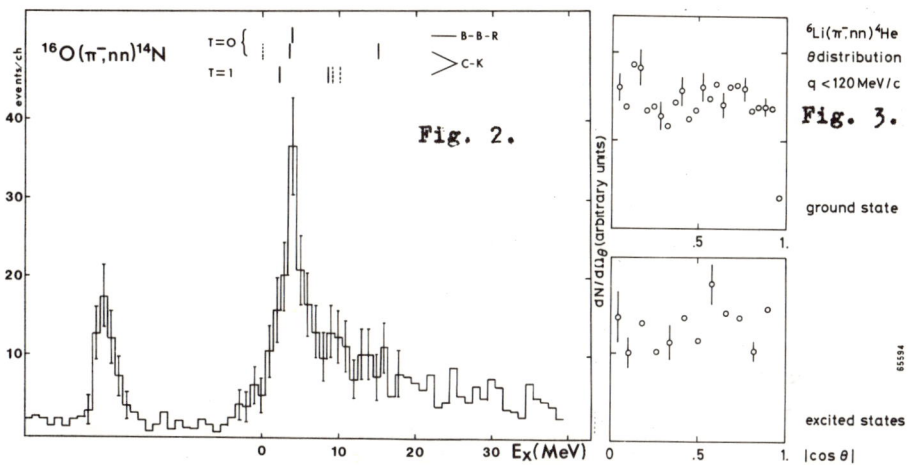

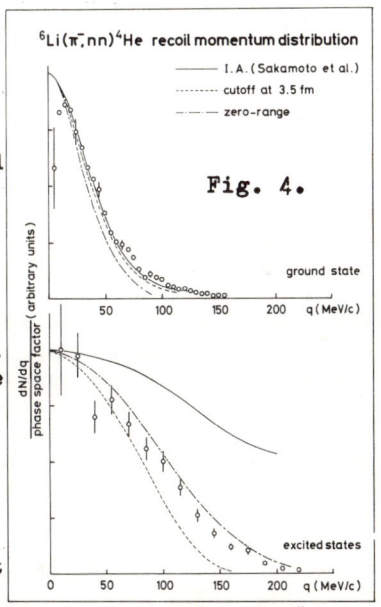

Fig.2. Excitation spectrum obtained with a heavy water target and 6 m TOF-pathes. The resolution for the peak at 4 MeV is 2-3 MeV. Theoretical predictions are indicated like in fig.1. The ground state (L=2) and the first excited state (T=1) in N-14 have been predicted, but are not visible in the spectrum.

Fig.3. Distribution of the angle Θ (see text) for the reaction on Li-6 leading to the ground state (top) and excited states (bottom) in He-4. The isotropic distributions indicate L=0 in both cases.

Fig.4. Recoil momentum distribution in the reaction on Li-6 leading to ground state and excited states in He-4. Our experimental results are compared with simple impulse approximation calculations by Sakamoto et al. 3). The dashed and solid curves are calculated with and without cut-off at R=3.5 fm, respectively. The dot-dashed curve is calculated with zero-range approximation.

1. S. Cohen and D. Kurath, Nucl. Phys. **A141**, 145 (1970)
2. V. V. Balashov, A. N. Boyarkina and I. Rotter, Nucl. Phys. **59** 417 (1964)
3. Y. Sakamoto, P. Cüer and F. Takeutchi, Phys. Rev. **C11**, 668 (1975)

Pion Production on Li^6 Via the Reaction ^{6}Li(p,dπ$^+$)He^5 at E_p = 800 MeV

J. Hudomalj-Gabitzsch, J. Clement, W. Dragoset, R. Felder,
G. S. Mutchler, T. M. Williams, and G.C. Phillips
Rice University, Houston, Texas 77001

E. V. Hungerford, M. Warneke, L. Pinsky, and J. C. Allred
University of Houston, Houston, Texas 77004

ABSTRACT

The external proton beam of the Los Alamos Meson Physics Facility (LAMPF) was used to study π$^+$ and π$^-$ production on light elements at 800 MeV incident proton energy. Although the Li^6 nucleus is not the simplest nucleus to test various reaction mechanisms for pion production, the fact that it can be well represented by a two-body system such as d-He^4, makes it possible to apply a similar general approach for study of three-body nuclear reactions as in the case of pion production on hydrogen and deuterium.[1,2]

Assuming d-He^4 configuration of Li^6, the pion production can be explained as proceeding mainly via the reaction pp→dπ$^+$ leaving the n-He^4 as a spectator. (Fig. 1) If this reaction mechanism is valid, the deuteron angular distribution from the reaction ^{6}Li(p,dπ$^+$)He^5 should have similar features to the deuteron angular distribution from the two-body reaction pp→dπ$^+$. We have measured both angular distributions in a kinematically complete experiment, by detecting the deuteron in the magnet arm, in coincidence with a pion in the TOF arm. A detailed description of the experimental technique is given in Ref. 1.

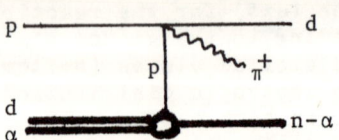

Fig. 1. Feynman diagram for the reaction ^{6}Li(p,dπ$^+$)He^5.

The very preliminary analysis of the data taken at five d-π$^+$ angle pairs show a strong enhancement in the deuteron momentum and n-He^4 spectator momentum spectra as predicted by the kinematics. However, there are broad structures in the spectra that could possibly be associated with other final-state interactions and a four-body break-up. Furthermore, since the charge of the pion could not be determined by measuring only its time-of-flight, there can also be some contribution from the reaction ^{6}Li(p,dπ$^-$)Li^5.

REFERENCES

1. J. Hudomalj-Gabitzsch, T. Witten, N. D. Gabitzsch, T. Williams, G. S. Mutchler, J. Clement, and G. C. Phillips Rice University; E. V. Hungerford, L. Y. Lee, M. Warneke, T. W. Mayes, and J. C. Allred, University of Houston, Phys. Lett. 60B, No. 2, 215 (1976).

2. E. V. Hungerford, J. C. Allred, K. Koester, L.Y. Lee, and B. W. Mayes, University of Houston; T. Witten, J. Hudomalj-Gabitzsch, N. Gabitzsch, T.M.Williams, J. Clement, G. S. Mutchler, and G. C. Phillips, Rice University. Abstracts of Contributed Papers, Sixth International Conference on High Energy Physics and Nuclear Structure, Santa Fe and Los Alamos, June 9-14, 1975.

IMPROVED ANALYSIS OF COULOMB-NUCLEAR INTERFERENCE EXPERIMENT FOR PIONS ON ^{16}O*

M. B. Johnson and M. D. Cooper
University of California, Los Alamos Scientific Laboratory
Los Alamos, New Mexico 87545

In this talk we propose a new procedure for analyzing Coulomb-nuclear interference experiments, in which the elastic scattering angular distribution of projectiles on nuclei is measured in the angular region where the Coulomb and nuclear amplitudes are comparable in size. In contrast to the conventional procedure,[1-3] detailed dynamical assumptions are not required in order to extract both the Bethe phase and the forward nuclear amplitude from the data. We test the new scheme by applying it to the ^{16}O data of Mutchler et al.,[1] who provide good quality angular distributions over the angular range $5° \leq \theta \leq 11°$ for π^+ and π^- incident at energies near the (3-3) resonance. Our basic results are: (1) the real part of the $\pi - ^{16}O$ forward nuclear amplitude vanishes at $T_\pi = 178 \pm 4$ MeV, a result substantially higher in energy than previously reported,[1] and (2) the extracted Bethe phase is in accord with the prediction of West and Yennie.[4]

Our method is based on two propositions. The first is that the data should be analyzed to extract the forward amplitude $f_N(\theta)$, where $f_N(\theta)$ is defined in terms of the complete elastic amplitude $F(\theta)$ and a Coulomb amplitude $f_C(\theta)$ by

$$f_N(\theta) = F(\theta) - f_C(\theta) . \tag{1}$$

This aspect of the analysis scheme parallels that proposed in Ref. 5 for analysis of total cross-section data and should therefore be reliable even for heavy nuclei and for low-energy incident pions, where the influence of the Coulomb interaction is very great. All three of the amplitudes in Eq. (1) depend on the charge of the pion, and we will indicate by a superscript + the amplitude for π^+ and by − that for π^-.

The second point is that knowledge of $f_N^+(0)$ and $f_N^-(0)$ determines directly the purely strong amplitude $f_S(0)$. Numerous theoretical investigations have been carried out to illuminate the connection between $f_N^\pm$ and f_S, beginning with Bethe[6] for proton scattering from nuclei and extending to the present time. For N = Z nuclei, all such theories are consistent with the following characterization: the dominant influence of Coulomb interaction on the nuclear amplitude is by the relation

$$f_N(0) = e^{i\phi} f_S(0) , \tag{2}$$

where ϕ is a complex number satisfying

---
*Work performed under the auspices of the U.S. Energy Research and Development Administration.

$$\phi \equiv \phi^+ = -\phi^-, \quad (3)$$

and, of course, where isotopic spin invariance implies

$$f_S(\theta) \text{ is independent of charge.} \quad (4)$$

Equations (2), (3), and (4) imply

$$f_S(0) = [f_N^+(0) f_N^-(0)]^{1/2} \quad (5)$$

$$\phi = -i \log [f_N^+(0)/f_S(0)]. \quad (6)$$

Table I shows the real and imaginary forward amplitudes $f_N^\pm(0)$ for several energies obtained by parametizing $f_N^\pm(\theta)$ as a function of angle and energy and fitting the parameters to the data of Ref. 1. The extracted values for $f_S(0)$ and Reϕ, obtained from Eqs. (5) and (6) are also shown. A very small imaginary part of ϕ was extracted. Also shown in the table are the values of Reϕ predicted by West and Yennie[4]

$$\text{Re}\phi_{WY} = \gamma C + \gamma \log \tfrac{2}{3} k^2 [r_S^2 + r_C(N)^2 + r_C(\pi)^2] + 2\sigma_0 + \text{Im} \frac{\gamma k}{3} \frac{r_C(N)^2}{f_S(0)} \quad (7)$$

where γ is the familiar Coulomb parameter,[7] C is Euler's constant, σ_0 the $\ell = 0$ point Coulomb phase shift, k the pion momentum, r_S a strong interaction radius, and $r_C(N)$ and $r_C(\pi)$ respectively the nucleus and pion rms charge radii. The last two terms in Eq. (7) arise because f_C was taken to be the Coulomb amplitude for a point source.[7] Fäldt and Pilkuhn[8] have proposed an alternative semiclassical theory for ϕ which gives a slightly different expression.

Table I Amplitudes (in fm) and the Bethe phase. Experimental errors on amplitudes are about ±0.25 fm and on phases about 10%.

Energy	Ref_N^+	Imf_N^+	Ref_N^-	Imf_N^-	Ref_S	Imf_S	Reϕ (expt.)	Reϕ (theory)
160	-0.5	8.1	1.4	8.1	0.5	8.2	0.12	0.13
180	-1.3	8.5	1.2	8.5	-0.1	8.6	0.14	0.13
200	-2.5	8.6	0.5	8.6	-0.6	8.7	0.16	0.14

REFERENCES

1. G. S. Mutchler et al., Phys. Rev. C11 (1975) 1873.
2. F. Binon et al., Nucl. Phys. B33 (1971) 42.
3. M. L. Scott et al., Phys. Rev. Lett. 28 (1972) 1209; G. S. Mutchler et al., Phys. Rev. C9 (1974) 1198.
4. G. B. West and D. R. Yennie, Phys. Rev. 172 (1968) 1413.
5. M. D. Cooper and M. B. Johnson, Nucl. Phys. A, to be published.
6. H. A. Bethe, Ann. Phys. 3 (1958) 190.
7. A. Messiah, Quantum Mechanics (Wiley and Sons, Inc., New York, 1961).
8. G. Fäldt and H. Pilkuhn, Phys. Lett. 40B (1972) 613; Phys. Lett. 46B (1973) 337.

STRONG ABSORPTION EFFECTS IN PION-NUCLEUS TOTAL CROSS SECTIONS[*]

W. A. Friedman, K. W. McVoy, J. E. Sedlak
Physics Department, University of Wisconsin, Madison, WI 53706

Recent pion total cross section data, taken at Brookhaven by Carroll et al.[1] for a wide range of nuclei, exhibit maxima clearly associated with the N*(3,3), which systematically (a) broaden, (b) increase in asymmetry, and (c) shift downward in energy as the A of the target is increased. We have investigated the extent to which these effects result primarily from the short pion mean free path near the (3,3) resonance, and whether any sensitivity to additional nuclear effects remain.

An optical model approach seems best adapted to a study of absorption effects, but in place of a direct computation of optical cross sections we have employed an eikonal approximation as suggested by Ericson and Hufner.[2] We employ the local optical potential $V_{\pi A}(r) = -(4\pi/2\omega) f_{\pi N}(0) \rho(r)$, neglecting recoil, and defining $f_{\pi N} = (Z f_{\pi p} + N f_{\pi n})/A$, we consider $\sigma_{\pi A} \equiv 1/2 (\sigma_{A\pi+} + \sigma_{A\pi-})$, to compare with the charge-averaged cross sections given by the BNL measurements. Employing a square density of spherical shape and radius R, we obtain by standard arguments the following simple expression:

$$\sigma_{\pi A} = 2\pi R^2 \, \text{Re}\{1 + 2/(2i\delta_o)^2 [(1-2i\delta_o)\exp(2i\delta_o) - 1]\}, \qquad (1)$$

where $2\delta_o = (\alpha + i)R/\lambda_o$ is the $\ell=0$ optical phase shift, $\alpha = \text{Re}[f_{\pi N}(0)]/\text{Im}[f_{\pi N}(0)]$ and $\lambda_o = 1/\sigma^T_{\pi N} \rho$ is the pion mean-free path. This expresses $\sigma^T_{\pi A}(E)$ directly in terms of the "absorption parameter" $R/\lambda_o(E)$, and has the expected limits $A\sigma_{\pi N}$ and $2\pi R^2$ for $R/\lambda_o \ll 1$ and $R/\lambda_o \gg 1$, respectively. It is primarily this absorptive "saturation" of $\sigma_{\pi A}$ at $2\pi R^2$ across the entire resonance region which accounts for the broadening of its peak as R (and so R/λ_o) increases, as originally suggested for ^{12}C by Ericson and Hufner.[2]

The square-density approximation, however, is found to underestimate the cross section, especially at resonance. This defect is largely rectified by the use of a more realistic density distribution as is indicated in Fig. 1, where square and gaussian shapes for ^{12}C are compared. The effect is even more pronounced for heavier nuclei where even the tails of the diffuse densities become opaque at resonance so that the nuclei appear to "swell" near the N* in a manner noted previously by Barshay et al.[3] This indicates that a significant fraction of the absorption occurs in this diffuse surface region. In fact, we find it to occur on the average at a density of only 0.04 nucleons/fm^3 (about 1/4 of the central density), suggesting that even total cross sections are quite sensitive to surface properties.

We have used realistic density distributions obtained from electron scattering data in eikonal calculations for several nuclei spanning values of A from 4 to 208. For these realistic densities it is helpful to use the concept of screening, and to express $\sigma_{\pi A}$ in terms of the function $A_{eff}(E)$ which represents the net number of unscreened nucleons on the bright side of the nucleus,

$$A_{eff}(E) = \int d^3r \rho(r) \exp(-1/2 \, \sigma_{\pi N} \int_\infty^z \rho(b,z')dz'). \quad (2)$$

We then obtain the total cross section from

$$\sigma_{\pi A} = \sigma_{\pi N}[\cos(\sigma_{\pi N}\alpha \, \partial/\partial\sigma_{\pi N}) - \alpha \sin(\sigma_{\pi N}\alpha \, \partial/\partial\sigma_{\pi N})]A_{eff}(E,\sigma_{\pi N}). \quad (3)$$

Note that for pion energies at 180 MeV we have $\alpha=0$ and $\sigma_{\pi A}=\sigma_{\pi N} A_{eff}$.

While the qualitative features of peak broadening are reproduced by our eikonal calculations, especially for $A \lesssim 60$, the downward shift for the heavier nuclei is not well reproduced. An optical model solution using the same potential does provide a substantial downward shift. In this optical calculation it appears that relative to the eikonal calculation, $\sigma_{\pi A}$ is enhanced for E < 180 MeV and is diminished for E > 180 MeV. This arises from systematic changes in the contribution of the peripheral impact parameters, and is due, presumably, to the change in sign of the real part of the optical potential, a change to which our calculation is insensitive (see Eq. (3)). Considering this effect we would expect the eikonal calculations near 180 MeV to be the most reliable, and indeed, at this energy our calculations, with realistic densities reproduce the experimental values to within 10%; poorest values result at the two mass extremes. Furthermore, it is noted that for $E_\pi = 180$ MeV, $\sigma_{\pi A}$, as a function of A, is well reproduced by $\sigma_{\pi A} = 13 \, A^{2/3}$ fm^2. If we take $\sigma_{\pi A} = 2\pi R^2$ and $R_\pi = r_\pi A^{1/3}$, we find the value $r_\pi = 1.44$. This value reflects the "swelling" discussed above, since for medium and heavy nuclei, the values of r_0 from electron scattering range from 1.2 to 1.3.

*Supported in part by National Science Foundation.

1. A.S.Carroll, et al., BNL preprint 1976.
2. T.E.O. Ericson, and J.Hufner, Phys.Letters 33B (1970) 601.
3. S. Barshay, C.B.Dover, J.P.Vary, Phys.Rev.C11 (1975) 360.

Figure 1: Calculated total pion-nucleus cross sections

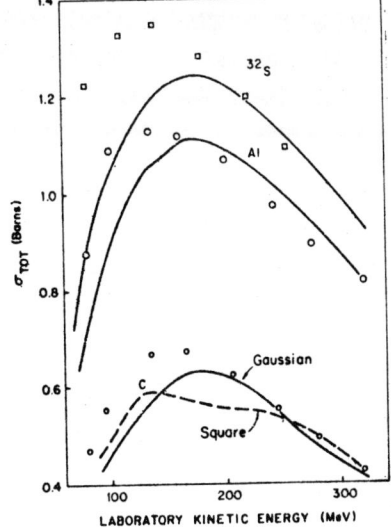

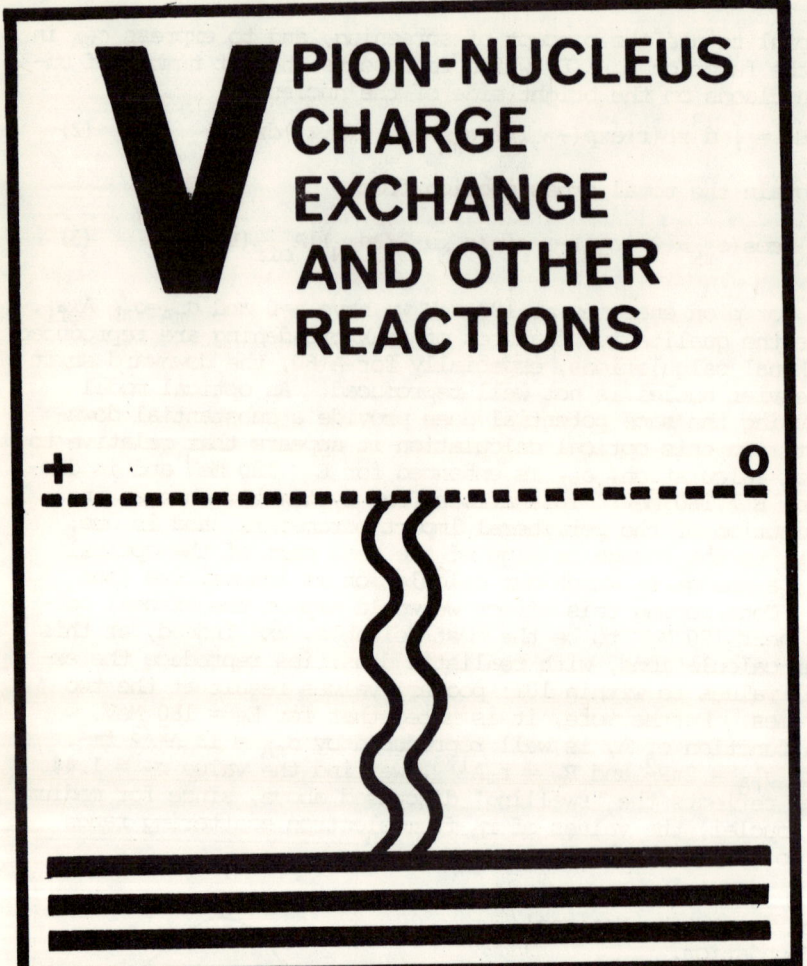

PION INDUCED INCLUSIVE REACTIONS †

Peter D. Barnes
Carnegie-Mellon University, Pittsburgh, Pennsylvania 15213

INTRODUCTION

In recent years there has been extensive discussion in the literature of inclusive reactions of the type

$$\left.\begin{array}{c}p\\ \pi\end{array}\right\} + A \rightarrow X + B + \text{anything}$$

where $X = \gamma$, π, p, d, ^{3}He, t, α, etc. Here X is the only detected object in the final state. This word <u>inclusive</u> covers a multitude of sins. In some experiments such as nucleon knockout reactions it may mean that measurement of an induced activity gives a cross section summed over a few final states of the residual system. On the other extreme are measurements of single nucleon emission in which case not even the total number of nucleons removed from target nucleus is established. The measured cross section is a sum over all possibilities. This afternoon I will tend to leave the discussion of the very specific inclusive reactions to the next two speakers and attempt to summarize what studies of the more inclusive pion induced reactions are teaching us about the π-nucleus interaction.

Of course, by its very definition a study of inclusive reactions does not sound very promising as a way to pin down detailed nuclear physics questions. I will try to show you that we now have a qualitative understanding of what is going on and that the results are at least sensitive to pion-nuclear interaction parameters that we desire to know. Unfortunately, extracting quantitative information is tied up in believing the details of a complicated reaction calculation the credibility of which is not yet fully established.

Figure 1 is a schematic particle energy spectrum which one might expect to observe in an inclusive reaction. The small direct reaction peaks and the quasi-elastic peak (which is important in the forward hemisphere) will not be treated here. The bulk of the cross section is in the evaporation and pre-equilibrium regions. The former has been well studied over many years. Of more recent interest have been attempts to calculate the pre-equilibrium region using, for example, pre-equilibrium models first suggested by Jim Griffin[1] (1966) and Intra-Nuclear Cascade models[2] (INC) developed at Oak Ridge, Brookhaven and Los Alamos using Monte Carlo techniques.

I can't go into the details here, but Fig. 2 indicates the viewpoint of these INC calculations. The nucleus is treated as a galaxy of nucleons with the density of finite nuclear matter. An incident projectile propagates through the system a distance characterized by its mean free path, scatters off a target nucleon, continues on at reduced energy until it scatters again and so forth. The recoiling nucleons each propagate in the appropriate direction and suffer

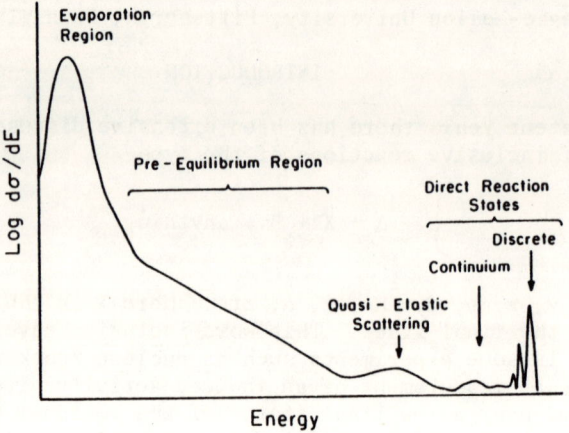

Figure 1. Schematic particle energy spectrum for an inclusive reaction.

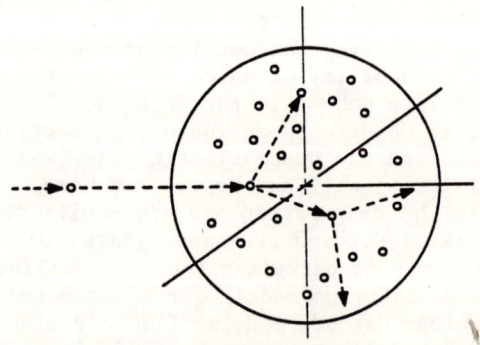

Figure 2. Schematic representation of the Intra-Nuclear Cascade calculations.[2]

their own interactions. Some nucleons penetrate the surface and escape, thus contributing to the nucleon energy spectra of Fig. 1. At each vertex in the figure, the appropriate two body on-shell scattering information is used to predict the energy and angle changes. In typical calculations, the propagation is stopped when the energy of a nucleon has fallen below e.g. the one particle separation energy. Subsequent particle emission is described in terms of standard evaporation models.

This is the language in which much of the inclusive pion induced reaction data is discussed. Before describing the results I would just like to address the question: How well does this treatment work with, for example, proton induced reactions? I will then review some stopped pion data where pion annihilation is guaranteed and some inflight pion data where annihilation does not necessarily happen. In these three cases information comes from two types of measurements, a) particle energy spectra, and b) gamma ray energy spectra. The first is very useful because it identifies specific objects leaving the target nucleus with a measured energy distribution. These spectra contain information on the <u>total energy</u> available for redistribution in the nucleus. Because of their inclusive character, they do not indicate how many nucleons are being removed from the target nucleus. Analysis of gamma ray spectra on the other hand permits identification of the final nucleus, and thus multiple nucleon removal is clearly established but the energy distribution of these nucleons is unknown.

II. P + Ni - A CASE STUDY

As a case study for testing the INC model, the next few slides show results for 100 MeV proton bombardment of $^{58}_{5}$Ni as obtained at the University of Maryland cyclotron.[3] Figure 3 shows the observed particle spectra. Notice that the p, d, and t spectra are very similar while the α spectra is falling very rapidly. Angular distributions are shown in Fig. 4. Notice that the very broad quasi-elastic peak in the proton spectra shows up at angles forward of 90° after which the slopes become very similar to each other and to the α spectra. Figure 5 compares the angle integrated proton spectra to the INC calculation and to a pre-equilibrium model calculation. (σ is effectively a normalization parameter). In both cases the shape is reasonably well predicted, although the absolute normalization is not established. The INC model underestimates the cross section by a factor 2-3. The authors[3] have also studied the induced gamma ray spectra and found the multiple "alpha particle" removal lines that have been much discussed in the literature. Table I shows that the <u>relative</u> strengths of these lines are well reproduced by the INC calculations but that the absolute cross section, e.g., for single nucleon removal is off by a factor of three to five. From their success in predicting these relative strengths, the authors conclude these "alpha removal" lines arise from sequential removal of the four nucleons.

Figure 6 shows similar success obtained by the CMU group[4] at

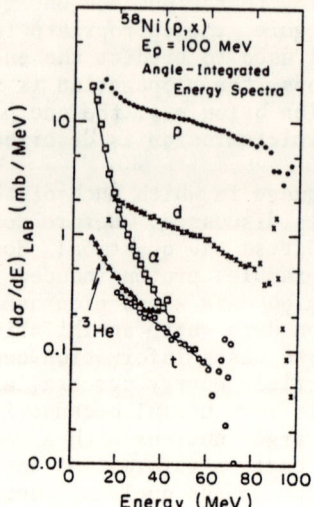

Figure 3. Angle Integrated Particle Emission Spectra observed in the bombardment of ^{58}Ni with 100 MeV protons.

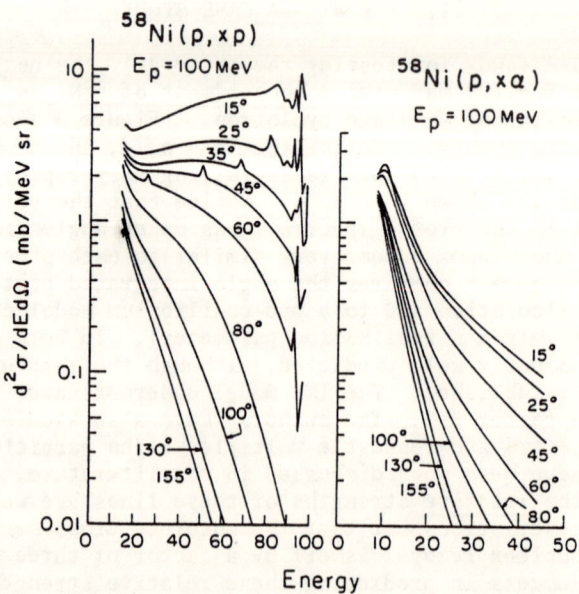

Figure 4. Angle dependence of the observed proton and alpha particle spectra (see Fig. 3).

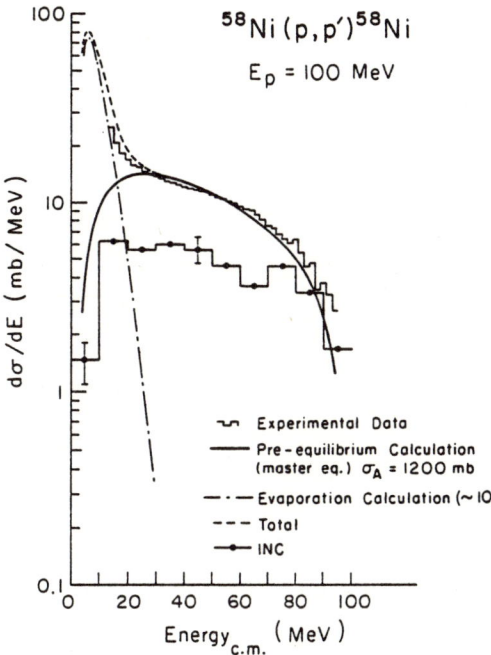

Figure 5. Comparison of the observed angle integrated proton spectra to an INC calculation and a pre-equilibrium model calculation.

TABLE I

100 MeV Proton Induced Gamma Ray Yields

Wall et al.

	^{58}Ni		^{56}Fe	
	Exp	Calc	Exp	Calc
$\sigma\ (1\alpha)$	19 mb	93 mb	12 mb	29 mb
$\dfrac{\sigma\ (2\alpha)}{\sigma\ (1\alpha)}$	0.34	0.36	0.26	0.32
$\dfrac{\sigma\ (3\alpha)}{\sigma\ (1\alpha)}$	0.05	0.013	-	0.004
$\dfrac{\sigma\ (1N)}{\sigma\ (1\alpha)}$	-	-	2.88	2.72
$\dfrac{\sigma\ (1p)}{\sigma\ (1\alpha)}$	0.48	0.28	0.4	0.33

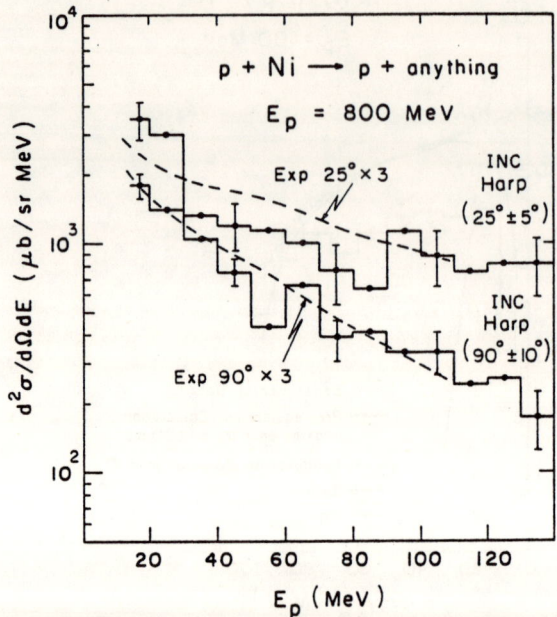

Figure 6. Comparison of proton emission spectra[4] for 800 MeV proton bombardment of Ni with INC calculations.

LAMPF for 800 MeV proton bombardment of Ni. We have used the INC code of Harp et al.[5] and find again that the shapes are well reproduced but that the normalization is again off by a factor of about three.

We conclude from the above that the INC calculation is reasonably successful in reproducing the proton induced inclusive reaction data except for the absolute normalization.

III. PION INDUCED REACTIONS

As the large meson facilities have come into operation over the past several years, pion induced inclusive reaction data similar in quality to that just shown has become available. Figure 7 shows the particle spectra obtained[4] in 235 MeV pion bombardment of Ni. Again the p, d, t and ^{3}He spectra are very similar in shape while the limited α spectrum again has a steeper slope. We will return to this point at the end of the talk.

In setting up an INC calculation for pions, one has to deal with the new phenomena of pion annihilation in which 140 MeV of extra energy is introduced into the system. This occurs presumably through the (π,2N) absorption mechanism. What is the probability that π annihilation will take place and what is the energy distribution of the reaction products?

The mean free path for nuclear pion absorption (annihilation) as a function of energy is shown in Fig. 8. This estimate is based on the survey and analysis of pion absorption as observed in Bubble Chamber and emulsion data and reported in a contributed paper submitted to this conference by Doss, Dytman and Silbar.[6] Notice that over a broad energy range the mean free path is about 4f which is comparable to the nuclear radius. This means that at e.g. 100 MeV pion annihilation is neither guaranteed nor excluded. Therefore INC calculations, which are very sensitive to the total energy available, must treat very carefully the question of how Δ's are formed and how they propagate in the nucleus.

How is the energy distributed in an annihilation process? We assume that the (π,2N) reaction is the dominant mode for π absorption. Figure 9 shows stopped pion, coincident nucleon energy spectra obtained by Lee et al.,[7] Nordberg et al.,[8] and Calligaris et al.[9] Notice that the more recent Lee data and the peaked distributions reported in the earlier work are in disagreement. This is a serious experimental discrepancy that leads to further confusion as I will soon show you.

IV. ANALYSIS OF GAMMA RAY DATA

Analysis of gamma ray spectra is very useful in establishing the yield distribution for multiple nucleon removal from the target nucleus. Figure 10 shows two distributions obtained for stopped pions by Engelhardt et al.[10] on light targets and by Ebersold et al.[11] on the much heavier ^{165}Ho. These distributions differ in both the shape and relative importance of neutron versus proton removal. In light targets neutrons and protons contribute

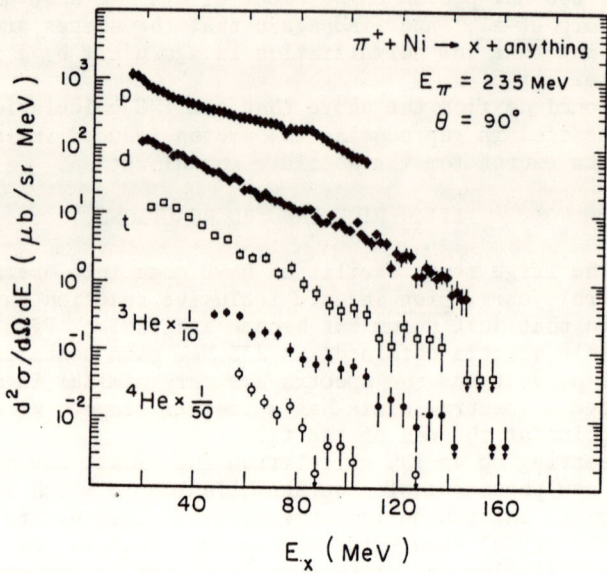

Figure 7. Particle emission spectra observed in the bombardment of Ni with 235 MeV pions.

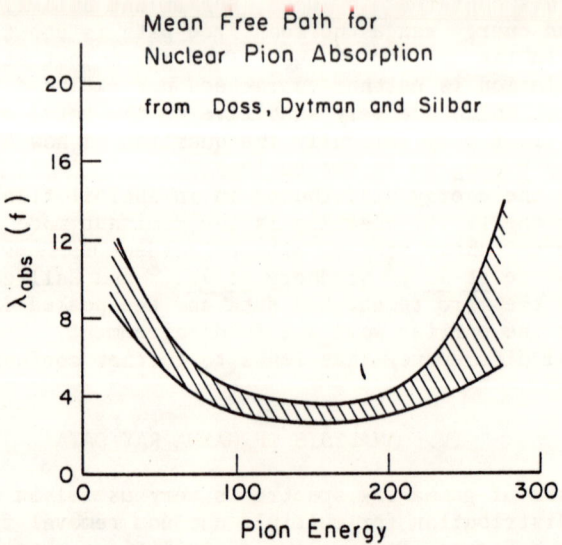

Figure 8. Mean free path for nuclear pion absorption as function of pion energy (adapted from Doss, Dytman, and Silbar).[6]

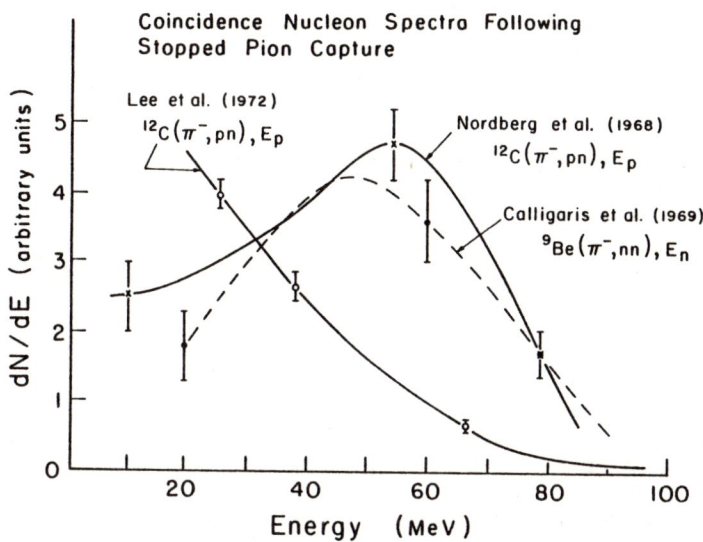

Figure 9. Comparison of experimental spectra[7-9] observed for the $(\pi,2N)$ process for pions stopping in ^{12}C and ^{9}Be.

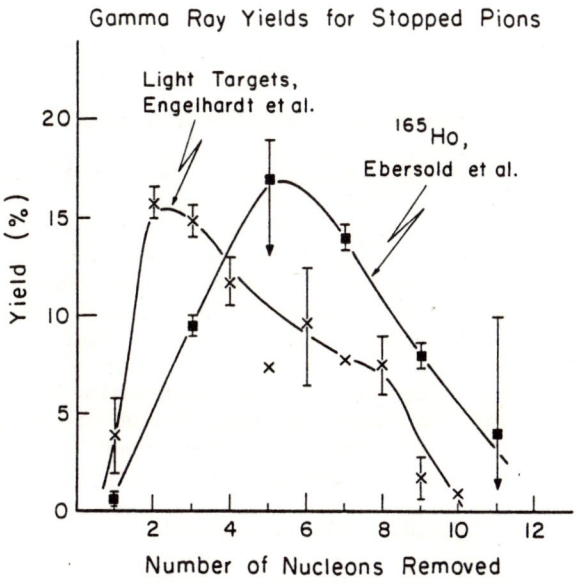

Figure 10. Comparison of multiple nucleon removal yield distributions as determined from gamma ray spectra for pions stopping in light targets and in ^{165}Ho.

about equally, while in ^{165}Ho only neutron removal is observed. In the case of inflight π induced gamma ray data on Ni, the ANL, LAMPF, Utah State, Iowa State and Texas A.&M. collaboration[12] report an average removal of 5.3 nucleons for both 100 and 220 MeV pions. The shapes of similar light target distributions have been calculated quite successfully by Zaider and Ashery and will be discussed in the following paper. Locher and Myhrer[13] have interpreted the ^{165}Ho data as a (π,2N) reaction followed by a (N,XN) reaction. By folding together the experimental (π,2N) energy spectra with the known (N,XN) cross sections, they predict the distribution shown on Fig.11. Notice that agreement is reasonable except when the Lee data[7] is used.

V. ANALYSIS OF PROTON SPECTRA

In these pion induced reactions we assume that the incident pions will primarily interact in p wave with nucleons and form deltas. It is clear from the above discussion of mean free paths that calculations of proton energy spectra will be very sensitive to the propagation properties of real deltas in nuclei. These deltas can decay back to the nucleon plus pion channel with probably Γ(E)

$$\pi + N \rightarrow \Delta \rightarrow \pi + N$$

followed by subsequent Δ formation. Alternatively the Δ's can scatter from a nucleon with cross section, σ, and generate two energetic nucleons

$$\pi + N + N \rightarrow \Delta + N \rightarrow N' + N''$$

giving the (π,2N) absorption mechanism. The probability Γ(E) of the decay and the cross section σ for the scattering are the two essential new ingredients to the INC calculations. Figure 12 compares the proton spectra[4] resulting from 235 MeV bombardment of Ni with the INC calculations of Harp et al.[5] Although there is qualitative agreement at low energies, the data falls off much more rapidly than the calculations. It would appear that pion absorption is overemphasized in the calculations. Joe Ginocchio at LASL has been studying this problem and reports[14] that Γ(E) was underestimated and the ΔN scattering cross section, σ, overestimated in the previous calculation.[5] By giving Γ the correct energy dependence and including a form factor in the one π exchange description of the ΔN scattering vertex, he gets the preliminary results shown on Fig. 13. These calculations are in much better agreement with the data.

V. COMPLEX PARTICLE EMISSION

We now turn to the question of complex particle emission in these inclusive reactions. The number of deuterons and tritons detected at 90° with 100 MeV of energy is far too great to be understood in an evaporation model. The most likely explanation is a

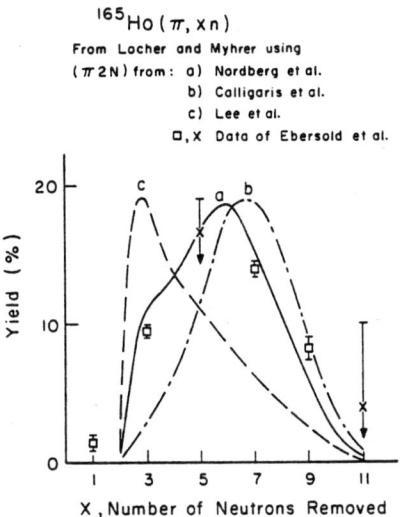

Fig. 11. Comparison of the ^{165}Ho yield distribution (see Fig. 10) with the calculations of Locher and Myhrer.[13]

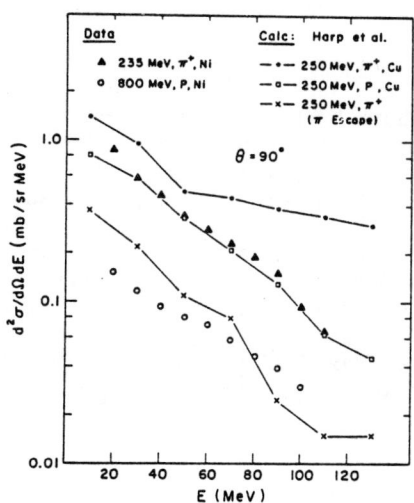

Figure 12. Comparison of pion induced proton emission spectra[4] to the calculation of Harp et al.[5]

pickup mechanism in which protons emerging from the nuclear surface occasionally pick up one or two nucleons if they have the same velocity. We test this hypothesis in the following simple-minded way. Let $\sigma^A(p_n)$ be the cross section for emissions of a complex particle of mass A, and momentum per nucleon p_n, and $P(p_n)$ be the probability of picking up the nucleon of the same velocity, p_n. Then we might expect:

$$\sigma^{A+1}(p_n) \propto \sigma^A(p_n) \, P(p_n)$$

This scaling law can be tested with the data of Fig. 7 by calculating the ratios

$$W_{A+1}(p_n) \equiv \frac{\sigma_{exp}^{A+1}}{\sigma_{exp}^A} \simeq P(p_n)$$

In this exercise we suppress the different nuclear structures of the complex particles which will introduce uncertainties of at least a factor of two. These cross section ratios are plotted in Fig. 14 for the CMU Ni data. Although the data cover four orders of magnitude in cross sections, the ratios are characterized by a single universal curve $P(p_n)$. Note, however, that the α data has a greater slope than expected, suggesting a different mechanism at work. To check the consistency of this interpretation, we compare results for Ni and Ag targets in Fig. 15 for both pion (235 MeV) and proton (800 MeV) bombardment. The shape of the curves is similar in all cases. However, the proton measurements give ratios a factor of two above the pion results.

We have mentioned earlier that for the 100 MeV proton + Ni data, the alpha spectra may just be evaporation spectra. Similar alpha particle spectra have been observed recently by Comiso et al.[15] for pions stopping in a carbon target. The shapes of these two spectra are compared on Fig. 16. Comiso et al.[15] find that the yield for ^{12}C is one α emitted per stopped pion when integrated over all energies.

In summary, the quality of this type of data has improved significantly over the past two years. There has been a corresponding improvement in the INC calculations. Most of the gross qualitative features appear to be understood although there are some puzzles when you look at detailed cases as I believe we will learn in the next few talks. Prediction of absolute cross sections continues to be a difficult problem.

In preparing this material, I have benefitted greatly from papers and discussions with a number of people: H. Engelhardt, S. Wall, J. Schiffer, and J. Ginocchio to name a few.

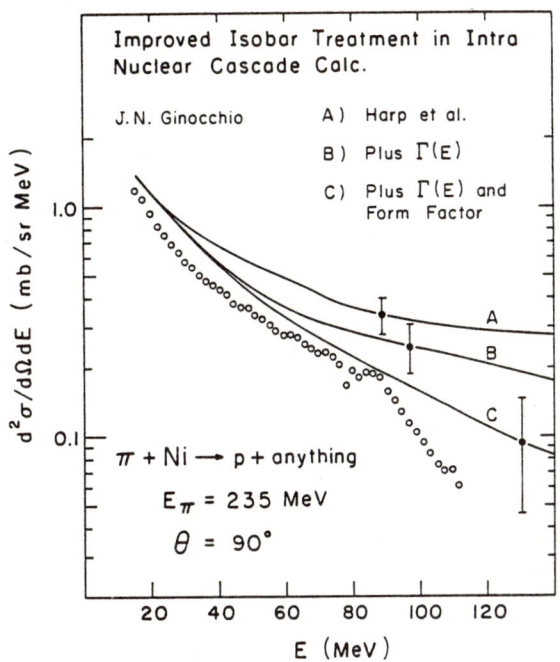

Fig. 13. Comparison of the modified INC calculations of J. Ginocchio with the data of Fig. 12.

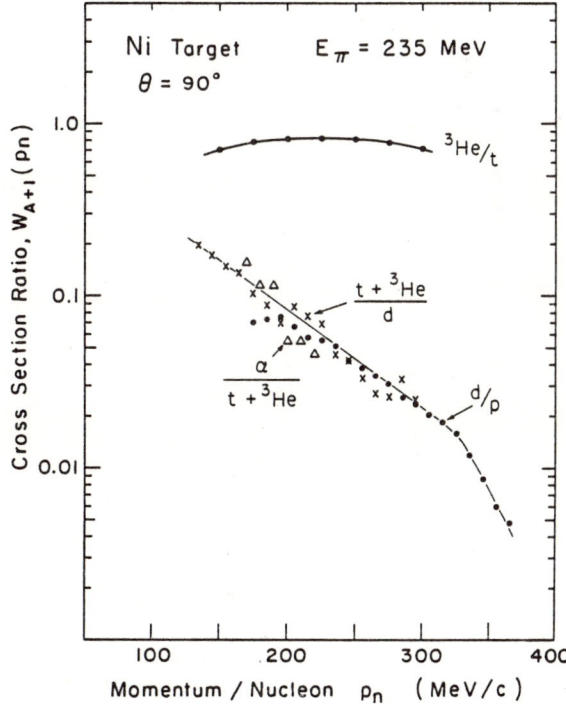

Fig. 14. Comparison of the complex particle emission cross section ratios (for mass A+1 to mass A) as a function of momentum per nucleon, p_n. The solid line is drawn to guide the eye and may be identified with the function $P(p_n)$, see text.

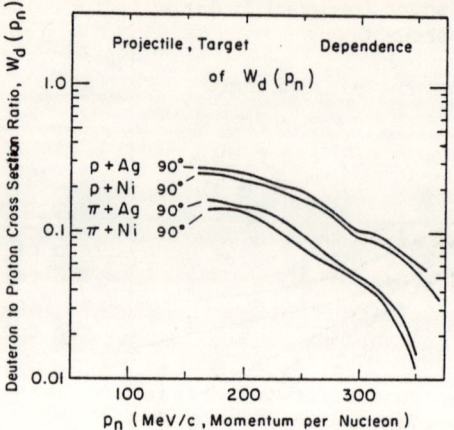

Figure 15. Comparison of the experimental deuteron to proton cross section ratios, $W_d(p_n)$, for both proton and pion projectiles and for Ni and Ag targets.

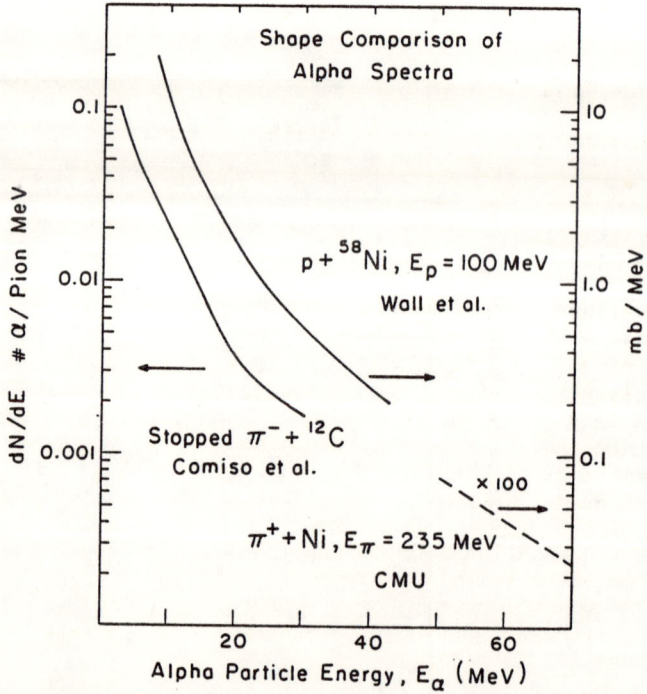

Figure 16. Shape comparison to α particle spectra from stopped pion absorption[15], 100 MeV inflight protons,[3] and 235 MeV inflight pions.[4]

† This work supported by the U. S. Energy Research and Development Administration.

REFERENCES

1. J. Griffin, Phys. Rev. Letters 17, 478 (1966); M. Blann, Phys. Rev. Letters 21, 1357 (1968).
2. H. W. Bertini, Phys. Rev. C6, 631 (1972); N. Metropolis et al., Phys. Rev. 110, 204 (1958); K. Chen et al., Physics Review C4, 2234 (1971).
3. J. R. Wu, C. C. Chang, H. D. Holmgren, N. S. Wall, J. P. Didelez, C. Butterfield, Proc. of the Second International Conf. on Clustering Phenomena in Nuclei, Nat. Tech. Information Ser. Report No. ORO-4856-26 (1975) p. 360.
4. J. F. Amann, P. D. Barnes, M. Doss, S. A. Dytman, R. A. Eisenstein, J. Penkrot, and A. C. Thompson, B.A.P.S. 19, 1007 (1974), Phys. Rev. Letters 35, 1066 (1975) and to be published.
5. G. D. Harp, K. Chen, G. Friedlander, Z. Fraenkel, J. Miller, Phys. Rev. C8, 581 (1973).
6. K. G. R. Doss, S. A. Dytman, and R. R. Silbar, "A Compilation of Pion Absorption Data", contributed paper submitted to this conference.
7. D. M. Lee, R. C. Minehart, S. E. Sobottka, and K. O. H. Zioch, Nucl. Phys. A182, 20 (1972).
8. M. E. Nondberg, K. F. Kinsey, and R. L. Burman, Phys. Rev. 165, 1096 (1968).
9. F. Calligaris, C. Cervigoi, I. Gabrielli, and F. Pellegrini, Nucl. Phys. A126, 209 (1969).
10. H. P. Engelhardt, C. W. Lewis and H. Ulbrich, Nucl. Phys. A258, 480 (1976).
11. P. Ebersold, Thesis ETHZ (1975) unpublished, and P. Ebersold, B. Aas, W. Dey, R. Eichler, J. Hartmann, H. J. Leisi and W. W. Sapp, Phys. Lett. 53B, 48 (1974).
12. H. E. Jackson, D. G. Kovar, L. Meyer-Scheitzmeister, R. E. Segel, J. P. Schiffer, S. Vigdor, T. P. Wangler, R. L. Burman, D. M. Drake, P. A. M. Gram, R. P. Redwine, V. G. Lind, E. N. Hatch, O. H. Otteson, R. E. McAdams, B. C. Cook, R. B. Clark, Phys. Rev. Letters 35, 641 (1975).
13. M. P. Locher and F. Myhrer, Helv. Phys. Acta, 49, 123 (1976).
14. J. N. Ginocchio, "Pion Induced Reaction in the Isobar Model", paper contributed to this conference; B.A.P.S. 21, 66 (1976); and private communication.
15. J. Comiso, T. Meyer, F. Schlepuetz, and K. O. H. Ziock, Phys. Rev. Letters 35, 13 (1975).

C. Cesare (U. of Trieste): There is no contradiction between Nordberg and Lee results in (π^-, np) spectra following π^- capture in light nuclei (^{12}C). Norberg et al. did not identify the charged particles which they assumed to be only protons. Now we know that a large percentage of deuterons of rather high energy emerge from the process. What is α evaporation process?

P. Barnes: My reference to "α evaporation" was mainly to point out the very strong energy dependence of the alpha emission spectrum in pion capture, and to point out that similar results are obtained in cases of proton bombardment, e.g. 100 MeV p on 58 Ni. Whether an evaporation model description of such spectra requires a preformation of alpha particles is an open question.

R. E. Segel (Argonne-Northwestern U.): We have been making gamma-ray measurements at Indiana at similar proton energies to those at Maryland. For ^{58}Ni we find as a preliminary result that α removal cross section is 2-3 times that reported by the Maryland group and thus our results are in better agreement with the calculations. However, in comparing the yields of all observable even-even nuclei we observe significant differences between the experimental results and those of a pre-equalibrium-evaporation code.

NUCLEON CHARGE EXCHANGE AND $(\pi,\pi N)$ REACTIONS

R.R. Silbar*
Theoretical Division, Los Alamos Scientific Laboratory
Los Alamos, NM 87544

ABSTRACT

Experimental data and theoretical estimates of $(\pi,\pi N)$ cross sections are discussed. It is found that the π^- to π^+ ratios in experiments where the residual A-1 nucleus is observed are strongly dependent on the probabilities for charge exchange of the recoil nucleon. These ratios are sometimes dramatically influenced by nuclear structure effects.

INTRODUCTION

I have been asked to briefly review recent developments in $(\pi,\pi N)$ reactions. In view of the time limitations, I will restrict my attention to considerations of charge exchange effects on ratios of these knockout cross sections, the topic in which most of the recent activity has occurred. Subjects such as the size of individual knockout cross sections[1] or angular distributions of the recoils from quasielastic scattering[2] will not be covered here.

FAILURE OF IMPULSE APPROXIMATION

There has been a long-standing awareness that the impulse approximation does not work for $(\pi,\pi N)$ reactions. The ratio R_n of the π^- and π^+ cross sections for neutron knockout would, in such a picture, be that of the free $\pi^- n$ and $\pi^+ n$ cross sections. Near the (3,3) resonance, then, $R_n^{IA} = \sigma(\pi^- n)/\sigma(\pi^+ n) \approx 3$. In the late 60's, however, the ratio for production of the β^+ emitter ^{11}C from a ^{12}C target bombarded by π^- and π^+ was found equal to 1.0±0.1, very different from the impulse approximation prediction.[3]

Naturally the number 1 is very mysterious and this provoked many theoretical proposals for why the ratio is so reduced from the impulse

---

*Work supported by U.S. Energy Research and Development Administration.

approximation value.[4] I will not go into this older material here. Such an experimental mystery demands experimental verification. In the last year, two groups - one working at LAMPF[5] and another in Leningrad[6] - have found the ratio R_n is not 1 but about 1.6 at the resonance. That is by itself a less mysterious number. But it was also found that the ratio, as a function of pion energy, rises from a number less than 1 at T_π below 100 MeV to about 1.8 at 300 MeV.

SIMPLE NUCLEON CHARGE EXCHANGE MODEL

The deviation of R_n from R_n^{IA} and the behavior of R_n as a function of T_π can be ascribed to the possibility that the recoil nucleon from the πN collisions can charge exchange as it leaves the nucleus.[4] If P is the probability for that charge exchange, then

$$R_n = \frac{\sigma_n^-}{\sigma_n^+} = \frac{\sigma(\pi^- n \to \pi^- n)(1-P) + \sigma(\pi^- p \to \pi^- p)P}{[\sigma(\pi^+ n \to \pi^+ n) + \sigma(\pi^+ n \to \pi^0 n)](1-P) + \sigma(\pi^+ p \to \pi^+ p)P} \quad (1)$$

$$\simeq (9-8P)/(3+6P)$$

near the resonance. A probability $P \simeq 0.2$ is all that is required to get agreement with experiment at that energy. More to the point, P is a decreasing function of T_π because the nucleon charge exchange cross section falls off rapidly with energy. This then produces the energy behavior of R_n compared with experiment in Fig. 1. That such a simple model should be so successful is remarkable.

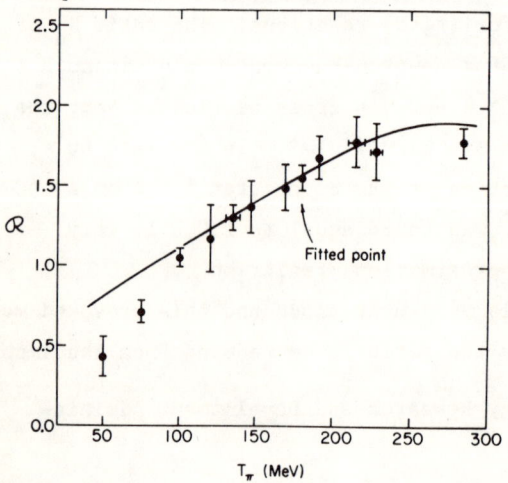

Subsequently the same model was shown to agree about as well for $R_n(T_\pi)$ for ^{14}N, ^{16}O, and ^{19}F, also measured by activation techniques.[7] More recently, Monahan

Fig. 1. Energy dependence of the predicted ratio $R = \sigma^-/\sigma^+$ from Eq. (1). Data taken from Ref. 5.

and Serduke[8] showed it also is in rough agreement with ratios for neutron and proton knockout cross sections for ^{58}Ni and ^{60}Ni, measured by observation of the deexcitation gammas of the residual nucleus.[9]

But is the success of this model spurious? For ^{64}Zn there is an indication that R_n is <u>larger</u> than the impulse approximation prediction.[10] Moreover, a recent argument by Karol[11] that the simple nucleon charge exchange model appears to predict a too large cross section, σ_{up}, for the two-step process, ^{11}B $(\pi^+,\pi^0 n)^{10}$C. Here the pion first charge exchanges on a neutron and then the subsequent recoil proton charge exchanges on the way out. If the (unmeasured) ^{11}B$(\pi^+,\pi N)^{10}$B cross section were about 50 mb, a typical value, then the model predicts a ^{10}C cross section about 5 times larger than observed.[3]

MODIFIED NUCLEON CHARGE EXCHANGE MODEL

Stimulated by these apparent contradictions, the following modifications of the model were worked out in collaboration with J.N. Ginocchio and M.M. Sternheim.[12]

Consider the cross section σ_n^- for removal of a neutron by a π^- incident, say, on a ^{12}C nucleus. For the direct process, only an effective number of neutrons, N_{eff}, can be knocked out without leaving the residual ^{11}C in a particle-unstable state (and hence not observed in the experiments being discussed). The possibility of charge exchange further <u>depletes</u> the final ^{11}C product. Let P_1^d be the probability for this, involving transitions of particle-stable states of ^{11}C to states of ^{11}B, stable or unstable. On the other hand, charge exchange of a struck proton can <u>enhance</u> the cross section for ^{11}C production. In this case any proton may be involved, but the charge exchange transition must be to a particle-stable state. Let P_2^e be that probability. Then,

$$\sigma_n^- \propto N_{eff}\sigma(\pi^- n \to \pi^- n)(1-P_1^d) + Z\sigma(\pi^- p \to \pi^- p)P_2^e . \qquad (2)$$

Similar formulas can be written for other knockout cross sections, introducing also a Z_{eff} and other probabilities $P_i^{d,e}$.

The simple nucleon charge exchange model[4] is recovered from equations like Eq. (2) by setting $N_{eff}=N$, $Z_{eff}=Z$, and all the $P_i^{d,e}$ equal to the semiclassical charge exchange probability P estimated there.

The reason for introducing different charge exchange probabilities, $P_i^{d,e}$, is that nucleon charge exchange strongly favors analog transitions. This is because the pn→np reaction is sharply forward peaked, i.e., involves small momentum transfers on the average. But non-analog transitions necessarily involve rearrangements of the space-spin part of the nuclear wave function (since the isospin part changes), and this generally requires a large momentum transfer to occur.

In what follows we will estimate the various $P_i^{d,e}$ by assuming that a given transition, if analog, goes with probability P, the semiclassical probability, and, if non-analog, with probability zero. The particular P_i^d or P_i^e is then a weighted average over the states involved. In fact, the weights in this average can be taken directly from nuclear spectroscopic factors, known either from theoretical shell model wave functions or from experiments involving single-nucleon pickup. Moreover, the effective nucleon numbers can also be estimated in terms of spectroscopic factors.

NUMERICAL RESULTS

Considering first the case of T=0 targets at resonance, the modified model gives

$$R_n = (9-8P)/(3+8P). \tag{3}$$

This is exactly the formula given years ago by Hewson.[13] The difference in R_n from the simpler model prediction, Eq. (1), is small and easily accomodated within the free parameter β hidden in the estimation of P.[4] Thus, the success of the model for the T=0 targets ^{12}C, ^{14}N, and ^{16}O is basically unchanged.

For T≠0 targets the situation is different. At resonance, the neutron and proton knockout ratios for T=1/2 turn out to be

$$R_n = \sigma_n^- / \sigma_n^+ = (9-7\nu P)/(3+15\nu P) \qquad (4a)$$

$$R_p = \sigma_p^+ / \sigma_p^- = (9-8\tfrac{1}{2}P)/(3+3\tfrac{1}{2}P) \qquad (4b)$$

and, for the ratio considered by Karol,[11]

$$R_{up} = \sigma_{up} / \sigma_n^+ = 2\nu'P/(3+15\nu P). \qquad (4c)$$

Here ν and ν' are appropriate ratios of sums of spectroscopic factors. Note the factor of ν' in Eq. (4c), which will reduce R_{up} from the value $2P/(3+6P)$ predicted by the simpler model. Also note that the proton knockout ratio R_p is independent of nuclear structure, i.e., does not involve ν or ν'.

Table I shows results for several T=1/2 nuclei and compares with experimental ratios when available. The ratio R_n for ^{19}F is somewhat larger than experiment but that for ^{31}P may be smaller. Not shown is the ratio R_{up} for ^{11}B, which is 0.024, some four times smaller than the prediction of the simpler model. Thus, for values of the (as yet unmeasured) $^{11}B \rightarrow {}^{10}B$ cross section of the expected size, the small $^{11}B \rightarrow {}^{10}C$ cross section is not in disagreement with the nucleon charge exchange model.

Table I Knockout ratios for T=1/2 targets

	^{11}B	^{19}F		^{31}P
ν	0.245	0.303		0.240
ν'	0.245	0.100		0.221
P	0.18	0.24	0.35	0.28
R_p	2.06	1.77	1.43	1.66
R_n	2.37	2.05	1.80	2.13
R_n exp		1.68 ± 0.11[a]		2.6 ± 0.5[c]
		1.52 ± 0.05[b]		
		1.78 ± 0.15[c]		

[a] Ref. 7 [b] Ref. 14 [c] Ref. 10

For targets with $T \geq 1$, we find formulas very similar to Eqs. (4). The proton ratio is again largely independent of nuclear structure and increases slowly with increasing neutron excess. Results for several nickel isotopes are shown in Table II.

Table II Knockout ratios for Ni isotopes

	^{58}Ni	^{60}Ni	^{62}Ni	^{64}Ni
R_p	1.6	1.8	1.9	2.0
R_p^{exp}	1.0±0.3	1.7±1.1	---	---
ν	0.22	0.17	0	0
R_n	2.0	1.8	3.9	3.9
R_n^{exp}	1.6±0.4	1.1±0.4	---	---

Experimental ratios from Ref. 9. These numbers may contain large background from secondary (n,2n) reactions (Ref.15).

Of special interest is the dramatic jump predicted for R_n in going from ^{60}Ni to ^{62}Ni. This results from the parameter ν going to zero at this point, which in turn is a statement that the isobaric analog states involved in the nucleon charge exchange have moved above the neutron emission threshold at this isotope.

This dramatic jump in R_n can occur in many other light and medium weight nuclei. It is a _qualitative_ prediction of the model and one which should be easily verifiable by experiments done in the near future.

SUMMARY AND CONCLUSIONS

So far the nucleon charge exchange model is in good semi-quantitative agreement with all experimental data on the knockout ratios. The following general features of the model (some of which have been pointed out in other approaches[16]) are to be tested in future experiments:

1) Ratios of knockout cross sections for $T \neq 0$ targets can be quite different from those for $T=0$.
2) The proton knockout ratio R_p is expected to be smooth and largely independent of nuclear structure.
3) In contrast, the neutron knockout ratios R_n can jump by a factor of two between neighboring nuclides.
4) There is a characteristic energy dependence for the deviations from the impulse approximation ratios.
5) It is not yet clear whether the charge exchange mechanism can account for all of these deviations.
6) It would be very interesting to check other consequences of nuclear charge exchange for $(\pi,\pi N)$ reactions by observations of the outgoing pion, or nucleon, or both.

Indeed, if the general picture presented here holds up as new data becomes available, there is a challenging possibility that $(\pi,\pi N)$ reactions might be used to extract new nuclear structure information.

REFERENCES

1. J. Hüfner, H.J. Pirner, and M. Thies, Phys. Lett. 59B, 215 (1975).
2. R.R. Silbar, Phys. Rev. C 11, 1610 (1975); R.R. Silbar and D.M. Stupin, ibid. 12, 1089 (1975).
3. D.T. Chivers et al., Nucl. Phys. A126, 129 (1969).
4. For references to many of the proposed mechanisms, see M.M. Sternheim and R.R. Silbar, Phys. Rev. Lett. 34, 824 (1975).
5. B.J. Dropesky et al., Phys. Rev. Lett. 34, 821 (1975).
6. L.H. Batist et al., Nucl. Phys. A254, 480 (1975).
7. N.P. Jacob and S.S. Markowitz, Phys. Rev. C 13, 754 (1976).
8. J.E. Monahan and F.J.D. Serduke, Phys. Rev. Lett. 36, 224 (1976).
9. H.E. Jackson et al., Phys. Rev. Lett. 35, 641 (1975).
10. H. Plendl and A. Richter (private communication).
11. P.J. Karol, Phys. Rev. Lett. 36, 338 (1976).
12. R.R. Silbar, J.N. Ginocchio, and M.M. Sternheim (to be published).
13. P.W. Hewson, Nucl. Phys. A133, 659 (1969).
14. P.J. Karol et al., Phys. Lett. 44B, 459 (1973).
15. J. Schiffer (private communication).
16. D. Robson, Ann. Phys. (N.Y.) 71, 277 (1972).

Lon-Chang Liu (Brooklyn College-CUNY): The importance of taking into account nucleon charge exchange should not be overemphasized. We know that the distortions of the incoming pion, of the outgoing pion as well as of the ejected nucleon are all very important. Nucleon charge exchange is only one aspect of such final state interactions.

If you do a detailed 3-body kinematic analysis for (π, π'p) as I did once, you will find that the momentum of the ejected nucleon is very small in some phase space region and very large in other regions. Consequently, analyzing only total cross sections is equivalent to making an average of complicated final state interactions. For this reason, the value of Ru obtained in your analysis may not be very meaningful. Only detailed theoretical analysis of coincidence-counter experimental results can provide reliable nuclear structure information.

R. Silbar: I agree that the role of nucleon charge exchange in (π, πN) can only be established with certainty in experiments which observe either the scattered pion, or, better, the recoil nucleon, or best, both. At the moment an experiment is in progress at LAMPF which looks at recoil protons. If the quasielastic events can be sorted out, the (π^-, p) cross section might be a useful way to settle the question.

Schiffer (Argonne National Lab): It almost seems as if this discussion were taking place in 1950. The pion is a low-momentum projectile and we have a good idea how the nucleus responds to such momentum transfers. The response function, to electron scattering for instance, contains the various multiple giant resonances. The effect of charge exchange and other final state interactions in the presence of the giant resonances and their isospin structure could give rise to π^+/π^- asymmetries. I don't see how such quasi-free calculations can be quantitatively meaningful when we know these effects to be important.

R. Silbar: I have calculated the excitation of the giant dipole resonance. It is very small (1-2 mb).

H. Pirner (Universitat Heidelberg): In our calculation of absolute (π,πN) cross sections done in Heidelberg we find that the following basic effects are important. (They also effect the π^-/π^+ ratio.) First, for low nucleon energies a compound nucleus may be formed. Second, inelastic reaction channels play an important role in the final state interaction of the fast nucleous. Last, the long mean free path of the nucleon together with the fact that it is a peripheral reaction allow the nucleon to escape without final state interaction. Therefore we have become very skeptical about the nucleon charge exchange as only explanation.

($\pi,\pi n$) REACTION ON LIGHT NUCLEI

Paul J. Karol[*]
Carnegie-Mellon University, Pittsburgh, PA 15213

ABSTRACT

A brief survey of experimental activation cross sections for knockout reactions by pions, near the (3,3) resonance and by protons on light nuclei is presented in conjunction with an exegetical comparison to recent theoretical calculations.

INTRODUCTION

Nucleon removal via simple (quasi-free) knockout induced by a fast projectile (electron, proton) has served to explore some of the basic structural features (nucleon removal energies, momentum distributions) of complex nuclei. Necessarily then, the mechanism for nucleon removal reactions must be well characterized in order to justify a unique structural interpretation of experimental data. The puzzling ($\pi^\pm,\pi^\pm n$) results on ^{12}C have seriously challenged the validity of the simple knockout reaction mechanism. Recent explanations have been predicated on the importance of a final state charge exchange interaction.

DISCUSSION

Comparison of high-energy (x,xn) reaction cross sections on a given nucleus should reflect mostly the magnitude of the quasi-free x-n scattering cross section. Consequently, the observation[1] that the ratio $R(\pi^-/\pi^+)$ for neutron removal by 190 MeV π^-'s and π^+'s is 1.5-1.6 instead of 3.0 as anticipated from free-particle cross sections has prompted a modified reaction pathway in which the charge-exchange of the struck nucleon with the residual nucleus plays a crucial part. Sternheim and Silbar[2] originally proposed a linear transport model in which the final nucleon charge exchange probability P altered the π^-/π^+ ratio to $R(\pi^-/\pi^+) = (9-8P)/(3+6P)$, and were able to account for ($\pi,\pi n$) reactions on ^{12}C, ^{14}N, ^{16}O and ^{19}F for 100 MeV $\leq T_\pi \leq$ 300 MeV with only a single normalization point. Karol[3] pointed out that the implied strength of the simple charge exchange hypothesis, as embodied in $0.2 \leq P \leq 0.5$, seemed to overestimate the "knight's move" reaction $^{11}B(\pi^+,\pi^0 n)^{10}C$ by a large factor. A more recent model[4] presupposes the necessary presence of isobaric analog states (IAS) for the final state charge exchange to occur appreciably. The ($\pi,\pi n$) reaction predictions on light nuclei again agree with experiment with the additional concordance that the $^{11}B(\pi^+,\pi^0 n)^{10}C$ cross section will be

[*]Chemistry Department

small. The latter, however, perhaps reflects the importance of isobaric analog states in the charge exchange but is no confirmation of the implied strength of the final state interaction. Further evidence cited in support of the charge-exchange-IAS hypothesis is agreement with recent $R(\pi^-/\pi^+)$ determinations[5] on ^{58}Ni, ^{60}Ni. However, the agreement with ^{60}Ni is only within a factor of 1.8 which should be compared to the original anomalous results on light nuclei of $R(\pi^-/\pi^+) = 1.6$ vs. expected 3.0 or a factor of 1.9 in agreement.

A rough estimate of the magnitude of the charge exchange probability P, assuming the linear transport treatment to be correct, may be arrived at by considering a $T = 1/2$ nucleus like ^{11}B with no analog state restrictions. The (p,n) cross section is given approximately by $\sigma(p,n) = \sigma_R P$ for protons of $30 < T_p < 100$ MeV. For ^{11}B, $\sigma(p,n)$ determined by activation measurements[6] between 50 and 100 MeV is ~10 mb from which one may infer that P is only ~0.05, not 0.2-0.5.

REFERENCES

1. B. J. Dropesky et al. Phys. Rev. Lett. **34**, 821 (1975).
2. M. M. Sternheim and R. R. Silbar, Phys. Rev. Lett. **34**, 824 (1975).
3. P. J. Karol, Phys. Rev. Lett. **36**, 338 (1976).
4. R. R. Silbar and M. M. Sternheim, Phys. Rev. Lett. (to be published).
5. H. E. Jackson et al. Phys. Rev. Lett. **35**, 641 (1975).
6. L. Valentin, Nucl. Phys. **62**, 81 (1965).

GAMMA RAYS FOLLOWING THE INTERACTION OF 70 MeV PIONS WITH S-D SHELL NUCLEI

M. Zaider, D. Ashery, S. Cochavi, S. Gilad, M. A. Moinester,
Y. Shamai and A. I. Yavin
Tel-Aviv University, Ramat-Aviv, Israel, and C.E.N. Saclay, France

ABSTRACT

The interaction of 70 MeV positive and negative pions with s-d shell nuclei was studied with the prompt gamma technique. The results are discussed in terms of cascade-evaporation calculations.

The prompt-γ technique[1,2] was used in order to study the interaction of 70 MeV pions with Na, Mg, Al, Si, P, S, K, and Ca targets. The measurements were carried out at C.E.N. de Saclay, using the secondary pion beam of the linear electron accelerator. The experimental arrangement was similar to that described in Ref. 1. Great attention was paid to the accurate measurement of count losses in both pion -and gamma- detection systems. This was done by using the output of a pulser in order to simulate, during the beam bursts, coincident "gammas" and "pions" of known rates. Information about dead-time losses was obtained by processing these simulated events through the electronic system of the real events. After correcting for those losses, the cross sections were found to be independent of pion-beam intensity. (For more details see Ref. 2.). A partial list of the measured cross sections is presented in the Table.

The measured cross sections for producing various residual nuclei were compared with theoretical predictions based on cascade-evaporation calculations, done with the ISOBAR[3] and EVA[4] codes (IE). Some modifications[5] were performed in the EVA code in order to make meaningful comparison with the data. The following are the main results and conclusions of this comparison:
1) Good agreement is found between the measured and calculated cross sections for reactions involving the removal of more than three nucleons. This agreement indicates that evaporation processes (simulated by the EVA code) are the main mechanism for the production of these nuclei.
2) The calculations also reproduce well the cross sections for reactions leading to the removal of one and two "α" particles. Since these calculations do not include any clustering effects, this agreement indicates that the experimental results may be understood without resorting to any additional[6] interaction mechanism. As may be seen from the Table, the cross section for "t" removal from a $T=\frac{1}{2}$ target nucleus is about as large as for "α" removal from the adjacent T=0 target nucleus. This suggests that the final nucleus, rather than the removed particle, is mainly responsible for the observed strength.

3) For the removal of a few (up to about three) nucleons the calculated cross sections are generally larger than the experimental ones. This indicates that the intranuclear cascade process (presumably responsible for these reactions) is not yet well understood.
4) Various ratios of cross sections for one nucleon removal with positive and negative pions disagree with both quasi free mechanism predictions and cascade-evaporation calculations. It is interesting to note that the large asymmetry found in the ratio $(\pi^-,\pi^-p+\pi^0n)/(\pi^+,\pi^+n+\pi^0p)$ for Si^{28} and S^{32} is well reproduced by the IE calculation. It appears that nucleon evaporation, charge exchange, and threshold effects account for this large asymmetry at $E_\pi = 70$ MeV.

REFERENCES

1) D. Ashery et al., Phys. Rev. Lett. 32, 943 (1974)
2) M. Zaider, Ph.D. Thesis, Tel-Aviv University (unpublished)
3) Harp et al., Phys. Rev. C8, 581 (1973)
4) Dostrowski et al., Phys. Rev. 116, 683 (1959)
5) M. Zaider and D. Ashery, the following paper
6) V.G. Lind, Preprint

TABLE
Measured cross sections (in mb) for emission of some light ions from s-d target nuclei.

Removed Equiv. Part.	Target	Na	Mg	Al	Si	P	S	K	Ca
p	π^+	14.6	15.1	8.9	4.3	34.8	21.3	19.3	-
	π^-	20.4	10.2	14.1	14.5	46.1	29.5	31.2	-
n	π^+	-	9.8	5.7	-	-	9.6	-	-
	π^-	-	2.7	3.2	-	-	-	-	-
t	π^+	14.4	3.1	16.8	6.0	16.9	-	Large	-
	π^-	17.0	0.3	15.9	-	22.0	-	13.7	-
α	π^+	-	20.2	13.4	24.4	-	21.1	-	24.7
	π^-	-	12.4	19.0	29.3	-	29.6	-	23.0

Funsten (William and Mary U.): Single nucleon emission from the decay of the pion excited intermediate states (inelastic pion scattering, giant resonance excitation) would produce unity π^+/π^- ratios. Our results on ^{39}Ca and ^{39}K from $\pi^\pm$ on a Ca target at SREL using the γ ray technique do not give unity ratios.

Schiffer (Argonne National Lab): Two comments and one question: You showed that the residues from $\pi + {}^{27}$Al and $\pi + {}^{32}$S were almost identical in magnitude. You seem to understand this, I don't, I find it remarkable.

You showed results with 80 MeV ^{3}He and 70 MeV pions on ^{27}Al that are almost identical, I find this also remarkable, while you seem to understand it.

Would you expect the 100 mb α-particle cross section that Yavin just showed us for individual targets from such calculations?

M. Zaider: No comment.

^{4}He(p,d)^{3}He REACTION AT Ep = 770 MeV

Pierre Couvert
Centre D'Etudes Nucleaires De Saclay, France

I shall speak to you about the neutron pick-up ^{4}He(p,d)^{3}He experiment done at a proton incident energy of 770 MeV at the Synchrotron "Saturne" of Saclay by the following co-workers: T. Bauer, A. Boudard, H. Catz, A. Chaumeaux, P. Couvert, M. Garçon, J. Guyot, D. Legrand, J. C. Lugol, M. Matoba, B. Mayer, J. P. Tabet, and Y. Terrien.

All the calculations I am presenting here about these experimental data have been performed by Alain Boudard, Michel Garçon and Yves Terrien from Saclay.

The experiment was performed by using the energy-loss spectrometer "SPES I" which was also used for pion production experiments as E. Aslanides and Y. Le Bornec told you this morning. This facility was used without any significant modification, except the introduction of a time-of-flight measurement to separate the outgoing deuteron from the lower energy tritons having the same magnetic rigidity. This time of flight measurement was useful at backward angles especially, where the pick-up cross-section we measured becomes small.

The experimental cross section angular distribution can be seen on Fig. 1. Statistical errors are very small (less than 5% except for large angles). The absolute normalization is known within ±15%.

What kind of calculations have been done on these data? Let us start with the classical one nucleon transfer model treated in the distorted wave approximation (DWBA). It is well known that the cross-section is proportional to the square of a six-dimensional integral

$$(1) \quad B_{\ell L} = \int d\vec{r}_{pn} d\vec{r}_{n^3He} \, \chi_d^{-*}(\vec{r}_d) \, D_L^*(\vec{r}_{pn}) \, \psi_\ell(\vec{r}_{n^3He}) \, \chi_p^+(\vec{r}_p)$$

where $\vec{r}_p = \vec{r}_{pn} + \frac{3}{4} \vec{r}_{n^3He}$, $\vec{r}_d = \vec{r}_{n^3He} + \frac{1}{2} \vec{r}_{pn}$

$\psi_\ell(\vec{r}_{n^3He})$ is the wave function of a neutron in the ^{4}He nucleus ($\ell=0$ for a neutron in a S state).

$D_L(\vec{r}_{pn})$ is the product $V_{pn} \Phi_d$ where V_{pn} is the neutron-proton interaction and Φ_d the internal wave-function of the deuteron. L=0 or 2 (S and D-states of the deuteron).

To have a rough idea of the physics involved, let us consider the plane-wave approximation. In that case, the integral reduces to the product of the Fourier transforms of ψ and D, and in the case of a neutron in a S-state in ^{4}He, it can be easily shown that

$$(2) \quad \frac{d\sigma}{d\Omega} \propto \left[D_o^2(\vec{Q}) + D_2^2(\vec{Q}) \right] \psi_o^2(\vec{P})$$

where $\vec{P} = -\vec{k}_b + \frac{B}{A} \vec{k}_a$, $\vec{Q} = \vec{k}_a - \frac{a}{b} \vec{k}_b$

Due to the high incident energy, P and Q vary from 2 to 3.5 fm^{-1} in the angular range we have studied. At those high momenta, the quantity into brackets is almost constant (see Fig. 2), which means that the <u>cross section is mainly sensitive to the form factor of the neutron</u>. Notice also that the D-state of the deuteron is predominant in the q-region involved.

That main feature of the reaction remains true in the DWBA approximation. The result of the introduction of a distorted wave function is that the quantity $(D_0^2+D_2^2)$ should no longer be taken at the fixed value Q, but in a small (1 to 1.5 fm^{-1}) window around Q. However, due to the constancy of $D_0^2+D_2^2$ in the region considered, it has been taken as a constant. That "local energy approximation" already used in a previous paper (1) is equivalent to writing $D_L(\vec{r}_{pn}) = D_L(Q)\delta(\vec{r}_{pn})$ and allows us to use the existing zero-range approximation code DWUCK. One problem remains: what to choose for the neutron form-factor ψ_0? Indeed, we do not know the neutron form factor at high momentum. However, we know the proton form-factor from electron scattering data and we have made the assumption that protons and neutron have the same wave-function in the ^{4}He nucleus.

Fig. 1 presents the results of such a DWBA calculation in absolute normalization for two evaluations of the neutron form-factor. The first one (curve I) is a parametrization of $\psi_0(\vec{r}_n{}^3\mathrm{He})$ given by LIM (2): the overall normalization of calculated cross sections is good, but the shape is very bad. The second evaluation of the form factor was done in the following way. We started with ^{4}He and ^{3}He wave-functions of the form

$$(3) \quad \Phi_A = N \prod_{i=1}^{A} e^{-r_i^2/2R^2} \prod_{j<i} (1 - e^{-r_{ij}^2/a^2})$$

the parameters of which were adjusted to reproduce the data of electron scattering from ^{3}He and ^{4}He nuclei (R=1.15 fm, a=0.85 fm. for ^{3}He; R=1.20 fm, a=0.77 fm. for ^{4}He). Then, the overlap of these two wavefunctions was computed, namely

$$(4) \quad \psi(\vec{r}_4) = \int \Phi_{3_{\mathrm{He}}}(\vec{r}_1,\vec{r}_2,\vec{r}_3) \Phi_{4_{\mathrm{He}}}(\vec{r}_1,\vec{r}_2,\vec{r}_3,\vec{r}_4) \delta(\frac{\vec{r}_1+\vec{r}_2+\vec{r}_3+\vec{r}_4}{4}) d\vec{r}_{123}$$

The result of the DWBA calculation corresponding to that neutron form factor is the curve II of Fig. 1: the shape of the calculated curve is rather good but the normalization is too low.

An alternative description of pick-up reaction at intermediate energy has been given by Tekou (3) in the case of the ^{12}C(p,d)^{11}C reaction. The spirit of that formalism is to introduce the multiple scattering theory to take account of the distortion in the entrance and exit channels. However, there are some basic problems which are unsolved in that theory, related to the fact that particles are different in the entrance and exit channels. Nevertheless, we have checked that, with the same ^{3}He and ^{4}He wave-functions as used before (Eq. 3), the calculated pick-up angular distribution is roughly similar to the one calculated in the DWBA formalism (same normalization; different shape).

I shall finish my talk with a more complex aspect of the pick-up process, I mean the possibility of transfering N^* resonances existing in nuclei. Starting from the description of baryon resonances in nuclei given by S. Jena and L. S. Kisslinger (4), R. Schaeffer (5) has developed a computer code that has been used to compute form-factors of N^* in ^{4}He. Table 1 gives the N^* considered and their orbital momentum in ^{4}He (only one possible value since ^{4}He has a spin 0). The result of the calculations of the ^{4}He(p,d)^{3}He reaction considered as a transfer of N^* described in the DWBA fortualism is presented in Fig. 3. One can see that the contribution to the cross section of N^* transfer look very small. (Only the more important contributions are shown.) However, it has to be pointed out that the recoil of ^{3}He in the ^{4}He nucleus has been neglected (the model of Jena and Kisslinger is a no-recoil model). That could underestimate the N* probabilities in ^{4}He by a large amount. On the other hand, we intend to repeat the experiment at a higher energy which should favour the transfer of N^* with respect to the neutron transfer.

REFERENCES

(1) S. D. Baker et al., Phys. Lett. 52B (1974) 57
(2) T. K. Lim, Phys. Lett. 44B (1973) 341
(3) A. Tekou, submitted to Nuclear Physics
(4) S. Jena and L. S. Kisslinger, Ann. of Physics 85 (1974) 251
(5) R. Schaeffer, private communication
(6) E. Rost, Nuclear Physics A249 (1975) 510
(7) E. Rost and J. R. Shepard, Physics Letters 59B (1975) 413

TABLE I

Mass (MeV)	$J\pi$	ℓ_{N*} in ^{4}He	$D(3\text{fm}^{-1})$ Ref. (6)
1440	$1/2^+$	0	22
1525	$3/2^-$	1	33
1550	$1/2^-$	1	10
1670	$5/2^-$	3	19
1688	$5/2^+$	2	36
1700	$1/2^-$	1	18

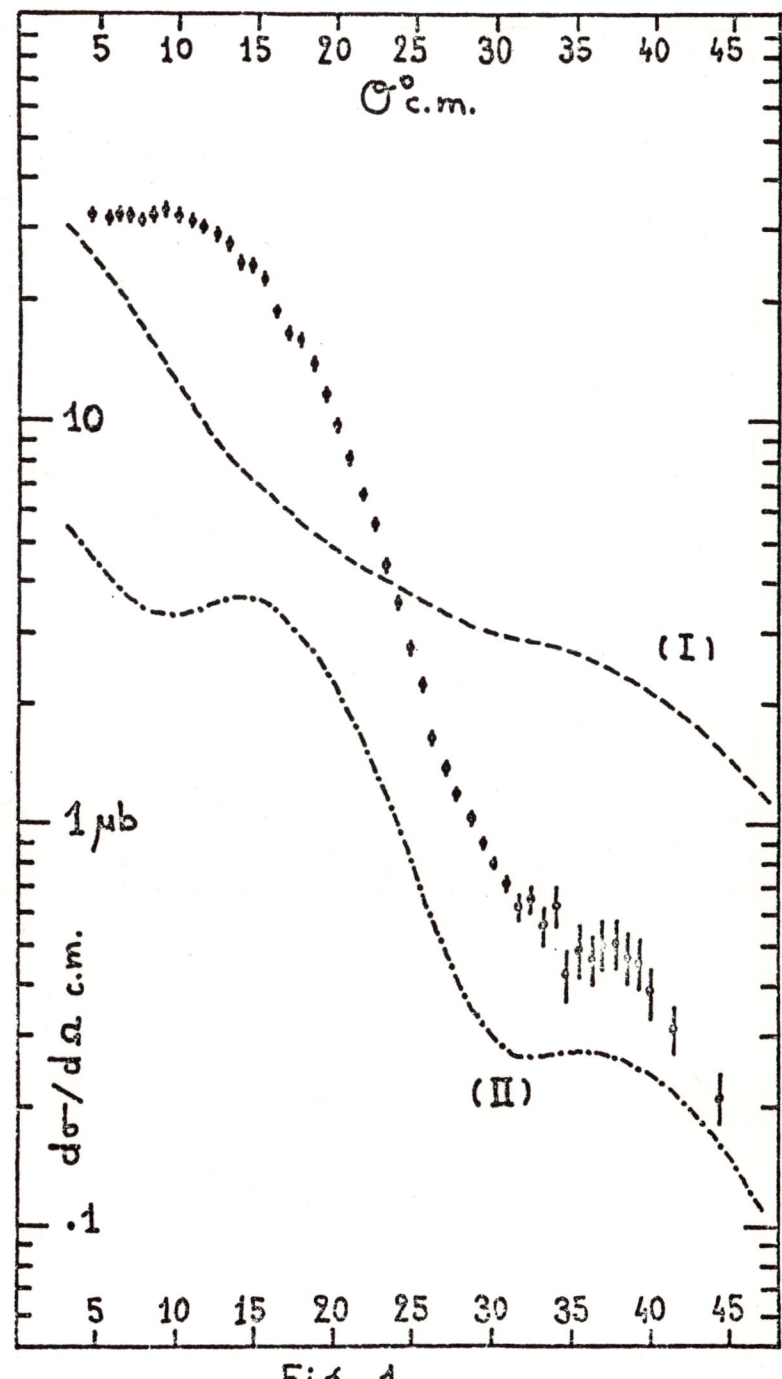

Fig. 1

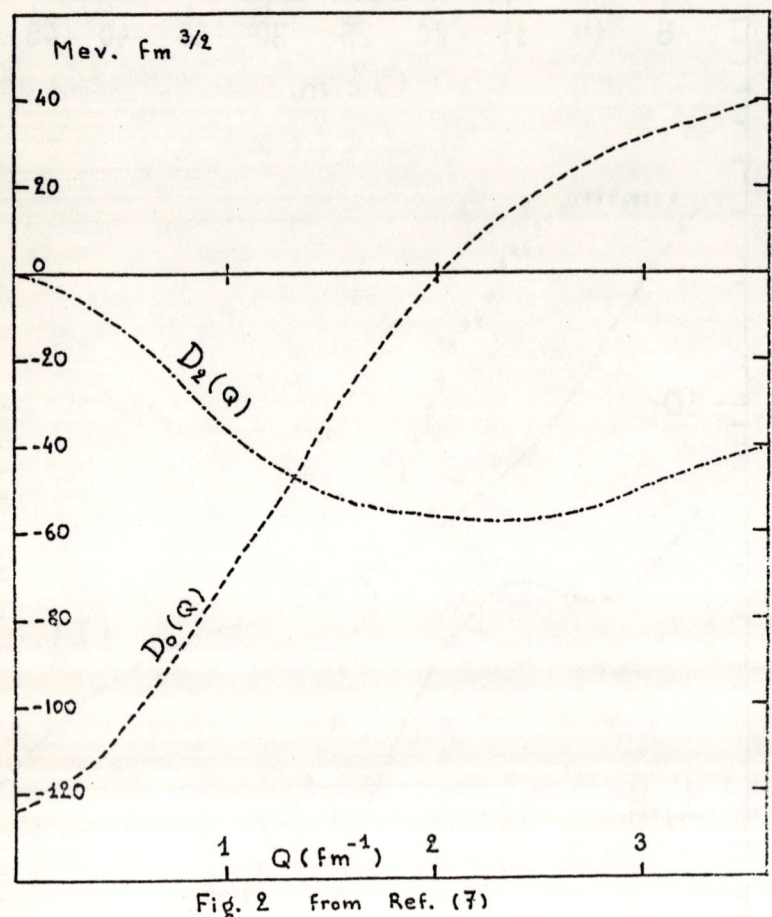

Fig. 2   from   Ref. (7)

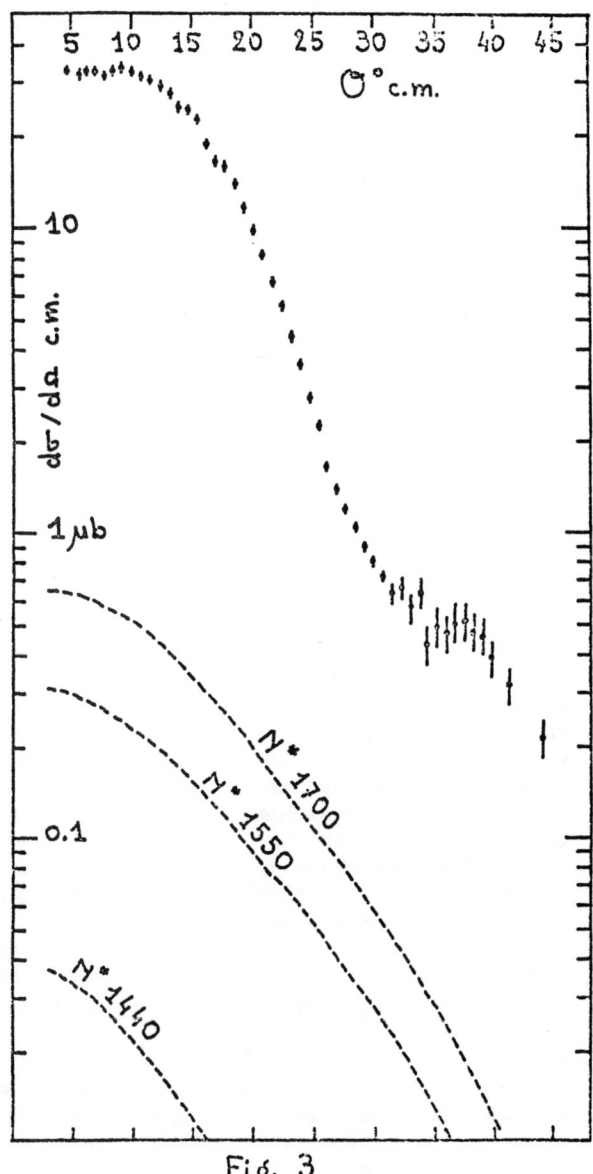

Fig. 3

PION CHARGE EXCHANGE AT REST

M. D. Hasinoff and M. Salomon
University of British Columbia, Vancouver, B.C., Canada V6T 1W5

A. Reitan[*]
University of Manitoba, Winnipeg, Man., Canada R3T 2N2

ABSTRACT

We present a description of charge-exchange absorption of stopped negative pions by complex nuclei, and estimate some branching ratios for this reaction.

GENERAL FEATURES

When stopped negative pions are absorbed by a nucleus from an atomic orbit, one of the reactions that may occur is the (π^-,π^0) charge exchange. To our knowledge, this reaction has previously been observed only for ^{1}H and ^{3}He,[1,2] although it has been looked for in several other cases.[3]

As far as light nuclei are concerned the reaction $A(\pi^-,\pi^0)$ at rest is energetically allowed only for ^{1}H, ^{3}He, ^{6}Li, ^{14}N and ^{27}Aℓ, most targets being acceptable from there on. However, the conservation of angular momentum and parity strongly suppresses the reaction for most targets that are possible from the point of view of energy conservation. Since the π^0, if emitted at all, has very low energy with respect to the nuclear system as a whole, it will largely be emitted in an s-wave in this system. Therefore, to lowest order only those charge-exchange absorptions are allowed which have a non-zero matrix element involving the s-wave part of the π^0 wave function. For absorption from an S-orbit this then requires the initial and final nuclei to have identical spins and parities; when the pion is absorbed from a P-orbit the nuclei must have opposite parities and a difference in spins of 0 or ±1 (0 → 0 transitions being forbidden), etc.

In very light nuclei, charge-exchange absorption from S-orbits is allowed for ^{1}H and ^{3}He, whereas in the case of ^{6}Li the reaction is not allowed for S- nor for P-orbits. For nuclei with mass number A < 100 there are only some ten cases for which the P-orbit charge-exchange reaction (including transitions to known excited states) is allowed; absorption from D- and higher orbits is, however, not so severely restricted. We also find that as the nuclear mass number increases the first-forbidden transitions, i.e. those that proceed via p- or higher waves in the overall system, become less suppressed compared to the allowed ones.

Considering transitions to excited final states in medium- to heavy-mass nuclei there is a large number of targets for which

---

[*]Permanent address: University of Trondheim, N-7000 Trondheim, Norway.

absorption from S-orbits is allowed, but in that case the branching ratio is of course strongly reduced due to the low population of negative pions. Even so, we find that for most targets S-orbit absorption may be the most likely channel for the charge-exchange reaction.

THEORY

In the charge-exchange reaction $A(\pi^-,\pi^0)B$ we picture the initial nucleus A as consisting of virtual particles C and p; the π^- then scatters with charge-exchange on the proton p, and the resulting virtual neutron n gets reabsorbed by C to form the final nucleus B. The transition rate thus depends on the matrix elements for the processes in the three vertices, as well as on the bound-state wave function for the absorbed pion, the calculational problem being to a large extent that of handling the angular-momentum couplings properly. We consider s-, p- and d-waves in the pion-nucleon vertex, using the impulse approximation, for the allowed transitions these πN partial waves are then transformed into overall s-waves through the Fermi motion of the virtual particles.

RESULTS

We have applied our model to charge-exchange absorption from nS, 2P and 3D orbits in a number of nuclei in the mass range A = 40-130. The 2P charge-exchange branching ratio is typically of the order 10^{-6}-10^{-5}, in agreement with preliminary TRIUMF data on Cu. As for the absorption from S orbits, among the abundant excited states in complex nuclei there is in most cases at least one for which such absorptions can proceed as allowed transitions. The corresponding branching ratios are not easily estimated, due to the unknown level populations. However, for the 2S level it only takes a population of about 0.01% (the corresponding number in muonic atoms being 2-3%[4]) for the absorption from this level to compete successfully with that from the 2P level. The 3D branching ratio is typically only of the order 10^{-11} - 10^{-10}.

REFERENCES

1. V.T. Cocconi et al., Nuovo Cim. 22, 494 (1961).
2. P. Truöl et al., Phys. Rev. Lett. 32, 1268 (1974).
3. V.I. Petrukhin et al., Nucl. Phys. 54, 414 (1964).
4. G. Backenstoss, Ann. Rev. Nucl. Sci. 20, 467 (1970).

NUCLEAR CORRELATIONS IN THE $^{13}C(\pi^+,\pi^0)^{13}N$ REACTION

E. Oset*

S.U.N.Y. at Stony Brook, Stony Brook, New York 11794

ABSTRACT

Pauli and short range correlations are studied and shown to be important in the $^{13}C(\pi^+,\pi^0)^{13}N$ reaction near the resonance, helping to reduce appreciably the discrepancies between the experiment and previous theoretical results.

The reaction $^{13}C(\pi^+,\pi^0)^{13}N$ in the region of the resonance[1] shows a flat integrated cross section while previous theoretical studies[2] show a pronounced minimum at $T_\pi \simeq 200$ MeV of about a factor five smaller than the experiment. One can think that the strong absorption shown in these models will be somehow reduced by a proper consideration of Pauli and short range correlations. The process of charge exchange takes place mainly on the surface, where the low density plus the fact that the nucleons are kept apart by correlations will increase the effective mean free path, leading to a reduction of the absorption. We have made some estimates of the effect of these correlations by using Glauber theory of multiple scattering. This approximation succeeds in reproducing elastic pion scattering in this region up to the first diffraction maximum and the differences beyond that point have little effect on the integrated cross section (the one we analyze for the charge exchange case).

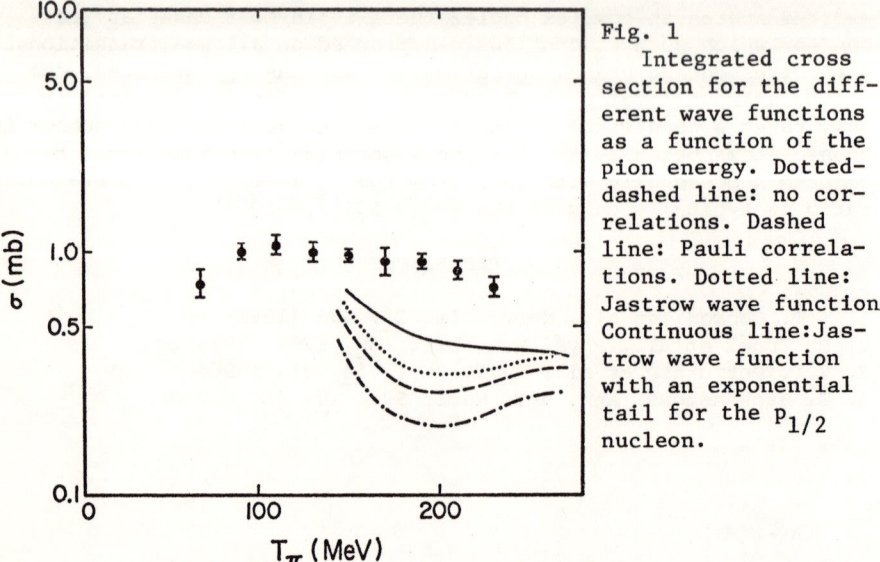

Fig. 1
Integrated cross section for the different wave functions as a function of the pion energy. Dotted-dashed line: no correlations. Dashed line: Pauli correlations. Dotted line: Jastrow wave function. Continuous line: Jastrow wave function with an exponential tail for the $p_{1/2}$ nucleon.

* On leave from University of Barcelona. Supported by GIFT(Spain)

The amplitude for our process is then given by
$$F(q)=\frac{ik}{2\pi}\int d^2\underline{b}\ e^{iq\underline{b}}<^{13}N|\sqrt{2}\Sigma_j\Gamma^{(v)}(\underline{b}-\underline{s}_j)\tau_j^+\prod_{\ell\neq j}(1-\Gamma^{(s)}(\underline{b}-\underline{s}_\ell))|^{13}C>$$
where $\Gamma^{(s)}$, $\Gamma^{(v)}$ are the isoscalar and isovector parts of the profile corresponding to the amplitude for πN scattering
$$f(q)=f^{(s)}(q)+f^{(v)}(q)\ \underline{\theta}\cdot\underline{\tau}\qquad (\underline{\theta},\underline{\tau}\ \text{isospin of pion and nucleon})$$

We have checked the consequences of considering the following wave functions in the nuclear matrix element of the amplitude:
1. Noncorrelated wave function, product of single particle harmonic oscillator wave function, filling the levels $s_{1/2}$ and $p_{3/2}$ with an odd nucleon in $p_{1/2}$ for ^{13}C. The nucleus of ^{13}N is the isobaric analog of ^{13}C, so it is given by $|^{13}N> = T^+\ |^{13}C>$.
2. Slater determinant of the single particle wave function to take into account Pauli correlations.
3. Jastrow wave function that takes into account additional short range correlations by choosing an appropiate correlation function.

The calculation of the matrix element of the amplitude with the non correlated wave function involves the diagonal matrix elements of the profile $\Gamma(\underline{b}-\underline{s})$ and only the odd neutron is allowed to undergo the charge exchange. The incusion of Pauli correlations allows the core neutrons of ^{13}C to contribute to the charge exchange through the non diagonal matrix elements of the profile. Short range correlations introduce a new degree of complexity and one has to deal with a cluster expansion to evaluate these matrix elements. We have used the F.A.H.T. cluster expansion[3] especially adapted to the process of multiple scattering[4]. The results can be seen in fig. 1, where the effects of correlations clearly tend to improve the agreement with the experiment. We have also considered the effects of changing the shape of the surface by taking a wave function with an exponential tail for the odd nucleon[5]. The combination of short range correlations with this particular shape of the tail strongly reduces the absorption as can be seen in the figure. The reaction is then very sensitive to details of nuclear structure.
In order to further improve the agreement with the experiment it is important to take these details into account at the same time that one looks for more accurate microscopic descriptions of the process.

The author is particularly thankful to Professor G. E. Brown for the suggestion of the work and valuable remarks along its realization and acknowledges some interesting discussions with Dr.M.Thies.

References
1. Y. Shamai et al. , Phys. Rev. Lett. 36 (1976) 82.
2. W.R. Gibbs, B.F. Gibson, A.T. Hess, G.H. Stephenson Jr., Phys. Rev. Lett. 36 (1976) 85.
3. J.W. Clark, P. Westhaus, Journ. Math.Phys. 9 (1968) 131, 149.
4. R. Guardiola, E. Oset, Nucl. Phys. 234A (1974) 458.
5. W.R. Gibbs, J.C. Jackson, W.B. Kaufmann, Phys. Rev. C9 (1974) 1340.

THE ELASTIC CHARGE EXCHANGE OF A PION ON ^{13}C
NEAR THE 3 3 RESONANCE REGION

S.FURUI

Institute of Physics, University of Tokyo, Komaba, Tokyo 153, Japan

The experimental result[1] of the elastic single charge exchange cross section of π^+ on ^{13}C is puzzling, since neither the energy dependence nor the absolute value near the 3 3 resonance region do not agree with theoretical predictions[2~4]. We studied the scattering in a simple eikonal model of Feshbach et al.[5] and tried to find characteristic features of this scattering process.

The eikonal model of Feshbach et al. is related to the optical limit of the Glauber's model[5]. The scattering amplitude can be written in the form,

$$F(\vec{k}',\vec{k}) = -\frac{1}{4\pi} \int d\vec{r}\, e^{-i\vec{k}'\cdot\vec{r}} \langle\chi|U(\vec{r})e^{i\int_{-\infty}^{z}(\sqrt{k^2-U(\vec{b},z')}-k)dz'}|\chi\rangle e^{ikz}$$

where $|\chi\rangle$ specifies the isospin state of the initial $\pi^+ + {}^{13}$C channel and the final $\pi^0 + {}^{13}$N channel. The potential $U(\vec{r})$ has the form,

$$U(\vec{r}) = 2E(V_c(\vec{r}) + V_p^0(\vec{r}) + t\cdot\tau V_p^1(\vec{r}))$$
$$= U_0(\vec{r}) + t\cdot\tau U_v(\vec{r})$$

where the suffix c denotes the core nucleons and p denotes the p-shell valence nucleon. The last scattering of the pion on the nucleus can be elastic $\pi^0 \to \pi^0$ or charge exchange $\pi^+ \to \pi^0$. We separate the amplitude to two corresponding terms.

$$F_d(\vec{k}',\vec{k}) = -\frac{1}{4\pi}\int d\vec{r}\, e^{-i\vec{k}'\cdot\vec{r}}(t\cdot\tau)U_v(\vec{r})e^{i\int_{-\infty}^{z}(\sqrt{k^2-U_0(\vec{b},z')}-k)dz'}e^{ikz}$$

$$F_{cc}(\vec{k}',\vec{k}) = -\frac{1}{4\pi}\int d\vec{r}\, e^{-i\vec{k}'\cdot\vec{r}} U_0(\vec{r})\int_{-\infty}^{z}dz'\, e^{-i\int_{z}^{z'}(\sqrt{k^2-U_0(\vec{b},z'')}-k)dz''}$$
$$\times [i(t\cdot\tau)U_v(\vec{b},z')/2\sqrt{k^2-U_0(\vec{b},z')}]e^{i\int_{-\infty}^{z'}(\sqrt{k^2-U_0(\vec{b},z'')}-k)dz''}e^{ikz}$$

$$= -\frac{1}{4\pi}\int d\vec{r}\, e^{-i\vec{k}'\cdot\vec{r}} U_0(\vec{r})\int_{-\infty}^{z}dz'[i(t\cdot\tau)U_v(\vec{b},z')/2\sqrt{k^2-U_0(\vec{b},z')}]$$
$$\times e^{i\int_{-\infty}^{z}(\sqrt{k^2-U_0(\vec{b},z')}-k)dz'}e^{ikz} \quad (1)$$

$$(t\cdot\tau) = \langle 0|t\cdot\tau|+\rangle = \sqrt{2}$$

$$\frac{d\sigma}{2\pi d(\cos\theta)} = |F_d(\vec{k}',\vec{k}) + F_{cc}(\vec{k}',\vec{k})|^2$$

where we retained terms linear in $U_v(\vec{r})$.

The integral

$$i\int_{-\infty}^{z} U_v(\vec{b},z')/2\sqrt{k^2-U_0(\vec{b},z')}\,dz' \simeq e^{i\int_{-\infty}^{z} U_v(\vec{b},z')/2\sqrt{k^2-U_0(\vec{b},z')}} - 1$$

corresponds to the phase shift due to the charge exchange before the final scattering at $\vec{r}$. The pair distribution of the valence nucleon and the core nucleon is explicit in this term.

We obtained the pion-nucleus potential by the impulse approximation employing the McKinley's[6] first phase shift of the pion-nucleon scattering. The nuclear density was replaced by the simple harmonic oscillator with the spring constant a = 1.6 fm.

The absolute value of the cross section is unfortunately 2-3 times smaller than the experiment. The ratio of the total cross section of the elastic SCX on ^{13}C and that of the elastic on ^{12}C agrees with the experiment near the 3 3 resonance region. In fig.1 we show our results with that of the DWIA (2), that of the Glauber's model[3] (3) and that of the Watson's multiple scattering series[4] (4).

The scattering amplitude in DWIA can be formally written as

$$\int d\vec{r}\, e^{-ik'z}(t\cdot\tau)U_v(\vec{r})\psi^{(+)}(\vec{k}',\vec{r})$$
$$+\int d\vec{r}'\int d\vec{r}\, [e^{ik'z'}U_o^*(\vec{r}')\tilde{G}_o^{(-)}(\vec{r}',\vec{r})]^*(t\cdot\tau)U_v(\vec{r})\psi^{(+)}(\vec{k},\vec{r}) \quad (2)$$

where $\tilde{G}_o^{(-)} = \lim_{\varepsilon\to 0}[E-i\varepsilon-K-U_o^*]^{-1}$

in momentum space. F_d and F_{cc} correspond to the two terms of eq.(2) respectively. One can regard the amplitude of eq.(1) as that of the elastic scattering in the I=3/2 channel. The effective potential in F_{cc} does not have the usual form

$$V(\vec{r})_{\alpha\beta} = \sum_i f(\vec{r})_i A_{\alpha\beta}$$

with r-independent factors $A_{\alpha\beta}$[5].

We remark that the interference of F_d and F_{cc} are constructive for the pion kinetic energy $T_\pi > 150 MeV$. The isovector correlation of the valence nucleon and the core nucleon was ignored[7]. The study is under way.

REFERENCES
1. Y.Shamai et al., Phys.Rev.Lett. 36(1975) 82.
2. N.Auerbach and J.Warszawski, Phys.Lett. 45B(1973) 171.
3. A.Reitan, Nucl.Phys. B68(1974) 387.
4. W.R.Gibbs et al., Phys.Rev.Lett. 36(1975) 85.
5. W.W.Bassichis et al.,Ann.Phys. (N.Y.) 68(1971)462.
   H.Feshbach and J.Hüfner, Ann. Phys.(N.Y.) 56(1970) 268.
6. J.M.Mc Kinley, Rev.Mod.Phys. 35 (1963) 788.
7. J.M.Eisenberg and A.Gal, Phys. Lett. 58B(1975) 390.

fig.1 The elastic single charge exchange cross section of ^{13}C.

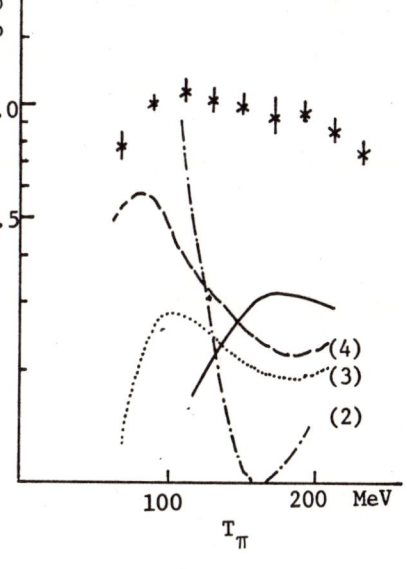

fig.1

PION CHARGE EXCHANGE IN THE (3,3) RESONANCE REGION

N. Auerbach, Department of Physics and Astronomy, Tel-Aviv University, Tel-Aviv, Israel.
J. Warszawski, Institut de Physique Nucleaire, B.P.1, 91 Orsay, France.

ABSTRACT

The $C^{13}(\pi^+,\pi^0)N^{13}$ reaction in the resonance region is studied using first order pion-nucleus optical potentials. In addition, model calculations of two-step processes are presented.

The recent measurements[1] of the $C^{13}(\pi^+,\pi^0)N^{13}$ reaction to the ground state of N^{13} exhibit a flat cross-section as a function of the pion energy in the range 30 MeV < T < 220 MeV. Results of previous calculations based on the first order optical model[2,3] showed that the (π^+,π^0) cross-section in C^{13} decreases as the energy of the pion approaches the resonance region.

We extended these calculations to include a series of optical model potentials each potential corresponding to a different off-shell extrapolation of the pion-nucleon t-matrix.

The general expression of the potentials used was:

$$V_{opt}(r) \sim (\alpha\rho(r) - \beta\underline{\nabla}\cdot(\rho(r)\underline{\nabla}) + \gamma\nabla^2\rho(r) - \delta\rho(r)\nabla^2) \qquad (1)$$

where $\rho(r)$ is the nuclear density and the coefficients $\alpha,\beta,\gamma,\delta$ are energy dependent functions of the pion-nucleon phase-shifts. Different off-shell models[4] for the t-matrix lead to different sets of functions.

The calculations were performed using a coupled-channels method in which the elastic, single charge exchange, and double charge exchange channels are included. In addition to the investigation of various types of optical potentials we also examined several models for the $p_{\frac{1}{2}}$ neutron density in C^{13}. All calculations irrespective of the potential or density used showed a sharp decrease of the (π^+,π^0) cross-section in the resonance region. (Of all the models used the largest cross-section we obtained at the resonance was about 0.2 mb).

In order to investigate the sensitivity of these results to variations of various parameters entering the theory we made several additional studies, still remaining in the framework of the first order optical potential.

(a) Because of strong absorption in the resonance region the (π^+,π^0) reaction takes place mainly at the surface. If for some reason the radius of the $p_{\frac{1}{2}}$ neutron density in C^{13} is considerably larger than predicted by nuclear models then this would reduce the effects of absorption. We have checked this point by increasing the radius and reducing the binding of the $p_{\frac{1}{2}}$ neutron. The (π^+,π^0) cross-section became flat as a function of energy but the magnitude increased only very slightly. A 30% increase in the radius led to about 10% increase in the cross-section.

(b) We reduce, quite arbitrarily, the absorption in the elastic channel. In order to get a (π^+,π^0) cross-section of about 0.6 mb we had to reduce the imaginary part of the potential by 70% which caused a reduction of the elastic cross-section by more than a factor of 2.
(c) We added to the potential the previously neglected spin-flip term and calculated its contribution using the DWIA. In the resonance region this term contributed less than 0.02 mb to the (π^+,π^0) cross-section.

Non-Direct Mechanisms. In view of these results it seems natural to go beyond the first order optical potential and to include second order processes. The possibility that such non-direct mechanisms may play a role in charge exchange has been mentioned in ref. 2. and recently second order effects were studied[6] using a Pauli correlation function and applying the Eikonal approximation.

Rather than using closure we consider here two step mechanisms in which the incident pion excites in the first step a specific intermediate state in C^{13} which in the second step is deexcited by charge exchange leaving the nucleus in the ground state of N^{13}. We performed a coupled channels calculation which included the elastic channel (1), the pion charge exchange to the g.s. of N^{13} (2) and an inelastic channel in C^{13} (3). As a first guess we used as intermediate states the giant isovector monopole and the giant dipole states. The coupling between channels 1 and 3 (V_{13}) was estimated using collective model transition densities. For the V_{23} coupling we used the same form as for V_{13} but reduced by a spectroscopic factor (S) to take into account the fact that only a small component of the giant state can lead to the g.s. of N^{13} via charge exchange of a single nucleon. The calculation showed that the monopole had almost no effect on the cross-section. In the case of the dipole the calculation yielded a less than 50% increase in the cross-section when we used a rather large S = 0.2 value. In an extreme case when absorption was reduced by 30% and S = 0.3, the cross section increased to 0.35 mb.

Although the inclusion of the dipole as an intermediate state had a noticeable effect on the (π^+,π^0) cross-section it was not sufficient to yield agreement with experiment. It is still possible however that a more careful treatment which will include additional intermediate states will result in an increased (π^+,π^0) cross-section in C^{13}.

REFERENCES
1. Y. Shamai et al., Phys. Rev. Letters 36, 82 (1976).
2. N. Auerbach and J. Warszawski, Phys. Letters 45B, 171 (1973).
3. D. Tow and J.M. Eisenberg, Nucl. Phys. A237, 441 (1975).
4. L.S. Kisslinger, Phys. Rev. 98, 761 (1955).
   H.K. Lee and H. McManus, Nucl. Phys. A167, 257 (1971).
   L.S. Kisslinger and F. Tabakin, Phys. Rev. C9, 188 (1974).
   G.A. Miller, Phys. Rev. C10, 1242 (1974).
5. E.C. Bartels and A.K. Kerman, MIT Report 2098-286, 1966.
6. J. M. Eisenberg and A. Gal, Phys. Letters 58B, 390 (1975).

PION-CARBON 12 WAVEFUNCTIONS AND INELASTIC SCATTERING*

W. R. Gibbs, A. T. Hess and G. J. Stephenson, Jr.
Theoretical Division, Los Alamos Scientific Laboratory
Los Alamos, New Mexico 87545

ABSTRACT

We have examined the optical model wave functions for a Kisslinger potential with free and fit parameters. The fit parameters produce much smoother wave functions and predict inelastic scattering to the first 2^+ state an order of magnitude smaller than the free parameters.

INTRODUCTION

Elastic π-^{12}C data at 50 MeV have been notoriously difficult to fit either with optical models or multiple scattering approaches. Recently, Cooper and Eisenstein[1] were able to obtain a good fit with parameters which apparently do violence to the s-wave π-N scattering amplitude.[2] Being curious as to the source of this improvement, we have examined the optical model wave functions produced by a Kisslinger potential using free parameters, using free parameters modified by the angle transformation necessary to obtain decent fits to π-^{4}He scattering,[3] and using the fit parameters. The wave functions were calculated with a modified version of ABACUS.[4] The general features persist from partial wave to partial wave; to illustrate these features we present the real and imaginary parts of the s-wave in figure 1. The two most striking features of the wave function from the fit parameters are the complete removal of the sharp kink at the nuclear surface and the strong suppression of the real part in the interior. The wave function from the angle-transformed parameters shows some of the suppression but does not remove the rapid variation near the surface and, as might be expected, does not significantly improve the fit to the elastic scattering.

INELASTIC SCATTERING

In view of the fact that the fit parameters decrease the magnitude of the scattering wave function in the surface and fact that the excitation of the collective 4.43 MeV 2^+ state is surface-peaked, we would expect a marked decrease in the inelastic cross section. We have calculated this cross section following the prescription of Hess and Eisenberg[5] and the results are presented in figure 2. There is a clear order of magnitude difference in the predictions over most of the angular range as well as a clear difference in shape. Precision measurement of this inelastic cross section would help immensely in determining if the particular choice of parameters made by Cooper and Eisenstein are truly reflecting some physics left out of the usual prescription.

*Work performed under the auspices of the U. S. ERDA.

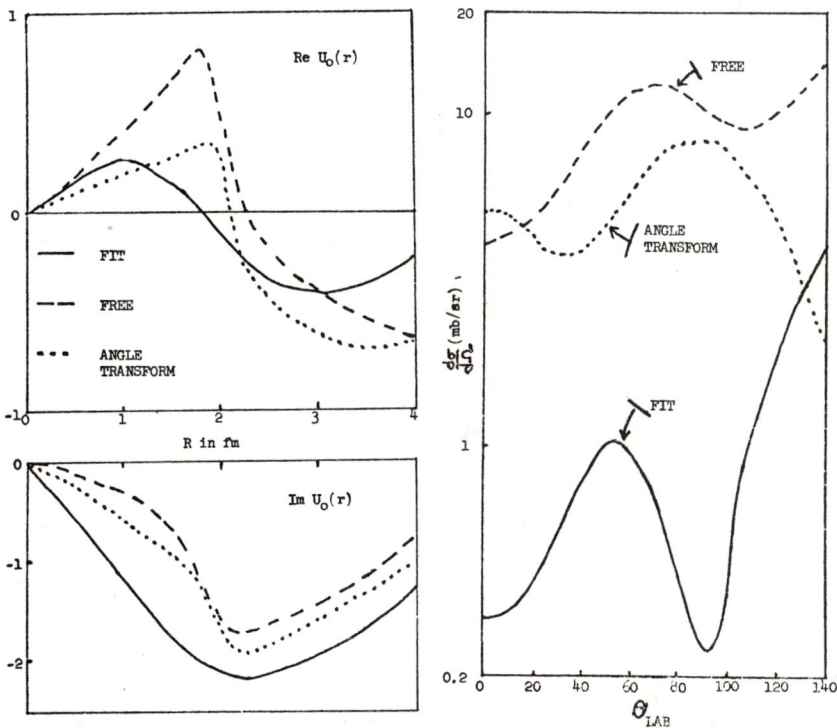

Fig. 1. Distorted wave functions.

Fig. 2. Inelastic cross sections to the 2^+, 4.43 MeV state of ^{12}C, as described in text.

REFERENCES

1. M. D. Cooper and R. A. Eisenstein, submitted to Phys. Rev. Lett.
2. A similar result was noted by Miller for the 30 MeV data; G. A. Miller, Nucl. Phys. A224, 269 (1974).
3. B. F. Gibson, W. R. Gibbs, A. T. Hess, G. J. Stephenson, Jr., and W. B. Kaufmann, Phys. Rev. C13, (to be published).
4. E. H. Auerbach and M. Sternheim, Brookhaven National Laboratory Report 12696 (1968); W. B. Jones, Ph.D. dissertation, University of Virginia, Charlottesville, Va., 1970.
5. A. T. Hess and J. M. Eisenberg, Nucl. Phys. A241, 493 (1975).

CROSS-SECTION FOR THE DOUBLE CHARGE EXCHANGE REACTION $\pi^+ + {}^4He \rightarrow \pi^- + 4p$ AT PION ENERGIES OF 98, 135, 145 AND 156 MEV

I.V. Falomkin, V.I. Lyashenko, G.B. Pontecorvo and
Yu. A. Shcherbakov
Joint Institute for Nuclear Research - Dubna

M. Albu, T. Angelescu, O. Balea, A. Mihul, F. Nichitiu,
A. Seraru
Institute for Atomic Physics - Bucharest

F. Balestra, R. Garfagnini, G. Piragino
Istituto di Fisica dell'Università - Torino
Istituto Nazionale di Fisica Nucleare - Sezione di Torino

C. Guaraldo, R. Scrimaglio
Laboratori Nazionali di Frascati, Italia

ABSTRACT

Cross-sections for the double charge exchange of positive pions on ^{4}He at various energies are presented. For measurements a high-pressure helium streamer chamber has been used. The experimental results are compared with theoretical model calculations based on pair correlations or meson exchange currents.

FORMALISM AND RESULTS

In this paper we present results of cross-section measurements for the double charge exchange of positive pions at pion energies 98, 135, 145 and 156 MeV. These measurements are a continuation of our previously reported work (1). For this experiment we used a high-pressure helium streamer chamber (2), which was both a target and a detector. The chamber was surrounded by a scintillation counter hodoscope, and was triggered by a secondary pion, passing through any one of the hodoscope counters. The tracks of the protons and the pion were easily identified due to a difference between their ionization densities. In order to evaluate the probability of background processes the experiment was carried out both with a ^{3}He and a ^{4}He filling of the chamber. In all about $1.9 \cdot 10^8$ pions passed through the chamber-target and 270 000 pictures were obtained. One elastic pion scattering event was obtained from about 100 pictures and one double charge exchange event was identified for every 10^4 pictures. Double charge exchange events were identified by submitting all candidates to a kinematical test. Background measurements with helium-3 were taken into account (1). Background processes (five-prong stars imitating the double charge exchange) were present both in helium-3 and in helium-4 with the same probability corresponding to about 30 μb. Fig. 1 presents all available experimental cross sections for the double charge exchange of positive pions on ^{4}He at different energies. Presented also are results of calculations for the reaction $\pi^- + {}^4He \to \pi^+ + 4n$ (3). The calculations were based on the pair correlation model. The cross sections for this reaction with positive pions shouldn't differ greatly. The same Fig. 1 shows the energy dependence of the total cross section calculated in ref. (4). As we see at low energies there is a considerable difference between the experimental and calculated cross section values for both models. At higher energies we see experimental data are near to meson exchange currents calculations.

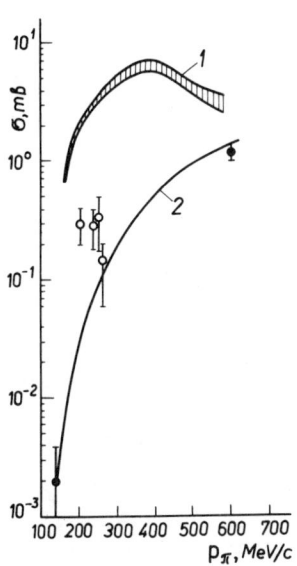

Fig. 1 Cross section of double charge exchange of π^+-mesons on ^{4}He versus pion lab. sys. kinetic energy-experimental points and calculated curves (3). The first and last points are from ref.(3) (at 40 and 485 MeV). The lower curve (2) is from exchanging currents model (4) calculations.

REFERENCES

1. I.V. Falomkin et al. Nuovo Cim. 22A, 333 (1974).
2. I.V. Falomkin et al. Nuovo Cim. 5, 757 (1972).
3. F. Becker, C. Schmit. Nucl. Phys. B18, 607 (1970).
4. J.F. Germond, C. Wilkin. Lett. Nuovo Cim., 13, 605 (1975).

PION DOUBLE CHARGE-EXCHANGE ON ^{4}He[*]

A. T. Hess, W. R. Gibbs, B. F. Gibson and G. J. Stephenson, Jr.
Theoretical Division, Los Alamos Scientific Laboratory
Los Alamos, New Mexico 87545

ABSTRACT

We have examined the reaction $^4\text{He}(\pi^\pm,\pi^\mp)4(^p_n)$ by considering two single π-N charge-exchange processes with intermediate off-shell propagation. Separable π-nucleon t-matrices having off-shell form factors were used along with fully antisymmetrized wave functions. Both angular distributions and total cross sections were calculated and compared to the existing data.

INTRODUCTION

In various experimental searches for the tetraneutron[1,2], data were obtained for the double charge-exchange reaction $^4\text{He}(\pi^-,\pi^+)4n$. Also, experiments[3,4] have been performed in which data were obtained for the double charge-exchange reaction of incident π^+ on ^{4}He. However, there has been some difficulty[5] in obtaining good agreement between theory and experimental results; discrepancies as large as 3 orders of magnitude exist. Some of this difficulty may be due to the improper treatment of antisymmetry of the nuclear wave functions or to the problems involved in a complete treatment of the five-body final phase space, but we believe the major difficulty lies in the effective π-nucleon t-matrix. A recent publication[6] considers the contribution to the process from the π-π scattering amplitude in which the incident pion scatters from a pion in the clouds which surround the nucleons. In this work, we report on our attempt to perform a complete calculation of the contribution to the double charge-exchange reaction arising from two π-N charge-exchange scatterings, which provides the obvious competitor to that mechanism of Ref. 6.

FORMALISM

We have used a double-scattering amplitude consisting of two single-scattering π-N charge-exchange amplitudes with intermediate off-shell propagation. Separable s- and p-wave amplitudes having off-shell form factors were used. Spin-flip was included and gives rise to single and double spin-flip contributions to the reaction. Fully antisymmetrized nuclear wave functions were used, incorporating the effects of Pauli principle supression in the final state of four identical nucleons. The integral over the final four nucleon phase space was done by Monte Carlo techniques. (Eq. 1).

$$\int \delta(\sum_{i=1}^{4}\vec{k}_i + \vec{k}' - \vec{k})\delta(\sum_{i=1}^{4}\frac{k_i^2}{2m} + \omega' - \omega + E_B)d\vec{k}_1 d\vec{k}_2 d\vec{k}_3 d\vec{k}_4 \quad (1)$$

[*]Work performed under the auspices of the U. S. ERDA.

RESULTS

We have calculated the 20° differential pion cross section for 140 MeV incident pions as a function of the outgoing pion kinetic energy and have compared the results to the Kaufman[2] data. A reasonable fit is obtained. We find that the Pauli suppression is significant and is, as expected, clearly most important near the high-energy end of the pion spectrum, where there is little energy to be shared by the four nucleons. We have also calculated the 0° pion cross section for outgoing pions as a function of the incident pion kinetic energy. Our results are low (~ a factor of 3) when compared to the Gilly data. Finally, in Fig. 1 we present our results for the total integrated pion double-charge-exchange cross section along with the two existing data points. Since our calculation should be most reliable for $E_\pi \leq$ 200 MeV, it is interesting to note that we agree with the data of Ref. 2 at 140 MeV and disagree by several orders of magnitude with the data of Ref. 4 at 100 MeV. While we feel that the results which we have obtained at low pion energies are reasonable, there are problems at the higher energies. These may result from neglecting multiple-scattering on the two spectator nucleons, ignoring the pion-nucleon partial waves higher than $\ell=1$, or neglecting N-N final state interactions. A more complete calculation awaits our ability to treat these additional features.

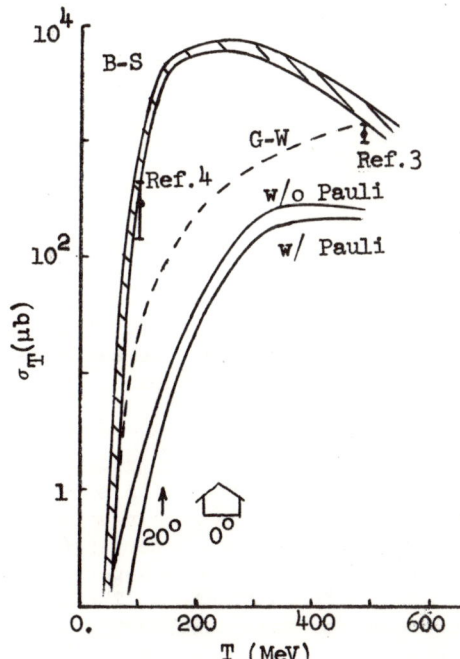

Fig. 1. Total cross section for ^{4}He(π^+,π^-)4p. The arrows indicate the energies of the single angle data.

REFERENCES

1. L. Gilly, et al., Phys. Lett. 1, 335 (1965).
2. L. Kaufman, et al., Phys. Rev. 175, 1358 (1968).
3. N. Carayannopoulos, et al., Phys. Rev. Lett. 20, 1215 (1968).
4. I. V. Falomkin, et al., Nuovo Cimento 22, 333 (1974).
5. F. Becker and C. Schmit, Nucl. Phys. B18, 607 (1970).
6. J. F. Germond and C. Wilkin, Nuovo Cimento Lett. 13, 605 (1975).

POSSIBLE EVIDENCE FOR SHORT-RANGE 4-N CORRELATIONS FROM $(\pi,\gamma X)$ REACTIONS

C. E. Stronach, J. H. Stith, and C. M. Dennis
Virginia State College, Petersburg, Va. 23803

B. J. Lieb and W. F. Lankford
George Mason University, Fairfax, Va. 22030

H. O. Funsten and W. J. Kossler
College of William and Mary, Williamsburg, Va. 23185

H. S. Plendl
Florida State University, Tallahassee, Fla. 32306

V. G. Lind
Utah State University, Logan, Utah 84321

ABSTRACT

Studies of prompt γ-ray spectra from pion reaction experiments on ^{32}S and ^{40}Ca performed with a Ge(Li) detector at the SREL synchrocyclotron indicate large cross sections for equivalent-alpha-particle removal, and for ^{3}He removal with shell transfer of the odd neutron. These phenomena are discussed in terms of the evidence they provide for existence of short-range 4-N correlations.

EQUIVALENT ALPHA REMOVAL

The observation of 2p-2n removal from ^{32}S and ^{40}Ca with surprisingly large cross sections/yields (Table I) from pion reactions near the $\Delta(1232)$ resonance and π^- absorption at rest, determined from prompt nuclear de-excitation γ-ray measurements, is superficially suggestive of direct reactions on 2p-2n clusters. These values are considerably larger than those for most neighboring

Table I. Selected Cross Sections (mb) and Yields (%) from $(\pi,\gamma X)$ Reactions on ^{32}S and ^{40}Ca.

Equiv. Removal		σ(220-MeV π^-)	σ(190-MeV π^+)	Yield (0-MeV π^-)
^{32}S	1α	68 ± 14	56 ± 11	6.79 ± .44
	2α	25 ± 5	32 ± 7	2.90 ± .20
	3α	14 ± 3	14 ± 4	1.57 ± .22
^{40}Ca	1α	90 ± 19	76 ± 12	5.92 ± .68
	2α	<72 ± 15	<53 ± 11	<2.10 ± .24
	3α	65 ± 15	61 ± 13	2.21 ± .34
	4α	13 ± 3	23 ± 5	2.65 ± .23

non-equivalent-alpha-removal daughter nuclei, which is not very startling because the total strengths of these reactions are seen in the cascades through the 2^+ states. Several of these levels are found to be Doppler broadened, however, with mean recoil momenta of several hundred MeV/c. These suggest that reactions involving removal of energetic heavy aggregates may be present, such as absorption on clusters, or cascade processes, including final-state pickup, rather than boil-off of successive particles in random directions.

Tentative results from studies of the interaction of 190-MeV π^+ with Ni show considerably larger cross sections for daughter nuclei corresponding to multiple equivalent-alpha removal compared with other even-even daughters, some of which are closer to the maximum-stability line than those corresponding to equivalent-alpha removal. This appears to confirm an earlier result of Jackson et al.[1]

EQUIVALENT ^{3}He REMOVAL

Examination of the cross sections and yields for states of nuclei corresponding to equivalent ^{3}He removal from ^{32}S and ^{40}Ca seems to provide yet stronger evidence for π reactions with surface 4-N clusters. Table II shows large cross sections for production of the $f_{7/2}^-$ single-particle states of the daughter nuclei. These levels also have large (d,p) spectroscopic factors[2]. This suggests an initial reaction with an alpha cluster, followed by recapture of a neutron, the final-state interaction being a stripping process. These results are in sharp disagreement with the predictions of statistical calculations, both for production of $\Delta A = 3$ daughters, and for the relative excitation of the individual states of ^{37}Ar.

Table II. Cross Sections (mb) and Yields (%) for Equivalent ^{3}He Removal from ^{32}S and ^{40}Ca.

Ex. State	σ(220-MeV π^-)	σ(190-MeV π^+)	Yield (0-MeV π^-)
$\pi + {}^{40}\text{Ca} \to {}^{37}\text{Ar}^*$			
1410 keV $1/2^+$	5.8 ± 1.4	4.9 ± 2.1	.26 ± .17
1611 keV $7/2^-$	29 ± 6	36 ± 7	1.86 ± .16
$\pi + {}^{32}\text{S} \to {}^{29}\text{Si}^*$			
1273 keV $3/2^+$	<45 ± 10	<44 ± 9	<5.24 ± .22
2028 keV $5/2^+$	19 ± 7	8 ± 4	.96 ± .30
2426 keV $3/2^+$	------	.9 ± .4	--------
3067 keV $5/2^+$	------	-----	--------
3624 keV $7/2^-$	6.2 ± 2.2	6.8 ± 2.8	1.56 ± .24

1. H. E. Jackson et al., Phys. Rev. Lett. 35, 641 (1975).
2. P. M. Endt and C. van der Leun, Nucl. Phys. A214, 1 (1973).

NEGATIVE PION ABSORPTION AT REST LEADING TO ^{11}Be BOUND STATES

B. Coupat, D.B. Isabelle, P.Y. Bertin
Lab. Physique Corpusculaire, Université de Clermont, France

A. Gérard, J. Miller, J. Morgenstern, J. Picard, B. Saghai, P. Vernin
DPhN/HE, CEN Saclay, France

ABSTRACT

The capture rate probabilities for negative pion at rest were measured for the $^{12}C(\pi^-,p)^{11}Be_{Bound}$ and $^{13}C(\pi^-,np)^{11}Be_{Bound}$ by an activation technique.

The study of negative pion absorption at rest by nuclei can provide information about effects such as short range correlations and clusters. In particular the absorption followed by emission of a single proton could be described as the interaction of the pion with one member of a proton pair producing a neutron sticking to nuclear matter[1] while the second member is ejected from the nucleus. But momentum conservation in the final state implies that the neutron has a rather large momentum then its sticking probability must be small. It is surely interesting to obtain results on reactions with a sticking neutron (π^-,p) and without sticking neutron (π^-,np) leaving the nucleus in the same final state.

By an activation method the two reactions $^{12}C(\pi^-,p)^{11}Be^*$ and $^{13}C(\pi^-,np)^{11}Be^*$ have been studied. The ^{11}Be residual nucleus is a β^- emitter with a 13.6 s half life and a 11.5 MeV energy end point. With such a method we measure only the probability for pion absorption leading to bound states of the residual nucleus. The interest of ^{11}Be is that only the ground and first excited (0.32 MeV) states are bound.

These experiments were performed with the Saclay pion beam facility[2]. In a first stage the production rate of ^{11}Be was measured in a natural carbon target (98.9 % ^{12}C, 1.1 % ^{13}C). The experimental set up has already been described[3].

A probability of $(4.5 \pm 0.8) \cdot 10^{-4}$ per stopped pion was found for the reaction $C_{natural}(\pi^-,p)^{11}Be_{Bound}$.

The second experiment was performed with a 98.1 % enriched ^{13}C target using a pulsed pion beam, the target activity being measured between pulses. A probability of $(1 \pm 0.1) \times 10^{-2}$ per stopped pion was found for the reaction $^{13}C(\pi^-,np)^{11}Be_{Bound}$. This is, to our knowledge, one of the first measurement of a (π^-,np) process for which the final state is known. Combining those two results we found that the probability for $^{12}C(\pi^-,p)^{11}Be_{Bound}$ is equal to $(3.4 \pm 0.8) \times 10^{-4}$.

As an activation method was used in this experiment the residual ^{11}Be activity can also be due to the reaction $^{12}C(\pi^-,\gamma p)^{11}Be$, but the probability associated to this process must be very small.

This is based on two facts : first the $(\pi^-,\gamma p)$ probability is much smaller than the total radiative capture probability which is known to be of the order of 2 %, second the proton spectrum for $^{14}N(\pi^-,p)$ was measured in an other experiment displaying a proton concentration at an energy corresponding to non radiative absorption, with an identical absorption probability.

In this experiment the ratio of the (π^-,np) to the (π^-,p) probabilities on C leading to the same residual nucleus ^{11}Be can be given for the first time and is found to be of the order of 30.

REFERENCES

1. J. Hüfner, Phys. Rep., 216, 1 (1975)
2. P.Y. Bertin et al., Internal Report, CEA DPhN/HE 71/3
3. B. Coupat et al., Phys. Let., 55, 286 (1975)
4. P.Y. Bertin, J. Phys., 36, C5, 13 (1975)

QUASI-FREE $(\pi,\pi N)$ SCATTERING AT THE (3,3) RESONANCE

V. E. Herscovitz, Th. A. J. Maris, P. M. Mors and C. Schneider
Instituto de Fisica
Universidade Federal do Rio Grande do Sul
Porto Alegre, Brasil

A quasi-free reaction is in practice only informative if its matrix element is to a good approximation factorizable in a (modified) matrix element for a free collision and a (distorted) momentum distribution[1]. The validity of this approximation is closely related to the validity of the WKB-approximation for the distortion of the incoming and outgoing waves, because both approximations demand that the main contributions are governed by momentum components near to the assymptotic momenta of the particles. A factorizable quasi-free $(\pi,\pi N)$ cross section could give information on the resonance propagation in the nuclear surface[2], with well-defined kinematics. Therefore we have investigated the quality of the WKB approximation for $(\pi,\pi N)$ scattering in the resonance region by comparing it with an extensive partial wave analysis. This note is a preliminary report of the work.

As an example we have taken the 200 MeV $(\pi,\pi p)$ reaction, knocking out a 2s proton of ^{40}Ca, giving a 0^+-state in ^{39}K; the separation energy is 11 MeV. The gaussian optical potentials have been taken purely imaginary for the pions and complex for the proton, with strengths following from the known total cross sections and forward scattering amplitudes of these particles for their scattering on free nucleons at the relevant energies.

The full curve of Fig. 1 shows the undistinguishable results for $|g'_{2s}(K=0)|^2$, i.e., the value of the distorted momentum distribution for vanishing recoil of the residual nucleus, in dependence of the pion scattering angle which in this case determines the kinematics. For a constant reduction factor simulating the distortion, this function would be a constant. The quite unexpected complete agreement between the WKB and partial wave results demand at least 25 angular momenta for each particle; the dotted curves represent the results for 15 and 20 angular momenta per particle. The 180 degree point of the curve was also calculated with $\ell \leq 30$ and the result differed negligibly from the corresponding $\ell \leq 25$ point.

We have calculated $|g'_{2s}(K)|^2$ for various kinematical situations and have also investigated the probability distribution for the spatial location in the nucleus of the quasi-free event. In all these cases a striking agreement between the WKB and partial wave calculations was found.

REFERENCES

1. G. Jacob and Th. A. J. Maris, Rev. Mod. Phys. **38** (1966) 121 and **45** (1973) 6.
2. L. C. Liu, Nucl. Phys. **A223** (1974) 523.

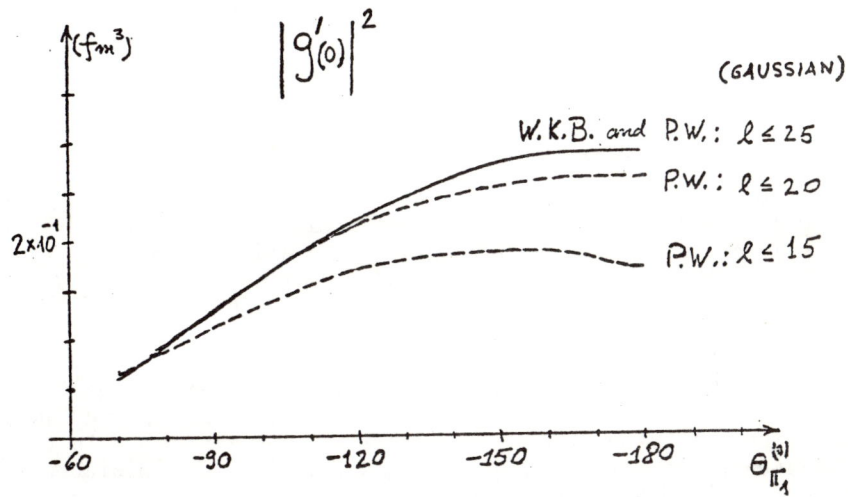

Figure 1: WKB and partial wave results for the distorted momentum distribution of 2s protons in ^{40}Ca, at vanishing momentum, in dependence of the pion scattering angle.

MODIFICATIONS OF CASCADE-EVAPORATION CALCULATION FOR
INTERPRETATION OF PROMPT-GAMMA TYPE EXPERIMENTS.

M. Zaider and D. Ashery, Tel-Aviv University, Ramat-Aviv, Israel

ABSTRACT

Modifications of the cascade evaporation codes ISOBAR-EVA related to the interpretation of prompt-γ type experiments are presented. The modified calculations are found to be in good agreement with experimental results from pion induced reactions.

The ISOBAR-EVA cascade-evaporation (IE) computer code[1] has frequently been used in the last few years to interpret and understand the results obtained in prompt-γ type experiments. In these experiments the prompt-γ spectra, resulting from the interaction of projectiles with nuclei, is observed, yielding cross sections for the production of specific final nuclei. The IE calculation (based on Monte Carlo simulation technique) consists of two stages: First, the intranuclear cascade interactions are simulated (ISOBAR), and a few nucleons are typically ejected from the original nucleus. Next, evaporation processes occur (EVA) until a particle-bound residual nucleus is reached. The spectrum of residual nuclei (total cross sections for producing specific final products) was usually used when comparing calculations with the experimental results. This is inconsistent with the fact that in a gamma type experiment, only a fraction of such cross sections (related only to the gamma transition which is actually observed) is measured. A procedure for calculating this fraction, using known branching ratios and the distribution of excitation energy in the final nuclei as predicted by the IE code, was developed.

Distortions of the calculated excitation energy distribution of the residual nuclei, due to the pairing energies δ used in the evaporation code, were also considered. The EVA code was modified so that only nuclei with excitation energy below the particle emission threshold were left at the end of the calculation. The effect of the above modifications is presented in Figure 1: the unmodified IE results (curve a), the calculations including modifications due to γ branching ratios (curve b) and the final results when corrections due to the pairing energy effects are also included (curve c). It can be seen that the agreement between these calculations and the experimental cross sections[2] obtained from the bombardment of a S^{32} target with 70 MeV negative pions is very good. It is gratifying that such an excellent agreement between theory and experiment is achieved with this first order calculation.

It is also intersting to note that the IE calculations do not include any normalization factor.

References:

1) Harp et al., Phys. Rev. C8, 581 (1973), Dostrowski et al., Phys. Rev. 116, 683 (1959).

2) M. Zaider et al., preceding paper.

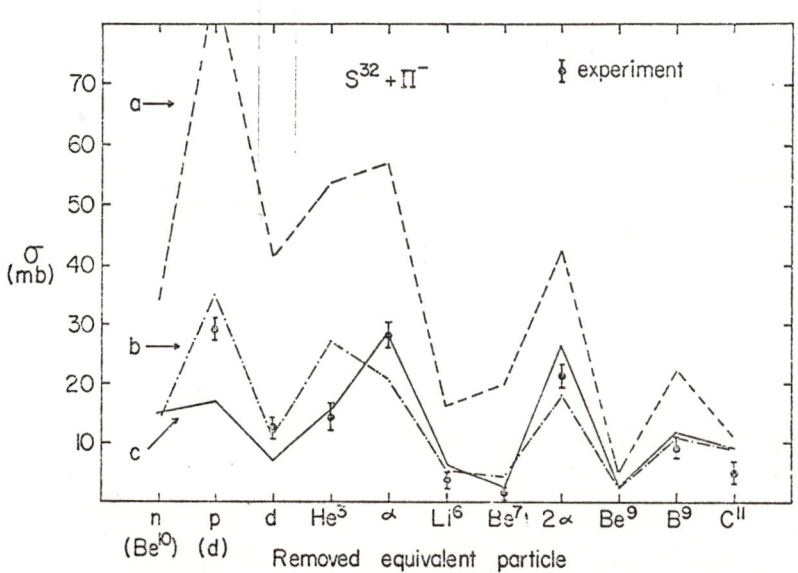

Fig. 1 - Measured and calculated prompt-γ cross sections for the interactions of 70 MeV π^- with a S^{32} target.

PION INDUCED REACTIONS IN THE ISOBAR MODEL*

J. N. Ginocchio
Theoretical Division, Los Alamos Scientific Laboratory
Los Alamos, New Mexico 87545

ABSTRACT

It is shown that the proton spectrum measured in pion induced reactions on nuclei can be reproduced in the isobar model using a phenomenological form factor which reproduces the pion production cross section in proton-proton collisions.

INTRODUCTION

In order to study the deep inelastic reactions induced by pions on nuclei, an intranuclear cascade model of pion-induced reactions was developed by G. D. Harp et al.[1] In this model the path of the pion is followed in the nucleus until it collides with a nucleon. For pions with energy below 330 MeV, the pion and nucleon form an isobar, Δ, with spin and isospin equal to 3/2; that is

$$\pi + N_1 \to \Delta . \tag{1}$$

The isobar is then followed until it either decays

$$\Delta \to \pi + N_1' \tag{2}$$

or scatters with a nucleon to two nucleons,

$$\Delta + N_2 \to N_1' + N_2' . \tag{3}$$

In either case each resulting particle is followed, its collisions and trajectory being determined using Monte Carlo techniques, until it either leaves the nucleus or is absorbed.

PION ABSORPTION

For pions the only mechanism for absorption in the isobar model is the two step one in which (1) is followed by (3). Originally the cross section for (3) was taken to be that given by a static one pion exchange model with a form factor of unity. A sensitive test of the adequacy of this model is given by the nucleon spectrum at 90° for pion induced reactions. The high energy part of this spectrum can be produced only by the pion absorption part of the reaction. A comparison of a calculation with this model of the proton spectrum at 90° for 235 MeV π^+ on Ni showed[2] that the calculation produced too large a cross section for energetic protons (>100 MeV) compared to the experimental results.

However this model has been improved with respect to the

*Work performed under the auspices of the U. S. ERDA.

propagation of the pion in the nucleus. Most importantly the cross section for (3) has been recalculated to include a phenomenological form factor[3] which was determined by reproducing the pion production cross section in nucleon-nucleon collisions by means of the inverse reaction

$$N_1' + N_2' \to \Delta + N_2 \to \pi + N_1 + N_2 \quad . \tag{4}$$

Using this new cross section the calculated proton spectrum agrees well with the measured spectrum.

REFERENCES

1. G. D. Harp, K. Chen, G. Friedlander, Z. Fraenkel, and J. Miller, Phys. Rev. $\underline{C8}$, 581 (1973).

2. J. F. Amann, P. D. Barnes, M. Doss, S. A. Dytman, R. A. Eisenstein, J. Penkrot, and A. C. Thompson, Phys. Rev. Lett. $\underline{35}$, 1066 (1975).

3. E. Ferrari and F. Selleri, Nuovo Cimento $\underline{27}$, 1450 (1963); E. Moniz, private communication.

A REMARK ON PION CAPTURE IN HEAVY NUCLEI

M.P. Locher and F. Myhrer[*]
SIN, 5234 Villigen, Switzerland

Recent experiments on γ-rays following π^- capture in heavy nuclei[1,2] indicate that emission of neutrons dominates and that high spin states are excited in the residual nucleus. We calculate the yield of neutrons from the (π^-,xn) reaction on ^{165}Ho with a two-nucleon absorption mechanism followed by (nucleon,xn) reactions using experimental inputs only[3]. For heavy nuclei the latter reaction proceeds mainly through the formation of highly excited compound states[4] which subsequently decay, mostly by the emission of neutrons. We show that the experimental observations[1,2] are completely consistent with this two-step model. In particular, the model implies that the yield curve $Y(x)$ for the (π^-,xn) reaction on heavy nuclei is closely connected to the shape of the energy spectrum of fast nucleon pairs emitted in pion absorption on light nuclei[5].

The energy spectrum $f(E)$ of the two nucleons originating from the elementary capture process $\pi^-(NN) \to NN$ on two bound nucleons is taken from experiments on light nuclei[5] to incorporate the effects of Fermi motion. The second step involves the interaction of the energetic nucleon pair with the recoiling (excited) nuclear core. We shall approximate this process by free nucleon-nucleus cross-sections[3,6] $\sigma_E(N,xn)$. Our model is therefore

$$Y(x) \propto \int_0^{m_\pi} dE\, f(E)\, \sigma_E(N,xn) \tag{1}$$

In Fig. 1 we compare the calculated neutron yield of pion capture in ^{165}Ho[3] with experiment[1]. Since for each x the $\sigma_E(N,xn)$ has a sharp maximum as a function of energy E[6], the observed peaking at $x = 6$ excludes, in a two-step model, any significant contributions from intermediate nucleons with energies different from 60 ± 25 MeV.

The observation of the high spin states in the residual nuclei[1,2] implies a minimal number of fast neutrons from a simple angular momentum argument (see Ref. 3).

---

[*] Present address: Theory Div., CERN, 1211 Geneva 23, Switzerland.

Figure 1: Neutron yield Y(x) versus numbers of emitted neutrons x from the reaction ^{165}Ho$(\pi^-,xn)^{165-x}$Dy* [1,3].

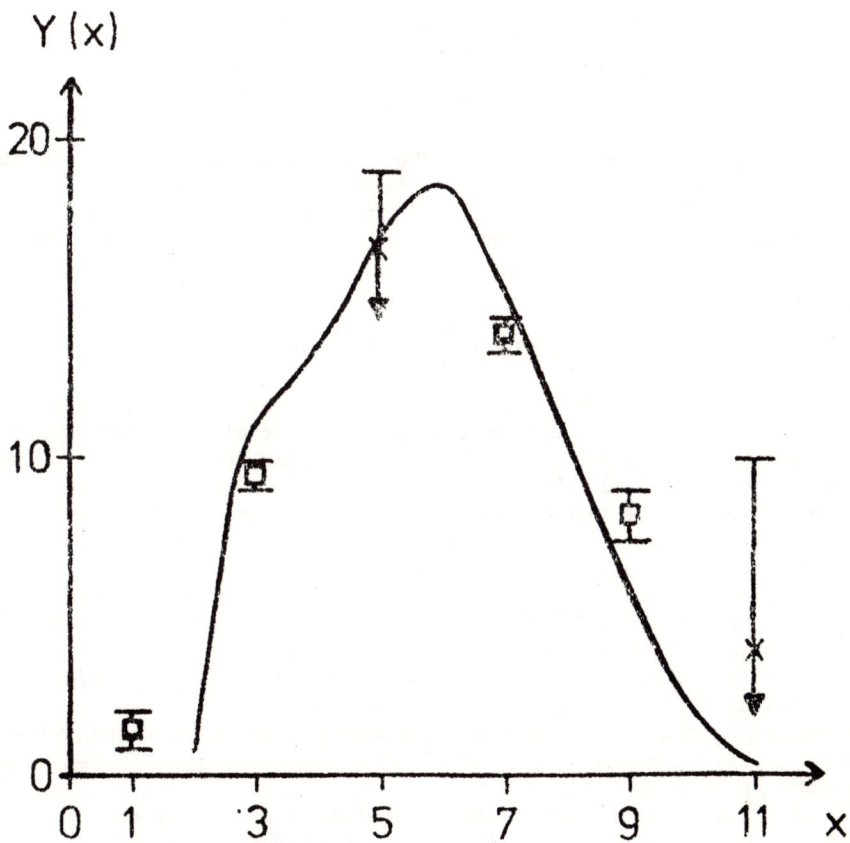

REFERENCES

1. P. Ebersold et al., Phys. Letters 58B, 428 (1975).
2. V.S. Butsev et al., Dubna preprint JINR, E6-8535 (1975).
3. M.P. Locher and F. Myhrer, Helv. Phys. Acta, in press.
4. N. Metropolis et al., Phys. Rev. 110, 185 (1958).
5. M.E. Nordberg et al., Phys. Rev. 165, 1096 (1968).
6. C.L. Rao et al., Can. J. Chem. 41, 2516 (1963).

A PROPOSED MEDIUM ENERGY DATA LIBRARY (MEDL)

E. D. Arthur and R. J. Barrett
Los Alamos Scientific Laboratory, University of California
Theoretical Division
Los Alamos, New Mexico 87545

ABSTRACT

The formation of a computerized compilation and bibliographic effort which would serve the data needs of the medium-energy community is discussed. It is shown that there will be a need in the near future for this type of effort for which no program exists. The important phases of the program and the types of data to be covered are briefly described.

INTRODUCTION AND SCOPE

Medium energy physics is now recognized as a separate and distinct discipline constituting a sizable number of researchers and facilities, representing a considerable dollar investment. The advent of new experimental facilities promises to greatly increase the amount of available medium-energy experimental data. For example, our estimates, based on searches of the Nuclear Science Abstracts and experimental proposals submitted to major medium-energy facilities, indicate that the rate of publication for π-nucleus data will increase from the present 500-1000 data points/year to 15-30 K data points/year in the near future.

The establishment of the medium-energy data library, proposed for funding in FY78, would serve the medium-energy community by reducing duplication of effort and through aiding the determination of future research goals. Its benefits to basic and applied researchers are briefly as follows.

   A. Basic Experimenters and Theorists - An easily accessible compilation of data would aid in the assessment of existing experimental data to compare theoretical predictions or to plan future experiments.

   B. Applied Researchers - The data base would provide a complete and orderly grouping of data pertinent to a particular problem, for example, π-radiotherapy.

The computerized effort would concentrate on the input areas shown in Fig. 1. Along with the compilation efforts, lines of communication would be established with facilities and groups throughout the world to ensure the completeness of the library and its widest distribution. Publications of bibliographical and compiled data would be distributed, and eventually specialized services available upon request would be offered.

343

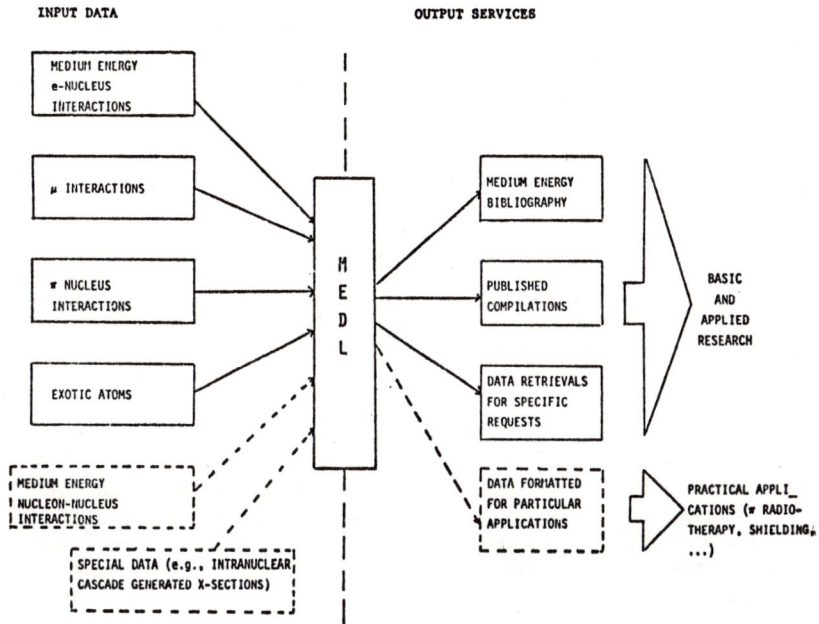

Fig. 1. A brief overview of the MEDL.

A COMPILATION OF PION ABSORPTION DATA

K.G.R. Doss*, S. A. Dytman*
Physics Dept., Carnegie-Mellon Univ., Pittsburgh, PA 15213

R. R. Silbar
Los Alamos Scientific Laboratory, Los Alamos, NM 87545

ABSTRACT

We have collected published data on the absorption of π^+ and π^- by nuclei at energies 20 to 300 MeV. We determine the absorption cross section per nucleon in a simple model and compare the results with the $\pi^+ d \to pp$ total cross section and with available theory.

TEXT

In the last 25 years, a large amount of data for total absorption cross sections has been published, which we summarize here.[1] Experiments done in bubble chambers and in cloud chambers each represent about half the data, so much of it is for carbon and almost all for light nuclei. The main ambiguity in the data is an inability to detect an uncharged pion in the final state, mislabelling an elastic or quasielastic charge exchange as absorption. Most groups have ignored this contribution. From the sparse supply of data existing[2], the charge exchange cross section seems to be about 10-20% of the absorption cross section which would make σ_{abs} 20-40% smaller.

The data is presented in terms of the absorption cross section per nucleon, σ_{abs}, defined in terms of the mean free path for absorption, λ_a, through

$$\lambda_a^{-1} = \rho_0 \, \sigma_{abs} , \qquad (1)$$

where ρ_0 is the uniform sphere density, $\rho_0 = 3A/4\pi R^3$, consistent with electron scattering measurements. With a shadowing correction for the effective number of nucleons, N_{eff}, the total nuclear absorption cross section is related to the mean free path by

$$N_{eff} = 2\pi\rho_0 \int b\,db\,dz \, \exp\left\{[(R^2-b^2)^{\frac{1}{2}}+z]/\lambda_a\right\} \qquad (2)$$

$$\sigma_{abs}^{tot} = \sigma_{abs} N_{eff} = \pi R^2 \, [1 - \tfrac{1}{2}(\tfrac{\lambda_a}{R})^2 + \tfrac{\lambda_a}{R}(1+\tfrac{\lambda_a}{2R})e^{-2R/\lambda_a}]. \qquad (3)$$

The extracted σ_{abs} are shown in Fig. 1 with the data[2] marked where the charge exchange subtraction was done most consistently.

Figure 2 compares a smooth representation of the compiled data with three times σ_{abs} for the deuteron.[3] (The factor of three compensates for the low deuteron density). The overall shapes are similar but the deuteron data falls off faster at higher energies. We also show in Fig. 2 the predictions of Beder and Bendix[4].

*Presently visitors at LAMPF: MS 831; Box 608; Los Alamos, NM 87545

REFERENCES

1. A bibliography is available upon request.
2. G. A. Blinov et al, Soviet Phys. JETP 35, 609 (1959)
   H. Hilscher et al, Nucl. Phys. A158, 602 (1970).
3. C. Richard-Serre et al., Nucl. Phys. B20, 413 (1970).
4. D. S. Beder, Can. J. Phys. 49, 1211 (1971); D. S. Beder and P. Bendix, Nucl. Phys. B26, 597 (1971).

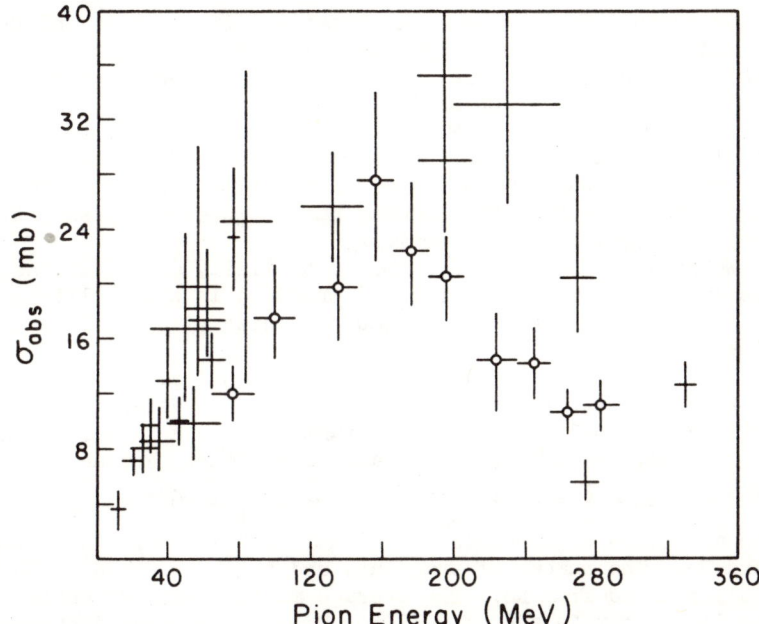

Figure 1. Absorption cross section per nucleon. ○ - Blinov et al

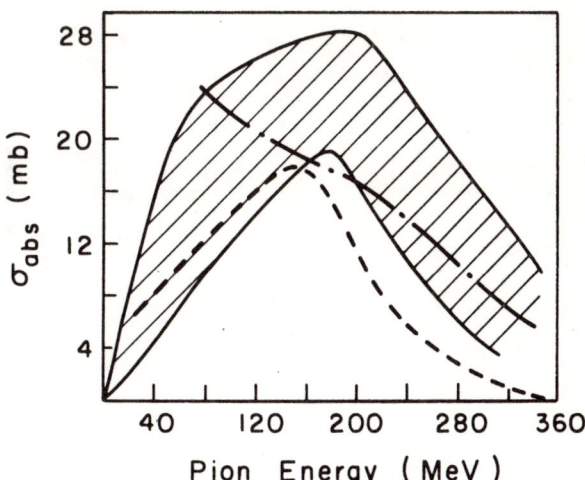

Figure 2. Cross hatches: smooth representation of all data.
--- $(\frac{3}{2})\sigma(\pi^+ d \to pp)$, Richard-Serre et al.
-··- Beder et al.

PARTICLE AND FRAGMENT MULTIPLICITIES FOR He AND Ne NUCLEI*

A. S. Kanofsky[††], R. C. Allen[+], and G. Lazo[+]
Dept. of Physics, Lehigh University[+]
Bethlehem, Penna. 18015
and
Brookhaven National Laboratory[†]
Upton, L. I., N. Y. 11973

We present in this paper a comparison of particle and fragment multiplicities resulting from collisions of pions of 2 GeV/c momentum on He and Ne nuclei. It was also attempted to obtain visible tracks with Argon as the streamer chamber gas, in order to obtain another A number. With Argon it was necessary to increase the voltage greatly and to add Ne or He gas in order to see streamers. The Argon was 10% by volume in the final mixture. Therefore, the data with Argon in the chamber has not been analyzed yet since it requires systematic corrections.

We show in figs. 1 and 2 the multiplicity distributions of pions (minimum ionizing particles) produced in pion collisions with nuclei, He and Ne, and also the multiplicity distributions of nuclear fragments.

The mean multiplicities $\bar{M}$, are shown in fig. 3.

With only two points, it is impossible to determine the exact functional dependence on A. For fragments and a linear dependence, $\bar{M} = C_1 + C_2 A$, we find $C_1 = .975$ and $C_2 = .0813$. For an $A^{2/3}$ dependence $\bar{M} = C_1' + C_2' A^{2/3}$, we find $C_1' = .613$ and $C_2' = .275$.

There are large differences in the multiplicity behavior between Ne and He. The number of fragments and pions produced in He is greatly reduced from Ne. The average multiplicity of nuclear fragments is 2.6 for Ne and only 1.3 for He. The angular distributions of the fragments are discussed in a later paper.

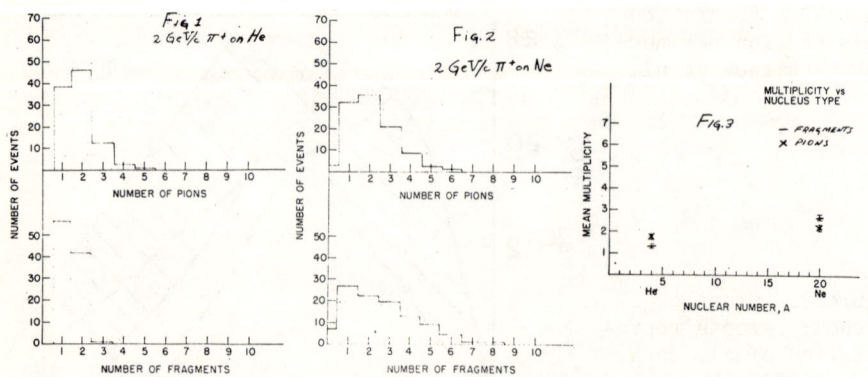

*Supported in part by the U.S.E.R.D.A. under contract E(11-1)-2894.

VI. FEW BODY SYSTEMS AND PION NUCLEUS PHYSICS

Pion Induced Reactions on Light Nuclei

F. G. Binon

Maître de Recherches au F. N. R. S. (Belgium)
Scientific Associate in the E. P. Division at CERN (Geneva)

INTRODUCTION

The title of this contribution is rather misleading for what I am going to consider here in fact is pion scattering of He^4 in the energy range ~20 to ~280 MeV.

The initial idea of the Organizers of this Conference was that I should give a critical review talk of all recent experiments on pion incuced reactions on light nuclei (A = 2, 3, 4). Since no new results in that area were submitted for presentation at this meeting, apart from the Dubna data on He^3 and He^4 which will be presented later in this session by Y. Shcherbakov, I thought that it would perhaps be a good idea to compare the experimental results on He^4 obtained in different laboratories. He^4 is by far the most studied light nucleus at the present time, particularly from the experimental point of view, and a rather extensive set of data is now available in the energy range extending from ~25 to about 300 MeV. It is a fact also that many theoreticians are working on those data (see, in particular, the contributed papers to this conference). Therefore, it is a good time to look at the coherence of those experimental results.

I do not intend to compare the data with the many theoretical calculations on He which have already been done. I think this is not the role of an experimentalist and, in these hard times we are living now, I don't want to take the bread out of the theoretician's mouths! Besides, certain theoretical contributions to the subject will be presented in other sessions.

My talk is divided into two main parts:
i) Presentation of the available data on He^4 and critical comparison
ii) Phase-shift analysis of $\pi^- - He^4$ scattering data

$\pi - He^4$ EXISTING DATA-COMPARISON

I. Total cross sections

Three sets of data are now available:
i) C. Wilkin et al. [1] The measurements were made with pions of both sign at the following kinetic energies: 110, 146, 189, 225, and 262 MeV.
ii) IISN-IPN Collaboration [2] The measurements were made with negative pions only, except at 110 MeV where both π^+ and π^- cross sections were measured. The π^- cross sections were measured at twelve energies between 67 and 285 MeV.

iii) LAMPF (preliminary) [3] Results are given for pions of both signs at the following energies: 51, 61, 75, 85, 95, and 105 MeV.

The results of the two first experiments are shown in Fig. 1. The agreement is good at the highest energies but at 110 and 146 MeV the results of the first experiment are higher by about 10% than those of the second one. This difference corresponds to nearly 10 standard deviations in the quoted experimental errors, which means that there is a real disagreement between the two experiments at those energies. However, it is interesting to note that the difference between π^+ and π^- cross sections at 110 MeV is, within the experimental errors, nearly the same in both experiments. That difference can be largely understood in terms of Coulomb distortion effects [1].

The preliminary data of LAMPF on He^4 are given on Fig. 2 together with the results of the two other experiments at energies below 120 MeV. The preliminary results of LAMPF given here are those which were presented last year in Santa Fe [3]. Though in agreement with the results of the IISN-IPN collaboration in the energy region around 100 MeV, there is a growing discrepancy between these two sets of data when the energy decreases. However during last year, a new and more complete analysis of the LAMPF data has been made which shows that the two last mentioned experiments are essentially in agreement with each other [4].

The IISN-IPN collaboration has also measured elastic differential cross sections (see below). This allows to subtract the elastic and Coulomb-nuclear interference contributions to the "partial total cross section" measured by each transmission counter used in the experimental set-up. The slope at zero solid angle of the least square fit line to the corrected data is a measure of the total inelastic differential cross section at $0°$. The results are shown in Fig. 3 together with the values of the total elastic differential cross section at $0°$. One sees that the total inelastic differential cross section at $0°$ has a constant value of about 25.2 mb/sr between 100 and 285 MeV. This is very different from the C^{12} case where a minimum is observed at about 160 MeV. [5].

II. Elastic differential cross sections

Five sets of data are now available:
i) M. Nordberg et al. [6]—π^- at 24 MeV
ii) M. Block et al. [7]—$\pi^{\pm}$ at 50, 58, and 65 MeV
iii) K. Crowe et al. [8]—π^- at 51, 60, 68, and 75 MeV
iv) Dubna [9]—$\pi^{\pm}$ at 68, 98, 135, 145, 156 MeV, and π^- only at 120, 174, and 208 MeV
v) IISN-IPN Collaboration [2]—π^- only at 110, 150, 180, 220, and 260 MeV

It is a well known fact that the experiments (ii) and (iii) disagree. They give essentially different results in the forward and backward directions. The integrated elastic differential cross section is always greater in the case of experiment (ii).

A comparison of the Dubna data with the results of experiments (ii) and (iii), at 68 MeV, shows that the Dubna results are in good agreement with those of K. Crowe et al. [8] and accordingly (see above) in disagreement with the data of M. Block et al.

Fig. 4, a, b, c, and d, show how the Dubna results compare with those of the IISN-IPN Collaboration. Though both experiments give essentially the same result in the energy region around 110 MeV, one immediately notices that the filling of the first dip differs a little, but significantly, around 150 MeV, differs nearly by one order of magnitude at about 180 MeV and by more than one order of magnitude around 220 MeV, the depth being always longer in the latter experiment.

I couldn't understand where such large differences could come from until I went back to the literature and found finally part, if not all, of the explanation in the article by I. Falomkin et al. [10] on the Dubna results in which the authors say (page 173): "Angular distributions in the laboratory system have been constructed with bins of 5°. For regions with a small number of events, larger bins have been used in order to reduce the statistical fluctuations!" Looking at Fig. 4 in Ref. 10, one can see that the width of their bins at the position of the first minimum sometimes exceeds 15°. Furthermore, on the same page 173 [10], one reads that "For elastic events, corrections for geometrical and landscape efficiency have been made." No other corrections being explicitly mentioned, one must conclude that no correction has been made, in particular, for the finite angular acceptance of the system (bin width). So, the numbers quoted in their tables do not correspond to differential cross-sections, as they say, but to integrated differential cross sections over bins of variable width. Looking back at Fig. 4, b, c, and d, taking into account the previous remarks, it becomes rather clear why not only the two experiments disagree at the position of the first minimum but also why the discrepancy grows when the dip becomes deeper. Nevertheless, I guess that not all the discrepancy can be accounted for by this correction only.

Of course the same remarks apply as well to the Dubna data on He^3 which have been taken in similar conditions. It might well be the explanation why R. Landau couldn't obtain a satisfactory fit to those data (see contributed paper by R. Landau).

In the article by Y. Shcherbakov et al. [9], the authors claim that the pion electromagnetic radius can be deduced from a phase shift analysis of their data ($\pi^{\pm}$). They give the following result: $\sqrt{<r^2>_\pi} = (0.83 \pm 0.17)$ fm. It is perhaps good to remind that the original proposal by M. Sternheim and R. Hofstadter [11], concerning the extraction of the pion electromagnetic radius from $\pi - He^4$ scattering data, is based on a precise measurement of the difference between π^+ and π^- differential cross sections in the region of the dip, that difference being linked directly to the pion form factor. The poor angular resolution of the Dubna experiment in the vicinity of the dip certainly washes out most, if not all, of the electromagnetic effects and I think it is most likely a hazard if the value they obtain for $\sqrt{<r_\pi^2>}$ is close to the value one expects it should be. I shall come back to this point later.

Phase Shift Analysis of π^- - He^4 Scattering Data

The IISN-IPN Collaboration data have been fitted with the following expression:

$$\frac{d\sigma}{d\Omega} = \left| |f_c| + f_N e^{2i\delta} \right|^2$$

where f_c is the Coulomb amplitude, with pion and He^4 form factors included, 2δ is a "standard Bethe phase" and f_N is the pure nuclear scattering amplitude. [2].

More specifically, the following phenomenological expression was used for f_N:

$$f_N(t) = \frac{k}{4\pi} \sigma_{Tot} [i + \rho] \exp [-R_s'^2 |t|/6] \prod_{j=1}^{N} (1 - t/t_j)$$

where ρ and R_s' are two real parameters, t_j are complex parameters and N is the number of dips showing up in the differential cross section.

I haven't the time here to enter into any detail about the significance and the value of these parameters. The values are given in Ref. [2] and a more complete discussion of this type of analysis will be found in a forthcoming paper [12]. The only thing I want to mention is that a more complete analysis of the combined data of K. Crowe et al. [8] and the IISN-IPN Collaboration [2] has been made with that formula and the result is that the fits are good at all energies and the parameters show a smooth behaviour as a function of energy. [12].

The point I want to make is that it is easy to "reconstruct" the phase shift parameters δ_ℓ and η_ℓ starting from the above expression for f_N. To see this, one first notices that since $|t| = 2k^2(1-\cos\theta)$, $\exp[-R_s'^2 |t|/6] = \exp(-z) \cdot \exp(z \cos\theta)$ where $z = R_s'^2 k^2/3$. But $\exp(z \cos\theta) = I_0(z) + 2\sum_{n=1}^{\infty} I_n(z) \cos(n\theta)$ where I_n are (modified) Bessel functions [13]. On the other side, $\cos(n\theta)$ can be expressed as a polynomial in $\cos\theta$, n being the largest power. The product factor $\prod_{j=1}^{N}$ being itself a polynomial of degree N in $\cos\theta$, it results that f_N can be expressed as a series of powers in $\cos\theta$.

On the other hand, the expression in partial waves of the nuclear amplitude leads to:

$$f_N(\cos\theta) = \sum_{\ell=0}^{\infty} (2\ell + 1) A_\ell P_\ell (\cos\theta)$$

where k, $A_\ell = [\eta_\ell \exp(2\delta_\ell) - 1]/2i$ and $P_\ell (\cos\theta)$ are polynomials in cos θ (Legendre polynomials). Thus, the partial wave expansion of the nuclear amplitude gives essentially another series of powers in cos θ.

The term by term identification of those two expressions is tedious but straightforward. It allows to express δ_ℓ and η_ℓ as functions of the parameters ρ, R'_s, and t_i.

Before I show you the results for the phase-shifts, I must draw your attention to the fact that the fit of the scattering data with the phenomenological expression for f_N, doesn't give the sign of Im t_i. This sign ambiguity is closely linked to the well known ambiguities encountered with the partial wave method. [12]. But what actually happens for He_4 is that for T ≤ 110 MeV, N = 1 (only one dip) and the negative solutions for Im t_1 always lead to solutions for η_0 and η_1 which are greater than 1, i.e. outside the unitary circle. So only the positive solutions for Im t_1 have to be taken into account at energies below or equal to 110 MeV. This is no more true for T > 110 MeV where both sign solutions for Im t_1 must be considered. This situation is illustrated in Fig. 5 where the two solutions for Im t_1 are shown at energies above 110 MeV only. By using the argument of continuity for the behaviour of Im t_1 versus T, it is most likely, though not proven, that either all values of Im t_1 have to be taken with the positive sign (full line) or Im t_1 is negative for energies above about 200 MeV and positive below (dashed line). The "reconstructed" phase-shifts corresponding to all Im t_1 positive (full line on Fig. 5) are given in Fig. 6. This choice for the sign of Im t_1 is the one which gives the smoothest behaviour for δ_ℓ and η_ℓ in the complex plane. Only the first four partial waves are shown for the sake of clarity.

Though the parameters δ_ℓ and η_ℓ are not determined with high precision in this analysis, in particular at the highest energies, their trajectories in the complex phase show at least a common tendency to cross the imaginary axis somewhere between ~160 and 220 MeV, i.e. in the energy region of the π-N 3.3 resonance. This similar behaviour of the various partial waves is what one expects to observe for π-N scattering on a nucleon inside the nucleus. This conclusion is further supported by the fact that the position of the first minimum in the differential cross section is energy independent and corresponds to $\theta_{cm} \approx 75°$. Simple kinematical arguments allow us to understand, at least qualitatively, why the minimum observed at $\theta_{cm} \approx 90°$ in π-N scattering is displaced towards lower angles when the nucleon is bound inside a nucleus. This argument gives at least some experimental support to the ideas which will be developed later in this session by F. Lenz concerning the formation and propagation of a Δ33 resonance inside the nucleus.

To finish, I would like to stress the practical interest of the phenomenological expression I have given above for f_N. In fact, I think that it is more than just ome more empirical formula which provides good fits to the data. Its physical content is richer than it appears at first sight. I have no time here to develop this idea further but it will be explained in the forthcoming paper [12]. In order to show the practical interest of the empirical formula, I shall use it as a "theoretical laboratory" to see what happens to the phase shifts when one changes the depth of the first dip in the differential cross section. Let me consider one particular energy, namely 110 MeV. The best fit to the data of Ref. [2] gives $\rho = 0.56$, $R'_s = 1.69$ fm, Re $(1 - z) = 0.704$, Im $(1 - z) = 0.185$, ($\overset{s}{t}_1 = 2k^2(1 - z_1)$) [2]. By changing the value of Im $(1 - z_1)$ one simulates a change of depth of the dip in the differential cross section. The results are given in the Table below for the three first partial views.

Im $(1 - z_1)$	S (L = 0)		P (L = 1)		D (L = 2)	
	δ	η	δ	η	δ	η
1	8.2°	.62	19.7°	.93	5.0°	.91
0.5	44°	.50	23°	.61	6.3°	.87
0.185	-15°	.76	25°	.55	6.4°	.80
0.100	-7.6°	.88	25.5°	.53	6.1°	.78
0.01	-1.0°	1.0	25.5°	.50	5.5°	.76
0	-0.3°	1.0	25.4°	.50	5.4°	.76

One sees how strongly the phase shifts depend on the filling of the dip, in particular the S-wave phase shifts. A good knowledge of the differential cross section at the position of the dip is therefore absolutely necessary if one wants to extract correct phase-shift parameters from the data. This brings me back to the end of the first part of my talk where I doubted very much the relevance of the Dubna data to determine the electromagnetic radius of the pion.

References

1. C. Wilkin et al, Nucl. Phys. $\underline{B62}$, 61 (1973).
2. F. Binon et al, Phys. Rev. Lett. $\underline{35}$, 145 (1975).
3. G. Burleson et al, 6th Intern. Conf. on H.E.P. and Nucl. Struct. Santa Fe and Los Alamos (1975). Abstracts of Contributed Papers, Paper I.D.6.
4. G. Burleson and M. Cooper, Private communication.
5. F. Binon et al, Nucl. Phys. $\underline{B17}$, 185 (1970).
6. M. Nordberg et al, Phys. Lett $\underline{20}$, 692 (1966).
7. M. Block et al, Phys. Rev. $\underline{169}$, 1074 (1968).
8. K. Crowe et al, Phys. Rev. $\underline{180}$, 1349 (1965).
9. Y. Shcherbakov et al, DUBNA preprint (1975), Submitted to Nuovo Cimento.
10. I. Falomkin et al, Nuovo Cimento $\underline{21A}$, 168 (1974).
11. M. Sternheim and R. Hofstadter, Nuovo Cimento $\underline{38}$, 1854 (1965).
12. F. Binon et al, To be published.
13. M. Abramowitz and I. Stegun Ed. Handbook of Math. Tables Dover Pub. (1967).

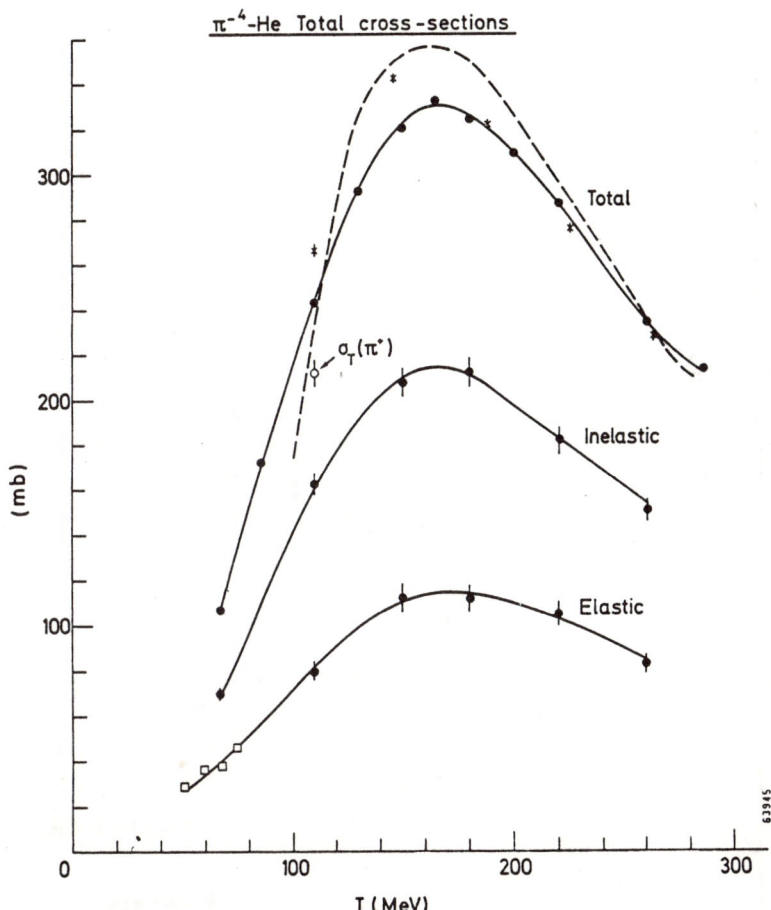

Figure 1 - π^--He^4 total cross section, total elastic cross-section and total inelastic cross section versus π^- energy in the laboratory. x: Wilkin et al [1]; o IISN-IPN Coll [2]; ▫ integrated elastic diff. cross section of K. Crowe et al [8].

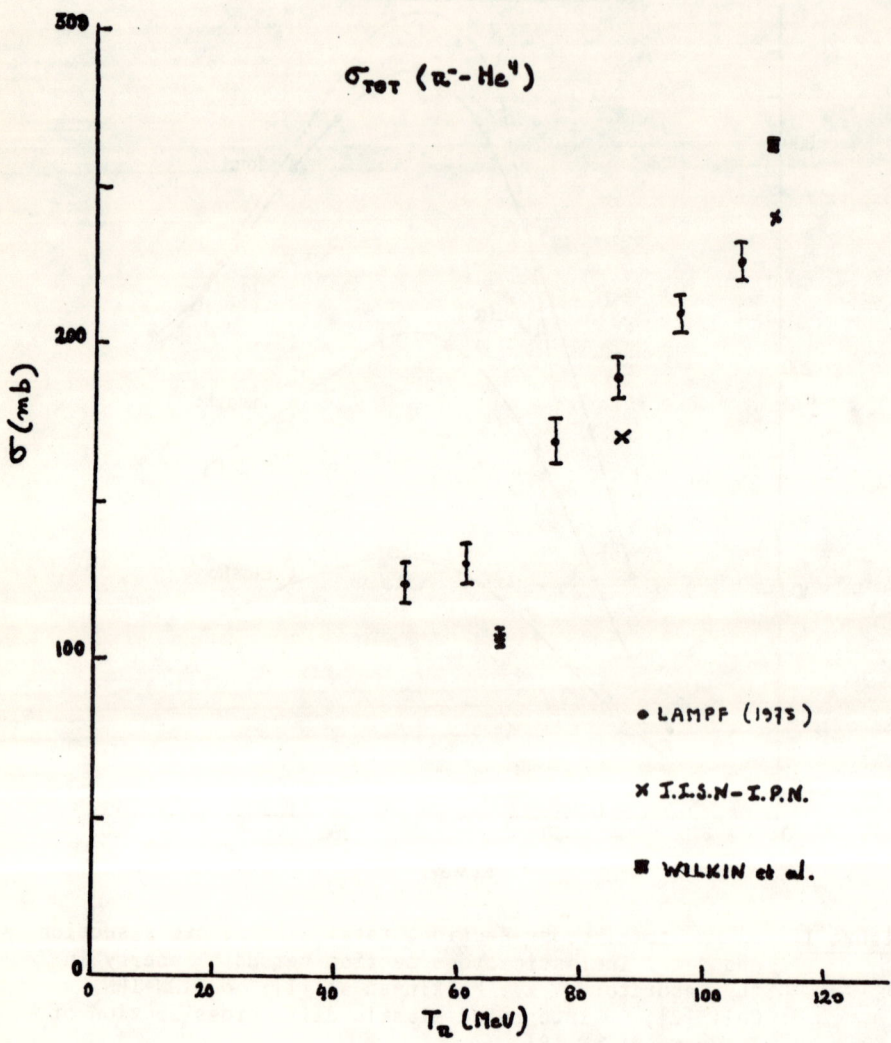

Figure 2 - π^--He^4 total cross section versus π^- energy in the laboratory. ◻ C. Wilkin et al [1]; x: IISN-IPN Coll [2]; o LAMPF [3].

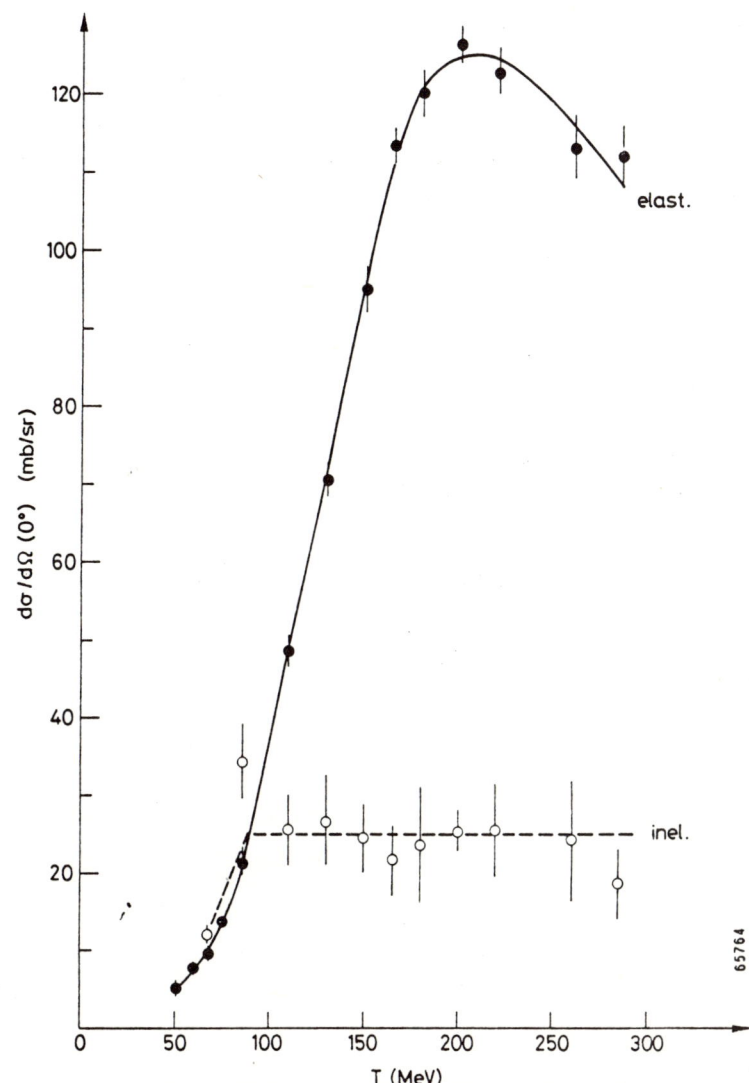

Figure 3 - π^--He^4 elastic and inelastic differential cross sections at 0° versus pion kinetic energy.

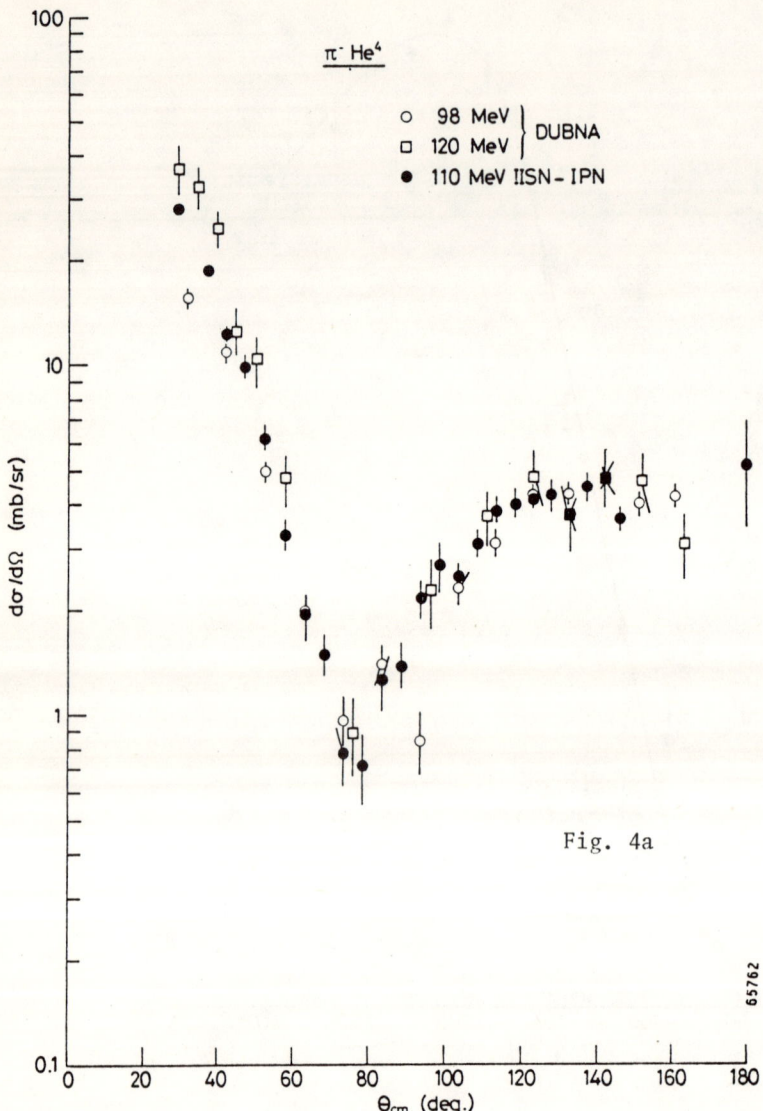

Figure 4 - a,b,c,d-Comparison of the measured elastic differential cross sections for π^--He scattering at DUBNA [9] and by the IISN-IPN Coll [2].

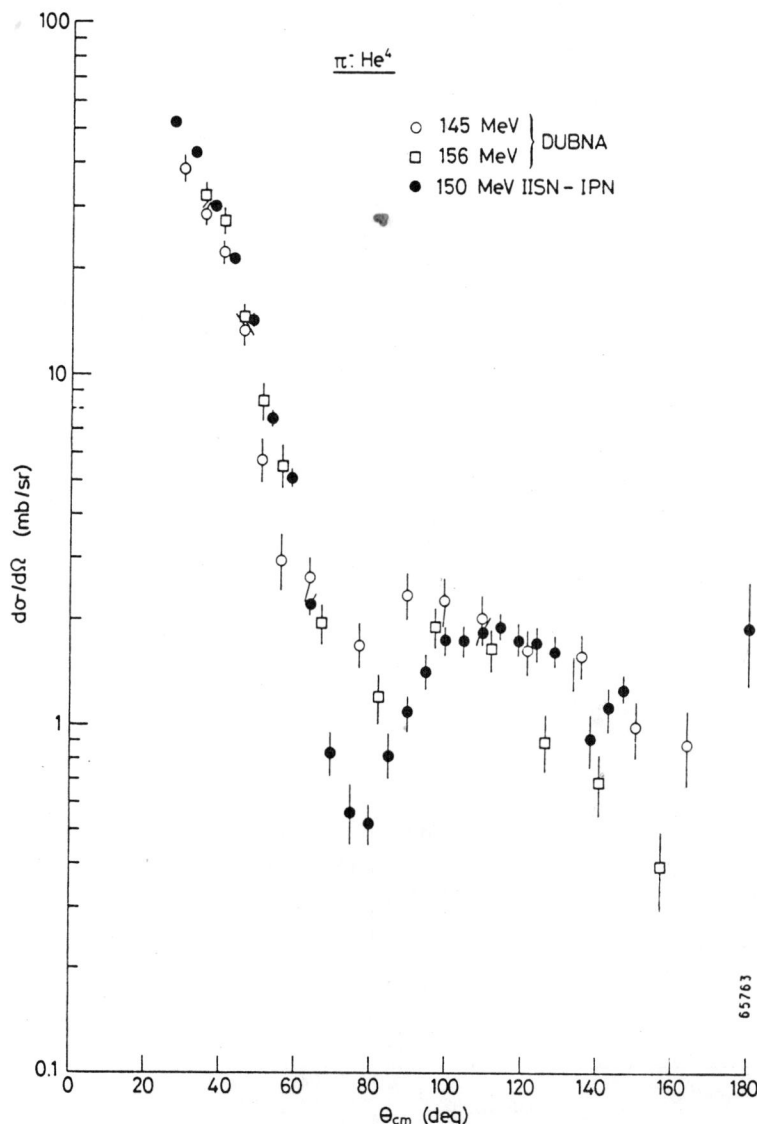

Fig. 4b

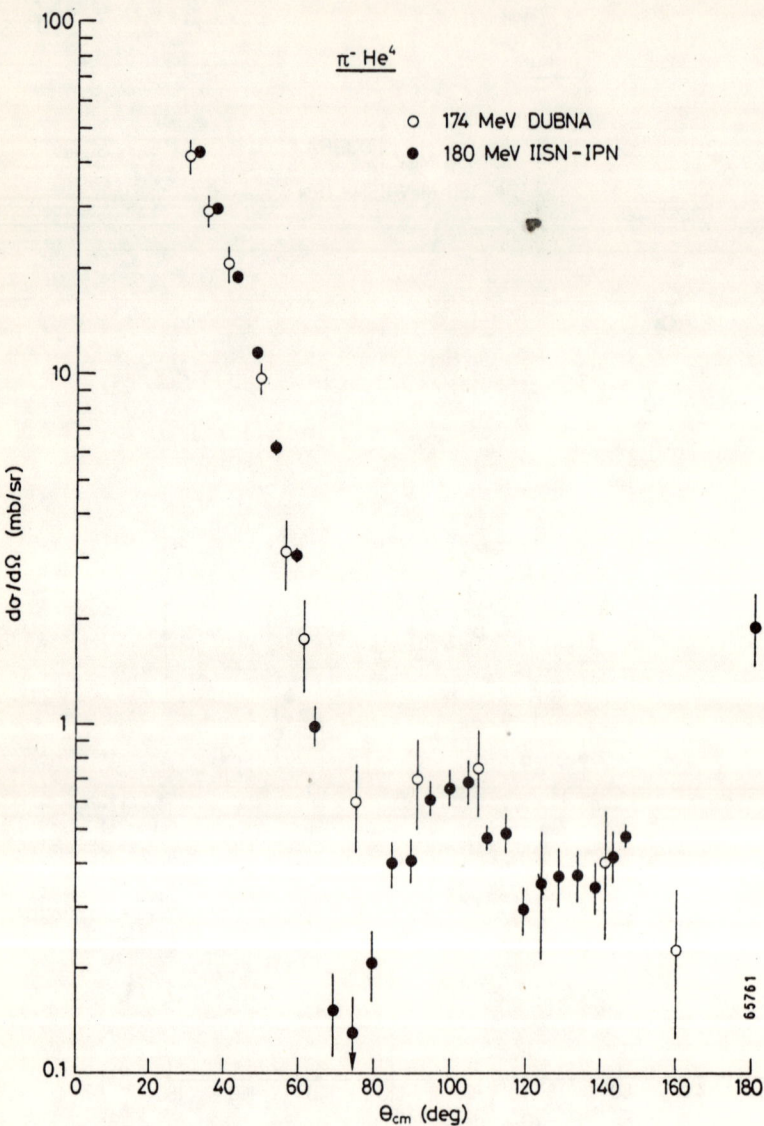

Fig. 4c

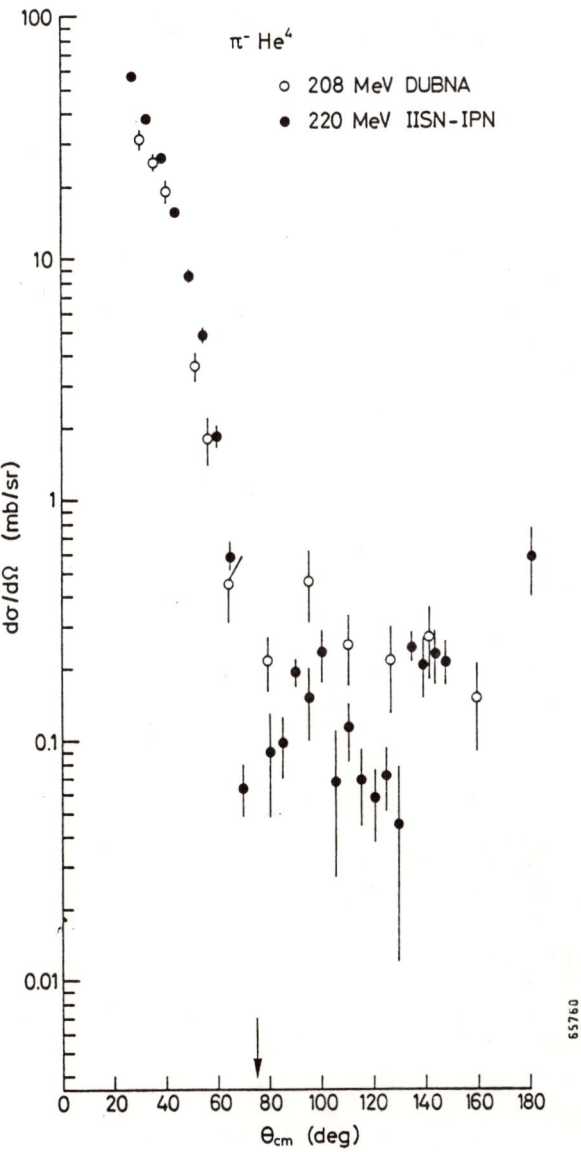

Fig. 4d

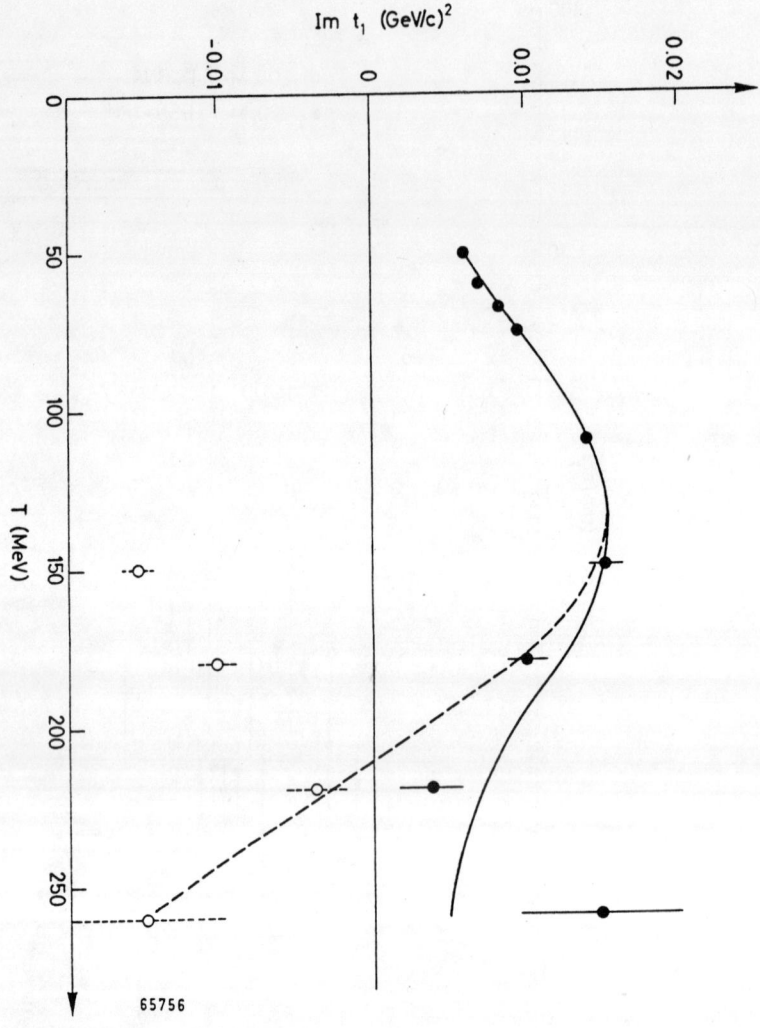

Figure 5 - Im t_1 (see text) versus pion kinetic energy in the laboratory.

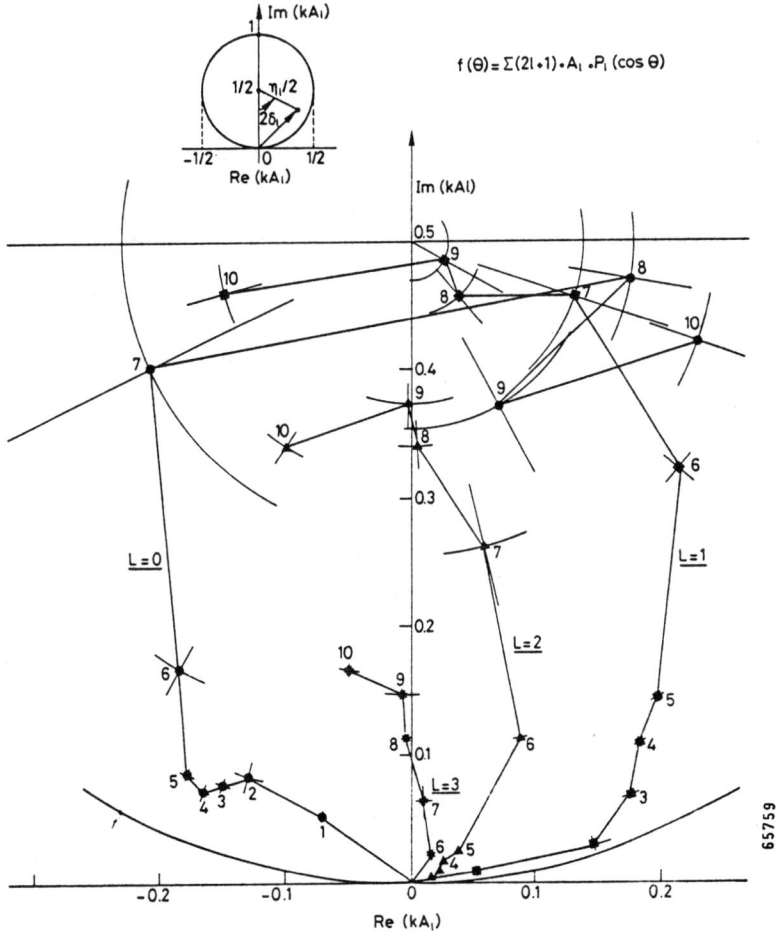

Figure 6 - Results of a "reconstructed" phase-shift analysis when all Im t_l are taken positive (see text). 1: 24 MeV. 2: 51 MeV. 3: 60 MeV. 4: 68 MeV. 5: 75 MeV. 6: 110 MeV. 7: 150 MeV. 8: 180 MeV. 9: 220 MeV. 10: 260 MeV.

Carl Shakin (Brooklyn College): L. C. Liu and I have checked your method of treating the Coulomb-Nuclear interference by doing an exact phse-shift calculation using a nuclear amplitude obtained from an optical potential that fits the π-^{4}He data. We find that the West-Yennie method can induce errors of 10-20 percent in the extracted values of the forward nuclear amplitude.

Binon: Since I'm a pure experimentalist, I used the only existing formalism that was available at the time the analysis was made. Now, I don't see where the West-Yennie method could be wrong. I checked out that the "I.R." contributions which W-Y showed to be negligible at high energies, are also negligible in our energy domain.

Martin Cooper (Los Alamos Sci. Lab): At what energy does the real part of the nucleon amplitude goes through zero?

F. Binon: Somewhere between 160 and 220 MeV. I'm afraid we can't be much more precise at the present time, in particular if the preceding argument is true.

Scattering of pions on ^{3}He and ^{4}He
in the Δ_{33} resonanse region

Yu.A.Shcherbakov

Joint Institute for Nuclear Research, Dubna

Abstract

Some experimental data on $\pi^{\pm}$-He^3, $\pi^{\pm}$-He^4 elastic and double charge scattering π-He^4 are presented. The comparison of the results with an optical model, Glauber Theory, FDR calculations, pair correlation and meson exchange current theories are considered.

Elastic scattering of charged pions by ^{3}He has been experimentally studied in the energy region 68-208 MeV /1/ at the Dubna cyclotron using a helium streamer chamber. The measured differential cross sections are shown in Fig.1.

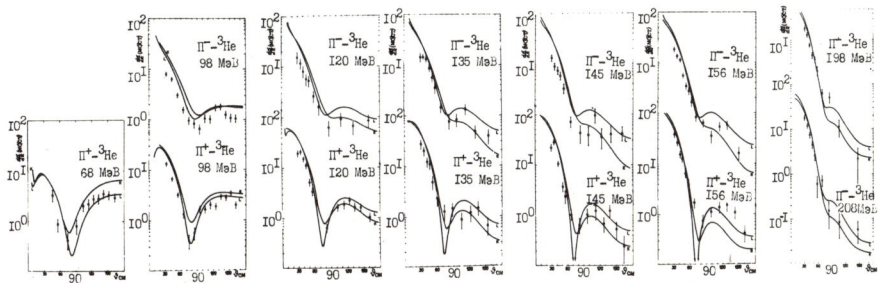

Fig.1. π-He^3-differential elastic cross sections.

Recently new experimental results have been obtained for elastic π^-He^3 differential cross sections also at 68 and 174 MeV. An interesting feature observed in these angular distributions is the fact that the minimum occurs at approximately the same angle ($\theta_\pi \sim 75°$ in cms) over the whole resonance region. The dips and secondary maxima are more pronounced for π^+He^3 reactions compared with π^-He^3 ones. The dip position remaines fixed also in the case of pion elastic scattering by He^4 it can be seen from Fig.2, where our experimental results are given in the energy interval 68-208 MeV. The same effect was observed in differential cross sections measured at 110, 150, 180, 220 and 280 MeV /3/. A quite different situation occurs for heavier nuclei: e.q. in the differential cross sections of elastic $\pi \cdot C^{12}$ scattering the dip position is shifted from 60° to 40° in the energy interval 70-280 MeV /4/. Rather than in the angular variable, the dip position

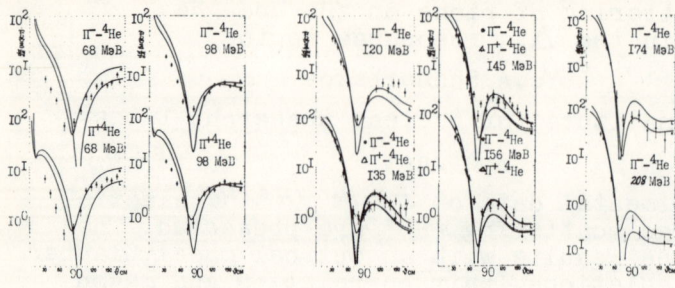

Fig.2. $\pi^{\pm}$He^4 elastic differential cross sections.

(first minimum) is a constant in the transferred momentum in this case. Different features of the π-He^4 and π-C^{12} differential cross sections can be explained qualitatively as follows. In the lightest nuclei (as He^3 and He^4) the multiple scattering plays a rather unimportant role and the gross structure of differential cross sections is determined by the elementary π-N amplitude alone. Therefore in the resonance region, where the pion-nucleon P-wave dominates, the dip position is given essentially by the zero of the Legendre polynomial $P_1(\cos\theta)$ being slightly modified by the distructive interference occuring between the pion-nucleon S- and P-waves as well as the kinematical transformation of the π-N amplitude from the pion-nucleon centre-of-mass system to a pion-nucleus one. On the other hand in the case of a π-C^{12} interaction - the multiple scattering becomes more important and the angular distributions are of a diffractive nature. The minimum is determined rather well by the first zero of the Bessel function $j_1(qa)$, where a-characterizes the effective nuclear dimension and q- denotes the transferred momentum.

The optical model calculations /5,1,2/ performed utilizing the Kisslinger type and Laplacian potentials describe qualitatively well the experimental data (Fig. 1 and 2). The dip position is predicted correctly in the case of both He isotopes. The optical potentials used contain spin-flip and charge-exchange terms and the difference between π^+ and π^--He^3 differential cross sections. However there are discrepancies between the calculated and experimental results in the small angle region. The discrepancies tend to increase with decreasing pion energy. The total elastic cross section are discribed still worse. As can be seen from Fig.3 only the high energy part of the $\sigma_{el}^{\pm}$(He^4) is more or less in agreement with the optical model predictions (with the Kisslinger potential), whereas the optical model completely fails to discribe $\sigma_{el}^{\pm}$(He^3) (see Fig.4). Recently some more optical model calculations were done by Landau /6/ using a different parametrisation of πN amplitude. The approach developed by Tabakin, Phatak and Landau and refined by Londergan, McVoy and Moniz

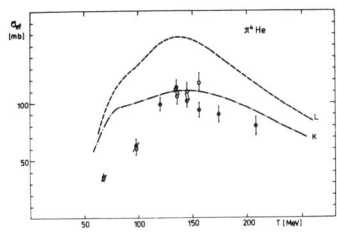

Fig.3. $\pi^{\pm}He^4$-total elastic cross sections.

is based on the assumption that the pion-nucleon interaction can be described by a separable potential. The off-energy shell behaviour of the resulting optical potential is physically more acceptable than that of Kisslinger and Laplacian potentials. The Landau's results discribe better the forward peak of our experimental differential cross sections for πHe^3 and πHe^4 elastic reactions at 98 MeV. More comparisons are needed to prove the theory definitely. At higher energies the Landau's calculations and our results both give good descriptions of experimental π He-data (see Fig.5,6). It was also proposed in ref. /6/ to use the elastic pion-He^3 scattering data for extracting information about the magnetic form factor of He^3 nucleus at $q^2<2fm$. As it can be seen from Fig.6, the magnetic formfactor influences the cross sections mainly in the dip region.

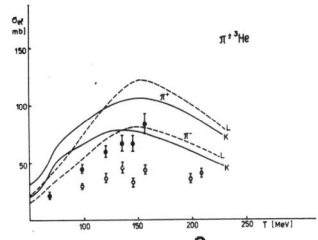

Fig.4. $\pi^{\pm}He^3$-total elastic cross sections.

Details of the nuclear structure were introduced in the optical model in another way by Mach /7/. In constructing the He^3 forfactor, semimicroscopic ground shell wave functions were used containing S'- and D-state admixtures. As it can be seen from Fig.7, comparing the dotted line (admixture parameters $P_{S'}=P_D=0$) with the remaining ones ($P_D=-0,3$, $P_{S'}=0$ - solid line, $P_D=0$, $P_{S'}=\sqrt{0,02}$ - dashed line and $P_D=0,3$, $P_{S'}=0$ - dot dashed line), the S' and D-state admixtures are not in a position to improve an agreement between

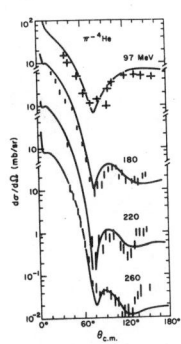

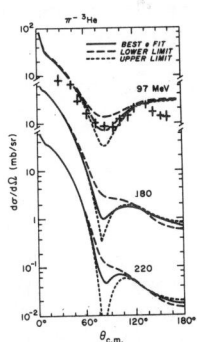

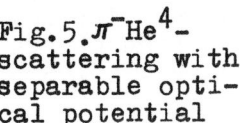

Fig.5. π^-He^4-scattering with separable optical potential

Fig.6. π^-He^3-scattering with different $Fm(q^2)$.

the experimental and the calculated results in the small angle region (the Laplacian optical potential was used). The admixtures influence the cross sections also, especially in the dip region. The main difference

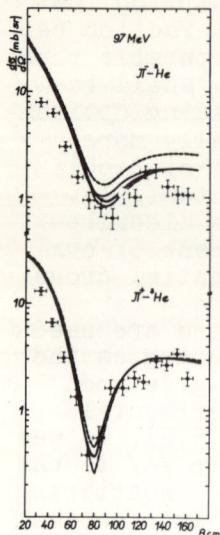

Fig.7. S' and D admixtures to 3He ground state.

between the two forementioned techniques consists in the different treatment of a pion exchange current. The extent is not known to which the spin formfactor of He^3 nucleus is affected by the exchange currents. Therefore the magnetic form factor values extracted from elastic π-He^3 data would not probably be very reliable at the present status of the theory.

The π-He^3 and π-He^4 differential cross sections were subject to analyses also on the basis of the Glauber model /8/. Although the model is used here rather far from its expected range of applicability the calculations can give us some idea about the magnitude of effects which are usually neglected in the optical model. Among these effects the nuclear recoil correlation and all the spin-flip and charge exchange effects produced by the spin and isospin dependent part of the π-N amplitude were studied in detail. In spite of the fact that the eikonal approximation proved to be a rather bad one for energies $E_\pi <$ 200 Mev, the differential cross sections are rather well reproduced in the energy interval 100-200 MeV (see full lines in Fig. 8 and 9). This situation is probably a result of delicate cancellations occuring in the Glauber model amplitude. Remarkable discrepancies were observed only in π-He^4 differential cross sections at large angles. If the nuclear recoil correlation is neglected (dashed line in Fig.9), a quite different (and worse) result is obtained. The Glauber model breaks down completely for $E_\pi <$ 100 MeV. Backward elastic scattering is believed to contain important information on the details of both the reaction mechanism and the nucleus structure. Recently the pion elastic scattering by He^4 was measured (Frascati-Turin-Dubna collaboration) in the angular interval 170-180° for energies 65-80 MeV. The results are compared (Fig.10) with optical model predictions (curve "Mach potential" means - Kisslinger potential with Fermi motion correction). The optical model seems to reproduce the measured data qualitatively well.

Experiments on pion elastic scattering by the lightest nuclei can also stimulate a new development of dispersions technique /9/. The pion-nucleus forward dispersion relations can yield predictions of Ref(0,w) as well as estimations of the pseudoscalar pion-nucleus coupling constant $f_{\pi He^3 H^3}$. A comparison of the calcu-

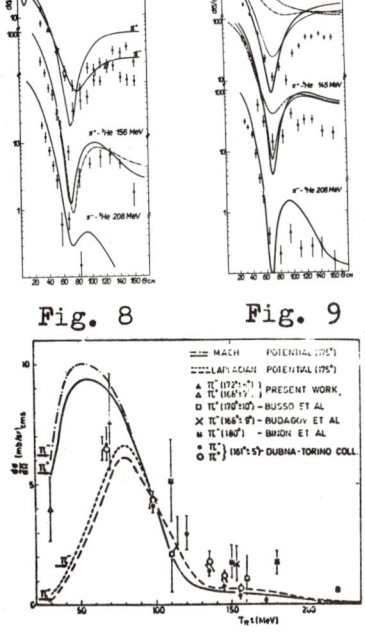

Fig. 8   Fig. 9

Fig. 10. He^4-backward scattering

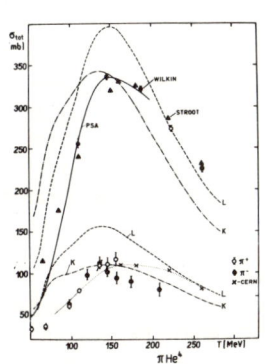

Fig. 11

lated Ref(0,w) with results of our preliminary phase-shift analysis /10/ shows that in the case of He^4 the theory predicts a bigger value of Ref(0,w) in the region w ~ 100 MeV and the experimental Ref(0,w) reach zero at smaller energy than the theoretical one. An analogous situation occurs also for C^{12} /4/. There are two possible sources of this discrepancy. Firstly the experiments on elastic scattering are not sufficiently accurate and Ref(0,w) is determined with a rather large error. Secondly the dispersion relation predictions of Ref(0,w) are based on experimental values of σ_{tot} for all energies and also the value of Imf(0,w) in the nonphysical region. Existing experimental σ_{tot} values /3 and 11/ differ somewhat for $T_\pi <$ 150 MeV. These results are shown in Fig. 11 with σ_{el}(T) from paper /2/. Our estimated values of σ_{el}(T) are in rather good agreement with the forementioned experiment /3/ for $T_\pi <$ 150 MeV and smaller values are obtained for $T_\pi >$ 150 MeV. New experiments are now being performed in order to discover the source of discrepancies. Similarly new experiments of σ_{tot}(T) are needed for $T_\pi >$ 150 MeV. In order to obtain more reliable values of Ref(0,w) (see Fig.12 - black dots data from /3/, a and b - curves from /9/ and /14/) it would be very useful to develop some physically sound models for calculating Imf(0,w) in the nonphysical region. The fact that the value of Imf(0,w) in the nonphysical region can influence remarkably the quantity Ref(0,w) was demonstrated* by Dubnička /12/ and by Dubnička and Meshcheryakov

---
* In ref. /13/ it was erroneously referred to our experimental data, which are not contained in our paper /2/.

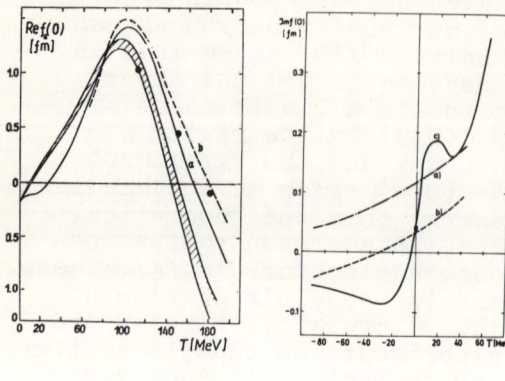

Fig. 12   Fig. 13

/13/. As a result of our energy-dependent phase-shift analysis we obtained (Fig.13) Imf(0,w) which is essentially negative in the nonphysical region unlike the parametrisation used by other authors /9, 14/. New measurements of the total π-He^4 cross section would be of great value. It would make possible a more precise evaluation of the dispersion intergral as well as to increase the accuracy of the phase-shift analysis. Such experiments should also confirm the difference observed in ref. /3,11/ between the total cross sections of π^+ and π^- on He .

As was mentioned earlier, elastic scattering $\pi^\pm$-He^3 experiments enable us to estimate the pion-He^3 coupling constant. Comparing this quantity with the pion nucleon coupling constant $f_{\pi N}$, we can get some idea about the role of the one nucleon exchange mechanism in pion-nuclear reactions. The elastic scattering information can be exploited in two ways. The simpler one consists in extrapolation of the differential cross section (according to Chew /16/) in the pole at $\cos\theta = -19,56$. However the accuracy of the existing data does not allow us to proceed in this way /17/. The accelerated convergence method /18/ essence of which is to move the pole position nearer to the physical region (Z=-5,95) gives also no reliable results. Even if the experimental accuracy is very high, serious difficulties would remain how to estimate the contribution from the near by lying nd-cut.

Recently Nichitiu and Mach /19/ have applied the Chew-Low /20/ extrapolation method in order to estimate the pseudoscalar pion-nucleus coupling constant. The value $f_{\pi He^3 H^3}$ =0,101±0,018 was obtained which is somewhat bigger than $f_{\pi np}$ =0,088, However the procedure used to obtain pion-He P-wave phase shift is not very convicing, The results of extrapolations in the case of π-He^3 and π-N -reactions are shown in Fig.14. Forward dispersion relations provide an alternative tool for extracting the coupling constant value. Total cross sections of π^--He^3 and π^+-He^3 reactions are an important incredient of this theory. At the present moment there are only single measurements of $\sigma^{\pm}_{tot}(He^3)$ /21/ in

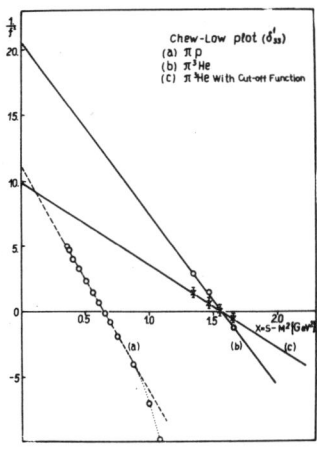

Fig. 14

rather small intervals of energy (120-240 MeV). Coupling constant estimation based on this experiment yields value $f_{\pi He^3 H^3}$ = =0,071±0,012. Finally the dispersion relation calculations performed by Kopeliovich /22/ using the Goldberger-Treiman relation result in value $f_{\pi He^3 H^3} \approx 0,16$. It is difficult to make some definite conclusions from these calculations since all of them are of very preliminary nature. It is interesting to note that the peak value of σ_{tot}^+ (He^3) (341±14 mb at 150 MeV) given by Spence /21/ is rather high compared to maximal values obtained for σ_{tot}^+ (He^4)(~343 mb, ref./11/, ~321 mb, ref. /4/) whereas the simple impulse approximation predicts $\frac{\sigma_{tot}^+ (He^4)}{\sigma_{tot}^+ (He^3)}$=1,14.

On the other hand the ratios:

$\frac{\sigma_{el}^+ (He^4)}{\sigma_{el}^+ (He^3)}$ =1,4±0,2; $\quad \frac{\sigma_{el}^- (He^4)}{\sigma_{el}^- (He^3)}$ =2,14±0,25;

$\frac{\sigma_{el}^+ (He^3)}{\sigma_{el}^- (He^3)}$ =1,90±0,27

extracted from our experiments are in a rather good agreement with the impulse approximation predictions (1,3:2,5:2,0).

The main attention was paid in this contribution to the problem connected with total and elastic cross sections of pion scattering by He^3 and He^4. Of great interest are also inelastic and absorption processes, single and especially double-charge-exchange reactions. The last reactions are probably sensitive to nucleon correlations, isobar exitations and to meson exchange currents.

The existing results of total cross section measurements of reaction $\pi^+ + He^4 \to \pi^- + 4p$ are shown in Fig. 15. Besides our new results (black circles) the results of Karajanopoulos and Block are given here. A comparison with model calculations (1 - pair corralation model /24/, 2-meson exchange current /25/) indicates that towards high energies the model based on exchange currents can give a quite good description of experimental results. However a more definite conclusion can be drawn only when more precise experiments will be available.

At the end of this brief review I would like to express my opinion that a further examination of

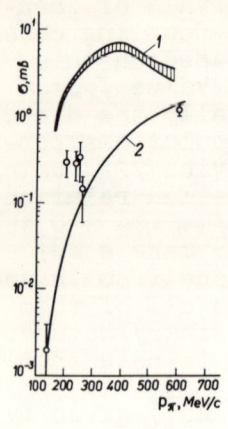

Fig. 15

pion nuclear reactions with very light nuclei can give us a deeper insight into the nuclear structure as well as into the details of pion-nucleon interactions. In the nearest future more quantitative results of both the theory and the experiment will be needed and we may hope that a meson factory programms would make this a reality.

I am grateful to R.Mach for stimulating discussions in preparating this paper.

References

1. Yu.A.Shcherbakov et al. Nuovo Cimento 31A,262 (1976)
2. Yu.A.Shcherbakov et al. Nuovo Cimento 31A,249 (1976)
3. F.Binon et al. BNL 19692 (1975).
4. F.Binon et al. Nucl.Phys., B17, 168 (1970).
5. R.Mach et al. Phys. Lett., 53B, 133 (1974).
6. R.H.Landau. Phys.Lett. 57B, 13(1975); R.H.Landau, S.C.Phatak, F.Tabakin, Ann. of Phys. 78, 299 (1975), J.Londergan, K.McVoy, E.Moniz Phys. Lett. 45B, 195 (1973).
7. R.Mach Nucl.Phys., A258, 513 (1976).
8. R.Mach et al. Contributed paper, this meeting
9. T.Ericson, M.Locher. Nucl.Phys., A148,1(1970).
10. M.Kulyukin et al. JINR, E1-6534, Dubna (1972).
11. C.Wilkin et al. Nucl.Phys., B62, 61(1973).
12. S.Dubnička. JINR, E2-6765, Dubna (1972).
13. S.Dubnička, V.A.Meshcheryakov JINR, E2-9399 (1975).
14. C.J.Batty et al. Nucl.Phys., B67, 492 (1973).
15. W.E.Meyerhoff, T.A.Tombrello. Nucl.Phys., A109, 1(1969); A206, 1(1973).
16. G.F.Chew. Phys.Rev., 112, 1380(1956).
17. O.V.Dumbrais et al. JINR, E2-1992, Dubna (1973).
18. R.E.Cutkosky, B.B.Deo Phys.Rev., 174, 1859 (1969).
19. R.Mach, F.Nichitiu. JINR, E1-9048, Dubna (1975).
20. G.F.Chew, I.Low. Phys.Rev. 101, 1570 (1957).
21. C.B.Spence, P.D.Thesis. Unpublished, Virginia (1975)
22. B.Z.Kopeliovich Nucl.Phys. (Soviet) 18, 1157 (1973).
23. G.Sqier et al. RL-73-037 (1973).
24. F.Becker, C.Schmidt. Nucl.Phys., B18, 607 (1970).
25. J.F.Germond, C.Wilkin. Lett.Nuovo Cimento, 13, 605 (1975).

Carl Shakin (Brooklyn College): There is important multiple scattering in the case of ^{4}He. However the dip in the Born term does not shift upon iteration of the series.

We can also make a perfect fit to the ^{4}He data at 110 MeV. This involves an integration over the Fermi motion and careful attention to off-shell effects and binding effects.

Shcherbakov: I would study with interest results of your calculations. I would be interested to know in which energy range your approach works.

Considering the multiscattering, I may say that we tried optical potentials and Glauber theory and both calculations were in fairly good agreement.

L. C. Liu (Brooklyn College - CUNY): This morning, Binon has mentioned that the filled dip in your 210 MeV π-^{4}He data in comparison with their results at 220 MeV is probably due to the poor resolution in your experiment. Is that comment correct?

Shcherbakov: I do not know how it is even possible to speak about an energy resolution because at that energy statistics gives the energy at the dip of order 20-50% and data are of essential preliminary nature. I am really surprised at the good agreement between our data and Binon et al. But if you take our low energy data, say 150 MeV, you may very well see that for π^+ and π^- He^3 scattering the resolution that we have chosen, 5-10%, is quite sufficient to see striking difference in π^+ and π^- scattering on He^3. This demonstrates clearly the presence of a spin-flip amplitude contribution.

A. N. Mitra (Univ. of Delhi): I have a question concerning the relevance of the magnetic form factor of H^3 in a pion-scattering amplitude. Does the connection come about through the PCAC effect?

Shcherbakov: I think that is a question better to address to Rubin Landau. He has studied the situation with form factors more or less in a detailed way.

R. Landau (Oregon State U.): I would like to indicate that an analysis of your ^{3}He and ^{4}He data has now been made (the full data were not originally available) and will be briefly presented tomorrow afternoon. The basic physics is that 1) these type of data definitely lead to the conclusion that a finite range πN interaction must be used to understand these data within a multiple scattering framework (i.e., the Local/Laplacian or Kisslinger potentials just won't do any longer), 2) the ^{3}He,^{4}He data are now at the level where some fundamentally new or important information can be examined. It is, in fact, data of this sort which has the potentiality of yielding some of the new information from the use of pion, which were discussed yesterday. As a matter of clarification, the π^- ^{3}He elastic cross sections are very sensitive to spin flip scattering, especially in the dip region. The spin flip scattering is of course sensitive to the spin distribution of neutrons within

^{3}He which, finally, is related to the magnetic moment form factor of ^{3}He. Thus, arises the sensitivity to magnetic moment distribution.

F. Binon (I.I.S.N.-Belgium): You claimed in your preprint that you could extract the pion radius value in the analysis of your data. Do you think that your poor resolution at the position of the first slip allows you to really get this number?

Shcherbakov: If you read once more our papers you will see that we extracted the pion radius essentially from K. Crowe et al., data. But in our other publications using our experimental data we have shown how to chose the most probable solution for the strong amplitude. Concerning resolution, in the streamer chamber it is possible to measure the scattering angles with a precision of 1°, but to extract r_π we need good statistics. The main difficulties in the interpretation of experimental data we have explained in our publications. Nevertheless my opinion is that the value $r_\pi = 0,8\pm 0,8$ fm. from energy dependent and independent analysis shows that at the present time we have no arguments to speak about some anomalies from πHe^4 for $F_\pi(q^2)$. Now it is necessary to measure r_π on the level 1%, but I think this we cannot reach in studying πHe scattering.

FADDEEV CALCULATIONS OF πD SCATTERING

A.W. Thomas

Theory Division, CERN - Geneva

ABSTRACT

We summarize the present status of the "Faddeev" calculations of πD scattering, with emphasis on what has been learnt about common approximation methods (for π-nucleus as well as πD). Some space is devoted to a discussion of the theoretical work which remains, including a suggestion of co-operation between theorists on a "homework" problem. Finally, we give examples of the interesting phenomena which one hopes to investigate through good πD experiments. Suggestions are made as to which experiments would be most useful.

INTRODUCTION

In this review we summarize recent progress in the understanding of the πD system based on integral equations of the Faddeev type. In addition we make suggestions as to the future course of these investigations. Such a review seems particularly appropriate now, in view of the recent proliferation of three-body calculations of πD elastic scattering. Moreover, there is a growing interest in this very fundamental system among experimental groups at the meson factories as well as CERN and Saclay. Hopefully our discussion will be of assistance in turning this interest into real proposals.

We restrict consideration to pion kinetic energies less than about 250 MeV, which are most easily amenable to present machines. It is also doubtful, because of the enormity of the calculation, that the Faddeev approach could play more than a guiding role above this energy.

For ease of reading, the text is divided into four sections. The first deals primarily with formal problems. Section B should be of wide interest, since we summarize there the present status of the Faddeev calculations, with emphasis on what has been learnt about the validity of various widely used approximations. Sections C and D should also be of quite general interest, since they describe not only what remains to be done in the three-body calculations, but also what physics may be learnt, and where one should look.

A. FADDEEV EQUATIONS ?

The main formal problem with "the Faddeev equations" for πD scattering is that they are like kangaroo feathers - there is no such thing ! Even at 50 MeV a pion is significantly relativistic, and non-relativistic potential theory is not applicable. Nevertheless, as was realized rather early, Faddeev's equations (involving only t-matrices, not potentials)[1] provide a guideline for generalization to the relativistic case [2,3,4].

It is standard practice to use the Blankenbecler-Sugar (BBS)* techniques to reduce the three-body Bethe-Salpeter equation to an integral equation in three variables per particle rather than four $[\vec{K}$ not $(k_0,\vec{k})]$[2], while guaranteeing two and three-body unitarity and Lorentz invariance. In the usual separable approximation, corresponding to bound-state or resonance dominance [5], this reduces to a set of coupled one-dimensional integral equations after angular momentum decomposition. From the numerical viewpoint these equations are no more difficult than their non-relativistic equivalents.

There are, however, a number of theoretical objections to these equations [6]. The non-linearity of the three-body propagator "$(\sum w_i)/(s-(\sum w_i)^2)$" [where w_i is $(\vec{k}_i^2+m_i^2)^{\frac{1}{2}}$, and s the invariant mass squared] is of some concern, because one cannot separate the energy of a non-interacting spectator. That is, two-body scattering in the three-body system does not reduce to the two-body BBS equation ! This is the "clustering problem". An apparent way of avoiding this problem for πNN, is to rely on the rather large mass of the nucleons to linearize the relativistic propagator (i.e., $s^{\frac{1}{2}}+\sum w_i \simeq 2\sum w_i$). The resulting relativistic Schrödinger equation (with propagator $[s^{\frac{1}{2}}-\sum w_i]^{-1}$) [7], lies at the heart of the theory of π-nucleus optical potentials based on the KMT approach [8,9]. However, unless one is very careful about defining the two-body relative momentum in terms of variables in the three-body centre-of-mass (c.m.), this does not solve the problem. Indeed, making the non-relativistic choice $(m_N\vec{k}_\pi-m_\pi\vec{k}_N)/(m_N+m_\pi)$ leads to a violation of three-body unitarity. The importance of this effect should increase with energy - as explained in detail in Ref. 7. Fortunately we shall see in §B that in practice the numerical error is at the one percent level in the resonance region.

From the formal three-body viewpoint, Aaron et al. (AAY) have solved the clustering problem [10]. More particularly, they show that one can demand that the two-body D-function depends only on the invariant mass of the interacting pair, and use two- and three-body unitarity to determine both the three-body propagator (as given above) and the two-body D-function. The latter is identical to the solution of the two-body BBS equation in the separable approximation. Another technical improvement from AAY, is the "special vector", which enables one to write p-wave two-body interactions as a Lorentz invariant three-vector dot product. In the πD system, where the P_{33} interaction is dominant, this "correction to the recoil correction" can be important [11].

Lest we give the impression that there are no theoretical problems, we record some of the most significant ones. First, every practical formulation makes the implicit assumption that the two-body t-matrices are independent of the invariant masses of

*To save confusion, abbreviations of author's names (e.g., BBS, AAY, etc.) are underlined, whereas other abbreviations are not (e.g., MSS, FSA, etc.).

the scattering particles. (See Ref. 3 for the clearest statement of this fact!) Second, there is a disagreement over the consistency of the way the special vectors of **AAY** have been used in practice (see Ref. 11). (This is related to the "angle-transformation problem" in π-nucleus scattering.) Third, all the calculations reviewed here make either the non-relativistic choice for the pair relative momentum, or some improvement along the lines suggested by **AAY**. However, it has been suggested recently that the Wightman-Garding choice is preferable for several theoretical reasons[12]. Finally, there is the standard three-body problem of determining the importance for the πD results of off-energy-shell changes in the two-body t-matrices. (This is not the same as problem one above.)

B. THE PRESENT SITUATION

Before discussing the most topical energy region [near the (3,3) resonance], we shall comment briefly on the πD scattering length. (The intermediate region (0-100) MeV, where there is only one three-body calculation[7], will be mentioned in § C.) There are a number of non-relativistic Faddeev (and non-Faddeev) calculations at zero energy, which agree quite well with each other and experiment. (This is reviewed in Ref. 13, but the interested reader is also referred to Fäldt's caution over a naïve interpretation of the Faddeev results[14].) No further progress can be expected unless this difficult experiment is repeated with greater precision and we have a better value of the πN isoscalar scattering length. From the zero energy results, we recall only that the contribution of pion absorption is quite significant[15].

Next we summarize those results which can be taken as firmly established in πD scattering near the (3,3) resonance, which is currently under the most active study[11,16-19]. Essentially every three-body calculation performed so far (in this region) has omitted all but the P_{33} (πN) and $^3S_1-^3D_1$ (NN) interactions. This leads to typically five coupled integral equations (in one variable) for each value of total angular momentum. Because of the complexity of these calculations we accept as firmly established only those results reported by at least two independent groups.

A major advantage of the three-body calculations is that they allow one to sum the multiple scattering series (MSS) exactly, without the need for some fixed scatterer approximation (FSA). Everyone is agreed that the Faddeev MSS converges. Indeed if NN rescattering is omitted (e.g., the triple scattering term "TS" involving a πN, NN and πN collision) three or four iterations suffice[18,19]. The inclusion of NN rescattering slows the convergence, and significantly alters the final answer[16,17,18]. This slow convergence is particularly frustrating since the single scattering (SS) result is typically only 20% wrong. An attempt to "unitarize" the SS result (which amounts to summing the MSS with a δ-function for the pion propagator) led to no significant improvement[13].

The effect of a correct treatment of three-body kinematics in the SS term is well understood - with the "resonance energy" being raised [16]. [As Brayshaw has established there is actually no πD resonance pole - rather a short cut corresponding to the πN resonant interaction in the three-body system [20].] The MSS broadens this resonance and tends to move it down in energy. Unlike the zero energy situation, "TS" does not significantly counteract the energy shift in SS. Near resonance it seems to mainly reduce the imaginary part of the partial wave amplitudes, and hence reduce cross-sections. This is very noticeable at backward angles.

Several groups have made detailed comparisons of their results with the FSA [18,19]. One could summarize the findings bluntly by saying that the FSA is never good - even in the forward direction. Woloshyn et al. find that the FSA gives total cross-sections 20% too high, as well as greatly enhanced differential cross-sections at 180° [18]. Mandelzweig et al. (MGE) find a very slow convergence to the static limit as they increase the nucleon mass [19]. Brayshaw agrees qualitatively with the FSA calculations of Kaufmann and Gibbs, but concludes that quantitatively the FSA is not good [17].

Following the discussion in §A, one would like to know the importance of the non-linear form of the three-body propagator. MGE show that replacing it by the linear form gives 1 % changes in partial wave amplitudes. At this conference, Rinat and Thomas have reported these changes to be only a few percent [11], with a correspondingly small effect on $d\sigma/d\Omega$ *. This is pleasing, since it not only suggests that the restriction of Ref. 7) to below 100 MeV was unnecessary, but also that the inherent lack of unitarity of the π-nucleus momentum-space optical model should have no practical importance. It is even more revealing to notice that the non-relativistic results of Myhrer and Koltun differ by only 10% from the fully relativistic results of Woloshyn et al. (e.g., at a πD c.m. energy of 110 MeV, $e^{i\delta_1} \sin\delta_1$ is 0.38+0.51i and 0.38+0.47i respectively) [16,18].

Perhaps those questions which remain open are even more interesting. A strong dependence on the deuteron wave function was reported by MGE [19]. The inclusion of short-range repulsion seems to reduce $d\sigma/d\Omega$ considerably at all angles. Backward scattering is increased by the inclusion of the deuteron d-state - as expected from its higher momentum components [11,19]. [Note that this is not the case at all energies. In fact, at 50 MeV it has been shown that increasing P_D lowers $d\sigma/d\Omega$ at 180° [7].]

The question of off-shell effects arises in any theory which involves unobservable extensions of the two-body scattering amplitudes. Disturbingly large effects at all angles have been reported by MGE when the πN range was approximately halved (without changing the πN phase shifts) [17]. Unfortunately this group has not demanded that their πN t-matrix satisfy the two-body BBS equation. This results in a much greater freedom than would be permitted in a complete dynamical description. (Similar reservations must be held about other reports of off-shell effects - e.g., in pion production.) Until a complete investigation is carried

*The observed differences could be caused by the lack of d-state (P_D=0%) in the full calculation (cf., P_D=4% in TH-MV).

out, with phase shift equivalent interactions which satisfy the two-body BBS equation, this question will remain undecided.

Finally we note that while the qualitative agreement of the three-body calculations is good, there are considerable quantitative differences. For example, there is disagreement over whether the largest p-wave phase shift $(J=2^+)$ goes through $0°$ or $90°$. In terms of partial wave amplitudes (f_{LJ}) this is only a matter of whether $\text{Im}(kf_{LJ})$ is above or below 0.5. Nevertheless such disagreement must not be tolerated much longer.

C. FUTURE THEORETICAL WORK

The immediate aim must be to clean up those areas of present disagreement. In this regard, it may be useful to agree on some basic "homework" problem (in the philosophy of Bethe [21]), involving relativistic effects, spin and isospin, which each group should calculate. It is essential, in view of past experience in simpler three-body systems (e.g., the triton), that there be some such independent check on accuracy. Then one would like to answer those open questions detailed at the end of §B.

Before making detailed comparison with experiment, a number of relatively uninteresting (but probably significant) corrections must be made. Any naïve estimate of the effects of πN partial waves other than P_{33} suggests at least 10% corrections. There is little difficulty in including such effects provided one can use perturbation theory. Only one three-body calculation has included these other πN channels. For 50 MeV $π^+d$ scattering they were shown to produce very strong interference effects – mainly from P_{13} [7]. At that energy such corrections can be made to first order with little inaccuracy.

The importance of NN scattering through the $^3S_1 - ^3D_1$ channel leads one to ask about the effect of other NN channels. For 180 MeV πD scattering, the NN t-matrix is needed at typically 100 MeV (c.m. energy). At this energy there are at least four NN partial waves with phase shifts as large as $^3S_1$.

Once these effects have been calculated and agreed upon (one year is minimal), one can think of extracting some really interesting new information from good πD experiments. There is considerable interest now in the N* (or ΔΔ) content of the deuteron (and complex nuclei), with an upper limit as high as (3-4)% suggested [12]. It has also been proposed recently that ρ-exchange should play an important role in modifying the pion rescattering process in nuclei [22]. Finally, while the effect of pion absorption on the πD scattering length is now fairly well understood, the same is not true at positive energy. In view of the strength of absorptive processes in complex nuclei, any insight gained from πD would be useful. This brief list of topics for future study is not exhaustive, but should serve to indicate the sort of information one hopes to glean from good πD data.

D. WHAT EXPERIMENT ?

Unless one is clever enough to find a measurement sensitive

to only the phenomenon of interest, the phenomena mentioned in §C will produce small effects. How small is unknown, since they have not yet been calculated, but Weise has suggested that ρ-exchange can produce significant corrections in backward π-nucleus scattering [22]. Because of the possibility of doing reliable calculations with recoil (and MSS) effects in the πD system, this seems the most likely place to establish such a mechanism. As a first test of the calculation it should, of course, fit in the forward direction. Thus, in order to see ρ-exchange effects (as well as the deuteron isobar content) we need good πD data ([3-5]%) from the forward peak to 180°, at perhaps five energies through the resonance region.

There are several reasons for expecting pion absorption to produce small effects in the resonance region. Experimentally, true absorption is less than 3% of the elastic plus break-up cross section. Using his boundary condition formalism, and introducing absorption through a pole in the P_{11} channel (cf., Afnan and Thomas at zero energy) [15], Brayshaw has shown that it gives very small effects on the 0+ amplitude in this region [17]. On the other hand, Beder has shown phenomenologically that absorptive effects can be very large at low energy (near 50 MeV) [23].

At present the Faddeev theory fits the 50 MeV π^+D data without absorption [7]. However, the calculation is very sensitive to the poorly known πN input at this energy, and small changes in this input, or improved πD data could easily change this. Thus there is a real interest in seeing good data in the (0-100) MeV region - once again over a full angular range. We also observe that since Coulomb corrections provide theoretical difficulties in this region [7,13], it would help to have both $\pi^\pm$ data at each energy.

SUMMARY

At present there are three calculations of πD cross-sections in the resonance region which include some relativistic effects [11,18,19]. Of these, two include spin and isospin. One of these omits NN rescattering, which has been established as an important correction. The third uses a simple deuteron wave function, which has also been suggested insufficient. No one has included other πN partial waves (except near 50 MeV [7]), or NN rescattering in other than $^3S_1-^3D_1$. More fundamentally, no two groups have used the same input to produce identical answers !

Nevertheless (as described in detail in §B) a number of things have already been learnt. The implications for π-nucleus scattering theory of the failure of the fixed scatterer approximation, and and of a correct treatment of the kinematics of the πN collision in the three-body system, has been stressed often [7,13,16-18] (the latter seems necessary to explain recent low energy π^+ ^{12}C and $\pi^\pm$ ^{4}He data [24]). The surprising accuracy of replacing the non-linear BBS propagator by the linear form [11,19] is good news for the π-nucleus work based on a relativistic Schrödinger equation.

It is to be hoped that the groups carrying out Faddeev calculations of πD scattering will co-operate on a common test problem, and will also work towards answering the open questions described in §C. If in addition, relatively uninteresting but necessary ingredients (e.g., πN interactions other than P_{33}, etc.) are included, then there is much to be learnt from good πD experiments. In the higher energy region one would hope to pin down effects like ρ-exchange and the N^* content of the deuteron. At lower energy, one expects to see the effects of pion absorption. To be somewhat more optimistic, we may learn something I have not even considered ! In any case, good experiments on this system which is fundamental to π-nucleus physics, will not be wasted effort !

REFERENCES

1. L.D. Faddeev, JETP(Sov.Phys.) 12, 1014 (1961).
2. R. Blankenbecler and R. Sugar, Phys.Rev. 142, 1051 (1966).
3. V.A. Alessandrini and R.L. Omnès, Phys.Rev. 139, B167 (1965).
4. D. Freedman, C. Lovelace and J. Namyslowski, Nuovo Cimento 43A, 258 (1966).
5. C. Lovelace, Phys.Rev. 135, B1225 (1964).
6. J.L. Basdevant and R.L. Omnès, Phys.Rev.Lett. 17, 775 (1966).
7. A.W. Thomas, Nuclear Phys. A258, 417 (1976).
8. A.K. Kerman, H. McManus and R.M. Thaler, Ann.Phys. 8, 551 (1959).
9. R.H. Landau, S.C. Phatak and F. Tabakin, Ann.Phys. 78, 299 (1973).
10. R. Aaron, R.D. Amado and J.E. Young, Phys.Rev. 174, 2022 (1968).
11. A.S. Rinat and A.W. Thomas, contribution to this conference on "Covariant Calculations of πD Scattering", and to be published.
12. H.J. Weber, "Relativistic (NN^*) Wave Function of the Deuteron", U.of Virginia Preprint (1976), and Ref. 12 therein.
13. A.W. Thomas, Proc.Int.Conf. on Few-Body Problems in Nuclear and Particle Physics, Université Laval (Québec 1975), p. 287.
14. G. Fäldt, U.of Lund preprint LUTP1974-13 (1974).
15. I.R. Afnan and A.W. Thomas, Phys.Rev. C10, 109 (1974); A.W. Thomas and I.R. Afnan, Phys.Lett. 45B, 437 (1973).
16. F. Myhrer and D.S. Koltun, Nuclear Phys. B86, 441 (1975).
17. D.D. Brayshaw, Phys.Rev. C11, 1196 (1975).
18. R.M. Woloshyn, E.J. Moniz and R. Aaron, Phys.Rev. C13, 286 (1976).
19. V.B. Mandelzweig, H. Garcilazo and J.M. Eisenberg, Nuclear Phys. A256, 461 (1976) and errata.
20. I.R. Afnan, invited paper on "Meson Reactions with Few-Body Systems", VII Int. Few-Body Conf., Delhi (1975).
21. G. Baym and C. Pethick, Ann.Rev.Nucl.Sci. 25, 27 (1975).
22. G.E. Brown and W. Weise, Phys.Reports 22C, 279 (1975); and W. Weise, private communication and to be published.
23. D. Beder, Nuclear Phys. B34, 189 (1971).
24. R.H. Landau and A.W. Thomas, Phys.Lett. (to be published), and contributions to this conference.

Eisenberg (Tel-Aviv): How were off-shell effects overestimated by Brayshaw and by Mandelzweig and co-workers?
Thomas: Brayshaw's formalism simultaneously incorporates two and three-body effects in the "interior region" in a way which is difficult to separate.
    Mandelzweig et al. do not seek a two-particle amplitude which satisfies a two-particle Blankenbecker-Sugar equation (contenting tehmselves with an onshell fit to the phase shift).

Yu. A. Shcherbakov (JINR-DUBNA): You have demonstrated how many troubles you have in this new method. But I am pleased that there already exists a theory suitable to apply to experiment (πd). May we have some hope to have some results in the near future with a theory for 4-body scattering (π^3He for example)?
Thomas: Certainly some very brave people have already made calculations on the four nucleon system at low energy. However, it seems very unlikely to me that (al least in the next few years) one could carry out a reliable calculation of π^3He scattering within a complete 4-body theory.

E. Hadjimichael (Fairfield Univ.): What prescription is followed in determining the form factors in the separable potential used in the Faddeev method.
A. Thomas: My personal preference is to choose a phenomenological form for the form factors, with a number of free parameters (ranges and strengths). These parameters are then adjusted so that the corresponding (on-shell) two-body t-matrix (obtained from the Blankenbecker-Sugar equation) reproduces the experimental phase shifts. That this idea is not accepted (yet) by everyone; it was discussed in answer to Eisenberg's earlier question.

Kopaleishvili (Tbilisi State Univ.): What can you say about the importance of taking into account the transformation from the two-body c.m. frame to the three-body c.m. frame?
Thomas: Such a transformation is of course included in the fully relativistic calculations (references 13, 18 and 19). The importance of omitting such a transformation has not been explicitly tested, to my knowledge. That it may not be too significant is suggested by a comparison of the curves CV and TH-MV in Ref. 13, since in the latter case, as well as using the linear form of the propagator, all kinematic transformation factors "f_i" of Ref. 7 were omitted.

K. W. McVoy (U. of Wisconsin): 1. Would experimental information on the polarization of the recoil deuteron be useful as a check on the accuracy of the calculations? 2. One way of viewing the "resonance" in πd or in general πA scattering is to use the fixed-scatterer approximation to look upon it as the interaction of a pion with a "nuclear crystal". In this limit one sees that, since the pion has a (3,3) resonance relative to each nucleon, the entire system will have A states, which split to form an "energy band" of overlapping resonances - all of which occur, in fact, in each πA partial wave. This is just the structure Moniz obtains in his isobar shell model. One would expect it to persist in proper 3-body calculations. Do the 3-body πd amplitudes show two poles in each partial wave?

Thomas: 1. Although T_{20} is sensitive to the % D-state, it is not clear that it is more sensitive to calculational details than the cross section itself. Indeed a computation by J. Alexander and myself using my phase shifts, in addition to the Glauber phases of Hoenig and Rinat, which are qualitatively different, gives almost identical $d\sigma/dt$ and T_{20}! This indicates the difficulty of ever making a phase shift analysis of πD scattering, and suggests the importance of finding more sensitive polarization parameters.

2. Most 3-body calculations are numerical. Some analytical work by Brayshaw indicates the presence of cuts rather than poles.

MULTIPLE SCATTERING CALCULATIONS IN πd ELASTIC SCATTERING*

Erasmo M. Ferreira
Stanford Linear Accelerator Center, Stanford, California
and Pontificia Universidade Catolica, Rio de Janeiro

From a theoretical point of view, the πd system is in a privileged position as compared to other π-nuclei systems, as the Faddeev equations provide the basis for an exact formulation of three body problems. As we have heard in the report by A. Thomas, efforts have been made to solve Faddeev equations for the πd system, and the success obtained with these first results is stimulating. Unfortunately, the attempts to obtain solutions of Faddeev equations face limitations of practical nature, as soon as the energy goes above the very low energy limit (let us say 100 MeV), due to the large number of coupled angular momentum states involved.

At this point the multiple scattering method comes into play. At energies which are not very low, the rather simple multiple scattering calculations, without appeal to model-dependent calculations, are able to give a fairly good description of the πd elastic scattering process.

Applications of the multiple scattering method to evaluate πd scattering processes have been made by several authors.[1] Technical details of the computations are not uniformly treated in the several papers, however, which makes it difficult to develop a critical feeling for the value and the limitations of the method. Some of the effects which are or are not accounted for in some of these computations (such as the fermi-motion dependence of the amplitudes, presence of D wave component in deuteron wavefunction, nucleon recoil) may have important consequences in the results, such as in large angle scattering. Besides, in each application only one or a limited range of values of the energy have been considered, and, comparing the results, we note that the performance of the calculations varies strongly with the energy.

The existing data are scarce, and of low statistics, and must be used in their entirety as a whole if one wishes to learn about the applicability of the method. We discuss here results which, although they are not a complete analysis, represent an effort in this direction.

We find that special attention must be given to a point which has been overlooked in most calculations, which is that of the indetermination in the values of the kinematical variables entering in the evaluation of the two particle amplitudes. One deals essentially with off-the-energy shell matrix elements of two-body transition operators. In general, these matrix elements are not known, and the values to be used have to be guessed, following some chosen prescription, from the on-the-energy values which are obtained from direct two-body experiments.

KINEMATICAL AMBIGUITIES. SOME SELECTED PRESCRIPTIONS.

The kinematical arbitrariness which is characteristic of this calculation can lead to very different predictions for the processes studied. We specify below three ways that can be taken to solve the indeterminacy. For easy future reference we call them prescriptions A, B, and C.

i. <u>Prescription A. Faddeev equations and reduction from three body to two body operators</u>. The exact three body amplitude for πd scattering given by

---
*Work supported by the Energy Research and Development Administration.

Faddeev equations can be expanded in terms of the two-body collision operators, in the form of a multiple scattering series. In the explicit evaluation of the terms of the expansion, care must be taken when expressing the matrix elements of operators defined in three-particle Hilbert space in terms of the usual two-body matrix elements. We can then see that the value of the energy parameter to be used becomes uniquely determined. Attention to this point has been called by Thomas.[3]

Let the three particles be labeled by the indices 1, 2, and 3 with momenta $\vec{p}_1$, $\vec{p}_2$, $\vec{p}_3$ in the lab system of reference. Let us select a pair (2,3), and treat the particle 1 separately. We define the new momentum variables

$$\vec{K} = \vec{p}_1 + \vec{p}_2 + \vec{p}_3, \quad \vec{k}_1 = (m_3 \vec{p}_2 - m_2 \vec{p}_3)/(m_2 + m_3)$$

$$\text{and } \vec{q}_1 = [(m_2 + m_3)\vec{p}_1 - m_1(\vec{p}_2 + \vec{p}_3)]/(m_1 + m_2 + m_3) \tag{1}$$

where $\vec{K}$ is the total momentum of the system, $\vec{k}_1$ is the internal momentum in the (2,3) pair relative to its center of mass, and $\vec{q}_1$ is the momentum of particle 1 with respect to the center of mass of the whole system. Defining $\mu_1 = m_2 m_3/(m_2 + m_3)$ and $M_1 = m_1(m_2 + m_3)/(m_1 + m_2 + m_3)$, the kinetic energy in the c.m. system ($\vec{K} = 0$) can be written

$$H_0 = (k_1^2/2\mu_1) + (q_1^2/2M_1) \tag{2}$$

The selected pair of particles can be any of the three possible choices, and new (not independent) sets of variables can be defined for each case. Each choice is called a channel. Let us call v_1 the potential acting between particles 2 and 3, v_2 the potential acting between 1 and 3, and so on. An important concept is that of the channel Hamiltonian

$$h_\alpha = (k_\alpha^2/2\mu_\alpha) + (q_\alpha^2/2M_\alpha) + v_\alpha \tag{3}$$

where there appears interaction only between the two particles forming the pair in channel α. The channel resolvent is

$$g_\alpha(z) = (z - h_\alpha)^{-1} \tag{4}$$

Channel operators depend on the relative coordinates of only two particles, and their matrix element between three free particle states can be expressed in terms of operators defined in the two-body Hilbert space. Let us call

$$\hat{h}_\alpha = k_\alpha^2/2\mu_\alpha + v_\alpha \tag{5}$$

the two-body Hamiltonian in channel α, and

$$\hat{g}_\alpha(z) = (z - \hat{h}_\alpha)^{-1} \tag{6}$$

the corresponding two-body resolvent. We can then reduce a three-body channel matrix element writing

$$\langle \vec{k}_\alpha \vec{q}_\alpha | g_\alpha(z) | \vec{k}'_\alpha \vec{q}'_\alpha \rangle = \delta(\vec{q}_\alpha - \vec{q}'_\alpha) \langle \vec{k}_\alpha | \left[\left(z - \frac{q_\alpha^2}{2M_\alpha} \right) - \hat{h}_\alpha \right]^{-1} | \vec{k}'_\alpha \rangle$$

$$= \delta(\vec{q}_\alpha - \vec{q}'_\alpha) \langle \vec{k}_\alpha | \hat{g}_\alpha \left(z - \frac{q_\alpha^2}{2M_\alpha} \right) | \vec{k}'_\alpha \rangle \tag{7}$$

The displacement in the value of the argument of the resolvent is very important here.

The three body transition matrix $T(z)$ is written as a sum $T = T_1 + T_2 + T_3$, where T_1, T_2, and T_3 satisfy the coupled equations $T_1 = t_1 + t_1 g_0 (T_2 + T_3)$ and similarly for T_2 and T_3. Here $g_0(z) = (z-H_0)^{-1}$ is the resolvent for three free particles, and

$$t_\alpha(z) = v_\alpha + v_\alpha g_\alpha(z) v_\alpha \qquad (8)$$

are channel α transition operators acting in the three particle Hilbert space, and satisfying a reduction relation analogous to Eq. (7).

The Faddeev version of the multiple scattering series is obtained in an obvious way by iterating the coupled integral equations written above. For the elastic scattering of particle 1 by the (2,3) bound pair the transition operator $T(z)$ can be expanded in the form of a multiple scattering series

$$T(z) = t_2(z) + t_3(z) + t_2(z) g_0(z) t_3(z) + t_3(z) g_0(z) t_2(z) + \ldots \qquad (9)$$

where the interpretation of the terms is the usual one, and all operators are defined in the three-particle Hilbert space. The reduction to matrix elements of two body operators is made with the appropriate shift corresponding to the energy of the particle which, in each term, does not participate in the process. Let E be the value of the total kinetic energy of the particle-deuteron system in the center-of-mass system, $\vec{P}$ the nucleon (particle 1) lab momentum, and $\vec{p}(\vec{p}')$ the initial (final) meson (particle 3) momentum in the lab system. For the term with particle 2 as spectator,

$$\langle \vec{P}', -\vec{P}', \vec{p}' | t_2(E) | \vec{P}, -\vec{P}, \vec{p} \rangle = \delta(\vec{q}_2' - \vec{q}_2) \delta(\vec{K}' - \vec{K}) \langle \vec{k}_1' | \hat{t}_2 \left(E - \frac{q_2^2}{2M_2} \right) | \vec{k}_1 \rangle \qquad (10)$$

where $\vec{K}(\vec{K}')$ is the total initial (final) momentum of the three particles, $\vec{q}_2$ ($\vec{q}_2'$) is the initial (final) momentum of the spectator with respect to the center of mass, $\vec{k}_1$ ($\vec{k}_1'$) is the initial (final) momentum of the meson relative to the center of mass of the interacting meson-nucleon system, and M_2 is given by $M_2 = m_N(m_N + m_\pi)/(2m_N + m_\pi)$.

The argument of the two body transition operator $\hat{t}_2$ then reads

$$E - q_2^2/2M_2 = E - [(2m_N + m_\pi)\vec{P} + m_N \vec{p}]^2 / [2m_N(m_N + m_\pi)(2m_N + m_\pi)] \qquad (11)$$

In the evaluation of the double scattering terms, one introduces complete sets of three free particle states between the operators, and the reduction to the two-body operators takes place in a manner analogous to that described above.

ii. <u>Prescription B. The meson collides with an on-shell physical nucleon.</u> If fermi-motion effects are taken into account, for each value and each direction of the nucleon momentum inside the deuteron, a different value is used for the relative energy between the incident particle and the nucleon.

iii. <u>Prescription C. The spectator nucleon is treated as an on-shell physical nucleon.</u> Experiments in which there is a breakup of the deuteron, and where an identification can be made between the spectator and the nucleon which was hit by the incident particle, show that the spectator nucleon recoils with a momentum distribution which is, in good approximation, the same as expected from the deuteron wavefunction. We are thus led to the assumption that the spectator nucleon behaves from beginning to end as an on-shell particle. The nucleon which participates in the collision must then be treated as an unphysical particle in the initial and final states. To fulfill energy conservation, the energy of the participant nucleon is equal to the deuteron mass m_d minus the energy $m_N + P^2/2m_N$ carried by the spectator nucleon, where P is the fermi-motion momentum. Thus the participant nucleon behaves as having an effective mass m_{eff} such that

$$m_{eff} + P^2/2m_{eff} = m_d - m_N - P^2/2m_N$$

In a certain sense, prescriptions B and C exchange the roles of the spectator and the participant nucleons. At zero fermi momentum the two prescriptions almost coincide, as then $m_{eff} = m_d - m_N \approx m_N$.

Prescription B has been often used in multiple scattering calculations of πd processes.[1] Prescription C was used in the analysis of pion deuteron breakup scattering.[4]

While prescription B seems to be intuitively more appealing, and prescription C has some kind of experimental support, prescription A has a safer theoretical basis. As off-energy-shell matrix elements are not intuitive quantities, we should rather rely on the more formal approach. The nucleons are not free physical particles inside the deuteron, and prescription A tells us how to take partially into account the effect in our calculation of the presence of two particles in the target nucleus.

In Fig. 1 are shown the values of the kinetic energy (excluded rest masses) in the center-of-mass system of the two colliding particles, as a function of the fermi motion momentum. The relative energy depends not only on the magnitude, but also on the direction of the fermi motion momentum, and the lines drawn represent the average value over all directions for a fixed magnitude P of the momentum. In prescription B, the value plotted for the energy does not depend much on the value of the fermi momentum, and remains almost constant, while in cases A and C the variation is strong. We can thus expect that fermi-motion effects may be stronger in cases A and C than in case B. These predictions have been confirmed by our calculations, covering the interval of energies from zero up to about 400 MeV. A

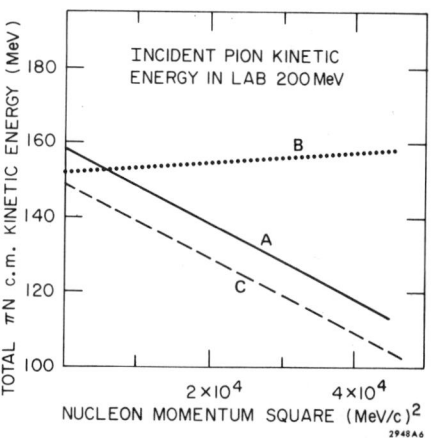

Fig. 1--Values of the total kinetic energy (rest masses excluded) in the πN c.m. system, according to prescriptions A, B, and C, against fermi-motion momentum squared. The energy values are averaged over all directions for a given magnitude of fermi momentum.

main observation is that fermi motion effects are extremely important for the correct evaluation of large angle scattering, because the strong cancellations which occur in the evaluation of the cross sections are sensitive to the proper account of the variation of the values of integrand as a consequence of these effects. A factor of up to four in the differential cross section can appear in the backward angles as the fermi motion effect is switched off.

We may expect that in the cases of prescriptions A and C the calculations are more sensitive to changes in the large momentum tail of the deuteron wavefunction than they are in case B.

OFF THE ENERGY SHELL BEHAVIOR OF AMPLITUDES

For each partial wave we must evaluate an off-shell amplitude $<k'|f_\ell(y)|k>$ where k, k' are the initial and final relative momenta of the colliding pair, and y is the energy parameter defined according to each of the prescriptions adopted. These three quantities are not related among themselves through the usual on-shell relations. Integration is made over all initial and final values of the nucleon momentum and the values of k, k', and y vary rather disconnectedly. We must define the matrix element as a function of these variables.

A simple recipe consists in writing $<k'|f_\ell(y)|k> = (kk')^{-\frac{1}{2}} \sin \delta_\ell(y) e^{i\delta_\ell(y)}$, where $\delta_\ell(y)$ is the physically measured πN phase shift at energy y. We have observed in the evaluation of πd cross section that, due to the smoothing caused by the integrations over k and k', it makes almost no difference to write $(kk')^{\frac{1}{2}}$ or simply k in the equation above.

Another possible specification for the off-shell extrapolation consists in defining a separable potential for each partial amplitude.[3]

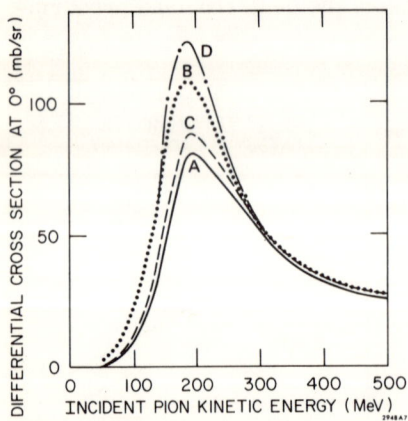

Fig. 2--Forward differential cross section for πd elastic scattering with Coulomb interaction switched off, comparing results obtained with kinematical descriptions A, B, C described in the text.
Curve D shows results obtained neglecting fermi-motion effects. The peak in Curve A is displaced about 6 MeV towards higher energies as compared to the other curves.

FORWARD SCATTERING

In forward scattering, as in the value of the total cross section, fermi motion effects are not so important, unless we are near the dominant and resonant wave. In the case of πd scattering near the P33 resonance the influence of the fermi motion effect can be about 35% in the forward cross section in the case of prescription C and 15% in the case of prescription B. This behavior can be seen in Fig. 2 where we plot the forward differential cross section for πd elastic scattering as a function of the meson incident energy, comparing the three different prescriptions and the usual calculation without account for fermi motion effects (prescriptions B and C coincide in this case). Of course the Coulomb interaction has not been taken into account.

We see in Fig. 2 that the position of the peak due to the P33 resonance is nearly the same in all cases, with a shift of about 6 MeV towards higher values of the energy observed in the case of prescription A. This is an important, although rather obvious, result as we should expect a displacement to occur in the position of the peak as a consequence of the shift in the value of the energy caused by the reduction from three-body to two-body operators. This result, which is shown in Fig. 2 for the nuclear (non-Coulomb) interaction for zero angle scattering, is also true of the total cross section, as the elastic πd scattering is almost completely forward. It is interesting to remark that larger shifts are expected to occur in the scattering by heavier nuclei.

We must call attention to the result, shown in the figure, that the values of the total and forward cross sections, evaluated with prescription A in the resonance region, are remarkably lower than the values obtained in the other two cases.

EXPERIMENTAL RESULTS AND THEORETICAL CALCULATIONS

It is hoped that the chronic scarcity of data on πd scattering will change soon, as already indicated by the recent experiment at 47.5 MeV by D. Axen et al.,[5] and the expected results of the measurements at 347 MeV/c (234.4 MeV kinetic energy) and 443 MeV/c (324.9 MeV kinetic energy) performed by a collaboration of the groups at the University of Virginia and at Los Alamos Scientific Laboratory.[6] There are reported experimental results on the elastic πd differential cross section for incident pions at 61 MeV,[7] 85 MeV,[8] 140 MeV,[9,10] 182 MeV,[11] 224 MeV,[12] 256 MeV,[13] 300 MeV,[14] and 330 MeV.[15] For large angle scattering, between 140 and 180 degrees in the laboratory system, there are results obtained by Schroeder et al.[16] at 375.7, 412.4, and 469.6 MeV. The work of Gabathuler et al.[13] also includes measurements of the backward cross section at 160 degrees lab scattering angle for incident pions of 141, 163, 185, and 208 MeV.

In Figs. 3, 4, 5, and 6 (see also Ref. 2) we confront those data with results of multiple scattering calculations, comparing the different prescriptions for the kinematical variables used in the evaluation of the two-body amplitudes. The calculations include double scattering terms, allowing for nucleon recoil, and including both the delta function and the principal value parts originated from the pole in the propagator. Corrections to the differential cross section arising from the double scattering terms never amount to more than 20 percent in the whole range of angles and of energies here considered. It is thus unnecessary to include fermi-motion dependence in the double scattering terms, which brings an important simplification in the numerical computations. The comparatively small contribution obtained for the double scattering terms makes us confident that higher order terms of the series can be neglected. The calculations account for fermi motion effects in the single scattering terms, and are made with Moravcsik wavefunction, with 7 percent d-wave component.

As shown in Fig. 3, the experimental results obtained at 47.5 MeV are reasonably well fitted by a multiple scattering calculation with the most usual treatment of the two-body kinematics, namely, prescription B. Fermi motion effects do not seem to contribute substantially to improve the quality of this theoretical curve. The other two prescriptions perform badly at this energy. However, we must remark that such observations should not be taken on their own as a basis of judgment about the method of calculation. In fact,

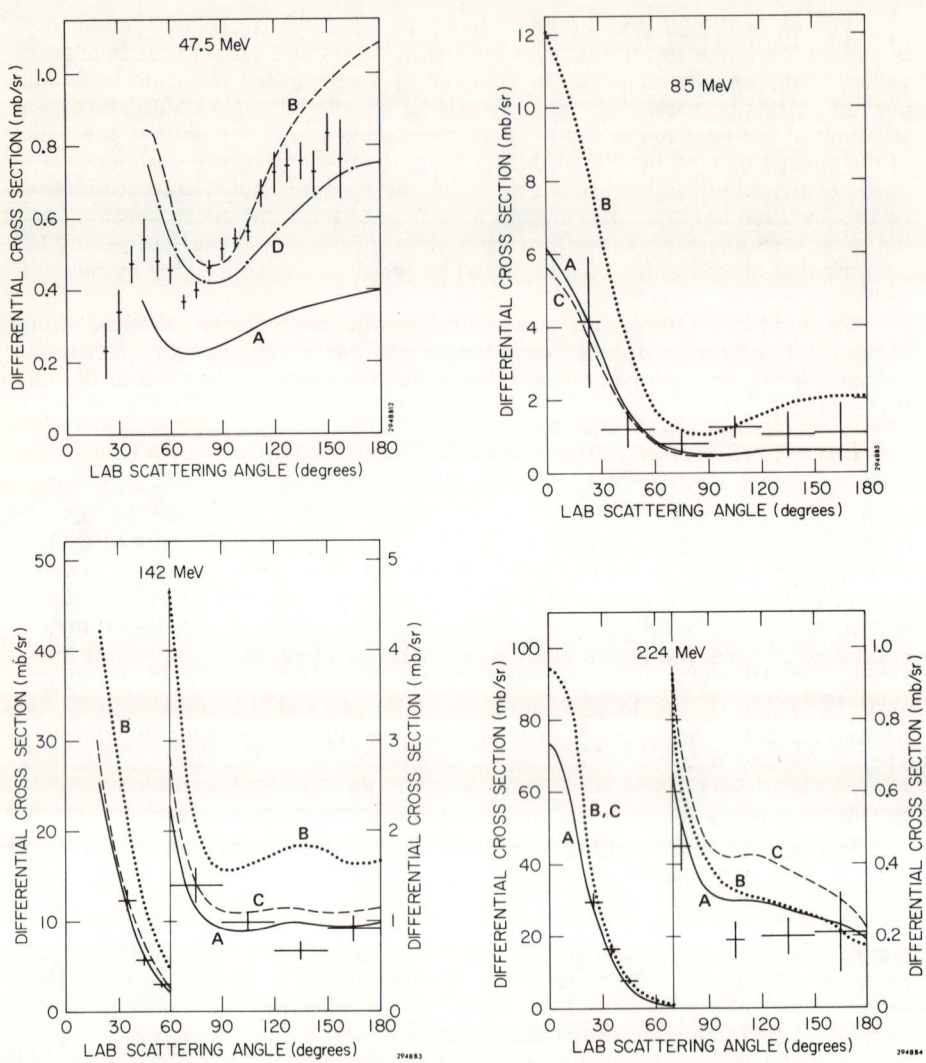

Figs. 3, 4, 5, and 6--Data on πd elastic scattering and theoretical curves representing results of multiple scattering calculations. Labels A (solid), B (dotted), and C (dashed) refer to the kinds of kinematical prescription described in the text. Curve D (dot-dashed) at 47.5 MeV is obtained without account for fermi motion. Notice the enlarged scale used for large angles at 142 and 224 MeV. The solid curves best fit the data, if 47.5 MeV is considered too low energy for this kind of theoretical calculation.

we can see that the situation becomes very different at slightly higher energies. Thus, at 85, 142, 182, 224, and 256 MeV, as exemplified in Figs. 4, 5, and 6 (see also Ref. 2), prescription A seems to describe the data better. From the inspection of the figures, we are led to believe that these theoretical calculations make sense at these energies. The poor results obtained with prescription A at 47.5 MeV should thus be taken as demonstrating that this energy is too low for a multiple scattering calculation involving only single and double scattering terms. Binding corrections, off-the-energy shell extrapolations, or complicated three-body mechanisms might play important roles at such low energies.

Above 230 MeV the experimental cross sections at large scattering angles gain a structure which is not reproduced by these theoretical calculations. At small angles up to about 70 degrees in lab system the calculated values are reasonable, but at large angles the calculations are wrong by a factor of about two. (See also Ref. 2 for the energy behavior of large angle cross sections.)

We may speculate on what may be the cause of this discrepancy. We notice that the strong reduction in the large angle experimental cross section, as compared to the calculated values, occurs suddenly as the energy goes above 230 MeV. At this energy some new dynamical phenomenon may have started to play a role. We may think for example that pion production and consequent reabsorption by the other nucleon may have started to contribute significantly. At these energies, which are above the threshold for pion production, this essentially three-body mechanism could eventually be responsible for a change in the dynamics of the process.

Another possible explanation for the observed discrepancy is that we may have entered in a range of momentum transfer where the effects of our insufficient knowledge of the deuteron structure may have started to affect the calculations. A change in the large momentum tail in the deuteron wavefunction may substantially change the value of the integral over internal fermi momentum in the expression of the differential cross section. As an example, we mention that the introduction of the d wave component in deuteron wavefunction causes an increase by a factor 2 in the calculated cross section at large angles in the energies of Schroeder experiment (375.7 MeV and over).

These effects due to changes in the deuteron structure or in the meson-nucleon interaction might be expected to be small at first sight. However we must notice that the value calculated for the πd differential cross section at large angles is several orders of magnitude smaller than the forward cross section, due to strong cancellations occurring in the integration procedure. The results obtained after such cancellations have a delicate and strong dependence on the quantities in the integrand.

The extreme sensitivity of the backward πd elastic cross section at large angles provides an excellent ground to study the deuteron structure and properties of the meson-deuteron and meson-nucleon interaction.

We find that more and more accurate experiments on πd scattering should be performed as soon as possible. The region of energies around and above 200 MeV should be carefully studied, as important changes in the process seem to take place in this region.

On the other hand, it is obvious that the theoretical effort must also be increased, both in the calculations with multiple scattering method and in direct solutions of Faddeev integral equations. A combination of the two methods, joining the nice features of each, may be an interesting and rewarding program of investigation.

REFERENCES

1. Multiple scattering calculations of πd elastic scattering: R. M. Rockmore, Phys. Rev. 105, 256 (1957); A. Ramakrishnan, V. Devanathan, K. Venkatesan, Nucl. Phys. 29, 680 (1962); H. N. Pendleton, Phys. Rev. 131, 1833 (1963); C. Carlson, Phys. Rev. C 2, 1224 (1970); D. S. Beder, Nucl. Phys. B 34, 189 (1971); W. R. Gibbs, Phys. Rev. C 3, 1127 (1971); R. L. Landau, Nucl. Phys. B 35, 390 (1971); J. M. Wallace, Phys. Rev. D 5, 1840 (1972); K. Gabathuler and C. Wilkin, Nucl. Phys. B 70, 215 (1974); E. Ferreira, L. P. Rosa, and Z. D. Thomé, Nuovo Cimento 20A, 277 (1974), and Nuovo Cimento Lett. 9, 707 (1974); M. A. Braun and V. B. Senyushkin, Sov. J. Nucl. Phys. 21, 147 (1975).
2. E. M. Ferreira, L. P. Rosa, and Z. D. Thomé, paper contributed to this conference (to be published).
3. A. W. Thomas, Proc. Int. Conf. on Few Body Problems in Nuclear and Particle Physics, Quebec, 27-31 Aug 1974 (Les Presses de L'Universite Laval, Quebec, 1975), p. 287.
4. E. M. Ferreira, L. P. Rosa, and Z. D. Thomé, Nuovo Cimento 21A, 187 (1974).
5. D. Axen, G. Duesdieker, L. Felawka, Q. Ingram, R. Johnson, G. Jones, D. Lepatourel, M. Salomon, W. Westlund, L. Robertson, Nucl. Phys. A 256, 387 (1976).
6. University of Virginia and Los Alamos Scientific Laboratory Collaboration - Preliminary Data, Los Alamos Report LA-6156-R, and private communication by J. McCarthy.
7. A. M. Sachs, H. Winick, and B. A. Wooten, Phys. Rev. 109, 1733 (1958).
8. K. C. Rogers and L. M. Lederman, Phys. Rev. 105, 247 (1957).
9. E. Arase, G. Goldhaber, and S. Goldhaber, Phys. Rev. 90, 160 (1953).
10. E. G. Pewitt, T. H. Fields, G. B. Yodh, J. G. Fetkovich, and M. Derrick, Phys. Rev. 131, 1826 (1963).
11. J. H. Norem, Nucl. Phys. B 33, 512 (1971).
12. J. L. Acioli, Ph.D. thesis, University of Chicago, 1968 (unpublished).
13. K. Gabathuler, C. R. Cox, J. J. Domingo, J. Rohlin, N. W. Tanner, C. Wilkin, Nucl. Phys. B 55, 397 (1973).
14. L. S. Dul'kova, I. B. Sokolova, and M. G. Shafranova, Sov. Phys. - JETP 35, 217 (1959).
15. G. Brunhart, G. S. Faughn, V. P. Kenney, Nuovo Cimento 24, 1162 (1963).
16. L. S. Schroeder, D. G. Crabb, R. Keller, J. R. O'Fallon, T. J. Richards, R. J. Ott, J. Trischuk, and J. Va'vra, Phys. Rev. Lett. 27, 1813 (1971).

A. Gal (Univ. of Virginia): A possible remedy for the insufficiency of single and double scattering terms in πd above the (3,3) resonance region may be provided by reflection terms. The pion loses energy in its first, large angle, collision, reaches the resonance and thus scatters back and forth for several cycles. This mechanism is more important above the resonance energy than below. For heavier nuclei these reflection terms can be included in a second order optical potential, as Keister has recently shown. However, his results indicate that the above effect is small, in contradiction to what Agassi and I found for π^- ^{4}He scattering.

Keister (CMU): I wish to clarify the point by Gal: Since I communicated my results to him, a mistake was found in the calculation, such that the multiple-reflection effects are actually quite large in nuclei within the fixed-scatterer approximation.

Ferrereira: The deuteron is quite different from larger nuclei with respect to contributions coming from the higher order terms of the multiple scattering series.

Carl Shakin (Brooklyn College): In a paper by Kowalski and Pieper, it was suggested that the FSA specification of the energy in the T matrix seemed to yield a better result for n-d scattering than an impulse approximation calculation using the Fadeev specification. It may be that the full solution of the 3-body equations tends to remove the shift in energy given in the impulse approximation term of the Fadeev multiple scattering series.

R. Amado (Univ. of Pennsylvania): I agree with Prof. Shakin. In an exact fixed scatterer model, that is for infinite mass, it can be shown that the total effect of the full multiple scattering series is just to shift the energy of the single scattering up to the "one shell" point. Of course the question here is whether the infinite mass case can teach us anything about the π-d system.

H. Garcilazo (Instituto Politecnico Nacional, Mexico): What is the influence of the double scattering term at energies above the resonance, and in particular at the energies of the new Virginia data?
E. Ferrereira: Less than 10% in all cases, even at large angles.

GRAPH SUMMATION METHOD IN πd PROBLEM

V.M.Kolybasov

Institute for Theoretical and Experimental
Physics, Moscow, USSR

The pion-deuteron interaction at low energy is interesting from two points of view. First, it is the simplest case of a pion-nuclear interaction, therefore it is the better example to study the accuracy and the application limits of various theories of pion scattering by nuclei. Secondly, comparing the theoretical results with data one gets a possibility to obtain a non-trivial information about pion-nucleon scattering lengths. Particulary, we mean the determination of the quantity b_o (the half-sum of the pion-proton and pion-neutron scattering lengths) which is now poorly known.

Pion-deuteron scattering at low energies was considered for the first time by Brueckner [1]. He has written the formula which took into account all order rescatterings but under an assumption that a nucleon has such a large mass that the recoil effects may be neglected. Really, this assumption is not quite good and (as it has been shown later [2]) the recoil effects strongly change the result. Present day calculations claiming a several per cent accuracy proceed either from Feynman graph summing method [2-6] or from solving the Faddeev equation with separable potentials [7-10].

Certainly, the essence of the matter is not in the formal method applied to solve the problem, but in the physical assumptions being made. Therefore, dividing the problem into the graph summation method and the method of Faddeev equations is rather conventional. However, at the first stage when understanding the nature of various effects is important rather than a high accuracy of calculations, the graph technique being quite evident has a number of obvious adventages.

Write the πN low energy scattering amplitude in the form

$$f_{\pi N} = b_0 + b_1 \vec{t}\vec{\tau} + [c_0 + c_1 \vec{t}\vec{\tau}](\vec{k}\vec{k}') \qquad (1)$$

where $\vec{t}$ and $\vec{\tau}$ are the pion and nucleon isospin operators, $\vec{k}$ and $\vec{k}'$ are the relative momenta before and after scattering. We shall use the following constants of the p-wave interactions: $C_0 = 0.208 \pm 0.008\ \mu^{-3}$, $C_1 = 0.180 \pm 0.005\ \mu^{-3}$, μ is the pion mass. Consider the two sets for the s-wave constants: the first is [11]

$$b_0 = -0.012 \pm 0.004\ \mu^{-1} = -0.017 \pm 0.006\ \text{fm}$$
$$b_1 = -0.097 \pm 0.007\ \mu^{-1} = -0.137 \pm 0.010\ \text{fm}$$
(A)

the second is [12]

$$b_0 = -0.005 \pm 0.002\ \mu^{-1} = -0.007 \pm 0.002\ \text{fm}$$
$$b_1 = -0.087 \pm 0.001\ \mu^{-1} = -0.123 \pm 0.002\ \text{fm}$$
(B)

Eq. (1) allows to take automatically into account the virtual pion charge exchange in multiple scattering.

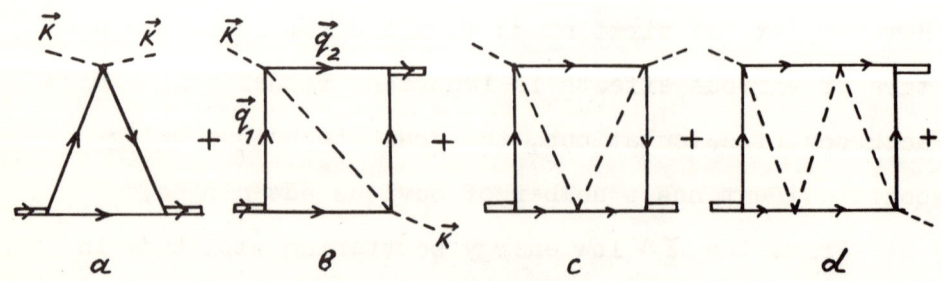

Fig. 1

The crucial point of the graph summing method in

πd scattering is that taking into account the s-wave part of interaction one discovers that the diagrams of fig. 1 lead (under certain assumption which will be mentioned below) to some modification of the Brueckner formula [2]:

$$a_{\pi d} = \frac{1}{1+\mu/m_d} \int d\vec{z}\, \psi_d^2(\vec{z}) \frac{1}{\delta} \Big\{ 2\tilde{b}_0 + 2\frac{e^{-\varkappa z}}{z^2}(\tilde{b}_0^3 - 2\tilde{b}_0^2\tilde{b}_1 - \tilde{b}_0\tilde{b}_1^2 + 2\tilde{b}_1^3)$$

(2)

$$+ 2\frac{e^{-\varkappa z}}{z}[(\tilde{b}_0^2 - 2\tilde{b}_1^2) - \frac{e^{-2\varkappa z}}{z^2}((\tilde{b}_0^2 - 2\tilde{b}_1^2)(\tilde{b}_0^2 - 2\tilde{b}_0\tilde{b}_1 - \tilde{b}_1^2) + 2\tilde{b}_1^4)]\Big\}.$$

Here $\psi_d(\vec{z})$ is the deuteron wave function, m_d is the deuteron mass, $\varkappa = \sqrt{2\mu \mathcal{E}_d}$, $\mathcal{E}_d$ is the deuteron binding energy, $\tilde{b}_i = (1 + \mu/m)b_i$, m is the nucleon mass,

$$\delta = 1 - \frac{e^{-2\varkappa z}}{z^2}(2\tilde{b}_0^2 - 2\tilde{b}_0\tilde{b}_1 - 3\tilde{b}_1^2) +$$

(3)

$$+ \left(\frac{e^{-2\varkappa z}}{z^2}\right)^2 [(\tilde{b}_0^2 - 2\tilde{b}_1^2)(\tilde{b}_0^2 - 2\tilde{b}_0\tilde{b}_1 - \tilde{b}_1^2) + 2\tilde{b}_1^4].$$

The diagrams including the nucleon rescattering one by another are not essential because they are of the order $\sqrt{\mu/m}\,(b_0/b_1)^2$ [2,4]. The eq. (2) shows that the main contribution is from the single and double scattering, therefore all further corrections should be introduced into those terms only.

The double scattering diagram of fig. 1b can be written in the form

$$M^{(2)} = C \int \frac{\varphi(\vec{q}_1)\varphi(\vec{q}_2) d\vec{q}_1 d\vec{q}_2}{(\vec{q}_1 - \vec{q}_2)^2 + \frac{\mu}{m}(q_1^2 + q_2^2) + 2\mu \varepsilon_d}, \quad (4)$$

C is a constant, $\varphi(\vec{q})$ is the deuteron wave function in the momentum representation. The term (μ/m) $(q_1^2 + q_2^2)$ takes into account the kinetic energy of nucleons in the intermediate state (the nucleons are not static). Deriving eq. (2) we have neglected such terms in the pion propagators. Really, the recoil effects are strong (of the order of $\sqrt{\mu/m}$) [2]. Therefore they essentially effect the results. Below this fact will be taken into account.

One must remember that about half a contribution from the double scattering term is due to the virtual exchange $\pi^-p \to \pi^0 n$ releasing the energy Q = = 3.3 MeV. Then in eq. (4) $\varepsilon_d \to (\varepsilon_d - Q) = -1.1$ MeV. Therefore the nonadiabetic effects turn to be suppressed and the variation of the final result is -0.002 fm.

Besides, it appears that even at zero energy of

the pion the p-wave part of the $\bar{\pi}N$ interaction gives essential contributions to the single and double scattering terms due to the interdeuteron motion of nucleons. However, these effects are of opposite signs and almost compensate one another.

The main results obtained by graph summing were later confirmed in the papers based on Faddeev equations. Those results are : a) the nucleon-nucleon rescattering diagrams are inessential; b) the main contribution is from the single and double scattering; c) the corrections to the Brueckner formula are large due to nonadiabatic effects; d) the p-wave part of $\bar{\pi}N$ interaction is essential in the single and double scattering terms separately but the total effect is almost compensated.

Besides the diagrams of fig. 1, one must consider the absorbtive processes when there is no intermediate pion. We shall accept the estimation of ref. [8] :

$$\alpha_{\pi d}^{(abs)} = -0.007 + i \cdot 0.006 \text{ fm}$$

Above we assumed the s-wave amplitudes of $\bar{\pi}N$ interactions to be constant. Generally speaking, it is a good approximation. But in the single scattering due to the strong compensation of the principle terms (as

$b_o \ll b_1$) the variation of the πN scattering amplitude becomes noticable [7,10]. The corresponding correction to the single scattering term is about -0.012 fm.

The numerical results for the pion-deuteron scattering length (including the contributions from separate terms) obtained with Gartenhaus-Moravcsic wave function and with two sets of s-wave πN scattering lengths are presented in the Table *).

Table.
(All the quantities are in fm)

Separate contributions	The act of πN parameters	
	(A)	(B)
Single scattering from eq.(1)	-0.036	-0.015
Double scattering from eq.(1)	-0.032	-0.025
Multiple scattering from eq.(1)	0.003	0.003
p-wave correction in the single scattering	0.006	0.006
Taking into account the variation of the πN amplitude for the single scattering term	-0.012	-0.012
Correction due to the nonadiabatic effects in the double scattering term (taking into account the energy release in the charge exchange $\pi^- p \to \pi^0 n$)	0.007	0.006
Taking into account the p-wave interaction in the double scattering	-0.007	-0.007
Absorbtive contribution	-0.007	-0.007
The total value of $\alpha_{\pi d}$	-0.078 ± 0.016	-0.051 ± 0.006
Estimations of b_o	$-(0.015^{+0.018}_{-0.015})$	$-(0.017^{+0.018}_{-0.015})$

The recent experimental measurement of the πd scattering length by the 1S level shift of the deuterium pionic atom [17] has yielded the following results:

$$\alpha_{\pi d}^{exp} = -(0.052^{+0.022}_{-0.017})\mu^{-1} = -(0.073^{+0.037}_{-0.024}) fm.$$

The problem can be reversed and the quantity b_o can be estimated by comparing the theoretical and experimental results. The corresponding values are given at the bottom of the Table. We see that within the errors the two values of b_o are in agreement with πd data. Nontrivial information about the quantity b_o is possible provided a large improvement of the accuracy of the 1S level shift of the deuterium pionic atom.

As to the improvement of the calculation accuracy, future activities is likely to be directed to solving the Faddeev equations but with realistic potentials instead of separable potentials.

References

1. K.A.Brueckner, Phys.Rev. 89, 934(1953).
2. V.M.Kolybasov, A.E.Kudryavtsev, JETP 63, 35(1972); Nucl.Phys. B 41, 510(1972).
3. V.M.Kolybasov, A.E.Kudryavtsev, JETP Letters 18, 527(1973).

4. A.E.Kudryavtsev, JETP 61, 490(1971).

5. P.Bendix, D.Beder, Phys.Lett. 49B, 140(1974).

6. V.M.Kolybasov, A.E.Kudryavtsev, Preprint ITEP-57 (1975).

7. N.M.Petrov, V.V.Peresypkin.Phys.Lett., 44B, 321(1973) Yad.Fiz., 18,791(1973); Nucl.Phys., A220, 277(1974).

8. A.W.Thomas, I.R.Afnan. Phys.Lett., 45B, 437(1973). I.R.Afnan, A.W.Thomas. Preprint FUPH-100 (Flinders University, 1974).

9. F.Myhrer, D.S.Koltum. Phys.Lett., 46B, 322(1973).

10. F.Myhrer, R.R.Silbar. Phys.Lett., 50B, 299(1974); F.Myhrer, Preprint (SIN, 1974).

11. V.K.Samaranayaka, W.S.Woolckock. Phys.Rev.Lett., 15, 936(1965).

12. D.V.Bugg, A.A.Carter, J.R.Carter. Phys.Lett., 44B, 278(1973).

13. V.M.Kolybasov, L.A.Kondratyuk, Phys.Lett., 39B, 439 (1972); Yad.Fiz., 18, 316(1973).

14. L.A.Kondratyuk. In "Interaction of high energy particles with nuclei", v.1, p.5, Atomizdat, Moscow, 1974.

15. G.Faldt. Preprint LUTP 1974-13(University of Lund, 1974).

16. V.M.Kolybasov, V.G.Ksenzov, JETP 72, No.1(1976).

17. J.Bailey, D.V.Bugg, U,Gastaldi et al., Phys.Lett. 50 B, 403(1974).

THE THEORY OF PION - ^{4}He SCATTERING

F. Lenz, SIN, Villigen

In my talk, I shall try to cover two main subjects. The first part deals with a purely theoretical problem. I shall show how, within the multiple scattering formalism, the dynamics of π-nucleon resonances, which are excited inside a nucleus, can be described. In the second part, I shall present the results of a calculation of π-^{4}He scattering and discuss in this special example the type of information about the dynamics of a Δ-nuclear system which can be obtained from studying π-nucleus scattering. I shall further try to indicate the relation of our calculation to other theoretical analyses of π-^{4}He scattering. The talk is based on a calculation of π-^{4}He scattering which has been performed in collaboration with M.Hirata and K.Yazaki.

The quantitative approaches developed so far to describe π-scattering on finite nuclei are all based on some version of the Watson-multiple scattering formalism. Within this formalism, the basic quantity to be calculated is the transition matrix τ describing the scattering of a pion on a bound nucleon

$$\tau = v + v \frac{\alpha}{E - H_A - T_\pi} \tau \qquad (1)$$

v is the π-N potential, H_A the nuclear Hamiltonian, T_π the kinetic enrgy operator of the pion and α projects onto the antisymmetrized intermediate nuclear states.

The general idea of the multiple scattering formalism is to relate, in an approximate way, τ with the t-matrix for pion scattering on a free nucleon

$$t = v + v \frac{1}{E - T_r} t \qquad (2)$$

where T_r is the kinetic enrgy operator of the π-N relative motion. The simplest and most widely used approximation is the static approximation in which τ is identified with t calculated at some energy $\bar{\varepsilon}(E)$.

$$\tau(E) = t(\bar{\varepsilon}(E))$$

In this approximation the only signature of the resonance in the π-nucleon system is a strong damping in the π-nucleus system. In order to account for the special dynamics due to the resonance in the subsystem it is necessary to go beyond the static approximation. In a first step, I neglect all but the dominant 3-3 amplitude. This amplitude is assumed to be separable and reads, if expressed in Lab.-quantities[+]

$$\langle \vec{k}', \vec{K}' | t(E) | \vec{k}, \vec{K} \rangle = t_0 \, \delta^3(\vec{k}+\vec{K}-(\vec{k}'+\vec{K}')) \cdot g^+(\vec{k}') \, D^{-1}(E - \frac{(\vec{k}+\vec{K})^2}{2(M+E)}) \, g(\vec{k}) \qquad (3)$$

[+] The difference between π-nucleus CM and Lab. system is neglected only in the formal part but not in the calculations.

$\vec{k}_i^{(i)}, \vec{K}^{(i)}$ are the pion and nucleon momenta respectively in terms of which the relative momenta are approximately given by

$$\vec{\kappa} = \frac{M\vec{k} - E\vec{K}}{M + E},$$ (4)

with the relativistic pion energy E and the nucleon mass M. The following form for the vertex function g is assumed

$$g(\vec{\kappa}) = \frac{1}{\kappa^2 + a^2}(\vec{S}\vec{\kappa})$$ (5)

with the range parameter a (300 MeV/c). $g^+(\vec{K})g(\vec{K})$ contains via $(S^+\vec{\kappa})(\vec{S}\vec{\kappa})$ the projection operator on the spin 3/2 π-N channel. (The corresponding isospin operators are not shown.)

The function $R_f(E)$ in the expression for $D(E)$

$$D(E) = E - R_f(E) + i\Gamma_f(E)/2$$ (6)

is fitted to the 3-3 π-N amplitude while $\Gamma_f(E)/2$ is related by unitarity to the form-factors $g(\vec{K})$.

Introducing the kinetic energy operator for the motion of the π-N CM

$$T_\Delta = \frac{p^2}{2(M+E)}$$ (7)

the elementary amplitude can be written as

$$\langle \vec{k}', \vec{K}' | t(E) | \vec{k}, \vec{K} \rangle = t_0 \, g^+(\vec{\kappa}')\langle \vec{K}'+\vec{k}' | D(E - T_\Delta) | \vec{K}+\vec{k}\rangle g(\vec{\kappa})$$ (8)

In contrast to free π-nucleon scattering, the CM motion of the interacting π-N system is not free if the scattering occurs in the nuclear medium. The modifications due to the surrounding nucleons is formally described by the so called binding corrections to the elementary amplitude, i.e. by the difference in the propagators in eqs. (1) and (2). We describe approximately the influence of the other nucleons by a potential in which the CM of the interacting π-N system is supposed to move. This amounts to replacing the energy $E-T_\Delta$ for free π-nucleon scattering in (8) by

$$E - T_\Delta \rightarrow E - T_\Delta - V - H_{A-1}$$

i.e. by the energy which is available to the π-N relative motion if the scattering occurs in the medium. Expanding

$$D(E - T_\Delta - V - H_{A-1}) \approx D(E) - H_\Delta^A$$ (9)

the Hamiltonian

$$H_\Delta^A = D'(E)(T_\Delta + V + H_{A-1})$$ (10)

has been introduced. H_Δ^A describes the system of A-1 nucleons and the CM motion of the interacting π-N system, the degree of freedom to which we refer in the following as the Δ. H_Δ^A is thus the (unperturbed) Δ-(A-1)N Hamiltonian. The π-nucleus first order optical potential is given by the ground-state expectation value of the modified π-N amplitude and reads in the momentum representation

$$\mathcal{U}(\vec{k}',\vec{k}) = t_0 \sum_n \int d^3K' d^3K \, \Psi_n^*(\vec{K}') g^+(\vec{\kappa}') \cdot$$
$$\cdot \langle \vec{k}'+\vec{K}' | \frac{1}{D(E)-H_\Delta^A} | \vec{k}+\vec{K} \rangle g(\vec{\kappa}) \Psi_n(\vec{K}) \qquad (11)$$

where Ψ_n are the nucleon single particle wave functions. Instead of the momentum eigenstates for the πN CM motion we may as well use any complete set of Δ-states. Introducing the Δ-h (Δ-hole) states $|\Delta n^{-1}\rangle$ the optical potential (11) can be rewritten as

$$\mathcal{U}(\vec{k}',\vec{k}) = \sum_{n\Delta\Delta'} F_{\Delta'n}^* \langle \Delta'n^{-1} | \frac{1}{D(E)-H_\Delta^A} | \Delta n^{-1} \rangle F_{\Delta n} \qquad (12)$$

The π-Δh vertex function $F_{\Delta h}$ is determined by the Δ and nucleon wave functions and the vertex function g of the π-N t-matrix (5).

$$F_{\Delta n} = F_{\Delta n}(k) = t_0^{1/2} \int d^3K \, g(\vec{\kappa}) \Psi_\Delta^*(\vec{K}+\vec{k}) \Psi_n(\vec{K}) \qquad (13)$$

The graphical representation of the optical potential is shown in the following diagram.

$$\mathcal{U} = \text{---}\bigcirc\text{---} = F^+ G_0^* F \qquad (14)$$

Iterating the optical potential in a Klein-Gordon equation generates the π-nucleus T-matrix

$$T = \text{---}\bigcirc\text{---} + \text{---}\bigcirc\text{---}\bigcirc\text{---} + \ldots$$
$$= \text{---}\bigotimes\text{---} = F^+ G^* F \qquad (15)$$

which in (15) has been formally summed by introducing the dressed Δ-h propagator.

$$G^* = \Rightarrow + \succ\text{---}\prec + \succ\text{---}\bigcirc\text{---}\prec + \ldots$$
$$= \boxed{} = \frac{1}{D(E)-H_\Delta^A - \mathcal{U}} \qquad (16)$$

$\mathcal{X}$ is the Δ-h interaction generated by the pion propagation and is given by the Δ-h vertex function F and the free pion propagator g_0.

$$\mathcal{X} = \quad\text{[diagram]}\quad = F g_0 F^+ \qquad (17)$$

The main difference to the standard one pion exchange potential is that the intermediate pion may propagate on its mass shell

$$k^2 = p^2 = E^2 - m_\pi^2$$

This leads to an imaginary part in the matrix elements of $\mathcal{X}$ which is directly related to the Δ-h vertex function

$$-\text{Im}\,\mathcal{X} \propto F(p) F^+(p) \qquad (18)$$

Denoting the eigenstates of the full Δ-h Hamiltonian

$$H_\Delta^A + \mathcal{X} \qquad (19)$$

with α and its eigenenergies with ε_α the full Δ-h Greensfunction reads

$$G^* = \sum_\alpha \frac{|\alpha\rangle\langle\alpha|}{D(E) - \varepsilon_\alpha} \qquad (20a)$$

and therefore the partial wave T-matrix

$$T_\ell(E) = \sum_\alpha \frac{F_\alpha^* F_\alpha}{D(E) - \varepsilon_\alpha} \propto \sum_\alpha \frac{\Gamma_{\ell\ell}^\alpha/2}{E - R^\alpha + i(\Gamma_{\ell\ell}^\alpha + \Gamma_{in}^\alpha)/2} \qquad (20b)$$

I must discuss briefly some necessary modifications of the formalism presented so far. First, the scattering of pions by nucleons via the background amplitudes is described by a standard static optical potential. Furthermore Coulomb-scattering has to be taken into account if elastic scattering at small angles is to be described correctly. Both Coulomb and background - scattering add an additional term δT_ℓ to the partial wave amplitudes (20b) and modify the one pion exchange interaction $\mathcal{X}$ (17) which now has to be calculated using a distorted pion propagator instead of g_0.

An important modification is due to the effect of the exclusion principle on the intermediate nuclear states, as is formally described by the antisymmetrization operator $\mathcal{A}$ in eq.(1). To illustrate this effect, the following double scattering term is considered.

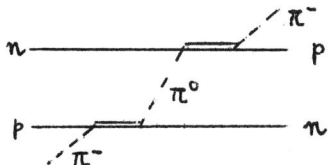

This term is contained in the iteration of the first order optical potential although it leads to a Pauli-forbidden state. Similar remarks apply to the corresponding spin flip second order terms.

We have eliminated these terms by replacing $\mathcal{Q}$ in eq. (1) by

$$\mathcal{Q} \rightarrow 1 - P_o \qquad (21a)$$

where

$$P_o = |0\rangle\langle 0| + \sum_{d.o.} |n\rangle\langle n| \qquad (21b)$$

The projection onto the ground state ($|0\rangle\langle 0|$) eliminates the double counting of those scattering events in which the nucleus is left in its ground state. The projection onto the doubly occupied s-orbitals eliminates the Pauli-forbidden states and is equivalent to taking into account the second order optical potential contribution arising from the Pauli-correlations. It can be shown that this modifies the one pion exchange interaction

$$\mathcal{H} \rightarrow \mathcal{H} - \delta\mathcal{H} \qquad (22a)$$

where $\delta\mathcal{H}$ is the "Fock-term" given by the following diagram

$$\delta\mathcal{H} = \qquad (22b)$$

Before I discuss the results of the calculations I shortly emphasize the main steps in the derivation of the formalism. As eq.(20) explicitly shows, pion scattering is described by the excitation of Δ-nuclear states (Δ-h states in the first order optical potential approximation). Thus, our final description is very similar to the isobar-doorway description of ref.1. Our derivation within the multiple scattering formalism may be considered as a microscopic justification of the isobar-doorway hypothesis.

Within the non-static multiple scattering formalism, these Δ-nuclear states describe the dynamics of the CM degree of freedom of the interacting π-N system. The Hamiltonian associated with this degree of freedom, the Δ-Hamiltonian, is determined by recoil and binding corrections to the elementary amplitude, cf. eq.(10). Pion multiple scattering adds to this Hamiltonian the retarded one pion exchange potential, eq.(19), which, to second order in the optical potential expansion, gets modified by the Fock-term, eq.(22a).

In the numerical calculations we have used Gaussian wave functions for the nucleons. The Δ-h Hamiltonian (20) has been diagonalized in an harmonic oscillator basis. The Δ-nucleus potential V, cf. eq.(10), has been chosen similar to the nuclear shell-model potential.

$$V = V_0 e^{-\mu r^2} \qquad (23a)$$

with

$$V_0 = -75 \text{ MeV}, \quad \mu = 0.25 \text{ fm}^{-2} \qquad (23b)$$

The main aspect in the following discussion of the results is the relation between the properties of the Δ-h states and pion scattering as is exhibited by eq.(20a). As a general result, I note that in all non-peripheral partial waves (pR ≥ ℓ) one Δ-h state dominates. As an example, the elastic widths corresponding to the 0⁻ Δ-h states at T_π=260 MeV are shown in Fig. 1 and compared to the elastic widths of the unperturbed (without pion multiple scattering) Δ-h states.

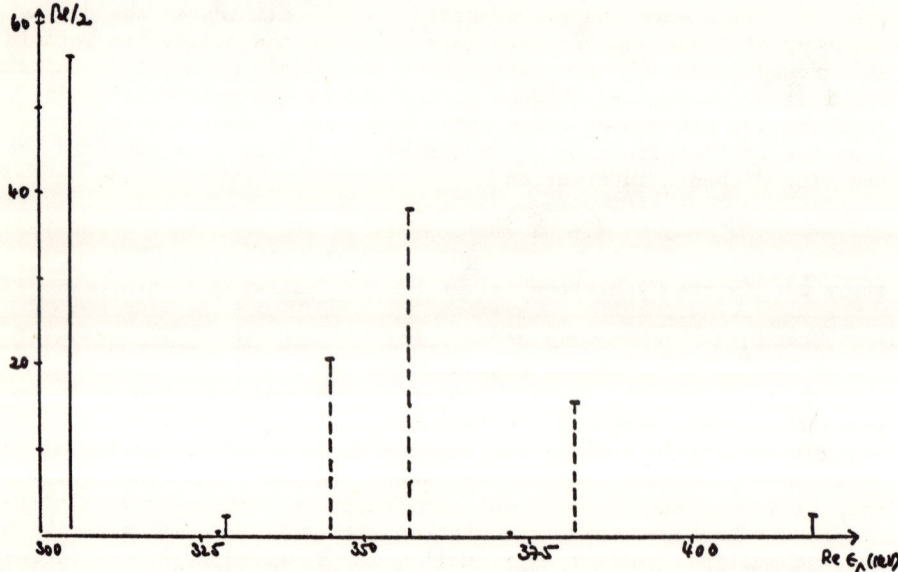

Fig. 1. Elastic half-widths of diagonalized (solid lines) and unperturbed Δ-h states. The abscissa denotes the real parts of the corresponding eigenenergies.

As the Figure shows, the effect of pion multiple scattering is to concentrate the transition strength to the π-nucleus ground state almost completely in one (diagonalized) Δ-h state. The reason for this "collectivity" can be understood in analogy to the collectivity in low energy nuclear physics. There it is the similarity between the residual interaction and, for example, the dipole operator which leads to the concentration of the dipole strength into one state.

In π-nucleus scattering it is the relation (18) between the imaginary part of the (residual) one pion exchange interaction and the pion emission and absorption operators F^+, F (13) by which one diagonalized Δ-h state is singled out. The coupling to the π-nucleus state is exhausted by this "giant" Δ-h state. +

Thus, the partial wave amplitudes (20b) are directly connected with properties of the corresponding collective Δ-h states. By studying different types of π-nuclear reactions (elastic, inelastic scattering, absorption and production) it should be possible to get detailed information about the structure of these dominant Δ-nuclear states and therefore about the dynamics of Δ-nuclear systems.

The type of information one can obtain from elastic scattering is specified by the following discussion. In Figure 2 are shown the experimental total and integrated elastic cross sections and the results of our calculation (solid lines).

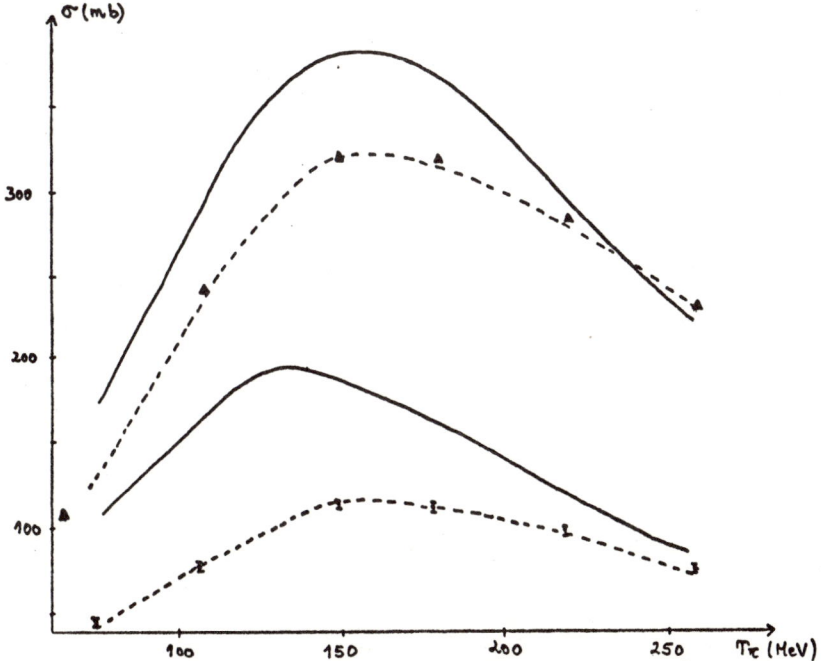

Fig. 2. Experimental [2] total (▲) and integrated elastic (✦) cross sections. The solid curves are calculated with $W_s = 0$ (25), the dashed curves are obtained from a fit of W_o.

+ It is problematic to speak about collectivity in the π-^{4}He system, where only one hole state is present. However, as preliminary calculations of π-^{16}O indicate, a similar phenomenon is observed in this more complex system. (See the talk of E.J.Moniz in this conference.)

The major failure of the calculation is the prediction of the integrated elastic cross sections which, at and below the resonance, are too large by about a factor of 2. Furthermore the ratio between elastic and inelastic cross sections is not even qualitatively reproduced. At the lowest energy for instance, a ratio

$$(\sigma_{ee}/\sigma_{in})_{th} \approx 1.6 \tag{24a}$$

is obtained in disagreement with the experimental value of

$$(\sigma_{ee}/\sigma_{in})_{exp} \approx 0.6 \tag{24b}$$

We have to conclude that, in the Δ-3N system, there are strong couplings to channels not contained in our calculation.

To account phenomenologically for these additional mechanisms we introduce into the Δ-nuclear Hamiltonian (20) an additional potential, the "spreading potential"

$$W_S = W_c \, e^{-\mu r^2} \tag{25}$$

and fit the (complex) strength W_o to total and integrated elastic cross sections at each energy. (The dashed lines in Figure 2 show the fit to σ_{tot} and σ_{el}.) The resulting energy dependence of W_o is shown in Figure 3.

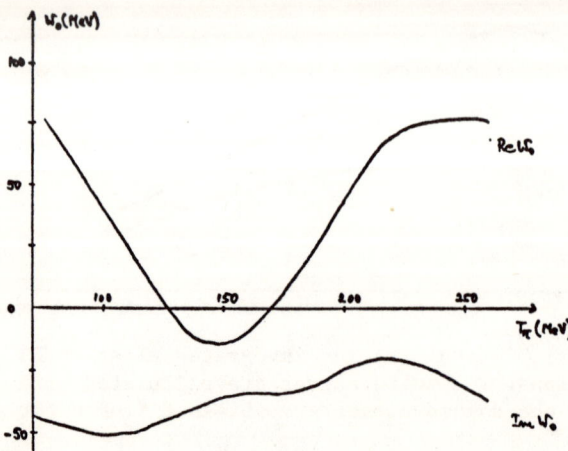

Fig. 3. Energy dependence of the strength W_o of the spreading potential W_S (25).

The real part of the spreading potential leads primarily to a shift in the energy dependence of the cross sections. By varying the imaginary part of W_0 it is possible to reproduce the experimentally observed values for σ_{el}/σ_{in}.

Having determined W_0 by this fit to the integrated cross sections, we have calculated the angular distributions. They are shown in Figure 4 in comparison with the experimental data [2,3].

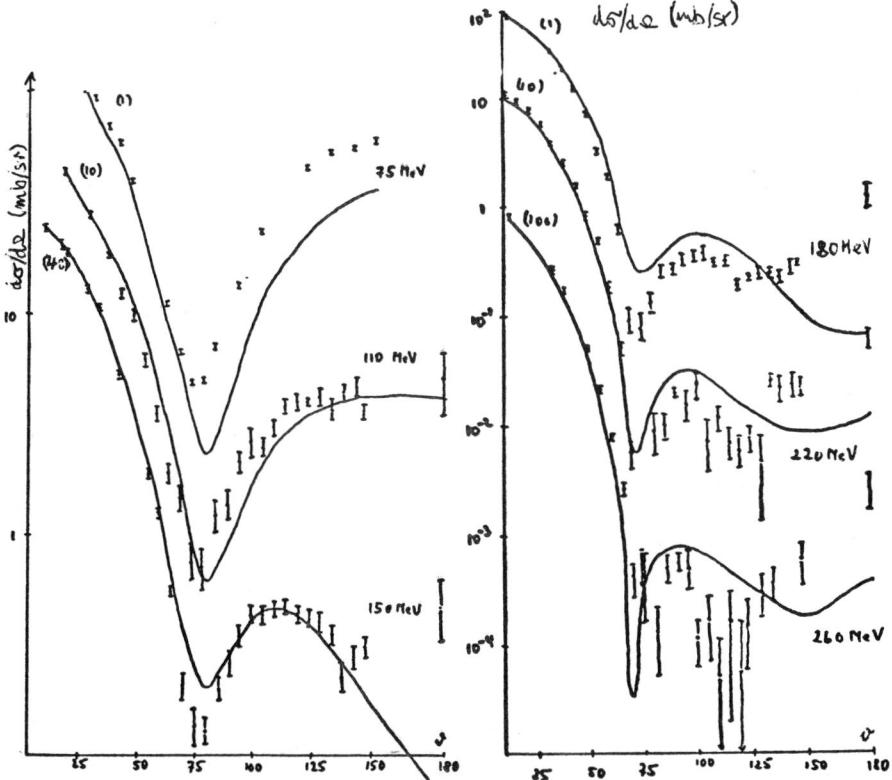

Fig. 4. Experimental and calculated differential cross sections. The cross sections have been divided by the numbers given in parentheses.

For not too high momentum transfers, quantitative agreement between theory and experiment is obtained. The position of the first minimum is correctly reproduced at all but the lowest energy. It can be shown that the failure in describing correctly the 75 MeV angular distribution is related to the behaviour of the π-^{4}He s-wave. At the lowest energy, this partial wave is practically not affected by the π-N 3-3 amplitude. We have indication that the discrepancy is due to neglecting π-^{4}He s-wave absorption. The discrepancies at high momentum transfers, especially in the region of the second (diffractive) minimum are partly related to the use of Gaussian nucleon wave functions. From this discussion of the angular distributions we conclude that, in the region of low momentum transfers, the coupling of the Δ-h states to the corresponding π-nuclear states is described correctly.

I can therefore focus the following qualitative discussion on the interpretation of the Δ-h eigenenergies. This will give us some indication of the physical origin of the spreading potential and furthermore allow us to relate our results with those of other calculations.

In the region below the resonance ($T_\pi \leq 150$ MeV) the scattering is dominated by the 1^+ partial wave which now will be discussed in detail. The possible Δ-h configurations are of the type (n is the radial quantum number)

$$nS_{3/2}\ s_{1/2}^{-1}$$
$$nD_{1/2}\ \text{"}$$
$$nD_{3/2}\ \text{"}$$

At the lowest energies the (on-shell) pion momentum is not large enough to cause an appreciable admixture of higher Δ-configurations, nS, nD, to the "ground state" configuration $0S_{3/2}\ s_{1/2}^{-1}$ in which most of the transition strength to the $|\pi_1 + 0\rangle$ state is concentrated. Therefore, we can replace the operators in eq.(16) for the Δ-h propagator (modified by Fock term and spreading potential), by the corresponding expectation values of the $0S_{3/2}\ s_{1/2}^{-1}$ configuration. The relation between the partial wave T-matrix and these expectation values reads

$$T_{1^+}(E) \propto \frac{\Gamma_{el}/2}{E - R^* + i(\Gamma_{el}/2 + \Gamma_{in}/2)} \quad (26a)$$

$$\Gamma_{el}/2 = -\text{Im}\langle \mathcal{H}\rangle = \left|F_{0S_{3/2}, S_{1/2}^{-1}}\right|^2 \quad (26b)$$

$$\Gamma_{in}/2 = \Gamma_f/2 + \text{Im}\langle \delta\mathcal{H}\rangle + \Gamma_{sp}/2 \quad (26c)$$

In a first order calculation, $\Gamma_{in}/2$ is simply given by the free half-width $\Gamma_f/2$ since every decay of the Δ is assumed to cause a brekup of the nucleus. In addition to the free decay, the Δ-h state may decay via an excitation of another Δ-h state. Consequently the π-nucleus partial wave amplitude is elastically broadened as compared to the π-nucleon amplitude.

The inelastic half width is reduced by $\text{Im}\langle\delta\mathcal{H}\rangle$ arising from the antisymmetrization of the intermediate nuclear states. For instance, the decay of the Δ via charge exchange or spin flip allowed in free π-nucleon scattering is forbidden if the nucleon is left in an $0s_{1/2}$ state. Therefore

$$\Gamma_{QF}/2 = \Gamma_f/2 + \text{Im} \langle \delta \mathcal{N} \rangle \qquad (26d)$$

has to be interpreted as the inelastic half-width which arises from the true nuclear breakups. The "spreading width"

$$\Gamma_{sp}/2 = - \text{Im} \langle W_s \rangle \qquad (26e)$$

accounts phenomenologically for the coupling of the Δ-h states to open channels different from the π-p-h (particle-hole) channels. The different contributions to the total width are shown in Fig.5.

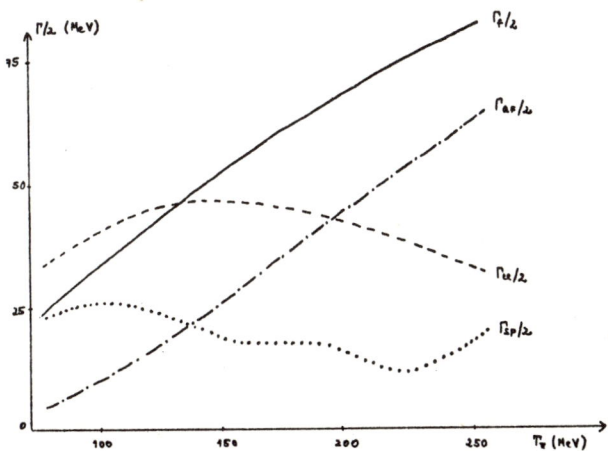

Fig. 5. Free half-width of the 3-3 πN amplitude and the different contributions, eq.(26), to the total width of the 1^+ π^4He partial wave amplitude calculated perturbatively.

Below 150 MeV the free width is quenched by 50% or more. Consequently, in the low energy region, the major contribution to the inelastic width is the spreading width. Thus the inelastic cross section is dominated by processes different from the nuclear breakup in π-nucleus quasifree scattering, in sharp contrast to the implicit assumption of a first order optical potential description.

It is now easily seen why it may not be necessary to account for these processes explicitly in the decription of elastic scattering if the influence of the Pauli-principle is not treated correctly. In the first order calculation of ref. 4) very good agreement with the experimental data has been obtained. Neglecting completely the influence of the Pauli-principle ($\Gamma_{QF}/2=\Gamma_f/2$, (26d)) increases the inelastic width by typically 20-25 MeV which is just of the order of magnitude of the spreading width (Figure 5). The calculation of ref.5) corrects for the double counting of the ground state transition (the term $|0\rangle\langle 0|$ in (21b)) by multiplying the optical potential with (A-1)/A. this accounts for about 40% of the quenching effect. The use of a Fermi-averaged amplitude increases further the free half-width. At 150 MeV, for instance, this increase is of the order of 10 MeV. Thus, due to the approximations, the result is again an overall increase of the inelasticity ($\Gamma_{QF}/2$ in (26d)) by about 20 MeV. The success of these first order calculations can therefore be explained as the result of the cancellation of quenching effects on the one hand and damping effects as described by the spreading potential on the other hand. The different models, although equally successful in describing elastic scattering, differ qualitatively in their predictions for the reactive cross section σ_{in}. While the first order descriptions imply that the reactive cross section is given by the quasi-free π-nucleus scattering cross section, our description predicts a major contribution to σ_{in} from more complicated processes (see below).

In a similar way, the different calculations can be related as far as the real part R^* (26a) of the Δ-h eigenenergies is concerned. In refs. 4,6) Fermi-motion and the binding energy of the nucleons have been taken into account. This amounts to assuming that the Δ moves freely inside the nucleus, i.e. the Δ-nucleus potential in the unperturbed Δ-h Hamiltonian (10) is implicitly assumed to be negligible. In our description, a repulsion in the spreading potential is needed,at all but one energy (Figure3), to compensate for the attractive Δ-nucleus potential (23).

We thus conclude that, in agreement with our calculation, the calculations mentionned above also indicate the presence of a strongly absorptive and repulsive coupling to other channels. In the first order calculations, a spreading potential does not have to be introduced explicitly to account for those couplings. Its effect is simulated by the approximations involved.

Among the possible processes which may be described by the imaginary part of the spreading potential, I mention the collision damping and the influence of absorption. At the lowest energies, the influence of the collision damping, i.e. the excitation of the nucleus via Δ-N collisions should be less important for phase-space reasons. Since no decrease in ImW_0 is observed at the lower energies (Figure3) the dominant process may be the true pion absorption, i.e. the process

$$\Delta N \rightarrow N \cdot N$$

Assuming an energy independent contribution from this process to Im W_0 of -45 MeV, the value of Im W_0 at T =75 MeV, we have calculated the absorption cross section. The results of this calculation are shown in Figure 6 together with experimental data obtained from cloud chamber measurements [7]. The experimental information is un-

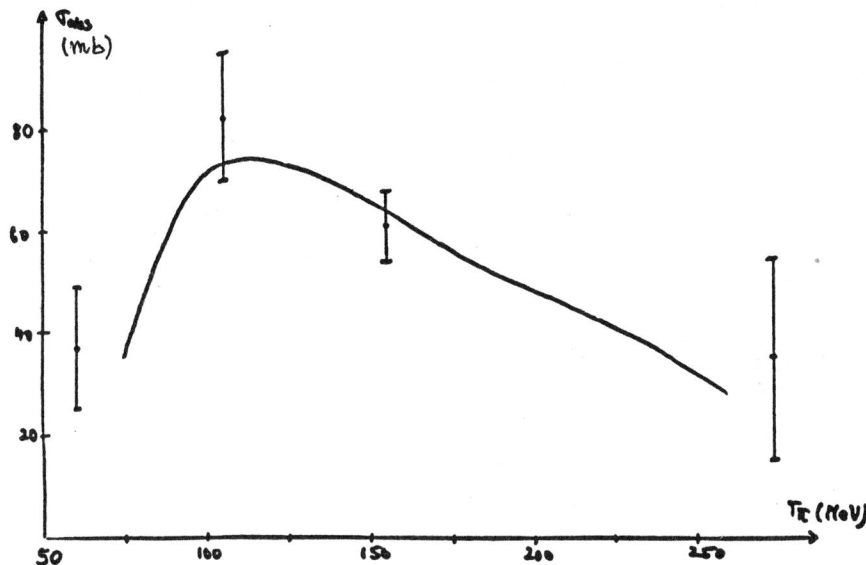

Fig. 6. Experimental [7] and calculated absorption cross sections as a function of the pion kinetic energy.

fortunately not unambiguous. At 153 MeV and 270 MeV pion kinetic energy, we have subtracted 20 mb to account roughly for the expected charge exchange cross section. At the lower energies, the charge exchange cross section can be expected to be small. Furthermore, we have extrapolated our results to the pion threshold and have obtained the correct order of magnitude for the width of the 2p state in pionic ^{4}He [8]. The interpretation of the observed damping of the Δ-h states as arising mainly from pion absorption is thus compatible with the existing, although poor, experimental data. It is tempting to attribute the observed repulsion of W_s to the dispersive effects of the pion absorption. This may be the case at the lower energies. However, the strong energy dependence in Re W_0 around the resonance (Figure 3) can most likely not be explained by this mechanism.

In concluding, I emphasize the following results of our analysis of π-^{4}He scattering. The scattering proceeds through the excitation of giant Δ-h states. The partial wave amplitudes are related directly to the properties of these states. Elastic scattering provides information about the coupling of these Δ-h states to the π-nucleus ground state and information about the eigenenergies of the Δ-h states. In the discussion of the eigenenergies we have shown that these giant Δ-h states couple strongly to channels different from the π-nucleus breakup channel. An important damping mechanism is most likely the true pion absorption. To resolve this problem of the physical origin of these damping mechanisms more experimental data specifying the major contributions to the reactive cross section are needed. More detailed information concerning the structure of the Δ-h states can be expected from a study of π-nuclear reactions like quasi-elastic scattering. Similarly, the theoretical study has to focus on the understanding of the basic couplings of the Δ-h states.

References.

1) L.S.Kisslinger and W.L.Wang, Phys.Rev.Lett. 30,1071, (1973), and preprint.

2) F.Binon et al. , Phys.Rev.Lett. 35,145,(1975).

3) K.M.Crowe et al. , Phys.Rev. 180,1349,(1969).

4) L.S.Celenza et al. , preprint.

5) R.H.Landau, Phys.Lett. 57B,13,(1975).

6) R.H.Landau and A.W.Thomas, preprint.

7) E.C.Fowler et al. , Phys.Rev. 91,135,(1953).
   M.S.Kozodaev et al. , JETP 11,300,(1960).
   Yu.A.Budagov et al. , JETP 15,824,(1962).

8) G.Backenstoss et al. , Nucl.Phys. A232,519,(1974).

Carl Shakin (Brooklyn College): We have performed a calculation of the Δ interaction in nuclear matter and find a strong damping width for the Δ which is also strongly density dependent. Since the π-nucleus interaction is largely peripheral, however, one expects that a calculation with the free π-N interaction is not unreasonable. Indeed we find we can fit the data from 110 MeV to 1 GeV with what is essentially the free π-N interaction using a covariant multiple scattering theory.

Lenz: In our calculations the density dependence of both, the damping of the Δ and the Pauli-corrections, is fully taken into account. At no place, have we used any typed of local density approximation which, obviously, would be hard to justify for a small system as ^{4}He. We find indeed a very strong variation of Pauli and dampling effects over the different partial waves, which reflects the strong density dependence. As I mentioned in the talk, one can understand qualitatively why one using a free pion nucleon amplitude may get a reasonable fit to the data. The success of such a description, however, does not imply that the corrections mentioned above are small. Rather, the small overall effect is the result of the cancellation of large Pauli corrections against large damping effects of the Δ.

T.-S. H. Lee (Argonne National Lab): It is a nice idea to introduce the Δ to describe the π-nucleus interaction. On the other hand, it is not trivial to develop many-body scattering theory, since it is a dynamical problem. Therefore, I would like to ask how do you determine the πNΔ vertex and the vertex in your 1-π exchange NΔ potential. On what physical grounds do you believe it?

Lenz: The Δ has not been introduced explicitly into the description of π-nucleus scattering. Rather, the starting point is the Watson multiple scattering formalism which actually is a many-body scattering theory. The many-body aspect gets lost only in the static approximation. In the non-static version of the multiple scattering formalism, the degree of freedom associated with the CM motion of the interacting πN system appears explicitly. Due to the presence of the pole in the 3-3 amplitude, the equation of motion of the πN-CM can be interpreted as that of the Δ. Consequently the πN→Δ vertex and the vertex in the one pion exchange potential are, both, determined by the form-factors of the 3-3 πN amplitude. Our final results depend only weakly on the assumed off-shell form of the amplitude when we include hard-core correlations.

L. Kisslinger (CMU): I believe that the strong differences between your results and the Isobar-Model results (discussed Monday) is the doorway space used. At low energy, on the wing of the resonance, the Δ-hole states might be reasonable and the collective state a reasonable approximation. At higher energies I believe that much more complicated states enter and that the Δ begin to be localized in the nucleus. A single state cannot give this description.

F. Lenz: As I mentioned in the talk, collective states appear only in the non-peripheral partial waves. In the peripheral partial waves pion rescattering or the Δ-h exchange interaction is less effective and therefore the coupling strength to the corresponding π-nuclear state is spread over several Δ-h states. With increasing pion energy the cross section is more and more dominated by the peripheral partial waves and therefore less influenced by the collective Δ-h states. This, however, does not imply that the "peripheral" Δ-nuclear states are of more complicated structure. Rather our calculations indicate that the collective states in the non-peripheral partial waves couple strongly to states of more complicated structure.

THE FEW-BODY PROBLEM AND PION-NUCLEAR PHYSICS[*]

B. F. Gibson[†]
Theoretical Division, Los Alamos Scientific Laboratory
Los Alamos, New Mexico 87545

ABSTRACT

Some of the pion-nuclear physics questions that pertain to pion-few-nucleon systems are explored. Those aspects of the problem which have relevance to the general study of pion-nuclear physics are emphasized, in particular the properties of the π-N interaction within a nucleus. Specific examples are restricted to elastic scattering from ^{2}H, ^{3}He, and ^{4}He and selected single and double charge-exchange reactions.

INTRODUCTION

The pion-few-nucleon problem is particularly attractive to those of us whose intuition is limited to systems in which the number of particles can be counted on the fingers of one hand. For these people, it was imperative that nature provide a distinct break between ^{4}He and ^{6}Li in the periodic table; their physics world is the nuclear 1s shell. For others, the cadre of real nuclei begins at the 1p shell and extends through the islands of stability beyond ^{208}Pb; these people are supposedly absent from today's session.

In addition to the simplicity and lack of certain many-body approximations, which for some make the few-body problem the only worthwhile endeavor, there are other legitimate and compelling reasons for our interest in few-body physics. This is particularly true of meson probes, and the π-d system may be one of the most interesting examples. Radiative π-d capture at rest[1] appears to be one of the most promising reactions for examining the low energy n-n interaction and hence speaking to the question of charge asymmetry in the N-N force. The reaction rate in this coincidence experiment is sufficient to provide the required statistics for a meaningful analysis, in contrast to the weak μ-capture reaction;[2] the uncertainty in the relatively feeble π-N interaction in the initial state has a very small effect upon the shape of the final-state spectrum. At higher energies, elastic π-d scattering appears to offer some hope of pinning down the elusive D-state probability in the deuteron.[3] As you are no doubt aware, the magnetic moment of the deuteron is no longer reconciled with a 7% D-state, since the ρ-π-γ coupling is now about 1/10 that used by Adler.[4] One can see in Fig. 1 that various experimental numbers can be interpreted as implying different answers to the question of what is the correct D-state probability in the deuteron. The estimates noted here tend to cluster in two regions: 4-5% and 7-9%. Each of these estimates can be called into question. Thus, Bertozzi was led to propose that one look at the polarization in elastic e-d scattering. Unfortunately, magnetic scattering and associated exchange current effects, which are not small,

---

[*]Work performed under the auspices of the U. S. ERDA.
[†]In collaboration with W. R. Gibbs.

1) Theoretical estimates ⇒ 3% < P_D < 10%.

2) μ_d = 0.8574 compared with $(\mu_p + \mu_n)$ w/o exch. cur. ⇒ $P_D \simeq$ 4%.

3) Local pot. fit to n-p scat. data (e.g. RSC) ⇒ PD ≃ 7%.

4) Separable pot. ^{3}H B.E. and a_2 calc. ⇒ $P_D \simeq$ 4-5%.

5) Certain p-d elastic scattering analysis ⇒ $P_D \simeq$ 7-9%.

6) Similar π-d elastic scat. analysis (filling min.) ⇒ $P_D \simeq$ 9%.

7) The π^+d → pp reaction analysis ⇒ $P_D \simeq$ 4-5%.

Fig. 1. A comparison of various estimates of the % D-state of the deuteron.

make the analysis difficult.[3,5] One can avoid the magnetic scattering problem by studying instead the polarization in elastic π-d scattering. Such an experiment is now underway at SIN;[5] the polarization of the recoiling deuteron is measured at θ_π = 180°, where only T_{00} and T_{20} are nonzero. The clear disadvantage of using a pion probe is that it interacts strongly with the target nucleons; therefore, multiple-scattering effects must be incorporated into the analysis. However, since the π-N angular distribution is peaked toward backward angles, single scattering will dominate even near 180° --- multiple-scattering corrections are not prohibitive. In Fig. 2 one can see that T_{20}, as a function of the incident pion kinetic energy, is sensitive to P_D. Hence, π-d scattering may indeed help to provide an answer to this interesting question, although quality Faddeev calculations may be required in the data analysis. In addition, we have the question of the N-N off-shell interaction --- a property that cannot be separated from 3-body forces in the interaction of 3 nucleons, be it in the triton binding energy, n-d scattering, or what have you. A probe differing from the two nucleons must be used. A pion (or possibly a K^+) appears to be the likely candidate.

Before one can proceed very far with pion probes, one must understand the basic π-N interaction. The pion-few-nucleon system appears to be well suited for studying such questions as the sensitivity to the π-N phase shifts, the role of effective amplitudes implied by frame transformations (or recoil factors or binding energy corrections), and the effect of varying the off-shell properties of the π-N interaction. A basic understanding of these points must be attained if meaningful analysis of pion interactions with nuclei in general is to be achieved. There are, of course, other questions which should be understood in their own right but which also have application to heavier nuclei: short-range N-N correlation effects and pion absorption are just two examples. In addition, there is the possibility of generating exotic charge states such as the trineutron and tetraneutron, exciting different giant resonances, and studying such large momentum transfer reactions mechanisms as the (p,π) process --- all of which make the pion-few-nucleon system an attractive laboratory.

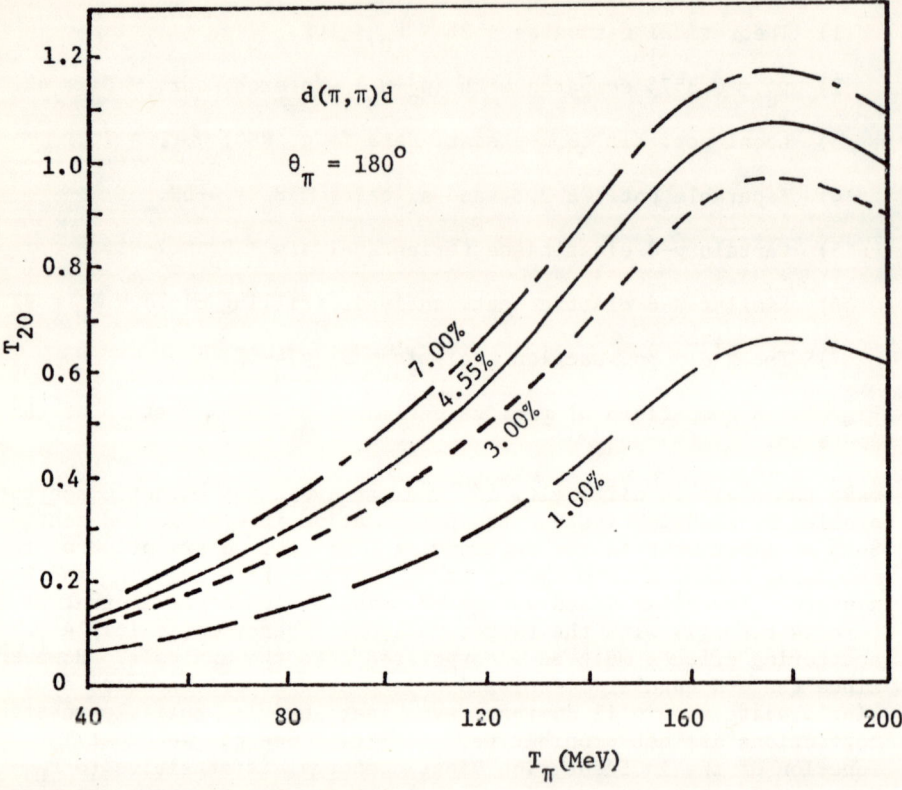

Fig. 2. Recoil deuteron polarization T_{20} as a function of pion kinetic energy at $\theta_\pi = 180°$ for various assumptions concerning the % D-state of the deuteron.

But, let us not, as is often the case with talks of this nature, peer too far into the future. There will be other conferences --- conferences built on greater experience, which will stand much closer to that future and be in a better position to forecast it. Let us instead look briefly at where we stand now and at what we might expect to be of interest or to accomplish over the next couple of years.

THEORETICAL ASPECTS

I will speak primarily about Hamiltonian approaches to pion-few-nucleon problems. I apologize for omitting the dispersion relation, N/D, and other such methods; but the Faddeev, optical model, and multiple scattering procedures can be used as a basis for understanding the pion-nucleus problem in a manner that can be more directly extended to complex systems. I shall view the pion-few-nucleon problem as playing a special role in pion-nuclear physics: one of testing our methods and sharpening our tools for more general pion-nucleus studies. I should also note that because the π-N interaction lends itself to a separable t-matrix approximation, often through a separable potential, I shall use that approximation almost exclusively.

In pion scattering, we approach relativistic velocities almost from the beginning. For T_π = 35 MeV, one finds already p/E = 0.6. Thus, one is forced to consider in most any calculation of interest (in addition to relativistic kinematics) the question of frames. This is a question which does not really arise in the Galilean-invariant, nonrelativistic problem. Because most pion-nucleus calculations are based on an underlying potential assumption, it can be reassuring to note that, since the dependence of the wave function upon the internal and center-of-mass variables factorizes[7] as

$$\langle \vec{Q}', \vec{q}' | \psi^{(+)}_{\vec{Q},\vec{q}} \rangle = (2\pi)^3 \delta^{(3)}(\vec{Q}' - \vec{Q}) \phi^{(+)}_{\vec{q}}(\vec{q}') ,$$

one can relate the half-off-shell t-matrix with arbitrary total momentum to that with zero total momentum:

$$T(E(Q,q), Q; \vec{q}', \vec{q}) = \frac{\omega(q') + \omega(q)}{E(Q,q') + E(Q,q)} \, t(\omega(q); \vec{q}', \vec{q}) .$$

where $E^2(Q,q) = \omega^2(q) + Q^2$. (The fully-off-shell relation is more complicated, but it can, under certain conditions, be approximated to a similar form.) In the π-d system, one is transforming between the π-N c.m. frame and the π-d c.m. frame. For relatively low energies (say, up to T_π = 200 MeV) one can show that p/E for the transformation is small. Thus, the transformation coefficient of the t-matrix is not so different from unity. An indication that this might be the case numerically can be inferred from the contribution to this conference by Rinat and Thomas,[8] although it is clear that it is to ones advantage to include the relativistically invariant dot products, etc. in the vertices as perscribed in Ref. 9 or 10.

Yet, before one is able to make serious use of this or any other theory in pion-nucleus physics, we must return to the grubby task of obtaining reliable and precise π-N phase shifts. Others before have emphasized the need. But I wish to make as specific an appeal as possible for 1-2% π-N data and the corresponding phase shifts. The data is needed from 30 MeV on up, especially emphasizing the small phase shifts, if one is to make meaningful low-energy pion-nucleus calculations. It is a tough assignment. But it must be undertaken.

I show in Fig. 3 two sets of phase shifts: one from the separable potentials used by Thomas in his recent π-d paper,[11] and one from a very recent analysis of the available π-N data below 100 MeV by Dodder.[12] The P_{33} and S_{31} phase shifts are not there, because most of the $\pi^+ p$ analyses agree reasonably upon these. Note the differences in the small phase shifts shown; I shall return briefly to the peculiar P_{11} phase shift again later. In the following sections I shall compare theoretical cross sections for various experiments using these and similar phase shifts. From those comparisons, I hope that it will be abundantly clear that, without precision π-N phase shifts, we are stuck for a quantitative understanding of the low-energy pion-nucleus problem.

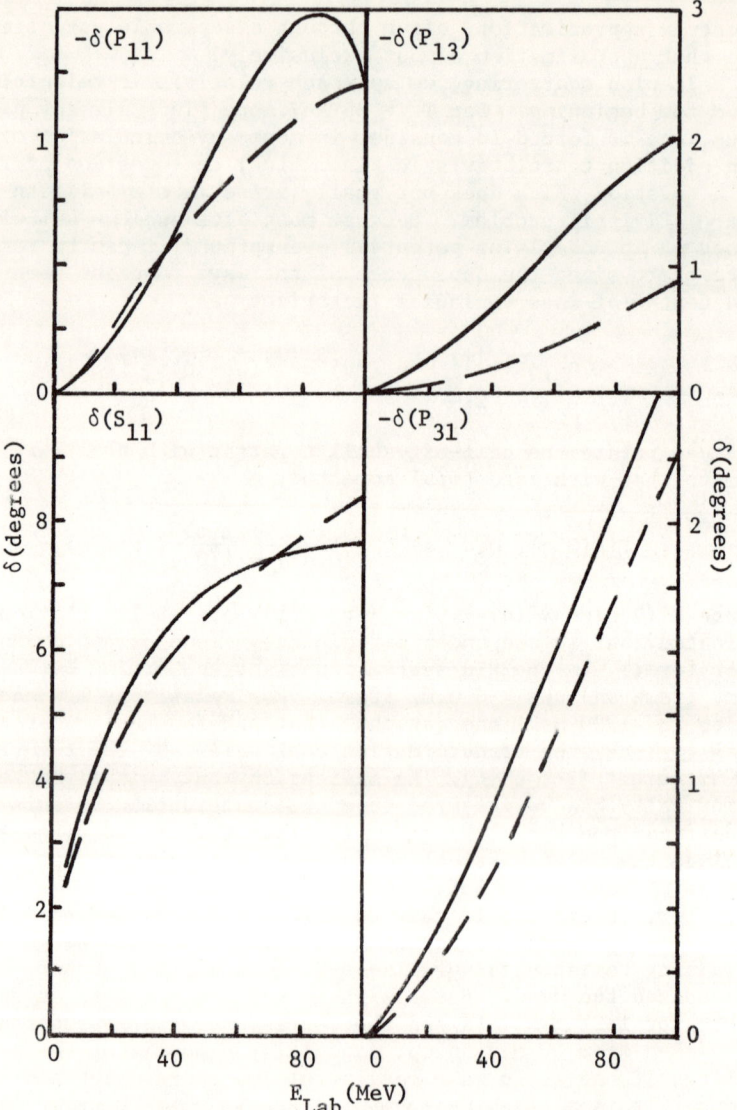

Fig. 3. π-N phase shifts from the analysis of Dodder (solid line) and the potentials of Thomas (dashed line).

This is, of course, not so true in the region of the (3,3) resonance, where that one amplitude dominates and there is little significant cancellation among various other amplitudes. But this energy region can be a relatively uninteresting one to examine for nuclear structure effects, because the nucleus becomes increasingly black as one approaches resonance.

As an aside, I would remark that the extrapolation of the measured phase shifts to zero energy in order to obtain the π-N scattering lengths can be tricky, because the radiative channels become non-negligible near threshold.

Before moving to specific experiments, it is necessary that I say a few words about the fixed-scatterer approximation.[13] Most people at this conference are familiar with the first-order optical model from low-energy nuclear physics. You are aware that this can be easily derived as an approximation to the Watson multiple-scattering series through use of the coherent approximation; i.e., under the assumption that the projectile does not excite the target nucleus from its ground state and therefore has an energy and corresponding velocity small compared with the Fermi properties of the target constituents. The opposite assumption, that the projectile has a high velocity and moves rapidly through the target before the constituent nucleons can move far, is the basis for the fixed-scatterer approximation. For nonlocal potentials, this is not the same as the closure approximation, although excitation of the nucleus is allowed in each of these cases (which one expects should happen when the energy of the projectile is large compared to the low-lying spectrum of the target). The target nucleons in the fixed scatterer approximation are fixed only in the sense that they move lethargicly compared to the pion projectile; the nucleons do absorb momentum and that effect is included approximately through an angle transformation discussed below. Consider a separable t-matrix of the form

$$t \sim \lambda_0(\omega) v_0(q) v_0(q') + \lambda_1(\omega) \vec{q} \cdot \vec{q}' \, v_1(q) v_1(q') \; ,$$

where $\omega^2 = k^2 + \mu^2$, having the form factors given by

$$v_\ell(q) = (k^2 + \alpha_\ell)^2 / (q^2 + \alpha_\ell^2) \xrightarrow[q \to k]{} 1 \; .$$

Here λ_0 and λ_1 contain the s- and p-wave phase shifts. Under the assumption of Galilean invariance, the $\vec{q} \cdot \vec{q}'$ is replaced by the relative momenta

$$\vec{q}_r \cdot \vec{q}'_r = (\vec{q} - \tfrac{\mu}{M} \vec{P}_i) \cdot (\vec{q}' - \tfrac{\mu}{M} \vec{P}_f)$$

$$\approx - q^2 \tfrac{\mu}{M} + (1 + \tfrac{\mu}{M}) \, \vec{q} \cdot \vec{q}' \; .$$

Thus, one obtains effective scattering amplitudes of the form

$$\bar{f}_0 \sim f_0 - q^2 \frac{\mu}{M} f_1$$

$$\bar{f}_1 \sim (1 + \frac{\mu}{M}) f_1$$

If, instead of the Galilean invariant perscription for adding velocities, we use the relativistic perscription, different effective amplitudes result including an induced d-wave. (One might expect that $\mu/M \to \omega/M$, but in actuality it is more like $\mu/M \to \mu^2/\omega M$.) Such an angle transformation also appears in the optical model work of people such as Mach, Landau, Tabakin, Miller, and others.[14] A brief history and more complete description of its application in the fixed-scatterer approximation can be found in Ref. 15 along with a discussion of binding energy shift and nucleon recoil effects.

The fixed-scatterer approximation is attractive for pion-nucleus problems because of the light mass of the projectile, because sophisticated wave functions containing correlations, deformation, etc. are easily incorporated into the calculation, and because the π-N interaction is strong in only a few partial waves. (It also has properties similar to an experiment in that, being a Monte Carlo calculation, small cross sections take a longer time to generate.) But, as the optical model is an approximation, so is the fixed-scatterer approach. Unfortunately, a true theoretical understanding of the error associated with either method is lacking. In the case of the optical-model approximation, one is seeking to understand the convergence of a series; i.e., terms beyond the first-order optical model where nuclear excitation, correlations, etc. contribute. In the fixed-scatterer approximation, the series (so to speak) has been summed exactly into an integral equation, and one is seeking to understand how well that integral equation corresponds to the real world. However, perhaps it is possible to sketch a picture indicating why the fixed-scatterer approximation may work for real nuclei, even near the (3,3) resonance when one might expect that the number of scatterings involved in the pion's transit of the nucleus is many and that the total transit time is so large that the fixed-scatterer approximation is poor. Consider the incomplete analogy of a tennis ball being volleyed against a wall by a fixed racquet strung at board strength. The closer the racquet to the wall, the more times the ball resonates between them before falling to the ground ---as a π^+ might resonate in the diproton system (see Ref. 10). However, if the racquet is strung loosely to simulate the weaker π^+-n interaction, the multiple scattering is severely inhibited, except at very short distances, which are supposedly prohibited by the N-N force in a nucleus. The same reduction in multiple scattering would occur in a four-cornered volley, as in π-^{4}He scattering, where two of the racquets were strung to represent the π^+-n interaction.

THE THREE-BODY SYSTEM

Let me now turn briefly to the π-d problem. Afnan and Thomas[16] showed some time ago, and Mizutani[17] has since confirmed, that absorption effects at zero energy can provide a sizeable correction to the amplitude, accounting for nearly all of the imaginary part. The increase in the cross section amounts to some 40%, although one must remember that the zero energy π-d cross section is abnormally small to begin with. Thomas was not so ambitious in his recently published 48 MeV π-d scattering calculations, where absorption due to the πNN ⇔ NN coupling was neglected.[11] Thomas argued that the approximation at that energy was reasonable; in view of the uncertainty in the π-N phase shifts, it is difficult to disagree. Certainly his treatment of low-energy π-d scattering is the most complete to date.

However, I would like to mention an alternate approach to the inclusion of absorption effects, one proposed by Koltun and Mizutani.[18] Theirs is intriguing because it has a rather straight forward extension to non-Faddeev calculations, in particular, to both optical-model and fixed-scatterer calculations. The Koltun-Mizutani prescription has more of a field-theoretic basis than that used by Afnan and Thomas; in practice it is similar in spirit to a DWBA calculation. An initial pseudo π-d problem is solved, just as in Afnan and Thomas but with a modified P_{11} potential --- a modification that removes (subtracts) the pion absorption and reemission contribution of the Born diagram. Recall that, as in the K^-p scattering problem where the πY coupled channel is open, the πNN ⇔ NN channel being open can make a

strongly attractive P_{11} interaction look as if it is repulsive, if one is careless in his interpretation of the small negative phase shift. The potential term described by the Born diagram is strongly repulsive; thus, the pseudo P_{11} potential is attractive. The resulting pseudo π-d scattering t-matrix is then corrected by the addition of an integral involving the operators and Green's function describing the pion absorption and reemission folded with the π-d wave functions for the pseudo problem. As I mentioned above, the numerical results[17] are very similar to those of Afnan and Thomas in the π-d scattering length calculation. However, the underlying theories do differ, and the Koltun-Mizutani procedure has a natural extension to optical-model and fixed-scatterer calculations for heavier systems. This is something that must be done, for it is known from experiment that such absorption processes can be important. It is even possible that the neglect of such absorption is responsible for making the 24 MeV π-^{4}He and 50 MeV π-C data difficult to fit, although absorption may result only in a renormalization or background effect and not alter the angular distribution significantly.

The πNN ⇔ NN reaction is interesting for another reason. Because of the large momentum transfer, the first order process (direct absorption on a single nucleon) is small in the πd ⇔ NN reaction, and the rescattering graphs dominate. (See Fig. 4.) Graph (a) provides only a small contribution to the cross section; graphs (b) and (c) permit the momentum transfer to be shared between the two vertices --- scattering and absorption. Thus, the reaction is sensitive to the off-shell properties of the π-N amplitude. In particular, at energies below 100 MeV, the reaction is effectively described as a second order process, and it can be used to determine the off-shell parameters α_ℓ of the previously described t-matrix. Hence, this reaction can be thought of as determining a form factor at the π-N scattering vertex, which accounts for

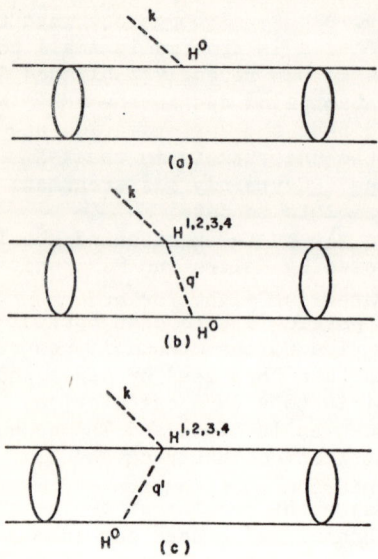

Fig. 4. First and second order graphs in the $\pi^+ d \to pp$ reaction.

the modification of the bare π-N interaction due to heavy meson exchange, pair diagrams, etc. The test of the parameterization comes in the use of that off-shell t-matrix in pion-nucleus scattering and reaction calculations where off-shell effects are important as they are in multiple scattering.

In Fig. 5, one can see selected results of the Goplen, Gibbs, and Lomon calculations: 1) for direct absorption by a single nucleon --- graph (a) of the previous figure; 2) for s-wave rescattering only --- graphs (b) and (c) with no p-wave π-N interaction --- an approximation sufficient only at the very lowest energies; 3) for the Kisslinger model which has zero range form factors (off-shell form factors identically one) for both s-wave and p-wave rescattering, a model which is clearly inadequate. In Fig. 6, one can see the quality of the fit to the data when the p-wave is included with

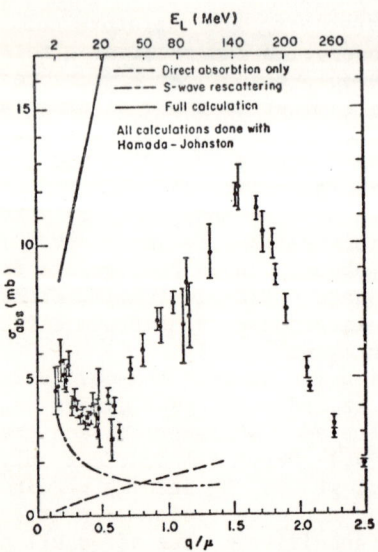

Fig. 5. Comparison of theoretical $\pi^+ d \to pp$ cross sections; the full calculation is for the Kisslinger model.

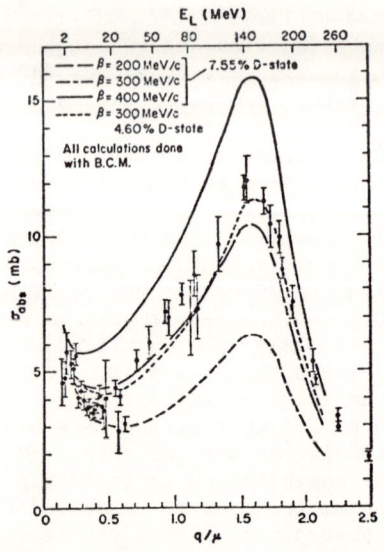

Fig. 6. Comparison of model $\pi^+ d \to pp$ results for various values of the p-wave off-shell parameter.

finite-range form factors to describe the off-shell π-N t-matrix. In
particular, the reaction can be seen to be sensitive to the p-wave
off-shell parameter, denoted as β in the figure. Although the fit is
not perfect, it is clear that, for this model of the π-N off-shell t-
matrix, a value of 300 MeV/c is a reasonable choice. Although the
reaction is not very sensitive to the s-wave parameter, a value of 500
MeV/c is indicated. In Goplen, Gibbs, and Lomon no form factor was
assumed for the absorption vertex. If one does assume a form factor
identical to that at the scattering vertex,[20] then the value of the
p-wave off-shell parameter can be increased to as much as 600 MeV/c.
This perhaps indicates the range of allowed parameter variation. However, this latter approximation of identical form factors is ad hoc,
and a value of 300 MeV/c appears to be in better agreement with the
π-d and π-^{4}He data, which we shall see in a moment. (As an aside,
recall from Fig. 1 that this particular reaction tends to favor a D-
state probability for the deuteron in the 4-5% range.)

Let us now turn to the need for the improved π-N phase shift data.
Thomas reemphasized this in his recent π-d potential calculation.[11]
But let us look at the situation when the off-shell dependence in the
calculation is fixed and only the phase shifts are varied.

At 47.5 MeV one does not expect to see much multiple scattering.
The nucleus is fairly transparent to pions in the 40-50 MeV region,
where the S_{31}, S_{11}, and P_{33} phase shifts are all about equal and only a
few degrees. Thus, π-d scattering is very sensitive to calcellations
among the larger amplitudes, and the smaller p-wave phase shifts reveal
their presence. Comparison between theory and data in this energy
range --- provided that the phase shifts are known --- will test most
of the basic assumptions of any model; e.g., the angle transformation
discussed previously in relation to the fixed-scatterer approximation.
Let us examine Fig. 7. Here we see the very different predictions, in
the fixed-scatterer approximation, for the phase shift parameterizations of Dodder[12] and Thomas.[11] A curve similar to the solid (Dodder)
curve results from the phase shifts of McKinley,[21] a much older parameterization whose qualitative features resemble in many ways those of
the Dodder analysis. In each case, the off-shell parameter α_1 was 300
MeV/c, and no energy shift was included in evaluating the phase shifts.
The large differences in the curves coming just from the small differences in the π-N phase shifts are apparent. In Fig. 8, one can see
that multiple scattering is not severe by comparing the curves for
α_1 = 300 MeV/c and 600 MeV/c. In addition, one can see the large effect of the angle transform; i.e., taking into account the momentum of
the struck nucleon. Although not shown in the figure, one should note
that treating the velocities relativistically in evaluating the angle
transform (in essence, using $\mu^2/\omega M$ instead of the Galilean invariant
μ/M) makes a difference of some 15-20% where the significant differences can be seen in comparing the other curves shown. But it should
be clear now that one can improve the model all day and not parallel
the data, if the π-N phase shifts are not known.

It is also discernible from present theoretical calculations that,
for the π-d system, the total cross section measurements do not differ-
entiate among models or phase shift parameterizations. In Fig. 9 are
plotted total cross section estimates utilizing the Thomas and McKinley

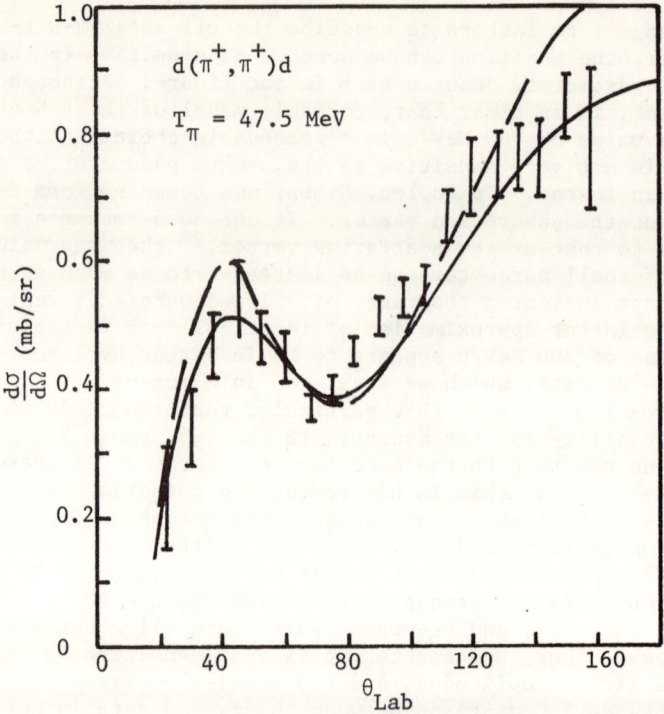

Fig. 7. Angular distribution of pions elastically scattered from the deuteron. The solid curve corresponds to Dodder's phase shifts; the dashed curve to Thomas'. In each case, α_1 = 300 MeV/c. Data are from Ref. 22.

phase shifts in the fixed scatterer approximation. There is little difference to be noted, as there is little difference if one removes the angle transformation from the fixed-scatterer calculations. There is also little difference if one compares with the results of Woloshyn, Moniz, and Aaron[10]---a Faddeev type calculation in which the s-wave and small p-wave π-N interactions were neglected. In contrast, at 142 MeV as shown in Fig. 10, one can see significant differences between the two phase shift parameterizations of Thomas and McKinley. The differing small phase shifts lead to varying back angle characteristics; neither curve agrees well with the data.[23] Differential measurements are required because the total cross section is determined by the forward scattering amplitude, and the forward scattering for the π-d system is essentially determined by the impulse approximation.

Before leaving the 3-body system, let me note that the important and interesting subject of nucleon-nucleon pion production has not changed significantly on the experimental side since the summary by Stephenson[24] at the Santa Fe conference last summer. However, there has been some slight progress on the theoretical side. An extension of the production operator model of Ref. 19 (for the pp → π⁺d transition)

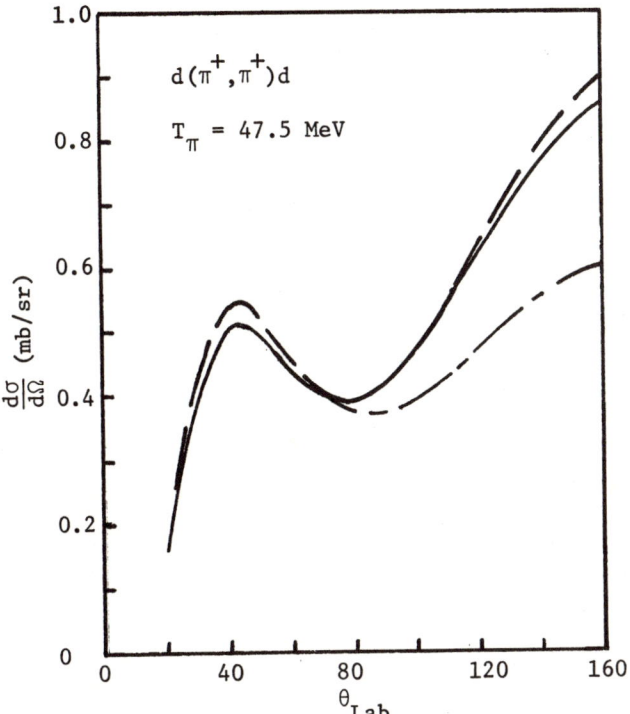

Fig. 8. Angular distribution of pions elastically scattered from the deuteron. All curves correspond to Dodder's phase shifts. The solid curve has $\alpha_1 = 300$ MeV/c; the dashed curve has $\alpha_1 = 600$ MeV/c; the dot-dashed curve has no angle transformation.

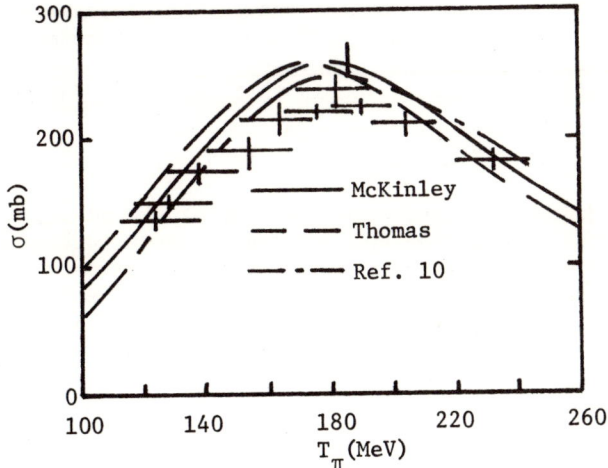

Fig. 9. Pion-deuteron total cross sections. Data are from Ref. 10.

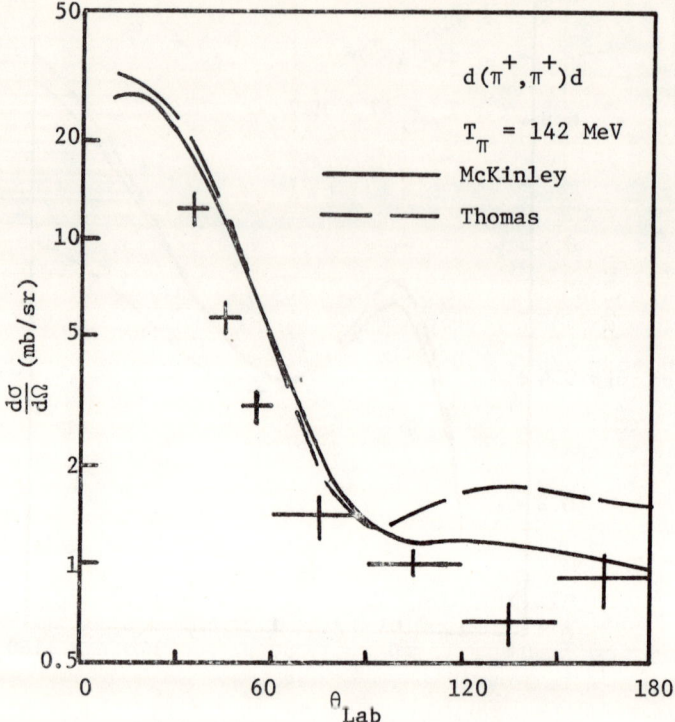

Fig. 10. Angular distribution of pions elastically scattered from the deuteron. The data are from Ref. 23; α_1 = 300 MeV/c.

was made to the continuum reaction; example π-N rescattering graphs are defined in Fig. 11 for the pp → pnπ⁺ production process, including final state n-p rescattering.[25] The n-p rescattering amplitude was modified to reflect the reduction in that amplitude due to the presence of the third particle:

$$\frac{b}{r}(e^{ik\, r_-} - e^{-\beta r}) f_{np}(E,\theta) \quad ,$$

where simple model estimates of b place it at around 1/5. Fitting that parameter to the zero degree 805 MeV data[26] (seen in Fig. 12), which is dominated by the (3,3) resonance, one also obtains very satisfactory fits to the data at other energies such as the 647 MeV data shown in Fig. 13, which has large n-p rescattering contributions.

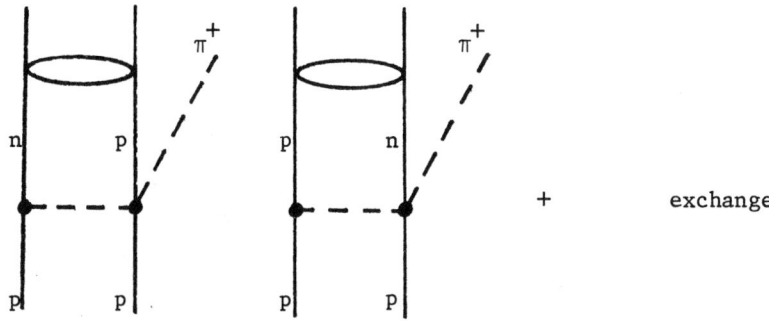

Fig. 11. Sample diagrams from the pp → pnπ reaction including n-p final state rescattering.

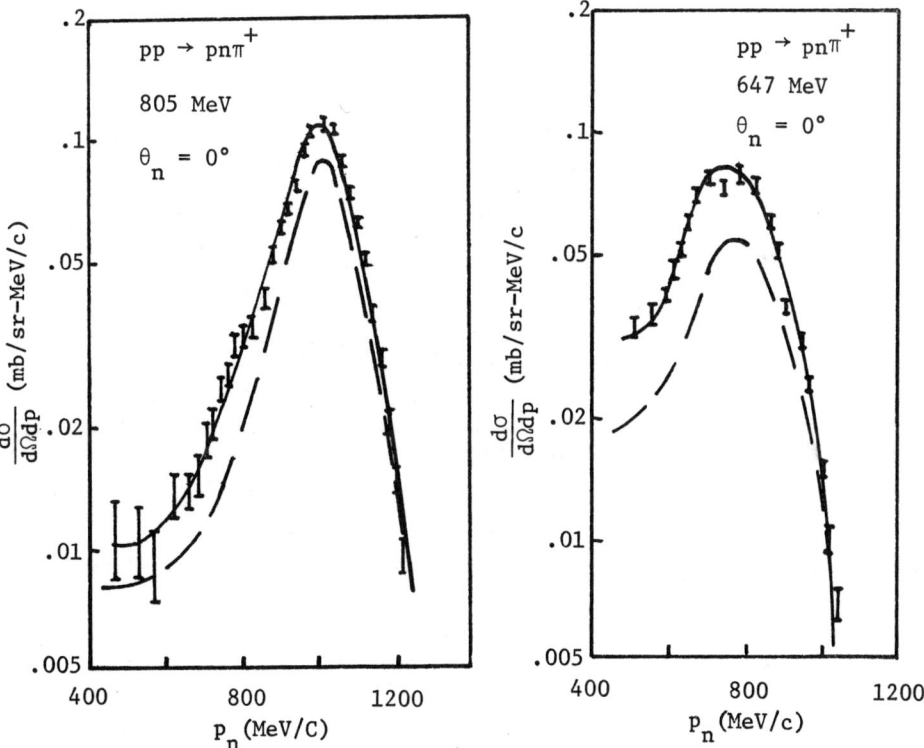

Figs. 12 and 13. Energy spectrum for forward neutrons; solid curve includes final state n-p interactions while dashed curve includes only the strong π-N (3,3) interaction.

THE FOUR-BODY PROBLEM

The pion-trinucleon problem is likely one of the more difficult to treat correctly of those which I shall discuss today. Both this and the π-^{4}He system will eventually receive proper treatment in terms of the Faddeev-Yakubovsky type equations, although optical-model, fixed-scatterer, etc. calculations must now suffice. Even here the difficulties are such as to make the π-^{3}He calculations less reliable than those for the π-^{4}He system. Besides the additional nucleon, the pion-trinucleon problem retains the complexity of spin and small wave function components found in the π-d problem. Even so, there are interesting effects to be noted.[27,28]

A comparison of π^+ scattering from the mirror nuclei or a comparison of π^+ and π^- scattering from ^{3}He (or ^{3}H) should help answer the question of convergence in the fixed-scatterer approximation. In either case, π^+-^{3}He scattering involves two dominant π^+p amplitudes whereas the other process involves only one such amplitude. If convergence is poor, one should find poor quality π^+-^{3}He fits to the data, while reasonable fits are obtained for either π^+-^{3}H or π^--^{3}He. Coulomb-nuclear interference effects should be interesting at low energy. Form factor effects will be significant at back angles.

In Fig. 14, one can see that at 50 MeV the various Coulomb-nuclear interference effects are the most interesting feature of the predictions. The repulsive π^+ Coulomb amplitude interferes destructively with the attractive strong interaction amplitude. This results in a second minimum inside the prominent p-wave minimum near 90°. If one looks at even lower energy, these two minima coalesce for the π^+-^{3}He case, where the charge of the target is two. For π^- scattering, the attractive Coulomb amplitude adds constructively with the attractive strong interaction amplitude, and there is only a single minimum.

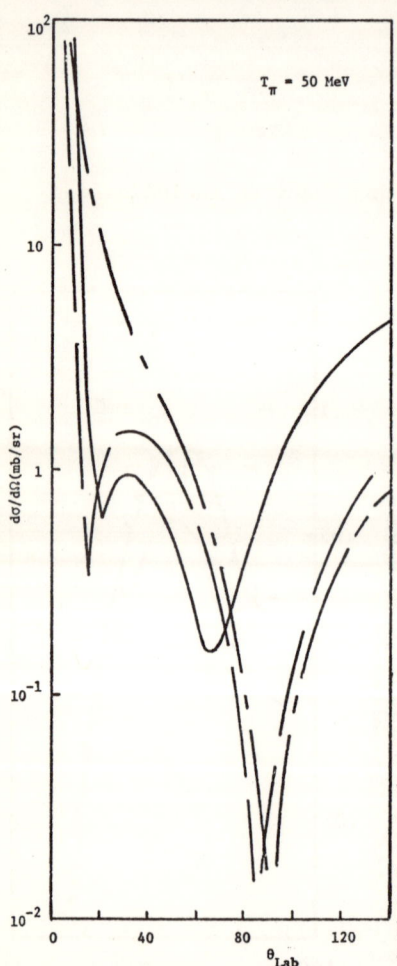

Fig. 14. Elastic scattering angular distributions at 50 MeV. The solid curve is π^+-^{3}He; the dashed curve is π^+-^{3}H; the dot-dashed curve is π^--^{3}He.

In Fig. 15 at 100 MeV, one can see only the remnants of the Coulomb-nuclear interference. The p-wave minima for the different scattering processes converge toward the same angle. They coincide at approximately 180 MeV, where the (3,3) resonance dominates. Note that, were it not for the differences in the ^{3}He and ^{3}H form factors, the π^+-^{3}He and π^--^{3}H curves would coincide except at forward angles, as would the π^--^{3}He and π^+-^{3}H curves.

In Fig. 16, one can see how the spin-flip cross section fills in the minimum in the non-spin-flip angular distribution. Landau, who has published very extensive optical model calculations for ^{3}He and ^{4}He, has suggested that one might use this feature of the π^--^{3}He cross section to study the spin-flip nuclear density[28] --- the corresponding magnetic form factor is studied in elastic electron scattering. I would caution that filling minima is not a good way to study anything. In particular, single-particle properties of the nucleus are better left to single-scattering probes such as one has in electron scattering, which can be carried out with precision at Bates.[29]

In addition to the elastic scattering possibilities, the trinucleons are also interesting because of the

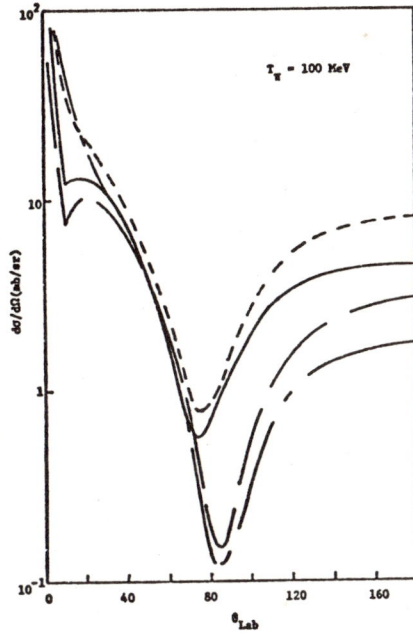

Fig. 15. Elastic scattering angular distributions at 100 MeV. The solid curve is π^+-^{3}He; the dashed curve is π^+-^{3}H; the dot-dashed curve is π^--^{3}He; the dotted curve is π^--^{3}H.

Fig. 16. Angular distribution of pions elastically scattered from ^{3}He. The solid curve is the sum of the non-spin-flip (dot-dashed curve) and the spin-flip (dashed curve) cross sections; α_1 = 300 MeV/c.

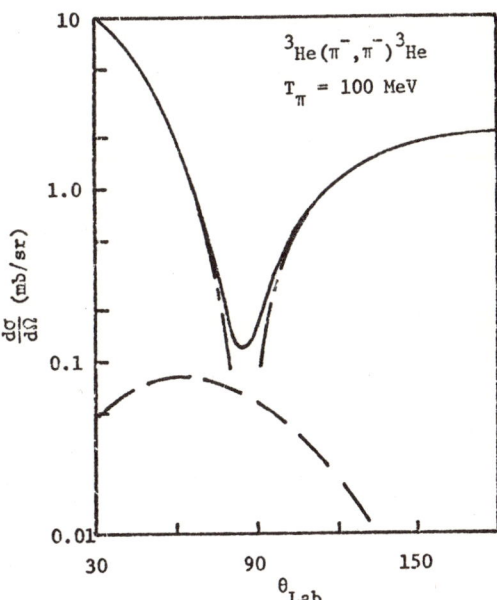

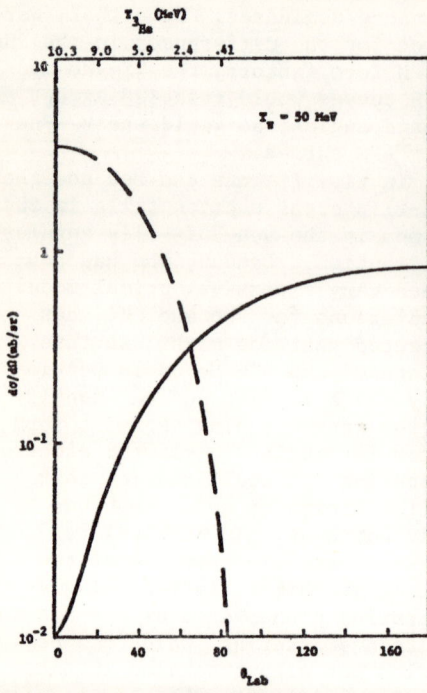

Fig. 17. Single-charge-exchange angular distributions: π^0 (solid curve) and recoil nucleus (dashed curve).

analog single-charge-exchange reaction ^{3}He$(\pi^-,\pi^0)^3$H or conjugate. Here, one can obtain partial angular distributions without the use of a π^0 detector. In particular, at low energy such as the 50 MeV shown in Fig. 17, where the backward peaking of the π^0 angular distribution results in a forward peaked spectrum for the recoiling residual nucleus, one can hope to obtain a large fraction of the angular distribution before encountering problems in extracting the recoiling nucleus from the target. As one studies higher energies, the π^0 angular distribution is more forward peaked, and the residual nucleus peak moves toward 90° where the nucleus picks up little kinetic energy in the collision. In heavier nuclei, such as ^{13}C, the analog transition is expected to exhibit a reduction because of the strong absorption due to multiple scattering, which markedly affects such monopole transitions. Several authors[25,30] have predicted that this would not be the case for the trinucleon analog transition (see Fig. 18). This prediction awaits experimental confirmation.

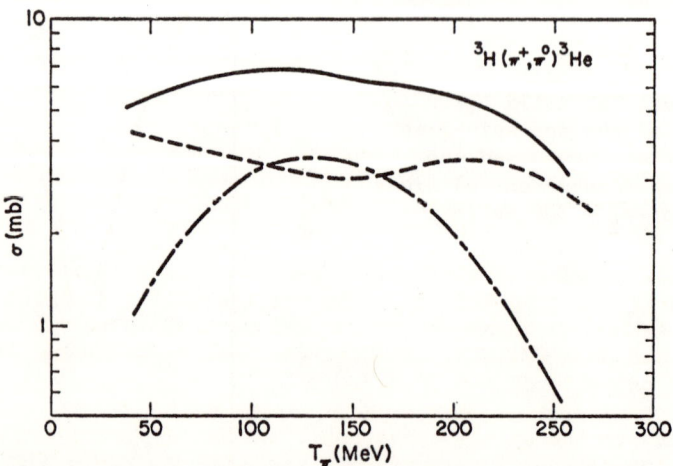

Fig. 18. Total cross section for the single-charge-exchange analog reaction. The solid curve sums the spin-flip (dashed curve) and non-spin-flip (dot-dashed curve) contributions.

THE FIVE-BODY PROBLEM

Since π-^{4}He elastic scattering should be simpler than the π-^{3}He problem, let me return to the point that fits to the low-energy data are not very significant in view of the present uncertainty in the π-N phase shifts. Both the non-relativistic[28] and relativistic optical-model[31] proponents, as well as the fixed-scatterer calculators,[15] have produced satisfactory fits to the excellent 110 MeV data of Binon, et al.[32] But let us look first at the 51 MeV data of Crowe, et al.[33], where the nucleus is more transparent. In Fig. 19, one can examine

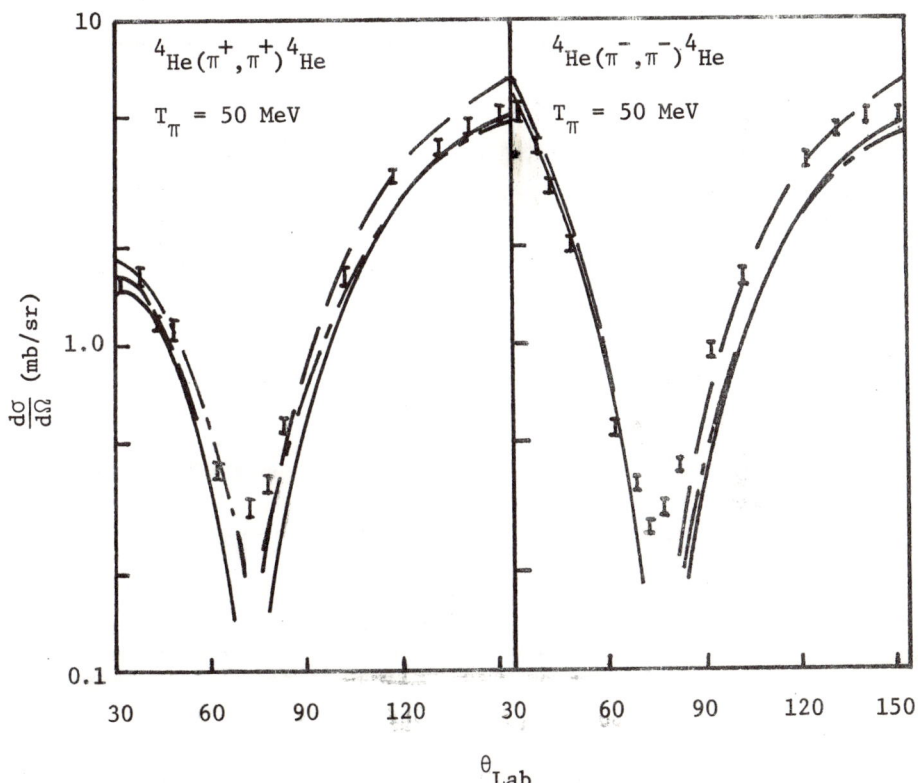

Fig. 19. Angular distributions of pions elastically scattered from ^{3}He; data are from Ref. 31. The solid curves correspond to McKinley's phase shifts; the dashed curve to Thomas'. Both curves have α_1 = 300 MeV/c and $\Delta E = -7$ MeV. The dot-dashed curves are for Thomas' phase shifts but with $\Delta E = -11$ MeV.

the fits to the data for the phase shifts of Thomas and McKinley, where the off-shell parameter is the same 300 MeV/c used in the π-d scattering discussion. Because multiple scattering is weak at this energy, the value of the off-shell parameter has little significance and the fits are essentially unaltered if one uses 600 MeV/c. It is clear that one needs better phase shifts. Otherwise, the model is not determined. With the McKinley phase shifts and an energy shift (binding energy subtraction) of 7 MeV, one can obtain reasonable fits to the existing data from 50 MeV through the 110 MeV data. However, at 50 MeV, one can obtain as good or a better fit with the Thomas phase shifts, if the binding energy subtraction is 11 MeV. Therefore, will the true π-N phase shifts please stand up?

If one looks instead at the 110 MeV data, the dependence upon the phase shifts has decreased, although the nucleus is not so black that it has entirely disappeared. (See Fig. 20) Even though the curves

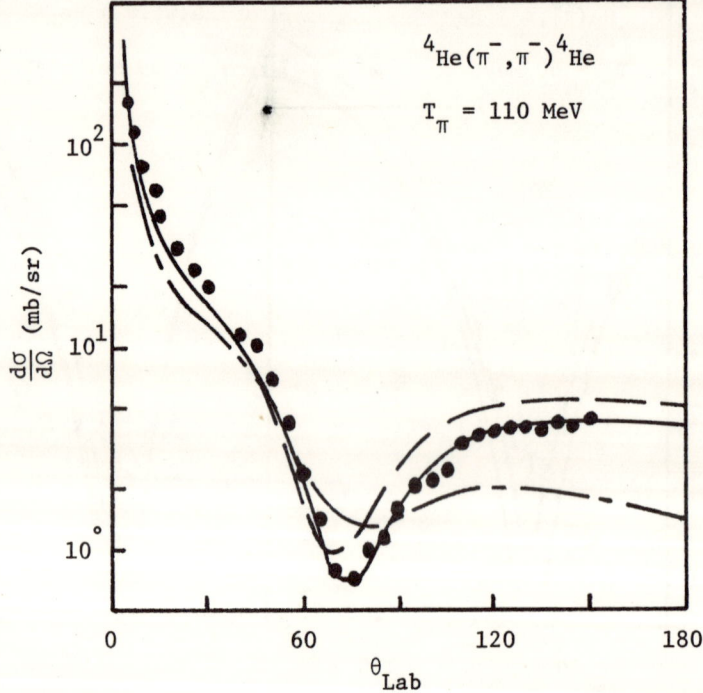

Fig. 20. Angular distribution of pions elastically scattered from ^{4}He; data are from Ref. 30. The solid curve corresponds to McKinley's phase shifts; the dashed curve to Thomas'. Both curves have α_1 = 300 MeV/c and ΔE = -7 MeV. The dot-dashed curve is for McKinley's phase shifts but with α_1 = 600 MeV/c.

for McKinley's and Thomas' phase shifts do not compare equally well with the data in Fig. 20, subtracting 9 MeV instead of 7 MeV when using the Thomas phase shifts will essentially reproduce the McKinley curve. Thus we again see a model ambiguity coming from the uncertainty in the phase shifts. On the other hand, multiple-scattering effects have become much more important, as one can see by comparing the two

McKinley curves having the off-shell parameters of 300 MeV/c and 600 MeV/c. For the latter, the minimum is not nearly deep enough nor are the back angle cross sections large enough. In fact, if the model is believable and if absorption is not a large factor at this energy, one might say that the p-wave off-shell parameter of 300 MeV/c is in very reasonable agreement with the πd → pp analysis of Ref. 19.

On the other hand, it is quite clear from comparison with all of the data between 51 MeV and 110 MeV that the angle transformation discussed above is required, if the fixed-scatterer calculations are to approach the data.[15] The minimum at approximately 75° is essentially stationary as a function of energy. (Recall that in the pion-trinucleon case this minimum moved.) It is the angle transformation that fixes this minimum in the fixed-scatterer approximation. It induces a large effective s-wave amplitude, one whose strength comes primarily from the large π-N p-wave amplitude. The resulting interference between the effective s-wave and p-wave amplitudes for equal numbers of protons and neutrons then produces the stationary minimum. In Fig. 21, one can see a comparison of the calculation without the angle transformation or the binding energy and recoil corrections, the calcula-

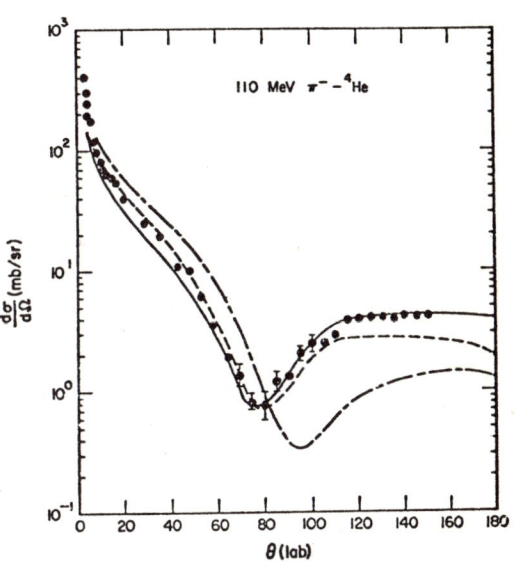

Fig. 21. Angular distribution for elastic scattering: dot-dashed curve is without angle transformation or energy correction; dashed curve is with angle transformation only; solid curve is the complete fixed-scatterer approximation.

tion with just the angle transformation, and the complete calculation. Each of these modifications to the basic fixed-scatterer approximation is sizeable, but one cannot ascertain how significant they are without precise π-N phase shifts with which to work.

Let me close this discussion of the five-body problem with brief mention of the pion double-charge-exchange reaction leading to a final state of four identical nucleons (4 protons or 4 neutrons). I speak of double charge-exchange on ^{4}He instead of the trinucleons because of the recent interest in that nucleus generated by the letter of Germond and Wilkin,[34] although it is interesting for either ^{3}He or ^{4}He because of the Pauli effects in the final state, in which all the nucleons are identical, and because short-range correlations can, in principle, be seen in this reaction, which cannot take place on a single nucleon. Germond and Wilkin suggested that one might find double charge-exchange on the pion cloud an important contribution to the cross section. An

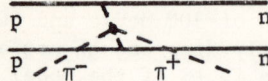
obvious competitor to this is the contribution to the double charge-exchange arising from the second order process involving two π-N charge-exchange scatterings. In a contribution to this conference, Hess, et al.[35] have estimated cross sections due to this latter process using separable s-wave and p-wave π-N amplitudes (which include spin-flip), fully antisymmetrized nuclear wave functions, and the complete 5-body phase space.

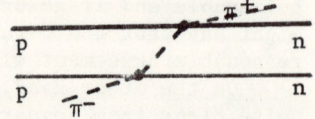

In Fig. 22, one can see the result for the 20° differential pion cross section at an incident pion kinetic energy of 140 MeV compared with the data of Kaufman, et al.[36] In Fig. 23, is a comparison of the calculated result for the 0° pion cross section for a fixed final pion kinetic energy of 176 MeV with the data of Gilly, et al.[37] The agreement with the data is not spectacular, due to neglect of higher partial waves, etc. but at 140 MeV the calculation is an improvement of 2300 in magnitude over the earlier result of Becker and Schmit.[38] A comparison of the total cross section predictions of Germond and Wilkin with those of Hess, et al. and Becker and Schmit is made in Fig. 24. The arrows indicate the energy regions of the single angle data mentioned above. In addition, the total cross sections of Falomkin, et al.[39] and Carayannopoulos, et al.[40] are indicated.

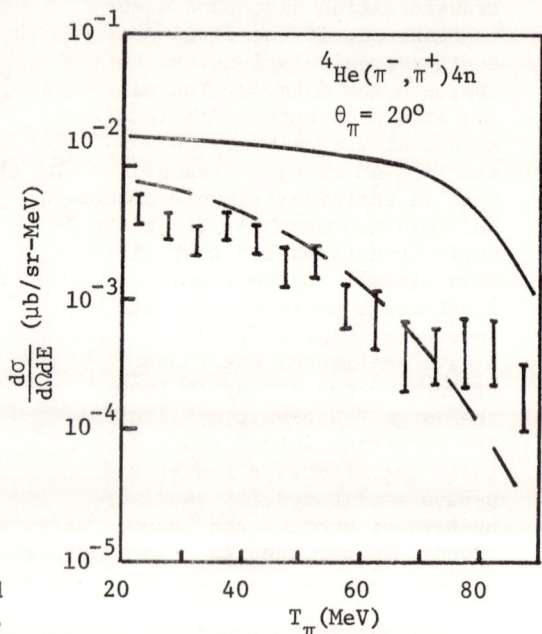

Fig. 22. Pion double-charge-exchange for initial T_π = 140 MeV. Solid curve is without Pauli correlations; dashed curve is with Pauli.

It is interesting to note for the experimentalists that the calculation which agrees reasonably well with the 20° data at 140 MeV disagrees with the 100 MeV total cross section point by several orders of magnitude. The calculations represented in Fig. 24 are qualitative, at best. However, the latest attempt at the double scattering estimate is a sizeable improvement over the previous work, and it appears that, if the Germond and Wilkin pion scattering is to be significant, it may well be at the higher energies --- above the resonance.

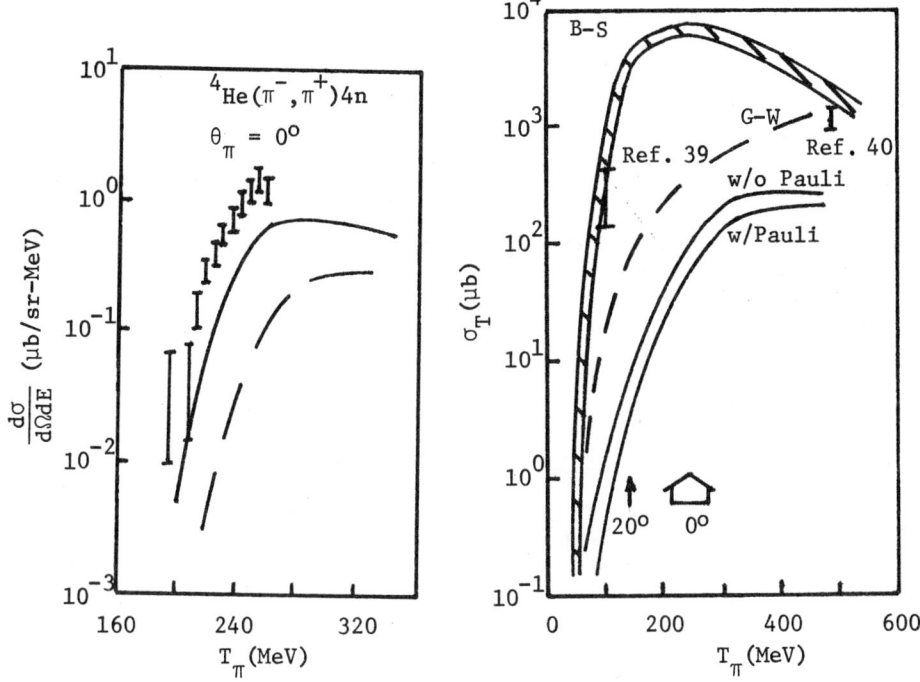

Fig. 23. Pion double-charge-exchange for final T_π = 176 MeV. Solid curve is without Pauli correlations; dashed curve is with Pauli.

Fig. 24. Pion double-charge-exchange total cross sections.

CONCLUSIONS

Let me conclude my remarks by reminding you that I have given a superficial and incomplete overview of the subject of pion-few-nucleon physics. I have not the time allotted to do it real justice. Besides, you have heard, or will hear, detailed presentations about the relativistic aspects, the field theoretic aspects, the isobar theory aspects, the Faddeev theory, the pion production reaction, the photoproduction reaction, the radiative pion absorption reaction, etc.

I trust that I have not left you with the impression that few-body physics is an isolated field of endeavor. It was my intent to provide some inkling of the physics of the pion-few-nucleon system whose understanding is important in the investigation of the more general subject to pion-nucleus interactions. The π-d problem is perhaps the single most important system for our study; here, we can achieve understanding of the model assumptions which will be applied to more complex nuclei. At low energy, π-d scattering is very sensitive to all of the phase shifts, because of amplitude cancellations. From it we also learn something of the energy regions where spin-flip and absorption are significant. The π-α system is almost as important. The fixed minimum in the angular distribution --- a property not common to simple diffractive scattering --- further tests our model assumptions. Here, we may also find pion absorption, double-spin-flip, or double-isospin-flip contributing significantly. Proper inclusion of Coulomb effects is easily examined in a comparison of π^+ and π^- scattering. Furthermore, multiple-scattering effects become large, and the off-shell properties of the π-N interaction come into play. The pion-trinucleon problem should prove more useful in testing the sophisticated aspects of of our models. It may also aid in our understanding of single-charge-exchange analog transitions.

One could continue the catalog. But let me stop and enunciate the hope that I have induced at least some of you to make the measurements and perform the analysis required to obtain the critically needed π-N phase shifts,[41] for at present, with the coming precision pion-nucleus data, we seem to have the cart before the horse.

ACKNOWLEDGEMENT

The speaker and his collaborator are deeply indebted to their colleagues A. T. Hess and G. J. Stephenson, Jr., with whom much of the research discussed here was carried out. The speaker also wishes to thank E. J. Moniz and D. S. Koltun for several lengthy discussions and enlightening comments and to acknowledge the helpful correspondence of A. W. Thomas.

REFERENCES

1. W. R. Gibbs, B. F. Gibson, and G. J. Stephenson, Phys. Rev. C11, 90 (1975).
2. W. R. Gibbs, B. F. Gibson, and G. J. Stephenson (to be published).
3. W. R. Gibbs, The Investigation of Nuclear Structure by Scattering Processes at High Energy, H. Schopper, Ed. (North-Holland Pub. Co., N. Y. 1975) pp. 205-25.
4. R. J. Adler, Phys. Rev. 141, 1499 (1966).
5. M. Chemtob, E. J. Moniz, and M. Rho, Phys. Rev. C10, 344 (1974).
6. W. Gruebler (private communication).
7. L. Heller, G. Bohannon, and F. Tabakin, Phys. Rev. C13, 742 (1976).
8. A. S. Rinat and A. W. Thomas (conference contribution).
9. R. Aaron, R. D. Amado, and E. J. Young, Phys. Rev. 174, 2020 (1968).
10. R. M. Woloshyn, E. J. Moniz, and R. Aaron, Phys. Rev. C13, 286 (1976).
11. A. W. Thomas, Nucl. Phys. A258, 417 (1976).
12. D. C. Dodder (private communication).
13. W. R. Gibbs, A. T. Hess, and W. B. Kaufmann, Phys. Rev. C (to be published).
14. R. Mach, Phys. Lett. 40B, 46 (1972) and Nucl. Phys. A258, 513 (1976); R. H. Landau, S. C. Phatak, and F. Tabakin, Ann. Phys. 78, 299 (1973); G. A. Miller, Phys. Rev. C10, 1242 (1974).
15. W. R. Gibbs, B. F. Gibson, A. T. Hess, G. J. Stephenson, and W. B. Kaufmann, Phys. Rev. C (to be published).

16. I. R. Afnan and A. W. Thomas, Phys. Rev. C$\underline{10}$, 109 (1974).
17. T. Mizutani, Thesis (Univ. of Rochester Tech. Report COO-2171-55).
18. D. S. Koltun and T. Mizutani (to be published).
19. B. Goplen, W. R. Gibbs, and E. L. Lomon, Phys. Rev. Lett. $\underline{32}$, 1012 (1974); see also B. Goplen, Thesis (LASL Report LA-5854-T).
20. W. R. Gibbs and B. F. Gibson (conference contribution).
21. J. McKinley, Rev. Mod. Phys. $\underline{35}$, 788 (1963).
22. D. Axen, G. Duesdicker, L. Felawka, Q. Ingram, R. Johnson, G. Jones, D. Lepatourel, M. Salomon, and W. Westland, Nucl. Phys. A$\underline{256}$, 387 (1976).
23. E. C. Pewitt, T. H. Fields, G. B. Yodh, J. G. Fetkovich, and M. Derrick, Phys. Rev. $\underline{131}$, 1826 (1963).
24. G. J. Stephenson, Jr., High Energy Physics and Nuclear Structure - 1975, AIP Conference Proceeding No. 26, pp. 55-62.
25. W. R. Gibbs, B. F. Gibson, G. Glass, and G. J. Stephenson (to be published).
26. G. Glass, M. E. Evans, M. Jain, R. Kenefick, L. Northcliff, C. Cassapakis, C. Bjork, P. Riley, B. Bonner, and J. Simmons, (to be published).
27. A. T. Hess and B. F. Gibson, Phys. Rev. C$\underline{13}$, 749 (1976) and LASL Report LA-6320-MS.
28. R. Landau, Ann. Phys. (N.Y.) $\underline{92}$, 205 (1975).
29. There is nothing implicitly wrong with the electron scattering results (see Ref. 28) which find $\langle r_{ch}^2(^3He)\rangle < \langle r_{mag}^2(^3He)\rangle$; see B. F. Gibson, International Conference on Photonuclear Reactions and Applications, B. L. Berman, Ed. (USAEC CONF 730301, 1973) pp. 373-383.
30. J. M. Eisenberg and V. B. Mandlezweig, Phys. Lett. $\underline{53B}$, 405 (1975); D. A. Sparrow, Phys. Lett. $\underline{58B}$, 309 (1975).
31. L. S. Celenza, L. I. Liu, and C. M. Shakin, "Covariant Calculation of Pion-^{4}He Elastic Scattering in the (3,3) Rosonance Region" (to be published).
32. F. Binon, et al., Phys. Rev. Lett. $\underline{35}$, 145 (1975).
33. K. Crowe, A. Fainberg, J. Miller, and A. Parsons, Phys. Rev. $\underline{180}$, 1349 (1969).
34. J. F. Germond and C. Wilkin, Nuovo Cimento Lett. $\underline{13}$, 605 (1975).
35. A. T. Hess, W. R. Gibbs, B. F. Gibson, and G. J. Stephenson, Jr. (conference contribution).
36. L. Kaufman, V. Perez,Mendez, and J. Sperinde, Phys. Rev. $\underline{175}$, 1358 (1968).
37. L. Gilly, M. Jean, R. Meunier, M. Spighel, J. Stroot, and P. Duteil, Phys. Lett. $\underline{19}$, 335 (1965).
38. F. Becker and C. Schmit, Nucl. Phys. B$\underline{18}$, 607 (1970).
39. I. V. Falomkin, et al., Nuovo Cimento $\underline{22}$, 333 (1974).
40. N. Carayannopoulos, J. Head, N. Kwak, J. Manweiler, and R. Stump, Phys. Rev. Lett. $\underline{20}$, 1215 (1968).
41. The need for better phase shifts is not so critical near the (3,3) resonance, where the nucleus becomes much blacker and less interesting --- except in the case of higher multipole, surface peaked reactions in which multiple-scattering absorption is not so important.

Carl Shakin (Brooklyn College): In the π-d absorption calculation, did one consider off-shell unitarity? That is, if you vary $v(k)$, do you recognize the fact that $\lambda_0(\omega)$ and $\lambda_1(\omega)$ are functionals of $v(k)$? Unless this is done you can get any result you wish for that cross section.
Gibson: The $\lambda_\ell(\omega)$ and $v_\ell(q)$ are in essence independent of each other in a manner similar to what one would find in dealing with an energy dependent separable potential.

A. N. Mitra (Univ. of Delhi): You mentioned about a substantial difference between the correction due to recoil effects brought about via Galilean invariance and a fuller relativistic treatment. It appears to me that if instead of using the pion mass 'μ' in the replacement $\vec{q} \to \vec{q} - \frac{\mu}{M}\vec{P}_i$, one uses the pion's relativistic energy ω_q, i.e. $\vec{q} \to \vec{q} - \frac{\omega_q}{M}\vec{P}_i$, the difference between the approaches would be considerably narrower, especially at the energies being considered. Do you agree?
B. Gibson: A relativistic perscription was used for adding the velocities in the actual calculations presented. This leads to a correction in the amplitudes more like $\frac{\mu^2}{\omega M}$ than $\frac{\omega}{M}$.

A. W. Thomas (CERN): Two comments: First, I was careful in the low energy πd calculations not to suggest that absorption would necessarily be small. In fact, in my talk the chance of getting at these effects was suggested as an important motivation for low energy πd experiments. Second, in relation to Brown's comment I remind you of the work of Hufner, et al., who have emphasized that the πN interaction is not easily defined as a short range process since nucleon exchange (as in Chew-Low) is a u-channel process. In addition, one should remember that a separable interaction with "range parameter" β, approximates the usual left-hand cut by a pole at $k=i\beta$. For comparison, a Yukawa potential with range parameter μ leads to a left-hand cut starting at $k=i\mu/2$. I hope this illustrates the difficulty of too simple an interpretation of the separable interaction parameter "β" as the mass of an exchanged particle.

L. Kisslinger (CMU): My impression is that the π-N data is not in quite as bad shape as the results which you have shown indicate. Bob Kelly, who does not seem to be at this session, has said that at least one or two of the analyses which you have shown should be rejected as it has used obsolete data. This reduces the spread in the results considerably.

COVARIANT N-N DYNAMICS AND πD SCATTERING*

D. D. Brayshaw
Stanford Linear Accelerator Center, Stanford, CA 94305

ABSTRACT

In this paper we show that ambiguities and approximations common to conventional three-body treatments of πD scattering can be largely eliminated. Thus, by employing a general dynamical framework which simultaneously describes the reactions NN → NN, NN ↔ NNπ, NN ↔ πD, NNπ ↔ πD, and πD → πD, uncertainties engendered by the relativistic reduction and off-shell extrapolation can be suppressed by demanding that NN elastic data (which are most sensitive) be correctly reproduced. With this constraint predictions for πD scattering are sensitive chiefly to the deuteron wave function, defining a model-independent nuclear probe.

INTRODUCTION

Three-body treatments of πD scattering possess a number of important advantages in that one may properly account for nucleon recoil and rescattering, and simultaneously describe breakup of the deuteron (with and without pion absorption). Such calculations are obviously useful in evaluating techniques of a more approximate nature applied to pion scattering on heavier nuclei. In addition, the rules governing Faddeev-like multiple scattering series provide a means of distinguishing the scattering pion from those involved in the nuclear force, eliminating frequent sources of double-counting. These advantages have prompted a number of calculations based on the non-relativistic Faddeev theory[1] and its covariant generalization.[2] The latter are clearly to be preferred, and in principle provide a means of testing the importance of specifically relativistic effects. However, the equations themselves are far from unambiguous, and have not been notably successful in many previous applications (particularly to mesonic systems[3]). Also, all Faddeev calculations to date share a common defect in that they are based on separable off-shell extensions; this is a considerable assumption for which there is little theoretical justification. Thus, while such calculations undoubtedly constitute a considerable improvement over earlier techniques, one must exercise some care in assessing the more subtle dynamical questions from this standpoint.

In this paper we point out that one may employ a more general three-body framework in order to suppress model-dependent effects arising from ambiguities in the relativistic reduction, off-shell behavior, and associated approximations. The essential idea is to consider the totality of <u>physical</u> NNπ states linked by the strong interaction (assuming conservation of angular momentum, parity, and isospin), and to describe them by a single covariant, unitary theory. In particular, one sector involves elastic NN scattering, since the pion may be absorbed before the nucleons separate. By requiring a detailed fit to the corresponding data (which are relatively clean and plentiful), parameters characterizing the unknown structure can be determined <u>empirically</u>, greatly reducing the uncertainties in the πD sector.

---

*Work supported by the National Science Foundation and the Energy Research and Development Administration.

DESCRIPTION OF NNπ SCATTERING

The suggested program would be quite impractical if one were limited to Faddeev-like calculations. This is because only modest off-shell variations can be produced in the separable approximation, and each change would require a new solution of the three-body equations. However, this author has defined an alternative set of equations by imposing generalized boundary conditions on the three-body system, and shown rigorously that the Faddeev amplitudes are included as a special case.[3,4] The covariant generalization of the resultant formalism has been highly successful in applications to the three-pion system,[3,5] and requires only a single computation in order to fit the relevant parameters. In fact, very few parameters are required to produce an excellent fit to NN scattering, since the minimal description constitutes an excellent first approximation to NN dynamics.[6]

As an illustration, we note that the $^1S_0$ NN state can be used to determine appropriate input to the 0^+ p-wave in πD scattering. The effectiveness of the NN description is shown in Table I, in which the singlet effective

TABLE I
$^1S_0$ Phase for 2-parameter NNπ Model

T_L (MeV)	δ_{mod} (deg.)	δ_{exp} (deg.)	T_L (MeV)	δ_{mod} (deg.)	δ_{exp} (deg.)
20	54.6	54.79 ± 0.52	160	20.8	20.04 ± 2.02
50	42.5	43.02 ± 0.98	200	15.0	13.96 ± 2.10
80	34.7	35.15 ± 1.34	240	10.2	8.45 ± 2.07
120	27.3	26.94 ± 1.77	280	6.1	3.43 ± 2.03

range parameters have been fitted; the tabulated points are a theoretical prediction of the model. Although additional parameters may be introduced to improve the agreement with the data of Ref. 7, this fit is sufficient to pin down the πD effective range expansion

$$\kappa^3 \cot \delta = -1/a_v + (r_v/2)\kappa^2 , \qquad (1)$$

and one obtains $a_v = 0.0647 \mu^{-3}$, $r_v = -47.8 \mu^{-1}$. Similar calculations of the remaining πD s- and p-waves at energies $T_L \leq 250$ MeV will be reported.

REFERENCES

1. I. R. Afnan and A. W. Thomas, Phys. Lett. 45B, 437 (1973); F. Myhrer and D. S. Koltun, Phys. Lett. 46B, 322 (1973).
2. R. M. Woloshyn, E. J. Moniz, and R. Aaron, Phys. Rev. C 13, 286 (1976).
3. D. D. Brayshaw, Phys. Rev. D 11, 2583 (1975), and earlier references therein.
4. D. D. Brayshaw, Phys. Rev. C (to be published).
5. D. D. Brayshaw, Phys. Rev. Lett. 36, 73 (1976).
6. D. D. Brayshaw, Stanford Linear Accelerator Center Report No. SLAC-PUB-1686 (1975).
7. M. MacGregor, R. Arndt, and R. Wright, Phys. Rev. 182, 1714 (1969).

Garth Jones (UBC and TRIUMF): 1) Re: $\frac{\sigma(pp \to pn\pi^+)}{\sigma(pp \to d\pi^+)}$. We are currently measuring these reactions in the 350 MeV range at TRIUMF using a magnetic spectrograph. Our statistical accuracy for $pp \to pn\pi$ is poor, at present, because we are looking at cH_2-C differences. Later this year, the situation will be improved when we install a liquid hydrogen target. However, we do not see a marked $pp \to pn\pi$ contribution relative to the $pp \to d\pi$ in this energy region. 2) Finally I would like to encourage the theorists to extend their investigations to include the d-wave pion component in the $pp \to d\pi$ reaction, as pion production left-right asymmetries using polarized beam depend sensitively on p-wave interference with the s- and d-waves. A preliminary look at recent data at 370 MeV suggests significant d-wave contribution even at this energy.

D. Brayshaw: The slide I showed contained only the 2^+ contribution to the $pp \to pn\pi$ and $pp \to d\pi$ cross sections. Other contributions are important, particularly for the latter, and the point is not that the former is relatively enhanced, but that one should expect a dip near 490 MeV. Although the other contributions are not negligible, I do not believe that the effect will be totally obscured.

A. W. Thomas (CERN): It seems that your physical interpretation contradicts what you yourself have emphasized in n-d scattering. That is, the two-body off-shell information is inextricably mixed with three-body forces (etc.) in the interior region. Thus, it is hard to understand how you can attribute these (apparently) necessary changes in the 2^+ amplitude to the off-shell behaviour of the P_{33} interaction. There may be far more interesting physical phenomena there.

D. Brayshaw: I have oversimplified somewhat in my presentation, and it is true that there is no unique explanation with regard to off-shell behavior vs. three-body forces. However, the effect does require a change in the overlap between the πD and $N\Delta$ channels, and the simplest way to understand this would be in terms of a modification in the off-shell (P_{33}) behavior.

PION PRODUCTION IN P-P INTERACTIONS AT 800 MeV[†]

R. D. Felder, J. Hudomalj-Gabitzsch, T. M. Williams, G. S. Mutchler,
J. M. Clement, K. R. Hogstrom, W. H. Dragoset, and G. C. Phillips

T. W. Bonner Nuclear Laboratories, Rice Univ., Houston, Tx. 77001

and

E. V. Hungerford, M. Warneke, B. W. Mayes, L. Y. Lee, and
J. C. Allred

University of Houston, Houston, Tx. 77004

ABSTRACT

The production of pions in p-p interactions was recently measured in a kinematically-complete counter experiment at 800 MeV. Kinematic conditions were chosen to examine in detail several pion production mechanisms.

INTRODUCTION

Measurements of single pion production in nucleon-nucleon reactions at medium energies provide some of the basic data necessary for a theoretical understanding of pion-nuclear physics. Pion production via the p-p interaction has been widely studied above 650 MeV for the reactions:

$$p + p \to d + \pi^+ \qquad (1)$$

$$p + p \to p + \pi^+ + n \qquad (2)$$

$$p + p \to p + p + \pi^0. \qquad (3)$$

The present experiment has measured each of the above reactions at kinematic conditions which allow particular production mechanisms to be investigated.

This experiment was recently performed at the Los Alamos Meson Physics Facility using an experimental apparatus similar to one previously described[1]. A magnetic spectrometer arm and a time-of-flight arm, each employing multi-wire proportional counters, were used to detect charged particles in coincidence and thereby completely determine the kinematics of three-body final states.

MEASUREMENTS

The cross section for reaction (1) was measured at five angle pairs with pion lab angles between 20° and 50°. For the same angles, reaction (2) was studied under kinematic conditions which permitted interactions between the final state nucleons. Preliminary data on reaction (2) are shown in Fig. 1, where the raw proton

[†] Work supported in part by US ERDA.

momentum distribution has been divided by the experimental momentum acceptance. The dashed curve is a plot of proton momentum vs. n-p relative energy as given by the scale on the right. The enhancement in the cross section near $E_{np} = 0$ is clear evidence for reaction (2)

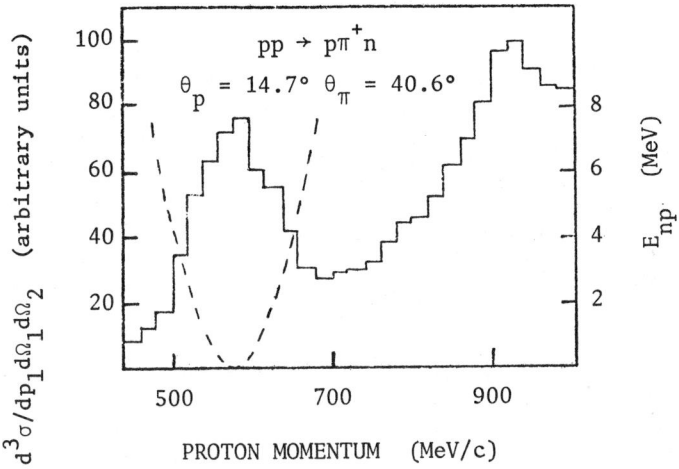

Figure 1. Proton Momentum Distribution

proceeding via the neutron-proton final state interaction (FSI). Similar data were recorded for reaction (3), in which the p-p FSI was observed by detecting both protons in the spectrometer arm.

A detailed investigation of pion production via the Δ^{++} resonance was also conducted for six angle pairs with pion lab angles between 15° and 45°. The broad peak near 930 MeV/c in Figure 1 is an example of pion production due to Δ^{++} formation. Kinematic conditions were chosen to observe the Δ^{++} mechanism in isolation, in interference with the Δ^{+} isobar, and at angles for measuring the angular distribution of the Δ^{++} decay products. Single-arm pion production was also measured by detecting pions in the spectrometer arm at lab angles between 15° and 60°.

The forthcoming analysis of this large body of pion production data recently measured at 800 MeV should provide interesting tests of the theoretical descriptions of pion production.

1. T. R. Witten et al., Nucl. Phys. A254, 269 (1975).

Silbar (LASL): You tantalized us with a statement that your experiment observed double pion production?
Felder: We have detected π^- in our pp pion production experiment; however, an estimate of the cross section for double pion production is not yet possible since significant backgrounds must still be removed.

PION PRODUCTION BY 800 MeV PROTONS FROM DEUTERIUM WITH A SPECTATOR NEUTRON

E. V. Hungerford, J. Lo, M. Warneke,
J. C. Allred, B. W. Mayes, L. Pinsky.
University of Houston, Houston, Texas 77004.

J. Clement, W. H. Dragoset, R. Felder, K. Hogstrom, J. Hudomalj-Gabitzsch, G. S. Mutchler, G. C. Phillips, T. Williams.
Rice University, Houston, Texas 77001.

ABSTRACT

Pion production induced by 800 MeV protons on deuterium has been investigated under conditions where the neutron in the deuteron target remains a spectator to the reaction. The cross section for this process is large and accounts for a large fraction of the total pion production in this kinematic region.

The University of Houston - Rice University collaboration is undertaking a series of investigations on pion production mechanisms induced by protons on very light elements (≤Li). These studies use the 600 MeV proton beam from the Space Radiation Effects Laboratory and the 800 MeV external proton beam facility at LAMPF. To date data have been taken with ^{1}H, ^{2}H, and ^{6}Li targets, with a magnetic spectrometer arm in coincidence with a time of flight arm. This experimental arrangement allows complete kinematic determination of the reaction products for a three body final state. Angular distributions and spectra are obtained for various possible reactions in interesting regions of phase space.

Pion production from deuterium is particularly interesting because at 720 MeV[1] inclusive pion production is less than the sum of the inclusive production of pions from protons on protons and protons on neutrons. Some of our pion production data from deuterium has been presented previously.[2] This paper concentrates on the most recent results at 800 MeV in which deuterons were emitted at angles less than 25°.

Analysis of this data is not yet complete, but at least for these angles, it was found that pion production mainly proceeds through the mechanism shown in Fig. 1. The process is simply the reaction p+p→d+π with the neutron in the target deuteron remaining as a spectator. An angular distribution for (dπ) angle pairs where the target neutron can remain at rest in the Lab system was obtained. In addition distributions where the neutron is required to remove momentum from the reaction process essentially maps out the Fermi distribution

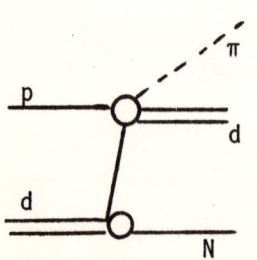

Fig. 1 Reaction Diagram

of neutron momenta in the target deuteron wave function.

In addition this reaction depends on the p+p→d+π amplitude in which the relative momenta of the nucleons in the final deuteron are sampled at large values. This amplitude is sensitive to the large momentum components and to the d state of the deuteron wave fundtion

REFERENCES

1. D. F. Cochran, et al. Phys. Rev. D6, 3085 (1972).
2. K. Hogstrom, et al. Bull. Am. Phys. Soc. 20, 1176 (1975).

COVARIANT CALCULATIONS OF πD SCATTERING

A.S. Rinat
CERN, Geneve and WIS, Rehovot, Israel

A.W. Thomas
CERN, Geneve

ABSTRACT

We report results of a covariant, unitary 3-body calculation including spin and isospin of πD scattering at 142 and 180 MeV. The major relativistic effects arise from a consistent choice of two-body relative momenta. Moreover, these effects tend to improve agreement with experiment.

DISCUSSION

Numerous attempts have been made to describe πD scattering for energies $E_\pi \lesssim 250$ MeV. Paramount amongst these are 3-body theories which attempt to incorporate relativistic effects while being manifestly covariant. We are aware of only one such calculation; in it spin and isospin have been neglected[2].

We report here the results of a calculation which differs from the mentioned one[2] in two respects i) spin and isospin are now included, ii) we make an alternative choice for the rleative momenta in two-particle vertices, such that the total energy is consistently conserved throughout.

We have solved coupled-channel equations[1] for E_π=142 and 180 MeV. The input in that energy region consists of the P_{33} πN and $^3S_1$ NN phases which are generated by t-matrices corresponding to separable driving terms[1]. (The $^3D_1$ NN admixture as well as smaller partial waves have been neglected).

Fig. 1 shows the πD elastic cross sections for these two energies. The curves 'CV' are the results of the full calculation. Those labeled TH are the results calculated in the approximation of Ref. (3) with a Galilean choice of relative momenta in two-body vertex functions but relativistic kinematics for the pion. Finally the curves TH-MV show the effect of properly defined relative momenta in a calculation, which is otherwise the same as TH. These momenta clearly exert a larger influence than other elements which render the description covariant. The agreement with experiment is best for CV.

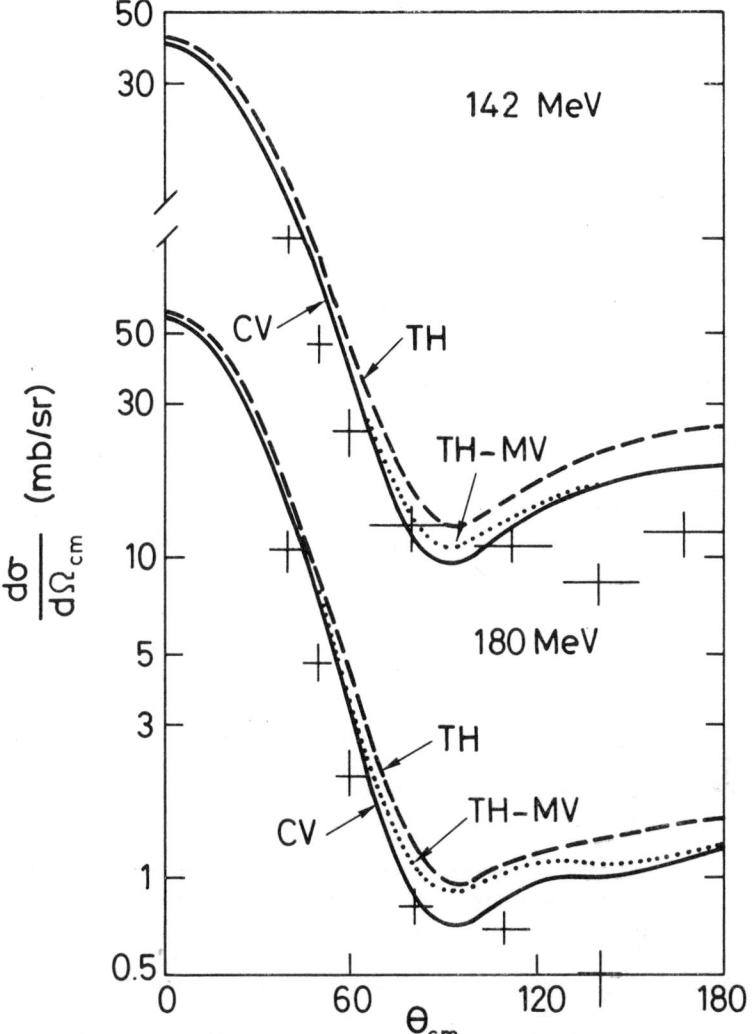

Fig. 1. Elastic πD cross sections for E_π=142 and 180 MeV. Curves labeled CV, TH, TH-MV, respectively correspond to a covariant calculation, to one with relativistic pion kinematics[3] and the same with properly defined relative momenta in vertices.

REFERENCES

1) R. Blancenbecler and R. Sugar, Phys. Rev. 142, 1031 (1966); D. Freedman, C. Lovelace and J. Namislowski, Nuovo Cim. 43A, 250 (1966); R. Aaron R.D. Amado and J.E. Young Phys. Rev. 174, 2022 (1968).
2) R.M. Woloshyn, E.J. Moniz and R. Aaron Phys. Rev., to be published.
3) I.R. Afnan and A.W. Thomas, Phys. Rev. C10, 109 (1974); A.W. Thomas, Nucl. Phys. A258, 417 (1976).

PION DEUTERON ELASTIC SCATTERING

E. M. Ferreira
Stanford Linear Accelerator Center and Pontifícia Universidade Católica
do Rio de Janeiro

L. P. Rosa and Z. D. Thomé
Universidade Federal do Rio de Janeiro, Brazil

In the present work we perform an extensive analysis of the existing experimental[1] data on pion-deuteron elastic scattering at low energies, in the framework of the multiple scattering series. We evaluate the single scattering and the double scattering terms, accounting for nucleon recoil and fermi-motion effects. The effects of the kinematical ambiguities in the use of the two-body πN amplitudes are studied. These ambiguities become more important when fermi-motion is taken into account, different choices of values of the effective meson nucleon colliding energy leading to predictions for the differential cross sections which are sometimes largely discrepant. We have used three different prescriptions (called A, B, C) for the resolution of these ambiguities.

A) The energy for the πN collision is determined by Faddeev equation.[2] If s is the center-of-mass kinetic energy for the πd system, and k is the nucleon momentum in the c.m. system, then the kinetic energy used in the πN center of mass when calling the πN amplitude is given by $s - k^2/2\mu$, where $1/\mu = 1/m_N + 1/(m_N + m_\pi)$.

B) The incident meson and the interacting nucleon are on the mass shell, with the physical values for their masses.

C) The spectator nucleon is on the mass shell, with kinetic energy $p^2/2m_N$. The interacting nucleon is off the mass shell, with momentum p, and energy given by the deuteron mass minus the spectator total energy.

To our knowledge, in the present work for the first time a calculation of the multiple scattering series is performed according to prescription A, which seems to be better founded theoretically. We are glad to note that the results obtained in case A give a much improved description of the data.

In the figures below we compare results obtained with different kinematical prescriptions. The solid curve in the figures refers to prescription A, while the dotted lines refer to B, and the broken line in Fig. 1 is obtained in case C.

Fermi-motion effects, which have been taken into account in all curves presented, improve substantially the description of the experimental results.

We see that the use of kinematical relations defined by the Faddeev equations brings a remarkable improvement to the multiple scattering calculation in the middle energy region.

At the higher energies and for large scattering angles, the Faddeev kinematics leads to values of the differential cross section which are small as compared to the data. The reasons for the discrepancy may perhaps be found in the deuteron structure, and changes in the large momentum tail of the wave function could improve the fitting.

At lower energies, as at 47.5 MeV, Faddeev kinematics <u>including fermi motion effects</u> predicts values for the differential cross section which are too low as compared to the data. The discrepancies may be due to the use of an inadequate set of phase-shifts, or to bad off-mass-shell extrapolations.

REFERENCES

1. Experimental data: 47.5 MeV, D. Axen at al., Nucl. Phys. A 256, 387 (1976); 85 MeV, K.C. Rogers et al., Phys. Rev. 105, 247 (1957); 142 MeV, E. G. Pewitt et al., Phys. Rev. 131, 1826 (1963); 182 MeV, J. H. Norem, Nucl. Phys. B 33, 512 (1971); 224 MeV. J. L. Acioli, Ph.D. thesis, Chicago University (1968); 256 MeV, K. Gabathuler et al., Nucl. Phys. B 55, 397 (1973); 234.4 MeV and 324.9 MeV, Preliminary Data, Los Alamos Report LA-6156-R; 375.7, 412.4, and 469.6 MeV, L. S. Schroeder et al., Phys. Rev. Lett. 27, 1813 (1971).
2. A. E. Thomas, Proc. Int. Conf. on Few Body Problems in Nuclear and Particle Physics, Laval U., Quebec, 27-31 Aug 1974, p. 287.

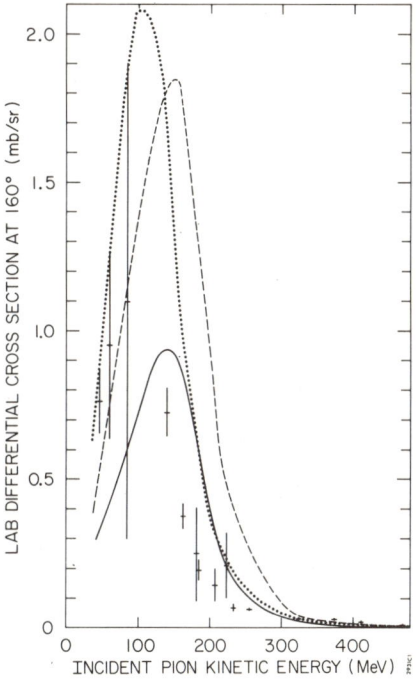

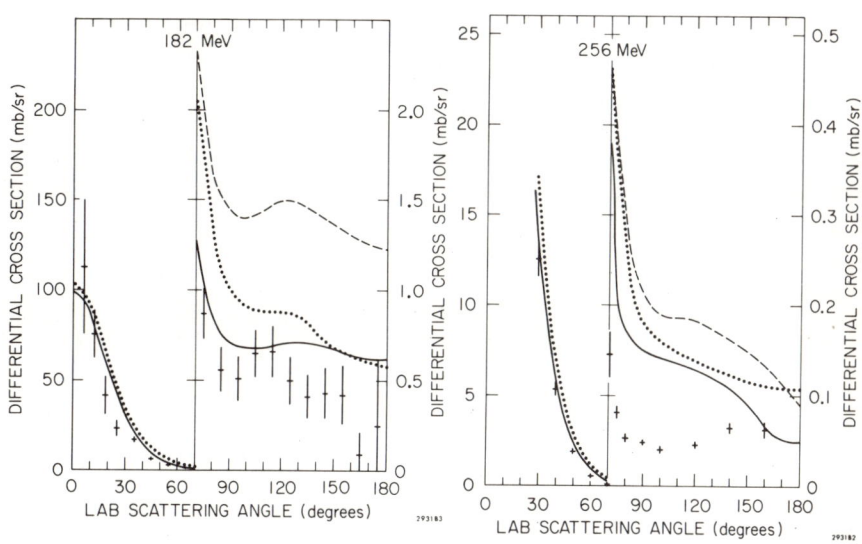

Figs. 1,2,3. Results of multiple scattering calculations using different prescriptions for the kinematics in πN collision. The solid lines show results for kinematics specified by Faddeev equation. For experimental points, see Ref. 1.

PION DEUTERON SCATTERING IN THE RESONANCE REGION

H. Garcilazo
Escuela Superior de Fisica y Matematicas,
Instituto Politecnico Nacional, Mexico 14 D. F.

ABSTRACT

We have calculated pion-deuteron elastic scattering in the impulse approximation, using Wick's three-body helicity formalism. We studied the dependence of the cross section on different forms of the off-shell pion-nucleon T-matrix.

INTRODUCTION

In a recent tree-body calculation of pion-deuteron elastic scattering using all angular momentum and all spin and isospin degrees of freedom, although keeping only the resonant 3,3 channel in the πN subsystem, it was found that for energies around the resonance, the impulse approximation contributes about 90% of the differential cross section[1]. Consequently, we have now undertaken to do a very complete calculation of the impulse approximation, by including all πN channels, and using a full relativistic angular momentum basis[2], as well as taking into account the analytic structure of the on-shell T-matrix.

THEORY

In order to calculate the πD amplitude, one needs the full off-shell πN T-matrix $t(p,p';s)$, where s is the invariant center of mass energy squared, which varies from a fixed value s_{max} up to $-\infty$. For a given channel characterized by isospin and angular momentum T and j, and initial and final helicities λ' and λ of the nucleon, we used the form

$$t^{\lambda\lambda'}(p,p';s) = g_{1-}(p,s)\tau_{1-}(s)g_{1-}(p',s)/g_{1-}^2(p_0,s) + (-)^{\lambda-\lambda'}$$
$$\times g_{1+}(p,s)\tau_{1+}(s)g_{1+}(p',s)/g_{1+}^2(p_0,s), \quad (1)$$

where $1\pm = j\pm 1/2$, and p_0 is the on-shell momentum given by

$$p_0^2 = (s-(M-\mu)^2)(s-(M+\mu)^2)/4s, \quad (2)$$

with μ and M the masses of the pion and the nucleon respectively. The form factors $g_1(p,s)$ were chosen of the form

$$g_1(p,s) = p^1/(\alpha^2 + p^2) s^2/s_0^2 + 1/2, \quad (3)$$

where α is a constant, and where the special form (3), ensures that the off-shell T-matrix (1), does not diverge at the point s=0, for which as we see from (2), $p_0 \to \infty$.

The on-shell T-matrix $\tau_1(s)$, is given in terms of phase shifts in the physical region $\sqrt{s} \gtrsim M+\mu$. In the unphysical region $M-\mu < \sqrt{s} < M+\mu$, we used fixed-t dispersion relations to construct it, while in the crossed region $\sqrt{s} \lesssim M-\mu$, we obtained it by using standard crossing relations.

RESULTS

In order to calculate our results, we constructed the on-shell πN amplitude, for all partial waves such that j=1/2 and j=3/2. We fixed the parameter s_0 in (3), at the threshold value $s_0 = (M+\mu)^2$. For the range parameters α in (3), we used the same value for all partial waves. We show in the figure, our results when we consider two different values of α, at three different energies. We see that going from a very big value of α to α =250 Mev/c, there are large variations in the cross section. The agreement between the data and our results for α =250 Mev/c, is good, which strongly supports our choice of the special form (3) for the form factors.

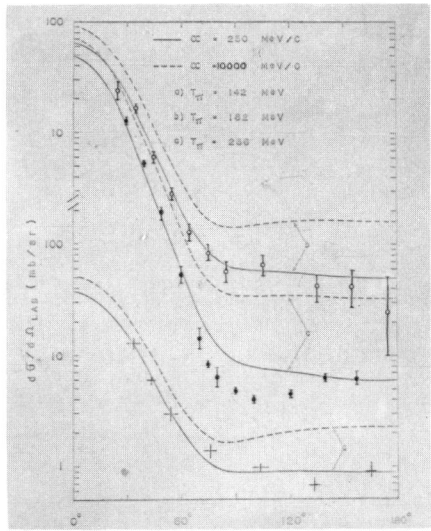

Fig. 1. Lab differential cross sections for πd scattering in the region of the 3,3 resonance.

1. V.B. Mandelzweig, H. Garcilazo, and J.M. Eisenberg, Nucl. Phys.A A256, 461 (1976).
2. G.C. Wick, Ann. Phys. (NY) 18, 65 (1962).

MULTIPLE SCATTERING ANALYSIS OF π^-He SCATTERING AT 1.12 GeV

Yukap Hahn
Physics Department, University of Connecticut, Storrs, Conn. 06268

ABSTRACT

Elastic scattering of negative pions by He nuclei at 1.12 GeV is analyzed by the multiple scattering theory. The contribution of the second order optical potential is found to be large in the region of the second diffraction maximum. The calculation is in fair agreement with the experiment for $q^2 \lesssim 0.4$ (GeV/c)^2.

INTRODUCTION

The multiple scattering theory of Watson is convenient in analyzing the pion scattering off nuclei at medium energies; in particular, the KMT formulation as modified by Feshbach and Hüfner[1] allows one to treat separately the dynamics of the target nucleus and the scattering of the pion projectile. Although the formalism contains many approximations, Feshbach et al[2] have applied it to the proton-He problem with moderate success. A slightly different approach was also tried[3] by constructing an effective inelastic channel, with somewhat different result. Mainly due to uncertainties in the available input parameters, these results are not conclusive. It is therefore of some interest to examine the pion scattering using the same theoretical procedures. We present here the application of the multiple scattering formalism to the π^-He scattering at 1.12 GeV, for which an experimental data is available[4]. This system has been analyzed earlier by Querrou[5], and more recently by Franco[6] using the Glauber diffraction theory and by Rule and Hahn[7] using the effective channel approach.

INPUTS

The required inputs for the multiple scattering theory are usually the pion-nucleon scattering amplitude and the He internal wave function. The on-shell amplitude is parametrized here as

$$t(\vec{q}) = a\,(i + \rho)\,\exp(-\beta q^2/2) \quad , \qquad (1)$$

where $a = k\,\sigma_{tot}/4\pi$ with $\sigma_{tot} = 35$ mb, and $\beta = 0.4 \pm 0.05$ f^{-2}. The parameter ρ is taken here to be approximately $\rho \approx -0.2$, but may well be determined from the πHe fit itself. The form (1) fits the spin and ispin averaged πN data for $q^2 \lesssim 0.6$ (GeV/c)^2. For the first order optical potential V_o, we used the single particle density for the He obtained from the electron scattering, while a Gaussian-type product wave function with a short-range dynamical correlation factor is used[2] in the evaluation of the higher order optical potentials.

RESULT OF THE CALCULATION

The spin and ispin averaged elastic amplitude for the πHe scattering is calculated by solving a set of coupled equations, one for the elastic and the other for inelastic. The coupling potential V_c is obtained from the pair correlation function in the separable approximation[2], while the propagation of the system in the intermediate states is described by the Green's function $\bar{G} = (\bar{E} + i\epsilon - K - \bar{V})^{-1}$ where K is the pion kinetic energy operator and $\bar{V}$ is the average distortion potential obtained from the third order optical potential. An average excitation energy of 30 MeV is included in $\bar{E}$. The set is solved by a partial wave analysis, with $\ell \lesssim 35$. The result is given in Fig. 1. First of all, the effect of the second channel is found to be very large; the V_o alone gives the cross section with the first minimum at the wrong q^2 and the second maximum too low by a factor of 5. The full two channel calculation agrees well with the experimental data of Combe et al[4] for the momentum transfer $q^2 \lesssim 0.4$ $(GeV/c)^2$, but tends to be flatter and lie above the experimental points for larger q^2. The present result differs from an earlier study[7], which employed the effective πN potential; further efforts are being made to clarify this discrepancy. The present study is also being extended to lower energies.

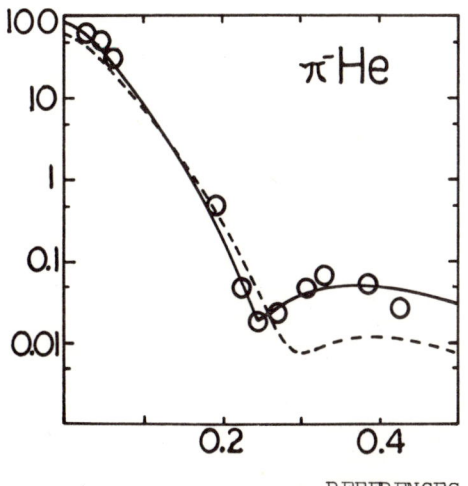

The π⁻He elastic cross section in mb/sr vs q^2 in $(GeV/c)^2$. The solid curve is the full two-channel result, while the dashed curve is for the single channel with V_o alone. The circles are the experimental data of ref. 4.

REFERENCES

1. H. Feshbach and J. Hüfner, Ann.Phys.(NY) **56**, 268 (1970).
2. H. Feshbach, A. Gal and J. Hüfner, Ann.Phys.(NY) **66**, 20 (1(71).
3. D. W. Rule and Y. Hahn, Phys. Rev. **C12**, 1516 (1975).
4. J. Combe et al, Nuovo Cimento **3A**, 663 (1971).
5. M. Querrou, Nuovo Cimento **3A**, 670 (1971).
6. V. Franco, Phys. Rev. **C9**, 1690 (1974).
7. D. W. Rule and Y. Hahn (to be published).

π-^{3}He, π-^{4}He ELASTIC SCATTERING AND π-^{3}He CHARGE EXCHANGE: AN IMPROVED CALCULATION

R. H. Landau[*]
Dept. of Physics, Oregon State University, Corvallis, OR 97331

Historically, studies of the three and four nucleon systems have helped lay much of the early foundation of nuclear physics. In the more recent past[1], we reported on a calculation of pion elastic scattering from 3He and 4He calculated with realistic nuclear form factors. Comparisons with the then-preliminary data on π- 4He were made and predictions concerning what may be learned by studies of π- 3He were described in detail. We now update these calculations; the preliminary 4He data have been published[2]; new 3He and 4He data have become available[3], improvements in the theoretical optical potential are being made[4], and other theoretical studies, using different approaches, have been reported[5]. In addition, we have extended our calculations of elastic scattering to also include the single charge exchange scattering from 3He (or 3H); the unified study of π- 4He, π- 3He elastic and charge exchange scattering is beneficial.

Our calculation is based on a first order optical potential in momentum space, which we write as the sum of products of nuclear n and p matter or spin form factors and πN off-shell spin flip and non-flip t-matrices. By making use of an optimal choice of nucleon momentum when making the factored approximation, it is possible to make the important Lorentz-angle transformation from the pi-nucleus to pi-nucleon C.O.M. quite reliable. The on- and off-shell πN matrices are related via a separable potential model which exactly reproduces the πN phase shifts, and the resulting optical potential is used in a relativistic Lippmann-Schwinger equation.

The four nuclear form factors for 3He are determined by forming appropriate linear combinations of the charge and magnetic form factors of 3He and 3H, thus including S, S' and D components of the wavefunction. Since there exists a rather large uncertainty in the magnetic form factors as determined by electron scattering, there is also a large uncertainty in spin form factors. In fact, since spin-flip scattering fills the minimum in π -3He scattering, pions may be useful in indicating if the magnetic size of 3He is as large as measured by electrons. For 4He we assume the nuclear spin form factors are zero and that the matter form factors are both equal to the experimental nuclear charge density with finite proton size removed.

The validity of our reaction theory is in a sense established by the quality of fit obtained to the scattering data from the relatively simple 4He nucleus. In this regard, we have two interesting new results. First, although the parameterless optical potential approach of Ref. (1) provided quite adequate fits to the earlier He data of Refs. (2) and (3), closer examination with more recent data (2),(3) indicates a most definite improvement in the quality of fit when a three body approach to recoil and binding corrections is used[4]. Secondly, the quality of the fits, at several energies, to the π^--4He data of Binon et al.[2] appears better than the fits obtained with the data of Shcherbakov et al[3]. However, if the latter's π^--4He and π^+- 4He data are plotted on the same gr

at each energy, the theory gives a much better fit to the combined data. (As far as the calculations go, there is little difference between π^+ and π^- - 4He scattering). Whether this is due to a deficiency in the theory or to the normalization of the data is not known.

When studying the $\pi^{\mp}$-3He data of Shcherbakov et al.^{3} we primarily concentrated on the π^- - 3He results since these hold the greatest promise of determining the interesting spin distribution. Unfortunately, if the difficulty in fitting the 4He data well indicates a deficiency in either theory or experiments, the π^- 3He results will be more difficult to interpret since now the π^+ 3He results cannot be combined with the π^- ones ($I \neq 0$). Nevertheless, we have found that the binding-recoil corrected optical potential does provide fairly good fits to the π^- 3He data at 98, 120 and 208 MeV. The π^+ fits are not quite as good. In addition, there does appear to be at least some indication that the magnetic (spin) distribution of 3He is smaller than indicated by electron scattering measurements. The availability of future experimental and theoretical results should help clear up this point.

Finally, our calculation of single charge exchange from 3He or 3H, calculated with the potential (1) in a distorted wave approximation, indicates: (1) the importance of binding and recoil corrections, especially scattering angles $\lesssim 70°$, (2) the importance of spin flip scattering^8, 3) a rather undramatic sensitivity to the nucleon spin distribution and 4) a total cross section which does not dip in the resonance region.

REFERENCES

1. R. H. Landau, Phys. Letters 57B (1975) 13; Ann. of Phys. 92 (1975) 205.
2. F. Binon et al., CERN Preprint (to Phys. Rev. Letters). See also (6).
3. Yu. A. Shcherbakov et al., Dubna-Torino preprint (to Nuovo Cimente).
4. R. H. Landau and A. W. Thomas, submitted to Phys. Letters, Nov. 1975; and contribution to this conference.
5. R. Mach, Nucl. Phys. A258 (1976)513; B. F. Gibson, W. R. Gibbs, A. T. Hess, G. J. Stephenson, Jr., and N. B. Kaufman, LASL, LA-UR-75-2154; D. A. Sparrow, U. of Colorado, COO-535-720; A. T. Hess and B. F. Gibson, LA-UR-75-1662.
6. R. Krowe et al., Phys. Rev. 180 (1969) 1344.
7. J. S. McCarthy et al., Phys. Rev. Letters 25 (1970) 884.
8. An obvious result since it is important for elastic scattering.

*Work supported in part by the Oregon State University Research Council (NSF Gu3662) and the O.S.U. Computer Center.

IS THERE AN ISOHELION ?

F. Nichitiu
Institute for Atomic Physics, Bucharest

Recently an energy dependent phase shift analysis of πHe^4 elastic scattering have been made[1] indicating a resonant behaviour for some partial waves. The structure of the P and D partial waves is sumarized in Table I.

	Max σ_{tot}^l	Max Im f_l	Max $\left\|\frac{df_l}{dT}\right\|$	Max σ_{el}^l	Max σ_{tot}^l	Re $f_l = 0$	min η_l	Max Im f_l	
$l=1$	16	18	108	118	123	146	148	148	T [MeV]
$l=2$	13	13	123	no	148	156	173	no	
	First			second		anomaly			

The first anomaly abserved in the interval of 10 to 20 MeV may be connected with the opening of the new inelastic channels ($\pi^0 He^3 p$, $\pi^0 H^3 n$). The second anomaly - the "resonant" region - suggest a compound nucleus formation with a non-negligible background.

We have attempt to extract the Breit-Wigner parameters from our P partial wave using various methods for adding the resonant amplitude to the background one[2,3]:

$$f = f_R + f_B + 2i f_R f_B \text{ or } S = S_R \cdot S_B \quad (1) \qquad f = f_R e^{2i\varphi} + f_B \quad (2)$$

where f_R is the B-W amplitude and f_B is an inelastic background. The resonant amplitude was parametrized as usual

$$f_R = \frac{0.5 \Gamma_{el}(E)}{E_R - E - \frac{i}{2}\Gamma_{tot}(E)} \quad \text{with} \quad \Gamma_i(E) = \delta_i \left(\frac{K}{K_R}\right)^3 \frac{2E_R}{E_R + E} \frac{(1 + (K_R R_i)^2)}{(1 + (K R_i))} \quad (3)$$

In our calculations we have used an unitary and also a non-unitary background amplitude inside the resonance region - for S_B unitary a constant background for each energy interval of 30 MeV and for S_B non-unitary an effective range expansion. Using an error of 8% for our P partial amplitude we have repeated the fit in many energy intervals.

A very good agreement for the whole energy interval tested (up to 180 MeV) is obtained by eq(1) with a non-unitary background across the resonance region (of course, the total amplitude is inside the unitary circle) and only a good χ^2 is obtained with unitary background from 65 to 170 MeV, with a better agreement for eq(1) than eq(2).

The value for the resonance mass is relatively stable E=4006-4012 MeV and the spread of the width over the methods are Γ_{el}=65-75 MeV and Γ_{tot}=135-160 MeV.

The factorization method for S matrix and the time decay ($\hbar/\Gamma$) shorter than the time for passing through the nucleus suggest that the πHe^4 resonance can be considerated as a nucleus (isonucleus) with a bound Δ_{33} resonance inside (binding energy of about 44 MeV) but near nucleus surface*) having a 1^+ level at 140 MeV and probable a 2^- level at about 160 MeV above πHe^4 mass. It is interesting to note that the mysterios narrow peak in the $He^4(\gamma,\pi^-p)$ reactions[4] take place at $\sqrt{s} \simeq 4030$ MeV just near presumbly 2^- level of our isohelion.

The background amplitude is due to scattering at small radial distance and multiple scattering, and it still contain the elementary Δ_{33} contribution - the reason why the non-resonant amplitude is large and strong depending on energy. In such cases the usual equations of adding the background to resonant amplitude may break down and we need a more elaborate methods than eq(1) or eq(2), but in any cases, there still is a fundamental question: how is a resonance correctly to be defined[5,6]?

References
1) L.Alexandrov,T.Angelescu,I.V.Falomkin,F.Nichitiu, Yu.A.Shcherbakov.Proc. Int. Conf. Few Body Problems in Nuclear and Particle Physics,Quebec,Canada,1974, p.348.
2) R.H.Dalitz,R.G.Moorhouse. Proc.Roy.Soc.Lond.A318,279, 1970.
3) C.Lovelace. Proc.Heidelberg Int.Conf.on Elementary Particles (Ed.H.Filthuth) p.79.
4) P.Argan,G.Audit,N.DeBotton,J.M.Laget,J.Martin, C.Schuhl,G.Tomas. Phys.Rev.Lett.29,1191,1972.
5) J.Hamilton,B.Tromborg. Partial Wave Amplitudes and Resonance Poles,Oxford at the Claredon Press 1972.
6) L.Fonda Fort. der Phys. 20,135,1972.
7) F.H. Heimlich et al. Nucl. Phys.A231, 509,1974.

---

*) The peak in the Δ_{33} resonance region observed in the $C^{12}(e,e')$ and $Li^6(e,e')$ reactions[7] can be explicated by the same surface formation of a bound Δ_{33} resonance with a width smaller than for the free Δ_{33} and equal approximately to that for electron scattering on bound nucleon (of about 75 MeV).

A LOWEST ORDER OPTICAL MODEL STUDY ON π-^{4}He SCATTERING*

T.-S. H. Lee
Argonne National Laboratory, Argonne, Illinois 60439

ABSTRACT

Within the framework of relativistic particle Hamiltonian theory, the lowest order π-^{4}He optical potential is constructed from the exactly soluble πN resonance model of Lee and Coester.[1] Compared with previous optical-model calculations, the main features of this study are: (1) the dependence of πN t matrix on the πN total momentum in the π-nucleus c.m. frame is rigorously determined by the relativistic particle Hamiltonian theory, (2) nucleon motion is treated exactly by folding the nonstatic πN t matrix into the ^{4}He wave function. Our calculations have shown that previously-used simplifying assumptions such as "angle transformation" and "fixed-scatterer approximation" could be unreliable to describe the scattering in the large momentum transfer region where the N-N correlation effects are expected to be important. The binding and Pauli effects on the πN interaction inside the nuclei are also under investigation by including the first-order corrections to the impulse approximation in the calculation.

---

*Work performed under the auspices of the U.S. Energy Research and Development Administration.

[1] T.-S. H. Lee and F. Coester, to be published.

MODEL FOR THE ABSORPTIVE COMPONENT IN LOW-ENERGY ELASTIC PION-DEUTERON SCATTERING

P. Goode and R. Rockmore
Rutgers University, New Brunswick, N.J. 08903

H. McManus
Michigan State University, East Lansing, Mich. 48824

ABSTRACT

We investigate the viability for low-energy pion-deuteron scattering of the R-matrix model for absorptive scattering suggested a long time ago by Brueckner.[1] In this model, the scattering of pions associated with the absorption process can occur through the absorption operator acting twice. The transition amplitude is separated into two parts associated with low and high nucleon momenta so that the adiabatic part largely identified with the impulse approximation may be removed. In the present work the absorption vertex is taken to have the customary Galilean-invariant form and no more than S-wave rescattering is considered; intermediate-state nucleon-nucleon interactions are taken into account in a simple two-parameter model.

REFERENCES

1. K. A. Brueckner, Phys. Rev. $\underline{98}$, 769 (1955).

PION-DEUTERON ABSORPTION*

W. R. Gibbs, B. F. Gibson, and G. J. Stephenson, Jr.
Theoretical Division, Los Alamos Scientific Laboratory
Los Alamos, New Mexico 87545

ABSTRACT

We have recalculated the $\pi^+ + d \to p + p$ reaction including a form factor at the absorption vertex in addition to the one previously included at the scattering vertex. We find no significant changes once the range parameter is adjusted to reflect the presence of the second form factor.

INTRODUCTION

The pion rescattering graph (Fig. 1) has long been known to dominate the $\pi + d \to N + N$ reaction. Goplen[1] showed that the use of a p-wave rescattering vertex of the form $\vec{q}\cdot\vec{q}'$ leads to much too large a cross section but that the introduction of a form factor at the vertex gave very reasonable results. In that work no form factor was included at the absorption vertex. Chew-Low type derivations of π-N elastic scattering tend to predict the two form factors to be of the same shape. For this reason, we have modified the code developed by Goplen[2] to include the absorption vertex form factor of the same form as that at the scattering vertex: $(q^2 + \alpha^2)^{-1}$.

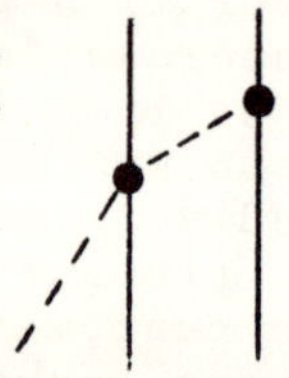

Fig. 1. Rescattering Graph.

TOTAL CROSS SECTION

In Fig. 2, we compare with the data our calculation of the total absorption cross section under the assumption that $\alpha = 600$ MeV/c. The general quality of the fit is similar to that obtained in Ref. 1 where only the scattering vertex was modified by a form factor. The N-N interaction utilized was the Lomon-Feshbach boundary condition model with a 4.6% D-state.

ANGULAR DISTRIBUTION

Recently, it has been suggested that this reaction may be sensitive to the deuteron D-state.[3,4] One problem with such a D-state determination is the possibility that the model dependence in the calculation will obscure the result. Since we have made what might seem to be a rather large variation in the model, it is interesting to see how large the effect is upon the angular distribution. In Fig. 3, we compare calculations for the quantity A from the $A + \cos^2\theta$ angular distribution with the data. The points designated with an X are from the recent data of Ref. 3. As may be seen, the general features are not altered significantly.

*Work performed under the auspices of the U. S. ERDA.

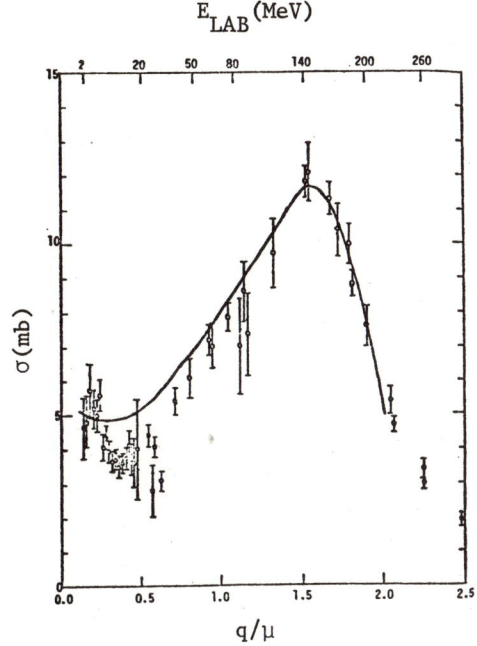

Fig. 2. Total absorption cross section.

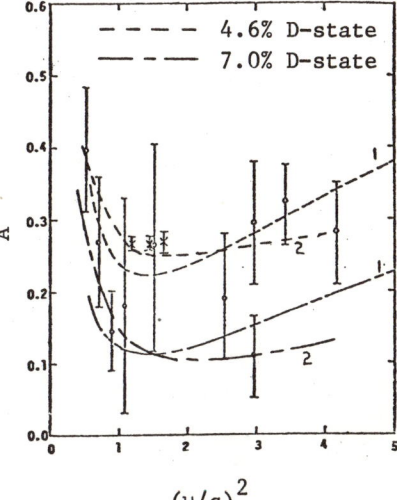

Fig. 3. Angular distribution coefficient "A". Both curves labeled "1" ("2") have a single (double) vertex form factor.

REFERENCES

1. B. Goplen, W. R. Gibbs, and E. L. Lomon, Phys. Rev. Lett. __32__, 1012 (1974).
2. B. Goplen, Thesis, LA-5854-T (Los Alamos Scientific Laboratory).
3. B. Preedom, et al., (submitted to Phys. Rev. Lett.).
4. W. R. Gibbs, Fourth International Conference on Polarization Phenomena in Nuclear Reactions, Zurich (August 1975).

INFLUENCE OF FORM FACTORS ON PIONIC DEUTERON DISINTEGRATION

M. Brack *
State University of New York, Stony Brook, N.Y. 11794

D. O. Riska
Michigan State University, East Lansing, Mich. 48824

W. Weise
University of Erlangen-Nürnberg, Erlangen, W-Germany

In a recent paper [1] we have presented a microscopic model for the process

$$\pi + d \longrightarrow p + p'. \qquad (1)$$

For the resonant p-wave absorption, we calculated the amplitude for (1) by summing Feynman graphs corresponding to the absorption on a single nucleon line (impulse approximation) and to the rescattering of both π and ρ mesons through the 33 resonance. For the s-wave πN rescattering we used the zero range $\pi\pi NN$ interaction of Koltun and Reitan[2]. Reid's soft core potential was used for the deuteron wave functions [3]. Without introducing form factors and using a rather strong ρ meson coupling constant $f_{\rho N} = f_\rho(1+\kappa_\rho)$ with $f_\rho = 1.0$ and $\kappa_\rho = 6.6$, we obtained a reasonable agreement with the experiment.

To investigate the effect of hadronic form factors, we include at each of the two vertices of the rescattered meson a monopole form factor:

$$F_{iNN}(k^2) = F_{i\Delta N}(k^2) = \frac{\Lambda_i^2}{k^2 + \Lambda_i^2} \ . \qquad (i = \pi, \rho) \qquad (2)$$

Here k is the transferred momentum. The form (2) is consistent with our use of a nonretarded meson propagator (see ref.1. A transfer of half the incident pion energy by the exchanged meson, as used by Goplen et al.[4], increased our results by less than 10%.) In the following we shall only discuss p-wave absorption which dominates the resonance region.

The dashed curves in Fig. 1 show the absorption cross section σ for the case where only a pion is rescattered. A cut-off mass of $\Lambda_\pi \simeq m_\rho$ already reduces the value of σ obtained without cut-off to less than the experimental value. This is in contrast to the findings of Goplen et al.[4] who needed cut-offs of less than 400 MeV in order to fit the experiment.

The inclusion of the ρ meson further cuts down the cross section, as already shown in ref. 1. If pion cut-offs below 1 GeV were used together with the ρ exchange, the peak value of σ would be less than half the experimental value. The solid line in Fig. 1 shows the result obtained with ρ exchange and setting $\Lambda_\pi = \Lambda_\rho = 1.5$ GeV (and $\kappa_\rho = 3.7$). This latter curve is not meant to be a fit; it rather defines some lower limits for the pion cut-off mass. (A variation of Λ_ρ between 1.5 and 2 GeV affected the results by less than 10%.)

* Work supported in part by USERDA Contract No. E(11-1)-3001.

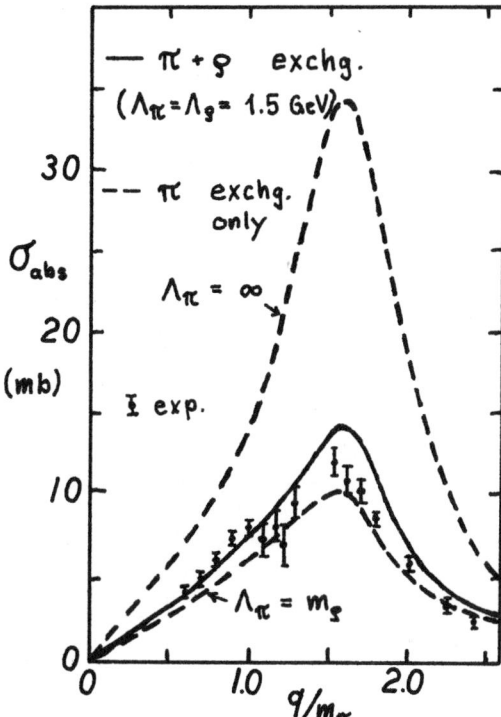

Fig. 1. Pion absorption cross section on deuteron as function of the incoming pion momentum q.

Our conclusions are thus opposite to those of ref. 4: After inclusion of the ρ meson, only very short ranged form factors give a reasonable agreement with experiment, especially for the larger values of κ_ρ = 5.5 - 6.6 found in recent analyses [5,6]. The low mass cut-offs of $\Lambda_\pi \simeq$ 700 MeV obtained from analyses of πN scattering data [7,8] seem to be incompatible with our calculations, even if we allow for an overall uncertainty in σ of 30 - 40% (mainly due to the uncertainty in the ρ meson coupling constant).

If pion cut-off masses lower than 1 GeV are to be believed, we have to look for further mechanisms which could enhance the cross section for the process (1), such as double rescattering or the exchange of higher mass bosons. Other improvements of the model, taking into account the momentum distribution of the deuteron, nonstatic vertex corrections, or the mass distribution of the ρ meson, will probably lead to minor corrections only.

Although many questions remain to be answered before we reach a quantitative understanding of the process (1), its extreme sensitivity to the pion cut-off energies may challenge new attempts to put the determination of form factors on a more rigorous theoretical basis.

REFERENCES

1. D.O.Riska, M.Brack and W.Weise, Phys. Lett. 61B, 41 (1976).
2. D.S.Koltun and A.Reitan, Phys. Rev. 141, 1413 (1966).
3. R.V.Reid, Ann. Phys. 50, 411 (1968).
4. B.Goplen, W.Gibbs and E.Lomon, Phys. Rev. Lett. 32, 1012 (1974).
5. F.Iachello, A.D.Jackson and A.Lande, Phys. Lett. 43B, 191 (1973).
6. G.Höhler and E.Pietarinen, Nucl. Phys. B95, 210 (1975).
7. K.Bongardt, H.Pilkuhn and H.G.Schlaile, Phys. Lett. 52B, 271 (1974).
8. W.Nutt and B.Loiseau, Nucl. Phys. B104, 98 (1976).

CALCULATION OF THE PION PRODUCTION REACTION ^{3}He$(p,\pi^+)^4$He

James H. Alexander
Dept. of Physics, University of British Columbia, Vancouver, B.C.

Harold W. Fearing
TRIUMF, Vancouver, B.C.

ABSTRACT

The differential cross section for the reaction ^{3}He$(p,\pi)^4$He has been calculated in a distorted wave impulse approximation model. Results are presented for incident proton lab energies of 415 and 716 MeV and centre-of-mass pion angles of 0° to 100°.

Recently there has been much interest in the reaction $A(p,\pi)A+1$ where the final nucleus is left in a definite state. These reactions are interesting because in principle they probe details both of π-nucleus interactions and of high momentum components of nuclear wave functions. Most data so far, except for some on deuterium, have been at threshold, though more medium-energy data are to be expected soon from the new meson facilities. In particular, the reaction ^{3}He$(p,\pi)^4$He has been studied and preliminary data are to be reported at this conference.[1] As the reaction $d(p,\pi)t$ was described fairly successfully in a DWIA calculation[2] the same model has been extended to ^{3}He$(p,\pi)^4$He.

The model used expresses the cross section for $A(p,\pi)A+1$ in terms of the cross section for $pp \to d\pi$ and a form factor which is basically a Fourier transform of the overlap of initial and final wave functions. Spin and antisymmetrization effects are included and distortion effects are put in via Glauber approximation. Details are as in Ref. 2.

The above model was used to calculate $d\sigma/d\Omega$ for ^{3}He$(p,\pi)^4$He at proton lab energies of 415 and 716 MeV for $0° \leq \theta_\pi^{cm} \leq 100°$. The effective momentum transfer in this range is 300-500 MeV/c at 415 MeV and 350-700 MeV/c at 716 MeV. It was found that the calculations were quite sensitive to the nuclear wave functions used. The form factor calculated in this model is somewhat similar to the charge form factor. Both Gaussian and Irving-Gunn S-state wave functions were tried. The former reproduce the charge form factor up to the first dip but are much too low beyond that, while the latter are too high over the first dip but reproduce the second maximum. As seen in the figure, the presence of higher momentum components in the Irving-Gunn wave functions leads to a cross section which decreases more gradually from forward direction than with the Gaussian wave functions.

The results shown here include distortion in the p^3He and the π^4He systems as in Ref. 2. Such distortions alter the shape of the cross sections to some extent, but primarily affect the normalization. An alternative way of putting in the distortion[2] or reasonable changes in the parameters can increase or decrease the overall normalization by factors of 2-3. The overall distortion effect is somewhat smaller at 716 MeV than at 415 MeV because of the smaller NN and πN cross sections at this energy.

At the time of this writing, only very preliminary data are available.[1] At 415 MeV both the normalization and shape of this data are reproduced well using the Irving-Gunn wave functions. At 716 MeV the shape is still in acceptable agreement with the data, but the normalization seems low by about a factor of 5. At both energies results with Gaussian wave functions decrease too rapidly, as would be expected, and give results sometimes as much as an order of magnitude too low at the larger angles.

Further refinements of the calculations are planned. Since the results are sensitive to the details of the wave functions, the calculations will be repeated using wave functions which better reproduce the charge form factors. Also as the D-states in the nuclear wave functions were significant in the $d(p,\pi)t$ calculations, serving to increase the cross sections at high momentum transfers,[2] they may be even more important in the $^3He(p,\pi)^4He$ calculations where the momentum transfer is greater. Results of these calculations will be reported at a later date.

REFERENCES

1. B. Tatischeff, private communication.
2. H.W. Fearing, Phys. Letters <u>52B</u>, 407 (1974); Phys. Rev. <u>C11</u>, 1210 (1975); Phys. Rev. <u>C11</u>, 1493 (1975).

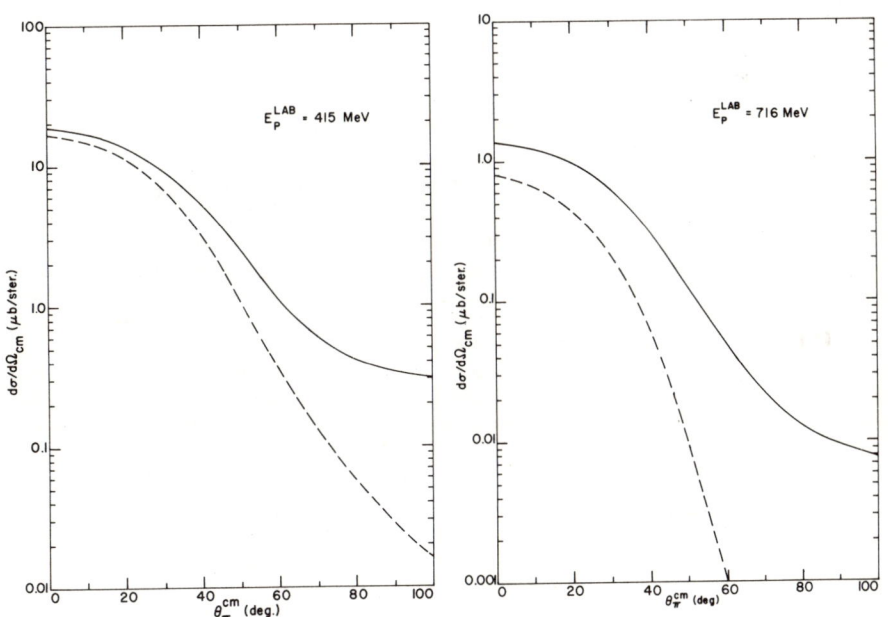

Fig.1. DWIA differential cross section for $^3He(p,\pi)^4He$ using Irving-Gunn (solid line) and Gaussian (dotted line) wave functions.

^{3}He(p, π^+)^{4}He REACTION AT 415 AND 716 MeV

B. Tatischeff, L. Bimbot, R. Frascaria, Y. Le Bornec
M. Morlet and N. Willis
Institut de Physique Nucléaire, B.P. n°1, 91406, Orsay, France

R. Beurtey, G. Bruge, P. Couvert, D. Garreta, D. Legrand
G.A. Moss and Y. Terrien
DPhN/ME, CEN-Saclay, B.P. n°2, 91190, Gif-sur-Yvette, France

Due to the high momentum components involved, the pion production induced by intermediate energy protons on light targets is expected to be an interresting reaction to investigate the role of Δ components. For very light nuclei (A $\leqslant$ 4), the S state component of the wave functions is rather wellknown, so that a comparison of experimental results from ^{3}He(p, π^+)^{4}He with classical calculations may reveal exotic components of the wave functions.

The differential cross sections for the ^{3}He(p, π^+)^{4}He reaction have been measured at 415 MeV and 716 MeV using the Saturne synchrotron proton beam[3] over a wide angular range (up to 100° CM). A liquid target (143 mg/cm^2), built by the IPN cryogenic service[4], has been used. The positive pions were detected by means of the SPES I facilities[5], essentially composed of a spectrometer and several drift chambers, scintillator counters and Cerenkov detectors. The beam intensity was monitored by a secondary electron chamber and calibrated with ^{11}C activity measurements from ^{12}C(p, pn)^{11}C. Very good reproducibility has been observed through different measurements at the same angle and same energy. Except for the smallest and the largest angles the statistical accuracy (plotted on the experimental results) was < 5%. An additional uncertainty, whose maximum value may be as large as $\pm$ 20%, is essentially due to the uncertainties on detectors efficiencies, solid angle and for a smaller part on target thickness, beam monitoring and muon contamination. The results are plotted in Fig.1 and Fig.2. When extrapolating the data, good agreement with the two previous measurements [1,2] is obtained.

Elementary calculations in the plane wave approximation have been done using one-nucleon and two-nucleon mechanisms. In the one-nucleon description which corresponds to a neutron stripping, the transition form factor corresponds to the neutron wave function in a zero range approximation. This function has been deduced from the nuclear density $\rho(r)$ corresponding to the elastic electron charge form factor $F(q)$. The computer code Piuck of Rost and Kunz has been used. For the two-nucleon mechanism calculation Ruderman's method, developped by Ingram [7], was followed. The experimental p(p, π^+)d cross sections of ref. 8 are introduced for the direct elementary process with the phase space term given by the Fearing's formula [9]. The energy of the incident proton for the elementary cross section is given by the Ingram's kinematical prescription. At Ep=415 and 716 MeV respectively this method gives an effective energy of 629 and 1040 MeV. For this last value the p(p, π^+)d cross-section is very small. We have considered the interaction of the incident proton on a nucleon moving in the same (and not opposite) direction, which allows us to fix the energy of the elementary process at the energy (616 MeV) corresponding to the maximum of the cross section. Harmonic oscillator wave functions for ^{3}He and ^{4}He nuclei are used with parameters adjusted in order to give good values of the charge root mean square radii (such wave functions do not

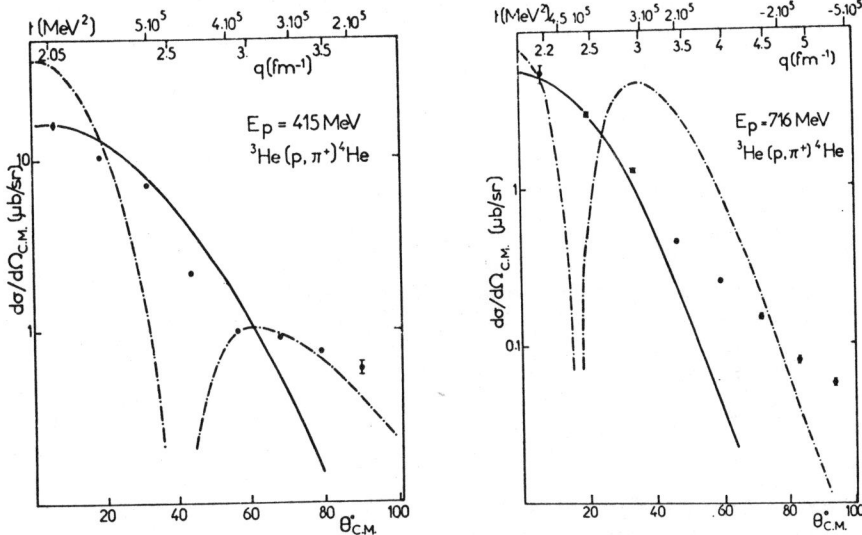

Fig. 1 The dotted-dashed lines correspond to the 1N mechanism and are multiplied by .2 Fig. 2
The full lines correspond to the 2N mechanism and are multiplied by 1.5

reproduce correctly the charge form factors at large momentum transfers.

The results corresponding to both calculations are shown in Fig. 1 and 2 (with a normalization factor). We note a deep minimum in the 1N case, at the zero of the Fourier transform of the neutron captured wave function $\varphi(r)$. In the 2N case the angular distribution, except for backward angles, reproduces rather well the experimental results. A change of 8% of one H.O. parameter does not change the angular shape but the normalization changes by a factor of 40%.

The 1N and 2N amplitude cannot be simply added especially due to the unknown relative phase. More complete calculations with S and D states giving a good description of both ^{3}He and ^{4}He nuclei are needed in order to study the role of isobars in the reaction mechanism.

References :
(1) K. Gabathuler et al., Nucl. Phys. B40(1972)32
(2) Yu.K. Akimov et al., Soviet Physics Jetp 14(1962) 512
(3) J. Thirion et al., Note CEA-N-1248
(4) S. Buhler, Proc. of the 8th Int. Cryogenic engineering conf. Berlin, May 70
(5) R. Beurtey, Brentwood Summer School, B.C. Canada, June 1975
(6) R.F. Frosch et al., Phys. Rev. 160 (1967) 874
(7) C.H.Q. Ingram et al., Nucl. Phys. B31(1971) 331
(8) C. Richard-Serre et al., Nucl. Phys. B20 (1970) 413
(9) H.W. Fearing, Phys. Lett. 52B (1974) 407 ; Phys. Rev. C11 (1975) 1210

SOFT-PION PRODUCTION IN N-N COLLISIONS

A.W. Thomas
CERN, 1211 Genève 23, Suisse

I.R. Afnan
School of Physical Sciences, Flinders University,
Bedford Park, 5042. Sth Aust., Australia.

ABSTRACT

The matrix element for s-wave pion production $nn \leftrightarrow \pi^- d$, is examined as a function of the external pion mass. Quite unexpectedly, in view of the well-known failure of soft pion theory for this reaction, it is found to be constant within a few per cent. Both the reason for this result and its implications are briefly discussed.

INTRODUCTION

It has been known for some time that soft-pion theory, despite its successes elsewhere, fails quite badly for s-wave pion production in $nn \leftrightarrow \pi^- d$ [1]. For real pions, the impulse approximation gives anomalously small results, owing to a "chance" cancellation of matrix elements leading to the S- and D-wave components of the deuteron. (A cancellation which led to suggestions of using this reaction to constrain P_D [2].)

More recently there has been a suggestion [3] that using PCAC one can put a model-independent constraint on the cross-section for μ-capture on nuclei leading to low-energy neutrinos. For such a process at small neutrino momentum, the cross-section is proportional to $|\underline{A}|^2 + |A_0|^2 + |\underline{V}|^2$, where A and V are the familiar axial-vector and vector currents. Using the standard PCAC arguments, the matrix element of A_0 is the amplitude for absorption of an s-wave "pion" of mass $\mu = m_\mu$ (the muon mass). Unfortunately the anomalous behaviour at the real mass ($\mu = m_\pi$), which we described above, makes one very dubious about the reliability of extrapolating from m_π to m_μ.

RESULTS

We have investigated this extrapolation by extending the model of Refs. 4 (and Ref. 2), with remarkable results -- summarized in Table 1. [The subscripts B-S and RSC refer to the Bryan-Scott (P_D = 5.38%) and Reid Soft Core (P_D = 6.47%) deuteron and $^3P_1$ wavefunctions, and M(μ) is the invariant matrix element for $nn \leftrightarrow \pi^- d$ as a function of the external pion mass "μ".] We notice the following:
i) As already stated, the sensitivity to P_D at $\mu = m_\pi$ is due to the difference in the single scattering contribution SS (i.e. no pion rescattering) in that case. ii) As μ decreases, this difference decreases (column four), and the SS contribution grows relative to the term in which the pion rescatters once, DS (see columns five and six). iii) The shift of importance from DS to SS is such that the invariant matrix element is essentially a constant (first two columns)!

Table 1

Variation of the invariant matrix element for
s-wave nn ↔ π^-d, with the external pion mass (μ).
$[\Delta \equiv \{|M|^2_{BS} - |M|^2_{RSC}\}/\{|M|^2_{BS} + |M|^2_{RSC}\}.]$

μ(MeV)	$|M(\mu)|^2/|M(m_\pi)|^2$			$\dfrac{M^{SS}(\mu)}{M^{SS}(\mu) + M^{DS}(\mu)}$	
	B–S	RSC	Δ	B–S	RSC
m_π	1.00	1.00	8.8%	16%	−3%
130.0	1.01	1.03	7.6%	21%	4%
120.0	0.98	1.03	6.5%	24%	10%
m_μ	0.97	1.04	5.0%	30%	20%
90.0	0.97	1.07	3.9%	39%	32%

SUMMARY

While investigations are continuing, we can make the following comments. First, the extrapolation of the matrix element of A_0 from m_π to m_μ can be carried out rather accurately (within ±5%) by assuming it is constant! Second, the close relationship of P_D to the OPEP tensor force suggests that there may be a conspiracy afoot. Further investigation of this phenomenon is clearly required -- particularly in the region $\mu \to 0$.

It is a pleasure to acknowledge conversations on this subject with M. Ericson and T.E.O. Ericson.

REFERENCES

1. M.E. Schillaci and R.R. Silbar, Phys. Rev. 185, 1830 (1969).
2. A.W. Thomas and I.R. Afnan, Phys. Rev. Letters 26, 906 (1971).
3. J. Bernabeu, T.E.O. Ericson and C. Jarlskog, Contributed paper (V.A.22) to the 6th Internat. Conf. on High-Energy Physics and Nuclear Structure, Santa Fe, 1975, and private communication.
4. A.E. Woodruff, Phys. Rev. 117, 1113 (1960).
   D.S. Koltun and A. Reitan, Phys. Rev. 141, 1413 (1966).
5. R.V. Reid Jr., Ann. Phys. (USA) 50, 411 (1968).
   R. Bryan and B.L. Scott, Phys. Rev. 177, 1435 (1969).

SLOW π- MESON ELASTIC SCATTERING ON NUCLEI

G.G.Bunatian, Yu.S.Pol

Joint Institute for Nuclear Research, Dubna, USSR; Lebedev Physical Institute of the USSR Academy of Sciences, Moscow, USSR

In order to describe the π-meson interaction with nuclei an effective quasipotential V_{eff} analogous to that for the bound states of π-mesic atoms is introduced. We used the most general type of the effective quasipotential V_{eff} including in particular the pair absorbtion of π-mesons, renormalization of the P-wave part of V_{eff}, consideration of the Fermi motion of nucleons. The dependence of the results on the nuclear density distribution form and on the type of kinematic transformation of the πN-scattering amplitude from πN system to pion-nuclei system was investigated. For low π-meson energies ($\leq$ 120 MeV) such potential with symmetrized Fermi distribution of nucleon density gives a sufficiently good description of elastic scattering differential cross-sections, all the results being weakly sensitive to possible uncertanties in V_{eff}. In the case of scattering of π-mesons with energies $\gtrsim$ 120 MeV through angles $>$ 40° only quantitative agreement with experiment is achieve. The fig. presents the results obtained for the ^{4}He nucleus. Different kinds of curve correspond to different enumerated above versions of V_{eff}.

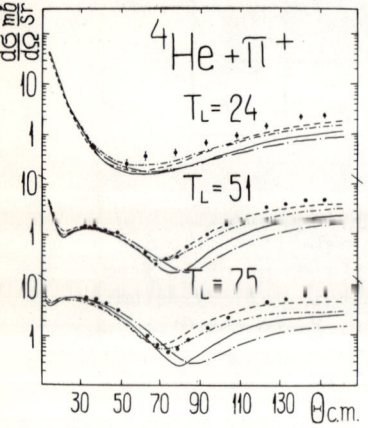

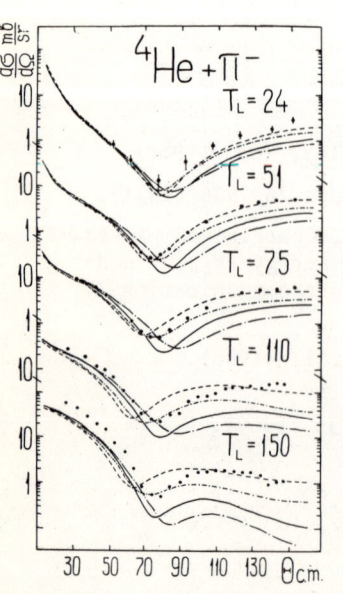

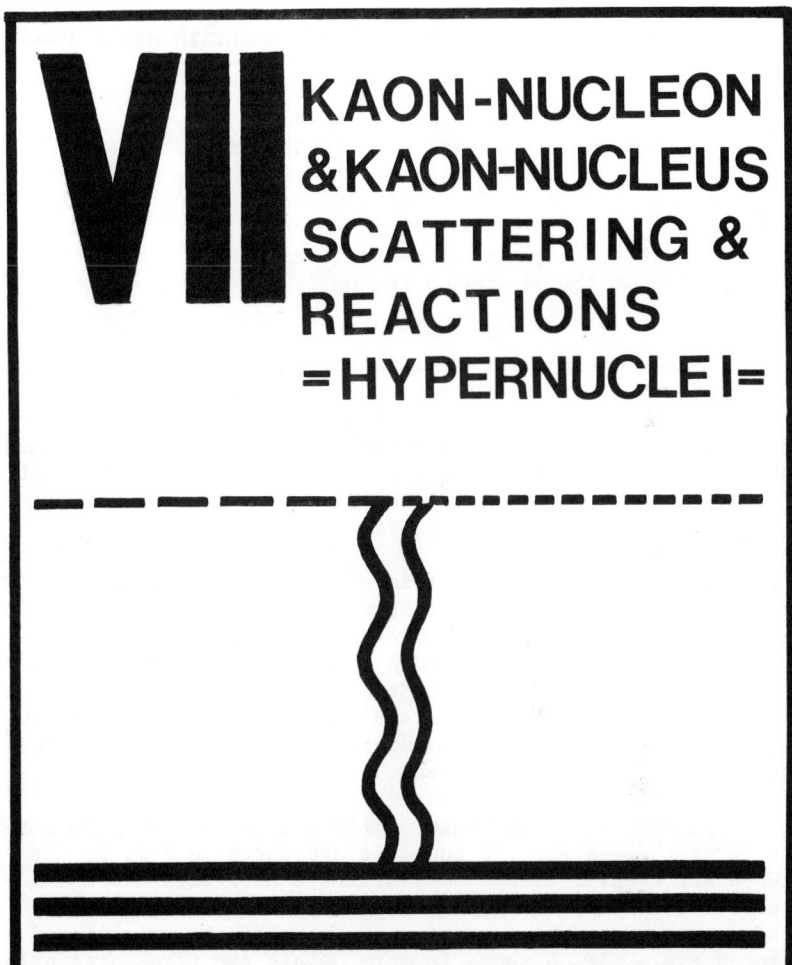

CURRENT STATUS OF KAON-NUCLEON ANALYSIS BELOW 3 GeV/c*

K. K. Li
Physics Department
Brookhaven National Laboratory, Upton, N.Y. 11973

ABSTRACT

A brief description of current results in the kaon-nucleon interaction in the resonance region is presented.

The low energy kaon-nucleon interaction in the resonance region has been extensively reviewed in the literature [1]. A brief description relating to the more recent results is presented. The results on the high precision total cross section measurements are shown in more detail as these have often revealed the first indication of the possible existence of resonances.

1. $\bar{K}N$ System:

It is well known that this system below about 2 GeV/c is very rich in structure. Fig. 1 shows the K^-p total cross section below about 3 GeV/c. The most prominent features are the peaks at about 400, 1000 and 1600 MeV/c, and a step and shoulder at about 800 MeV/c. Fig. 2 shows in more detail the recent results below about 1 GeV/c. It reveals further smaller structures at about 600 and 900 MeV/c. The statistical uncertainty of this type of measurement is typically better than 0.5%. The corresponding cross sections on deuterium are shown in Fig. 3 and 4.

The derived total cross sections [2] of the two pure I-spin states of the system, I=0 and I=1, are shown in Fig. 5. Many structures are clearly seen. Most of these have been confirmed by the results of detailed partial wave analyses to be resonances. Indeed, this gives a good picture of where most of the established Y^* resonances are. Fig. 6 shows in more detail the structures below 1 GeV/c. Apart from the known $\Lambda(1690)$, $\Sigma(1670)$ and $\Sigma(1770)$, there are evidences for several more new ones.

There are many detailed measurements of the cross sections, angular distributions and polarization of the elastic and the various inelastic channels. Many partial wave analyses [3] have been performed to search and study these resonance states. The most recent results are from the Rutherford-Imperial College (RL-IC) collaboration [4] in the c.m. energy range of 1480-2170 MeV, and from the College de France-Saclay (C-S) group [5] from 1950 to 2550 MeV. There are more recent data which were not included in the original RL-IC analysis and so can be used as a check on their predictions. One is from the study of the $K_L^0 p$ reactions. In particular, Fig. 7 shows the pure I=1 $\Sigma^0\pi^+$ production cross sections [6]. The dotted curve is the prediction from the RL-IC analysis and

*Work supported by the Energy Research and Development Administration.

the solid curve is the fitted result. The agreement is quite good, except perhaps at low energy. The other is from the low energy K^-p elastic polarization measurement from the BNL-Yale group. Fig. 8 shows their preliminary results [7] and the curves are the RL-IC predictions. Again, there is general agreement except at low energy. Also RL-IC predicted a rapid change in polarization at about 1050 MeV/c which is not shown in the data.

The parameters of the well-known Y^* resonances from these recent results are shown in Table I. They may change slightly with more and better data. The existence of these states is, however, considered established. Listed also are the names used by the Particle Data Group [8], and the energy of the corresponding structure in the total cross section [2,9].

There are many more probable or possible Y^* resonances observed that need further confirmation. For instance, Table II lists the parameters of some of the possible Y^* shown in Fig. 6. In paricular, the RL-IC analysis gives a possible $P_{1/2}$ resonance at 1738 MeV which might correspond to the $\Sigma(1715)$. The $\Sigma(1580)$ reported earlier [10] was subsequently observed by Litchfield [11] in an analysis of the reaction $K^-p \to \Lambda\pi$. However, the preliminary data reported [12] by the ANL-CMU group on the reaction $K_L^0 p \to \Lambda\pi$ do not require such a state.

2. KN System:

The main interest in the low energy KN reactions has been on the possible existence of the exotic resonances Z^*, because such states cannot be described by the conventional quark model for the baryons. As the system has strangeness + 1, inelastic effects become appreciable only at well above 700 MeV/c. The first indication of the possible existence of such states came from the total cross section measurement [13].

(i) I=1:

Fig. 9 shows the K^+p total cross section above about 1 GeV/c. Structures are seen at about 1900, 2150 and 2500 MeV. All subsequent studies, however, have essentially been on the lower structure at 1900 MeV. Fig. 10 shows more recent data in low energy, the elastic cross sections and the cross sections of the inelastic 1-pion and 2-pion production [14]. The structure can be described qualitatively by the sharp rise and the plateau of the 1-pion production and the fall of the elastic cross sections. Most recent partial wave analyses in this region agree qualitatively. Fig. 11 shows the Argand diagrams from the analyses of Cutkosky et al [15], Martin [16] and Adams et al [17]. They all show a small counter clockwise loop in the $P_{3/2}$ amplitude, but none of them favor a resonance interpretation. The result of the analysis of Arndt et al [18] using a two channel K-matrix formalism favored a $P_{3/2}$ resonance with a mass of 1787 MeV. However, recent K^+p polarization results cast some doubts on this conclusion. Fig. 12 shows the polarization results from the BNL-Yale group [7]. The prediction from Adams et al (in solid curve) shows fair agreement while that from Arndt et al (in dashed curves) seem to be too small. Thus the existence of the

Z_1^* is still an open question though recent results tend to favor a non-resonance interpretation. As there is a large non-resonance background, the detection is very difficult.

(ii) I=0:

Fig. 13 shows the K^+d total cross sections. The derived I=0 cross section [19] in Fig. 14 shows a shoulder at about 1700 MeV and a peak at about 1850 MeV. A broad peak at about 1750 MeV below the inelastic threshold is obtained after subtracting away the inelastic contribution as shown in Fig. 15. It is almost completely elastic.

Results from various partial wave analyses have been inconclusive, due mostly to the lack of polarization data. The result of Aaron et al [20] shows resonance behavior in the $D_{3/2}$ wave (Fig. 16) with a mass of 1830 MeV and a width of 100 MeV. The more recent works are from the BGRT collaboration [21] and from Martin [16]. Fig. 17 shows the three solutions from the BGRT results. Solution A showing no resonance behavior is incompatible with low energy polarization data. Solution C shows possible resonance behavior in the $P_{1/2}$ wave, and solution D favors a resonance in $P_{1/2}$. The results of Martin (Fig. 17) is somewhat in between solution C and D, and show counter clockwise loops in $P_{1/2}$ and $D_{3/2}$. Using solution D, the BGRT group obtained the resonance mass at about 1740 MeV, width of about 300 MeV and almost completely elastic.

The reaction $K_L^0 p \rightarrow K_S^0 p$, as first pointed out by London [22], offers interesting possibility in the Z_0 ambiguities as its amplitudes is a superposition of both I=0 and I=1 states of S=+1, and only I=1 state of S= -1. The results from different experiments using various combinations of Z_1 and Y_1 solutions are, however, somewhat inconclusive. The results from the Tel-Aviv Heidelberg collaboration [23] as shown in Fig. 18 using the BGRT Z_1 solution [24] and Y_1 solution from Langbein and Wagner [3] below 1.8 GeV and from Hemingway et al [3] above 1.8 GeV favors solution A (solid curve) rather than solution C (dashed curve). The ANL-CMU results [25] as shown in Fig. 19 using solution A and D with the BGT Z_1 solution [26] and Y_1 solutions from various groups seem to favor solution D in reproducing the backward peak. It can be seen that the method is very sensitive to the Y_1 solution. The BEGPR results [27] as shown in Fig. 20 using the BGRT Z_1 solution and the RL-IC Y_1 solution show solution D can reproduce the shape in the angular distribution better than solution A.

Perhaps the current experiment at BNL to measure the $K^+n \rightarrow K^+n$ polarization will distinguish these solutions more definitively.

Parameters of established Y* resonances

PDG	Mass (MeV) From σ_T	Wave	J^P	Mass M(MeV)	Full Width Γ (MeV)	Elasticity x	Dπ	Λπ
Λ(1405)		S_{01}	$\frac{1}{2}^-$	1405	40	0	1.0	
Λ(1520)	1518	D_{03}	$\frac{3}{2}^-$	1519	15	0.47		
Λ(1670)	1645?	S_{01}	$\frac{1}{2}^-$	1670	45	0.20		
Λ(1690)	1692	D_{03}	$\frac{3}{2}^-$	1690	60	0.24		
Λ(1815)	1820	F_{05}	$\frac{5}{2}^+$	1822	81	0.57		
Λ(1830)		D_{05}	$\frac{5}{2}^-$	1825	94	0.04		
Λ(1860)		P_{03}	$\frac{3}{2}^+$	1900	72	0.18		
Λ(2100)	2100	G_{07}	$\frac{7}{2}^-$	2110	250	0.30		
Λ(2350)	2350	*H_{09}	$\frac{9}{2}^+$	2370	170	0.14		
Λ(2585)	2585							
Σ(1385)		P_{13}	$\frac{3}{2}^+$	1385	35		0.12	0.88
Σ(1670)	1670	D_{13}	$\frac{3}{2}^-$	1670	50	0.08		
Σ(1765)	1770	D_{15}	$\frac{5}{2}^-$	1774	130	0.41		
Σ(1915)	1912	F_{15}	$\frac{5}{2}^+$	1920	130	0.05		
Σ(2030)	2025	F_{17}	$\frac{7}{2}^+$	2040	190	0.24		
Σ(2250)	2255	{*G_{19}	$\frac{9}{2}^-$	2211	77	0.02		
		*D_{15}	$\frac{5}{2}^-$	2267	97	0.07		
Σ(2455)	2455							
Σ(2620)	2620							

*Seen in partial wave analysis only recently. Listed here because they account for the structures in total cross section.

TABLE I

RESULTS OF THE FITS DESCRIBED IN THE TEXT. J IS THE SPIN AND X THE ELASTICITY OF A RESONANCE

Isospin	Laboratory Momentum (MeV/c)	Mass (MeV/c²)	Width (MeV/c²)	Height (mb)	(J+1/2)x
0	685	1646 ± 7	20	1.3	0.04
0	784	1692 ± 4	38	12.4	0.48
0	875	1735 ± 5	28	6.3	0.29
1	546	1583 ± 4	15	2.8	0.06
1	602	1608 ± 5	15	2.4	0.06
1	657	1633 ±10	10	1.4	0.04
1	737	1670 ± 4	52	6.5	0.23
1	833	1715 ±10	10	7.0	0.30

TABLE II

REFERENCES

1. See, for example, the review by P. J. Litchfield, London Conference (1974).
2. The procedure for unfolding the Fermi motion and calculating the Glauber-Wilkin shielding correction is described in:
   R. L. Cool et al., Phys. Rev. $\underline{D1}$, 1887 (1970).
   R. J. Abrams et al., Phys. Rev. $\underline{D1}$, 1917 (1970).
3. See, for example,
   (a) R. Armenteros et al., Nucl. Phys. $\underline{B14}$, 91 (1969).
   (b) J. K. Kim, Phys. Rev. Lett. $\underline{27}$, 356 (1971).
   (c) B. Conforto et al., Nucl. Phys. $\underline{B34}$, 41 (1971).
   (d) W. Langbein and F. Wagner, Nucl. Phys. $\underline{B47}$, 477 (1972).
   (e) A. T. Lea et al., Nucl. Phys. $\underline{B56}$, 77 (1973).
   (f) R. J. Hemingway et al., Nucl. Phys. $\underline{B91}$, 12 (1975).
   (g) P. Baillon and P. J. Litchfield, Nucl. Phys. $\underline{B94}$, 39 (1975).
4. Rutherford Lab.-Imperial College collaboration, Rutherford Lab. Report RL-75-182 (Dec. 1975).
5. A. de Bellefon et al., Nucl. Phys. $\underline{B90}$, 1 (1975).
   A. de Bellefon et al., Paper F2-02 submitted to the Palermo International Conference on High Energy Physics (June 1975).
6. Bologna-Edinburgh-Glasgow-Pisa-Rutherford collaboration, Rutherford Lab. Report RL-76-016. See also references for other data points given in this report.
7. M. E. Zeller, talk given at the Winter Meeting of the Brookhaven High Energy Discussion Group, 1975.

8. Review of Particle Properties, Particle Data Group, Rev. of Mod. Phys. 48, Part II (April 1976).
9. D. V. Bugg et al., Phys. Rev. 168, 1466 (1968).
   T. Bowen et al., Phys. Rev. D2, 2599 (1970).
10. K. K. Li, Baryon Resonances-73, p. 283 (1973). Purdue University, West Lafayette, Indiana.
    A. S. Carroll et al., Particles and Fields 1975 APS/DPF, Seattle, p. 376 (1975), University of Washington, Seattle, Washington.
11. P. J. Litchfield, Phys. Lett. 51B, 509 (1974).
12. G. Keyes et al., Am. Phys. Soc. Bull., p. 646 (April 1976).
13. R. L. Cool et al., Phys. Rev. Lett. 17, 102 (1966).
14. U. Casadei et al., CERN/HERA 75-1 (1975).
15. R. E. Cutkosky et al., Nucl. Phys. B102, 139 (1976).
16. B. R. Martin, Nucl. Phys. B94, 413 (1975).
17. C. J. Adams et al., Nucl. Phys. B66, 36 (1973).
18. R. A. Arndt et al., Phys. Rev. Lett. 33, 987 (1974).
    P. H. Steinberg, New Directions in Hadron Spectroscopy, p. 352, (1975), Argonne National Laboratory.
19. A. C. Carroll et al., Phys. Lett. 45B, 531 (1973).
20. R. Aaron et al., Phys. Rev. D7, 1401 (1973).
21. Bologna-Glasgow-Rome-Trieste collaboration, Nucl. Phys. B71, 138 (1974).
22. G. W. London, Phys. Rev. D9, 1569 (1974).
23. R. T. Ross, Palermo Conference rapporteur talk, Rutherford Lab. Report RL-75-115 (1975).
24. G. Giacomelli et al., Nucl. Phys. B20, 301 (1970).
25. Y. Cho et al., Phys. Lett. 60B, 293 (1976).
26. W. Cameron et al., Nucl. Phys. B78, 93 (1974).
27. Bologna-Edinburgh-Glasgow-Pisa-Rutherford collaboration, Rutherford Lab. Report RL-76-015 (1976).

David D. Brayshaw (SLAC): Unless the scattering is almost elastic, one should be cautious in making statements regarding "resonant" or "non-resonant" behavior based on Argand plots, "speed" or the like. In the A_1 3π system, for example, I have shown that it is quite possible to have a resonance pole without "resonant" phase behavior; while, in the πD 2^+ state I discussed yesterday, the phase goes through 90° and satisfies all the usual criteria, but there is no associated pole in the amplitude.

Li: No comment.

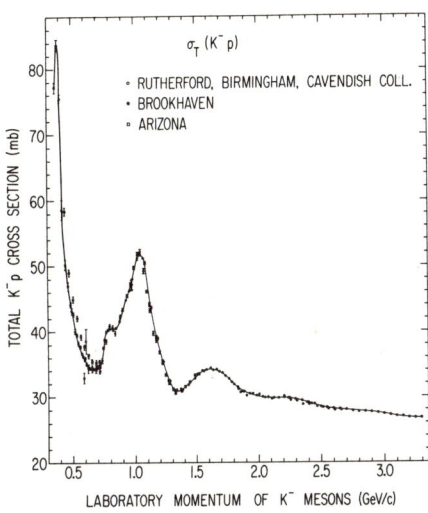

FIG. 1

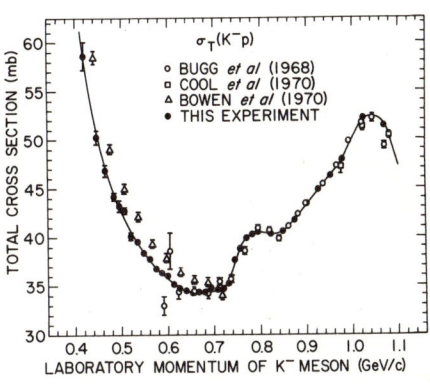

FIG. 2

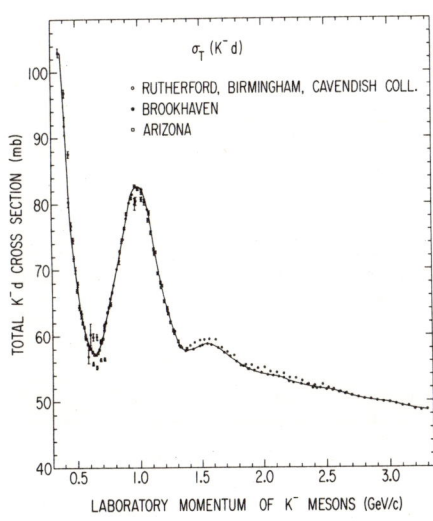

FIG. 3

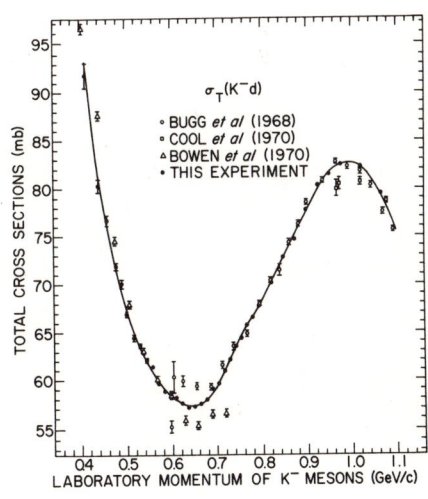

FIG. 4

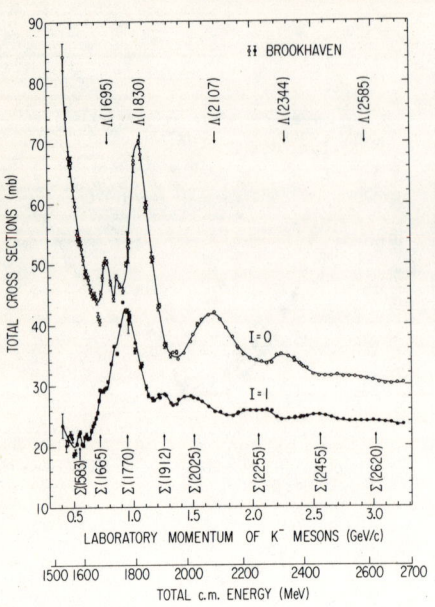

FIG. 5

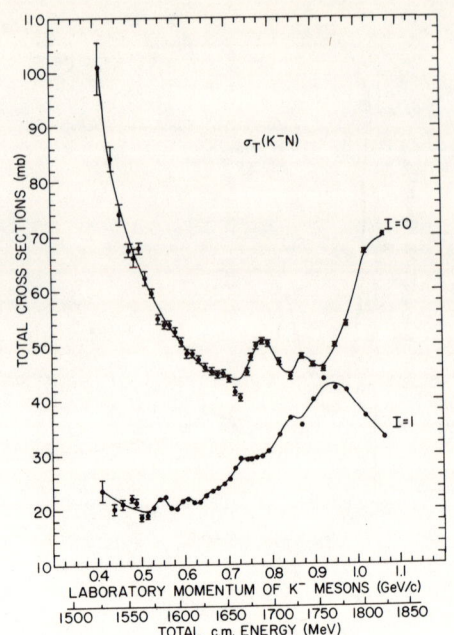

FIG. 6

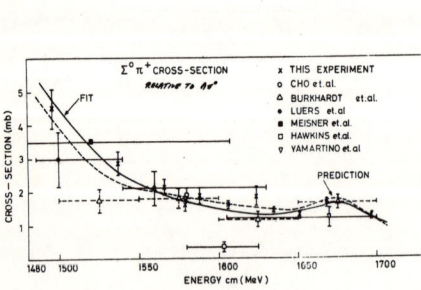

FIG. 7

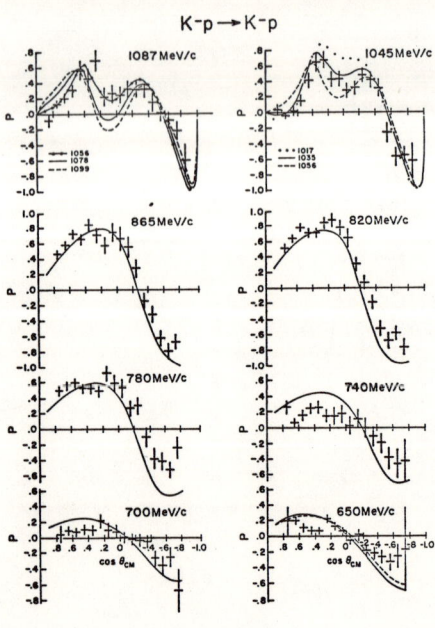

FIG. 8

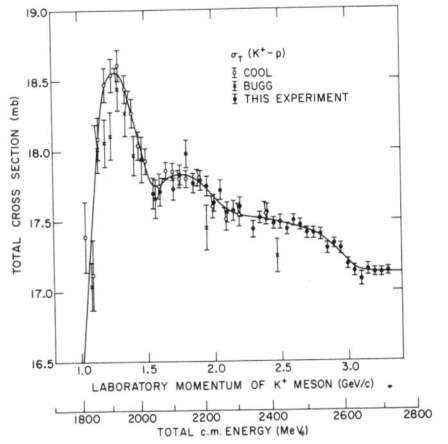

FIG. 9

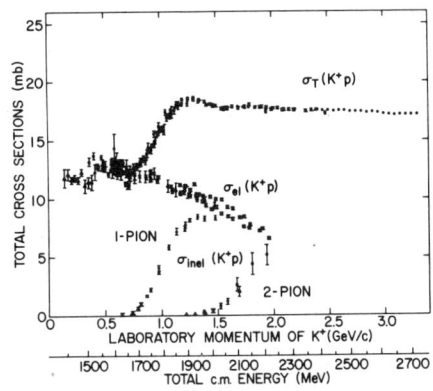

FIG. 10

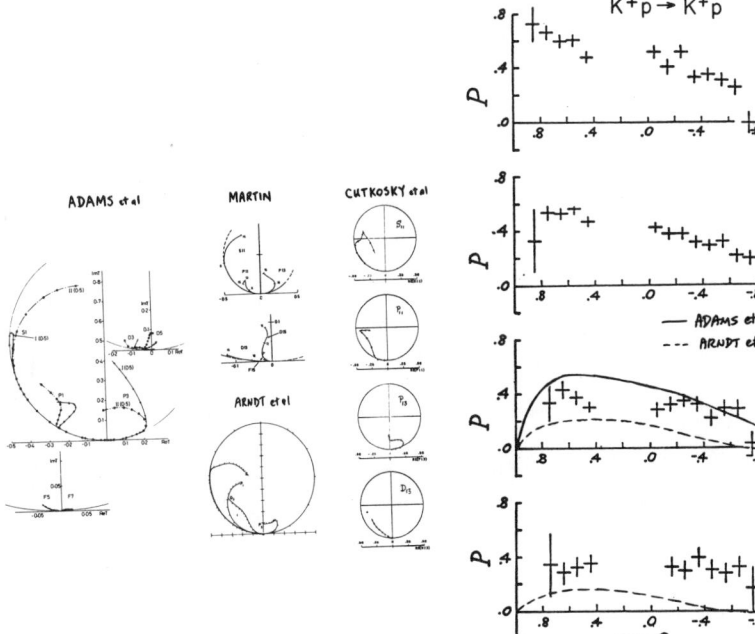

FIG. 11

FIG. 12

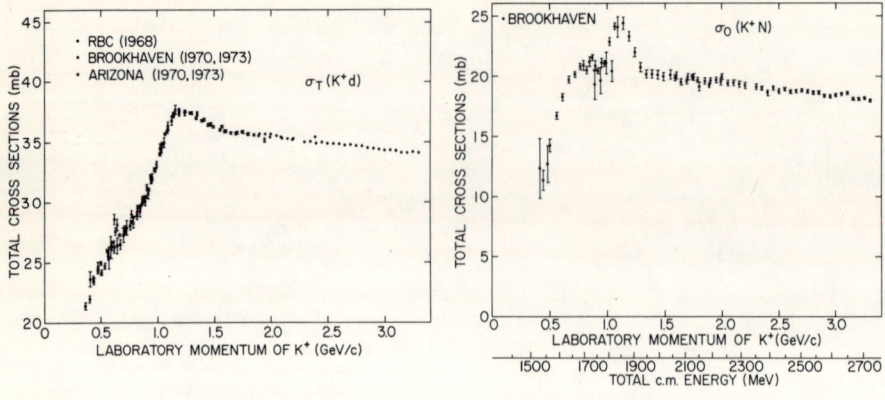

FIG. 13

FIG. 14

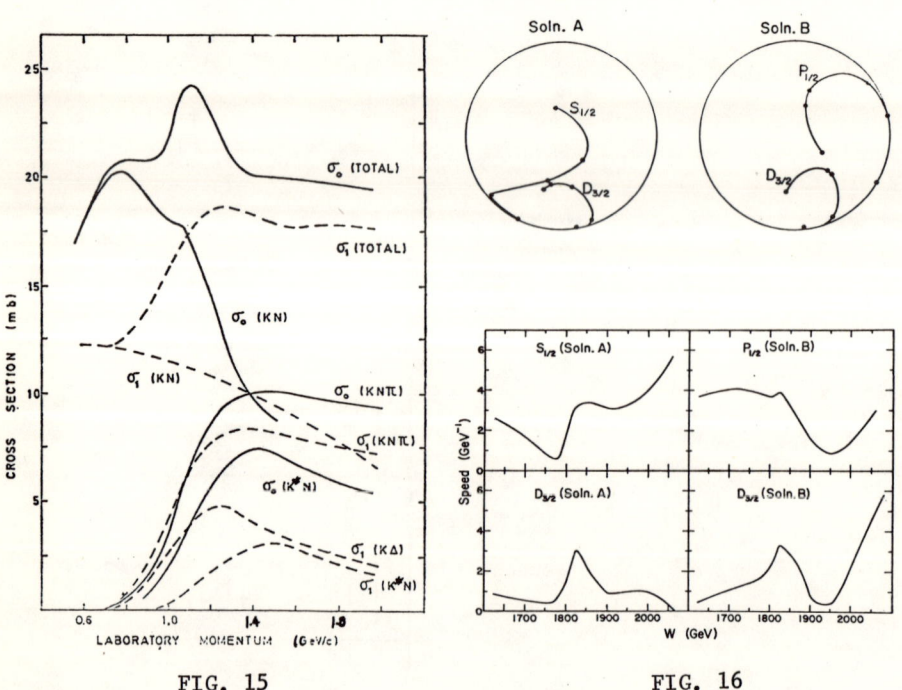

FIG. 15

FIG. 16

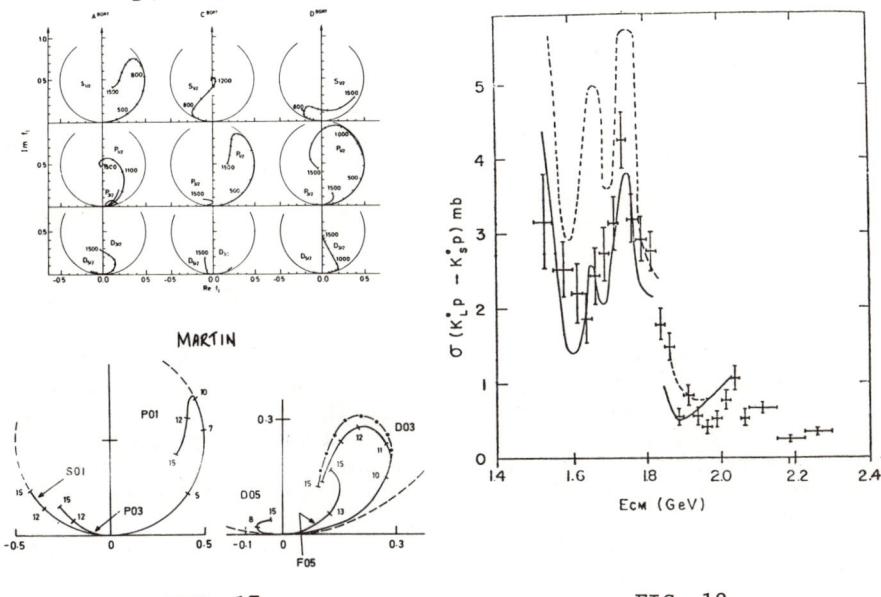

FIG. 17

FIG. 18

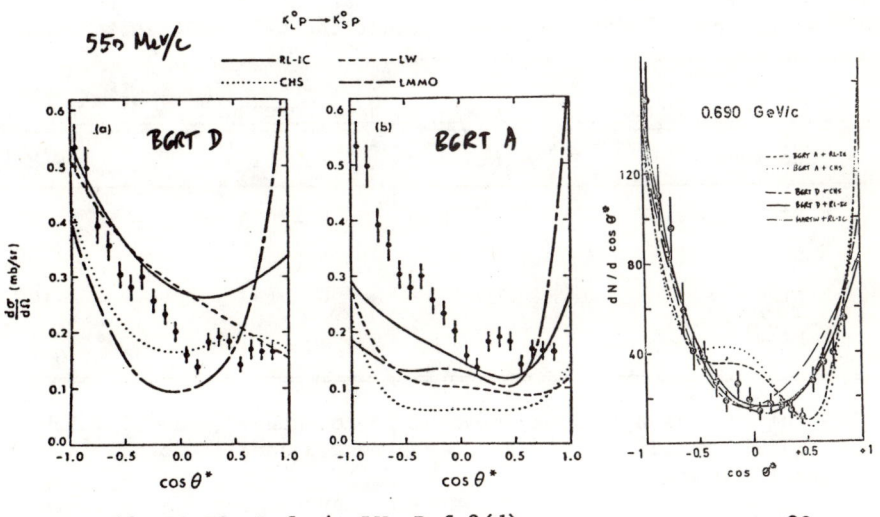

FIG. 19 RL-IC: Ref. 4, LW: Ref 3(d)
CMS: Ref. 3(a), LMMO: Ref. 3(e)

FIG. 20

THE OFF-SHELL $\bar{K}N$ T-MATRIX

M. Alberg
University of Washington, Seattle, Washington 98195

ABSTRACT

The on-shell formalism for the $\bar{K}N$ T-matrix is described. Phenomenological models of off-shell extrapolation are discussed. Several tests of off-shell behavior are reviewed.

INTRODUCTION

Theories of the kaon-nucleus interaction require a knowledge of the kaon-nucleon off-shell T-matrix. In principle, the $\bar{K}N$ interaction, both on and off-shell, could be derived from field theory, but our knowledge of $\bar{K}N$ dynamics is inadequate for this task. We must resort to phenomenological models which reproduce the experimentally determined on-shell T-matrix and provide a reasonable extrapolation to off-shell values. In this talk I will describe the models of the $\bar{K}N$ T-matrix that have been developed for use in the kaon-nucleus problem, so that the energy region of interest ranges from about 100 MeV below the $\bar{K}N$ threshold of 1432 MeV to about 100 MeV above it.

ON-SHELL T MATRIX

The K^- and $\bar{K}^0$ mesons are members of an isotopic doublet (I = 1/2) of strangeness S = -1. Near threshold, the $\bar{K}N$ channel is strongly coupled to two-body channels that include pions and the S = -1 hyperons Σ and Λ. The K^-N reactions of significant cross-section are shown in Table I.

Table I K^-N Reactions Near Threshold

reaction	Q (MeV)	reaction	Q (MeV)
$K^-p \to K^-p$	0	$K^-n \to K^-n$	0
$K^-p \to \bar{K}^0 n$	-5	$K^-n \to \Sigma\pi$	100
$K^-p \to \Lambda\pi$	180	$K^-n \to \Lambda\pi$	180
$K^-p \to \Sigma\pi$	100		

This energy region includes several Y* resonances, one of which, the Y_0^* (1405) of width Γ = 40 MeV, is strongly coupled to the $\bar{K}N$ system in the I = 0 state[1].

In the approximation of complete charge-independence it is appropriate to consider states of definite isospin. The $\bar{K}N$ T-matrix then splits into two symmetric submatrices, T^0 for the I = 0 state and T^1 for the I = 1 state[2].

$$T^0 = \begin{pmatrix} T^0_{11} & T^0_{12} \\ T^0_{21} & T^0_{22} \end{pmatrix} \begin{matrix} (\bar{K}N)_0 \\ (\Sigma\pi)_0 \end{matrix} \qquad (1a)$$
$$\;(\bar{K}N)_0 \;\;(\Sigma\pi)_0$$

$$T^1 = \begin{pmatrix} T^1_{11} & T^1_{12} & T^1_{13} \\ T^1_{21} & T^1_{22} & T^1_{23} \\ T^1_{31} & T^1_{32} & T^1_{33} \end{pmatrix} \begin{matrix} (\bar{K}N)_1 \\ (\Sigma\pi)_1 \\ (\Lambda\pi)_1 \end{matrix} \qquad (1b)$$
$$\;(\bar{K}N)_1 \;(\Sigma\pi)_1 \;(\Lambda\pi)_1$$

These on-shell T-matrix elements should be determined from experiment. The data includes differential cross-sections for the elastic channel and partial cross-sections for the other channels. Several analyses[1,3,4,5,6] have parameterized this data in terms of the real reaction matrix K

$$K = \left[T^{-1} + ik \right]^{-1} \qquad (2)$$

in which k is a diagonal matrix of channel momenta. The on-shell T-matrix elements are then found from the K-matrix elements given by one of these analyses.

MODELS FOR THE OFF-SHELL T-MATRIX

Theoretical descriptions of the $\bar{K}N$ off-shell T-matrix have been proposed by two groups: Krzyzanowski, Wrzecionko and Wycech in Warsaw, and Henley, Wilets and myself in Seattle. The models have many features in common. Both make the assumption that a separable $\bar{K}N$ potential exists. A separable form of the T-matrix follows, which is reasonable for energies at which the $\bar{K}N$ interaction is dominated by a single resonance. This condition is satisfied near threshold in the I = 0 state because of the Y_0^*. Both analyses use the Yamaguchi[9] form of the separable potential

$$V^I_{ij}(k,k') = \lambda^I_{ij} \, v_i(k) \, v_j(k') \qquad (3)$$

$$v_i(k) = \frac{1}{k^2 + \beta_i^2}$$

in which i,j = 1, 2 for I = 0 and i, j = 1,2,3[8] or 1[7] for I = 1. The parameter β_i corresponds to the inverse range of the force in the i'th channel. Substitution of this potential in the Lippmann-Schwinger equation

$$T = V + VGT \tag{4}$$

yields a solution for the T-matrix

$$T_{ij}^I(E;k,k') = v_i(k) A_{ij}^I(E) v_j(k') \tag{5}$$

in which $A_{ij}^I(E)$ is a complex function of the potential strengths λ_{ij}^I, their ranges, and the energy denominators which appear in G for the different channels.
The T-matrix is on-shell when

$$k^2 = 2\mu_i(E - M_i) \tag{6a}$$

and

$$k'^2 = 2\mu_j(E - M_j) \tag{6b}$$

in which μ_i is the reduced mass and M_i is the total mass in channel i. The T-matrix is half-off-shell if only one of these equations is true, and fully-off-shell if neither is satisfied.

Wycech and his collaborators fit the on-shell values of this T-matrix to the analysis of B.R. Martin[3], while we use Kim's analysis[6]. Wycech et al also constrain their T-matrix to fit the Y_o^* below threshold. The two sets of parameters for the $\bar{K}N$ potentials in the I = 0 state are compared in Table II.

Table II   Parameters of the I = 0 $\bar{K}N$ potential

potential model	Warsaw	Seattle
β_1 (fm^{-1})	5.42	5.56
β_2 (fm^{-1})	4.05	5.56
λ_{11}/β_1 (MeV)	-220	-326
$\lambda_{12}/\sqrt{\beta_1\beta_2}$ (MeV)	-207	-112
λ_{22}/β_2 (MeV)	+132	-412

In our model the ranges of the forces in the $\bar{K}N$ and $\Sigma\pi$ channels are set equal to each other. In both models these forces are found to have very short ranges, about .2 fm. This result agrees with the field-theoretic model in which the $\bar{K}N$ force is due to particle exchange. Conservation of angular momentum and parity require the exchange of vector particles with $J^P = 1^-$, so the lightest particle that can be exchanged is the $\rho(750 \text{ MeV})$, which corresponds to a force of short range (<.3 fm).

A comparison of the on-shell behavior of the T^o_{11} matrix element in both models is shown in Fig. 1. In each case the Y^*_o resonance is reproduced below threshold, with energies and widths that are within the experimental errors. The differences in the curves reflect the sensitivity of the models to the phenomenological analyses of the 2-body data.

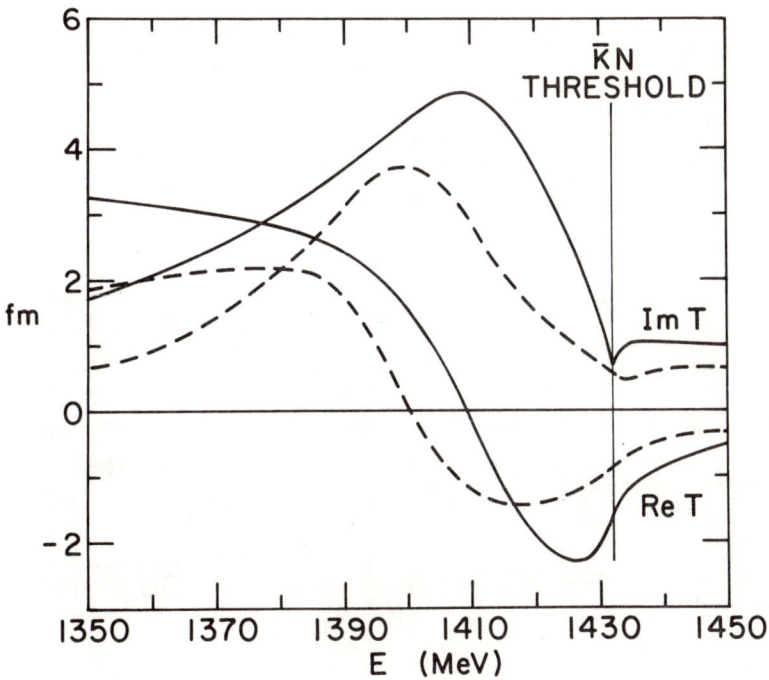

Fig. 1. The matrix element T^o_{11}. The smooth curves were calculated by Alberg et al; the dashed curves were calculated by Krzyzanowski et al.

Off-shell values of the T-matrix are needed in the theory of the kaon-nucleus interaction. For example, in our calculation[7] of the kaon-nucleus optical potential, equation (5) is solved in the presence of the nuclear medium. Energy denominators of the form $E - M_1 - k^2/2\mu_1$ are replaced by $E - M_1 - K^2/2M_1 - V_N - V_K - k^2/2\mu_1$, in which $\vec{K}$ is the center-of-mass momentum of the K^-N pair, and V_N and V_K are optical potentials for the nucleon and kaon respectively. $A^I(E)$ now behaves like $A^I(E + \Delta E)$. A typical value of ΔE in the surface of the nucleus would be $\Delta E = 10(1 + i)$ MeV. The T_{11}^o matrix element which is off-shell by ΔE is compared with T_{11}^o on-shell in Fig. 2.

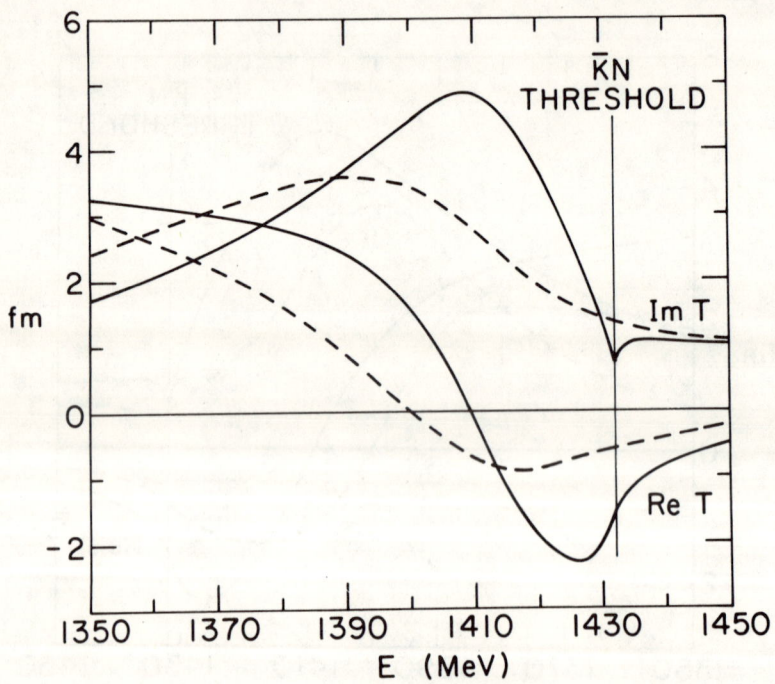

Fig. 2. The matrix element T_{11}^o calculated on-shell (smooth curves) and off-shell (dashed curves).

$$T_{11}^o \text{ (on-shell)} = v(k)\, A_{11}^o(E)\, v(k) \qquad (7a)$$

$$T^O_{11} \text{ (off-shell)} = v(k) \, A^O_{11} (E + \Delta E) \, v(k) \qquad (7b)$$

$$k^2 = 2\mu_1 (E - M_1)$$

A striking difference in the on-shell and off-shell behavior is seen. The Y^*_o resonance is broadened and shifted downward in energy. This illustrates how the off-shell extrapolation of the T-matrix changes the effect of the resonance in our calculation of the optical potential.

The models of the $\bar{K}N$ T-matrix which I have discussed are unfortunately very similar. Separable potentials which do not have the Yamaguchi form should be investigated. One would not expect, however, that the off-shell behavior of the T-matrix would be very model-dependent, because the range of the $\bar{K}N$ force is so short.

TESTS OF THE OFF-SHELL T-MATRIX

In principle we should be able to test the $\bar{K}N$ T-matrix off-shell by a comparison of K^--nucleus experiments with theory. For example, K^--nucleus elastic and inelastic scattering, and non-mesonic K^- absorption, include off-shell effects. I have referred to one area in which off-shell effects have been examined, namely K^--mesic atoms, for which a K^--nucleus optical potential has been constructed from the elementary $\bar{K}N$ interaction. We have carried out a Brueckner type of many-body calculation of the optical potential which is dependent on the off-shell behavior of the $\bar{K}N$ T-matrices. Shifts and widths of kaonic energy levels have been calculated from this optical potential. In Table III a comparison of on-shell and off-shell calculations is made.

Table III   Shifts and widths of the 3 d level in ^{31}P

	E (keV)	$-i\Gamma/2$ (keV)
experiment[10]	−0.33 (±.08)	−i 0.72 (±.06)
on-shell	−0.52	−i 0.82
off-shell	−0.55	−i 0.71

The difference between the two calculations is significant, especially in the widths. However, other effects, such as non-locality of the optical potential, and variations in the parameters of the nuclear matter distribution also cause changes of the same order of magnitude in the shifts and widths. It is difficult to make a quantitative study of off-shell effects by means of a complicated many-body system.

Another test of off-shell effects might be made in a 3-body

system, for which the Faddeév equations explicitly depend on the off-shell 2-body T-matrices. Two calculations have been made in such systems. Myhrer[11] studied the K^-d system, while Revai[12] treated low-energy K^--nucleus scattering as a 3-body problem composed of the kaon, a bound nucleon, and the residual nucleus.

Myhrer calculates the K^-d scattering amplitude as a function of the binding energy of the deuteron. He assumes that the $\bar{K}N$ interaction is dominated by the Y_o^*, and writes a separable T-matrix

$$T(E; k,k') = v(k) A(E) v(k')$$

in which the Breit-Wigner type amplitude is given by

$$A(E) = \frac{\lambda}{[E + \Delta E - E_Y] + \frac{i\Gamma}{2} \frac{k_1}{k_Y} \left|\frac{v(k_1)}{v(k_Y)}\right|^2}$$

$$k_1^2 = 2\mu_1(M_1 + \Delta E - M_2) \qquad k_Y^2 = 2\mu_2(E_Y - M_2)$$

in which ΔE is the binding energy of the deuteron, E_Y and Γ are the energy and width of the Y_o^*, and $v(k)$ has the Yamaguchi form. Myhrer uses this T-matrix in the $\bar{K}N$ multiple scattering series to calculate the K^-d scattering amplitude. He finds that in order to get the correct sign of the real part of this amplitude at threshold he must have a unphysically large binding energy of 20 MeV for the deuteron: This shows that more than just the binding energy must be put into ΔE.

Revai examines the scattering of a low energy kaon on a bound nucleon. He uses separable potentials to describe the interaction of the bound nucleon with the kaon and the residual nucleus. The kaon interacts only with the bound nucleon. The potentials are taken as Yamaguchis of appropriate ranges and strengths to fit the $\bar{K}N$ scattering length and the nucleon binding energy. Revai solves the Faddeév equations for the K^--(bound nucleon) scattering length as a function of the binding energy of the nucleon. His results are shown in Fig. 3.

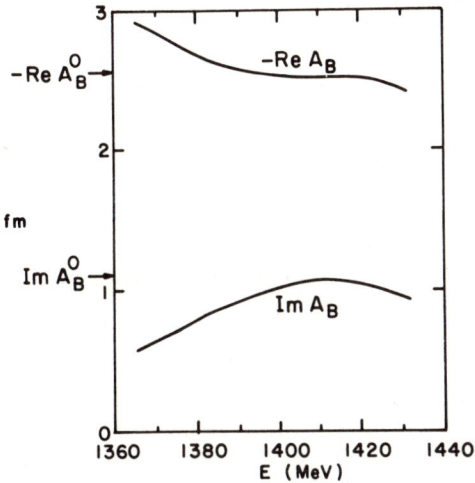

Fig. 3. The kaon-(bound nucleon) scattering length, from Revai[12].

Revai claims that A_B^0, the free $\bar{K}N$ scattering length multiplied by kinematic factors, is a good approximation to A_B, the kaon-(bound nucleon) scattering length. His conclusion is qualitatively correct, but quantitative calculations should use the off-shell calculation. Both Myhrer and Revai examine only the threshold behavior of the off-shell effects, and a careful study of the off-shell $\bar{K}N$ T-matrix should include non-zero values of the kaon momentum.

CONCLUSION

Off-shell effects in the $\bar{K}N$ T-matrix appear to be significant in several problems. Only one type of separable potential model for the off-shell T-matrix has incorporated the full multichannel nature of the $\bar{K}N$ interaction. Alternative theoretical descriptions should be developed and compared with the existing calculations. A simple binding energy correction does not seem to be sufficient to describe off-shell behavior. Kaonic atom data should be complemented by other experimental tests of off-shell effects.

REFERENCES

1. Y. A. Chao, R. W. Kraemer, D. W. Thomas and B. R. Martin, Nucl. Phys. B56, 46 (1973).

2. R. H. Dalitz and S. F. Tuan, Ann. Phys. (N. Y.) 10, 307 (1960).

3. B. R. Martin and M. Sakitt, Phys. Rev. $\underline{183}$, 1345 (1969).

4. A. D. Martin and G. C. Ross, Nucl. Phys. $\underline{B16}$, 479 (1970).

5. J. Thompson, Proc. of the Duke Conf. on Hyperon Resonances, (1970).

6. J. K. Kim, Phys. Rev. Lett. $\underline{19}$, 1074 (1967).

7. M. Alberg, E. M. Henley and L. Wilets, Phys. Rev. Lett. $\underline{30}$, 255 (1972); Ann. Phys. (N. Y.) $\underline{96}$, 43 (1976).

8. W. Krzyzanowski, J. Wrzecionko and S. Wycech, Acta Phys. Pol. $\underline{B6}$, 259 (1975).

9. Y. Yamaguchi, Phys. Rev. $\underline{95}$, 1628 (1954).

10. G. Backenstoss, J. Egger, H. Koch, H. P. Povel, A. Schwitter, and L. Tauscher, Nucl. Phys. $\underline{B73}$, 189 (1974).

11. F. Myhrer, Phys. Lett. $\underline{45B}$, 96 (1973).

12. J. Revai, Phys. Lett. $\underline{33B}$, 587 (1970).

Th. A. J. Maris (Univ. of Rio Grande do Sul): You remarked that in principle the off-shell matrix element could be calculated from field theory. It seems to me that the off-shell matrix element is not a priori defined. In field theory the LSZ technique gives a time-ordered product of interpolating fields, which (up to a renormalization) are only uniquely defined on the mass shell. Field transformations change the off-shell matrix elements without changing the Physics.

J. V. Noble (Univ. of Virginia): I would like to comment on the use of Faddeev equations in K-d scattering. Although it is true that the input to a Faddeev equation is the off-shell 2-body t-matrices, and that the 3-body amplitude will then be 2 and 3 body unitary, I must emphasize that unitarity is not dynamics. Unless the 2-body t-matrix satisfies a Lippmann-Schwinger equation, the Faddeev equation will not be equivalent to the Schroedinger equation with a Hermitian Hamiltonian, and therefore will represent different dynamics. Now, I note that in Myhrer's calculation the K-N propagator is an ad hoc form designed to reproduce a resonance, rather than being drived from the vertex functions via a Lippmann-Schwinger equation. A similar thing was done years ago in the 3-α problem by Duck, and he ran into similar difficulties to those found by Myhrer. Thus I would suggest that the problem in K-d is not the theory but its application.

Myhrer (CERN): A comment to Noble's remark. The $Y_0^*(1405)$ is a resonance in the $\pi\Sigma$ channel, but how to treat this in the K^-p channel is an open question. A separable potential approach like Revai's gives a bound state pole in the K^-p I=0 channel. The coupling to the $\pi\Sigma$ introduce a width to this bound state. Instead of starting from a potential in the Lippmann-Schwinger equation to generate the K^-p t-matrix, one can explicitly construct a resonance in the K^-p t-matrix. The consequences are that in a K^- bound nucleon or K^-d Faddeev calculation the latter approach gives different results compared to Revai's results.

Alberg: Your remarks correctly describe Revai's calculation, but not ours. Our I=0 $\bar{K}N$ channel does not contain a bound state, The resonance appears only when the coupling to the $\Sigma\pi$ channel is turned on.

Segel (Argonne-Northwestern): While there is a well established optical potential for use in low energy nuclear reaction studies, this potential tells us little about either nuclear structure or the nucleon-nucleon interaction. Is there any reason to be more optimistic about the kaon-nucleus situation and to feel that the establishment of an effective kaon-nucleus interaction will lead to the obtaining of fundamental information?

Alberg: The presence of the Y_0^* distinguishes the kaon-nucleus interaction from the nucleon-nucleus problem. The enhancement of the K^-p interaction should enable us to extract information about the low-density part of the proton distribution from kaonic atom data.

J. Law (Univ. of Guelph): In answer to Segel's question as to what one can learn from Kaonic work. I would like to indicate the line Deloff and I have taken. While M. Alberg has concentrated on the $\bar{K}N$ force, we have also concentrated on the nuclear structure aspect. We've fitted $\bar{K}N$ potentials to the complex scattering lengths. The resonance is there in our potentials. By using a multiple scattering formalism it is possible to separate the force aspect from the nuclear structure aspects. In fact we find a 2ℓ-moment dependence of the nuclear distribution on the kaonic atom level shifts.

G. A. Miller (Univ. of Washington): A major difficulty that occurs in the low energy proton nucleus problem is the strong influence of the compound nucleus. This makes a detailed theoretical analysis difficult. However this difficulty does not occur for K^- nucleus scattering at low energies.

L. Kisslinger (CMU): It is difficult to derive theoretical low energy potentials. Have you solutions which can fit some of the higher resonances so they can be used for in-flight experiments? Even at low energy the higher resonances could play an important role due to binding-Fermi motion effects.

M. Alberg: We have not attempted to fit higher resonances for kaon experiments above threshold.

A. N. Mitra (Univ. of Delhi): This is just a comment on Prof. Kisslinger's questions on the role of D-wave, etc., resonances which could be important in this analysis. It is, however, necessary to recognize the limitations of a Yamaguchi parametrization (such as we have heare) which allows only one resonance (in a definite J-state) at a time. To have more than one resonance one must be prepared to put in more Yamaguchi terms corresponding to the additional J-states. Off-shell, etc., considerations are of course not going to help in this regard.

M. Alberg: Our present analysis uses only s-wave potentials to describe the Y_o^*. Of course resonances in other angular momentum states could be described by additional separable potential terms.

M. K. Banerjee (U. of Maryland): I do not know much about K^-N system, but it seems to me that there are two kinds of forces. These can be ω exchange between K^- and N giving rise to strong short ranged local attractive potential. K^-N can combine to form Λ or Σ. This mechanism will give rise to repulsive (!) factorable separable potential. Has any one tried a combination of a local and factorable separable potentials.

Mary Alberg: This would be an interesting approach. To my knowledge no one has tried it.

NEW DATA ON THE $(K^-, \bar{\pi})$ REACTION

K. Kilian
Max-Plank-Institut für Kernphysik, Heidelberg

ABSTRACT

Population of hypernuclear states has been observed in recoilless Λ production via the strangeness exchange reaction (K^-, π^-) on ^{9}Be, ^{12}C, ^{16}O, ^{32}S and ^{40}Ca.

Spectra with pronounced peaks and nearly no background have been obtained. New information about the Λ-nucleus interaction can be extracted.

INTRODUCTION

The strangeness exchange reaction (K^-, π^-) has become an excellent tool to produce and investigate hypernuclei [1]. Performing this reaction on bound neutrons

$$K^- + (n + R) \to (\Lambda + R)^* + \pi^- \qquad (1)$$

hypernuclei are produced whose excitation levels show up in the K^-, π^- energy loss spectrum. Advantage may be taken of the fact, that the Λ particle can be produced without recoil [2]. For K^- momenta between 300 and 1000 MeV/c and for π^- detected at 0^0 the Λ recoil is small ($\leqslant$ 100 MeV/c) compared to the Fermi momenta of the target nucleons ($\sim$ 250 MeV/c) and there is a high probability of forming a bound hypernuclear system were the Λ stays within the well of the residual nucleus. Moreover the complexity of the formed states is drastically reduced. One expects, that a Λ particle replaces a neutron in the nucleus in a one step process without otherwise changing its wave function.

It is of general interest to compare such nuclear and hypernuclear states which have the same configuration. The Λ particle can be considered as a probe which "marks" one of the initially indistinguishible neutrons in the manybody system of the nucleus. One might be able to learn what interaction an individual particle feels in the multi particle system of the nucleus.

EXPERIMENT

Since the last review on hypernuclear experiments [1] there exist now new data on recoilles production of $^{16}_{\Lambda}$O $^{32}_{\Lambda}$S and $^{40}_{\Lambda}$Ca [3]. The latter two hypernuclei being the heaviest known so far. The (K^-, π^-) reactions were studied at a K^- momentum of 900 MeV/c, using a separated K^- beam at the CERN PS. About 3500 K^- per burst hit the targets. In order to achieve reasonable yields rather thick targets of 4.8 g/cm² H₂O, 9.6 g/cm² S and 6.3 g/cm² Ca were used. Therefore the experimental resolution of about 3 MeV was entirely determined by energy straggling in the targets. The energy loss between the incoming K⁻ and the outgoing π⁻ at 0° was determined using a focussing magnetic double spectrometer shown in figure 1. The target

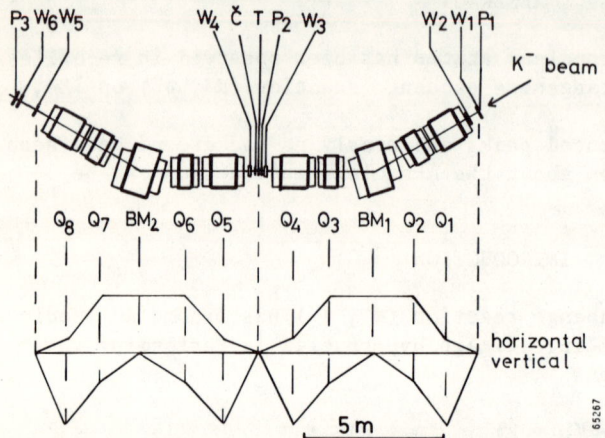

Fig. 1. Magnetic double spectrometer.
BM : bending magnets,
Q : quadrupoles,
p : plastic scintillators,
W : multiwire drift chambers,
T : target,
C : **liquid hydrogen** Cerenkov counter.

position was in the common focal plane between the two identical spectrometer parts. Since only standard beam transport elements were used, the optical aberrations were taken into account by applying higher order corrections during the evaluation of the exact particle momenta. For this purpose the space coordinates and the corresponding angles in all focal planes have been measured. We used twenty four planes of multiwire drift chambers, which allowed the particle coordinates to be determined with an accuracy of 0.4mm fwhm. The final energy resolution for the spectrometer after "software" corrections was < 1 MeV.

The (K⁻, π⁻) reaction was identified by requiring in three ways that the interaction took place in the target. First the time of flight between the entrance and exit of the spectrometer was measured with plastic scintillators. For K⁻ the time of flight is 68 ns ($\beta = 0.87$) and 60 ns for π⁻($\beta = 0.98$). Therefore 64 ns are expected for the (K⁻, π⁻) reaction. Corresponding to a time resolution of 0.6 ns fwhm, events with (64 ± 0.3 ns) have been considered for the final analysis. The main background in this sample results from decays of K⁻ mesons, where the timing window accepts decays occuring within ± 0.6m around the target. The total number of decays in this region is more than four orders of magnitude larger than the expected reaction rate, but only a small fraction of these decays yields particles which are accepted in the phase space of the second spectrometer part. These are predominantely events with big decay-angles ($>4°$ for $K\rightarrow 2\pi$) occuring in the region behind the target focus.

Most of the events with big angles can be rejected by requiring secondly that the measured vertical coordinate at the target position is in agreement with that coordinate reconstructed from the measured positions and angles at the entrance and exit of the spectrometer. The high accuracy for the reconstruction of the (K^-, π^-) vertex which guarantees the rejection of the K^- decays in flight is a big advantage of this focussing spectrometer design. In general, in the momentum region below 1 GeV/c the multiple scattering limits the accuracy of the vertex reconstruction unless one is using a focussing design and chambers and targets positioned in focal planes.

A liquid hydrogen Cerenkov counter of 3.5 cm thickness (less than 0.3 g/cm^2) was placed just behind the target. Due to a refractive index of n = 1.11 the threshold velocity for Cerenkov light emission in liquid hydrogen is $\beta = 0.9$ so that the π^- mesons produce light while the slower passing K^- do not. By requiring as a third condition a Cerenkov signal in this counter, the remaining K^- decay events in the region behind the Cerenkov radiator are ruled out. Liquid hydrogen has the advantage over all other radiator materials with the correct refractive index, that it does not produce any (K^-, π^-) reaction background in the investigated hypernuclear spectra.

RESULTS

The new results for the (K^-, π^-) reaction on ^{16}O, ^{32}S and ^{40}Ca are shown in figure 2. In addition reevaluated data on $^{9}_{\Lambda}Be$ and $^{12}_{\Lambda}C$ are given for comparison which are combined with some recent measurements. By more stringent requirements in the reconstruction of the trajectories also in the reevaluated data the background could be eliminated almost completely.

The spectra are plotted as a function of the Λ binding energy B_Λ. $B_\Lambda = 0$ corresponds to a Λ which is just unbound on the residual nucleus groundstate. The hypernuclear excitation energy E_{ex} is just the difference between the B_Λ of the groundstate and the actual Λ binding energy. Since the groundstates in hypernuclei heavier than $^{14}_{\Lambda}N$ are not known experimentally, the indicated groundstates in $^{16}_{\Lambda}O$, $^{32}_{\Lambda}S$ and $^{40}_{\Lambda}Ca$ are calculated by extrapolating the known binding energies of the very light hypernuclei [4]. Under the chosen kinematical conditions a population of the groundstate is not expected.

In all spectra with the exception of $^{40}_{\Lambda}Ca$, a strong transition to a narrow state ("peak") with a width comparable to the resolution (~ 3 MeV) is seen. In addition at higher excitations a broad "bump" with a width of the order of 10 MeV is observed. The $^{40}_{\Lambda}Ca$ shows only one broad distribution but the sharp rise at lower excitations may indicate that in $^{40}_{\Lambda}Ca$ the narrow peak and the broad bump overlap in energy. Relevant quantities for the spectra are summarized in table 1.

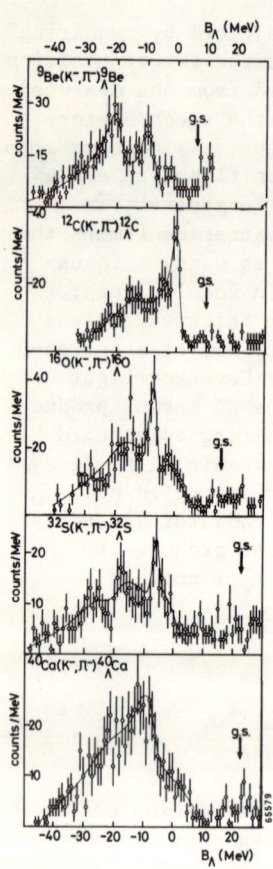

Fig. 2. Spectra of the (K^-, π^-) reaction on ^{9}Be, ^{12}C, ^{16}O, ^{32}S, and ^{40}Ca, as a function of the Λ binding energy B_Λ. $B_\Lambda = 0$ corresponds to a zero relative energy between the Λ particle and the core nucleus groundstate.

For all choosen targets it can be seen that the hypernuclear configurations populated are well concentrated in a limited interval of the excitation energy and that the 0^0 cross section integrated over the whole spectrum, is always about 1 mb/sr (see table 1). This is a strong argument for the assumption that we observe only contributions of the one step process in the (K^-, π^-) reaction. The mean free path λ of the K^- and π^- in the nucleus is about 1 to 2 fermi. Only the very narrow ring at the nuclear surface where the (K^-, π^-) path goes approximately 1 to 2 fm in nuclear matter can contribute to the 0^0 single step reaction. Further interactions of the incoming K^- or outgoing π^- would increase the excitation energy of the hypernucleus and most probably change the (K^-, π^-) angle. Simple geometrical considerations show that the volume of this ring depends only on the mean free path and not on the radius of the sphere. Therefore the cross section for the one step reactions which is expected to be proportional to this volume should in fact be identical on all nuclei. For the interpretation of the measured spectra as corresponding to target like excitations it is crucial that we really deal with a single step reaction alone where the reaction took place on a single neutron without any additional interaction.

In a quantitive interpretation of the spectra in the particle hole (1p, 1h) picture [5,6] the excitation energies which can be reached by a one step (K^-, π^-) reaction on a nuclear target are given by the sum of the excitation energy of the formed Λ and the energy of the created neutron hole. The recoil free strangeness exchange selects Λ-particle neutron hole pairs in the same spin-space configuration. The situation is schematically shown in figure 3 where e.g. a $1S_{1/2}$ neutron hole is created together with a $1S_{1/2}$ Λ and so on up to the last bound neutron of the target which is transformed into a Λ in the same configuration. From all targetlike hypernuclear configurations the $1S_{1/2}$ (1p, 1h) configuration is expected to have the highest excitation energy E_{max} simply corresponding to the energy difference of the last

Table 1

Target nucleus	Ground state B_n (MeV)	Ground state B_Λ (MeV)	Narrow peak B_Λ (MeV)	Narrow peak Γ_{exp} (MeV)	Narrow peak $V_{n\Lambda}$ (MeV)	Narrow peak $\sigma(0°)$ (mb/sr)	Broad bump B_Λ (MeV)	Broad bump Γ_{exp} (MeV)	Broad bump $V_{n\Lambda}$ (MeV)	Broad bump $\sigma(0°)$ (mb/sr)	Whole spectrum $\langle B_\Lambda \rangle$ (MeV)	Whole spectrum $\langle V_{n\Lambda} \rangle$ (MeV)	Whole spectrum $\sigma(0°)$ (mb/sr)
$^{9}_{\Lambda}$Be	2	7	-9	5	11	0.3	-21	11	23	0.8	-16	18	1.1
$^{12}_{\Lambda}$C	19	10	0	4	19	0.4	-12	13	31	0.7	-8	27	1.2
$^{16}_{\Lambda}$O	16	14*	(-1) / -7	7 / 4	17 / 23	0.2 / 0.4	-17	14	33	0.5	-11	27	1.3
$^{32}_{\Lambda}$S	15	20*	-6	5	21	0.4	-16 / -27	10 / 12	31 / 42	0.5 / 0.3	-15	30	1.2
$^{40}_{\Lambda}$Ca	16	21*	(1) / (-9)	2*5 / 5	15 / 25	0.2 / 0.2	(-22)	17	38	0.7	-14	30	1.1

The main parameters of observed resonances in the (K^-, π^-) reaction on different nuclei. B_Λ is the only directly measured quantity. $B_\Lambda = 0$ corresponds to zero relative energy between the Λ particle and the core nucleus ground state. The ground state energies with an asterisk are theoretical estimations. $V_{n\Lambda}$ is the difference in binding energy between the last neutron in the target nucleus and B_Λ: $V_{n\Lambda} = B_n - B_\Lambda$. The error in the relative yields amounts to about 30%. The numbers in parentheses represent a tentative description of the low-energy shoulder as weak narrow peaks. The decomposition is not significant statistically.

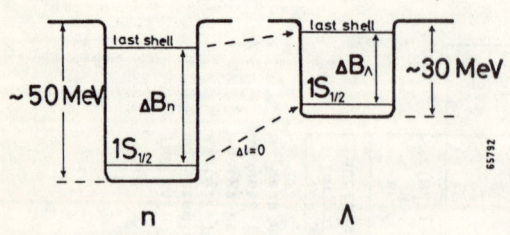

Fig. 3. Single particle energies for n and Λ in different potential wells. The highest excited targetlike (1p, 1h) configuration in the $1S_{1/2}$ shell differs from the hypernuclear groundstate which has a n-hole in the last shell by an energy ΔB_n.

neutron shell and the $1S_{1/2}$ shell, shown as ΔB_n in fig. 3. ΔB_n is experimentally determined [7] and is in good agreement with the excitation energies of the broad bumps in all measured spectra. The energetically lowest excitation is expected to correspond to a (1p, 1h) excitation in the last shell with an E_{min} value given by ΔB_Λ (fig.3). This interpretation was confirmed by the spectra of $^{12}_\Lambda C$ and $^{9}_\Lambda Be$[13]. In the $^{16}_\Lambda O$ spectrum even a splitting between the $P_{3/2}$ and $P_{1/2}$ configurations is indicated. A mixing between these two configurations by a residual Λ-nucleon interaction can be expected since the energy difference between these configurations is only some MeV. Mixing may shift the position of these two configurations and will enhance the intensity of the upper state. Following this line of argument, the sharp peak in $^{32}_\Lambda S$ would be assigned to the $D_{5/2}$ particle-hole excitation. Similarly to $^{16}_\Lambda O$, also in $^{40}_\Lambda Ca$ the major strength of the 2s, 1d shell would be concentrated in the $D_{5/2}$ particle-hole excitation it might be shifted and appears to be embedded in the broad bump. Simple calculations of ΔB_Λ assuming a Saxon-Woods potential, which reproduces the hypernuclear groundstates [4], yield for the energies of the last shell contribution in $^{16}_\Lambda O$, $^{32}_\Lambda S$, and $^{40}_\Lambda Ca$, E_{ex} = 11, 19 and 19 MeV, respectively. The disagreement between these values and the observed ones can be explained by mixing of (1p, 1h) configurations in the last shell.

Recently A. Bouyssy and J. Hüfner (BH) [8] have compared our data with model calculations in the framework of the DWIA (details in ref.6). They have extacted numerical results. Parts of the results are already shown in a contribution to this Conference [9]. In this analysis the neutron single particle energies were taken from experiments [7] and the single particle states of the Λ were computed from an energy independent shell model potential with a depth D_Λ and a spin orbit strength $U^{so}_{\Lambda\Lambda}$. The exact structure of the excited (1p, 1h) hypernuclear states was obtained by diagonalizing a residual Λ nucleon interaction.

The values D_Λ were determined so that they reproduced the measured Λ binding energy B_Λ for the narrow peaks. The resulting values for D_Λ are shown in fig.4. A mean value for D_Λ derived from this analysis on medium heavy nuclei is

$$D_\Lambda = 28 \pm 3 \text{ MeV for } r_0 = 1.1 \text{ fm} \qquad (2)$$

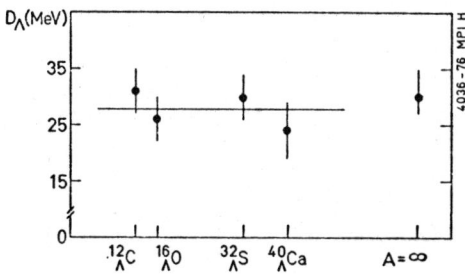

Fig. 4. Depth of the Λ-nucleus shell model potential ($r_0=1.1$fm) The value "∞" is the extrapolated value from the literature, 4.
Picture taken from ref. 8.

in good agreement with values obtained by extrapolating groundstate binding energies of light hypernuclei [4]. The concept of an average shell model potential for the Λ is supported by this agreement. Taking the D_Λ values found, the positions of hypernuclear groundstates can be predicted, e.g. B_Λ = (17 ±4)MeV for $^{40}_\Lambda$Ca. Fixing groundstate positions experimentally is an important task. From the energy splitting and intensity ratio of the split peak in the $^{16}_\Lambda$O spectrum a value for the Λ-spin orbit strength $U^{SO}_{\Lambda\Lambda}$ was deduced [9]. Relative to the neutron spin-orbit strength U^{SO}_{NA} the authors (BH) find

$$U^{SO}_{\Lambda\Lambda} = (0.0 \pm 0.3) U^{SO}_{NA} \qquad (3)$$

which means an SU(3) breaking one body spin orbit potential. Having fixed all parameters they calculate the full spectrum of the (1p, 1h) hypernuclear states and the cross section for each state. Fig. 5 shows the last result for $^{16}_\Lambda$O. The energies of the calculated states essentially fit to the experimental results. The calculated differential cross sections integrated over the energy show the experimentally verified A-independance and the correct magnitude.

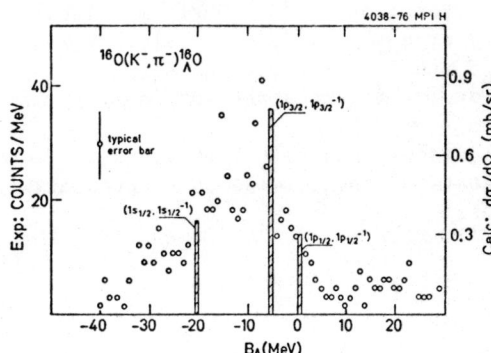

Fig. 5. Comparison of experimental (circles) and calculated (bars) spectra for the reaction ^{16}O (K⁻, π⁻)$^{16}_\Lambda$O*. The dominant configurations are indicated for each peak. The shaded areas correspond to the expected intensities.
Picture taken from ref. 8.

The broad structure at high excitations in the spectrum is attributed to the contributions of the (K⁻, π⁻) reaction on neutrons in the inner shells. The intensities of these bumps, however, are higher than the expected contribution, assuming a direct population in the

(K^-, π^-) reaction that should be proportional to the number of nucleons in the bound shells. For $^{16}_\Lambda O$ e.g. one expects the peak region to contain ~3 times the intensity of the bump. A 1:1 ratio is found (table 1). Also in the (BH) analysis this disagreement cannot be explained. Corrections by (2p, 2h) admixtures (<10%) and population of 1^- states (<20%) are too small. This discrepancy in the relative intensities may be resolved if one takes into account a continuous background due to quasi-free processes[14]. However, as discussed by A. Gal,[4] then the result for the Λ-spin orbit strength $U^{SO}_{\Lambda\Lambda}$ in eq. (3) may be questionable.

An alternative interpretation of the spectra in a collective picture was proposed about 10 years ago by Lipkin [10]. In this case, one assumes that, in the first approximation, all the particle-hole states are degenerate and form a strangeness analogue resonance, which splits due to the difference in the ΛN and NN interactions. (See also ref. 15.) In Table 1, the relevant parameter, $V_{n\Lambda} = B_n - B_\Lambda$, that measures the energy needed to transform a neutron into a lambda, and should be independent of the target nucleus in the model of Lipkin, is listed for different bumps. The peaks have been assigned to strangeness analogue resonances[11]. But starting with a strangeness analogue state that splits due to the symmetry-breaking effects one may have to take an average $<V_{n\Lambda}>$ which considers the whole resonance. Since a detailed theory is lacking, one can assume as a first approximation that the energy of the resonance $<B_\Lambda>$ is given by the center of gravity of the peaks. The $<V_{n\Lambda}>$ in Table 1 is obtained using this value of the excitation. The value of $<V_{n\Lambda}>$ is very constant for all measured targets, with the exception of ^{9}Be. It is premature to decide if the disagreement in the case of ^{9}Be is due to different B_n or that this target is too light to show collective effects. In the present sample, the neutron binding energies B_n are practically the same with the exception of ^{9}Be. Shell model calculations reproduce the constancy of $<V_{n\Lambda}>$ also very nicely. $<V_{n\Lambda}> \approx$ 24 MeV is found for $^{12}_\Lambda C$, $^{16}_\Lambda O$, $^{32}_\Lambda S$ and $^{40}_\Lambda Ca$ [12]. But the sample of targets studied is too small to decide if the agreement in $<V_{n\Lambda}>$ is accidental or not.

I want to aknowledge useful discussions with Professors A. Bouyssy and J. Hüfner and I am grateful for permission to present some of their unpublished results.

REFERENCES

1. B. Povh, Proc. 6th Internat. Conf. on High-Energy Physics and Nuclear Structure, Santa Fe and Los Alamos, N.M. 1975 (eds. R. Mischke, C. Hargrove and C. Hoffman) (Los Alamos, 1975), p. 173.
2. M.I. Podgoretski, Zh. Eksper. Teor. Fiz. $\underline{44}$ (1963) 695 (Engl. transl. : Soviet Phys. JETP $\underline{17}$, 470 (963)).
   H. Feshbach and A.K. Kerman, Preludes in Theoretical Physics (North Holland, Amsterdam, 1966) p. 260.
3. W. Brueckner, B. Granz, D. Ingham, K. Kilian, U. Lynen, J. Niewisch, B. Pietrzyk, B. Povh, H.G. Ritter and H. Schroeder to be published in Phys. Lett. B.
4. A.R. Bodmer, Proc. Summer Study Meeting on Nuclear and Hypernuclear Pysics, Brookhaven 1973, p. 64.
5. N. Auerbach and A. Gal, Phys. Letters $\underline{48B}$, 24 (1974).
6. J. Hüfner, S.Y. Lee and H.A. Weidenmüller, Phys. Letters $\underline{49B}$, 409 (1974). Nuclear Phys. $\underline{A234}$, 429 (1974).
7. G.J. Wagner, Lecture Notes in Physics, Vol. $\underline{23}$, Springer, Berlin (1973) p. 16.
   J. Källne and B. Fagerstrom, Proc. 5th Int. Conf. on High-Energy Physics and Nuclear Structure, Uppsala, 1973 (North Holland, Amsterdam, 1974), p. 369.
8. A. Bouyssy and J. Hüfner, to be published.
9. A. Bouyssy Contributed paper to this Conference.
10. H.J. Lipkin, Phys. Rev. Letters $\underline{14}$, 18 (1965).
    H.J. Lipkin, Proc. Summer Study Meeting on Nuclear and Hypernuclear Physics, with kaon beams, Brookhaven 1973 (Brookhaven National Laboratory, Upton, NY, 1973), p. 148.
11. J.P. Schiffer and H.J. Lipkin, Phys. ReV. Letters $\underline{35}$, 708 (1975).
    J.P. Schiffer, see these proceedings:Panel Discussion.
12. A. Bouyssy, private communication.
13. W. Brueckner, M.A. Faessler, K. Kilian, U. Lynen, B. Pietzyk, B. Povh, M.G. Ritter, B. Schürlein, H. Schröder and A.H. Walenta, Phys. Lett. $\underline{55B}$, 107 (1975).
14. A. Gal, see these proceedings:Panel Discussion.
    R.H. Dalitz and A. Gal, submitted to Phys. Lett. (May 1976).
15. Nguyen Van Giai, these proceedings, invited talk.

A. I. Yavin (Tel Aviv Univ.): 1) You quoted a cross section of 1.2 mb. How does this cross section compare with that of a free nucleon? 2) Can you show us a graph of the background for kaons and pions going all the way through. In otherwords, how does the spectrum look when you change the window up and down by 4 m/sec. (to look at (π,π) and (K,K) reactions)?

K. Kilian: 1) We find on our targets about 0.3 to 0.5 times the elementary differential cross section. 2) I do not have such a graph here. In order to get a realistic estimate for the background it is most convenient to switch the experiment simply to $(K^+\pi^+)$. No hypernuclear reaction is possible there, but scattering and decay processes are unchanged. We find a smooth background increasing to higher pion energies with its maximum outside of the hypernuclear energy region (on the right side of the ground state position indicated in the spectra). Less than 20% of the total rate in the shown spectra is due to background.

Marc Rayet (Brussels Univ.): There is an interesting piece of information which is lacking in that type of data, that is hypernuclear ground states. Would you propose a way of modifying your kinematics in order to get a chance to populate such ground states?

Kilian: An experiment is set up at Brookhaven and we are preparing one at CERN, where the $K^-\pi^-$ reaction will be measured at angles above 10 to 20°. The momentum transfer on the Λ should be sufficient enough to populate hypernuclear ground states.

Elie Aslanides (Strasbourg, France): Another way to look for hypernuclear ground states is to use the three body reaction $p+A \rightarrow p'+K^++B$. In a SACLAY-STRASBOURG-TORINO collaboration we are actually tuning up a spectrometer at Saclay to look simultaneously at eventual p-K pairs at 0°. The spectrometer used is a large band (±18%), high resolution (~5 10^{-4}) and large solid angle (~20 msr), QDD type spectrometer, SPES II.

HYPERNUCLEAR SPECTROSCOPY AND STRANGENESS ANALOG STATES

Nguyen Van Giai
Division de Physique Théorique[+],
Institut de Physique Nucléaire, 91406 Orsay, France

ABSTRACT

The 0^+ states of hypernuclei in the vicinity of closed shell nuclei are studied in the framework of a self-consistent particle-hole model. The symmetry breaking due to the difference between the interactions V^{NN} and $V^{\Lambda N}$ still allows for the persistence of some collectiveness. These semi-collective states will not be strongly seen because of their structure.

INTRODUCTION

It is the common belief that the study of the properties of hypernuclei is useful for obtaining information and better understanding about two different topics, namely i) the effective interaction between strange and non-strange baryons in a nuclear medium, and ii) the structure of nuclei. At the present time most of the observed hypernuclei are those containing only one Λ particle in addition to neutrons and protons and therefore the information one can get concerns mainly the Λ-nucleon interaction. The Λ particle is a $J^\pi = 1/2^+$ baryon with a mass rather close to the nucleon mass, but it does not obey the Pauli principle with respect to nucleons. This feature makes the Λ a very interesting and unique probe for nuclear structure.

For a rather long period most of the experimental data came from the measurements of the decay of hyperfragments in nuclear emulsion[1], thus limiting the available information mainly to the ground state energies and spins of light hypernuclei. In the recent years a number of experiments using K^- beams to produce excited hypernuclei through the strangeness exchange reaction[2-4] :

$$K^- + {}^A Z \rightarrow \pi^- + {}^A_\Lambda Z^*$$

[+]Laboratoire associé au C.N.R.S.

have stimulated a renewal of interest in the subject because
high lying states can thus be reached. Besides other spectroscopic information these and further experiments should provide
a test of the relevance of unitary symmetry in nuclear
systems through the observation of many fragmented
"strangeness exchange" states (SES) or a strongly collective
strangeness analog state (SAS) as suggested earlier by
Lipkin, Feshbach and Kerman[5-7]. We shall hear about the
present status of the strangeness exchange experiments with
K^- beams, especially on some s-d shell nuclei, in
Dr. Kilian's talk[8].

On the theoretical side a great deal of effort has
been devoted to the study of light hypernuclei and to the
extrapolation of binding energies to heavy and infinite
systems[9]. More recently there have been a few papers dealing with the interpretation of strangeness exchange
reactions on some p-shell nuclei[10-12]. This interpretation
is based on the assumption of the shell model for the
hypernucleus and the idea that the major part of the
breaking of unitary symmetry comes from the difference
$\Delta V_{\Lambda N}$ between the depths of the average potentials that
a Λ particle and a neutron feel. If this difference does
not depend much on the single particle states involved a
strongly collective state -the SAS- will exist, just as
in the case of isobaric analog states, even though $\Delta V_{\Lambda N}$
might be large. Dalitz and Gal[13] have also suggested
the existence of supersymmetric states in p-shell hypernuclei at excitation energies much lower than the SAS.

At the present time we have a fairly good microscopic description of the ground state properties of
ordinary nuclei by means of the Hartree-Fock (HF) approximation with effective nucleon-nucleon interactions.
Such calculations are able to reproduce reasonably well
a large body of data including ground state energies,
charge distributions and single particle spectra for a
large number of nuclei[14]. Our knowledge of the single
particle properties of a Λ inside a nucleus is not so good
mainly because of the lack of experimental information
concerning medium and heavy hypernuclei where the shell
model should be more valid. It is nevertheless possible to
use the HF model together with an effective Λ-nucleon
interaction and get the correct trend of the ground state
separation energies B_Λ from the light hypernuclei up to
extrapolated values in heavy systems. This has been done
by Rayet[15]. It seems then reasonable to use these self-
consistent average fields for the nucleons and the Λ to
describe excited hypernuclear states and especially SAS
in the framework of a particle-hole model. Such an approach
is very common in ordinary nuclear spectroscopy. For the
case of isobaric analog states the self-consistent
particle-hole model is quite appropriate as we shall see
in the next section. It also proves helpful in studying

the fragmentation of the SAS into several SES due to the breaking of unitary symmetry.

THE PARTICLE-HOLE MODEL AND UNITARY SYMMETRY

Before applying the particle-hole model we want to stress the fact that the model does not introduce any symmetry-breaking by itself as long as the consistency relation between the single particle energies and the matrix elements of the residual interaction is satisfied. If the SU(3) unitary symmetry is exact, then the ground state $|\phi\rangle$ of the nucleus $^A Z$ and the unitary analog state (UAS) $|UAS\rangle \equiv U_-|\phi\rangle/\langle\phi|U_+U_-|\phi\rangle^{1/2}$ obtained by the action of the U-spin lowering operator on $|\phi\rangle$ are two degenerate eigenstates of the Hamiltonian H. The state $|UAS\rangle$ contains a component where a neutron has been changed into a Σ (the Σ-analog state $|\Sigma\text{-AS}\rangle$) and a second component with a neutron changed into a Λ (the Λ-analog state $|\Lambda\text{-AS}\rangle$). Because of the large symmetry breaking the energy difference between the two components is large compared to the nuclear scale. We shall now concentrate on the Λ component. The state $|\Lambda\text{-AS}\rangle$ can be obtained by considering the operators:

$$\tilde{U}_- = \sum_i \Lambda_i^+ N_i, \quad \tilde{U}_+ = (\tilde{U}_-)^+, \quad \tilde{U}_3 = \frac{1}{2} \sum_i (N_i^+ N_i - \Lambda_i^+ \Lambda_i) \qquad (1)$$

where Λ_i^+ and N_i^+ create a Λ and a neutron respectively in a state i and the summations run over an arbitrary complete set of states. Then we have:

$$|\Lambda\text{-AS}\rangle \equiv \tilde{U}_-|\phi\rangle/\langle\phi|\tilde{U}_+\tilde{U}_-|\phi\rangle^{1/2} \qquad (2)$$

Let us now consider the case where $|\Phi\rangle$ is the ground state of a doubly closed shell nucleus and can be represented by a Slater determinant $|0\rangle$. We restrict the space of (1 particle)$_\Lambda$-(1 hole)$_N$ to the subspace spanned by the vectors:

$$|\lambda\nu\rangle \equiv \Lambda_\lambda^+ N_\nu |0\rangle \qquad (3)$$

where $\lambda \equiv (n_\lambda \ell_\lambda j_\lambda)$ and $\nu \equiv (n_\nu \ell_\nu j_\nu)$ are such that $n_\lambda = n_\nu$, $\ell_\lambda = \ell_\nu, j_\lambda = j_\nu$. The corresponding wave functions are denoted by $\varphi_\lambda(\Lambda)$ and $\varphi_\nu(N)$.

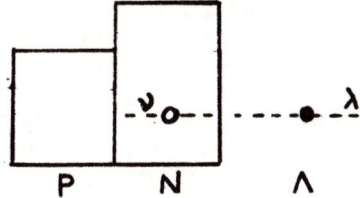

Fig. 1. The particle-hole configuration $|\lambda\nu\rangle$

The Hamiltonian can be written as :

$$H = T_N + T_P + T_\Lambda + V^{NN} + V^{PP} + V^{\Lambda\Lambda} + V^{NP} + V^{N\Lambda} + V^{P\Lambda} \qquad (4)$$

and its matrix elements in our subspace take the simple form :

$$\langle \lambda'\nu' | H | \lambda\nu \rangle = \delta_{\nu\nu'} \delta_{\lambda\lambda'} (E_0 + \varepsilon_\lambda - \varepsilon_\nu) - V^{\Lambda N}_{\lambda'\nu, \lambda\nu'} \qquad (5)$$

with the following definitions :

$$E_0 = \langle 0 | H | 0 \rangle$$

$$\varepsilon_\nu = \langle \nu | T_N | \nu \rangle + \sum_{\nu'} \tilde{V}^{NN}_{\nu\nu', \nu\nu'} + \sum_{\pi'} V^{NP}_{\nu\pi', \nu\pi'} \qquad (6)$$

$$\varepsilon_\lambda = \langle \lambda | T_\Lambda | \lambda \rangle + \sum_{\nu'} V^{\Lambda N}_{\lambda\nu', \lambda\nu'} + \sum_{\pi'} V^{\Lambda P}_{\lambda\pi', \lambda\pi'}$$

In the above equations the summations run over the neutron and proton states which are occupied in the $^A Z$ nucleus, $\tilde{V}^{NN}$ stands for the antisymmetrized matrix elements whereas V^{NP}, $V^{\Lambda N}$ and $V^{\Lambda P}$ are non-antisymmetrized matrix elements. We further assume that the sets of single particle states $\{\nu\}$, $\{\pi\}$ and $\{\lambda\}$ are such the wave functions having the same quantum numbers $(n\ell j)$ are practically the same although their energies ε_ν, ε_π and ε_λ may differ somewhat. This assumption is quite reasonable if the single particle states are HF states[15]. Neglecting also the mass differences between Λ, N and P we obtain :

$$\varepsilon_\lambda - \varepsilon_\nu = \sum_{\pi'} (V^{\Lambda P}_{\lambda\pi', \lambda\pi'} - V^{NP}_{\lambda\pi', \lambda\pi'})$$

$$+ \sum_{\nu'} (V^{\Lambda N}_{\lambda\nu', \lambda\nu'} - V^{NN}_{\nu\nu', \nu\nu'} + V^{NN}_{\nu\nu', \nu'\nu}) \qquad (7)$$

Now if the unitary symmetry is exact one would have $V^{NN} = V^{PP} = V^{\Lambda\Lambda} = V^{NP} = V^{N\Lambda} = V^{\Lambda P}$ and eq. (7) becomes :

$$\varepsilon_\lambda - \varepsilon_\nu = \sum_{\nu'} V_{\lambda\nu', \lambda'\nu} \qquad (8)$$

where $(n_\lambda \ell_\lambda j_\lambda) = (n_\nu \ell_\nu j_\nu)$ and $(n_{\lambda'} \ell_{\lambda'} j_{\lambda'}) = (n_{\nu'} \ell_{\nu'} j_{\nu'})$. This relation simply tells us that the Hamiltonian matrix (5) has an eigenvector $\propto \sum_\nu |\lambda'\nu'\rangle$ with an eigenvalue E_0. This particular eigenvector is a coherent superposition of particle-hole states and it is just the state $|\Lambda\text{-AS}\rangle$ defined by (2). We also note that the consistency relation (8) is satisfied if the same interaction V is used to calculate the energies ε_λ and ε_ν by means of HF calculations in the nuclei $^{A+1}_{\ \ \Lambda}Z$ and $^A Z$ respectively and at the same time is the particle-hole residual interaction.

In fact, the Λ-nucleon interaction is weaker than the nucleon-nucleon interaction, i.e., unitary symmetry is strongly broken. We assume for convenience that the interactions are still charge symmetric, although it is known that the Λ-nucleon interaction has an appreciable charge symmetry breaking component due to Λ-Σ mixing, and write:

$$V^{\Lambda N} = V^{NN} + \delta V$$
$$V^{\Lambda P} = V^{\Lambda N} \text{ and } V^{NP} = V^{NN}$$

In this case eq. (7) becomes:

$$\varepsilon_\lambda - \varepsilon_\nu = \sum_{\nu'} V^{\Lambda N}_{\lambda\nu',\lambda'\nu'} + \{\sum_{\pi'} \delta V_{\lambda\pi',\lambda\pi'} + \sum_{\nu'} (\delta V_{\lambda\nu',\lambda\nu'} - \delta V_{\lambda\nu',\nu'\lambda})\} \quad (9)$$

It is interesting to compare with the case of the isobaric analog state in a heavy nucleus where the equivalent of eq. (9) would be:

$$\varepsilon_\pi - \varepsilon_\nu = \sum_{\nu'}' V^{NP}_{\pi\nu',\pi'\nu} + \sum_{\pi'} (V^C_{\pi\pi',\pi\pi'} - V^C_{\pi\pi',\pi'\pi}) \quad (10)$$

where V^C is the Coulomb interaction and the summation on π' runs over all occupied proton states whereas $\sum_{\nu'}'$ only runs over excess-neutron states because of the Pauli principle. The term $\sum_{\pi'}$ in eq. (10) represents the Coulomb energy $\Delta E_C(\pi)$ of a proton in a state π and because of the long range nature of V^C the quantities $\Delta E_C(\pi)$ do not depend much on the particular state π: $\Delta E_C(\pi) \simeq \overline{\Delta E_C}$. Therefore the isobaric analog state $T_-|0\rangle$ still remains an approximate eigenstate of H but its energy will be shifted above the parent state energy E_0 by the Coulomb displacement energy $\overline{\Delta E_C}$.

In contrast to the Coulomb force responsible for the isospin symmetry breaking, the interaction δV is of short range and consequently one must expect the quantity in brackets in eq. (9) to depend in general on the particular state λ involved. I.e., we cannot assume it to be approximately constant as in the Coulomb case. Thus it seems unlikely that a fully coherent Λ-AS might emerge as an approximate eigenstate and it is more probable that the Λ-AS will be distributed over several particle-hole states. To what extent this fragmentation of the Λ-AS takes place depends of course on the detailed symmetry breaking interaction δV.

STRANGENESS EXCHANGE STATES AND Λ-AS IN CLOSED SHELL NUCLEI

The particle-hole model is applied to the description of the hypernuclear states obtained by changing a neutron into a Λ in the nuclei ^{12}C, ^{16}O, ^{32}S and ^{40}Ca. These nuclei have been used as targets in strangeness exchange reactions[2,3,8]. Their ground states $|\Phi\rangle$ can be described by a Slater determinant $|0\rangle$ to a good approximation.

The single particle basis $\{\varepsilon_\nu, \varphi_\nu\}$ for neutrons are obtained by performing HF calculations in the nuclei ^AZ using the effective Skyrme interaction. This velocity dependent as well as density dependent force (through the three-body term) is of the type

$$V^{NN} = t_0(1+x_0 P_\sigma)\delta(\vec{r}_1-\vec{r}_2) + \frac{1}{2}t_1(\vec{k}'^2\delta(\vec{r}_1-\vec{r}_2) + \delta(\vec{r}_1-\vec{r}_2)\vec{k}^2)$$
$$+ t_2\vec{k}'\cdot\delta(\vec{r}_1-\vec{r}_2)\vec{k} + iW_0\vec{k}'\cdot\delta(\vec{r}_1-\vec{r}_2)[(\vec{\sigma}_1+\vec{\sigma}_2)\times\vec{k}]$$
$$+ t_3\delta(\vec{r}_1-\vec{r}_2)\delta(\vec{r}_1-\vec{r}_3) \tag{11}$$

where P_σ is the spin exchange operator, $\vec{k}$ is the relative momentum operator and $\vec{k}'$ is the adjoint of $\vec{k}$. It depends on the six phenomenological parameters t_i, x_0 and W_0. Several sets of parameters have been determined by requiring for each set a good overall description of total energies and charge radii for spherical nuclei ranging from ^{16}O to ^{208}Pb [14]. Among these parameter sets we have selected the interaction SIII which seems to give rise to a correct non-locality in the nuclear average field (effective mass $m^*/m = 0.76$ in nuclear matter) and therefore it leads to a single particle level density which is correct not only near the Fermi surface but also for the deep lying states.

The single particle basis $\{\varepsilon_\lambda, \varphi_\lambda\}$ for the Λ particle have been calculated by Rayet[15]. They are obtained by performing HF calculations in the hypernuclei $^{A+1}_\Lambda Z$. The effective nucleon-nucleon interaction is also the force SIII whereas the effective Λ-nucleon force is taken of the simple form

$$V^{\Lambda N} = - U_0(\pi\mu^2)^{-3/2} e^{-r^2/\mu^2} \tag{12}$$

with $\mu = 1.044$ fm corresponding to two pion exchange. This form for $V^{\Lambda N}$ is certainly an oversimplification but is has the advantage that only one adjustable parameter appears, namely the strength U_0. The value of this parameter has been determined by fitting the experimental value of $B_\Lambda(^5_\Lambda He) = 3.12$ MeV in a variational model which treats properly the relative Λ-^4He motion[15,16] since the HF model would bring in sizeable center-of-mass effects in such a light system. The value adopted in the present calculations is $U_0 = 230$ MeV fm^3. In Table I we show the values of the depth V_Λ of the Λ-nucleus potential which come out of the HF calculation for various $^{A+1}_\Lambda Z$ systems.

Nucleus	$^{13}_\Lambda C$	$^{17}_\Lambda O$	$^{33}_\Lambda S$	$^{41}_\Lambda Ca$	infinite matter
V_Λ [MeV]	37	35	41	37	35

Table I. The depth V_Λ of the Λ-nucleus potential

The effective interaction (12) gives a purely central
Λ-nucleus potential. A spin-orbit or tensor component in
$v^{\Lambda N}$ would give rise to a spin-orbit part in the Λ-nucleus
potential. However, little is known about the spin-orbit
splitting of the single particle states of the Λ in the
nucleus. In a contribution to this Conference, Bouyssy
has analyzed the relative cross-sections and the difference
between the excitation energies of the first two 0^+ states
seen in the reaction $^{16}O(K^-,\pi^-)^{16}_\Lambda O$ [8]. Since these states
correspond mostly to the configurations $(1p1/2)_\Lambda - (1p1/2)_N^{-1}$
and $(1p3/2)_\Lambda - (1p3/2)_N^{-1}$, one can get some information on the
spin-orbit strength V^Λ_{LS} of the Λ-nucleus potential. Because
of the large uncertainties in the experimental data it is
only possible to deduce that V^Λ_{LS} and V^N_{LS} (the neutron-
nucleus spin-orbit strength) have the same sign and that
$|V^\Lambda_{LS}| \lesssim \frac{1}{3}|V^N_{LS}|$. In the following we shall characterize the
Λ spin-orbit splitting by the quantity $a^\Lambda_{so} = (\varepsilon^\Lambda_{n\ell j_<} - \varepsilon^\Lambda_{n\ell j_>})/$
$(\varepsilon^N_{n\ell j_<} - \varepsilon^N_{n\ell j_<})$ and use it as a free parameter instead of
adding a two-body spin-orbit term to the interaction (12).
We have checked that the results of the particle-hole
diagonalization are not strongly sensitive to the values of
a^Λ_{so} if we keep a reasonable range of variation $-0.3 < a^\Lambda_{so} < 1.0$.
The results presented here correspond to the value
$a^\Lambda_{so} = 0.3$.

In Table II are shown the particle-hole energies
$\varepsilon^\Lambda_{n\ell j} - \varepsilon^N_{n\ell j}$ in different nuclei. In $^{12}_\Lambda C$ there is a large gap

Nucleus	$^{12}_\Lambda C$	$^{16}_\Lambda O$	$^{32}_\Lambda S$	$^{40}_\Lambda Ca$
1s1/2	18.9	19.0	20.2	20.8
1p3/2	14.5	16.1	19.1	19.5
1p1/2		11.9	15.7	17.2
1d5/2			16.2	17.4
2s1/2			9.3	12.0
1d3/2				13.0

Table II. The particle-hole energies $\varepsilon^\Lambda_{n\ell j} - \varepsilon^N_{n\ell j}$ (in MeV).

between the 1s1/2 and 1p3/2 configurations. The off-
diagonal matrix element is about 1.3 MeV so that there
will be little mixing between the two states. The situation
is similar in $^{16}_\Lambda O$ where the 1p1/2 state is well separated
from the others and will not couple to them. In all four
nuclei the configurations corresponding to a neutron hole

in the highest or the two highest subshells have always the lowest energy and are well separated from the other configurations. However, in $^{32}_{\Lambda}S$ and $^{40}_{\Lambda}Ca$ the average distance between the configurations containing a neutron hole in one of the low lying subshells is of the order 1.-1.5 MeV and one may expect some configuration mixing. This explains for the results shown in Table III. The quantities ω_i are the excitation energies of the 0^+ states $|\Psi_i\rangle$ in the

Nucleus	ω_i[MeV]	$|S_i|^2$	wave function					
			1s1/2	1p3/2	1p1/2	1d5/2	2s1/2	1d3/2
$^{12}_{\Lambda}C$	11.5	0.46	−0.21	0.98				
	17.7	0.54	0.98	0.21				
$^{16}_{\Lambda}O$	12.8	0.12	−0.08	−0.15	0.99			
	17.5	0.32	−0.32	0.94	0.11			
	22.1	0.56	0.94	0.30	0.12			
$^{32}_{\Lambda}S$	20.4	0.16	0.05	0.02	0.02	0.03	0.998	
	26.8	∼0.	−0.08	−0.06	0.92	−0.37	0.	
	27.6	0.20	−0.08	−0.33	0.33	0.88	−0.02	
	30.7	0.13	−0.57	0.79	0.09	0.21	0.	
	33.8	0.51	0.81	0.51	0.19	0.19	−0.06	
$^{40}_{\Lambda}Ca$	20.3	0.18	0.06	0.01	0.01	0.02	0.987	0.14
	21.3	0.06	−0.04	−0.06	−0.13	−0.11	−0.14	0.97
	25.3	∼0.	−0.08	−0.04	0.87	−0.47	∼0.	0.06
	26.0	0.14	−0.09	−0.43	0.40	0.80	−0.03	0.10
	28.3	0.13	−0.59	0.75	0.12	0.27	0.01	0.07
	31.2	0.49	0.79	0.50	0.22	0.24	−0.08	0.11

Table III. Excitation energies ω_i, spectroscopic factor $|S_i|^2 = |\langle\Psi_i|\Lambda\text{-AS}\rangle|^2$ and wave function amplitudes for 0^+ states.

hypernucleus $^A_{\Lambda}Z$. The ground state values of B_{Λ}, defined as the difference between the ground state energies $E_0(^{A-1}Z)$ and $E_0(^A_{\Lambda}Z)$ of the nucleus ^{A-1}Z and the hypernucleus $^A_{\Lambda}Z$, are also calculated in this model. They are 12.5, 14.7, 22.3 and 23.2 MeV for $^{12}_\Lambda C$, $^{16}_\Lambda O$, $^{32}_\Lambda S$ and $^{40}_\Lambda Ca$, respectively. The spectroscopic factors $|S_i|^2 = |\langle\Psi_i|\Lambda\text{-AS}\rangle|^2$ are a measure of the collectiveness of the states $|\Psi_i\rangle$ (except in $^{12}_\Lambda C$ where the two states have comparable spectroscopic factors although they are of a non collective character). The wave function amplitudes are also shown in Table III.

The general trend of the results is that, except for $^{12}_\Lambda C$, the highest state has several components on configurations containing a neutron hole in low lying states and the main amplitudes have the same sign so that its spectroscopic factor is the largest although it remains far from 1. In other words in spite of the strong breaking of unitary symmetry and the short range character of the interaction δV some coherence in this highest state still survives.

It has been suggested recently by Schiffer and Lipkin[12] that the relevant quantity to look at in the case of the strangeness exchange reactions is not the B_Λ value of the excited state but rather the Q-value of the reaction defined (to a constant) as $Q_{\Lambda N} = E_0(^A_\Lambda Z) - E^*(^A_\Lambda Z)$. If $Q_{\Lambda N}$ varies smoothly throughout the periodic table this would be an indication that the concept of SAS is valid. If we calculate $Q_{\Lambda N}$ for the highest and most collective 0^+ state in the four nuclei considered here, the values are -22.2, -22.4, -23.5 and -24.0 MeV in $^{12}_\Lambda C$, $^{16}_\Lambda O$, $^{32}_\Lambda S$ and $^{40}_\Lambda Ca$ respectively. It seems that these states are good candidates for being the approximate Λ-AS. However, these states will not be strongly seen experimentally because they involve deep lying neutron holes which cannot be easily created since the incoming K^- are mostly absorbed before they can interact with 1s or 1p neutrons in a medium or heavy nucleus. Thus the low lying 0^+ states (neutron hole in the highest shell) will be more favored even though their spectroscopic factors are comparatively smaller.

We have already mentioned the fact that the results do not depend much on the value of a^Λ_{so}. In particular, the position and the semi-collective character of the highest state are not affected when we vary a^Λ_{so} from -0.3 to 1. However, the present results are certainly sensitive to the choice of the parameters of the interactions V^{NN} and $V^{\Lambda N}$. A comparison with the experimental spectra[8] shows that the calculated levels are systematically to low. This could be due to the choice of the Skyrme interaction SIII which may not reproduce well enough the non-locality of the nucleon-nucleus potential. One could for instance choose a V^{NN} interaction corresponding to a neutron effective mass $m^*/m \simeq 0.6 - 0.7$. Such a choice would still be acceptable but would push up the levels by a few MeV. A velocity-dependent interaction $V^{\Lambda N}$ leading to a Λ effective mass $m^*_\Lambda/m < 1$ could also make the Λ single-particle spectrum less dense and more similar to the neutron spectrum. This would favor the collectiveness effect for the high lying levels. Finally, a more complete treatment should include the effects of the widths of the single particle states and especially of the deep ones. These effects may render the collectiveness less evident and in any case would give to the highly excited states a large width.

THEORETICAL ANALYSIS OF THE FORMATION OF UNITARY ANALOG STATES

We consider now the formation of UAS in strangeness exchange reactions $^A_Z(K^-, \pi^-)^A_\Lambda Z^*$. We use the broken SU(3) symmetry model described above. Using the standard multiple scattering approach so familiar in π-nucleus physics, we obtain a set of linearized Klein-Gordon equations which couple the different channels:[17]

$$(-\nabla^2 + M_i^2) \chi_i = \sum_{j=1}^{3} 2E_j V_{ij} \chi_j \qquad (13)$$

where $\chi_1 = |K^-, \phi\rangle$, $\chi_2 = |\pi^-, \Lambda\text{-AS}\rangle$ and $\chi_3 = |\pi^-, \Sigma\text{-AS}\rangle$. In Eq.(13) M_i and E_i are the meson masses and total energies in each channel, and the potentials V_{ij}^i are related to the nucleon distribution $\rho(r)$ (normalized to 1) by:

$$2E_i V_{ij}/A = b_0^{(ij)} k_i^2 \rho + b_1^{(ij)} \nabla\cdot\rho\nabla + b_2^{(ij)} \nabla^2\rho + b_4^{(ij)} \nabla^2(\nabla^2\rho) \qquad (14)$$

The parameters $b^{(ij)}$ can be obtained from the K^-N two-body data.[18]

To study the cross-section corresponding to a particular Λ hypernuclear state $|\Psi_i\rangle$ we simply replace in Eq.(13) V_{12} by $V_{12} S_i$ where S_i is the spectroscopic factor defined before. In Table IV are shown forward cross-sections calculated at $p_K = 390$ MeV/c, using only the ρ term for the diagonal as well as the non-diagonal potentials. These results must be considered as preliminary since it appears that the $\nabla\cdot\rho\nabla$ and $\nabla^2\rho$ terms are important for elastic scattering[18] and the cross-sections are sensitive to the input parameters $b^{(ij)}$. The angular distributions which come out of this calculation are strongly forward peaked. This is in agreement with the experimental analysis presented by Bonazzola et al. in a contribution to this conference.

The alternative approach to calculate the cross-sections makes use of the distorted-wave impulse approximation (DWIA). The forward cross-section of the process $|\phi\rangle \to |\Psi_i\rangle$ becomes the product of $d\sigma/d\Omega$ $(K^-N \to \pi^-\Lambda; \theta=0°)$ times a factor which depends on the momentum transfer and the absorption in the entrance and exit channels. The DWIA reproduces fairly well the energy-integrated forward cross-sections but has difficulty to explain the relative cross-sections of the different states.[11]

Nucleus	ω_i [MeV]	$d\sigma/d\Omega$ ($\theta=0°$)
$^{12}_\Lambda C$	11.5	1.6
	17.7	1.9
$^{16}_\Lambda O$	12.8	0.4
	17.5	1.1
	22.1	1.9

Table IV. Calculated forward cross sections (in mb/sr) for the different 0^+ states.

CONCLUSION

We have tried to show that in spite of apparently unfavorable circumstances due to the short-range character of the unitary symmetry breaking part of the baryon-baryon force (as opposed to the long-range nature of the isospin violating Coulomb force in the case of isobaric analog states) some collectiveness may still survive and supports the concept of unitary analog state as an approximate eigenstate of the hypernuclear system. Being built mainly on deep neutron holes this semi-collective state seems hard to produce experimentally. On the theoretical side many questions are still unanswered. The coupled-channel treatment of the strangeness exchange reaction must be improved by including the surface terms (see Eq.(14)) which seem to be important.[18] One important point is to describe the widths of the strangeness exchange states since these widths can be quite large. The problems of the spin-orbit potential and the effective mass of a Λ inside a nucleus are also open. The observation of Σ-analog states in Σ-hypernuclei would be extremely interesting. There is also the possibility that the strangeness exchange reactions can excite other states than the analog states. In a contribution to this Conference Hoshi and Fujita show that the L=1 transitions can be appreciable even in the forward direction. Further measurements including angular distributions and using heavier target nuclei will help to answer some of these questions.

ACKNOWLEDGMENTS

It is a pleasure to thank L. Kisslinger for arousing my interest in this subject and for his collaboration. I am indebted to M. Rayet for his help in the hypernuclear HF calculations and for many discussions. Thanks are due to A. Bouyssy, J. Hüfner and B. Povh for interesting discussions.

REFERENCES

1. For a recent compilation, see M. Juric, G. Bohm, J. Klabuhn, U. Krecker, F. Wysotzki, G. Coremans-Bertrand, J. Sacton, G. Wilquet, T. Cantwell,

F. Esmael, A. Montwill, D. H. Davis, D. Kielczewska, T. Pniewski, T. Tymieniecka and J. Zakrzewski, Nucl. Phys. B52, 1 (1973).

2. M. A. Faessler, G. Heinzelmann, K. Kilian, U. Lynen, H. Piekarz, J. Piekarz, B. Pietrzyk, B. Povh, H. G. Ritter, B. Schurlein, H. W. Siebert, V. Soergel, A. Wagner and A. H. Walenta, Phys. Lett. 46B, 468 (1973).

3. G. C. Bonazzola, T. Bressani, R. Cester, E. Chiavassa, G. Dellacasa, A. Fainberg, D. Freschi, N. Mirfakhrai, A. Musso and G. Rinaudo, Phys. Lett. 53B, 297 (1974); Phys. Rev. Lett. 34, 683 (1975).

4. B. Povh, in Proceedings of the Santa-Fe Conference on High-Energy Physics and Nuclear Structure, Ed. D. E. Nagle et al., AIP, New York p. 173 (1975).

5. H. J. Lipkin, Phys. Rev. Letters 14, 18 (1965).

6. H. Feshbach and A. K. Kerman, in Preludes in Theoretical Physics, Ed. A. De-Shalit, H. Feshbach and L. Van Hove, North Holland, Amsterdam, p. 260 (1966).

7. A. K. Kerman and H. J. Lipkin, Annals Phys. 66, 738 (1971).

8. K. Kilian, these Proceedings.

9. For a review, see A. Gal, in Advances in Nuclear Physics, Ed. E. Vogt and M. Baranger, vol. 8 (1975).

10. N. Auerbach and A. Gal, Phys. Lett. 48B, 22 (1974).

11. J. Hüfner, S. Y. Lee and H. Weidenmüller, Nucl. Phys. A234, 429 (1974); A. Bouyssy and J. Hüfner, Heidelberg preprint (1976).

12. J. P. Schiffer and H. J. Lipkin, Phys. Rev. Letters 35, 708 (1975).

13. R. H. Dalitz and A. Gal, Phys. Rev. Letters 36, 362 (1976).

14. See, for instance, M. Beiner, H. Flocard, Nguyen Van Giai and Ph. Quentin, Nucl. Phys. A238, 29 (1975).

15. M. Rayet, preprint (1975) and private communication.

16. R. H. Dalitz and B. W. Downs, Phys. Rev. 111, 967 (1958).

17. L. S. Kisslinger, Phys. Rev. 157, 1358 (1967).

18. L. S. Kisslinger, these Proceedings.

V. N. Fetisov (P. N. Lebedev Inst. of Physics, Moscow, U.S.S.R.):
My question is: What can you say about the widths of excited states
of hypernuclei (analog- or quasi-analog and other states)? It may
be strongly connected with correct interpretation of observed quasi-
elastic peaks.
N. V. Giai: A calculation of the widths of hypernuclear states does
not exist yet. These widths can be of the order of several MeV
especially for the states built on deep hole configurations. A
correct estimate is certainly needed for a better understanding of
the experimental data.

G. N. Epstein (Michigan State Univ.): Are you happy about the
Gaussian form you use for the ΛN interaction? Would your results
change much if you used some other form?
N. V. Giai: The Λ-nucleus potential and the Λ-N particle-hole
matrix elements depended very little on the detailed readial form
of the interaction but rather on its strength. One could use for
example a zero-range interaction and get essentially the same
results.

E. Rost (Colorado):Have you considered the odd-A nuclei? In particular, can you comment on whether the data on ^{9}Be reflect a splitting in J, in T, or just an accident?
N. V. Giai: I have so far only applied the $(particle)_\Lambda$-$(hole)_N$
model to nuclei with closed shells or subshells.

A. Gal (Univ. of Virginia): (i) It seems strange, in view of the
strong nuclear absorption, that the deep neutron shells are
strangeness-exchanged in your calculation whereas the valent neutron-
shell excitation is not identified by you with the observed Q~20 MeV
narrow excitations. A way out of this difficulty would be to normalize the strength of the ΛN effective interaction to D_Λ~34 MeV, which
you have used. This may have the gross feature of a shift of your
calculated spectrum to higher excitation energies, thus leading to
the identification I have alluded to. (ii) A general remark about
the HF calculations involved: the deformation allowed in these is
radial only. Shape deformations may be more pronounced. This was
the conclusion Bassichis and I reached in our 1970 P.R. calculation,
although the nuclear interaction there was defective, not leading to
saturation.

N. V. Giai: Concerning (i), the calculated excitation energies can
be shifted up by the procedure you suggest, and also by using a
neutron effective mass lower than the value m*/m=0.76 adopted here,
as discussed in the written version of my talk. Concerning (ii),
I believe that shape deformation is a tiny effect for hypernuclei
which are close to spherical nuclei. Indeed, for ordinary nuclei
in the vicinity of a magic nucleus the spherical Hartree-Fock
solution is a very good approximation to the deformed solution.

M. Rayet (Brussels Univ.): In connection with A. Gal's comment, I may add a more precise statement to justify the use of a spherical Hartree-Fock formalism in calculating the Λ single particle energies and wave functions. In my Hartree-Fock calculations with a Skyme potential, radial compression turns out to be only about 2% in $_\Lambda O^{12}$ (this goes up to 6% with a softer potential like Brink and Boeter's). If one likes the view that shape distortion is only a small effect compared to the overall radial compression (as was shown years ago by Ho and Volky for example), one way be confident that a spherical Hartree-Fock calculation will still show the essential features of the Λ motion inside closed shell nuclei.

AN OVERVIEW OF HYPERNUCLEAR PHYSICS*

Herman Feshbach
Laboratory for Nuclear Science and Department of Physics
Massachusetts Institute of Technology
Cambridge, Massachusetts   02139

The motivations for the study of hypernuclei are three-fold. Hypernuclei are a new form of matter whose properties are just beginning now to be uncovered. From the understanding of these properties it will be possible to determine the nature of the force acting between the hyperons, such as the Λ and the Σ, with nucleons, one of the fundamental forces of nature. And finally the hyperon acts as a probe of its host nucleus, and this furthers our study of the properties of nuclei [1].

Up to recently only the binding energy of the light hypernuclei up to $^{15}_{\Lambda}N$, the binding energy of the "emulsion nucleus", with the average mass number, A, lying between A=40 and A=100, were known. From the study of stopped negative kaons a radiative transition had been observed in either $^{4}_{\Lambda}H^*$ or $^{4}_{\Lambda}He^*$, (an identification has recently been made, See contribution to this conference by Pniewski) while highly excited states had been found in $^{12}_{\Lambda}C$ and $^{14}_{\Lambda}N$. This forms a rather meager data base, far poorer than the nuclear data available before World War I. It was only after information regarding the low lying excited states of the nucleus and some of the electromagnetic properties of the ground and excited states was developed that it became possible to develop the great syntheses represented by the nuclear shell and rotational models. There is now hope that we are at the beginning of a rapidly increasing growth of data concerned with the properties of hypernuclear systems. Kaon beams of increased intensity, together with spectrometers with increased resolution will soon become available. Progress will be made through the opportunities presented for the formation of excited hypernuclei via the recoiless (K^-,π^-) reaction.

The formation of such states in which a Λ replaces a neutron have been demonstrated by Bressani's [2] and Povh's groups [3]. The work of the latter group in which the excitation of two levels in $^{9}_{\Lambda}Be$, and one in $^{12}_{\Lambda}C$, $^{16}_{\Lambda}O$ and $^{32}_{\Lambda}S$ were described to this conference by Kilian. Some controversy regarding the broad peak seen in $^{40}_{\Lambda}Ca$ remains. As described by Gal to this conference, Dalitz and Gal have proposed a "quasi-free formation" reaction which also generates a π^- in the final state. The amplitude for this process can interfere with the amplitude for the recoiless process. The superposition of a non-resonant reaction amplitude together with a resonance contribution is often observed in reactions, e.g., those involving isobar analog states, and its presence in the (K^-,π^-) is not surprising. It however poses an additional task, that of decomposing the observed cross section into these two components. Their differing kinematic

---

* This work is supported in part through funds provided by ERDA under Contract E(11-1)-3096.

properties, their differing dependence upon the kaon energy should
make such a separation possible. In any event the improved experimental facilities shortly to become available will be very helpful.
They are needed in any event to determine whether there are additional structure peaks or valleys, (there are several candidates in
the present data) and to determine the widths of the resonances.
The first of these requires improved statistics while the second
requires improved resolution as well. Improved experimental facilities will also make it possible to measure pion angular distributions.
The forward π^- observations now being made are very selective in
terms of the hypernuclear states that can be excited. At other
angles other excitations may become more prominent and it may become
possible to determine the hypernuclear ground state energy.

It is important to understand the nature of the reaction mechanism governing the (K^-,π^-) reaction. It is a completely one step
process occuring in the surface of the target nucleus? Hüfner et al.
[4] discuss the following two step processes which they estimate
might make appreciable contributions to the reaction.

$$K^- + p \rightarrow \pi^- + \Sigma^+ \qquad \Sigma^+ + n \rightarrow \Lambda + p \qquad \text{(a)}$$

$$K^- + n \rightarrow \pi^- + \Sigma^0 \qquad \begin{cases} \Sigma^0 + p \rightarrow \Lambda + p \\ \Sigma^0 + n \rightarrow \Lambda + n \end{cases} \qquad \text{(b)} \qquad (1)$$

$$K^- + p \rightarrow \pi^0 + \Lambda \qquad \pi^0 + n \rightarrow \pi^- + p \qquad \text{(c)}$$

These reactions are to occur within nuclei so that reaction (1a) is
more accurately written:

$$K^- + (N,Z) \rightarrow \pi^- + (N, Z-1, \Sigma^+)$$

$$(N, Z-1, \Sigma^+) \rightarrow (N-1, Z, \Lambda) \qquad (2)$$

In any event one should look for processes like (2a) or (1b) in which
Σ hypernuclei can be formed in a one-step process by examining the
π^- spectrum for appropriate values of the K^- energy. To see how
important two step processes are one should look for the production
of π^+ which can only occur for K^- projectiles by a two-step mechanisms such as:

$$K^- + p \rightarrow \Sigma^- + \pi^+ \qquad \Sigma^- + p \rightarrow \Lambda + n \qquad \text{(3a)}$$

$$K^- + n \rightarrow \Sigma^- + \pi^0 \qquad \Sigma^- + p \rightarrow \Lambda + n, \; \pi^0 + p \rightarrow \pi^+ + n \qquad \text{(3b)}$$

$$K^- + p \rightarrow \Lambda + \pi^0 \qquad \pi^0 + p \rightarrow \pi^+ + n \qquad \text{(3c)}$$

It is of course always possible to produce π^+ from π^- by double
charge exchange but the cross section for this process is small.
Reaction (3a) is very similar to (1a) but (3b) and (3c) do differ
from (1b) and (1c). However if a substantial cross section is found
for π^+ production it would indicate a substantial contribution of the

two step process to the π^- production. The detection of the π^+ particle would seem to be relatively easy.

If a one step process dominates the (K^-,π^-) reaction, the indicated method of analysis is the DWBA. In order to apply the DWBA it is necessary to perform experiments in which the K^- is elastically scattered by nuclei. These experiments are essential for the quantitative interpretation of kaon reactions. The influence of the Y^* resonances will clearly be important but it would also be interesting to see if the (K^-,π^-) resonance channel has any impact on the elastic scattering. For the exit channel one will need to know the π^- - hypernucleus scattering. Here the effects of the various (π^-,Λ) resonances will need to be understood. Hopefully from the insight into the behavior of particle resonances inside nuclei gained from the analyses of the π^- - nucleus elastic scattering in which the Δ resonances play an important role it will prove possible to at least approximately describe the π^- - hypernucleus scattering. Eventually when the reaction mechanisms are sorted out it will become possible to study the π^- - hypernucleus reaction via the (K^-,π^-) reaction.

Some controversy has been associated with the question of whether or not the resonance or bound state formed in the (K^-,π^-) reaction is a collective state or simply a doorway state formed by the substitution of the Λ for a neutron. There is a substantial difference from the isobar analog case where the isospin symmetry breaking force is the long range Coulomb force. This potential has comparatively small non-diagonal nuclear matrix elements and qualitatively can be thought of as simply a constant, adding the Coulomb energy to the nuclear energy for the state in question but otherwise not affecting the nuclear wave function. In the case of hypernuclei, the symmetry breaking force the range is of the order of the range of nuclear forces so that it is no longer possible to carry over the above discussions for the Coulomb force, to hypernuclei. The present evidence in light nuclei (see Giai's report to this conference) except for the case of $^9_\Lambda Be$ (see Gal's report to this conference) would indicate that the resonances reported by Kilian are not of the collective type simply because the level spacing is too large. However, the possibility remains that such collective states can be formed in heavier hypernuclei as pointed out by Rayet [5].

An earlier suggestion that in the heavier nuclei, the Λ wavefunctions would differ substantially from the corresponding neutron wavefunctions is not borne out by Rayet's Hartree-Fock calculations. His calculated spectra analyzed for the energy separation of the major shells give the following results:

Table I

Hypernucleus	$^{13}_\Lambda C$	$^{17}_\Lambda O$	$^{41}_\Lambda Ca$	$^{91}_\Lambda Zr$	$^{209}_\Lambda Pb$
Calculated	12 MeV	12	10	8	6
$40/A^{1/3}$	17.5 MeV	15	11.7	8.9	6.75

Note that the energy separation of the major shells is given by the first line of the table. Moreover the corresponding Λ and neutron wave functions show large overlap as indicated by Table 2.

Table 2

Wave Function Overlap

single particle states	$1s_{1/2}$	$1p_{3/2}$	$1h_{11/2}$
Nucleus			
Oxygen	.996	.982	
Lead	.993		.989

The major difference between the Λ and neutron wavefunctions is in their exponential tail reflecting their differing binding energies. The possibility of forming collective Λ - nucleon states is clearly a consequence of these results. On the other hand, the possibility of exciting the $1s_{1/2}$ Λ state when the doorway state is formed by replacing a neutron $ns_{1/2}$ state by a Λ would appear to be small since this Λ state will be pure to within a few percent.

I will not comment on Gal's remarks regarding the "supersymmetric" state in $^9_\Lambda Be$ except to emphasize their importance.

Let me conclude this section which has been concerned mostly with the (K^-, π^-) reaction by pointing to another recoil-less method for forming hypernuclei - namely, the radiative capture of the K^- particles in flight, the (K^-, γ) reaction. If the incident kaon has a momentum of 600 MeV/c in the laboratory system then the free process $K^- + p \rightarrow \Lambda + \gamma$ will leave the Λ at rest if the γ propagates in the forward direction. This gamma ray would have an energy of 600 MeV. There would be essentially no background. The main problem would be the detection of this gamma ray, eventually with good energy resolution. I am told this is difficult.

Consider next the nature of the ΛN potential. This will contain all the forms familiar to us from the nucleon-nucleon potential such as the central, spin-spin, tensor, spin-orbit, and exchange potentials. These are however some significant differences:

(1) The spin orbit term is written:

$$V_{SO} = \upsilon_N(r) \vec{\sigma}_N \cdot \vec{L} + \upsilon_\Lambda(r) \vec{\sigma}_\Lambda \cdot \vec{L} \qquad (4)$$

where $\vec{L}$ is the orbital angular momentum, and $\vec{\sigma}_N$ and $\vec{\sigma}_\Lambda$ are the spin operators for the nucleon and Λ respectively. V_{SO} can be rewritten as follows:

$$V_{SO} = \frac{1}{2}(\upsilon_n + \upsilon_\Lambda)(\vec{\sigma}_N + \vec{\sigma}_\Lambda) \cdot \vec{L} + \frac{1}{2}(\upsilon_N - \upsilon_\Lambda)(\vec{\sigma}_N - \vec{\sigma}_\Lambda) \cdot \vec{L} \qquad (5)$$

The second term is usually not present in the nucleon-nucleon potential. No hard information on the strength of these forces exists. Gal, Soper and Dalitz [6] from their analysis of p-shell hypernuclei estimate that the first term has a magnitude of the same order as the nucleon-nucleon spin orbit potential. They conclude that the second term may be present but is considerably smaller.

(2) Since the elementary exchange process which generates a major contribution to the Λ-N potential requires the exchange of 2 pions, three body potentials may be important in hypernuclei. See Figure 1.

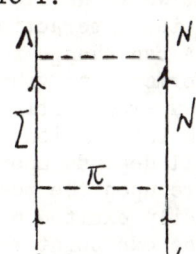 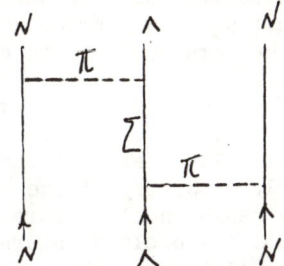

These are quite complicated in nature and I will not attempt their description. Bhaduri, Loiseau and Nogami's [7] calculations of these three body potentials have been used by Gal, Soper and Dalitz in their discussions of p-shell nuclei. A review was given by Dalitz in his 1973 BNL lectures [1]. We cannot say much regarding the strength of the 3 body potentials particularly because the large value 13.59 ± 0.15 MeV obtained for $B(\Lambda)$ of $^{15}_{\Lambda}N$ has disturbed the fit Gal et al. [6] obtained for the p-shell hypernuclei. That fit, which was based on the $p_{3/2}$ hypernuclei, predicted a value of 11 MeV for $B(\Lambda)$ of $^{15}_{\Lambda}N$. They are now engaged in obtaining a new fit and we await their results with interest. The expectation is that there will be a considerable reduction in the importance of the three body potential.

(3) The value of $B(\Lambda)$ for $^4_{\Lambda}He$, 2.39 ± 0.03 MeV is found to be greater than that of $^4_{\Lambda}H$, 2.04 ± 0.4. If the only isospin symmetry breaking potential were the electrostatic potential, the value of $B(\Lambda)$ for $^4_{\Lambda}H$ would be greater than the value for $^4_{\Lambda}He$. The empirical result indicates the presence of attractive charge symmetry breaking potential of the form

$$V_{CSM} = \tau_3 \left[A(r) \vec{\sigma}_\Lambda \cdot \vec{\sigma}_N + B(r) + C(r) S_{\Lambda N} \right] \quad (5)$$

where $S_{\Lambda N}$ is the tensor potential operator. The existence of V_{CSM} can be inferred from the effect of the isospin mixing of the Λ and Σ° [8]. It is equivalently possible to regard the physical Λ as a mixture of a pure isospin zero Λ and the Λ plus $\pi°$. Using only the B term in (5), the strength of B can be determined from the $B(\Lambda)$ mentioned above. Dalitz finds this strength to be of the order of a few MeV.

I shall not attempt to summarize the thorough discussion Gal, Soper and Dalitz have made of the p-shell nuclei in view of the revisions now being made. I should mention that in their earlier calculations they find that the strength of the central component of the ΛN potential is a few percent greater than the triplet potential. One should also mention the problem of the predicted binding energy of $^5_\Lambda$He has not been resolved. The predicted value of 5.5 MeV is considerably greater than the experimental value of 3.12 ± 0.02 MeV.

We turn next to what can be learned regarding nuclei from study the properties of hypernuclei [9]. In this context we think of the Λ as acting like a probe which as a consequence of its interaction with the core nucleus can change the latter. The Λ can change the latter's radius, its rate of rotation if it is deformed, its vibration frequency, and for example the energy gap in superconducting nuclei. It can change its electromagnetic and weak interaction properties. These effects can be considerable. It all depends upon the polarizability of the core nucleus and therefore upon the position of the nearby nuclear core states which can be readily excited by the residual Λ - nucleon interaction. As an example one can point to the results of Dalitz, Soper and Gal for the p-shell hypernuclei. In the case of the ground J=1⁻ state of $^8_\Lambda$Li one finds considerable mixing of the core ^{7}Li ground state with the nearby excited state of ^{7}Li at 0.48 MeV. Clearly the nature of the nuclear wave functions of the two ^{7}Li states involved is being probed. This effect should be more pronounced when the core of the hypernucleus is in an excited state since there is then a greater density of nuclear states. A number of additional examples for which experimental data is as yet not available for the most part will be discussed below.

Compressibility is an important nuclear property for which we have as yet only rough values. The following indicates the possibility that the properties of hypernuclei may yield information on this important number. Because of the presence of the Λ the semi-empirical mass formula for the energy becomes

$$E = E(A) + E_A(\Lambda) \qquad (6)$$

where the first term gives the nuclear energy for a nucleus of mass A. The second term in a Fermi gas model has the form:

$$E_A(\Lambda) = \frac{a}{r^2 A^{2/3}} - \frac{b}{r^3} \qquad (7)$$

where a and b are empirical constants, r is related to the nuclear radius R, $R = r A^{1/3}$. The first term in E(Λ) represents the kinetic energy of the Λ in the nucleus, which is inversely proportional to R^2. The second term is the contribution of the potential energy assuming that the Λ - nucleon interaction involves only a two body potential. If the three body potential is important, $E_A(\Lambda)$ will contain a term proportional to A. In the absence of evidence to the contrary we shall assume that the three body term does not modify the empirical form of E(Λ), but of course this issue is of great interest in itself and will be settled once the values of B(Λ) for a sufficient number of hypernuclei with substantially differing values of A have been measured. With this hypothesis the values of a and b can be determined as follows [10].

$$-E_A(\Lambda) \xrightarrow[A\to\infty]{} B_\infty = 27 \text{ MeV} = \frac{b}{r_o^3} \quad (8)$$

while

$$\frac{a}{r_o^2 A^{2/3}} = E_A(\Lambda) - E_\infty(\Lambda) = B_\infty - B_A(\Lambda) \quad (9)$$

These estimates hold when r takes on its equilibrium value r_o. That value is determined by the condition:

$$\frac{E}{r} = 0 \quad \text{at} \quad r=r_o \quad (10)$$

If r'_o is the value of r for the unperturbed core nucleus the fractional change

$$\xi \equiv \frac{r_o - r'_o}{r'_o}$$

is found to be:

$$\xi \simeq -\frac{B_\infty + 2B_A(\Lambda)}{r'^2_o E''(A)} = \frac{B_\infty + 2B_A(\Lambda)}{KA} \quad (11)$$

where K is the nuclear "compression modulus". For K=150 MeV and $^{15}_\Lambda$N, ξ=.026.

The question now arises as to how this change in radius might be observed. One possibility is to compare the value of B(Λ) for hypernuclei with mirror core nuclei. However in order to do so an estimate of the effect of the charge symmetry breaking potential is needed. In a Fermi gas model that energy,

$$\Delta E(\text{CSB}) = -\frac{T_z}{A} \frac{b}{r_o^3} V_o \quad (12)$$

where b is the range of V_{CSB} and V_o is its depth. This result should be compared with the term in the Coulomb energy in the same model, neglecting change.

$$E_c = \frac{3}{5} \frac{e^2}{r_o A^{1/3}} \left[\frac{A^2}{4} + AT_z + T_z^2 \right] \quad (13)$$

which is linear in T_z. We see that ΔE(CSB) becomes very much smaller than that term as A increases, although it is larger in the $^4_\Lambda$H and $^4_\Lambda$He case. Using the value of V_o determined by that case (see earlier discussion) which may be an underestimate for its value in more massive nuclei, we find that the cross-over point, which is of

interest in itself, will occur early in the p-shell. In any event,
for sufficiently large A one can neglect ΔE(CSB) opening up the possibility of determining r_o and therefore K.

Odd A hypernuclei will differ substantially from odd A nuclei
primarily because of the Pauli principle. The nucleon added to an
even nucleus can only occupy an orbit at the Fermi momentum. The Λ
can occupy an orbit ranging from the deepest bound orbit through all
the possible single particle orbits so that it is a much more effective and versatile probe. Consider a vibrational nucleus where the
vibrational phonon is 2^+. The lowest energy levels of such a nucleus
are shown in the figure. Two possible single particle orbitals are

```
                      _____ 0,2,3,4,6      1p3/2
                                              _____
                      _____ 0,2,4

                      _____ 2

   1s1/2              _____ 0
  _____

    Λ              core nucleus                 Λ
```

shown for the Λ. Needless to say an additional nucleon could not be
put into these orbits. We now consider the effects of the Λ on the
core assuming, using perturbation theory language, that the Λ will
not change its orbit because of the Λ-nucleus interaction. Under
these circumstances, the $1s_{1/2}$ Λ can only induce ΔJ=0 transitions
which would then involve a two phonon transition of say the zero
phonon state, to the second excited level of the core. On the
other hand, the $1p_{3/2}$ Λ will be able to induce a ΔJ=0 and ΔJ=2. The
latter permits a one phonon transition coupling for example the 0^+
two phonon state to the 2^+ three phonon and the 2^+ one phonon state.
It is clear that the vibrational spectrum based on the ground state
of the hypernucleus will differ significantly from that based on the
excited state in which the Λ is in a $1p_{3/2}$ orbital. From observation
of these spectra it would be possible to determine the one and two
phonon matrix elements of the core nucleus which can be compared with
results obtained by inelastic alpha particle scattering.

When the core of the nucleus is deformed the Λ can change its
deformation. We can calculate that change using the Inglis cranking
formula. Using deformed harmonic oscillator orbitals the change in
the moment of inertia, Δ, we obtain the irrotational flow result when
the Λ is in the ground state,

$$n_x = n_y = n_z = 0:$$

$$\Delta \mathcal{J} = \frac{9}{8\pi} (5M_\Lambda R_o^2) \beta^2$$

when R_o is the average radius and β, the deformation parameter
$\Delta R/R_o = 0.95\beta$. However Λ could be in an excited state. For an
arbitrary $n_x = n_y$ and n_z we obtain:

$$\Delta \mathcal{J} = M_\Lambda R_o^2 \left[(n_y + n_z + 1) \left(\frac{45 \beta^2}{8\pi} + \ldots \right) \right.$$

$$\left. + (n_z - n_y) \sqrt{\frac{4\pi}{5}} \left(\frac{4}{3\beta} + \frac{4}{3} \sqrt{\frac{5}{4\pi}} + \ldots \right) \right)$$

indicating the possibility of substantial changes in the rotational spectrum built upon an excited state of the Λ.

Another possibility of probing the core nucleus comes from the Coriolis coupling of the Λ and the core rotator nucleus. If the ground state band of the core is a 0,2,4,... band then the hypernucleus will, if the Λ is placed in a $1s_{1/2}$ orbit, have a K = 1/2 band. Here we should be concerned with the value of the decoupling parameter. However the Λ might be in a deformed p orbit with a magnetic quantum number of 1/2 or 3/2. The latter would give rise to a K=3/2 band which would interact via the Coriolis coupling with K=1/2 band. The moment of inertia would be modified. The relevant parameter is the ratio

$$|\langle \chi_{3/2} | J'_+ | \chi_{1/2} \rangle|^2 / [\varepsilon(3/2) - \varepsilon(1/2)]$$

$\chi_{1/2}$, $\chi_{3/2}$ are the internal wave functions for the K bands. We here have the opportunity for determining experimentally the matrix element in the numerator when the extra particle, the Λ, is in various orbits.

When the core is a superconducting nucleus, some special effects may be observable because of the possible presence of a three-body Λ-nucleon potential. It is clear that such a force will modify the size of the gap and thus test the quasi-particle description. But as important, because of the correlated nucleon pairs in a superconducting nucleus, a particular component of the three body potential will be most effective. One may also expect anomalies in B(Λ) for these nuclei.

Doorway states such as the giant dipole resonance, or isobar analog states can serve as a core nuclear states of hypernuclei. Presumably one should be able to detect these by means of their special modes of decay, such as the unique proton energy in the case of the decay of an isobar analog state. There could also be evidence for their formation in the (K^-, π^-) reaction at appropriate K^- momenta.

The recoil-less reaction leading to the formation of a hypernuclear resonant state can also provide nuclear information. At the very least the cross section for the process can be related via the simple DWBA to the probability that the final core nuclear wave function was present in the original target. Similar information is available from the (d,p) reaction with the difference that in the hypernuclear case one must assess the impact of the Λ - neutron hole interaction not present in the (d,p) case. A similar relationship to the (p,2p) reaction exists for the case when the neutron is removed from a deep state. A more complex speculation suggests the possibility a 2p-1h collective states in which the particles are the

Λ and proton and the hole is a neutron level. In terms of operators on the parent nuclear wave function this would be obtained by the action of the operator U_{T_-}.

The presence of the Λ will also affect the electromagnetic properties of the core. Since the Λ is neutral, the electric multipole moments, both static and transition, are of obvious interest as it is in the case of the valence neutron. But there is a very important difference, the possible occupancy of orbits by the Λ is not blocked by the presence of the core nucleons. Hence one can observe the induced polarizability of the core nucleus by the Λ in a variety of orbits.

When the Λ is in the intense electric field of a heavy core nucleus the polarizability of the Λ may also play a role.

Since the Λ does have a magnetic moment it is necessary to include its contribution to the magnetic properties of the hypernucleus. Two aspects can enter. On the one hand the value of μ_Λ might change, for example, be quenched because of the presence of the surrounding nucleons, which provide the possibility of Λ-nucleon exchange currents. And again the core can be polarized by the presence of the Λ and thus change its electromagnetic properties.

Because the Λ has zero isospin, the zero isospin magnetic dipole nuclear transition which is approximately forbidden, will in hypernuclei have an additional contribution. The excited core nucleus can decay virtually to its final state via the lambda-nucleus interaction with the Λ moving to an excited state. The Λ would then radiate magnetically.

The exploitation of many of the phenomena discussed above depends upon the observation of excited states of hypernuclei and the transitions between them. It is too early to be certain that these observations will be possible. We await the results of the attempts to do so with great interest.

REFERENCES

[1] See Proceedings of the Summer Study Meeting on Nuclear and Hypernuclear Physics with Kaon Beams, ed. by H. Palevsky BNL 18335, Brookhaven National Laboratory (1973).
[2] G. C. Bonazzola et. al. Physics Letters 53B, 297 (1974) Phys. Rev. Letters 34, 683 (1975).
[3] W. Brueckner et. al. submitted to Physics Letters March, 1976.
[4] J. Hüfner, S. Y. Lee and H. A. Weidenmuller Nuclear Physics A234, 429 (1974).
[5] M. Rayet, Self Consistent Calculations of Hypernuclear Properties with the Skyrme Interaction - to be published.
[6] A. Gal, J. M. Soper and R. H. Dalitz, Ann. of Phys. (N.Y.) 63, 53 (1971), 72, 445 (1972).
[7] R. Bhaduri, B. Loiseau and Y. Nogami, Nucl. Phys. B3, 380 (1967).
[8] R. H. Dalitz and F. von Hippel, Phys. Letters 10, 153 (1964).
[9] See 1973 discussion by author in Reference [1].
[10] See A. Bodmer in Reference [1].

Segel (Argonne-Northwestern): We have seen the very nice work from CERN. Kaon beams 100 or more times as intense as the CERN beam with a much lower contamination, are now being designed. Thus many of the experiments that you have described which now may appear to be very difficult are likely to soon become quite feasible.
Feshbach: No comment.

MODEL FOR LOW ENERGY KAON-NUCLEON INTERACTION IN THE I = 0 STATE*

S. C. B. Andrade and E. M. Ferreira**
Pontifícia Universidade Católica, Rio de Janeiro, Brazil

Unitary Padé approximants were recently used[1] to build a model for the low energy kaon-nucleon interaction. The model is based on a Lagrangian

$$\mathcal{L}_{int} = \{-ig_{N\Lambda K}(\bar{p}\gamma^5 \Lambda K^+ + \bar{n}\gamma^5 \Lambda K^0) + h.c.\} + \{-ig_{N\Sigma K}(-\bar{p}\gamma^5 \Sigma^0 K^+ + \sqrt{2}\bar{p}\gamma^5 \Sigma^+ K^0 + \bar{n}\gamma^5 \Sigma^0 K^0 - \sqrt{2}\bar{n}\gamma^5 \Sigma^- K^+) + h.c.\} + \lambda\{\bar{p}p\bar{K}^+ K^+ + \frac{1}{2}(\bar{p}n\bar{K}^0 K^+ + \bar{n}n\bar{K}^+ K^+ + \bar{p}p\bar{K}^0 K^0)\} \quad (1)$$

written in terms of nucleon, hyperon, and kaon fields, and consisting of the usual Yukawa terms ($g_{N\Lambda K}$ and $g_{N\Sigma K}$) and of the contact interaction (coupling constant λ) acting directly only on the I = 1 state of the KN system. The amplitudes are calculated starting from the first diagonal Padé approximant to each partial wave amplitude. This procedure guarantees the S-matrix unitarity, and the model achieved remarkable success when applied to the I = 1 state of the kaon-nucleon system.

As an extension of that work, we have now applied the model to study the KN interaction in the isospin I = 0 state. To obtain the [1,1] Padé approximant we must evaluate all Feynman diagrams up to fourth order. The Yukawa coupling constants occur in the final expressions only through the effective coupling constant

$$g^2 = (g^2_{N\Lambda K} - 3\eta g^2_{N\Sigma K})/4\pi$$

with $\eta \simeq 0.97$. Once the scattering length $a_s^{(0)} = -0.3f$ in the I = 1 state is given, λ is fixed in terms of an effective coupling G^2 defined by the combination

$$G^2 = (g^2_{N\Lambda K} + \eta g^2_{N\Sigma K})/4\pi$$

and then G^2 can be determined by fitting the other K^+p scattering data. It is commonly accepted that the value of G^2 lies somewhere in the interval from 10 to 20. Thus g^2 becomes the only free parameter for the description of the I = 0 interaction. The present knowledge of the values of the Yukawa coupling constants indicates that $0 < g^2 < G^2$.

We have obtained for the I = 0 state an s-wave scattering length of about -0.4 fermi, and positive $p_{1/2}$ and negative $p_{3/2}$ phase shifts, with typical scattering lengths $a^{(0)}_{p_{1/2}} = +0.08$ and $a^{(0)}_{p_{3/2}} = -0.02$ (fermi)^3 for $G^2 = g^2 = 15$. These results may be compared to the solutions of phase shift analysis by B. C. Wilson et al.[2] Their preferred sets present also negative s-wave shifts while both $p_{1/2}$ and $p_{3/2}$ phases are positive.

Figure 1 shows the values of the differential cross section for elastic $K^+n \to K^+n$ scattering, as extracted from results of K^+d scattering experiments,[2] together with the predictions from our model. The differential

---
* Work partially supported by Energy Research and Development Administration.
**Present address: Stanford Linear Accelerator Center, Stanford, CA 94305.

polarization observed[3] in $K^+d \to K^0pp$ at 600 MeV/c is well reproduced by our calculations, as shown in Fig. 2. Figure 3 shows the experimental results on K^+d total cross section,[4] together with our results.

Some of the noted discrepancies can be attributed to the way K^+p and K^+d experiments have been analyzed to produce the I=0 data, but we must also point out the deficiencies of the model, which does not include pions explicitly and represents all kinds of short range forces by the contact term in the Lagrangian.

On the whole, however, we think that the model reaches relative success. Without free parameters, we have predicted the correct sign of the s-wave amplitude, and the approximate magnitude of the K^+n cross section, which are the most reliable data available at the present in the I=0 state of the KN system.

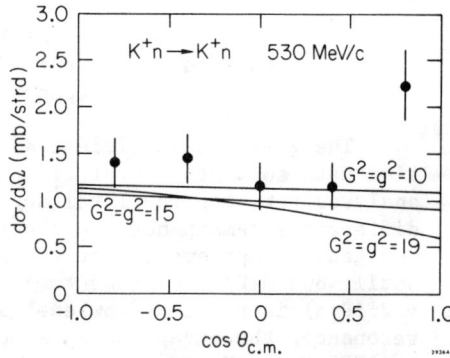

Fig. 1--Predicted differential cross section for $K^+n \to K^+n$ at 530 MeV/c for some values of G^2 and g^2. Data from K^+d experiments.[2]

REFERENCES

1. S. C. B. Andrade, E. M. Ferreira, and L. Ye Chang, Nucl. Phys. B **87**, 485 (1975).
2. B. C. Wilson et al., Nucl. Phys. B **42**, 445 (1972); ibid., **56**, 346 (1973); ibid., **71**, 138 (1974).
3. A. K. Roy et al., Phys. Rev. **183**, 1183 (1969).
4. A. S. Carrol et al., Phys. Lett. B **45**, 531 (1973).

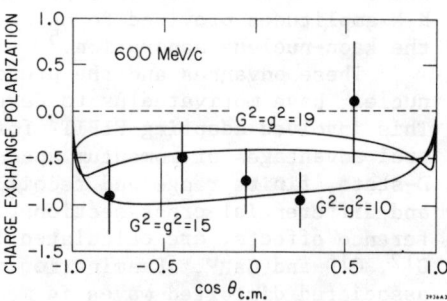

Fig. 2--Differential polarization for $K^+n \to K^0p$ compared to experimental data.[3]

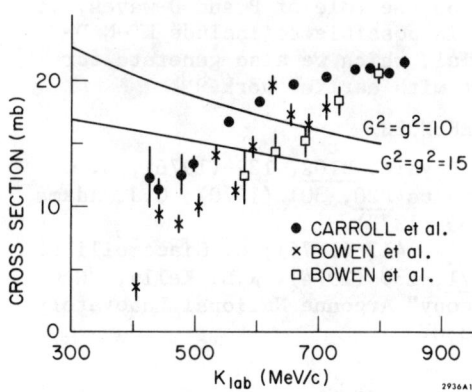

Fig. 3--Calculated cross section in the KN, I=0 state, compared to experiments.[4]

K^+-NUCLEUS ELASTIC SCATTERING[†]

R. A. Eisenstein
Physics Department, Carnegie-Mellon University,
Pittsburgh, PA. 15213

F. Tabakin
Physics Department, University of Pittsburgh
Pittsburgh, PA. 15260

There exists sufficient K^+-nucleon cross-section and polarization data so that several groups have provided impressive phase shift analyses for both the $I = 1$[1] and $I = 0$[2] states. Very significant differences remain between the resultant amplitudes, but there is now general agreement that the S_{11} is repulsive (and remains elastic until ~ 800 MeV/c lab momentum), and that the combination $\delta(P_{11}) + 2\delta(P_{13})$ is small at low energies. Most analyses yield no $Z_1^*(P_{13})$ resonance; the exception is Arndt et al.[3] who predicted a Z_1^* state at 1787 MeV. However, recent preliminary BNL polarization experiments favor the non-Z_1^* solutions.[4] Further tests via spin-rotation experiments have been suggested by Cutkosky et al.,[1] who also show that the non-Z_1^* solutions have appreciable low energy P-waves, in contrast to Arndt's result. Indeed, the BGRT and the Martin[2] amplitudes have appreciable $I = 1$ and $I = 0$ P-waves, with P_{01} and D_{03} being the only remaining resonance possibilities. Significant $I = 1$ and $I = 0$ D-waves are also present. Preliminary examination of the K N amplitudes provided to us suggests that P-waves might dominate the kaon-nucleus absorption.[5]

These advances and the proposals to scatter K^+ mesons from nuclei[6] have motivated us to develop an optical potential for kaons. This involved adapting PIPIT[7] for kaons, which offers, among the several advantages of momentum space methods, a easy incorporation of D-state, finite range and recoil effects. Total, elastic, inelastic and differential cross sections, along with Coulomb-nuclear interference effects, are calculated for 0.4 to 1.2 GeV/c K^+ incident on C^{12}, O^{16} and Ca^{40}. Examination of the absorption coefficients and associated distorted waves is made to gain insight into the degree and localization of absorption. The sensitivity to various phase shift uncertainties, with emphasis on the role of P and D-waves, is determined for various nuclei. It is possible to include K^+-N D-waves in an $\vec{r}$ space optical potential, which we also generate for convenience and to make contact with earlier work.[8]

REFERENCES

1. R.E. Cutkosky et al., Nuclear Physics B102, 139 (1976); G. Giacomelli et al., Nuclear Physics B20, 301 (1970); C.J. Adams et al., Nuclear Physics B66, 36 (1973).
2. B.R. Martin, Nuclear Physics B94, 413 (1975); G. Giacomelli et al. (BGRT), Nuclear Physics B71, 138 (1974). R.L. Kelly, "New Directions in Hadron Spectroscopy" Argonne National Laboratory Report ANL-HEP-CP-75-58 page 330.

3. R.A. Arndt et al., Phys. Rev. Letters 33, 987 (1975).
4. M.E. Zeller et al. BAPS 21, 70 (1976).
5. G. Keyes, private communication.
6. P.D. Barnes, R.A. Eisenstein, W.R. Wharton and E.V. Hungerford, "The Measurement of the $K^{\pm}$ Elastic Scattering from Selected Nuclei at 650-1000 MeV/c". Research proposal submitted to Brookhaven National Laboratory.
7. R.A. Eisenstein and F. Tabakin, PIPIT: A Momentum Space Optical Potential Code for Pions", contribution to this conference.
8. S. Wycech, in "Proceedings of the Summer Study Meeting on Nuclear and Hypernuclear Physics with Kaon Beams", (1973) page 307; A. Deloff and J. Law, Phys. Rev. C10, 1688 (1974).

† Work supported by the ERDA and the NSF.

ANGULAR DISTRIBUTION FOR THE (K^-, π^-) REACTION ON ^{12}C AND ^{27}Al

G.C. Bonazzola, T. Bressani, E. Chiavassa, G. Dellacasa,
M. Gallio, A. Musso and G. Rinaudo

Istituto di Fisica Superiore dell'Università, I 10125 Torino, Italy
Istituto Nazionale di Fisica Nucleare, Sezione di Torino

Recent experiments[1,2,3] have demonstrated the usefulness of the strangeness exchange reaction (K^-, π^-) on nuclei to study the production and the properties of hypernuclei.
The interpretation of the experimental data is not unique. In particular it is not clear whether an independent particle model[4,5] or a collective one[6] is more adequate to describe the hypernuclear states. Theoretical speculations were up to now based on the observed excitation energy spectra, which were the only reported result. The knowledge of the angular distributions for the different final states would be essential for the determination of the quantum numbers and for a comparison with different models.

Since our spectrometer allowed the simultaneous measurement of π^- emitted from 0° to 20° in the lab., we reanalysed our data on ^{12}C and ^{27}Al and we present here the relative angular distributions, without integrating the spectra over the entire solid angle as done before. We developed a reconstruction program for the angles in the interaction point which takes into account the complicated shape of the magnetic field in the target region and the multiple Coulomb scattering in the target (~ 5 gr/cm^2). The precision on the reconstruction of the emission angle is $\pm 1°$ (fwhm). The figure shows the observed ecitation spectra for $^{12}_\Lambda C$ (a), $^{27}_\Lambda Al$ (b) and the corresponding angular distributions (c and d). In our kinematics ($p_{K^-} = 390$ MeV/c) the momentum transfer varies between 40 MeV/c (at 0°) and 100 MeV/c (at 20°).

In the spectrum of $^{12}_\Lambda C$ peak 1 can be interpreted as due to the ground state formation and peak 2 as due to the formation of excited states in which a p-shell neutron in ^{12}C is replaced by a Λ in a p-orbital. The angular distributions confirm this interpretation since the production of the excited states has a maximum at 0°, according to the picture of a replacement of a neutron with a Λ without changing the angular momentum and spin. The production of the ground state has an opposite trend, in agreement to the fact that it is necessary to change a p-shell neutron into a s-shell Λ.

The interpretation of the spectrum of $^{27}_\Lambda Al$ is not so simple

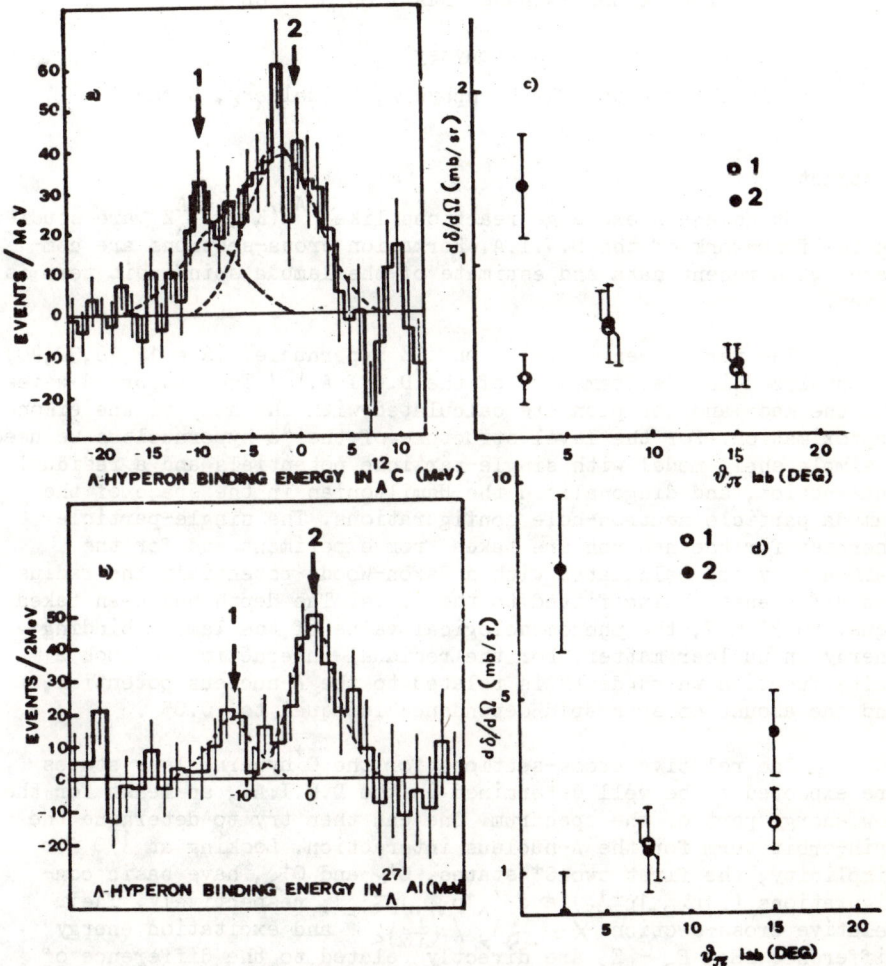

since none of the observed peaks cannot be interpreted as due to the formation of the ground state, but are both due to excited states. The shape of the angular distribution suggests that peak 2 is due to a replacement of a neutron with a Λ in the same shell, whereas peak 1 involves a change of shell.

References:
1) G.C. Bonazzola et al., Phys. Lett. 53B, 297 (1974)
2) G.C. Bonazzola et al., Phys. Rev. Lett. 34, 683 (1975)
3) W. Brückner et al., Phys. Lett. 55B, 107 (1975)
4) N. Auerbach and A. Gal, Phys. Lett. 48B, 22 (1974)
5) J. Hüfner et al., Phys. Lett. 49B, 409 (1974)
6) H.J. Lipkin, Phys. Rev. Lett. 14, 18 (1965).

STRANGENESS EXCHANGE REACTION ON NUCLEI

A. Bouyssy

Max-Planck-Institut für Kernphysik, Heidelberg, Germany

Abstract

Strangeness exchange reactions like $^A Z(K^-,\pi^-)^A_\Lambda Z$ are studied in the framework of the D.W.I.A. Formation cross-sections are compared with recent data and estimate of the lambda spin-orbit term is given.

The very recent data[1] on $^A_\Lambda Z$ hypernuclei (A = 12,16,32,40) are analyzed in the framework of the D.W.I.A.[2] The distorted waves for the kaon and the pion are calculated with the help of the eikonal approximation. For the level structure of the $^A_\Lambda Z$ hypernucleus we used a simple shell model with single-particle potentials and a residual interaction, and diagonalized the Hamiltonian in the space of the lambda particle neutron-hole configurations. The single-particle energies for the neutron are taken from experiment and for the lambda they are calculated with a Saxon-Woods potential, the radius and diffuseness being fitted to the r.m.s. The depth has been taken equal to 30 MeV, the phenomenological value of the lambda binding energy in nuclear matter. For the residual interaction we took a delta function which depth is related to the Λ-nucleus potential, and the amount of spin-spin dependence is equal to -0.05.

The relative cross-sections for the 0^+ hypernuclear states are expected to be well determined in the D.W.I.A., at least for the low energy part of the spectrum. One can then try to determine the spin-orbit term for the Λ-nucleus interaction. Looking at ^{16}O for simplicity, the first two 0^+ states, $0^+_<$ and $0^+_>$, have basic configurations $(_\Lambda 1p_{1/2} n 1p^{-1}_{1/2})$ and $(_\Lambda 1p_{3/2} n 1p^{-1}_{3/2})$, respectively. Their relative cross-section $X = \left(\frac{d\sigma}{d\Omega}\right)_> / \left(\frac{d\sigma}{d\Omega}\right)_<$ and excitation energy difference $\Delta E = E_> - E_<$ are directly related to the difference of spin-orbit terms for the lambda and the neutron, $\Delta V^{LS} = V^{LS}_\Lambda - V^{LS}_n$. A non-zero value for ΔV^{LS}, which means SU(3)-breaking one-body spin-orbit potential, is necessary to reproduce both characteristics, as can be seen on fig. 1, from which one can deduce $|V^{LS}_\Lambda| \lesssim \frac{1}{3}|V^{LS}_n|$ (we took $V^{LS}_n = +20$ MeV). Similar conclusions can be obtained from an analysis of the other hypernuclei.

Our results are in rather good agreement with the data, but, as shown on fig. 2, in the case of ^{16}O, there is a shift between the excitation energies of the observed peaks and the calculated ones, increasing with the mass number A. Moreover, in the high-energy part of the spectrum, the cross-sections are systematically smaller than the observed intensities of the bumps, but here, rescattering of the pion with excitation of 2p-2h configurations, as well as Σ-conversion, could be important. These problems are now under investigations.

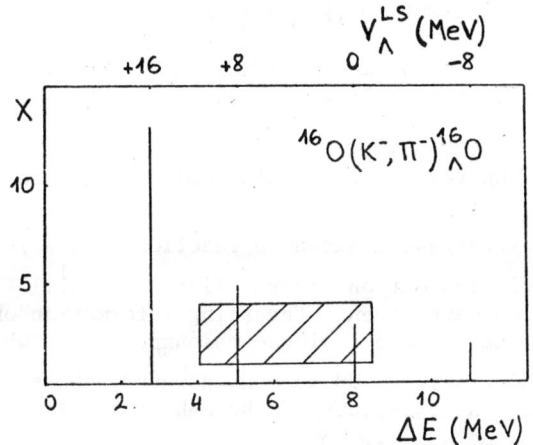

Fig. 1

Relative cross-section for the first two 0^+ states versus their excitation energy difference, for different values of the Λ-nucleus spin-orbit potential. Shaded area is deduced from data of ref. 1.

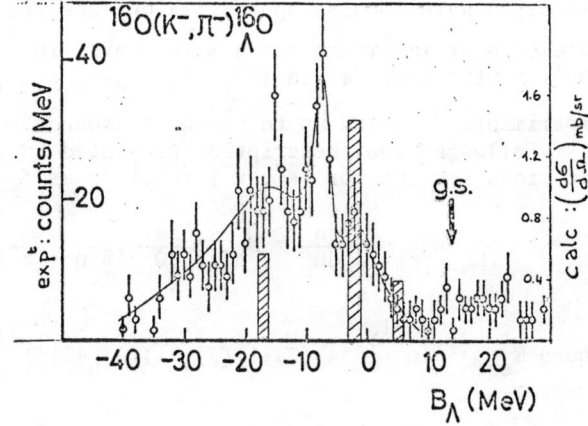

Fig. 2

Shaded areas correspond to the formation cross-sections (scale on the right) and binding energies (lower scale) of the calculated 0^+ hypernuclear states. ($V_\Lambda^{LS}=0$) Data are from ref. 1.

References

1) K. Kilian, New Data on the (K^-,π^-) reaction. Invited talk at this Conference.
W. Brückner et al., CERN-Heidelberg collaboration, CERN-preprint submitted to Physics Letters
2) J. Hüfner, S.Y. Lee and H.A. Weidenmüller, Nucl.Phys. A234(1974)429.

STRANGE GIANT RESONANCE IN (K⁻, π⁻) REACTION

N. Hoshi
Department of Physics, University of Tokyo, Tokyo, Japan

T. Fujita
Department of Physics, University of British Columbia, Canada

In recent years, the strangeness exchange reaction (K⁻, π⁻) on nuclear targets has been carried out on several light nuclei[1]. The experiments on this reaction have given interesting information on the possible existence of the strange analogue resonance (SAR) which was discussed by Kerman and Lipkin[2]. Up to now, however, those experiments do not indicate the existence of the SAR states as analysed by Hüfner, Lee and Weidenmüller[3].

In this paper, we point out an important contribution to (K⁻, π⁻) reaction from strange giant resonance in which one unit of angular momentum L is transferred. In analogy with the giant dipole resonance in usual nuclei, we call the resonance as strange giant dipole resonance (SGDR) for the L=1 case. According to a simple distorted wave impulse approximation, one finds that the SGDR becomes most important for q ∼ 100 MeV/c in ^{40}Ca, where q indicates the momentum transfer in (K⁻, π⁻) reaction. It is noted that in the experiments reported up to now q is around 80 ∼ 100 MeV/c also[4].

Following the prescription presented in ref. 3, we describe the cross section for (K⁻, π⁻) reaction as follows,

$$\frac{d\sigma}{d\Omega} = N_{eff} \left(\frac{d\sigma}{d\Omega}\right)_{K^-n \to \pi^-\Lambda}$$

where $N_{eff}^{(j_\Lambda j_n^{-1})J} = (2J+1)(2j_\Lambda+1)(2j_n+1) \begin{pmatrix} j_\Lambda & j_n & J \\ 1/2 & -1/2 & 0 \end{pmatrix}^2$

$$\times |<n_\Lambda \ell_\Lambda| \tilde{j}_J(r) |n_n \ell_n>|^2$$

The function $\tilde{j}_J(r)$ is the spherical Bessel function in the absence of any distortion. The distortion is taken into account with the help of the eikonal approximation.

We calculate N_{eff} in the reaction ^{40}Ca(K⁻, π⁻)$^{40}_\Lambda$Ca, taking into account the L=1 transition. Fig. 1 shows our calculated results for N_{eff}. Dashed blocks indicate the contributions due to the L=1 transitions, while white blocks show those of the L=0 case.

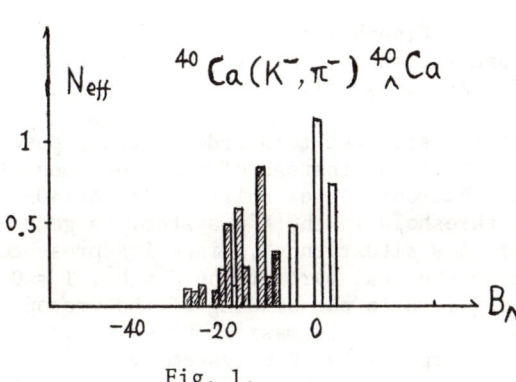

Fig. 1.

From Fig. 1, one can see that the contributions due to L=1 are large enough to be comparable with the L=0 case and should correspond to SGDR. Although our calculations are based on some crude approximations, they suggest that the SGDR gives much interesting information on nuclear structure as well as on the properties of hypernuclei.

In this paper, we do not report the contributions due to the L=2 transition. According to our preliminary calculation, the effect is found to give a contribution which is roughly one-fourth of the L=1 case. However, it should be noted that one would like to see how much the L=2 transition contribute to (K$^-$, π^-) reaction and how the strange giant quadrupole resonance are made.

REFERENCES

1. G. C. Bonazzota et al., Phys. Lett. $\underline{53B}$(1974)297, Phys. Rev. Lett. $\underline{34}$(1975)683.
   W. Brückner et al., Phys. Lett. $\underline{55B}$(1975)107.
2. A. K. Kerman and H. J. Lipkin, Ann. Phys. $\underline{66}$(1971)738.
3. J. Hüfner, S. Y. Lee and H. A. Weidenmüller, Nucl. Phys. $\underline{A234}$ (1974)429.
4. B. Povh et al., private communication.

THE Λ(1405) IN NUCLEAR MATTER AND IMPLICATIONS FOR KAONIC ATOMS

J. M. Eisenberg*
Department of Physics and Astronomy, Tel-Aviv University,
Tel-Aviv, Israel.

It is well known[1,2] that the simplest zero-order optical potential for kaonic atoms implies repulsion instead of the experimentally observed attraction for the hadronic level shifts. The Λ(1405) resonance, some 27 MeV below threshold in the K^-p system, is generally believed to account for this situation[3,4], since its presence implies a change in the sign of the real part of the $J = \frac{1}{2}^-$, $T = 0$, channel amplitude at that energy, while the binding of the proton and the kinetic energy of the K^-p center of mass each serves to reduce the effective internal energy of the K^-p system by some 5 MeV to 15 MeV. We here wish to point out that a dynamic effect of the Λ(1405) in nuclear matter tends to raise the resonance position by a comparable amount, and therefore further boosts the effective change of sign in the real part of the optical potential.

The effect in question is shown as an inset in the figure and involves the dissociation of the Λ(1405) into a Σ-hyperon and pion, the latter forming a Δ(1236) J=T=3/2 resonance with a nucleon in the medium. The modification of the pion propagator in nuclei leads to a shift in the position of the resonance and a (generally small) change in its width. To evaluate these we consider the Λ(1405) self-energy,

$$\Sigma_\Pi(E) = -\frac{iG^2}{\pi} \int d^4k \; \frac{1}{E-k_o-M_\Sigma-\frac{k^2}{2M_\Sigma}+i\varepsilon} \; \frac{1}{k^2+\mu^2-k_o^2+\Pi(k,k_o)} \; \frac{1}{(k^2+\Lambda^2)^2} ,$$

where G is the effective Λ(1405) → Σπ coupling constant, M_Σ and μ are the two final-state masses in this decay, Λ is its vertex cut-off parameter, and $\Pi(k,k_o)$ is the pion self-energy in the medium. For the free particle $\Pi \rightarrow -i0^+$, whence G^2 is easily evaluated in terms of the width, $\Gamma = -2\text{Im}\Sigma(M_\Lambda) \cong 40$ MeV, from the above equation. In the medium, we take $\Pi(k,k_o) \cong -4\pi\rho f_{\pi N}(k,k_o)$ for nuclear density ρ and a πN amplitude given by the theory of Chew and Low[5], and estimate $\Sigma_\Pi(E)$ by considering $\Pi(k,k_o)$ to be reasonably slowly varying for approximate k_o-integration in which the pion pole dominates. The k-integration is then performed numerically, yielding results for the shift Δ in the position of the Λ(1405) as shown in the figure. The parameter Λ proves insensitive (within the line thick-

---

* Work supported in part by the U.S. National Science Foundation, the Israel Center for Absorption in Science, and the Israel-U.S. Binational Science Foundation.

ness of the curves for $3\mu \leq \Lambda \leq 9\mu$), but the πN cutoff parameter α in $\Pi(k,k_o) \sim (1+k^2/\alpha^2)^{-2}$ is more critical.

Relevant nuclear densities in kaonic atoms are[4] $\rho/\rho_0 \sim 1/8$ for central densities $\rho_0 \cong 0.16$ fm^{-3}. For these values, the effective position of the resonance is raised by some 5 MeV to 7 MeV, so that this effect appears to be comparable to the kinematic ones in bringing about a sign reversal in the kaonic-atom optical potential.

REFERENCES

1. G. Backenstoss et al., Phys. Letters 38B, 181 (1972).
2. T.E.O. Ericson and F. Scheck, Nucl. Phys. B19, 450 (1970).
3. S. Wycech, Nucl. Phys. B28, 541 (1971).
4. W.A. Bardeen and E.W. Torigoe, Phys. Rev. C3, 1785 (1971).
5. G.F. Chew and F.E. Low, Phys. Rev. 101, 1570 (1956).

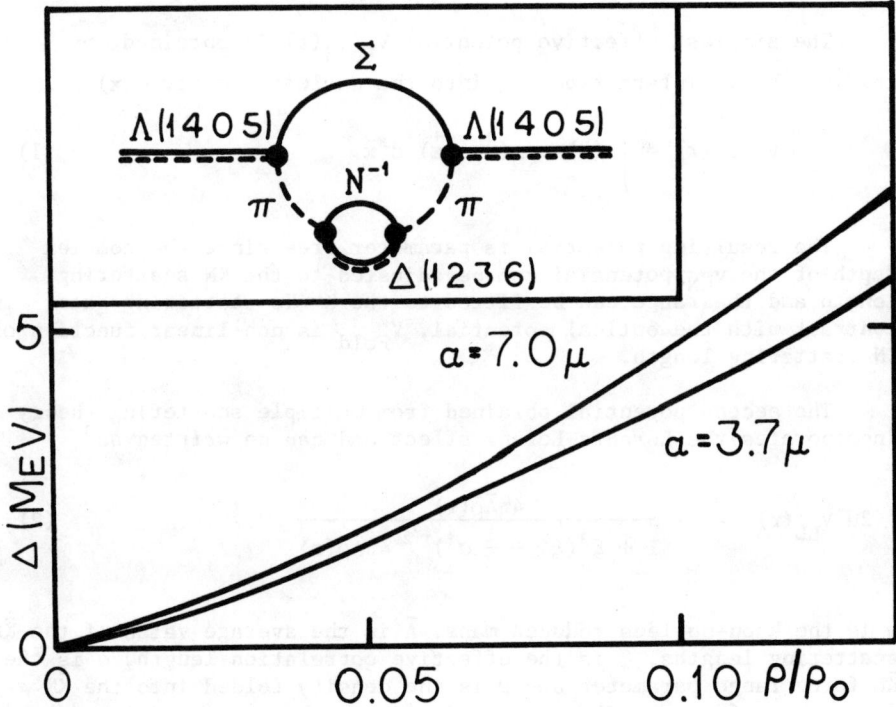

Fig. 1. Shift Δ in the position of the $\Lambda(1405)$ in nuclear matter due to the dynamic effect shown in the inset.

KAONIC ATOMS LEVEL SHIFTS: POTENTIAL APPROACH[†]

A. Deloff[*] and J. Law
University of Guelph, Guelph, Ontario, Canada, N1G 2W1

ABSTRACT

Kaonic atoms level shifts and widths have been calculated using two different forms of effective kaon-nucleus potential.

INTRODUCTION

The kaonic atoms level shifts and widths can be calculated in two ways. The first method[1] is to calculate the kaon-nucleus scattering length directly from multiple scattering theory and relate[4] this quantity to the complex level shift. The second method[2,3] seeks instead an effective kaon nucleus potential which used with Klein-Gordon equation yields the shifts and the widths. We would like to present the results obtained by means of the potential model.

The simplest effective potential $V_{Fold}(r)$ is obtained[2] by folding the $\bar{K}N$ interaction v_{KN} into the nuclear density $\rho(x)$

$$V_{Fold}(r) = \int \rho(x) \, v_{KN}(\vec{r} - \vec{x}) \, d^3x \,. \tag{1}$$

The resulting potential is parameter free since the complex depth of the v_{KN} potential can be adjusted to the $\bar{K}N$ scattering length and the range can be fitted to the $\bar{K}$-^{4}He scattering data. In contrast with the optical potential, V_{Fold} is non-linear function of $\bar{K}N$ scattering length.

The second potential obtained from multiple scattering theory incorporates the Lorentz-Lorenz effect and can be written as[3]

$$2\mu \, V_{LL}(r) = - \frac{4\pi \bar{A} \tilde{\rho}(r)}{1 + \xi^3 (\xi^2 + \frac{1}{2} \sigma^2)^{-\frac{1}{2}} \, 4\pi \bar{A} \tilde{\rho}(r)} \tag{2}$$

μ is the kaon-nucleus reduced mass, $\bar{A}$ is the average value of the $\bar{K}N$ scattering lengths, ξ is the effective correlation length, σ is the $\bar{K}N$ force range parameter and $\tilde{\rho}$ is the density folded into the $\bar{K}N$ force shape factor. The parameters ξ and σ have been adjusted to the lowest levels shifts and widths for light nuclei. As seen from

[†] Supported by National Research Council, Canada.
[*] On leave of absence from Institute for Nuclear Research, Warsaw.

(2) V_{LL} is also non-linear in $\bar{A}$.

The level shifts (ϵ) and widths (Γ) have been obtained from the potentials (1) and (2) using the method described in ref. 4 and the results are given in table I.

Table I: Kaonic atoms level shifts and widths.

element	level	-ε (eV)			Γ (eV)		
		experiment	V_{Fold}	V_{LL}	experiment	V_{Fold}	V_{LL}
^{10}B	2p	208± 35	234	161	810± 100	963	627
^{11}B	2p	167± 35	245	158	700± 80	948	675
^{12}C	2p	590± 80	723	452	1730± 150	2088	1775
^{12}C	3d	-	-0.16	0.13	0.98±0.19	1.04	0.61
^{31}P	3d	330± 80	266	326	1440± 120	1895	1315
^{31}P	4f	-	-0.56	0.46	1.97±0.33	3.12	1.88
^{32}S	3d	550± 60	457	381	2330± 200	3067	2101
^{32}S	4f	-	-1.13	0.97	3.25±0.41	6.35	3.70
^{35}Cℓ	3d	770±400	844	903	3800±1000	4441	3318
^{35}Cℓ	4f	-	-1.59	1.63	5.69±1.50	11.39	6.73
^{64}Zn	4f	-60±300	285	420	1700± 400	2556	1850
^{27}Aℓ	3d	130± 50	63	88	490± 160	619	412
^{28}Si	3d	240± 50	112	197	810± 120	1209	761
^{63}Cu	4f	240±220	194	239	1650± 720	1650	1250
^{58}Ni	4f	180± 70	98	226	590± 210	1448	971
^{58}Ni	4f	260± 90	98	226	1340± 140	1448	971
^{59}Co	4f	80± 50	59	152	980± 150	1044	701
^{107}Ag	5g	500±130	184	262	2420± 510	1817	1393
^{114}Cd	5g	250±120	204	423	3000± 350	2710	1954

The agreement with experiment is quite good for the lighter elements. The long tail of V_{Fold} gives Γ_{up} systematically too large, which is not the case for V_{LL} as V_{LL} becomes repulsive for large r. For heavier elements hyperfine effects obscure the comparison with experiment.

REFERENCES

1. A. Deloff and J. Law, A simple level shift formula for kaonic atoms. (1976, sub. to Phys. Rev. Lett.).
2. A. Deloff and J. Law, Phys. Rev. 10C, 1688 (1974).
3. A. Deloff, The Lorentz-Lorenz effect in kaonic atoms. (1976, sub. to Phys. Lett.).
4. A. Deloff, Phys. Rev. 13C, 730 (1976).

KAONIC ATOMS LEVEL SHIFTS: MULTIPLE SCATTERING APPROACH[†]

A. Deloff[*] and J. Law
University of Guelph, Guelph, Ontario, Canada, N1G 2W1

ABSTRACT

A multiple scattering formalism is used to obtain the koan-nucleus scattering length directly in terms of the $\bar{K}p$ and $\bar{K}n$ scattering lengths and the 2ℓ moments of the nuclear density. The coulomb corrected kaon-nucleus scattering length is related to the complex level shifts of a kaonic atom.

INTRODUCTION

The multiple scattering set of equations[1] can be solved for the scattering amplitude, by assuming the fixed scatterer approximation, closure and that the t-matrices are derivable from separable potentials. The kaon-nucleus scattering length A_ℓ (uncorrected for Coulomb effects) is obtained as[2]

$$[(2\ell+1)!!]^2 A_\ell = \left[\frac{ZA_p}{1+A_pG} + \frac{NA_n}{1+A_nG}\right] \langle r^{2\ell} \rangle \left[1 + X\left(1+A\delta_{\ell 0} - AU_{2\ell}\right)\right]. \quad (1)$$

$A_{p,n}$ are $\bar{K}p$, $\bar{K}n$ scattering lengths (corrected for a kinematic factor[3]). $\langle r^{2\ell} \rangle$ is the 2ℓ- moment of the nuclear density $\rho(r)$ (normalised to unity);

$$X = \frac{G}{A} \frac{[Z A_p(1 + A_nG) + N A_n(1 + A_pG)]}{[1 - G[(Z-1)A_p + (N-1)A_n] + A_pA_n(1-A)G^2]}, \quad (2)$$

$$U_{2\ell} = \frac{4\pi \xi^3}{\langle r^{2\ell} \rangle} \int d^3r \, \rho^2(r) \, r^{2\ell}. \quad (3)$$

ξ is the two nucleon correlation length[3]. G is an effective propagator which depends on the shape of the separable potential and the mean nucleon separation; Z, N, A have their usual meanings. A_ℓ can be corrected for Coulomb effects[2] (A_ℓ^c). The kaonic atom complex level shift is related to A_ℓ^c by[4]

$$-\varepsilon_\ell - \frac{i\Gamma_\ell}{2} = \delta E_\ell = -|B_\ell|[2/(2\ell+1)]^{2\ell+2} A_\ell^c [r_B^{2\ell+1}(2\ell+1)!]^{-1}. \quad (4)$$

[†] Supported by National Research Council, Canada.
[*] On leave of absence from Institute for Nuclear Research, Warsaw.

$|B_\ell|$ is the Klein-Gordon point charge coulomb binding energy and r_B the Bohr radius. The calculated shifts and widths are given in Table I, using a Fermi nuclear density. The agreement with experiment is quite good for the lighter elements. For the heavier ones, hyperfine effects would obscure the comparison.

Table I: Kaonic atom level shifts and widths.

element	level	a (fm)	c (fm)	$-\epsilon$ (eV) experiment	$-\epsilon$ (eV) theory	Γ (eV) experiment	Γ (eV) theory
^{10}B	2p	0.48	2.20	208±35	141	810±100	690
^{11}B	2p	0.48	2.17	167±35	145	700±80	738
^{12}C	2p	0.44	2.39	590±80	455	1730±150	1944
^{12}C	3d			-	0.08	0.98±0.19	0.50
^{31}P	3d	0.56	3.21	330±80	348	1440±120	1343
^{31}P	4f			-	0.33	1.97±0.33	1.83
^{32}S	3d	0.59	3.20	550±60	629	2330±200	2199
^{32}S	4f			-	0.74	3.25±0.41	3.87
^{35}Cℓ	3d	0.57	3.33	770±400	1022	3800±1000	3340
^{35}Cℓ	4f			-	1.32	5.69±1.50	6.81
^{64}Zn	4f	0.55	4.32	-60±300	538	1700±400	1625
^{27}Aℓ	3d	0.52	3.07	130±50	90	490±160	412
^{28}Si	3d	0.59	3.01	240±50	200	810±120	814
^{63}Cu	4f	0.50	4.29	240±220	319	1650±720	1065
^{58}Ni	4f	0.56	4.19	180±70	277	590±210	908
^{58}Ni	4f			260±90	277	1340±140	908
^{59}Co	4f	0.569	4.08	80±50	193	980±150	664
^{107}Ag	5g	0.5029	5.377	500±130	365	2420±510	1026
^{114}Cd	5d	0.5848	5.27	250±120	635	3000±360	1481

REFERENCES

1. M. Goldberger and K. M. Watson, Collision Theory (John Wiley, 1964).
2. A. Deloff and J. Law, Hadronic Atoms (1976, in preparation)
   A. Deloff and J. Law, A simple level shift formula for Kaonic Atoms (1976 sub. to Phys. Rev. Lett.).
3. A. Deloff, The Lorentz-Lorenz effect in Kaonic atoms (1976 sub. to Phys. Lett.).
4. A. Deloff, Phys. Rev. C13, 730 (1976).

APPLICATION OF FORWARD DISPERSION RELATION TO KAON ^{12}C SCATTERING

K. ARAI, I. ENDO and M. KIKUGAWA
Physics Department, Hiroshima University, Hiroshima, Japan

Encouradged by successful use of forward dispersion relation (FDR) in pion nucleus scattering,[1] we apply FDR to Kaon ^{12}C scattering. Such an analysis is of special interest because we can expect that characteristic features of nuclear system with strangeness quantum number are exhibited without referring to any specific models. Assuming that i) the total cross sections of both K$^+$ and K$^-$ scatterings tend to the same (constant) limiting value at high energies, ii) contributions from all the poles on the complex ω(lab. total energy of kaon) plane are represented by a single effective pole at $\omega=\omega_p$, iii) from threshold ($\omega=m_k$) to $\omega=\omega_1$ ($\omega_1=m_k+5$ Mev), complex scattering length approximation is valid and iv) that the imaginary part of K$^-$C scattering amplitude below threshold is obtained by analytic continuation of the one slightly above threshold, we write the FDR for $f^{(\pm)}=\frac{1}{2}\{f(K^-C)\pm f(K^+C)\}$ as follows.

$$\mathrm{Re}\, f^{(-)}(\omega) = \frac{2\gamma\omega}{\omega^2-\omega_p^2} + \frac{2\omega}{\pi}\int_{\omega_0}^{\omega_1}\frac{\mathrm{Im}\, f^{(-)}(\omega')}{\omega'^2-\omega^2}d\omega' + \frac{\omega}{4\pi^2}\int_{\omega_1}^{\infty}\frac{k'\{\sigma_T^{K^-}(\omega')-\sigma_T^{K^+}(\omega')\}}{\omega'^2-\omega^2}d\omega' \quad (1)$$

$$\mathrm{Re}\, f^{(+)}(\omega) = \mathrm{Re}\, f^{(+)}(m_k) + \frac{2\gamma\omega_p}{\omega^2-\omega_p^2}\frac{k^2}{\omega_p^2-m^2}$$

$$+ \frac{2k^2}{\pi}\int_{\omega_0}^{\omega_1}\frac{\omega'\,\mathrm{Im}\, f^{(+)}(\omega')}{(\omega'^2-\omega^2)k'^2}d\omega' + \frac{k^2}{4\pi^2}\int_{\omega_1}^{\infty}\frac{\{\sigma_T^{K^-}(\omega')+\sigma_T^{K^+}(\omega')\}}{(\omega'^2-\omega^2)}\frac{\omega'}{k'}d\omega' \quad (2)$$

where $K=\sqrt{\omega^2-m_k^2}$ and the principal part should be taken in each integrals.

We performed the integration (1) and (2) using reported K±C total cross section[2,3] fitted with a polynomial above 830 MeV and by assuming linearly rising from $\omega=\omega_1$ up to 830 MeV. For K$^-$C scattering length a_- we adopted the value determined in the analysis of K-mesic atom.[4] Effective pole residue γ and K$^+$C scattering length have been varied so as to give best fitting Re f to experimental value at 1.65, 2.01 and 2.26 GeV/c.[5] The obtained values are $\gamma=11.5\pm2.9$ and $a_+=0.65\pm0.35\, m_\pi^{-1}$. Comparison with experimental values is shown in Table I.

Table I

k (Gev/c)	FDR (Gev^{-1})	Re f(-) Experiment (Gev^{-1})	FDR (Gev^{-1})	Re f(+) Experiment (Gev^{-1})
1.68	21.5	21.0 ±4	-16.1	-7.05 ±4
2.01	21.5	22.9 ±4.5	-20.4	3.7 ±5
2.26	20.3	21.2 $^{+8}_{-9.5}$	-24.4	-11.1 $^{+8}_{-9.5}$

We observed that the value of γ is stable enough under variation of ω_0. This behavior suggests that the value does not depend much on the specific model employed for Im f(±) below threshold.

Unlike π nucleus scattering, γ is slightly larger than coherent sum of elementary hyperon poles $\gamma=7.4$ which is obtained by using $g_\Lambda^2/4 =10.8$ and $g_\Sigma^2/4\pi=3.8$(SU(3) values)[6] and by taking into account the nucleon-hyperon mass difference.

REFERENCES

1. T. E. O. Ericson and M. P. Locher, CERN 71-14; Spring School on Pion Interactions at Low and Medium Energies.
2. D. V. Bugg et al., Phys. Rev. 168, 1466 (1968).
   R. J. Abrams et al., Phys. Rev. D4, 3235 (1971).
3. S. P. Denisov et al., Nucl. Phys. B61, 62 (1973).
4. R. Seki, Phys. Rev. Lett. 29, 240 (1972).
5. B. Gobbi et al., Phys. Rev. Lett. 29, 1278 (1972)
6. H. Pilkuhn et al., Nucl. Phys. B65, 460 (1973).

BLACK SPHERE MODEL FOR THE LINE WIDTHS OF KAONIC
AND ANTIPROTONIC ATOMS

W.B. Kaufmann* and H. Pilkuhn
Institut fur Theoretische Kernphysik, Universitat Karlsruhe
Federal Republic of Germany

ABSTRACT

It is shown that the line widths of K^- and $\bar{p}$ atoms are adequately described by the WKB approximation for a totally absorbing nucleus.

The line widths of kaonic and antiprotonic atoms are easily calculated in the WKB approximation,

$$\Gamma = (n\pi)^{-1} E_n \exp(-2I), \quad I = \int_R^{r_1} K(r)dr, \quad E_n = mZ^2\alpha^2/2n^2 \quad (1)$$

where E_n is the unperturbed atomic binding energy, r_1 is the inner turning point of the unperturbed problem and R is the new turning point near the nuclear surface which is produced by the strong optical potential. The formulation (1) corresponds to total absorption, but the energy shifts can be included if some reflexion at R is included.

We have shown that a good description of the widths of all nuclei heavier than Be is obtained by taking the imaginary wave number $K = ik$ from the unperturbed problem and chosing R such that a fixed number of nucleons N is at distances greater than R[1]. For kaons, we find $N = 3.0$ and for antiprotons $N = 1.2$. This description sheds little light on the microscopic aspects of absorption, but it emphasizes the role of the nuclear surface, and explains to some extent the success of perturbation theory calculations. It should be particularly useful for antiprotonic atoms where practically nothing is known about the elementary $\bar{p}p$ interaction. For circular orbits, one gets approximately

$$\Gamma \sim \frac{2}{\pi} E_n \, n^{-5\lambda+n-1} \, y^{2\lambda} \, e^{2(n+1/4-y/n)}, \quad (2)$$

$$y = Z\alpha mR, \quad \lambda^2 = \ell(\ell+1) = n(n-1). \quad (3)$$

1) W. Kaufmann and H. Pilkuhn, Karlsruhe University preprint (1976).

* On sabbatical leave from Arizona State University, Tempe, Arizona 85 281, USA.

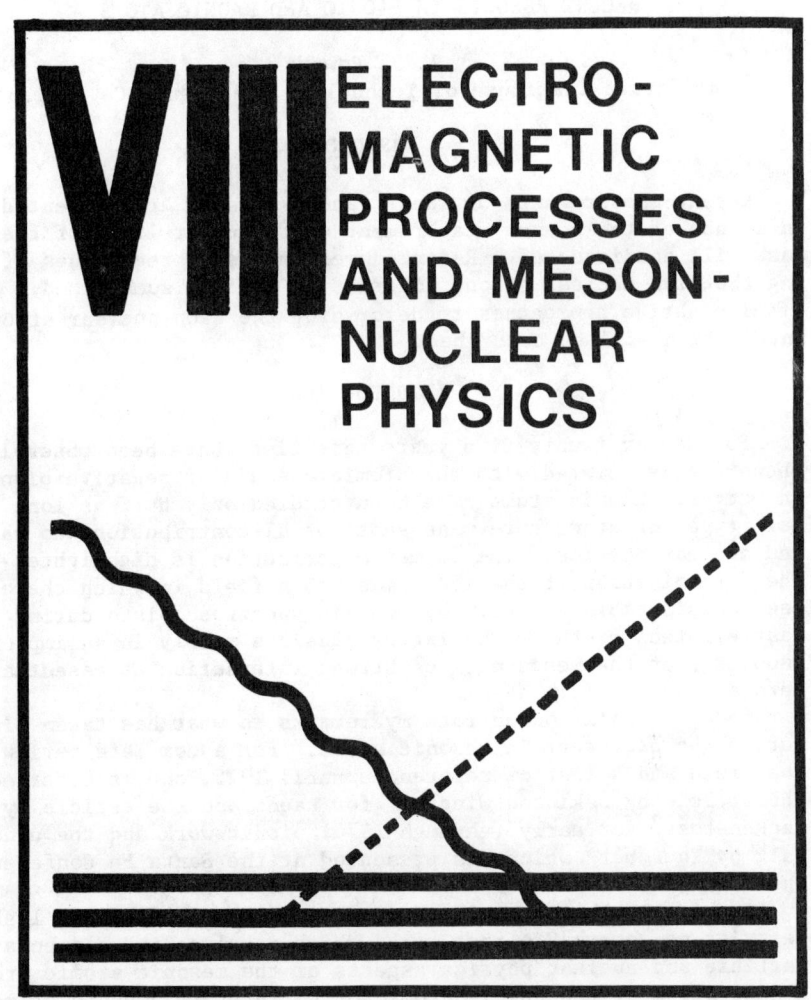

VIII ELECTROMAGNETIC PROCESSES AND MESON-NUCLEAR PHYSICS

RECENT RESULTS IN PIONIC AND KAONIC ATOMS

R. J. Powers
California Institute of Technology, Pasadena, Ca. 91125

ABSTRACT

Recent measurements of the π^- and K^- masses are presented. The influence of the former measurement on the upper limit of the ν_μ mass will be discussed. Recent developments in techniques of studying the pion-nuclear strong interaction will be summarized. A couple of alternative approaches to describing the kaon-nuclear strong interaction will be described.

INTRODUCTION

For nearly twenty-five years scientists have been observing phenomena associated with the atomic cascade of negative pions in matter. Kaonic atoms have been studied only half as long. Yet both types of atoms have made substantial contributions to particle and nuclear physics. The former contribution is highlighted by the determination of the meson masses, a field in which there has been considerable activity by mesonic spectroscopists during the past eighteen months. The latter consists mainly in an improved knowledge of the meson-nucleus strong interaction at essentially zero energy.

Today I shall concentrate my remarks to what has taken place during the past year in mesonic atoms. For a complete review of the field and a list of references until 1975, one is referred to the article by Seki and Wiegand[1] for kaons and the article by Backenstoss[2] for early (through 1970) pionic work and the updated talk by Tauscher[3] which was presented at the Santa Fe Conference in 1975.

In view of the specialized nature of this conference, I shall restrict my remarks to that work, which involves the elementary particle and nuclear physics aspects of the mesonic atomic problem. A discussion of the purely atomic or solid state aspects of mesonic atoms, can be found in the first three references.

MESONIC ATOM INFORMATION

Before I begin the discussion of recent results, I would like to take a few minutes to review the observables (and their sensitivities) available to the mesonic spectroscopist. Once the meson has penetrated the last electron cloud during its atomic cascade towards the nucleus, the information that this probe can reveal will depend on its distance from the nucleus. If the meson has little overlap with the nucleus, it will view the nucleus as simply a source of a very strong electric field. Under such conditions

the energies of mesonic transitions can be used for determining
such meson properties as its mass and polarizability. Indeed the
definitive measurements of the π^- and K^- mass come from mesonic
atoms. The basic assumption that is made is that the electromagnetic interaction is completely understood and hence all radiative
effects are calculable. Recent results in muonic x rays seem to
indicate that this is possible to at least 30 ppm[4,5].

In addition, even electromagnetic properties of the nucleus
itself are detectable even though there is negligible overlap of
the meson with the nucleus. For example, the spectroscopic quadrupole moment of ^{165}Ho has been obtained with 3% accuracy from the
hyperfine structure of $5g \rightarrow 4f$ transitions at SIN.[6]

When there is appreciable overlap of the meson with the nucleus
strong interaction effects begin to emerge. First there are the
direct effects. Not only are the energy levels shifted relative
to the electromagnetic eigenvalues but also the possibility of
strong absorption by the nucleus modifies the transition probabilities of the atomic transitions and becomes visible as a broadened
line width. I categorize these effects as direct because they
can be directly related to an effective meson-nucleus interaction
as these effects result from a modified Hamiltonian.

But there are also indirect effects. Long before the meson
energy levels are measurably shifted by the strong interaction,
absorption can take place and manifests itself in a reduced intensity for radiative transitions from the affected state.

A second but much less common indirect effect can occur if a
resonance occurs between a nuclear energy level and the energy
difference between two mesonic states, one of which is affected by
the strong interaction. If the resonance condition permits the
mixing of atomic states where appreciable absorption occurs, the
intensity of the radiative transitions from the mixed mesonic
states will be diminished. Examples of such resonances have already been reported for both pionic[7] and kaonic[8] atoms recently.
We shall discuss this work in more detail below.

What should be stressed here is that in the region of the
nucleus it is the strong interaction itself that is being studied.
Unlike muonic atoms and electron scattering where the interaction
is so well known that very minute details of the nuclear charge
distribution can be revealed, our understanding of the pion-nuclear
interaction is at best semi-phenomenological making the extraction
of matter distribution parameters still somewhat tenuous. Fortunately there are several ways of looking at the problem. It is
not inconceivable that now that precise low energy pion scattering
data is becoming available, it will be possible to disentangle
the matter properties from the strong interaction. The scattering
data is more suited to probing the interaction by virtue of its
greater degrees of freedom—namely, variable momentum transfer;
whereas pionic atoms provide only a limited number of pieces of
information, albeit more precise and hence ideally should be reserved for probing the moments of the matter distribution. This

idea of complementarity of information is already quite successful in the field of nuclear charge distributions where electron scattering data provides basically the shape of the charge distribution and muonic atom energies provide very precise determinations of a limited number of radial charge moments.

Today I propose first to discuss what has been learned recently about mesonic properties. Then we will turn to the question of the meson-nucleus interaction and the possible extraction of nuclear matter parameters.

MESON PROPERTIES FROM MESONIC ATOMS

Recently an experiment performed in Leningrad[9] reduced the uncertainty in the pion mass by nearly a factor of four. I personally do not find it surprising that this most recent work was performed with a crystal spectrometer. What is amusing is that the experiment was performed not at one of the highly touted meson factories but at a long existing accelerator using an ingenious trick (Fig. 1).

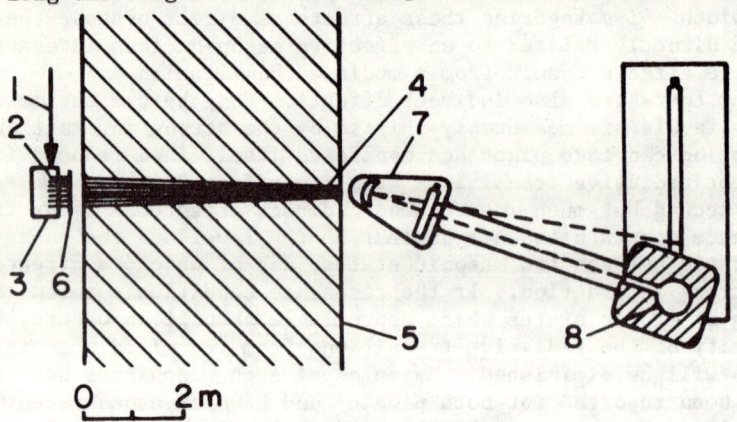

Fig. 1 -- Experimental set-up of crystal spectrometer. 1 indicates the proton beam of the 1 GeV accelerator at Gatchina incident on copper disks (2) which serve as the principal pion producer. Many escaping pions stop in the target (Ca or Ti)(3) producing pionic x rays. After having passed through the collimator 4, which contains numerous copper plates to block direct view of the production target, the x rays are diffracted by the curved crystal 7 into a detector (8). Calibrations are performed at position 6 which sits on the beam side of 5 meters of shielding (5).

The pion production target (Cu) was interleaved with the pionic x-ray target (Ca or Ti) so that effectively copious numbers of π^-s were able to stop in the material of interest. By viewing the pionic x-ray production target through a 5m long collimator which screened the spectrometer crystal from direct view of the pion production target, Marushenko et al. were able to gather sufficient data in 10 hours to attain an accuracy of 12 parts per million.

A summary of the most recent determinations of the π^- mass is shown in Table 1. There is very excellent agreement between the CERN[10] value which was obtained with a Ge(Li) detector and the two crystal spectrometer experiments performed at Berkeley[11] and at Leningrad (Gatchina).

Table I Pion Mass

π Atoms	Used	Value (keV)	Ref.
I	$5g \to 4f$	139569 ± 6	CERN[10]
Ti	$6h \to 5g$		
Ti,Ca	$4f \to 3d$	139566 ± 10	Berkeley[11]
Ti,Ca	$4f \to 3d$	139565.7 ± 1.7	Leningrad[9]

If we combine this most recent result of the π^- mass (139565.7 ± 1.7 keV) with the accepted value for the μ mass (105659.48 ± 0.35 keV)[12] with the most recent value[13] of the μ^+ momentum from π^+ decay at rest (29787 ± 5 keV/c) we can determine the rest mass of the muon neutrino assuming the mass equality of positive and negative pions from the CPT theorem. The result is

$$m_\nu^2 = 0.047 \pm 0.392 \text{ MeV}^2/c^4 \qquad (1)$$

From this it follows that with a probability of 90% the inequality

$$m_\nu < 0.66 \text{ MeV}/c^2 \qquad (2)$$

holds. In spite of the fact that the new value for the muon momentum is nearly a factor of three improvement over previous determinations, this fact has been more than compensated by the recent pion mass measurement. As a result at the present time the dominant source of error (96%) in our knowledge of the muon neutrino mass is the precision of the muon momentum measurement. This latest value for m_ν is quite compatible with the value[14] inferred from $K_{\mu3}$ decay which is of similar accuracy.

For completeness we point out that the kaon mass is now known to 34 ppm if one averages the kaonic x-ray results of Columbia[15] and CERN[3] reported at the Santa Fe Conference last year.

In the determination of the π^- and K^- masses, it has been tacitly assumed that there was no appreciable shift of the atomic levels due to a possible polarization of the meson by the nuclear field. Current experimental results on the polarizability of the proton[16] and theoretical results[17,18] seem to indicate that such effects should be at least an order of magnitude smaller than current experimental accuracies of mesonic atomic transition energies.

MESON-NUCLEUS INTERACTION

During the past couple years there has been considerable activity--both experimental and theoretical--concerning kaonic atoms. This is undoubtedly a reflection of the fact that strong interaction effect data with kaonic atoms have been available for only a few years. Now work is being carried on by groups at LBL, CERN, Brookhaven and RHEL.

On the other hand, after a flurry of experimental activity in the early seventies, the study of strong interaction effects in pionic atoms has been relatively subdued during the past year. This must be at least partially due to the fact that most of the remaining work requires intense pion beams and the two major meson factories at SIN and LAMPF have only begun to operate more or less continuously. However, at least three techniques are being used (or at least being discussed) to study strong interaction effects in pionic atomic states: a) directly, with pionic x rays (direct measurement of pionic transition widths and shifts); b) indirectly with missing pionic x rays (E2 resonance effects which attenuate radiative transitions; and c) without pions (proton elastic scattering near π^- production threshhold).

We shall first discuss the pionic atom work in order of decreasing "exoticism".

Pionic Atoms Without Pions

Recently G. T. Emery[19] suggested that by studying proton elastic scattering near the π^- production threshold, it might be possible to determine pionic binding energies for states not normally visible or for targets (such as ^{14}O) which are not available in sufficient quantities to perform a standard pionic x-ray experiment. One possibility would be to study the total cross section near threshold for $^{13}C(p, \pi^-)^{14}O$. In Fig. 2 we can see a plot of the possible energy dependence near threshold for experimental conditions presumably available at accelerators like the Indiana Cyclotron.

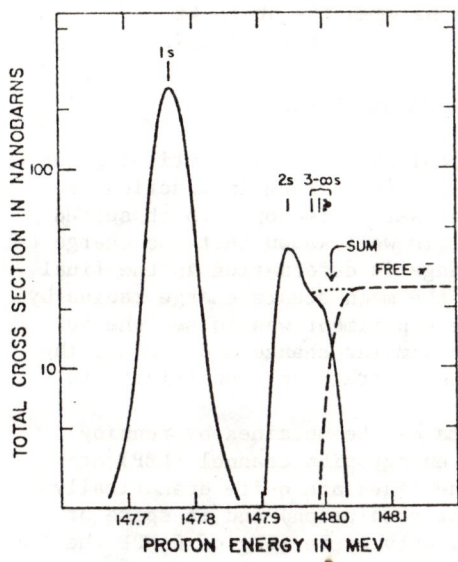

Fig. 2 Energy dependence near threshold for $^{13}C(p,\pi^-)^{14}O$ assuming "realistic" experimental conditions. The cross section for producing ^{14}O pionic atom states is given by the full line while that for producing a free π^- is given by the dashed line.

This technique—"pion-less" pionic atoms is based on a suggestion by Tzara[20], who has proposed observing near-threshold photoproduction of negative pions. The cross section magnitudes were estimated using the single proton emission branching ratio following pion capture[21] from the 1s state.

E2 Nuclear Resonance

Bradbury et al.[7] have reported the observation of E2 nuclear resonance in Cd which allows the determination of the pionic 3d state strong interaction shift. In ^{112}Cd and ^{111}Cd the energy difference between the 5g and 3d pionic states matches quite well nuclear energy levels which are reachable from the nuclear ground state by an E2 interaction. By measuring the relative attenuation of the 5g → 4f transitions in each isotope (each normalized to its own 6h → 5g intensity) one can infer the binding energy of the 3d state in a way that is fairly independent of atomic cascade calculations. In these particular isotopes the 4f → 3d transitions can also be measured directly and thus serve as a check for the technique. Although the technique is not one of great universality, it does make possible the observation of strong interaction effects in states not normally observable. This seems to be the case[22] in ^{110}Pd where the 4f to 3p energy difference is in resonance with the 2+ level at 373.8 keV. Preliminary analyses of pionic data[23] taken at LAMPF support these theoretical calculations.

Similar resonance effects have been observed in kaonic

Mo at LBL[8]. Here the resonance is between the 6h to 4f energy difference and the 2+ level in ^{98}Mo at 787.4 keV.

Direct Measurement of Pionic Transitions

At LAMPF the Caltech-Wyoming collaboration is continuing the studies[24] performed at SREL[25] looking for strong interaction isotope shifts for pionic x rays. One set of isotopes which seemed promising was Nd^{148} and Nd^{150}. It is well known that the charge distribution undergoes a rapid change in deformation as the final two neutrons are added increasing the mean square charge radius by 51.4 ± 0.6 mfm[26]. The goal of the experiment was to see whether the neutrons might be undergoing a similar change and whether the pionic 4f → 3d transitions would demonstrate any sensitivity to this difference.

Fig. 3 shows what type of data can be obtained by running approximately 24 hours at the low energy pion channel (LEP) at LAMPF using a Ge(Li) detector. The lines are quite dramatically broadened (three times the detector resolution) and in spite of the low yields (∼ 30%) and the low duty cycle (6%) of LAMPF the signal-to-noise is quite good.

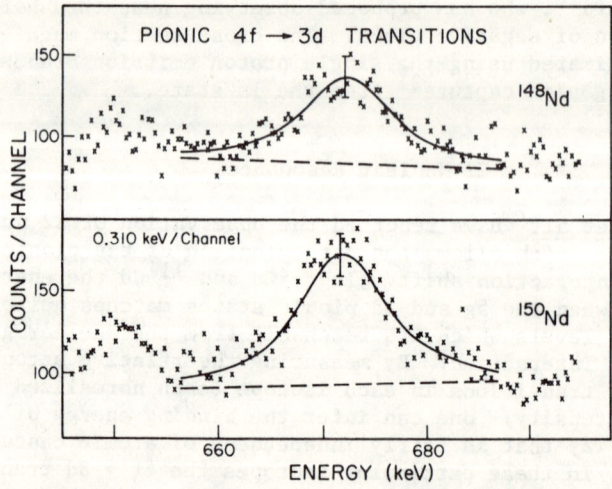

Fig. 3 4f → 3d pionic spectra in ^{148}Nd and ^{150}Nd, showing the fitted Lorentzians. The 9 keV broadening is easily observed with a Ge(Li) which had 3 keV resolution. The results can be found in Table III.

In the interpretation of these data we used an effective potential derived from the multiple scattering theory of Ericson and Ericson.[27] It has both local and nonlocal contributions.[28,29]

$$V_N = V_\ell(r) - (2\mu)^{-1} \nabla \cdot \alpha(r) \nabla \qquad (3)$$

In order to emphasize the dependence of this potential on the neutron, proton and matter densities denoted by ρ_n, ρ_p and $\rho_m = \rho_n + \rho_p$ we write

$$V_\ell(r) = -(4\pi/2\mu) \{[b_o \rho_m + b_1(\rho_n - \rho_p)] + iB_o \rho_m \rho_p\} \qquad (4)$$

Similarly, for the nonlocal part we use

$$\alpha(r) = \alpha_o(r)/[1 - \alpha_o(r)/3] \qquad (5)$$

where $\alpha_o(r) = -4\pi \{[c_o \rho_m + c_1(\rho_n - \rho_p)] + iC_o \rho_m \rho_n\}$

Now, in principle, the values for these parameters can be derived from multiple scattering theory. These values are given in Table II and are taken from Ref. 2 with slight corrections for the difference in the definition of the two nucleon parameters B_o and C_o.

Table II

Fitted optical potential parameters and neutron half-density radius from 4f→3d pionic data

	2 parameter fit	3 parameter fit	Theory	
b_0	-0.033	-0.033	-0.03	m_π^{-1}
b_1	-0.117	-0.117	-0.09	m_π^{-1}
B_0	0.076	0.076	0.044	m_π^{-4}
c_0	0.186 ± 0.002	0.219 ± 0.026	0.21	m_π^{-3}
c_1	0.139	0.139	0.18	m_π^{-3}
C_0	0.228 ± 0.022	0.249 ± 0.027	0.19	m_π^{-6}
Δc	0	0.095 ± 0.074		fm
χ^2/DF	25.7/18	24.2/17		

Using the above potential we fitted all pionic x-ray data which demonstrate appreciable strong interaction effects in the 3d level ($42 \leq Z \leq 60$) to the p-wave parameters c_o and C_o. There were two reasons for this restriction:
a) it was hoped that restricting the fit to a limited region of the periodic table would provide a potential more appropriate to Nd;
b) by fitting only the 3d data we avoided the problems associated with the fact that the absorptive s-wave parameter B_o is not well predicted by multiple scattering theory as the 3d state is very insensitive to the s-wave parameters.

The remaining parameter c_1 has only a weak influence on the fitted values given in Table III. As a result this parameter c_1 and the s-wave parameters b_o, b_1 and B_o were held fixed to the values determined in Ref. 28 and which were determined by a fit to all pionic x-ray data ($10 \leq Z \leq 83$).

Table III

Summary of pionic x-ray energies

Z	A	4f→3d Transition Energy (keV)		Width (keV)	
		Exp.	Fit	Exp.	Fit
42	96	323.2(2)	323.30	0.56(10)	0.581
49	115	442.8(5)	442.90	2.8(6)	1.927
50	118	460.3(6)	461.64	1.9(12)	2.250
55	133	561.0(9)	561.62	3.7(12)	4.480
56	138	582.99(27)	582.80	4.3(9)	5.004
57	139	604.0(9)	604.57	6.2(20)	5.781
58	140	627.31(29)	626.77	5.6(10)	6.673
58	142	627.50(33)	626.64	6.5(9)	6.519
60	148	671.68(32)	671.88	8.81(123)	8.032
60	150	671.61(22)	671.57	9.27(110)	7.757

It is interesting to note that because of the relative size of c_o and c_1 in the nonlocal potential, the relative influence in the potential of the neutron distribution ($c_o + c_1$) is nearly five times that of the proton distribution ($c_o - c_1$). The two parameter fit to the 3d data give quite excellent agreement between the experimental energies and the fitted energies and yield potential parameters that are typically within 15% of the values predicted by multiple scattering theory.

Bearing in mind that the pion potential is not completely under control, we investigated what sensitivity our data would have to the neutron distribution. First we assumed that both neutron and proton distributions could be represented by Fermi distributions with identical skin thicknesses but different half density radii which differed by an amount proportional to $A^{1/3}$ i.e.

$$c_n = c_p + (\Delta c) A^{1/3} \qquad (6)$$

The parameter Δc was allowed to vary simultaneously with c_o and C_o and yielded somewhat different potential parameters (See Table II for the results of the 3 parameter fit). It is interesting to note that Δc is not very much different from zero. This fit suggests that if there is a difference between the half density radii of the proton and neutrons distributions it is not much bigger than 0.5 fm. This is not too surprising a result and is not too informative since it describes an "average" behavior in this region of the periodic table.

A much more interesting result was found by examining the real part of the isotope shift between ^{148}Nd and ^{150}Nd. It was found that if we assume we know what the change in the proton distribution is (from muonic x rays), we find that the isotope shift

$$\Delta E_{150-148} = -0.07 \pm 0.35 \text{ keV} \qquad (7)$$

corresponds to a change in the equivalent radius of the neutron distribution of 52 ± 53 mfm as opposed to 51.4 ± 0.6 mfm for the protons. The error does assume that our knowledge of the strong interaction potential is approximately ($\sim 10\%$) correct. Admittedly some rather gross assumptions have been made in the interpretation. There might be appreciable deviations from a Fermi distribution for the neutrons at the surface where the interaction is the strongest.

On the other hand the real part of the isotope shift depends mostly on the real part of the nuclear potential, where one finds quite good agreement between multiple scattering theory and the fitted parameters. What is clearly needed is some good low-energy pion scattering data to help determine better the potential so that the pionic atom data can be used exclusively to determine the radial moments.

Help does seem to be on the way from the theoreticians. A contributed paper to this conference by F. Hackenberg et al.[30] suggests a mechanism by which the absorptive s-wave parameter can be made to agree with the phenomenological fits. They assume that the absorption can be treated as a scattering off-shell from one nucleon and an absorption by a second one.

Kaonic atoms

The bulk of the published kaonic strong interaction data appears in the review article by Seki and Wiegand.[1] Within the last few months some additional data has appeared from Rutherford[31]. A selected set of these data can be found in the contributed paper by Deloff and Law[32] which is being presented at this conference. With a few exceptions (mostly at low Z) the precision of the experimental shifts and widths is less than 30%. As a result no clear approach of interpreting these data (either theoretical or phenomenological) has yet emerged that would allow the extraction of nuclear matter parameters.

Generally the most successful approaches to interpreting kaonic data have been based on the assumption that the kaon-nuclear strong interaction can be described by a complex potential which strongly distorts the bound state kaon wave function inside the nucleus. As a result, as for the pion, the Klein-Gordon equation is solved numerically with this potential added to the Coulomb potential.

The point of divergence of theoretical approaches involves the way in which this potential is constructed. Some approaches have been purely phenomenological; others have tried to tackle the problem directly by making attempts to relate this effective potential to $\overline{K}N$ scattering lengths. A complete discussion of numerous attempts are described in Ref.'s 1 and 3. A paper by Deloff and Law[32] has been submitted to this conference which uses the potential approach with two different forms for the potential. One is obtained[33] by folding the $\overline{K}N$ interaction potential V_{KN} into the nuclear density. This approach is essentially parameter free as it uses the $\overline{K}N$ scattering length and K^--^{4}He scattering data to determine respectively the strength and the range of the $\overline{K}N$ interaction.

An alternative approach is to use a potential obtained from multiple scattering theory which incorporates the Lorentz-Lorenz[34] effect and can be written as follows:

$$2\mu V_{LL} = - \frac{4\pi \overline{A}\tilde{\rho}(r)}{1 + \xi^3(\xi^2 + \frac{1}{2}\sigma^2)^{-1/2} \, 4\pi\overline{A}\tilde{\rho}(r)} \tag{8}$$

where μ is the reduced mass of the kaon-nucleus system, $\overline{A}$ is the average value of the $\overline{K}N$ scattering lengths, ξ is the effective nucleon-nucleon correlation length, σ is the $\overline{K}N$ interaction range parameter and $\tilde{\rho}$ is the smeared density resulting from the convolution of the $\overline{K}N$ form factor with the nuclear density. In this latter approach the authors allowed ξ and σ to vary freely and in general found a better fit to experiment with these two free parameters than one gets by using the more conventional approach[34] of allowing $\bar{A}$ to vary and setting $\xi = 0$. The major advantage of this latter approach is that it is able to fit both strong interaction widths which have been measured directly as well as those inferred from intensity measurements.

Conclusions

For years pionic and kaonic atoms have been extending the promise of nuclear matter distribution information. Information from pionic atoms has always been hampered by the complexity of the interaction and by the fact that pionic atoms by themselves cannot produce for a given nucleus enough information to describe both the potential and to determine the matter distribution parameters. The real hope for this field lies in good low energy pion scattering data which can be used to remove some of the burden from the pionic spectroscopists. A theoretical explanation for the discrepancy of the fitted values and those obtained from multiple scattering theory for the absorptive s-wave parameter is needed.

On the other hand kaonic atom work suffers from a decided lack of experimental precision and lack of data. As a result there are almost as many theoretical approaches to the interpretation of the data as there are experimental data. Many theoretical approaches seem to give reasonable agreement with experiment but I believe that is more a reflection of the imprecision of the data than of the appropriateness of the interpretation.

Meanwhile the experimentalists have moved on to using antiprotons for probing nuclear matter distributions with the hope (perhaps naive) that the interpretation of these data will be simpler.

REFERENCES

1. R. Seki and C.E. Wiegand, Ann. Rev. Nucl. Sci. 25, 241 (1975).
2. G. Backenstoss, Ann. Rev. Nucl. Sci. 20, 467 (1970).
3. L. Tauscher, Proceedings of the Conference on High-Energy Physics and Nuclear Structure, Santa Fe 1975 (New York: American Institute of Physics), p. 541.
4. L. Tauscher et al., Phys. Rev. Lett. 35, 410 (1975).
5. M. S. Dixit et al., Phys. Rev. Lett. 35, 1633 (1975).
6. P. Ebersold et al., Phys. Lett. 53B, 48(1974).
7. J. N. Bradbury et al., Phys. Rev. Lett. 34, 303 (1975); Phys. Rev. Lett. 34, 1064 (1975).
8. G. L. Godfrey et al., Phys. Lett. 61B, 45 (1976).
9. V. I. Marushenko et al., J.E.T.P. Lett. 23, 80 (1976).

10. G. Backenstoss et al., Phys. Lett. 43B, 539 (1973); see also Ref. 3.
11. R.E. Shafer, Phys. Rev. D 8, 2313 (1973).
12. Particle Data Group, Rev. Mod. Phys. 48, S21 (1976).
13. M. Daum et al., Phys. Lett. 60B, 380 (1976).
14. A.R. Clark et al., Phys. Rev. D 9, 533 (1974).
15. S.C. Cheng et al., Nucl. Phys. A254, 381 (1975).
16. P. Baranov et al., Phys. Lett. 52B, 122 (1974).
17. M.V. Terent'ev, Yad. Fiz. 19, 1298 (1974).
18. T.E.O. Ericson and J. Hüfner, Nucl. Phys. 47B, 205 (1972).
19. G.T. Emery, Phys. Lett. 60B, 351 (1976).
20. C. Tzara, Nucl. Phys. B18, 246 (1970).
21. B. Coupat et al., Phys. Lett. 55B, 286 (1975).
22. M. Leon, Phys. Lett. 53B, 141 (1974).
23. M. Leon, private communication.
24. A.R. Kunselman et al., "Pionic 4f → 3d Transitions in ^{148}Nd and ^{150}Nd and the Neutron Isotope Shift", to be published.
25. D.A. Jenkins et al., Phys. Rev. 185, 1508 (1969).
26. R. Engfer et al., At. Data and Nucl. Data Tables 14, 509(1974).
27. M. Ericson and T.E.O. Ericson, Ann. Phys. (N.Y.) 36, 323 (1966).
28. D.K. Anderson et al., Phys. Rev. Lett. 24, 71 (1970).
29. D.K. Anderson et al., Phys. Rev. 188, 9 (1969).
30. F. Hachenberg et al., "S-Wave Pion Absorption by Nuclei", contributed paper this conference.
31. C.J. Batty et al., Phys. Lett. 60B, 355 (1976).
32. A. Deloff and J. Law, "Kaonic Atoms Level Shifts: Potential Approach", contributed paper this conference.
33. A. Deloff and J. Law, Phys. Rev. C 10, 1688 (1974).
34. A. Deloff, "The Lorentz-Lorenz effect in kaonic atoms", submitted to Phys. Lett. 1976.
35. R. Seki, Phys. Rev. C 5, 1196 (1972).

A. Bernstein (MIT): Before matter radii differences can be obtained one needs a calibration on the isovector parameters. Pb^{208} and Zr^{90} are suggested as possibilities.

Powers: Your point is well taken. Unfortunately nature has been rather cruel to the pion spectroscopist. The pionic strong shifts in both nuclei are fairly small. A measurement of the ^{208}Pb 5g→4f (575 keV) strong shift to 100 eV would determine it to only 7%. But perhaps in light of the uncertainty in matter distribution parameters this might be sufficiently precise to provide an interesting comparison. In ^{60}Ni, on the other hand, the pionic shifts are nearly five times larger.

Gilbert Shen (Fermilab): I believe the $K_{\mu 3}$ measurement of the neutrino mass represented a 90% confidence level.

Powers: That is correct.

PHOTOPRODUCTION OF PIONS IN LIGHT NUCLEI

C. Tzara
C.E.N. Saclay

The photoproduction of low energy pions on nuclei has been studied for a few years only, first at the Saclay electron linac, with the collaboration of a group from Louvain University, and at the Bates Linac. The $\pi^{\pm}$ and π^0 are all under investigation, and the experimental results, although some are still in a preliminary state, are the subject of many theoretical investigations.

The relation with the radiative capture of a π^- from a mesic atom orbit is evident when we consider

$$\gamma(AZ) \to \pi^+ (A, Z-1)$$

and $\pi^-(AZ) \to \gamma (A, Z-1)$

as merely related by transposition (ingoing particle = outgoing anti-particle, here the pion). However, important differences between these two reactions explain why they are generally considered in distinct categories. In radiative capture processes, one must know the partial capture rate from each atomic orbit and their respective population. For instance, in a medium nucleus, the s orbit is not even reached by the pion in the course of its atomic life. This poses difficult problems to the interpretation of the data. On the other hand, the technique used for this process allows at once a wide band of final nuclear states, whereas in the photoproduction studies, the excitation of the various final states is reached in a lengthy way by successive steps.

Apart from these practical differences, the basic physics is the same. What makes it so attractive? Let us consider the elementary process:

$$\gamma N \to \pi N$$

We know experimentally that under 20 MeV for charged pions and 2 MeV for the neutral one, the amplitude is of axial character and that the pion is produced in an s state. More specifically, it has the form $i\sqrt{2}\, eg\, \vec{\sigma}\cdot\vec{\epsilon}$. We know also that the πN scattering amplitude is small at these low energies.

The P.C.A.C. hypothesis unifies many aspects of the low energy pion-nucleon system. One of its conclusions is that the photo- and electroproduction amplitudes at threshold are porportional to the axial form factor of the nucleon.

Turning to the nuclear case, we can then understand either by using impulse approximation or by generalising the P.C.A.C. approach, that the photoproduction at threshold is directly related to the transition axial form factor of the nucleus. Thus photoproduction at and near threshold (and radiative capture from the S orbit) is on the same footing as the magnetic electron scattering and the muon capture, which are also mainly governed by the axial form factor. Approximately, the knowledge of one of these processes is sufficient to determine the two others. But for each of them, corrections must be made to this simple picture. As we shall now see in the case of ^{6}Li, the experimental precision now allows tests of these corrections.

My excuse for presenting first the results we have obtained at Saclay is that I must clear up the confusion created by our publication two years ago of the cross section for the reaction:

$$\gamma + {}^6Li \rightarrow {}^6He + \pi^+$$

which was found 60% lower than the theoretical estimates available at that time.[1] While calculations were redone, we improved our experimental setup to cure the deficiency of the primitive one.

Let me describe the procedure used to extract from the data the quantity of interest, the cross section which we express on a nucleus A as:

$$\sigma_A(k) = q/k \, |S_A|^2$$

q and k are the momenta of the pion and the photon in the C.M. reference frame. S_A is the amplitude for the process on the details of which I shall comment later. For the moment, it suffices to know that it splits in two parts:

$$|S_A|^2 = \frac{2\pi(Z-1)\alpha \, W/q}{\exp\{2\pi(Z-1)\alpha W/q\} - 1} \, |t_A|^2$$

t_A is almost constant near threshold when s wave production is allowed. W is the pion energy, Z the target charge.

The data obtained in an experiment are reaction yields versus the maximum energy of the photon spectrum:

$$A_A(k) = \int_{k_A}^{k} dk' \, \frac{dN}{dk'} \, \varepsilon \, \nu_A \, C_A \, \sigma_A(k')$$

where dN/dk' is the photon spectrum normalised to a given integrated energy, ε is the efficiency of the detecting system, ν_A the number of nuclei per unit area of the target, C_A a correction factor depending of the target, k_A the threshold in the laboratory.

Up to now we measure $A_A(k)$ relatively to the yield for a proton target (actually CH^2). The critical points in the measurement are the following: (1) Formation of a clean and reproducible Bremsstrahlung beam. It requires a good control of the electron beam and a good photon collimation. These aspects have been improved since our first experiment. The reproducibility of the end point is better than 30 keV. However, the absolute energy scale does not coincide with the nominal one as given by the fields in the electron beam handling system (Fig. 1). (2) Constitution of a detecting system of constant efficiency and low background linked to a powerful data recorder.

The principle of the detection is to record the e^+ from the $\pi^+ \to \mu^+ \to e^+$ decay, after each photon burst, in two lucite Cerenkov followed by a plastic scintillator. After a beam burst, the detectors receive the e^+ and parasitic pulses from neutron capture and the long lived ^{12}B produced in them by the reaction $^{12}C(n,p)^{12}B$. These backgrounds are greatly reduced by the requirement of a triple coincidence. We record the scintillator timing for each event in order to characterize the μ^+ decay time. During the burst, the detectors receive an intense electromagnetic shower. Their output, even microseconds after the burst, can be affected, and hence their detection efficiency would depend on the photon beam intensity and the nature of the target. The pulse height delievered by each detector during an event (triple coincidence) is recorded. A preliminary study of the dependence of the reaction yield and pulse height distributions versus the beam intensity fixes the limit above which these quantities begin to shift. During the data taking runs, the beam intensity is set under this limit. The efficiency ε is then constant and identical for the two targets.

In our original experiment, we used a worse collimation, two detectors in coincidence, and no recording of their pulse heights. In the light of our subsequent experience, we suspect that the efficiencies were different for the ^6Li and CH_2 targets and that our published value of the ^6Li cross section relative to that on the proton is uncertain.

One word about the normalization and the treatment of the data. The quantameter, which measures the energy in the photon beam, is linked to a leaky integrator whose time constant is equal to the μ^+ decay time. The time interval in which the events are recorded is locked to the sampling time of the quantameter integrator. The normalization is thus independent of the detailed shape of the beam burst.

The time distribution of the events after each burst shows in addition to the exponential muon decay a long lived component (^{12}B?). The events in a delayed wide interval (90 μsec) are recorded. From measurements below threshold, we know the contribution of this background in the useful time interval. At 1 MeV above threshold, it represents 1/3 of the yield for ^6Li and 1/20 of the proton yield. At increasing energies, it becomes soon negligible.

Finally, the correlated pulse height distribution in the two Cerenkov counters displays a definite separation between a low pulse region, the only one present below the reaction threshold, and a large pulse domain. The reaction yield is then studied as a function of the threshold set for the sum of the pulse heights in the two Cerenkov counters. At this stage, we obtain the reaction yield for the two targets as a function of nominal electron energy and of the counters threshold. The yields have been normalized to the same number of nuclei in the targets, and the same photon attenuation and positron losses in the target. An example is given in

Fig. 2.

The treatment of the data proceeds as follows: to minimize the theoretical input, one chooses the data between thresholds and 2 MeV above where the cross sections have the form:

$$\sigma_p(k) = \frac{q}{k} |E_{0+}(0)|^2 = \frac{q}{k} a_p$$

and

$$\sigma_{Li}(k) = \frac{q}{k} \frac{2\pi\gamma}{\exp(2\pi\gamma)-1} |E_{Li}(0)|^2 = \frac{q}{k} \frac{2\pi\gamma}{\exp(2\pi\gamma)-1} a_{Li}$$

The photon spectrum is that of Bethe-Heitler, apart from high energy tip, where it is smoothly extrapolated to the finite end point value determined by Jabbur and Pratt.[2] The following parameters: ΔE: the difference between the real and nominal energies a_{Li}/a_p : the ratio between the reduced cross sections are used to fit the experimental yields. a_{Li}/a_p and the χ^2 are found for various pulse heights threshold. One sees on Fig. 3 that the χ^2 are reasonable and a_{Li}/a_p stable versus the threshold value. We obtain

$$\alpha_{exp} = \frac{a_{Li}}{a_p}\bigg|_{exp} = 0.098 \pm 0.004$$

(error determined by the fitting procedure). The ratio of theoretical value of α to the experimental ones are: from Bergström[3] (set B wave function): α_{th}/α_{exp} = 1.02 from Delorme and Figureau[4] (same set): α_{th}/α_{exp} = 1.17; from Cannata, Lucas, Werntz[5]: α_{th}/α_{exp} = 1.17±0.14. (the error comes essentially from the theory). Note that in the case of radiative pion capture, these authors find from a recently published experimental value:[6]

$$R_{th}/R_{exp} = 1.24 \pm 0.43$$

(here the error comes essentially from the experiment). In other words, the two experimental approaches are in agreement. Once α_{Li}/α_p is obtained from the reaction yields in the lower energy bin (2 MeV above threshold) it should be possible to exploit the data at higher energies. For the proton cross-section, it is known that its variation is in first order linear in $k-k_t$, the energy above threshold. We suppose the same variation for the Li cross section. We know that here the continuum starts at 0,9 MeV, and that a resonance exists at 1,9 MeV. But phase space in the final state makes the cross section to the continuum extremely small, unless there is a resonance. In that case, the cross section resembles that of a bound state. Thus we retain this hypothesis. The two cross sections can then be written as:

$$\sigma_p = q/k \, a_p \{1 + \varepsilon_p(k-k_t)\}$$

$$\sigma_{Li} = q/k \, a_{Li} \frac{2\pi\gamma}{\exp(2\pi\gamma)-1} \{1 + \varepsilon_{Li}(k-k_t)\}$$

$$+ q'/k \, a_{Li}^* \frac{2\pi\gamma'}{\exp(2\pi\gamma')-1}$$

where a_{Li}^* is the parameter for the excitation of the level, and q' and $\gamma' = \alpha Z W'/q'$ are the pion momentum and Sommerfeld factor for this excitation. The fit to experimental points furnishes the following values:

$$\varepsilon_p = (11.5 \pm 3) \times 10^{-3} \text{ MeV}^{-1}$$

$$\varepsilon_{Li} = (23 \pm 26) \text{ " "}$$

$$a_{Li}^*/a_{Li} = 0.64 \pm 0.24$$

the last two quantities are completely correlated.
A calculation by Delorme and Figureau[4,7] gives:

$$\varepsilon_{Li} = 21 \times 10^{-3}$$

and

$$a_{Li}^*/a_{Li} = 0.68$$

in good agreement with ours. ε_p is known from extrapolation from higher energies or dispersion theories:

$$\varepsilon_p = 11 \times 10^{-3}$$

also in good agreement. A final remark: we have used a Bethe-Heitler Bremsstrahlung shape without producing a noticeable difference in a_{Li}/a_p and a_{Li}^*/a_{Li}.
Other nuclei under investigation at Saclay are:

$$\gamma + {}^{14}N \to \pi^+ + {}^{14}C \text{ or } {}^{14}C^*$$

where we found

$$a_{{}^{14}C}/a_p = (7 \pm 2.3) \times 10^{-4}$$

to be compared to a calculated value of 2.3×10^{-3} reported in a communication to this Conference by Figureau and Mukhopadhyay, and

$$a_{{}^{14}C}^*/a_p = (6 \pm 1) \times 10^{-2}$$

for the excitation to the levels situated around 7-8 MeV. The π^+ photoproduction shows in this case its power to detect special configurations in nuclear states.

$\gamma^{12}C \to {}^{12}B \pi^+$ and $\gamma^3 He \to {}^3H \pi^+$ measurements are still in a preliminary stage.

Recently, the physicists working at the Bates Linac have attacked the same technique by measuring the reaction

$$\gamma + d \to n + n + \pi^+$$

which is essentially governed, in the vicinity of threshold, by the scattering length of the two neutrons. They report their results in a communication to this Conference by Booth, Chasan, Bernstein and Bosted. By comparison with a simplified theory,[8] they claim that their results are compatible with $a_{nn} = 18 \pm 2$ in striking agreement with the adopted value: $a_{nn} = 16-17$. However, before drawing conclusions, one must use in the theory a more refined description of the deuteron and of the interaction hamiltonian, and extend the range of validity above two MeV. Considering that a 5% experimental uncertainty is now possible to achieve, it seems urgent for the theorists to improve the computation. See Fig. 4.

The experiments on π^- photoproduction were initiated at the Bates Linac on ^{11}B and ^{12}C. The only practical way to detect these reactions is to record the residual activity. In a paper submitted to this Conference, Bernstein, Paras, Turchinetz, Chasan and Booth report on their results on $\gamma, {}^{12}C \to \pi^-, {}^{12}N$. The normalization was done by comparing to the reaction $\gamma, {}^{14}N \to {}^{12}N, 2n$ and the background due to the two step reaction $\gamma, {}^{12}C \to {}^{12}B, p; {}^{12}C, p \to n, {}^{12}N$ obtained by extrapolating the data taken below threshold. As can be seen from the Fig. 5, the background is much larger than in π^+ photoproduction. The normalization procedure is also less direct. Despite these difficulties, the authors have succeeded to obtain a cross section which agrees near threshold with the theoretical ones. The variation above threshold seems faster than even the theoretical curve computed by Nagl and Überall who take in account s, p, d, pion waves, momentum dependent term in the interaction Hamiltonian, etc.[9]

Before attacking the question of the π^0 photoproduction, I would like to briefly summarize the status of the charged pion photoproduction. From the experimental point of view, the technique for studying the π^+ production has attained a good precision and sensitivity. It is possible to measure the cross section on the proton down to 0,2 MeV above threshold, and to determine cross sections on nuclei relative to the proton as low as 1/100, with a precision of the order of ~5%. However the range of nuclei is limited to $Z \lesssim 10$ above which the coulomb effect and the form factor render the cross section too small. More work remains to be done on d, ^3He, ^{12}C, ^{14}N. The measurement of the π^- production seems to be abandoned due to the experimental difficulties.

The theory suffers from uncertainties of various natures, which according to (3) attain 12%. For the moment, we are thus far from the hope that magnetic electron scattering, weak interaction and low energy pion interaction can be unified in a precise manner. On the other hand, some interesting problems such as the modification of the elementary amplitude in a nucleus are work being studied because the experiment is able to discriminate between various models.

THE NEUTRAL PION PHOTOPRODUCTION

The incentive for the experiments, started last summer at Saclay, was the simple (naïve?) following remark.

If we wish to obtain information about the $\gamma, p \to p, \pi^0$ and $\gamma, n \to n, \pi^0$ elementary amplitudes, we can use as targets a series of light nuclei whose structure enhances differently the production from the neutron and the proton. In particular, very near the threshold, the spin flip amplitude dominates. Thus the ^{3}He(t), where the two protons (neutrons) are coupled to S=0, acts essentially as a neutron (proton) target. We were of course aware of the fact that various effects could mar this picture: the momentum distribution of the nucleons, the coupling with the charged pion production.

Why do we need these elementary amplitudes? Especially at threshold, the various theories are in disagreement and the experimental knowledge is still imprecise. Extrapolation at threshold from the data gives:

channel	$n \pi^+$	$p \pi^-$	$p \pi^0$	
$Re\ E_0^+$	28,3±0,5	-31,8±2	-1,7±?	in $10^{-3}\ c/m_{\pi^+}$

from the general relation

$$\mu_{\pi^0 n} = -\frac{1}{\sqrt{2}}\{\mu_{\pi^+ n} + \mu_{\pi^- p}\} + \mu_{\pi^0 p}$$

one obtains:

$$Re\ E_{0+}\Big|_{\pi^0 n} = 0.8 \pm 1.5$$

Our hope was to improve this knowledge.

The experimental setup is derived simply from that used in π^+ photoproduction. The two arms of three counters are now set in coincidence, with lead converters in front of each. The observation of pairs generated by π^0 decay gammas is made during the beam burst, stretched to its maximum (20 μ sec., repetition rate: 1000/sec.).

The targets are liquid gas with thin windows. In addition to the precautions mentioned above for the π^+ measurements, one must avoid any interaction with a heavy element, whose coherent production would overwhelm the interesting counts. Charged pion production and subsequent charge exchange is negligible.

We have obtained preliminary data on proton and ^{3}He between 1 and 8 MeV and at 4 and 8 MeV on d. What is presented in Fig. 6a and Fig. 7a are the ratio between the yields on ^{3}He or d to the proton. A comparison with a theory requires the knowledge of:

1) The Bremsstrahlung spectrum. See above.

2) The efficiency of the π^0 detector as a function of the photon energy and π^0 emission angle. It is computed by a Monte Carlo procedure.

3) The differential cross section for photoproduction on the various targets.

Thus a theoretical yield is computed. What we have plotted in Fig. 6b and Fig. 7b are the ratios of the theoretical yields $Y(^3He/p)$ and $Y(d/p)$.

The theory we used is the simplest approximation, assuming:
- impulse approximation,
- frozen nucleons,
- E_o and M_1+ multipoles only,
- Pure s state in d and ^{3}He, and no mixed symmetry state in ^{3}He.

Three values for $E_o+(\pi^0)$ were used: $(0; 1; 1.5) \times 10^{-3}$.

The comparison between the experiment and this theory shows a general agreement. However, we have since discovered that the hydrogen target superficial mass could be less than thought by 40%. The ratios of experimental yields should then be decreased by 40%. The future experiments will be improved by a better knowledge of the target densities, and by a better rejection of subthreshold background.

At inspection of the theoretical curves, one may wonder if the sensitivity of such an experiment will ever be sufficient. The possible effects of the nucleon momentum distribution and of the intermediate $\pi\pm$ photoproduction render the prospect even worse. Delorme and Figureau[10] have computed that the cross section is, in the case of a ^{6}Li target, increased by a factor 3 when these effects are taken into account. Koch and Woloshyn[11], and communication to this Conference, find also large changes (an increase in the case of d, a decrease in the case of ^{3}He). As our data are in general agreement with a first order calculation, there seems to exist a large discrepancy between them and the more elaborate theoretical cross sections. This situation should induce new investigations about π^0 photoproduction.

REFERENCES

1. J. P. Deutsch, D. Favart, R. Prieels, B. Van Ostaeyen, G. Audit, N. de Botton, J. L. Faure, C. Schuhl, G. Tamas, C. Tzara, Phys. Rev. Letters 33, 316 (1974).
2. R. J. Jabbur, R.H. Pratt, Phys. Rev. 129, 184 (1963).

3. J. C. Bergström, I. P. Auer, R. S. Hicks, Nucl. Phys. A251, 401 (1975).
4. G. Delorme, A. Figureau, Private communication.
5. F. Cannata, C. W. Lucas, C. W. Werntz, Phys. Rev. Letters 33, 1316 (1974).
6. H. W. Baer, Y. A. Bistirlich, K. M. Crowe, N. de Botton, J. A. Helland, P. Truöl, Phys. Rev. C8, 2029 (1973).
7. G. Delorme, A. Figureau, Private communication.
8. C. Tzara, Nucl. Phys. A256, 381 (1975).
9. A. Nagl, H. Überall, to be published.
10. G. Delorme and A. Figureau, Private communication.
11. J. H. Koch, R. M. Woloshyn, Phys. Lett. 60B, 221 (1976).

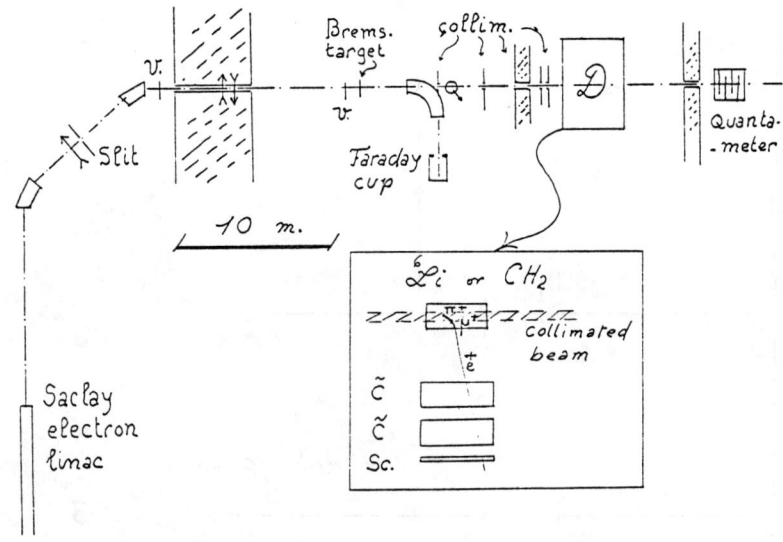

fig 1

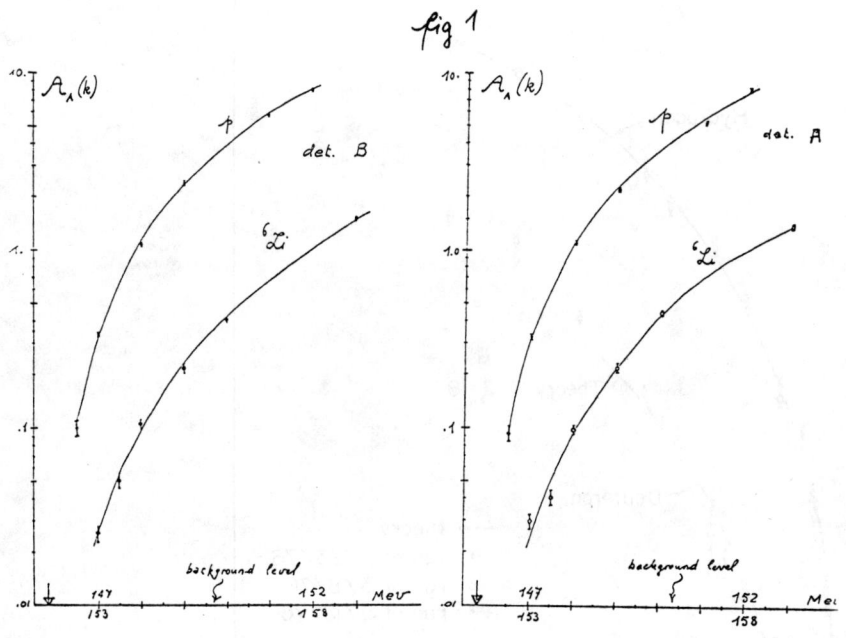

fig 2

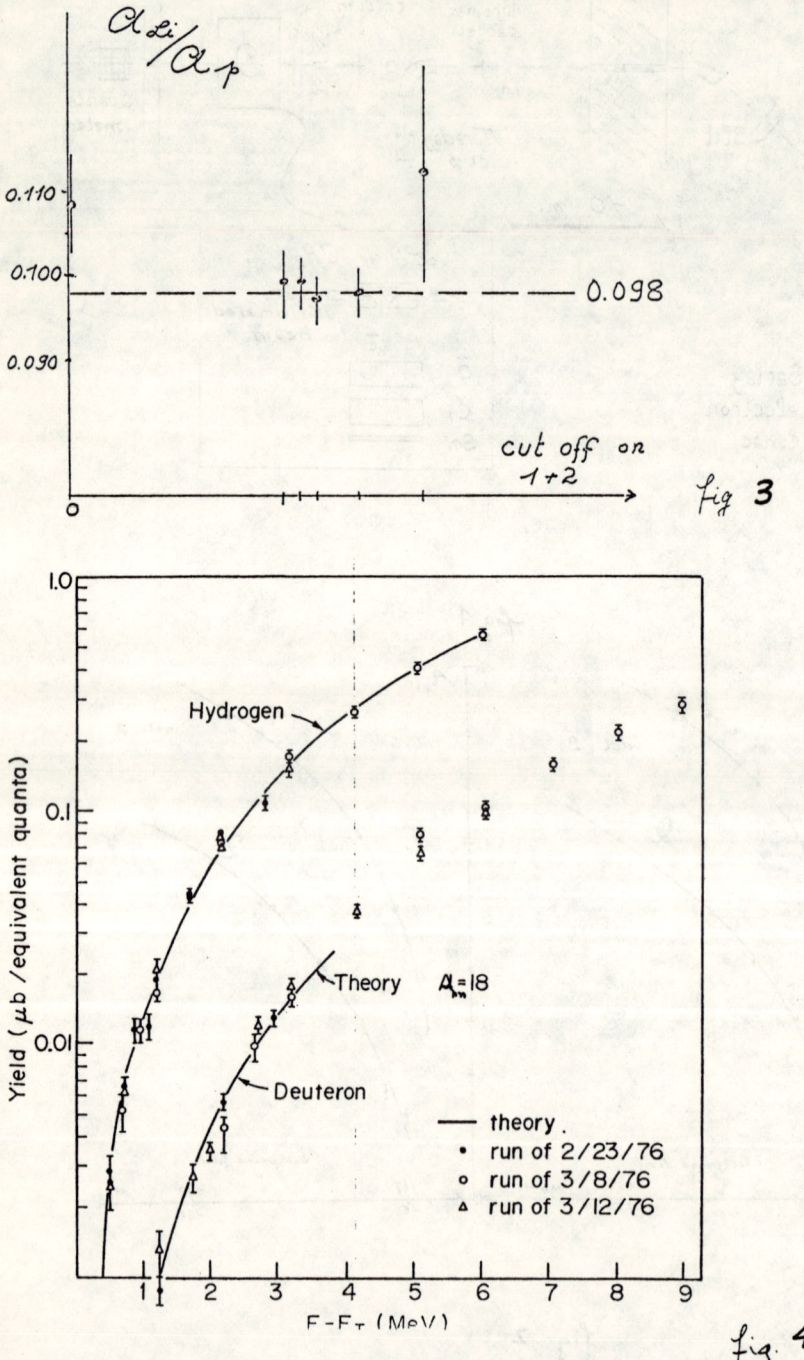

fig 3

fig. 4

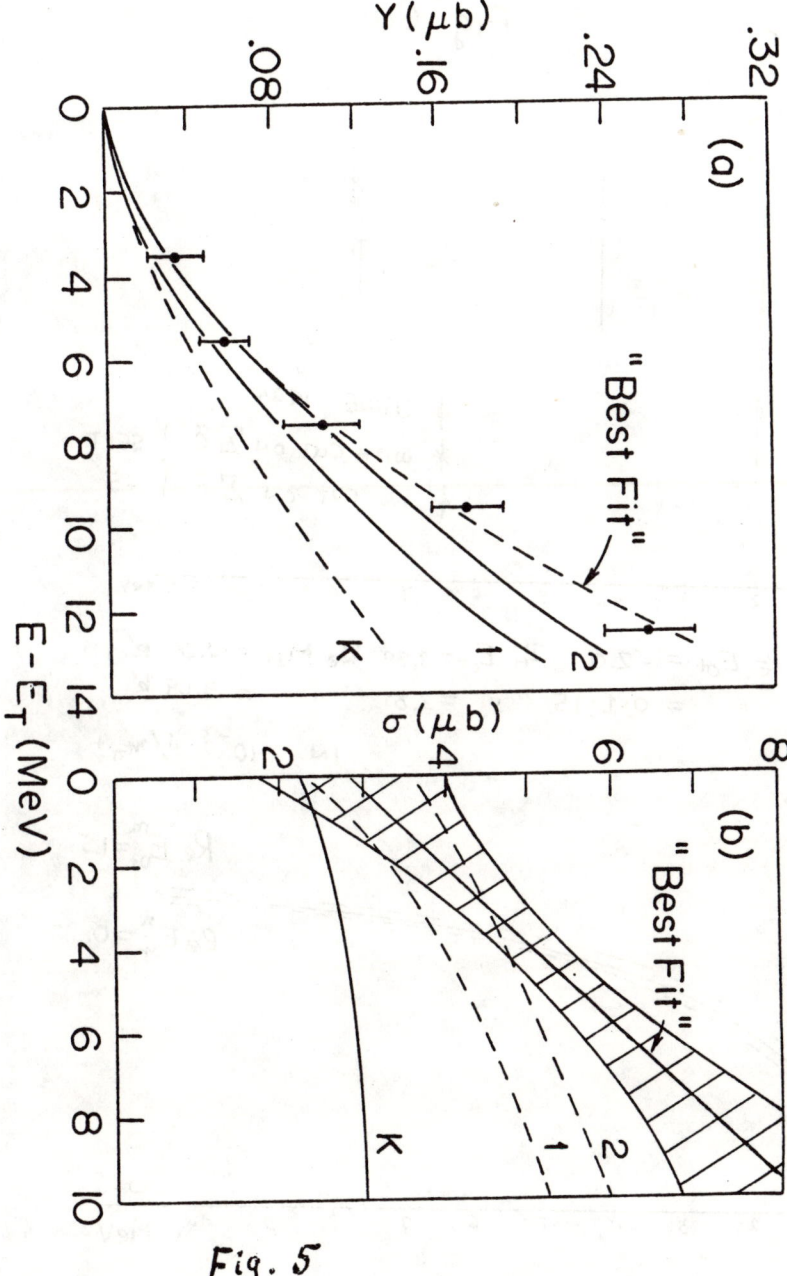

Fig. 5

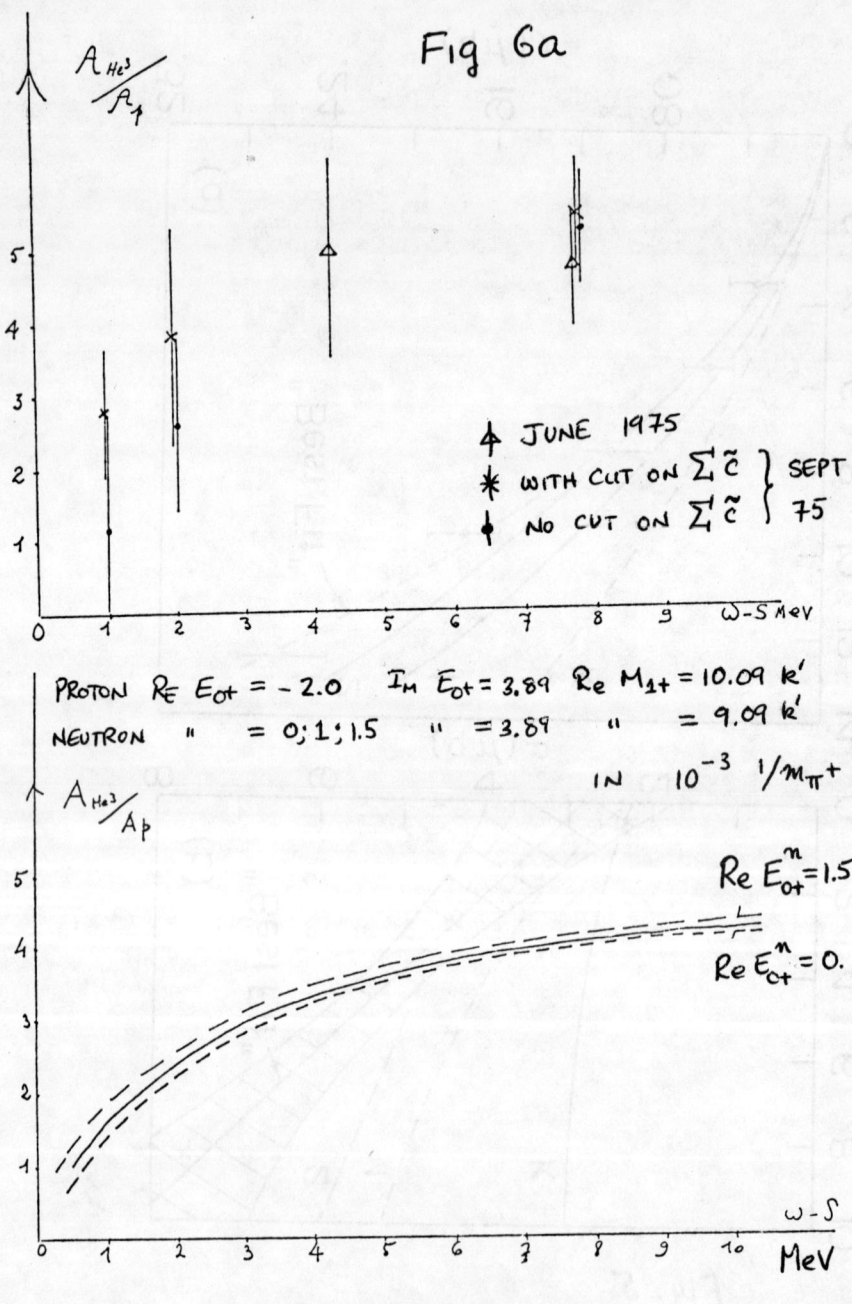

Fig 6a

Fig 6b

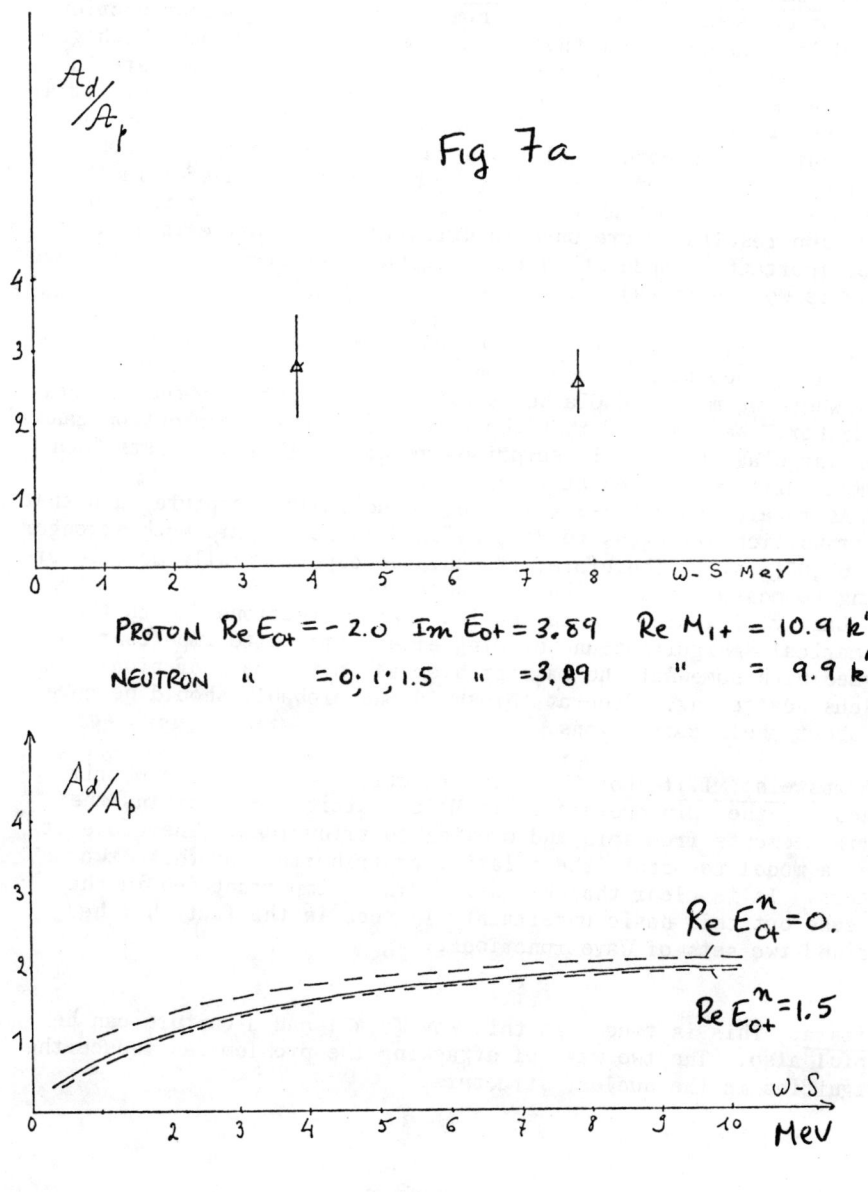

Fig 7a

PROTON $Re\ E_{0+} = -2.0$ $Im\ E_{0+} = 3.89$ $Re\ M_{1+} = 10.9\ k'$
NEUTRON " $= 0; 1; 1.5$ " $= 3.89$ " $= 9.9\ k'$

Fig 7b

J. Barry Cammarata (Stanford Univ.): Though there is now apparent agreement between theory and experiment on the ^{6}Li photoproduction, I would like to point out that the nuclear wave functions which give this agreement (Bergstrom et al., set B) give a µ-capture rate in ^{6}Li which is more than 2 standard deviations below the Deutsch et al. experimental value. The significance of this is evident from the facts that (1) in both (γ,π) and µ-capture reactions the Gamow-Teller operator is dominant and (2) the momentum transfer in the two processes is comparable. Thus to properly interpret the photoproduction results (where uncertainties in the theory exist) it is quite important to understand the µ-capture results (where the theory is well defined).

Anup Rej (Fysisk Institutt, Trondheim): We have made calculations on $\pi°$ photoproduction on deuterium near (3,3) resonance using a model where we modified Glauber's eikonal propagator by the Fresnel propagator. We observed that the double scattering correction can be as large as 20%, but it surprises me to see that Koch gets such enormous difference even at threshold.

As regards $\gamma N^{14} \to C^{14} + \pi$ reaction, we noted in µ capture case that the transition strengths to 2_1^+, 2_2^+ and 3^- states are much stronger than $6^-, 0^-$ (g.s.). Therefore, it would be experimentally more interesting to measure transitions to those levels.

Lastly, I would like to mention that corrections due to the kinematical ambiguities and binding effect are quite important and plagued with somewhat the same problem as in the case of pion-nucleus scattering. Even at threshold one probably should be careful about these corrections.

A. Bernstein (MIT): For the (γ,π) reaction one needs the matrix element of the spin operator. In M1 transitions one obtains the matrix elements from spin and orbital contributions. Therefore it takes a model to obtain the relative contributions of these two factors. It is clear that Bergstrom did an important job in the Li^6 case but this basic uncertainty is seen in the fact that he obtained two sets of wave functions.

C. Tzara: This is true. In this way (γ,π^+) and µ capture can be helpful also. The two ways of attacking the problem can reduce the ambiguities in the nuclear structure.

RADIATIVE PION CAPTURE IN NUCLEI †

Peter Truöl

Physik-Institut der Universität
Zurich, Switzerland

In recent years radiative pion absorption has been the subject of
intensive experimental and theoretical investigations. The observation of high-resolution photon-spectra has been shown to reveal a
considerable amount of information both on the capture process itself as well as on the structure of the recoil-nucleus. These data
were obtained at the Berkeley 184"-cyclotron and have been properly
reviewed on several occasions [1]. Since the Santa Fé conference in
1975 the first meson factory data have become available from Los
Alamos [2] and SIN [3]. It is these new results, which I will report on
here. The measurements deal with the following aspects

1) Search for bound and unbound states in the $A=3, T_z = 3/2$-system:
$^3H(\pi^-,\gamma)^3n$

2) Further investigation of giant magnetic dipole states in 1p-shell
nuclei: $^6Li(\pi^-,\gamma)^6He$; $^7Li(\pi^-,\gamma)^7He$; $^9Be(\pi^-,\gamma)^9Li$; $^{12}C(\pi^-,\gamma)^{12}B$

3) Isobaric analogs of giant spin isospin states in ^{16}O and structure of ^{18}N:
$^{16}O(\pi^-,\gamma)^{16}N$; $^{18}O(\pi^-,\gamma)^{18}N$

I. $^3H(\pi^-,\gamma)3n$

In radiative absorption in Tritium the trineutron excitation spectrum can be studied free from complications arising from other
strongly interacting particles in the final state. It was hoped that
a measurement of the photon spectrum would reveal evidence for
T=3/2 resonances or even a bound state of the trineutron,
if such states should exist contrary to negative results of previous
less sensitive experiments using other reactions [4]. The final analysis of a first measurement using the Berkeley-pairspectrometer [5]
at the extended low energy pion-channel at LASL with a target of liquid Tritium has recently been published [2]. Though the total number
of thousand events in the spectrum shown in Figure 1 may seem small,
a large experimental effort was necessary to collect them. It was
only possible through the rare coincidence of an excellent cryogenics group experienced in building Tritium targets, an accelerator
crew providing stable, high intensity pion beams, the willingness
of the laboratory leadership to deal effectively with the extreme
radiation safety hazards in handling such a large volume of
Tritium (23 cm^3; 57 kCi) and lastly the availability of a large

acceptance spectrometer. The spectrum displayed contains only
Tritium events, the backgrounds from the other hydrogen isotopes in
the target and the target container have been subtracted. The resolution (3 MeV at 130 MeV) and the acceptance of the spectrometer were
verified in frequent runs with an identical hydrogen target cell. The
absolute branching ratio was found to be 4.5 ± 0.8 %. It is in fair
agreement with the predicted value [6] of 6.5 ± 2.0 %. The latter was
obtained using the impulse approximation with a complete Hamiltonian
obtained from $\pi^- p \to n\gamma$ [1], realistic Tritium wavefunction and a final
state interaction treated in the Amado model. The large error is
mainly due to the uncertainty of the total 1s-pion absorption. Since
the radiative rate can in general be calculated to approximately
± 10 %, it may be used vice-versa to predict the width $\Gamma_{1s}(^3H)$ of the
pionic 1s-level, using the experimental branching ratio. The result
is Γ_{1s} = 1.02 ± .18 eV. The shape of the spectrum is well represented
by the theoretical curve, which doesn't assume any resonance. It can
be concluded, that except for a small excess of events at excitation
energies of 7 and 13 MeV no evidence for the existence of resonances
or a bound state, characterized by a line with energy greater than
127 MeV exists.

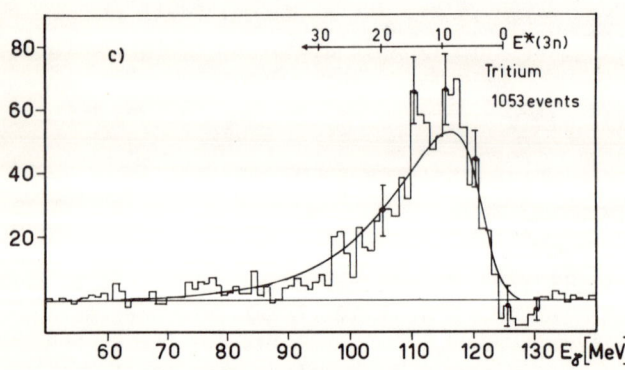

Figure 1: Photon spectrum from $\pi^- {}^3H \to nnn\gamma$ (1MeV bins) Solid curve: Theoretical calculation of Phillips and Roig, normalized to the total number of events upper scale: Tri-neutron excitation energy

II. The SIN-pairspectrometer

A new pairspectrometer has been constructed, assembled and tested at
SIN. The set up, shown in Figure 2, is rather similar to the one
used in previous experiments.

The use of three large multiwire proportional chambers as detectors
for the electron-positron pair resulted in an improvement of some
spectrometer properties:
1) the acceptance was doubled using a large window frame magnet and
   bigger detectors
2) the spatial resolution was improved to ±1mm in the critical horizontal coordinate with 3 wire planes per chamber wound 90° and
   ± 60° relative to this axis.

3) the smaller mass of the MWPC's reduces multiple scattering, which combined with 2) decreases the FWHM of the 129.4 MeV hydrogen calibration line to 1 MeV (2 MeV previously) (see also the 134 MeV ^{6}Li$(\pi^-,\gamma)^6$He(g.s.) line in Figure 2).

4) the faster data acquisition combined with the hard-ware and PDP 11/40 on-line pattern recognition feasible with MWPC's allow an early rejection of background events and consequently an on-line, however approximate energy determination for the surviving good events.

An example of such an on-line spectrum is shown in Figure 2 (insert).

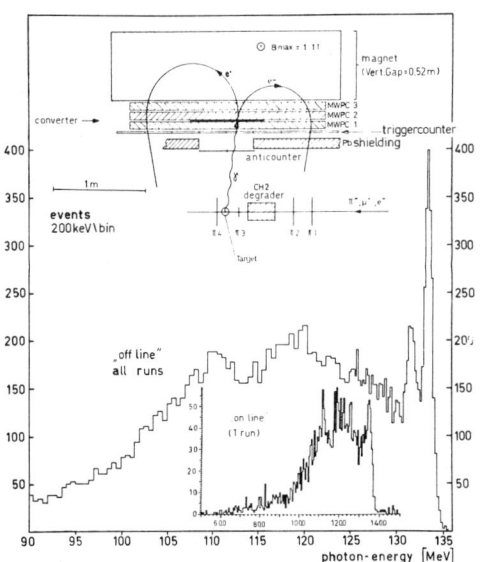

Figure 2:
Upper part: Sketch of the experimental set-up
Lower part: Photon spectrum from radiative pion absorption in ^{6}Li after off-line analysis. The insert shows a spectrum from one one-hour run as displayed on-line by the PDP-11/40

Preliminary results from a first series of measurements with several light nuclei as targets have been contributed to the conference[3]. They include a remeasurement of the photon-spectrum for ^{6}Li, ^{12}C and ^{16}O and new data for ^{7}Li, ^{9}Be and ^{18}O. Though no absolute branching ratios for the observed transitions to isolated nuclear levels are available at present, a number of interesting results have already emerged.

III. M 1 - states

A prominent feature of the existing data for 1p-shell nuclei[5,10] is the strong excitation of analogs of giant-M1 states in the target nucleus, notable examples being ^{6}Li, ^{10}B, ^{12}C and ^{14}N. In the new data M1-transitions are observed for ^{6}Li, ^{7}Li, ^{9}Be and ^{12}C. In the ^{6}Li-spectrum (Figure 2) the contributions to the ^{6}He ground-state ($J^\pi = 2^+$, E*=0, T=1, Eγ=133.9 MeV) and the first excited state (0^+, 1.8, 1, 132.1) are now nicely separated, where the ground-state transition has twice the strength of the first excited state.

This is in agreement with the Berkeley-result. In the ^{7}Li-spectrum (Fig. 3a) the upper line corresponds to the ^{7}He-ground-state ($3/2^-$, 0, 3/2, 126.6)-transition. The highest energy peaks in the ^{9}Be-spectrum (Figure 3b) represent two ^{9}Li-levels with ($3/2^-$, 0, 3/2, 124.5) and ($1/2^-$, 2.7, 3/2, 121.8). All three levels are analogs of known M1-transitions in the target nucleus.

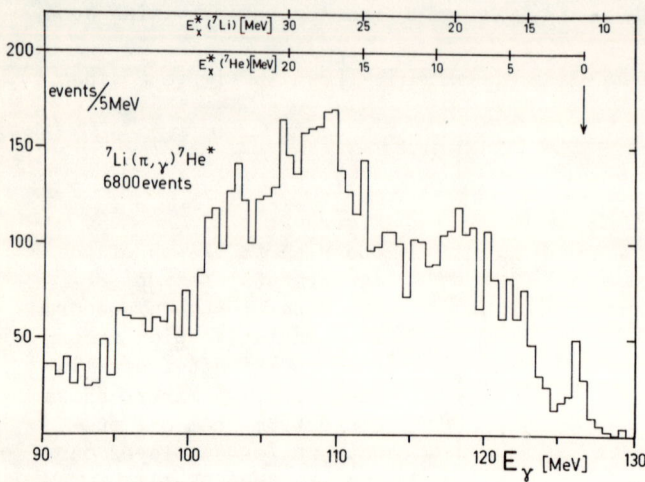

Figure 3a:
Photon spectrum from ^{7}Li(π^-,γ)^{7}He. The upper scales indicates the excitation energy in ^{7}He and in ^{7}Li. Arrow: ^{7}He ground-state

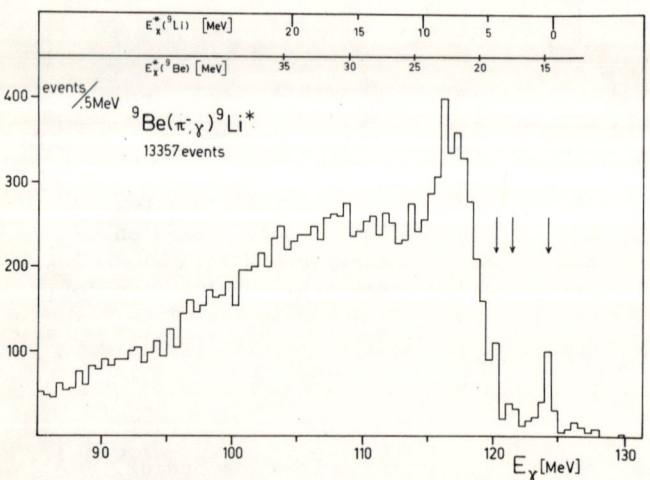

Figure 3b:
Photon-spectrum from ^{9}Be(π^-,γ)^{9}Li. Upper scales: Excitation-energy in ^{9}Li and ^{9}Be. The arrows point to known ^{9}Li-levels

In Table I the results for the ^{6}Li, ^{7}Li and ^{9}Be target are summarized. An approximate relative branching ratio is given, i.e. the contribution of the particular transition relative to the total radiative absorption fraction. A rough correction for the efficiency drop towards lower energies was made.

Table I Relative branching ratios

Target	Recoil	J^π;T	E_γ (MeV)	E_R^* (MeV)	E_T^* (MeV)	branch. (%)	ratio
^{6}Li	^{6}He	0^+;1	133.9	0.0	3.6	7.6	(7.0) [5]
^{6}Li	^{6}He	2^+;1	132.1	1.8	5.4	4.1	(3.4)
^{7}Li	^{7}He	$3/2^-$;3/2	126.6	0.0	11.3	1.7	
^{9}Be	^{9}Li	$3/2^-$;3/2	124.5	0.0	14.4	1.2	
^{9}Be	^{9}Li	$1/2^-$;3/2	121.8	2.7	17.0	0.7	

In the ^{12}C-spectrum - not shown here - the first high resolution radiative capture results [7] have been reproduced, but in addition the improved energy resolution allows one to separate the ^{12}B-ground-state from the first three excited states.

The branching ratios for these states have been determined by normalizing to the hydrogen-peak in the data from a CH_2 target, and can be compared to the prediction of Maguire and Werntz [8].

Table II Branching ratios to ^{12}B-bound states

J^π;T	E_γ (MeV)	E_T (MeV)	E_R (MeV)	absolute branching ratio exp. (%)	theor. (%)	multipolarity 1s	2p
1^+;1	125.0	0.0	15.1	.084 ± .006	.071 ± .010	E1	M1
2^+;1	124.1	1.0	16.0	.022 ± .003	.029 ± .004	M2	M1
2^-;1	123.3	1.7	16.6	.003 ± .002	.022 ± .003	E2	E1
2^+;1	122.4	2.6	17.2	.019 ± .003	.011 ± .004	M1	E1

The last column lists the dominant photon multipoles contributing to the rate from the two pionic orbits. In the theoretical calculation the radiative π-capture matrix-elements have been related in an model-independent way to photoabsorption, electron-scattering and β-decay matrixelements involving the same nuclear states or their analogs in the target nucleus. The agreement is satisfactory except for the 2^- state, which does seem to be only weakly excited. Similar results are given by Skupsky [9] however with a four times larger rate calculated for the 2^- state. It appears that the improved experimental data now allow a much more critical test of the nuclear wavefunctions employed in the calculations, since in comparing several levels the large uncertainties induced from the errors of the X-ray widths, yields and capture schedules can be overcome.

We can combine the new data for the 1p-shell nuclei with the Berkeley result for ^{10}B and ^{14}N [10]) to compare with the results from 180°-electron-scattering. In Figure 4 the values for Γ_γ^0/Γ_W given in the review of Fagg [11]) are contrasted with the relative branching ratios to the $T_z=1$ analogs of these transitions for radiative pion capture.

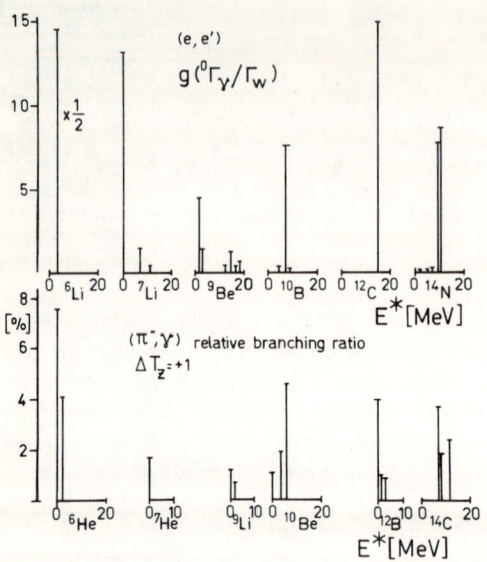

Figure 4: Upper part: M1-transition strength vs. excitation energy for several 1p-shell nuclei (from Fagg, ref. 11) $g=(2J^*+1)/(2J^0+1)$
Lower part: Relative strength (fraction of total rate) for $\Delta T_z=+1$ analogs of M1-states excited in radiative pion capture vs. excitation energy.

The concentration of isovector M1-strength into a few low-lying levels was predicted by Kurath [12]) for self-conjugate 4N+2 nuclei. The experimental confirmation in inelastic electron scattering was shown by Nang [13]) to be consistent with a small amount of SU(4)-super multiplett impurity present in the ground-state on which these transitions are built. Mukhopadhyay and Cannata [14]) have shown, that for reactions dominated by the $\vec{\sigma}\tau^+$-Gamow-Teller-Operator such as μ-capture - SU(4) symmetry implies, that the allowed strength for the transitions to the analogs of these M1-states is also concentrated into a few levels. Furthermore for 4N, 4N+1-nuclei such transitions would be absent in the limit of exact SU(4)-symmetry. Consequently the μ-capture rates [15]) predicted for e.g. ^{9}Be are much smaller than for neighbouring nuclei and more fragmented. The (π^-,γ) operator for 1s-capture is also of Gamow-Teller-Type. Even for the large amount of 2p-capture the important momentum dependent terms have a similar structure [8,16]). It is clear therefore, that the (π^-,γ)-reaction offers an attractive tool for the isovector M1-states. The observed pattern for ^{10}B and ^{14}N confirmed these expectations and explicit shell model calculations gave good agreement with the experimental branching ratios. Such detailed calculations do not exist for ^{7}Li and ^{9}Be, but it is apparent, that the transition rates in these nuclei are considerably smaller than in neighbouring nuclei, which reflects the SU(4) selection rule. However, the reduction is only a factor of 4 to 5, whereas in μ-capture one expects 60, because of the small SU(4) impurity in ^{9}Be. This fact remains to explained. The two ^{6}Li-^{6}He transitions have been a prefered testing ground for different theoretical approaches to radiative pion absorption. Excellent agreement with the (π^-,γ)-data and all weak and electromagnetic processes involving the same states or their analogs could be obtained with one set of wavefunctions. These analyses have been reviewed previously in great detail [1,22] and will not be discussed here.

IV Spin-Isospin-States

A further objective of (π^-,γ)-studies was the search for transitions to giant-multipole spin-isospin-states. These were discovered in ^{12}C [7] and the new data give identical results. A detailed comparison between theory and experiment can be made for ^{16}O. The new data is displayed in Figure 5 together with the calculated [16] branching ratios for the levels indicated at their expected positions.

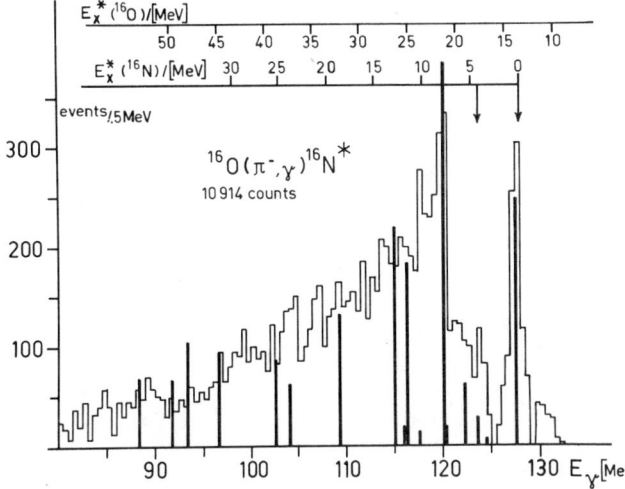

Figure 5:
Photon-spectrum from ^{16}O$(\pi^-,\gamma)^{16}$N*.
Solid lines: Predicted spin-isospin states with their strength normalised to the ground-state transition.
Dipole states $(E_\gamma(\text{MeV}),J^\pi)$ 115, 1$^-$; 116.5, 120.5, 122.1, 128, 2$^-$
Quadrupole states: 88, 1$^+$; 103, 96.5, 2$^+$; 109.5, 104, 93, 3$^+$

The latter have been normalized relative to the highest-energy peak in the spectrum near 128 MeV, which corresponds to an unresolved quartett of states including the 2$^-$-ground-state, expected [17] to contribute 86% of the peak. I have only displayed one of several similar calculations [18,19]. In ^{16}O no M1-states are expected, so the theory deals with giant-dipole and quadrupole spin-isospin states only. The quadrupole excitations ($J^\pi = 1^+$, 2^+, 3^+) are concentrated in in the region 30 to 50 MeV excitation energy, i.e. between 90 and 110 MeV in the photon spectrum. Between 115 and 128 MeV we find the giant-dipole states ($J^\pi = 1^-$, 2^-). Clearly present in both data and theory is the (2^-, 7.6, 1, 124.4)-MeV-state. The branching ratio seems to agree also, however as always in radiative capture studies it is not clear how much non-resonant-background from direct reactions of the type $^{16}O(\pi^-, \gamma n)^{15}N$ has to be subtracted. Some structure around 115 and 118 MeV could be taken as evidence for the stronger 1^- and 2^- transitions expected in this area. Similarly a speculative identification of the structure around 105 with 2^+, 3^+-quadrupole states can be made. In both cases the levels need to be shifted in energy. It is clear that further experimental work improving the statistical accuracy is needed here. Somewhat disturbing is the fact, that the absolute theoretical branching ratios overestimate the experimental ones by a factor of two to four. It is attributed to ground-state correlations in the target[16]. This conclusion is somewhat supported by the calculation of Szydlik and Werntz[17] for the ground-state transition, who use semi-phenomenological matrix-elements from weak and electromagnetic processes and obtain good agreement with the experimental value.

V. New features

Concluding this short review, some observations are listed, which at present cannot be reconciled with existing theoretical interpretations. ^6Li, ^7Li: Broad bumps appear in the spectra centered around $E\gamma$= 120 MeV(^6Li), 112 MeV(^6Li), 118 MeV(^7Li), 108 MeV(^7Li), corresponding to excitation energies (Recoil, Target) of (14,17.5), (22, 25.5), (8.5,20) and (18.5,30)MeV. The second peak in ^6Li is quite probably the analog of the T=1, $33F$-resonance seen both in ^6Li and ^6Be[20]. The others may be associated with $1\hbar\omega$ excitations[22]. ^9Be: A strong transition is observed, which even appears to be split[2], centered at $E\gamma$= 117 MeV(7.5 MeV) in ^9Li, 22 MeV in ^9Be. The ~10% **transition strength and 3-4 MeV width are both larger than for $1\hbar\omega$ excitations observed in other 1p shell nuclei.**
^{18}O: In Figure 6 the spectrum obtained with a
$D_2^{18}O$ target is shown. One identifies the upper peak at 124.5 MeV as the transition to the T=2, J^π= 0^-, 1^- or 2^- ^{18}N ground state. The spectrum looks surprisingly similar to the ^{16}O MeV-spectrum, with another peak appearing an excitation-energy of 7.5 MeV. The additional two neutrons do not seem to influence the distribution of radiative capture strength too much, 6.5% of all the strength goes to the ground-state compared to 8% for ^{16}O. Recently measured photo-proton and photo-neutron cross-section on ^{18}O[21] also show a peak at the equivalent energy of the upper peak. Since nothing is known about the level diagramm of ^{18}N or about the ^{18}O T=2 levels in this energy region, the data, when improved in statistical accuracy will yield first information on

the spectroscopy of this unexplored domain.

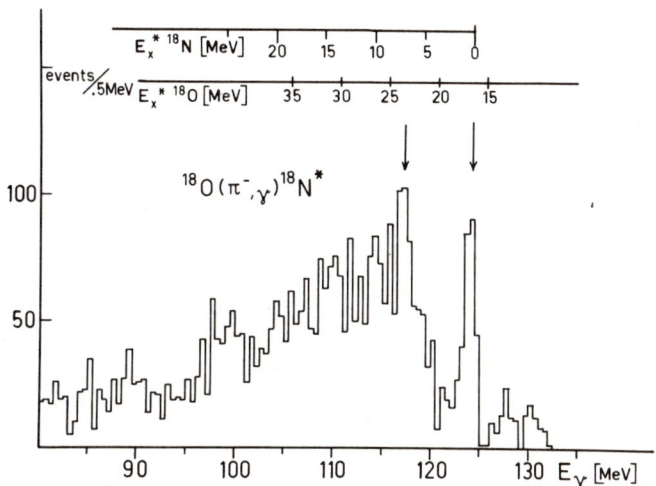

Figure 6: Photon spectrum from $^{18}O(\pi^-,\gamma)^{18}N^*$. Upper scales: Excitation energies in ^{18}O and ^{18}N. The events above 125 MeV are from the deuterium content of the target (also in ^{16}O, Figure 6)

† The work reported here was carried out by two groups;
At SIN: J.C. Alder, B. Gabioud, F. Hoop, C. Joseph, J. F. Loude, H.A. Medicus, N. Morel, A. Perrenoud, J.P. Perroud, D. Renker, H. Schmitt, G. Strassner, M.T.Tran, P. Truöl, B. Vaucher, H.v. Fellenberg, E. Winkelmann, C. Zupancic, Universities of Lausanne, München and Zürich, supported by SIN, Swiss Nationalfonds and Deutsche Forschungsgemeinschaft.

At LASL: H.W. Baer, J.A. Bistirlich, S. Cooper, K.M. Crowe, E.R. Grilly, J.P. Perroud, R.H. Sherman, F.T. Shively, P. Truöl, University of California, Case Western Reserve University, LASL; supported by US ERDA

1) H.W. Baer, K.M. Crowe, Proc. Int. Conf. on Photonuclear Reactions and Applications, Asilomar (1973)
   J. Eisenberg, Proc. V Int. Conf. on High-Energy Physics and Nuclear Structure, Santa Fé (1975)
   H.W. Baer, K.M. Crowe, P. Truöl, to be published in Adv. in Nucl. Physics Vol 9
2) J.A. Bistirlich et al., Phys. Rev. Letters 36 (1976), 942
3) J.C. Alder et al., contributions to this conference
4) S. Fiarman, S.S. Hanna, Nucl. Phys. A251 (1975), 1

5) see e.g. H.W. Baer et al., Phys. Rev. C8(1973), 2029
6) A.C. Phillips, F. Roig; Contr. to V. Int. Conf. on High-Energy Physics and Nuclear Structure, Santa Fé (1975)
7) J.A. Bistirlich et al., Phys. Rev. C5(1972), 1867
8) C.F. Maguire, C. Werntz, Nucl. Phys. A205(1973), 211
9) S. Skupsky, Nucl. Phys. A178(1971), 289
10) H.W. Baer et al., Phys. Rev. C12(1975), 921
11) L. Fagg, Rev. Mod. Phys. 47(1975), 683
12) D. Kurath, Phys. Rev. 130 (1963), 1525
13) P.T. Nang, Nucl. Phys. A185(1972), 413
14) N.C. Mukhopadhyay, F. Cannata, Phys. Letters 51B (1974), 225
15) N.C. Mukhopadhyay, Physics Letters 45B(1973), 309
16) J.D. Vergados, Phys. Rev. C12(1975), 1278
17) P. Szydlik, C. Werntz, Phys. Letters 41B(1974), 209
18) H. Ohtsubo, T. Nishiyama, M. Kawaguchi, Nucl. Phys. A224(1974), 164
19) J.D. Murphy et al., Phys. Rev. Letters 19(1967), 714
20) E. Ventura et al., Nucl. Phys. A219(1974), 157
21) B.L. Berman et al., Bull. Am. Phys. Soc. 21(1976), 68; and priv. communication.
22) J.D. Vergados, Nucl. Phys. A220(1974), 259

A. N. Mitra (Univ. of Delhi): This concerns the E_γ spectrum you showed in the reaction $\pi^- + H^3 \rightarrow (3n) + \gamma$. It appeared there was some (3n) structure in the neighborhood of the threshold, probably above. You were reluctant to discuss its significance, but there are reasons to expect some sort of structure in that region. The precise position, (i.e., whether it is below threshold - as a bound state, or above threshold - as a resonance) is probably model-dependent, but the more interesting point which is model-independent is its spectroscopic structure, viz, a $(1p)^2$ configuration. Do you see any way of measuring the spin? Probably this is hard, but it will resolve an interesting and long-standing controversy.

P. Truöl: Well, I would say that a measurement of the spin is just impossible. But if I understand the people at Los Alamos correctly, we are going to have another run so we'll get another chance to resolve the question of whether the state is there or not. But to determine the spin one must make a coincidence experiment with one or two of the neutrons and that to my mind is impossible.

THRESHOLD PHOTOPRODUCTION OF PIONS*

Justus H. Koch

Laboratory for Nuclear Science and Department of Physics
Massachusetts Institute of Technology
Cambridge, Massachusetts 02139

In parallel with the experimental talk of Tzara, I will focus my remarks on the nuclear photoproduction of pions near threshold and use as examples light targets for which data already exist. What can we learn from this reaction? First, this process is an excellent probe for the isobaric analog states of the nucleus. Since the produced particle is a pseudoscalar, we are probing the axial formfactor of the target nucleus. This provides an independent test for nuclear model wavefunctions derived from electron scattering. Second, pion photoproduction on nuclei, in particular in the threshold region, is very sensitive to the final state interactions of the produced pion with the residual nucleus. One can thus hope that these experiments will help to fill the gap in our knowledge of the π-nucleus interaction between π-mesic atoms and the low energy pion scattering experiments. And finally, nuclear photoproduction allows us to study the production mechanism itself, e.g. to investigate modifications of the free $\gamma + N \rightarrow \pi + N$ amplitude in the nuclear medium.

* Research supported in part by ERDA under contract E(11-1)-3069.

These three points of interest, nuclear structure, π-nucleus interaction and the production mechanism, also specify the theoretical inputs necessary for an analysis of pion photoproduction. Below, these three aspects will be discussed by using the microscopic approach to this reaction [1].

The production amplitude for a free nucleon can be obtained in several ways. One possibility is to evaluate the diagrams in Fig. 1 [2]. For charged pions, good agreement with experiment is obtained with the first three diagrams, but for π^0 photoproduction, the ω and N* contributions have to be included. For applications in the threshold region, we expand the resulting amplitude in powers of the momenta, keeping only the lowest multipoles. This is done in a general frame, where the initial nucleon has momentum $\vec{p}$ [see e.g. Ref. 3]:

$$M_{\pi\gamma} = C_1 \left[\vec{\sigma}\cdot\vec{\epsilon} + \xi \frac{\vec{\sigma}\cdot\vec{k}\ \vec{\epsilon}\cdot\vec{p}}{k^2} \right] + C_2\, \vec{\sigma}\cdot\vec{\epsilon}\ \vec{k}\cdot\vec{q} + \cdots \quad (1)$$
$$\equiv \vec{\epsilon}\cdot\vec{J}^5$$

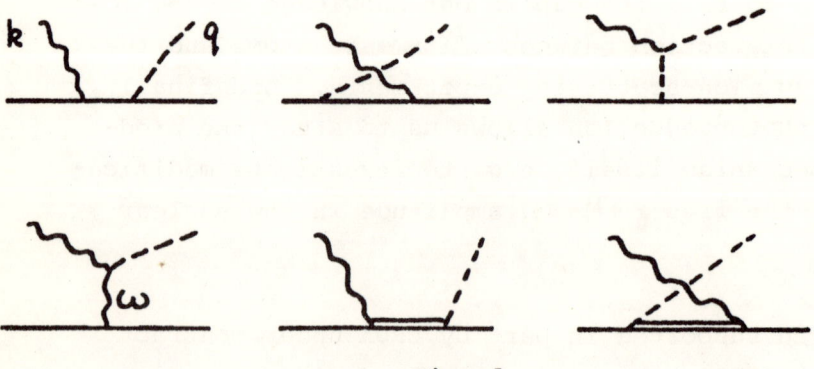

Fig. 1.

The C_i depend on the total energy and are operators in the nucleon isospin space

$$C_i = C_i^+ \delta_{\beta 3} + C_i^- \tfrac{1}{2}[\tau_\beta, \tau_3] + C_i^0 \tau_\beta \quad . \tag{2}$$

In the standard microscopic approach, this free single nucleon amplitude is used for each of the target nucleons. Just as in the impulse approximation in π-nucleus scattering, this raises the same questions: what is the size of the binding corrections, the off-shell uncertainties, etc.

Using the DWIA for the outgoing pion, the nuclear photoproduction amplitude becomes in this approximation

$$\langle \pi, f | M | \gamma, i \rangle = \langle f | \int d\vec{x}\, \varphi_\pi^*(\vec{x})\, \vec{\varepsilon} \cdot \vec{J}_{(\vec{x})}^5\, e^{i\vec{k}\cdot\vec{x}} | i \rangle . \tag{3}$$

Equation (3) is the basis for the following discussion of the nuclear structure and pion physics aspects in the nuclear γ, π reaction.

All the pion physics enters through the pion wavefunction $\varphi_\pi(\vec{x})$ which is distorted by the Coulomb and strong nuclear potentials. So far, our main information about the strong interaction part for low energies comes from π-mesic atoms and is summarized in the optical potential [4]

$$V_{\pi^-} = -\frac{4\pi}{2\mu}\left\{ a_0 \varrho + a_1(\varrho_n - \varrho_p) + A\varrho^2 + \vec{\nabla}\cdot\alpha(r)\vec{\nabla}\right\}, \tag{4}$$

$$\alpha(r) = b_0 \varrho + b_1(\varrho_n - \varrho_p) + B\varrho^2,$$

with the parameters chosen to fit the observed strong level shifts and widths. To illustrate the great importance of the final state interaction, we show in Fig. 2 the total cross section for $\gamma + {}^{12}C(g.s.) \to \pi^- + {}^{12}N(1^+1)$, keeping only the dominant $\vec{\sigma}\cdot\vec{\mathcal{E}}$ term in Eq. (1) and

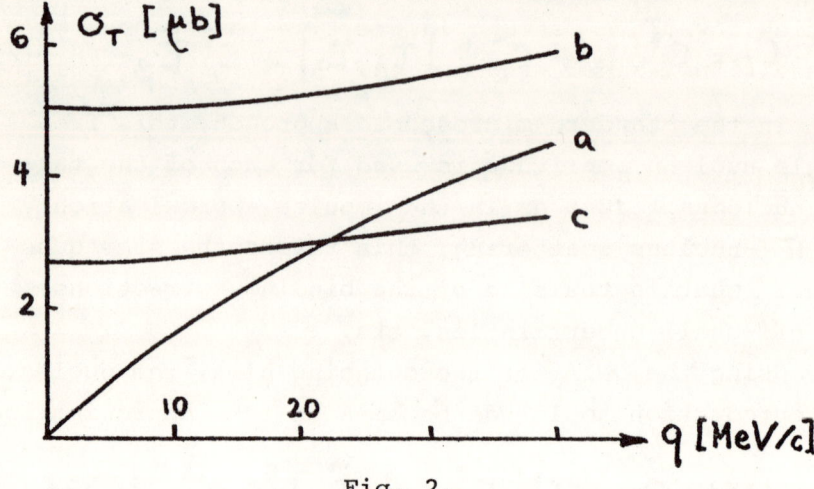

Fig. 2.

the pion s-wave. The nuclear Coulomb interaction changes the shape of the total cross section drastically, from a roughly linear increase for a plane pion wave, curve a, to the steplike shape of curve b. Inclusion of the strong π - nucleus interaction then reduces the height of this step by 1/2. As one increases the energy of the incident photon energy, a new step will appear as other nuclear final states become accessible. [5]. On the

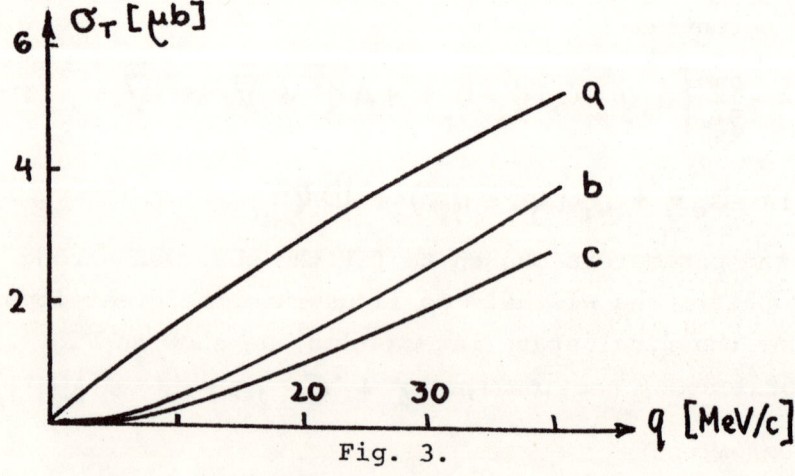

Fig. 3.

other hand, for $\gamma + {}^{12}C(g.s.) \rightarrow \pi^+ + {}^{12}B(1^+1)$,
Fig. 3, the Coulomb repulsion keeps the threshold cross section down. The strong interaction then causes only an additional small suppression. Figure 4 explains the

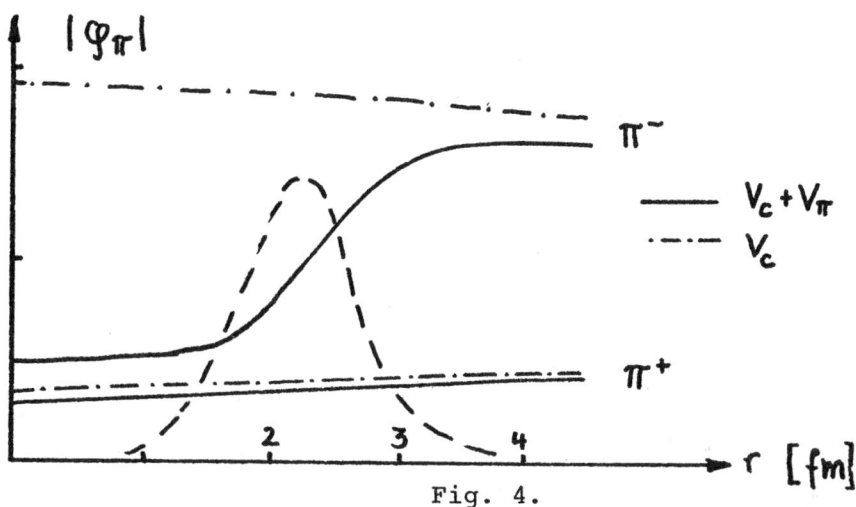

Fig. 4.

different behavior of the π^+ and π^- by showing the corresponding pion wavefunctions, φ_π. The strong slope of the π^--wavefunction explains why e.g. the $\vec{q}$-dependent terms in the production operator, Eq. (1), can cause important corrections, larger than one might expect from the small asymptotic pion momentum. Photoproduction of negative pions can therefore serve as a sensitive probe of the strong π-nucleus potential. This interaction is examined under unusual circumstances. While in π-mesic atoms or in elastic π-nucleus scattering the target is in its ground state, e.g. ${}^{12}C(0^+0)$, a different spin and isospin configuration is involved in charged pion photoproduction, e.g. ${}^{12}B(1^+1)$ or ${}^{12}N(1^+1)$.

Unfortunately, even though the nucleus is

quite transparent for a low energy pion, not the entire nuclear volume contributes to pion photoproduction. Due to the nuclear transition density (dashed curve in Fig. 4), the main contribution to the production process comes from the surface region of the target.

The qualitative examples given above have shown that for positive pions the total cross section is not very sensitive to the strong interaction part of the nuclear potential. For a discussion of the nuclear structure aspects, we therefore use a reaction where a π^+ is produced: $\gamma + {}^6Li(g.s.) \rightarrow \pi^+ + {}^6He(0^+1)$ [6-9]. The information we already have about the initial and final nuclear states, $|i\rangle$ and $|f\rangle$ in Eq. (3), comes from elastic and inelastic electron scattering, μ-capture and β-decay. Clearly the model wavefunctions used to evaluate the matrix element in Eq. (3) must also give reasonable fits to these processes, especially for momentum transfers near 140 MeV/c. Figure 5 shows four theoretical curves for the total photopro-

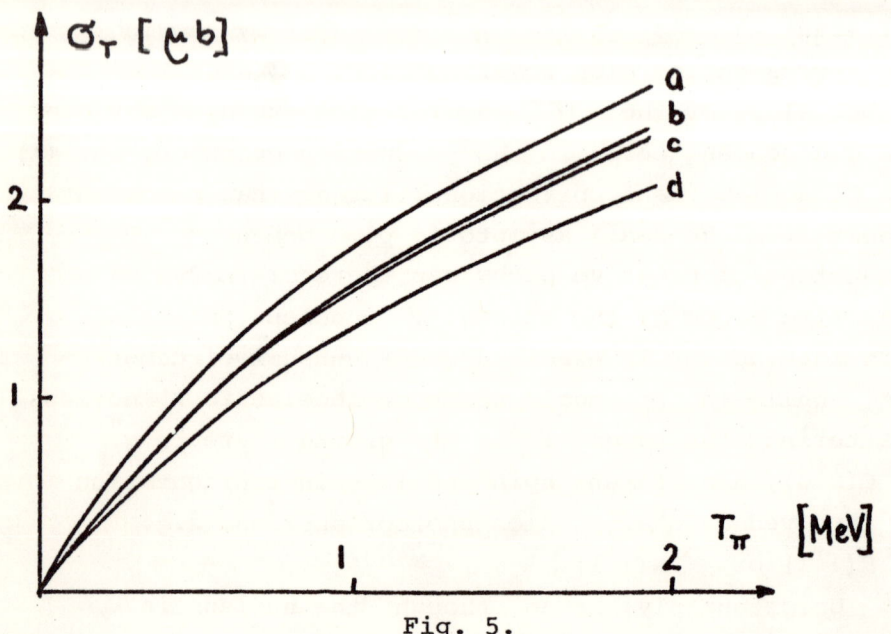

Fig. 5.

duction cross section. Curve a [6] is obtained by using
the nuclear wavefunctions of Donnelly and Walecka [10],
who restrict the model space to the harmonic oscillator
p-shell and then use the observed electromagnetic properties of ^{6}Li to fix all free parameters. Using the
same combination of states, but replacing the harmonic
oscillator by Woods-Saxon wavefunctions [9] yields the
cross section labelled c. Curves b and d are taken
from the work of Bergstrom [8], who uses a phenomenological p-shell transition density that involves more
free parameters than the harmonic oscillator model of
Ref. 10. While all these parametrizations give nice fits
to the inelastic electron scattering data, they yield
different predictions for the photoproduction cross section. The transverse M1 formfactor, which is measured in
inelastic electron scattering to the 0^+1 level of ^{6}Li
has two contributions, a convection current part and a
part proportional to the spin density. Therefore, fits
to the electron scattering data, which yield the same
total M1 formfactors, give different predictions for the
γ, π^+ reaction depending on how much of the M1
strength is concentrated in the spin part. The latest
photoproduction data for ^{6}Li shown by Tzara lie closest
to curves b,c, Fig. 5. For a more detailed comparison
of the data with the theory, it is necessary to evaluate
the full production amplitude, Eq. (1), with the various
nuclear models.

Finally, an important tool for investigating the
production mechanism itself is the photoproduction of
neutral pions [3]. While charged pion photoproduction
near threshold is dominated by the static Kroll-Ruderman
interaction, $\vec{\sigma} \cdot \vec{\varepsilon}$, the π° production amplitude vanishes in the limit of a static nucleon. Therefore,

neutral pion production can be used to study the non-static terms in the production operator, e.g. the momentum dependent terms in Eq. 2 which include terms due to the transformation of the single nucleon operator to the π -nucleus c.m. frame. Also, since the amplitude is an order of magnitude smaller than for charged pions, a two step production mechanism, whereby a charged pion is produced on one nucleon and then becomes a π^0 by charge-exchange scattering off a second nucleon, is important near threshold [3,11]. The π^0 thus plays a special role in that it is extremely sensitive to our model for the production mechanism itself. In addition, π^0 photoproduction has another unique feature: while the charged pions are produced incoherently and are therefore mainly probing the nuclear surface, a π^0 can be produced coherently, leaving the target in its ground state. Therefore, π^0 production is, in principle, also a probe for the pion wavefunction deep in the nuclear interior.

For a theoretical description of nuclear pion photoproduction, all three physical aspects -- nuclear structure, pion-nucleus interaction and production mechanism -- have to be taken into account. However, the preceding remarks have shown how each of these aspects enters on a different level near threshold for each of the three charge states of the pion: For a π^+, accurate nuclear structure information is most important, the π^- cross section is very sensitive to the strong π-nucleus interaction and the π^0 depends crucially on details of the production mechanism itself. The three charge states of the pion, together with the possibility of choosing specific nuclear transitions [12] and different targets therefore allow us to sort out the dependence on the various theoretical inputs and to focus our attention on particular aspects of the underlying physics.

References

[1] For a discussion of the elementary particle approach see M. Ericson and M. Rho, Physics Reports $\underline{5}$, (1972) 58.
[2] R. D. Peccei, Phys. Rev. $\underline{181}$ (1969) 1902.
[3] J. H. Koch and R. M. Woloshyn, Phys. Lett. $\underline{60B}$ (1976) 221.
[4] G. Backenstoss, Ann. Rev. Nucl. Sci. $\underline{20}$ (1970) 467; L. Tauscher and W. Schneider, Z. Physik $\underline{271}$ (1974) 409.
[5] A. Bernstein et al. (Paper VIII.6 contributed to this conference) have measured the $^{12}C(\gamma,\pi^-)^{12}N$ cross section for pion energies > 4 MeV. Nagl and Überall (contributed paper VIII.3) include higher partial waves for the pion and some of the momentum dependent terms in Eq. (1), which substantially improves agreement with the data.
[6] J. H. Koch and T. W. Donnelly, Phys. Rev. $\underline{C10}$ (1974) 2618.
[7] F. Cannata et al. Phys. Rev. Lett. $\underline{33}$ (1974) 1316.
[8] J. C. Bergstrom et al. Nucl. Phys. $\underline{A251}$ (1975) 401.
[9] J. B. Cammarata and T. W. Donnelly, Stanford Preprint ITP-524 (1976).

[10] T. W. Donnelly and J. D. Walecka, Phys. Lett. 44B (1973) 330.
[11] J. H. Koch and R. M. Woloshyn, Paper VIII.2 contributed to this conference.
[12] A. Delorme and N. Mukhopadhyay, Paper VIII.8 contributed to this conference.

B. Cammarata (Stanford Univ.): From low-energy theorems obtained from PCAC and current algebra the dominant part of the (γ,π) transition operator is given by the axial current operator. This operator has three parts: an axial, induced pseudoscalar and a tensor (second-class) term. In work done with Bill Donnelly, the influence of the tensor term (whose presence in the nuclear axial current is suggested by recent nuclear β-decay experiments) has been studied in the electroproduction and photoproduction of pions from nuclei at threshold. We find that these processes are very sensitive to the tensor term. In particular for the ^{6}Li$(\gamma,\pi^+)^6$He reaction we find that the tensor term reduces the theoretical cross section.
J. Koch: No comment.

PHOTOPRODUCTION OF NEGATIVE PIONS
AND THE GROUND STATE WAVE FUNCTION OF ^{12}C

K. Srinivasa Rao[+]

MATSCIENCE, The Institute of Mathematical Sciences,
Madras-600 020, INDIA

ABSTRACT

The energy dependence of the total cross section for the $^{12}C(\gamma,\pi^-)^{12}N$ reaction has been studied taking into account the presence of two-particle-two-hole (2P-2h) correlations in the ground state wave function, in the impulse approximation. The results are in good agreement with the experimental data in the first pion-nucleon resonance region.

THEORY

The recent experimental measurement[1] of the total cross sections for photoproduction of negative pions from ^{12}C leading to the ground state of ^{12}N, in the first pion-nucleon resonance region, reveals that the data obtained are in agreement with our phenomenological surface production model calculations.[2] In ref. 2, we assumed the ground state of ^{12}C to be spherical. However, it is known[3] that the ^{12}C ground state is better described as a deformed state and Rowe et al.[4] have obtained the same by doing a complete matrix diagonalization within the 1p-shell. In the absence of experimental data on the 2p-2h admixtures in the ground state of ^{12}C, as is available for ^{16}O, we make use of the model wave function of Rowe et al.[4]

We assume the ground state wave function of ^{12}C to be approximated[4] by:

$$0.739 |op-oh\rangle - 0.575 |(1p_{1/2}^2)_{J=0,T=1} (1p_{3/2}^{-2})_{0,1}\rangle$$
$$+ 0.35 |(1p_{1/2}^2)_{1,0} (1p_{3/2}^{-2})_{1,0}\rangle$$

and denote this be DGS. In the pure shell (PS) model case the strength of the op-oh state would be 1 and the 2p-2h admixtures would be absent. The ground state of ^{12}N is assigned to the 1p-1h configuration $(1p_{1/2})(1p_{3/2}^{-1})$, in the independent particle model (IPM). The Gillet-VinhMau[5] (GV) wave function of ^{12}N obtained in the Tamm-Dancoff approximation includes configuration mixing. The omission of 3p-3h correlations in ^{12}N is justifiable due to their insignificant contributions in an analogous situation.[6]

In Fig. 1, we plot the total cross section for $^{12}C(\gamma,\pi^-)^{12}N$ as a function of incident photon energy, in the impulse approximation, assuming volume production of pions. As was found[7] in the case of $^{16}O(\gamma,\pi^+)^{16}N$, the 2p-2h correlations play a dominant role in reducing the theoretical cross section values throughout the energy region. It is interesting to note the significant role of Final State Interactions in a study of the same reaction by Nagl and

---
[+]Present Address: Physics Dept., Catholic University, Wash. DC 20064

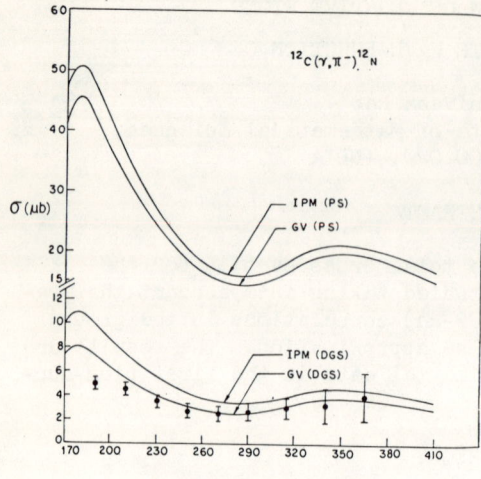

Überall[8] and in the study of $^{51}V(\gamma,\pi^+)^{51}Ti$ by Freed and Ostrander[9]. Further, theoretical work on this reaction is considered desirable.

The author is thankful to Professor Alladi Ramakrishnan for his interest in this work and is grateful to Prof. H. Überall for an interesting discussion.

Fig. 1. Total cross section for $^{12}C(\gamma,\pi^-)^{12}N$.

REFERENCES

1. V.D. Epaneshnikov et al., Sov. J. Nucl. Phys., <u>19</u>, 242 (1974).

2. K. Srinivasa Rao, V. Devanathan and G.N.S. Prasad, Nucl. Phys. <u>A159</u>, 97 (1970).

3. Y. Abgrall, et al., Nucl. Phys., <u>A131</u>, 609 (1969).

4. D.J. Rowe, et al., Phys. Rev. <u>C3</u>, 73 (1971) and J.C. Parikh, Private Communication.

5. V. Gillet and N. VinhMau, Nucl. Phys. <u>54</u>, 321 (1964).

6. A.K. Rej and T. Engeland, Phys. Letts. <u>45B</u>, 77 (1973).

7. K. Srinivasa Rao and V. Devanathan, Phys. Letts., <u>32B</u>, 578 (1970).

8. A. Nagl and H. Überall, this conference.

9. N. Freed and P. Ostrander, Pennsylvania State University preprint (1975).

Anup Rej (Fysisk Institutt, Trondheim): We already know from previous calculations on pion photoproduction and μ capture on ^{12}C, ^{14}N, ^{16}O that higher order deformations of the nucleus are important. This is what you learn from your calculation. But a rescattering calculation would tell you that the single scattering approximation for such big nuclei has little validity due to large rescattering contributions.

K. Rao: This is the first time that the effect of 2p-2h correlations in the ground state wave function of ^{12}C on pion photoproduction from ^{12}C is reported. The data for $^{12}C(\gamma,\pi^-)^{12}N$ have started coming in (Epaneshnikov et al.) and it is worthwhile to do theoretical work. As for multiple scattering corrections, we appear to be nowhere near the final answer and the studies on even deuterium are not without contradiction. What was presented here, is in my opinion a first step in the right direction. The effect of Final State Interactions has to be studied thoroughly, especially since the enormous reductions in the cross sections for $^{11}B(\gamma,\pi^-)^{11}C$ and $^{51}V(\gamma,\pi^+)$ Ti, reported by Handel and Weise and Freed and Ostrander, respectively, using optical potentials for pion-nucleus scattering, are rather disturbing.

G. Alberi (Istituto Fisica Teorica, Trieste): I would like to comment about the possibility of testing the presence of 2 particle 2 hole component in nuclei with high energy hadrons, where double scattering is quite important, both in the initial and in the final state.

J. Koch (MIT): How do your wave functions fit elastic scattering and μ^- capture data.

K. S. Rao: Since the same initial and final nuclear states are involved in both π^+ photoproduction and μ^--capture, the wave functions used in our studies of charged pion photoproduction from ^{12}C, play a significant role in muon capture too. However, the Cohen-Kurath wave functions give the best fit for muon capture, β-decay and electron scattering by ^{12}C. At present, Prof. N. Freed at the Pennsylvania State University and I, are studying charged pion photoproduction from ^{12}C using these wave functions.

SPIN-FLIP TRANSITION STRENGTH OF $^{12}C(\gamma,\pi^+)^{12}B$

K. Shoda, H. Ohashi and K. Nakahara
Laboratory of Nuclear Science, Tohoku University,
Tomizawa, Sendai 982, Japan

ABSTRACT

Energy distributions of π^+ emitted from $^{12}C(e,e'\pi^+)^{12}B$ reaction have been measured at $\theta=30°$, $60°$, $90°$ with the electron beam of $E_e=195$ MeV. The result is used to deduce the spin-flip transition strength between the ground state of ^{12}C and the residual states of ^{12}B. Such obtained strength is compared with the theory and is found in agreement with the generalized Helm model.

The cross section of photopion production from ^{12}C has been calculated by Überall's group[1,2]. They estimated it with the generalized Helm model fit to the transverse form factor measured with electron scattering. However experiments have never been made to compare with this kind of theory.

The cross section of the (γ,π^+) reaction should be analyzed from the spectra of π^+ obtained from $(e,e'\pi^+)$ reaction, because both the reactions relate with each other as shown by

$$\left(\frac{d^2\sigma}{d\Omega dE}\right)_{(e,e'\pi^+)} = \sum_i \int \left(\frac{d\sigma_{E_{Ri}}}{d\Omega}\right)_{(\gamma,\pi^+)} \cdot N_{h\nu}(E_e,E_\gamma) dE_\gamma , \qquad (1)$$

where $(d\sigma_{E_{Ri}}/d\Omega)_{(\gamma,\pi^+)}$ is (γ,π^+) differential cross section relate to the residual energy E_{Ri}, and also $N_{h\nu}$ is the virtual photon spectrum associated with an electron with energy E_e. The spectrum $N_{h\nu}$ shows a sudden increase at the end point energy, then the π^+ spectrum described by eq.(1) is to show a sudden increase at the position depending on E_e and E_{Ri} since smoothly varying cross section and large residual level spacing are expected for strong transitions in the present reaction.

An experiment has been made to measure the energy distribution of π^+ on the $(e,e'\pi^+)$ reactions with 195 MeV electron beam from Tohoku University linear accelerator. Particles were momentum analyzed with a 178.9° deflecting double foccussing spectrometer and detected with a ladder of 33 channels of Si(Li) solid state 3-coincidence system. All charged nuclear particles were stopped in the first detector of the coincidence system and rejected. The pulse height were well separated between π^+ and e^+ and discriminated with bias setting. The back ground of e^+ was corrected though it was very small. Correction for the life of π^+ was also made.

The result is shown in Fig. 1. The solid curve in the figure is theoretical result with the generalized Helm model calculated by the same method in refs. 1,2) and also with eq.(1) including the energy resolution for the experimental condition.

Besides the above comparison, $(d\sigma_{ERi}/d\Omega)(\gamma,\pi^+)$ was deduced from the present spectra by the least square fits with eq.(1) over about ±2 MeV around the pion energy of the expected strong transitions. These results are compared with the theoretical estimates in Table I where the recent theoretical result with the shell model is also shown[3].

As shown by Fig. 1 and Table I, the experimental values are in agreement with the generalized Helm model within factor $1/2 \sim 1$. The larger yield of π^+ found in the lower energy side in Fig. 1 may reflect the giant resonance tail in the reaction.

REFERENCES

1) H. Überall, B. A. Lamers, C. W. Lucas and A. Nagl, Phys. Lett. **44B**, 324 (1973).
2) F. Cannata B. A. Lamers, C. W. Lucas, A. Nagl, H. Überall, C. Werntz and F. J. Kelly, Can. J. Phys., **52**, 1405 (1974).
3) J. B. Seaborn, V. Devanthan and H. Überall, Nucl. Phys. A219, 461 (1974). V. Devanathan et.al., Proc.Int.Conf. on Nucl.Phys.Munich 1973 vol.1,p.674

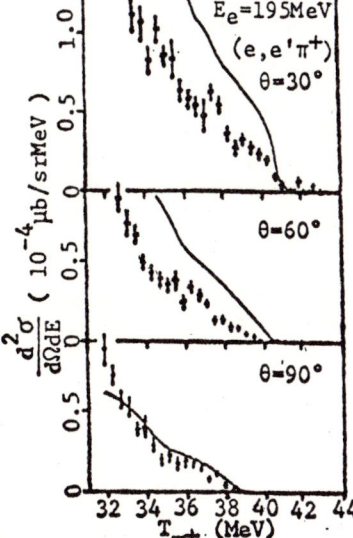

Fig.1. Energy distribution of π^+. Solid curve shows the theoreticl result (ref. 2).

Table I. Comparison of $(d\sigma_{ERi}/d\Omega)(\gamma,\pi^+)$

Level Number[a]	1	2	10	11
E_{Ri} in ^{12}B (MeV)	0	0.95	4.53	5.0
J^π	1^+	2^+	2^-	4^-
Main mode[a]	M1	E2,s	M2,si	M4
$\left(\dfrac{d\sigma_{ERi}}{d\Omega}\right)(\gamma,\pi^+)$ $\left(10^{-4}\dfrac{\mu b}{sr}\right)$ $\theta=30°$	1.1±0.2 (2.32)b) (8.2) c)	0.04±0.2 (0.14)b) (0.38)c)		
$\theta=60°$	0.08±0.02 (0.75)b) (3.3) c)	0.43±0.04 (0.36)b) (1.3) c)		
$\theta=90°$	0.01±0.07 (0.10)b) (1.0) c)	0.45±0.12 (0.45)b) (2.1) c)	0.83±0.13 (0.24)b) (1.6) c)	

a) Notations are used as in refs.1,2).
b) Theory (generalized Helm model,1,2).
c) Theory (shell model, 3).

THE $^{12}C(\gamma,\pi^-)^{12}N$ REACTION NEAR THRESHOLD

A. M. Bernstein, N. Paras and W. Turchinetz
Massachusetts Institute of Technology, Cambridge, Mass. 02139*

B. Chasan and E. C. Booth
Boston University, Boston, Mass. 02115*

In this paper we report a study of the total cross-section for the $^{12}C(\gamma,\pi^-)^{12}N$ reaction by observing the 16.3 MeV end point, 11 msec β^+ radioactivity of ^{12}N. Since there is only one particle stable bound state in ^{12}N, this technique measures the cross-section to that one state only.

The experiment was performed with a bremsstrahlung beam produced by electrons from the MIT Bates Linear Accelerator impinging on a Ta radiator. The cross-section as a function of photon energy must be extracted from a measurement of the yield,

$$Y(E_o) = \int_{E_T}^{E_o} \sigma(E) \phi(E_o,E) dE,$$

where E_o is the peak bremsstrahlung energy, E_T is the threshold energy, $\sigma(E)$ is the total cross-section and $\phi(E_o,E)$ is the bremsstrahlung spectrum function.

The yield for the $^{12}C(\gamma,\pi^-)$ reaction is shown in Figure 1a, along with the theoretical predictions. The calculation of Koch[1] uses the $\vec{\sigma}\cdot\vec{\epsilon}$ interaction with a coupling strength adjusted to fit the (γ,π^-) production on the neutron. Only s-wave pions, distorted in a Kisslinger-type optical potential are considered. Because of these approximations, this calculation should be accurate only in the near threshold region. The calculation of Koch[1] gives results which are in agreement with experiment up to 4 MeV above threshold, but are too small above that energy.

The calculation of Nagl and Uberall[2] uses the full interaction Hamiltonian[3] and includes s, p and d pion waves distorted in a Kisslinger-type optical potential. The differences between the curves in Figure 1a lie in different choices for the pion momentum in the coefficients of the interaction Hamiltonian. In curve 1, the asymptotic pion momentum is used. In curve 2, the local pion momentum is used.

A two parameter "best fit" to the data was made assuming a step at the threshold, which is the effect of the final state Coulomb interaction in the (γ,π^-) reaction, and a linear rise above threshold. This curve is shown in Figure 1. This does not imply that the data excludes curvature in the cross-section, but because of the limited number of data points, a fit with more unknown parameters is not justified.

The cross-sections versus energy are shown in Figure 1b. It can be seen that the best fit extracted for the cross-section at threshold is 2.9 ± 1.1 µb, which is consistent with all of the calculations. In the radiative pion capture in ^{12}C the theoretical predictions[4] (with errors of approximately 30%) are in agreement with experiment[5]. The (γ,π^-) calculations of Nagl and Uberall[2] have a larger slope than that of Koch[1] because p wave pions and the momentum dependent terms in the Hamiltonian were included. However, it can be seen that the slope of the best fit cross-section is somewhat greater than any of the theoretical curves. Further experimental and theoretical work on this reaction would be desireable.

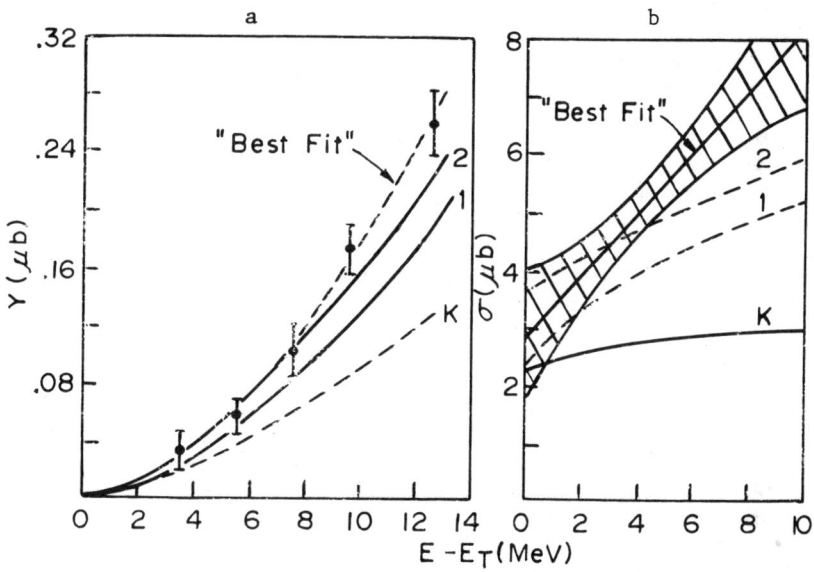

Fig. 1 (a) Yield and (b) cross-section versus energy above threshold. The shaded zone shows the error in the best fit.

1. J. H. Koch, Private Communication. J. H. Koch and T. W. Donnelly, Nucl. Phys. B64, 478 (1973) and Phys. Rev. 10C, 2618 (1974).
2. A. Nagl and H. Uberall, Private Communication and to be published, A. Nagl, F. Cannata and H. Uberall, Phys. Rev. C12, 1586 (1975)
3. G. F. Chew, M. L. Goldberger, F. E. Low and Y. Nambu, Phys. Rev. 106 1345 (1957), F. A. Berends, A. Donnachie and D. L. Weaver, Nucl. Phys. B4, 1 (1967).
4. W. Maguire and C. Wentz, Nucl. Phys. A205, 211 (1973).
5. H. W. Baer et al, Phys. Rev. C12, 921 (1975) H. W. Baer and K. M. Crowe, Conf. on Photonuclear Reactions and Applications, Asilomar, Calif., March 1973, B. L. Berman, editor, and references quoted there.

Threshold Photoproduction of π^+ Mesons in Deuterium

E.C. Booth and B. Chasan*
Boston University, Boston, Mass. 02215

A. Bernstein and P. Bosted[+]
Massachusetts Institute of Technology
Cambridge, Mass. 02139

The theory of pion photoproduction in complex nuclei is based on the physical picture of a photon interacting with an individual nucleon, the interaction amplitude being identical with the free nucleon case. This is the essential assumption of the impulse approximation, which is the basis for most calculations reported thus far.[1]

The validity of the impulse approximation is hard to establish in complex nuclei because calculations involve assumptions about nuclear wave functions and final state interactions which inevitably introduce theoretical undertainties. These are minimal for deuteron production particularly near threshold, where the dominant interaction amplitude is well known to be $\vec{\sigma}\cdot\vec{\varepsilon}$.[2] Here ε is the photon polarization and σ the nucleon spin operator.

We have studied threshold π^+ production from the deuteron using the bremsstrahlung beam of the MIT Bates Linear Accelerator. The excitation function for this reaction was obtained by detecting positrons emitted in $\pi\mu e$ decay. The 1.6 usec half life of the muon permitted detection of the positrons after the short (2 usec) beam pulses. The positrons were detected with a four-counter telescope consisting of two 1/16" plastic scintillators sandwiched between two thick (4 cm.) plastic Cerenkov counters. The background rate was found by measuring the D_2O yield just below the D threshold, then extrapolating into the energy region of interest by measuring the yield at 151 MeV with a carbon target. At higher energies the subtraction was made, along with the needed oxygen correction, by studying the yield from a BeO target. The background is the dotted line in Fig. 2.

This experiment is a measurement of a quantity proportional to the yield

$$Y(E_o) = \int_{E_T}^{E_o} N(E,E_o)\sigma_d(E)dE$$

where $N(E,E_o)$ is the bremsstrahlung shape function, E_T is the threshold energy and $\sigma_d(E)$ is the total deuteron photoproduction cross-section. In order to normalize our measurements we compared our results to $\gamma(P,n)\pi^+$. The solid curve in Fig. 1 is the hydrogen excitation function calculated with the well established threshold $\gamma(P,n)\pi^+$ cross-section.[3] The bremsstrahlung spectrum, which ultilizes the end point result of Jabbur and Pratt[4], is due to Mathews.[5] The fit to the experimental points involved a scaling of the "theoretical" excitation function, and shifting the energy scale. Thus the

---

*Partially supported by tne National Science Foundation

+Partially supported by the Energy Research and Development Administration

absolute energy scale was ultimately determined by fitting the proton excitation function.

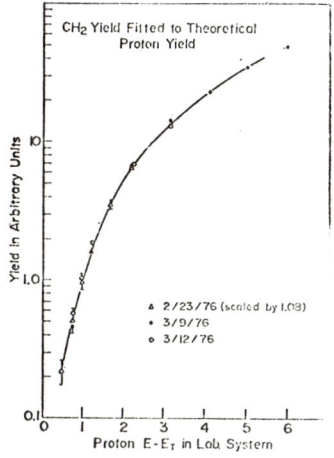

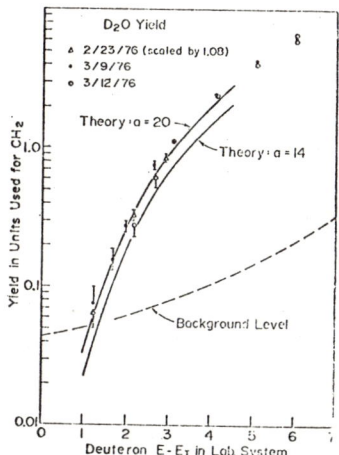

Fig. 1 Quadruple coincidences per quantameter unit for polyethene target with background subtracted.

Fig. 2. Quadruple coincidences per quantameter unit for heavy water with the background below threshold subtracted.

Fig. 2 shows the experimental excitation function for the deuteron. The solid curves were obtained from calculations of $\sigma_d(E)$ carried out by Tzara[6], and represent two values of the neutron neutron scattering length. They have been scaled by the factor determined by the proton fit, and are corrected for the difference in H and D target thickness. It can be seen that there is good agreement with the data for $a_n > 20$; $a_n \approx 22$ gives perhaps the best fit. Since the currently accepted value is[7] $a_n = 18\pm 1.5$ this experiment qualitatively supports Tzara's calculation, but gives a result slightly higher than theory. Small target thickness corrections are still to be made, but should not change the results by more than 5%.

REFERENCES

1. F.J. Kelly, L.S. McDonald and H. Uberall, Nucl. Phys. A139, 329 (1969).
   J.H. Koch and T.W. Donnelly, Nucl. Phys. B64, 478 (1973).
2. G.F. Chew, M.L. Goldberger, F.E. Low, and Y. Nambu, Phys. Rev. 106, 1345 (1957).
3. J. Deutsch, D. Favart, B. Van Ostoeyen, G. Audit, N. de Botton, J.L. Faure, Cl. Schuhl, G. Tamos, C. Tzara, Physical Review Letters, 33, 316 1974.
4. R.J. Jabbur and R.H. Pratt, Phys. Rev. 133, B1091 (1964).
5. J. Mathews, Private Communication.
6. C. Tzara, Nucl. Phys. A256 381 (1976).
7. W.O. Lock and D.F. Measday, Intermediate Energy Nuclear Physics (Methuen London, 1970) p. 50.

NEAR THRESHOLD π^0 PHOTOPRODUCTION FROM THE DEUTERON

J. H. Koch*
Massachusetts Institute of Technology, Cambridge, Ma. 02139

R. M. Woloshyn**
University of Pennsylvania, Philadelphia, Pa. 19174

ABSTRACT

Total and differential $d(\gamma,\pi^0)d$ cross sections are calculated for photon energies 0 to 5 MeV above threshold.

Neutral pions can be produced coherently from a nuclear target. π^0 photoproduction therefore provides a probe for the nuclear matter distribution and the pion wavefunction over the entire nuclear volume. The π^0 photoproduction amplitude for a single nucleon depends crucially on small non-static terms, which are absent in the well known Kroll-Ruderman amplitude. Thus, before proceeding to analyze π^0 photoproduction from a more complicated target, it is important to test our understanding of the elementary amplitude on a simple and well understood target such as the deuteron.

In a recent letter[1] we have presented the total (γ,π^0) cross sections for few-body targets in the threshold limit. Since experiments for photoproduction of low energy π^0's are now being performed[2], we present below a calculation of the total differential $d(\gamma,\pi^0)d$ cross section for photon energies from 0 to 5 MeV above threshold.

The elementary γ-N photoproduction amplitude expressed in terms of CGLN invariants in the two body c.m. frame is transformed to the γ-d c.m. frame. The invariants are evaluated by using the field theoretic pion, nucleon and ω-pole diagrams. A single pion rescattering of the produced pion is retained, which yields a large contribution, where first a charged pion is photoproduced on one nucleon and then charge-exchange scatters from the other target nucleon.

Fig. 1 shows the calculated reduced cross sections as a function of T_γ, the photon lab energy above threshold. The dash-dot line is the cross section for $p(\gamma,\pi^0)p$, while the solid and dashed lines represent $d(\gamma,\pi^0)d$ for the direct and direct plus rescattering mechanism. The deuteron is described by a Hulthen type wave function. The ratio of deuteron proton cross sections, which is measured in experiments, is plotted in Fig. 2. Fig. 3 shows the reduced angular distribution in the c.m. frame for the deuteron with (——) and without (----) the rescattering term. The rescattering changes the differential and total cross section quite drastically

---
*Work supported in part by the Energy Research and Development Agency.
**Work supported in part by the National Science Foundation.

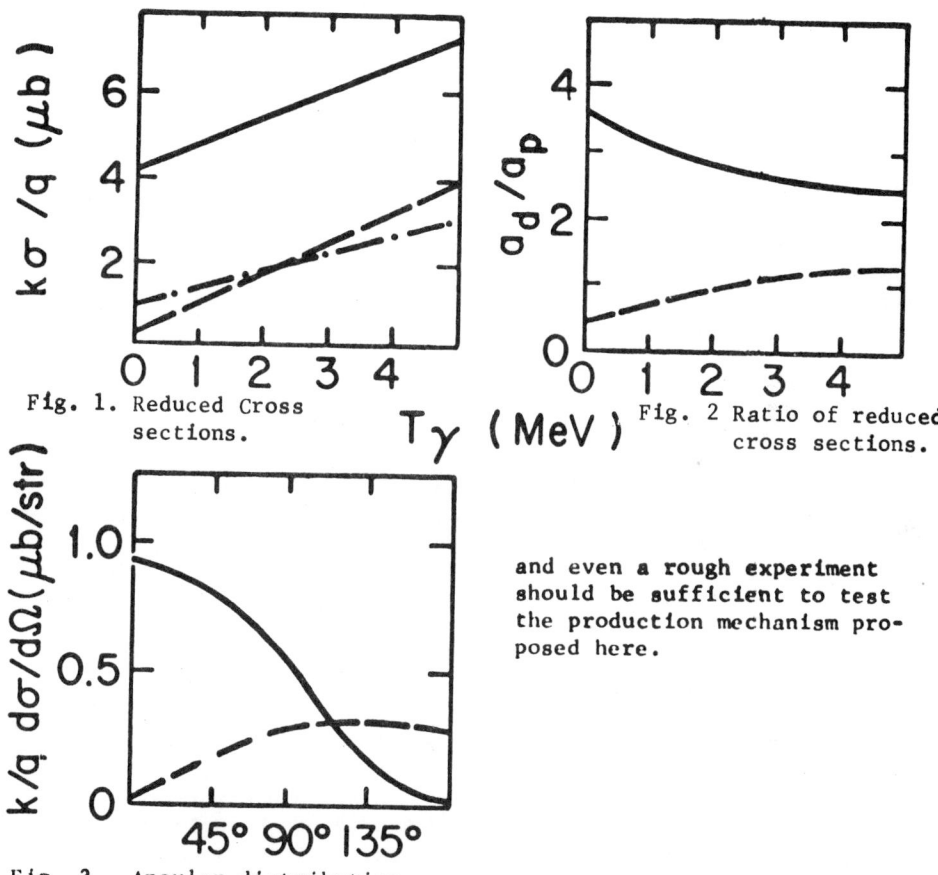

Fig. 1. Reduced Cross sections.

Fig. 2 Ratio of reduced cross sections.

Fig. 3. Angular distribution

and even a rough experiment should be sufficient to test the production mechanism proposed here.

REFERENCES

1. J. H. Koch and R. M. Woloshyn, Phys. Lett. <u>60B</u>, 221 (1976).
2. C. Tzara, private communication.

π^- PHOTOPRODUCTION IN ^{12}C

Anton Nagl* and H. Überall*+

Catholic University of America, Washington DC 20064

ABSTRACT

Charged pion photoproduction on nuclei is calculated taking into account the complete transition operator, Coulomb and nuclear distortion of the partial pion waves, and initial nuclear motion in impulse approximation. The calculated yield of the reaction $^{12}C(\gamma,\pi^-)^{12}N_{gd}$ is compared with MIT data of Bernstein et al., and earlier data of Epaneshnikov et al.

THEORY

Our formalism for calculating pion photoproduction from complex nuclei, based on the impulse approximation, uses the complete transition operator of the elementary pion photoproduction reaction on a nucleon[1], employs all contributing partial waves of a pion wave function distorted by an optical potential[2], and includes an accurate kinematical treatment which permits calculations down to threshold. Local pion momenta are used, and effects of initial nuclear motion and nucleon pair correlations are allowed for.

The formalism is applied to a calculation of the reaction

$$\gamma + {}^{12}C \rightarrow {}^{12}N_{gd} + \pi^- \quad (1)$$

whose cross section was measured[3,4] by observing the activity of the ^{12}N ground state (all excited states being particle-unstable[5]) after bremsstrahlung irradiation of ^{12}C. The nuclear model used is the Helm model[6]; another recent calculation[7] of (1) uses a shell model but includes no final-state interactions.

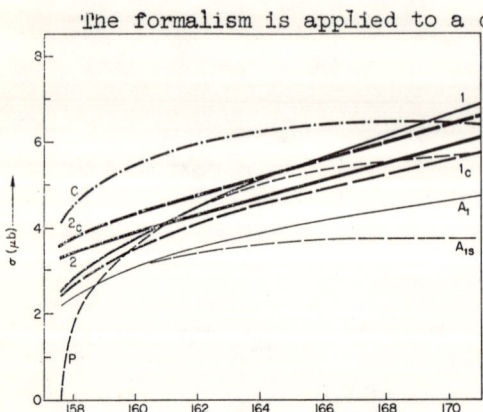

Fig. 1. Total cross section of Reaction (1) near threshold, as a function of photon energy.

---

*Supported in part by the National Science Foundation
+Also at Naval Research Laboratory, Washington DC 20375

Fig. 1 shows calculated cross sections near threshold as follows: P(C) uses plane-wave (or Coulomb-distorted) pions, all other curves use Coulomb-and strongly interacting pions. Curves 1 and 2 employ two different versions of local pion momenta, and 1_c, 2_c include pair correlation effects. Curves A_1 (and A_{1S}) use only the threshold interaction (and S waves only).

Figs. 2 compare our results, folded into the bremsstrahlung spectrum, with yield data of MIT[4] (triangles) and Russia[3] (circles). The theory is seen to be slightly low near threshold, and somewhat high at high energies where the plane wave seems to fit slightly better.

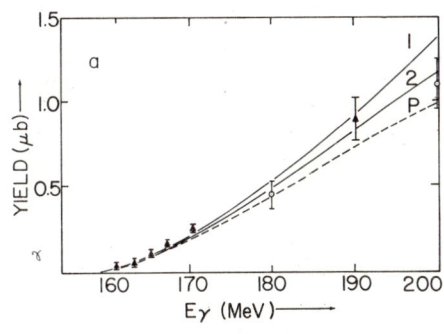

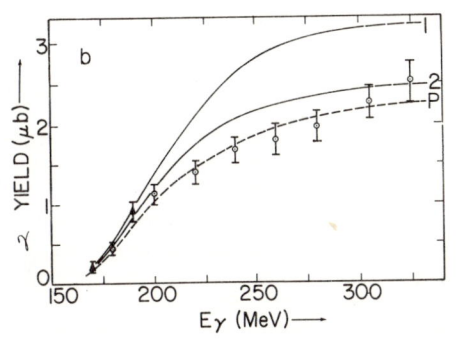

Fig. 2. Bremsstrahlung-folded yield of Reaction (1) from threshold to 200 MeV photon energy (a), to 300 MeV (b), compared with MIT data[4] (triangles) and Russian data[3] (circles)

REFERENCES

1. F.A. Berends, A. Donnachie, and D.L. Weaver, Nucl. Phys. **B4**, 1 (1967).
2. M. Krell and S. Barmo, Nucl. Phys. **B20**, 461 (1970).
3. V.D. Epaneshnikov et al., Sov. J. Nucl. Phys. **19**, 242 (1974).
4. A.M. Bernstein et al., Phys. Rev. Lett. (to be published).
5. K. Srinivasa Rao, V. Devanathan, and G.N.S. Prasad, Nucl. Phys. **A159**, 97 (1970).
6. H. Überall, *Electron Scattering from Complex Nuclei* (Academic, 1971).
7. K. Srinivasa Rao, this conference.

A PWIA Analysis of Charged Pion Photoproduction from ^{12}C

K.Baba, I.Endo, M.Fujisaki, S.Kadota and Y.Sumi
Department of Physics, Hiroshima University, Hiroshima, Japan

H. Fujii and Y.Murata
Institute for Nuclear Study, University of Tokyo, Tokyo, Japan

S.Noguchi
Department of Physics, University of Tokyo, Tokyo, Japan

A.Murakami
Department of Physics, Saga University, Saga, Japan

In a series of experiments on pion photoproduction from ^{12}C in the energy range between 300 and 850MeV, the following characteristic features have been revealed:[1,2]
(i) A singly-peaked structure is observed in the pion momentum spectra in the whole range of energy at both of lab angles 28.4° and 44.2°.
(ii) The position of the peak is lower by about 35MeV/c than the pion momentum corresponding to the free nucleon kinematics.
(iii) The width of the peak is substantially broader than our experimental resolution and is proportional to the photon energy.
(iv) There is no significant difference between π^+ and π^-; the π^+/π^- ratio is almost the same as that of the elementary processes.

Here we present consequences of our plane wave impulse approximation (PWIA) calculation applied to the photo-pion reaction on ^{12}C.

Following the prescription for analyzing (e,e'p) and (p,2p) reactions,[3] we write

$$\frac{d^2\sigma}{d\Omega_\pi dp_\pi} = \int dE_S d\Omega_N \frac{p_N^2}{p_\pi^* E_\pi E_{N'}} \frac{E_\pi^* + E_{N'}^*}{|p_\pi/E_\pi + (\vec{p}_{N'} \cdot \vec{p}_\pi)/p_{N'} E_R|} \left(\frac{d\sigma}{d\Omega_\pi^*}\right)^{free} Q(E_S, \vec{p}_N) f_\pi, \quad (1)$$

where $(E_N, \vec{p}_N)$, $(E_\pi, \vec{p}_\pi)$, $(E_{N'}, \vec{p}_{N'})$ are the 4-momenta of the target nucleon in the nucleus, the produced pion, recoiled nucleon, respectively, E_S is the separation energy, E_R is the energy of residual nucleus and the quantities with asterisk are those in the center-of-momentum system of the pion and recoiled nucleon. For the spectral function $Q(E_S, \vec{p}_N)$, we adopt a harmonic oscillator type one which is given by a simple shell model, and the relevant parameters are taken from the (e,e'p) experiment.[4] The elementary cross section $(d\sigma/d\Omega_\pi^*)$ is taken from the experimental data by Fujii et al.[5] A reduction factor f_π is introduced in (1) since the cross section is much reduced probably because of a large absorption probability of the produced pions in escaping the target nucleus. Numerical calculation has been performed by using a Monte Carlo method.

Apart from the absolute magnitude which can be fitted by f_π, the only parameter left free in our calculation, the calculated results show that the features (i) to (iii) above are quite well reproduced both for π^+ and π^- by using the same values for parameters in the spectral function as in the (e,e'p) reaction. Therefore if we regard

f_π as a purely phenomenological parameter we can get excellently good fits to the data as shown in Fig.1 where a few examples of our fit are given.

From these fits it is found that f_π is a rather slowly varying function of only p_π. Now if we assume that f_π is entirely due to pion absorption which takes place uniformly in the nucleus, we have[6]

$$f_\pi = \frac{3}{x^3} \left(\frac{x^2}{2} - 1 + (1 + x) e^{-x} \right), \qquad (2)$$

with $x=2R/\lambda$, where R is the nuclear radius and the mean free path of pion λ is given in terms of the nuclear density ρ and absorption cross section σ_{abs} as $\lambda=1/\rho\sigma_{abs}$. Values of σ_{abs} deduced from fitted f_π are plotted in Fig.2 together with corresponding average pion-nucleon total cross section $\sigma_T(\pi N)$. This figure clearly shows that there must exist a large absorption effect; indeed we need values of σ_{abs} even larger than those of $\sigma_T(\pi N)$ above the 3-3 resonance. This may suggest that some other nuclear effect, such as two-nucleon absorption, should be explicitly taken into account.

Although such a large absorption is left unexplained, our PWIA analysis performed entirely in parallel with quasielastic electron scattering is verified to be very suitable as well as useful in analyzing $(\gamma,\pi^\pm)$ experiments. At the same time we may say that $(\gamma,\pi^\pm)$ experiments are quite promising for the study of nuclear structure, particularly to investigate the behavior of neutron under the same experimental conditions as those for proton.

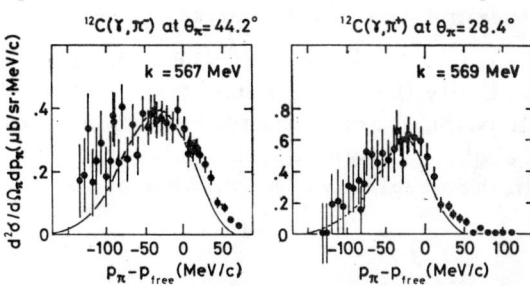

Fig.1. Typical examples of pion momentum spectra for the reactions $^{12}C(\gamma,\pi^\pm)$; errors are statistical only and p_{free} denotes the pion momentum calculated from free nucleon kinematics. Solid curves are the result of the present PWIA calculation (1).

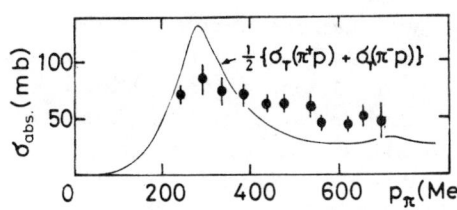

Fig. 2. Values of σ_{abs} calculated from the fitted values of f_π and by using (2). Smooth curve is the average pion-nucleon total cross section as a function of pion momentum p_π.

1. I.Endo et al., Phys. Letters 47B, 469(1973).
2. K.Baba et al., Preprint HUPD-7513(Hiroshima Univ.), 1975.
3. G.Jacob and Th.A.Maris, Rev. Mod. Phys. 38, 121(1966); 45, 6(1973).
4. H.Hiramatsu et al.,Phys. Letters 44B, 50(1973).
5. T.Fujii et al., Phys. Rev. Letters 28, 1672(1972); 29, 244(E)(1972).
6. S.Fernbach et al., Phys. Rev. 75, 1352(1949).

EXCITATION OF THE CARBON-14 GROUND STATE BY THE CHARGED PION PHOTOPRODUCTION AT THRESHOLD IN NITROGEN-14

A. Figureau
Institut de Physique Nucléaire (et IN2P3) - Université Lyon-1
43, Bd du 11 Novembre 1918, 69621 Villeurbanne (France)

N. C. Mukhopadhyay
S. I. N. Theory Group, CH-5234 Villigen (Switzerland)

The transition $^{14}C_{g.s.}$ ($J^\pi T = 0^+1$) $\xrightarrow{\beta^-}$ $^{14}N_{g.s.}$ (1^+0), besides having practical value, is theoretically quite interesting. Its anomalously long life-time (log ft ~ 9) is an unmistakable signature of the presence of tensor force [1] in the nuclear two-body interaction which causes the Gamow-Teller matrix element (GTM) to vanish in this case. The vanishing of the GTM has important consequences for other weak [2] and electromagnetic [3-5] interactions. Thus the inverse muon capture cross-section is small and dominated by the pseudoscalar form factor [2]. In the Kroll-Ruderman (KR) limit, cross sections for the 1s radiative pion capture and charged pion photoproduction at threshold (CPP) should be quite small, as also suggested experimentally in the former case [4].

CPP has the advantage, over muon and radiative pion capture in exciting selectively the ^{14}C ground state. Thus it should be a good probe to study the corrections to the KR limit, and pionic distorsions near threshold. Currently, the following experimental quantity, defined in the usual way, is available [5]:

$$a_{^{14}N \to ^{14}C_{g.s.}} / a_p = (7.0 \pm 2.3) \, 10^{-4} \qquad (1)$$

With this in view, we examine here the CPP and its sensitivity to the nuclear and pionic wave functions in the impulse approximation. We use, in this calculation, a general form [6] of the effective Hamiltonian which includes s, p and d pion partial waves and satisfies the requirements of Galilean and gauge invariance principles. Among the many nuclear wave functions that we consider are realistic ones of Cohen and Kurath [1], effective wave functions [3] obtained by fitting the M1 form factor from the reaction $^{14}N(e, e')^{14}N^*$ (2.31 MeV), and the unrealistic wave functions obtained in the LS and jj limits. Harmonic oscillator radial functions are used with oscillator parameters deduced from the charge radius of ^{14}N and the inelastic electron scattering experiments

For pion distorsion, we include the Coulomb and strong interaction effects, the latter being parametrized in the Krell-Ericson form.

Our calculations indicate that the CCP reaction populating $^{14}C_{g.s.}$ depend on the nuclear configuration mixing, the nuclear radius and the p-wave pion-nucleus optical potential, roughly in this order of decreasing importance. Non realistic nuclear models are easily shown to be inadequate to interpret the experimental result (1), which could even be used for a rough selection among the others. Thus, for illustration, wave functions of ref. (3) predict a ratio (1) which is 1/11 of the corresponding Cohen-Kurath estimate. This conclusion remains unaffected by a sensible variation of the oscillator parameter. With large theoretical errors, mostly due to the uncertainty in the pion distortion, we obtain the value $2.3 \cdot 10^{-3}$ for the ratio (1), in the Cohen-Kurath model with b = 1.68 Fm. This is much smaller than the value given in the same conditions by the KR limit of the effective Hamiltonian.

REFERENCES

1. H. J. Rose, O. Haüsser and E. K. Warburton, Rev. Mod. Phys. 40, 591, (1968)

2. N.C. Mukhopadhyay, Phys. Lett., 44B, 33, (1973)

3. N. Ensslin, W. Bertozzi, S. Kowalski, C. P. Sargent, W. Turchinetz, C. F. Williamson, S. P. Fivozinsky, J. W. Lightbody Jr. and S. Penner, Phys. Rev., C-9, 1705 (1974)

4. H. W. Baer, J. A. Bistirlich, N. De Botton, S. Cooper, K. W. Crowe, P. Truöl and J. D. Vergados, Phys. Rev., C-12, 921, (1975)

5. N. De Botton, in "Effets mésoniques dans les noyaux", Saclay (1975)

    J. P. Deutsch, Private communication

6. J. Delorme, A. Figureau, and M. Krell, to be published.

PHOTO- AND ELECTROPRODUCTION OF PIONS
AND KAONS NEAR THRESHOLD FROM NUCLEAR TARGETS[†]

J. B. Cammarata[*] and T. W. Donnelly[**]
Institute of Theoretical Physics, Department of Physics
Stanford University, Stanford, California 94305

ABSTRACT

Theoretical results for electroproduction of charged pions from light nuclei near threshold are presented as extensions of recent work on pion photoproduction. Possible effects of induced tensor second-class currents are examined. The production of kaons is discussed and compared with the pion problem.

DISCUSSION

In recent work[1] the photoproduction of charged pions from nuclei near threshold was examined in a general framework, though with specific application to the reaction ^{6}Li$(\gamma,\pi^+)^6$He. By concentrating on the threshold region (where the kinetic energy of the pion is only a few MeV) one may be assured that only s-wave pions contribute. This greatly simplifies the analysis and interpretation of experimental results. The same is true for the electroproduction reaction $^A_Z X(e,e'\pi^{\pm})^A_{Z\mp 1}X$ near threshold. Of course photoproduction is just a special case of electroproduction; namely, where the photon is on its mass shell. However, then the polarization vector for the electromagnetic field has only transverse components and only the transverse axial-vector multipoles are allowed: $\hat{T}_J^{el5}$ and $\hat{T}_J^{mag5}$, in our notation[2,3]. Thus in photoproduction in lowest order there are no monopole transitions. In contrast to this, in electroproduction the exchanged photon has both transverse and longitudinal (or scalar) polarization and so the time-like and longitudinal multipole projections of the axial-vector current, $\hat{M}_J^5$ and $\hat{L}_J^5$ (see Refs. 2, 3) are present as well as the two transverse multipoles. From previous work[4] on semi-leptonic weak interactions in nuclei and, specifically, from investigations of neutrino and anti-neutrino reactions we know that the $\hat{M}_J^5$ multipole is the only one which contains the possible effects of induced-tensor second class currents. Generally the transverse multipoles dominate over such effects; however, for monopole transitions $0^+ \rightarrow 0^-$ there can be no transverse contributions and the sensitivity to second class currents is greatly enhanced.

[†]Research supported by the National Science Foundation.
[*]Address after 1 September 1976: Department of Physics, Virginia Polytechnic Institute, Blacksburg, Virginia 24061
[**]Alfred P. Sloan Foundation Fellow.

In this contribution we present results for charged pion electroproduction from light nuclei (^{3}He, ^{6}Li, ^{11}B, ^{12}C and ^{16}O) leading to specific final nuclear states. We examine in detail the possibility of measuring experimentally the nuclear axial form factor over a wide range of momentum transfer and the effects of second class currents. Finally we discuss the extension of these ideas to kaon photo- and electroproduction from nuclei where the final hadronic state contains a kaon and a hypernucleus.

REFERENCES

1. J. B. Cammarata and T. W. Donnelly, Stanford Preprint ITP-524.
2. J. S. O'Connell, T. W. Donnelly, and J. D. Walecka, Phys. Rev. C6, 719 (1972).
3. J. D. Walecka, in Muon Physics, ed. V. W. Hughes and C. S. Wu (Academic Press, New York, 1972) p. 113.
4. T. W. Donnelly and J. D. Walecka, Phys. Lett. 41B, 275 (1972).

THE CHARGED PION PHOTOPRODUCTION ON ^{12}C NEAR THE 3 3 RESONANCE REGION

S. FURUI

Institute of Physics, University of Tokyo, Komaba, Tokyo 153, Japan

We performed an exploratory analysis of the charged pion photoproduction on ^{12}C for the incident photon momentum k_γ = 319 MeV/c $\sim$569 MeV/c. The spectrum of the emerging pion was obtained at angles in the laboratory system 28.4° and 44.2°.[1]

The photoproduction can be regarded as incoherent, and it may be allowed to separate the production step and the final state interaction. We write the number of pions emerging from the nucleus as the sum of the pions produced from each circular region whose impact parameter b and the z-coordinate is specified. In analogy with the Glauber's multiple scattering theory, the differential cross section can be written in the form[2]

$$d\sigma/d\Omega(\gamma \to \pi^i(\vec{q}))\{n^{(0)}(\vec{b},z) + n^{(1)}(\vec{b},z)\tilde{\lambda} + n^{(2)}(\vec{b},z)\tilde{\lambda}^2 + \cdots\}, \quad (1)$$

where

$$n^{(j)}(\vec{b},z) = [e^{-\sigma_{\pi \to \pi}T(\vec{b},z)}(\sigma_{\pi \to \pi}T(\vec{b},z))^j]/j!$$

is the effective number of scatterers. Assuming that the pion propagates along the straight path, we defined the effective forward pion-nucleon (π-N) scattering cross section in the nucleus by $\tilde{\lambda}\sigma_{\pi \to \pi}$. T(b,z) is the optical thickness along the path of the produced pion. Eq.(1) reduces to

$$d\sigma/d\Omega(\gamma \to \pi^i(\vec{q})) \ e^{-(1-\tilde{\lambda})\sigma_{\pi \to \pi}T(\vec{b},z)}$$

which is equivalent to the result derived by Adler et al.[3] and Sternheim and Silber[4] for the case that the backward π-N scattering is ignored.

The cross section $\sigma_{\pi \to \pi}$ includes the effective absorption and the charge exchange contribution.[3] Since the actual path of the pion is not a straight line, we replaced $\lambda\sigma_{\pi \to \pi}$ by the average of the forward scattering ($\theta < 90°$) on a nucleon with the Fermi motion.

Taking into account the π-N charge exchange, the differential cross section can be written in the form[3]

$$d\sigma/d\Omega(\pi^i) = \sum_{j=1,3} M_{ij} \ d\sigma/d\Omega(\gamma \to \pi^j)$$

where M_{ij} is a 3×3 matrix. Assuming the dominance of the 3 3 resonance in the π-N scattering, one can write M_{ij} by the linear combination of the integrals with the nuclear density $\rho(\vec{b},z)$,

$$\int d^2b \ dz \ d\phi \ e^{-(1-\tilde{\lambda})\sigma_{\pi \to \pi}T(\vec{b},z)} \rho(\vec{b},z) \ .$$

In the case that the pion propagates along the path which intersects with the incident beam direction by an angle θ, the optical thickness $T(\vec{b},z)$ becomes ϕ dependent. We replace the average over ϕ by the mean values for the smallest $T(\vec{b},z,\phi)$ and for the largest

$T(\vec{b}, z, \phi)$.

The pion photoproduction cross section on a nucleon in the nucleus $d\sigma/d\Omega(\gamma \to \pi^j)$ was calculated with the Fermi gas model. We approximated the elementary production cross section by the resonance contribution on N_{33}^*, N_{11}^* and N_{13}^*, parameterized by Walker[5] and the π-exchange Born contribution. Pauli principle and the Fermi motion were taken into account following Benz et al.[6]

The effective pion absorption cross section was parameterized as the ref.4). It has the maximum of about 30 mb near the pion kinetic energy $T_\pi \simeq 205$ MeV.

The final result is shown in Fig.1 and Fig.2 for the production angle $\theta = 28.4°$ and $44.2°$ respectively. At $k_\gamma = 319$ MeV/c which corresponds to the N_{33}^* production, agreement is good. $k_\gamma = 469$ MeV corresponds to the threshold of $\gamma+N \to \pi+N_{33}^*$ channel, N_{33}^* being regarded as a particle of mass 1.236 GeV. The part of the cross section suppressed by the Pauli principle at $k_\gamma = 469$ MeV/c is about 17μb/sr and 5μb/sr for $\theta = 28.4°$ and $44.2°$ respectively. Detailed comparison with the experiment must, however, await further study.

The cross section is sensitive to the effective pion absorption cross section. The parameterization employed by Adler et al.[3] gives the cross section near $k_\gamma = 319$ MeV/c too large.

The charge-exchange probability is of the order of several %. It is smaller than the model of Adler et al.[3] This is due to the fact that the backward pion-nucleon scattering is ignored. Since on shell scattering is assumed, it is small in our model.

I would like to thank Drs. H. Fujii and S. Noguchi for supplying me the new data before publication.

REFERENCES
1) I. Endo et al., Phys. Lett. 47B 469 (1973), H. Fujii and S. Noguchi (Private communication).
2) R.J.Glauber, in High Energy Physics and Nuclear Structure (Plenum, N.Y. 1970) p.207.
3) S.L. Adler et al., Phys. Rev. D9 2125 (1974).
4) R.R.Sternheim and M.M.Silber, Phys. Rev. C10 2215 (1974).
5) R.L.Walker, Phys. Rev. 182 1729 (1969).
6) P. Benz et al., Nucl. Phys. B65 158 (1973).

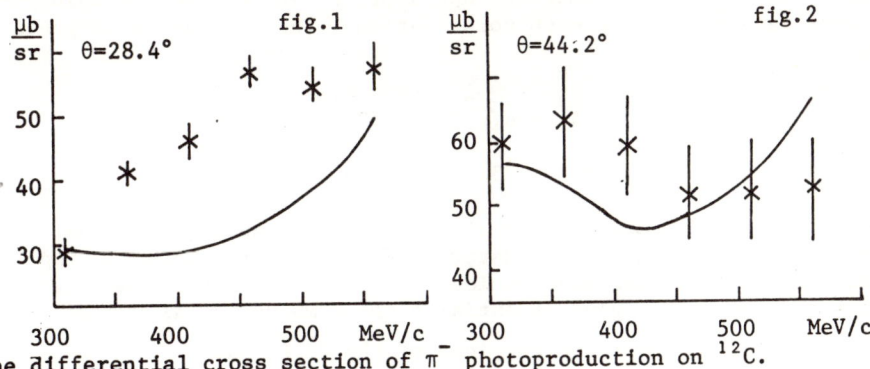

The differential cross section of π^- photoproduction on ^{12}C.

RADIATIVE PION CAPTURE IN ^{3}He*

W. R. Gibbs, B. F. Gibson, and G. J. Stephenson, Jr.
Theoretical Division, Los Alamos Scientific Laboratory
Los Alamos, New Mexico 87545

ABSTRACT

We have calculated the rate for radiative pion capture from a 1s atomic orbital in the $\pi^- {}^3\text{He} \to \gamma {}^3\text{H}$ analog transition. The nuclear model result utilizing the impulse approximation agrees essentially with the particle model result. Combining this rate with our estimate of the $\pi^- {}^3\text{He} \to \pi^0 {}^3\text{H}$ rate, we obtain a Panofsky ratio $P_3 < 2.5$.

INTRODUCTION

The Panofsky ratio

$$P_3 = \frac{\tau_x^{-1}(\pi^- {}^3\text{He} \to \pi^0 {}^3\text{H})}{\tau^{-1}(\pi^- {}^3\text{He} \to \gamma {}^3\text{H})} \qquad (1)$$

has been measured by two different groups[1,2] to be either $2.68 \pm .13$ or $2.28 \pm .18$. We have recalculated the radiative pion capture rate and estimated a value of P_3 which lies between the two experimental numbers.

THE TRANSITION OPERATOR

We make use of the non-relativistic capture operator[3]

$$O_i^- = \bar{A} \frac{2\pi i}{\sqrt{km_\pi}} \tau_i^- \vec{\sigma}_i \cdot \vec{\epsilon}[1 + \frac{1}{2M} (\vec{P}_p - \vec{P}_n) \cdot \hat{k}] \qquad (2)$$

where m_π and M are the charged pion and nucleon masses, k is the photon energy, and $\vec{P}_p$ and $\vec{P}_n$ are the proton and neutron momenta. The τ^- is the isospin lowering operator, $\vec{\sigma}_i$ is the nucleon spin operator, and $\vec{\epsilon}$ is the photon polarization. The constant $\bar{A}$ is given by

$$\bar{A}^2 = \alpha(g^2/4\pi)(1/2M^2)$$

$$= 23.0 \times 10^{-4} \text{fm}^2 ,$$

where $\alpha = 1/137.0$ and $g^2/4\pi = 14.28$. For the capture reaction, the correction to the static operator is $(1 + k/2M)$ for both the impulse approximation nuclear model calculation and the particle model.[4] Thus the two models give essentially the same numerical result.

*Work performed under the auspices of the U. S. ERDA.

NUMERICAL RESULTS

For capture from the 1s atomic orbital (point Coulomb wave function), one can easily show that

$$\tau^{-1} = 2\pi c (2\pi)^3 \int \delta^4 (\Sigma P_i^{(4)}) |\overline{M}|^2 \frac{d^3p}{8\pi^3} \frac{d^3k}{8\pi^3} \quad (3)$$

$$= (8c\overline{A}^2)/(a_3^B)^3 \frac{k}{m_\pi} \left|\frac{1 + k/2M}{1 + k/3M}\right|^2 \overline{F}^2 \quad (4)$$

where $\overline{M}$ contains the expectation of the operator as well as the pion wave function, c is the speed of light, $a_3^B = 101.7$ fm is the Bohr radius, $k \approx m_\pi(1 - m_\pi/6M)$, and $\overline{F}^2$ is a factor accounting for the π-N initial state rescattering and the Fourier transform of the overlap of the initial and final nuclear states. As noted in Ref. 1, the square of the overlap determined from β-decay is about 0.95. The magnetic spin-flip electrom scattering form factor at the appropriate momentum transfer is approximately 0.78. Initial state π-N scattering reduces the rate by about 5%. Thus $\overline{F}^2 \approx 0.55$, and

$$\tau^{-1} \approx 3.1 \times 10^{15} \text{ sec}^{-1}$$

The corresponding charge exchange rate is

$$\tau_x^{-1} = \frac{8}{9} \frac{c}{(a_3^B)^3} \frac{q_f}{m_\pi} (1 + \frac{m_\pi}{3M}) |a_1 - a_3|^2 \overline{F}^2 \quad (5)$$

where $|a_1 - a_3| = 0.262\ m_\pi^{-1}$, $q_f = 32$ MeV/c, and $\overline{F}^2 \approx 0.92$, since the momentum transfer is much smaller. Thus, for this rate we estimate

$$\tau_x^{-1} = 7.7 \times 10^{15} \text{ sec}^{-1} .$$

The resulting Panofsky ratio is $P_3 = 2.48$, with an uncertainty of at least 0.1 from the uncertainty in the nuclear structure extimates for the wave function factors. In addition, one might expect that the N* anomalous moment effects,[5] which appear to be sizeable in the π⁻p → γn rate and necessary[3] in order to obtain agreement with the measured Panofsky ratio P_1, might increase the ^{3}He radiative capture rate by as much as 7%. Thus, our estimate of the ^{3}He Panofsky ratio might be reduced to as low as $P_3 = 2.32$, in disagreement with the latest measurement.[1]

REFERENCES

1. P. Troul, et al., Phys. Rev. Lett. **12**, 1268 (1974).
2. G. A. Zaimidoroga, et al., JETP **21**, 848 (1965) and **24**, 621 (1969).
3. W. R. Gibbs, B. F. Gibson, and G. J. Stephenson, Jr. (to be pub)
4. M. Ericson and A. Figureau, Nucl. Phys. **B3**, 609 (1967).
5. R. D. Peccei, Phys. Rev. **181**, 1902 (1969).

The SIN-Pairspectrometer and Radiative Pion Capture in ^{12}C
J.C.Alder, B.Gabioud, F.Hoop, C.Joseph, J.F.Loude, H.A.Medicus, N.Morel, A.Perrenoud, J.P.Perroud, D.Renker, H.Schmitt, G.Strassner, M.T.Tran, P.Truöl, B. Vaucher, H.v.Fellenberg, E.Winkelmann, Č.Zupančič
Universities of Lausanne, München and Zurich

A 180° pairspectrometer for the detection of medium energy photons emitted in a variety of pion and muon-induced radiative processes has been assembled at SIN, aiming at a resolution of 1 MeV (at 130 MeV) and an efficiency of .5% with large solid angle. This instrument includes a window frame magnet with a usable field volume of 250x65x52 cm^3 and a maximum field of 11 kG. At the nominal strength of 8 kG the field was mapped with an accuracy of .3%. Photons are converted in a 90 μm gold foil into electron-positron pairs, which are detected by multiwire proportional chambers. Each chamber has a useful aperture of 213x48 cm^2 and is constructed with three anode and four cathode planes. With these detectors the acceptance (including conversion efficiency) is .35%, the solid angle is 1.4%. For the anodes, gold plated tungsten wires (20 μm Ø) are used which are spaced 2mm and are oriented under ±60° and 90° with respect to the magnet center plane. The cathode planes consist of Cu-Be wires (50 μm Ø), spaced 1mm, stretched horizontally and supported every 50 cm by nylon wires. The gap size is 8 mm. The chambers are sealed with 15 μm mylar foils against the air and the helium bag inside the magnet. Specific electronic amplifiers and a read-out system have been developed. With the 900 μV sensitivity of the amplifiers the chamber plateau extends from 4.6 to 5.5 kV, using the "magic gas" mixture. The read-out is performed by a PDP 11-40 through controllers, one for each plane of wires, which encode the address of the first wire and the size of every cluster of hit wires into 16 bit words [1]. The use of three planes of wires with different orientations facilitates the recognition of crossing trajectories and reduces ambiguities. The topology of good events is recognized on-line. High-energy photon triggers are separated from background events, are then reconstructed and a fast preliminary determination of energy is made. This energy measurement is based on an interpolation of tabulated energies in function of entrance angles and distances between entrance and exit point. The on-line resolution obtained this way is 2.5 MeV. In the off-line analysis, the chamber information is transformed into a system of independent coordinates, where the corrections to the on-line guess of the photon parameters (energy, conversion point and entrance angle) are expressed by power series. The coefficients of these expansions are determined by a fit to Monte Carlo events. With this method a final resolution of 1.0 MeV (FWHM) at 130 MeV is achieved, as is apparent from our first experimental results for radiative capture of pions in a CH$_2$-target, displayed in Fig. 1. The spectrum shows the $\pi^-p \to n\gamma$-line at 129.4 MeV,

the transitions to ^{12}B bound states between 121 and 126 MeV, and
strong spin-isospin states in ^{12}B superimposed on the quasifree
background, as observed previously by the Berkeley group [2]. At
lower energies, the $\pi^-p\to\pi^0n$-continuum is also visible. With our
resolution we are able for the first time to separate the contributions to the 4 lowest states of ^{12}B at 0, .95, 1.67 and 2.62 MeV.
The branching ratios, normalized relative to the 129.4 MeV line
based on (1.17+.06)% pions captured by the protons in CH_2 [2]), are
given in the table. With the exception of the 2$^-$ state, the agreement with the theoretical predictions of Maguire and Werntz [3]) is
good. For the giant resonance region we also confirm the Berkeley
results which showed that the transition strength is concentrated
into two levels, as was predicted for collective 2$^-$ and 1$^-$
states [4-6].

1) J.F.Loude et al., ISPRA Nucl. Electronics Symp. (1975)
2) J.A.Bistirlich et al., Phys. Rev. C5, 1867(1972)
3) C.F.Maguire, C.Werntz, Nucl. Phys. A205, 211(1973)
4) H.Ohtsubo et al., Nucl. Phys., A224, 164 (1974)
5) S.Skupsky, Nucl. Phys. A 178, 289 (1971)
6) F.J.Kelly, H.Ueberall, Nucl.Phys. A 118, 302 (1968)

E_γ	$E^*(^{12}B)$	$E^*(^{12}C)$	J^π,Γ	Branching ratios (x 10^{-5})		
(MeV)				This work	Berkeley	Theory
125.0	0.	15.1	$1^+,-$	84 ± 6 ⎫		71 ± 10 [3]) 76 [5]) /74 [4])
124.1	0.95	16.0	$2^+,-$	22 ± 3 ⎬ 91 ± 9		29 ± 4 37/40
123.3	1.67	16.8	$2^-,-$	3 ± 2 ⎭		22 ± 3 96
122.4	2.62	17.7	$1^-,-$	19 ± 3		11 ± 4 37
120.6	4.4	19.5	?,-	188 ±12	185 ±19	
117.6	7.4	22.5	?,0.9	175 ±19	159 ±16	
Pole				1620 ±90	1490 ±150	
Total				2120 ±120	1925 ±200	

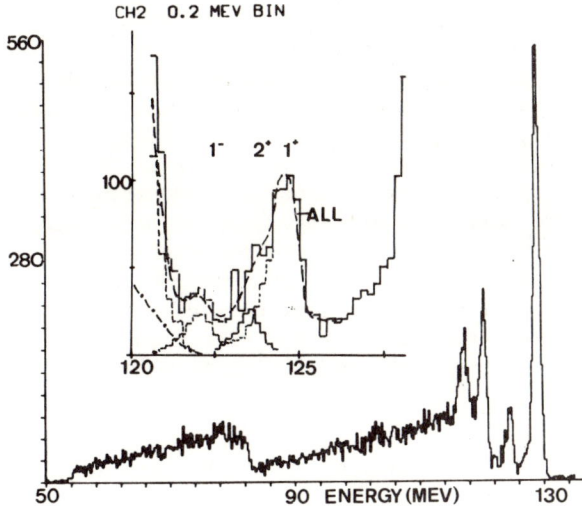

Figure 1

Photon spectrum from radiative pion capture in polyethylene. Insert: Section around 125 MeV enlarged with contributions of the three ^{12}B-bound states.

Investigation of 1p-Shell Nuclei ^{6}Li, ^{7}Li and ^{9}Be with Radiative Pion Capture *

J.C.Alder, B. Gabioud, C. Joseph, J.F.Loude, H.A.Medicus, N.Morel, A.Perrenoud, J.P.Perroud, D. Renker, H.Schmitt, G.Strassner, M.T.Tran, P.Truöl, H.v.Fellenberg, E. Winkelmann

Universities of Lausanne, München and Zürich

Previous studies of radiative capture in 1p-shell nuclei, namely the self-conjugate nuclei ^{6}Li, ^{10}B and ^{14}N [1,2], have demonstrated that this reaction is a useful tool to investigate the $\Delta T_Z=+1$ analogs of giant M1 states observed in 180° inelastic electron scattering. These states contribute a large fraction to the total observed radiative capture branching ratio. This observation was explained [2,3] by the fact that the transition operator is of Gamow-Teller type, even if most of the pions get absorbed from a 2p-state. A similar pattern is predicted for μ-capture [4], where the same few low-lying states should absorb more than 90% of the total allowed strength. The calculations [4] for μ-capture treat also other p-shell nuclei as ^{13}C and ^{9}Be, thus it is obvious to study the (π^-,γ)-reaction for these nuclei, too. We have measured the photon spectrum for ^{9}Be, ^{6}Li and ^{7}Li, detecting the high-energy photons in the pairspectrometer at SIN. ^{6}Li and ^{7}Li in addition to ^{9}Be were chosen, because for ^{6}Li previous measurements [1] were not able to clearly separate the 2^+ excited state of ^{6}He at 1.8 MeV from the ground state in the spectrum, and for ^{7}Li very little is known about the final state nucleus ^{7}He. [6] Our photon spectra are shown in Figures 1, 2 and 3. The general features of the ^{6}Li-spectrum are identical with the Berkeley result, though now the two ^{6}He states (133.9 and 132.1 MeV) are nicely separated with our resolution of 1 MeV (FWHM) at 130 MeV. Their relative contributions are $2^+(1.8)/0^+(0) = .50 \pm .05$. The nature of the broad peaks at 120 and 110 MeV is not clear at present, since they cannot be associated with known level structures. Similar peaks appear in the ^{7}Li spectrum around 118 and 108 MeV. The ^{7}He gound state, which is unbound by 440 keV with respect to the ^{6}He+n channel, is observed at $E_\gamma=126.6$ MeV. This state is the analog of the $J^\pi=3/2^-$, $T=3/2$ state at 11.25 MeV in ^{7}Li, which is excited by M1-transitions. In ^{9}Be, lines to the known ^{9}Li-levels $E^*=0$ MeV, $J^\pi=3/2^-$, $T=3/2$, $E_\gamma=124.5$ MeV and (2.69, 1/2$^-$, 3/2, 121.8 MeV) are observed, both analogs of M1 levels in ^{9}Be. The first one collects 39% of the allowed strength in μ-capture. About 54% are supposed to go to levels with (4.3, 5/2$^-$, 3/2, 120.2) and (5.4, 3/2$^-$, 3/2, 119.1) [4]. These levels are observed in the ^{7}Li(t,p)^{9}Li reaction. [5] We notice, however, that most of the strength besides the always present quasifree contribution is going to one broad or possibly two narrower resonances with $E_\gamma=116.5$ and 117.6 MeV ($E^*=8.0$ and 6.9 MeV) about which no previous information is available.

* work supported by Fonds National Suisse and Deutsche Forschungsgemeinschaft and SIN.

1) H.W.Baer et al. Phys. Rev. C8, 2029(1973)
2) H.W.Baer et al. Phys. Rev. C12, 921(1975)
3) J.D. Vergados, Nucl.Phys., A239 271 (1975)
4) N.C. Mukhopadhyay, Phys.Letters 45B, 309 (1973)
5) P.G.Young, R.H. Stokes, Phys. Rev. C4, 1597(1971)
6) F.Ajzenberg-Selove, T. Lauritsen, Nucl.Phys. A227, 1(1974)

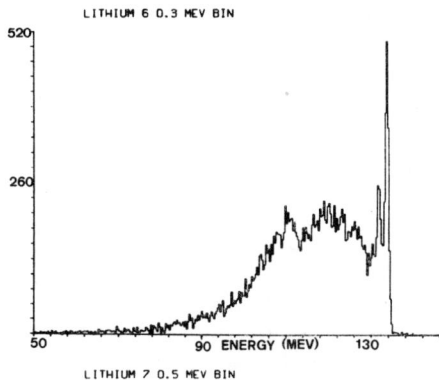

Figure 1:

Photon spectrum from ^{6}Li(π^-,γ)

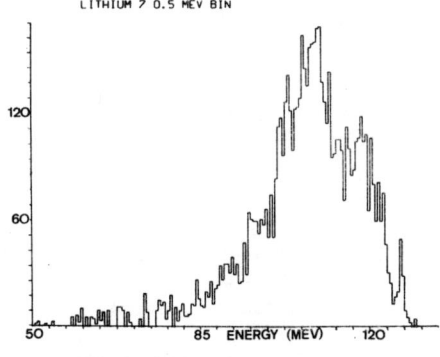

Figure 2:

^{7}Li(π^-,γ)

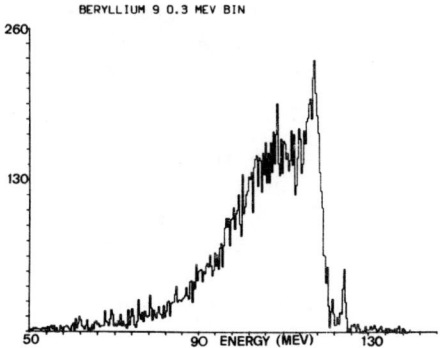

Figure 3:

^{9}Be(π^-,γ)

Radiative Pion Capture in Oxygen-Isotopes

J.C.Alder, B.Gabioud, C.Joseph, J.F.Loude, H.A.Medicus, N.Morel,
J.P.Perroud, A.Perrenoud, D.Renker, H.Schmitt, G.Strassner,
M.T.Tran, P.Truöl, H.v.Fellenberg, E.Winkelmann
Universities of Lausanne, München and Zurich *

Among the first nuclei suggested as possible targets for a radiative pion capture experiment was ^{16}O. Murphy et al.[1] predicted a strong probability to excite spin-isospin states in this reaction, preferentially the $T_Z=+1$ analogues of the $J^\pi=2^-$, 1^- dipole excitations, but a large fraction of the transition strength was expected for quadrupole excitations, too. Later calculations [2,3] have confirmed this behaviour in greater detail. The Berkeley group [4] has observed the photon spectrum and confirmed the dipole transitions experimentally. We have remeasured the photon spectrum with the high-resolution pair-spectrometer at SIN, to overcome some of the difficulties of the first experiment, such as low statistics and large inflight subtraction due to the thick target. Furthermore, our better resolution, 1.0 MeV (FWHM) at 130 MeV, would have allowed to observe possible fine structure not apparent in the first experiment. In addition to ^{16}O, we also stopped pions in ^{18}O, where we reach final states in ^{18}N, a $T_Z=+2$ nucleus, of which so far only the ground state was known. Both targets were available in the form of D_2O. We present in figures 1 and 2 the preliminary results for the photon spectra. The contribution from the plastic container of the target has been subtracted, but the deuterium contribution (less than 5% of total events for ^{16}O) is still included. This spectrum is known to have a peak near 130 MeV and then falls off smoothly to lower energies [5]. In the ^{16}O-spectrum we observe the transition to the 2^- ^{16}N-ground state (at $E_\gamma=128.0$ MeV) and to the 0^-, 1^-, and 3^- states which are within 400 keV above the ground state. However, the 2^- state is expected [6] to contribute 86% of the total rate into this region. The relative branching ratio which we observe is about 7%, in agreement with the Berkeley result, which gave .15% for this transition and 2.2% for the total rate. The second prominent peak at 120.4 MeV (Excitation energy in ^{16}N 7.6 MeV, in ^{16}O 21.0 MeV) can be associated with the 2^- state at this energy predicted [3] to have 34% of the total dipole transition strength. However very little is visible of other transitions, e.g. to 1^- at $E_\gamma=115$ MeV expected to contain 20% of the dipole strength and 2^- at 116.5 with 17%, and the quadrupole transitions to positive parity states with energies between 90 and 110 MeV.

For ^{18}O we find a spectrum which looks quite similar to that of ^{16}O. Clearly distinguishable is the transition to the ^{18}N ground state ($J^\pi=0^-$, 1^- or 2^-), with about 4% of the total rate, which we expect at 124.5 MeV. This state so far has only been seen in n-p and t-^{3}He-reactions. 7 MeV lower, we observe a further peak in the spectrum. The energy would correspond to an excitation energy of 23.5 MeV in ^{18}O. At this energy a prominent peak is observed in the recently measured photoproton and photoneutron cross-sections on ^{18}O.[7] We seem to observe the $T_Z=+2$ analog of this state here. A very preliminary analysis indicates, that the total radiative capture rate for ^{18}O is lower by 30% than the one in ^{16}O.

1) J.D. Murphy et al., Phys. Rev. Lett. 19, 714(1967)
2) H. Ohtsubo, T. Nishiyama, M. Kawaguchi, Nucl. Phys. A224(1974) 164
3) J.D.Vergados, Phys. Rev. C12, 1278(1975)
4) J.A.Bistirlich et al., Phys. Rev. C5, 1867(1972)
5) J.W.Ryan, Phys. Rev. 130, 1554(1963)
6) P.Szydlik, C, Werntz, Phys. Letters 41B, 209(1974)
7) B.L.Berman et al., Bull. Am. Phys. Soc. 21, 68(1976), priv. comm.

* work supported by SIN Fonds National Suisse and Deutsche Forschungsgemeinschaft.

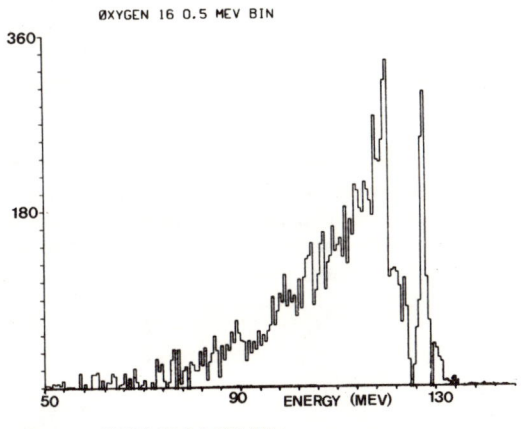

Figure 1:

Photon spectrum from ^{16}O(π^-,γ) (0.5 MeV bins)

Figure 2:

Photon spectrum from ^{18}O(π^-,γ) (0.5 MeV bins)

SEARCH FOR 2γ EMISSION IN π CAPTURE : A PROGRESS REPORT

J. Deutsch, D. Favart, P. Lipnik, P. Macq and R. Prieels
Institut de Physique Corpusculaire, University of Louvain
B - 1348 Louvain-la-Neuve, Belgium

INTRODUCTION

It has been pointed out by Ericson and Wilkin [1] that annihilation reaction of π^- virtual pion in nuclei $\pi^- + \pi^+ \to 2\gamma$ could lead to observable rates of 2γ emission in π-capture. Subsequent work by Bernabeu and Rho [2] has shown that second order radiative effects in the "normal" 1γ radiative π-capture could also lead to emission of two photons. However, this contribution was found to be negligible in light elements.

In an attempt to measure the 2γ/1γ branching ratio R in pion capture we started an experiment at the CERN Synchrocyclotron. Progress made so far allows us to present an upper limit on this ratio as measured in ^{12}C.

EXPERIMENTAL SET-UP

We have used a 150 MeV/c pion beam derived from the extracted proton beam incident on an external target. In doing so we were able to take advantage of the new performances of the CERN synchrocyclotron after completion of its improvement program. Details on the beam characteristics can be found elsewhere [3]. For beam monitoring, we used a conventional plastic detector telescope set-up. The beam was degraded in a carbon absorber, collimated in (6 x 6) cm^2 and stopped in a (10 x 10 x 3) cm^3 graphite target. High energy gamma rays were detected by a set of four lead-glass detectors and observed in coincidence with incoming beam-particle. The lead-glass detectors covered about 145° in the plane normal to the beam direction, as shown in fig. 1. Plastic scintillators located in front of the lead-glass detectors allowed us to reject the charged particles from the target. Rejection of coincidences between two adjacent γ-detectors which may originate from a single γ-ray, was achieved, partially at least, by lead shielding and anticounters located between the lead-glass detectors. However it appears, from the preliminary data obtained, that some spurious coincidences still remained.

RESULTS

The results presented here were obtained in an exploratory run with a total of about 2.5 10^8 π^--stop. The branching ratio R is deduced from the observed coincidence rate assuming a spatial correlation [4] between the two gamma-rays given by $d\sigma \simeq (1 + \cos \theta) d\cos \theta$ where θ is the angle between the two γ-rays.

A Monte Carlo program was used to estimate our detector efficiency in coincidence, as a function of θ, for the various pairs of detectors. Our results are summarized in table 1; we have indicated, for each pair of detectors, the angular region covered by the pair and its acceptance as estimated by the Monte-Carlo program.

CONCLUSION

At the present stage of the experiment, one cannot ascertain that the coincidences observed are due to true 2γ events. On the contrary, the events observed in adjacent detectors are probably generated by single γ, and our rejection efficiency for this type of coincidences needs to be improved. At large angle we observed coincidence from π^0-decay γ-rays due to inflight charge exchange of the π^- in the target (charge exchange for π^- at rest is energy forbidden in ^{12}C). The rate of coincidences between counters 1 and 4 is of the order of $10^{-3} \gamma\gamma/\pi$-stop whereas in a previous work [5] performed in similar conditions the observed coincidence rate was about ten times smaller. However the lack of details on the previous experiment (beam energy spread, geometry) does not allow us to draw any conclusion on this point. Non-adjacent detectors pairs are certainly less contaminated by these effects but other sources of instrumental effects may exist and at the present time, the data derived from these pairs of detectors can only be used to give an upper limit on the rate $(2\gamma/\pi\text{-stop}) \leq 3.10^{-5}$. Using the branching ratio $(1\gamma/\pi\text{-stop}) = (1.92 \pm 0.20)$ % measured in ^{12}C by Bistirlich et al [6], we deduce the following upper limit on R : $R \leq 1.5\ 10^{-3}$. Let us recall that the estimate for R is $R \simeq 10^{-4}$ [1].

TABLE 1

Detector pair	Angular region (degrees)	$\gamma\gamma$-rate observed per π^--stop	Acceptance assuming $(1+\cos\theta)$ correlation
(1-2) or (3-4)	20-65	5.10^{-8}	8.10^{-4}
(2-3)	20-45	6.10^{-8}	4.10^{-4}
(1-3) or (2-4)	55-100	1.10^{-8}	5.10^{-4}
(1-4)	95-145	8.10^{-7}	5.10^{-4}

REFERENCES

1. T.E.O. Ericson et C. Wilkin, Phys. Letters 57B (1975) 345
2. Bernabeu and M. Rho - private communication
3. Louvain Internal report, IPC-N-7602
4. Search for a new mode of π^- capture in nuclei, CERN PHIII-74/37
5. V.I. Petrukhin and Y.D. Prokoshkin, Nucl. Physics 54 (1964) 414
6. J.A. Bistirlich et al., Phys. Rev. C5 (1972) 1867

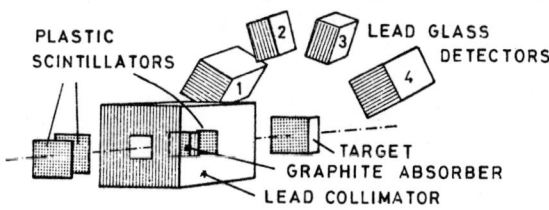

Fig. 1 : Sketch of the experimental set-up

THE TRIUMF π^o SPECTROMETER AND THE PANOFSKY RATIO IN HYDROGEN AND DEUTERIUM

M. Salomon, D. Berghofer, M.D. Hasinoff, R. MacDonald,
D.F. Measday, J. Spuller and T. Suzuki
University of British Columbia, Vancouver, B.C., Canada

J.K.P. Lee
McGill University, Montreal, Quebec, Canada

J.M. Poutissou, R. Poutissou and P. Depommier
University of Montreal, Quebec, Canada

ABSTRACT

We present the performance and preliminary results obtained with two large NaI crystals operated in singles and coincidence.

GENERAL FEATURES

Two large NaI crystals (18"ϕ × 20" long and 14"ϕ × 14" long) are used in singles and coincidence modes for the detection of high energy γ-rays. Their energy resolution at 130 MeV is 5% and 8% respectively. For a well collimated beam of monoenergetic electrons the energy resolution is 4% and 6.5%.

The timing resolution in both counters is 2.5 ns for γ-rays of energy larger than 20 MeV. This feature was used to separate neutrons and γ's by the time of flight method.

RESULTS

We have used this counters to detect γ-rays emitted when negative pions are stopped in several targets.

The Panofsky ratio in Hydrogen $P = \omega(\pi^-p \to \pi^o n)/\omega(\pi^-p \to \gamma n)$ was measured and a preliminary value of 1.56 ± .04 was obtained. We expect to reduce the error by a factor of 4 and the final value will be presented at the conference.

Similar measurements have been performed in deuterium, $^6Li, ^{12}C$, copper and cadmium targets. Singles and coincidence spectra where obtained and new limits for the charge exchange rate will be presented.

Experimental results are compatible with recent calculations by Hasinoff, Salomon and Reitan also to be presented at this conference.

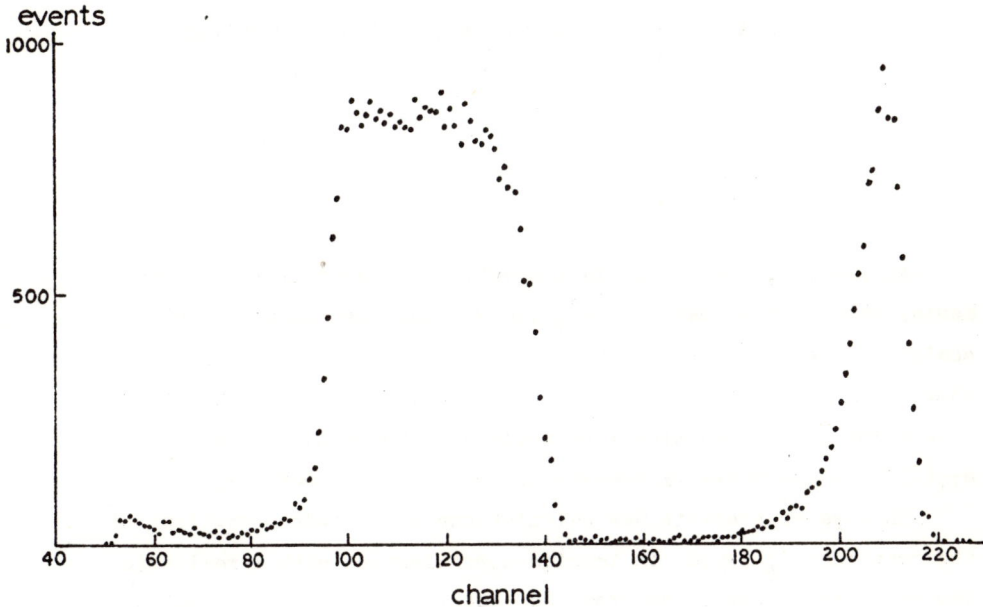

FIG. 1 γ-Ray Spectrum in the 18"φ × 20" long NaI Crystal for π⁻ Stopped in a LH$_2$ Target.

ELASTIC η- MESON PHOTOPRODUCTION FROM DEUTRONS

Yu.N. Krementzova, A.I. Lebedev

P.N. Lebedev Physical Institute, Moscow, USSR

At present, the available theoretical interpretation of experimental data on the reaction $d(\gamma,\eta)d$ suggests the essential isoscalar excitation of the nucleon resonance $S_{II}(1535)$ [1]. This conclusion contradicts the quark model as well as the analysis of the η - meson photoproduction from nucleons, the latter shows that $S_{II}(1535)$ is essentially excited by isovector photons [2].

This paper presents new calculations of the cross sections of the reaction $d(\gamma,\eta)d$ in impulse approximation with a realistic deutron form-factor taken from the description of the d-d scattering cross section with the admixture of the D-state $P_D = 6.5\%$ in the deutron wave function [3].

One can see from Fig.1 that if $S_{II}(1535)$ has an essential isovector excitation (for the ratio of the isoscalar nucleon amplitude to the proton amplitude, $T_o/T_p = 0.35$), than the cross section value is too small to explain the experimental data [1]. However, the cross section calculated for $T_o/T_p = 0.84$ (for an isoscalar character of the excitation) satisfactory agrees with the experiment.

It is worth noting that there exist remarkable differencies between the cross sections under small angles, and, especially, between the asymmetry coefficients , Fig.2, in the case of η - meson photoproduction by polarized photons for A2 and B2 variants of the nucleon amplitudes [4].

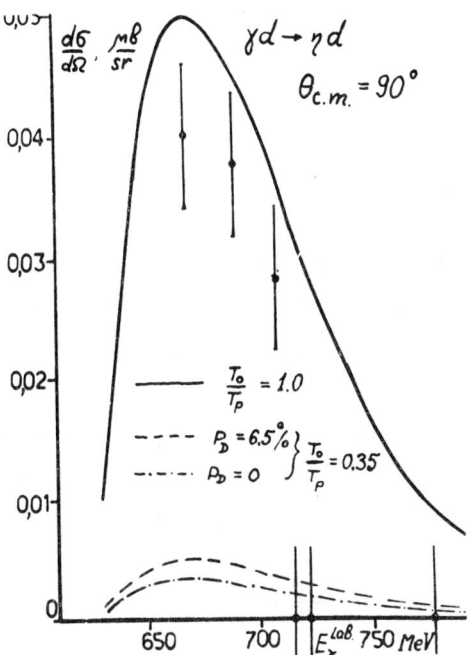

Fig.1

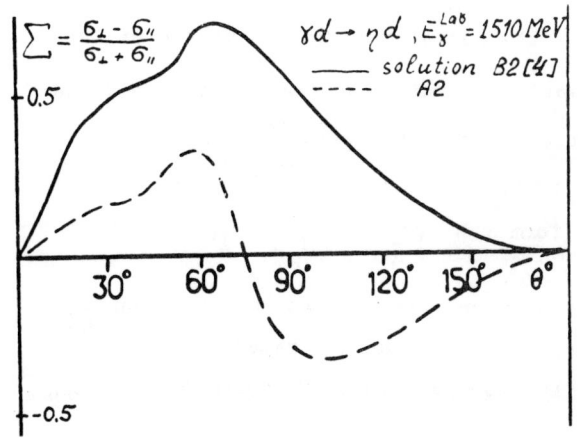

Fig.2

REFERENCES

1. R.L. Anderson, R. Prepost, Phys. Rev. Letters. 23, 46 (1(69).
2. R.L. Walker, Proc. 4th Int. Symp. on Electron and Photon Interactions, Liverpool, p.21 (1969).
3. F. Alberi et al., Nucl. Phys., B17, 621 (1970).
4. H.R. Hicks et al., Phys. Rev., D7, 2614 (1973).

PARTIAL η - MESON PHOTOPRODUCTION CROSS SECTION FROM Li^6

A.I. Lebedev, V.A. Trjasuchev, V.N. Fetisov

P.N. Lebedev Physical Institute, Moscow, USSR

Some information on the photoexcitation amplitudes of the nucleon resonances, important for the development of the hadron quark models [1], may be obtained from the experimental study of the η - meson photoproduction reactions, as we show it in this paper by considering the following reaction as an example :

$$\gamma + Li^6(J^P = 1^+, T = 0) \rightarrow Li^6(3.56 \text{ MeV}, J^P=0, T=1) + \eta \quad (1)$$

Calculations were performed in impulse approximation with nuclear wave functions of the oscillator shell models [2] as well as with the amplitudes of the process $p(\gamma,\eta)p'$ taken from the paper [3]. The change of the isospin of the nucleus Li^6 in the course of the reaction (1) causes an isovector character of the reaction amplitude. Required for calculations, isovector parts of the reduced widths, characterizing the photoexcitation of the nucleon resonances, can be written in the form : $\gamma^v = \frac{1}{2}\gamma^p (1 - \gamma^n/\gamma^p)$

The reduced widths of the resonance photoexcitation of proton, γ_p, and of neutron, γ_n, were defined following the paper [4].

In the Table we give the average values of the ratios γ_n/γ_p.

Resonances	$S_{11}(1535)$	$S_{11}(1700)$	$P_{11}(1470)$	$P_{11}(1750)$	$D_{13}(1520)$	$F_{15}(1688)$
$\lvert\gamma_n/\gamma_p\rvert$	0.33	1.0	0	1.0	0.68	0.062

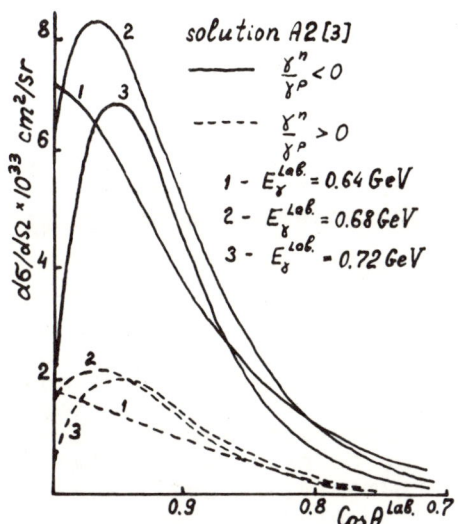

Fig.1

In Fig.1 we have plotted the -meson angular distributions in the region of the resonance $S_{II}(1535)$. The cross sections are given as the functions of the photon energy in Fig.2. One can see from these Figs.1,2 that the cross sections calculated either with the essential isoscalar amplitudes or with essetial isovector ones largely differ from one another. Hence, it should be expected that the experimental data on - meson photoproduction from Li^6 will give some valuable information on the nucleon resonance nature.

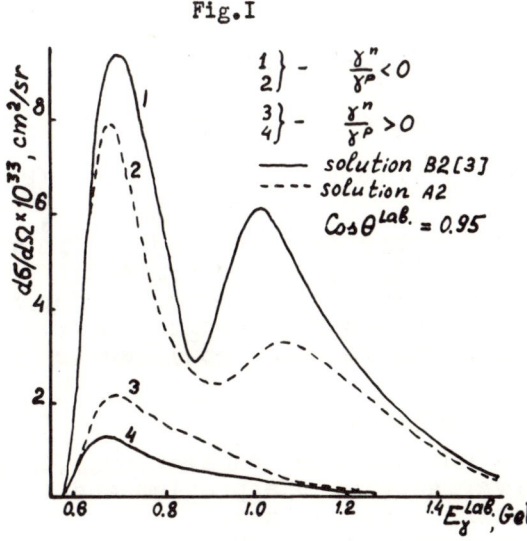

Fig.2

REFERENCES

1. R.L. Walker, Proc. 4th Int. Symp. on Electr. and Phot. Interactions, Liverpool, p.21 (1969).
2. T.W. Donnelly, J.D. Walecka, Phys. Lett. 44B, 330 (1973).
3. H.R. Hicks et al., Phys. Rev., D7, 2614 (1973).
4. Particle Data Group, Phys. Lett. 50B, 1 (1974).

INVARIANT IMPULSE APPROXIMATION AND PIONIC ATOMS

R. Mach[*)]
Joint Institute for Nuclear Research, Dubna, USSR

ABSTRACT

The impulse approximation is modified so that the Galilean invariance will not be violated. A method is suggested how to take the nuclear motion very accurately into account. As an example, characteristics of ^{4}He pionic atom are calculated.

IMPULSE APPROXIMATION

Interactions of pions with nuclei are usually studied in the frame of Watson's multiple scattering theory. An important constituent of this theory is the impulse approximation, which consists of $\tau(E) \sim t(\tilde{E})$. Here, t and τ are scattering matrixes describing the pion scattering by a free and a bound nucleon, respectively. The reaction energy is denoted as E. The choice of $\tilde{E}$ is usually made on the basis of intuitive arguments, e.g. $\tilde{E}=E$ [1] and $\tilde{E}=E-U$ [2], where U is the mean nucleon binding energy. A general drawback of the two choices is that they violate the Galilean invariance of the matrix elements $\langle n\vec{p}'|t(\tilde{E})|n\vec{p}\rangle$ which enter the Watson's series. We have

$$\langle n\vec{p}'|t(\tilde{E})|n\vec{p}\rangle = \int \langle n'|\vec{k}_1', \vec{k}_2, .., \vec{k}_A\rangle \langle \vec{p}\vec{k}_1'|t(\tilde{E})|\vec{p}\vec{k}_1\rangle \langle \vec{k}_1, .., \vec{k}_A|n\rangle d\tau$$

$$\langle \vec{p}'\vec{k}_1'|t(\tilde{E})|\vec{p}\vec{k}_1\rangle = (2\pi)^3 \delta(\vec{p}'+\vec{k}_1'-\vec{p}-\vec{k}_1)\langle \vec{q}_f|t(z)|\vec{q}_i\rangle, \qquad (1)$$

$$z = \tilde{E} - (\vec{p} + \vec{k}_1)^2/2(m + M), \qquad (2)$$

where $\vec{p}'$ and $\vec{p}$ is the initial and final pion momentum, respectively. These quantities are denoted by $\vec{q}_{i(f)}$ in πN c.m.s. Nuclear states are labelled as n' and n; m and M denote the pion and nucleon mass, respectively. The dependence on the total π-nucleus momentum (which survives in z when $\tilde{E}$ is chosen according to ref. [1-3]) can be avoided by postulating

$$\tilde{E} = (p^2 + p'^2)/4m + (k_1^2 + k_1'^2)/4M. \qquad (3)$$

The choice made in eq. (3) ensures the translational, time-reversal and Galilean invariance of the problem and it leads to several interesting consequences.

MOTION OF TARGET NUCLEONS

The fixed scatterer approximation is commonly used in evaluating the matrix element $\langle n\vec{p}'|t(\tilde{E})|n\vec{p}\rangle$, i.e. one neglects the dependence of $\langle \vec{q}_f|t(z)|\vec{q}_i\rangle$ on nucleon momenta ($\vec{k}_1' \sim \vec{k}_1 \sim 0$). The error involved is of order m/M ~ 1/7. This procedure can be substantially improved. Making use of eq. (3), the nucleon momenta $\vec{k}_1'$ and $\vec{k}_1$ can be approximated by effective

[*)] On leave from Institute of Nuclear Physics, Řež, ČSSR.

$$\vec{k}_1'^o = \frac{\vec{x}'}{A} - \frac{A-1}{2A}(\vec{Q}_i - \vec{Q}_f) \qquad \vec{k}_1^o = \frac{\vec{x}}{A} + \frac{A-1}{2A}(\vec{Q}_i - \vec{Q}_f) \qquad (4)$$

ones in the expression for $\langle \vec{q}_f | t(z) | \vec{q}_i \rangle$. We have

$$\langle n\vec{p}' | t(\tilde{E}) | n\vec{p} \rangle = (2\pi)^3 \delta(\vec{p}+\vec{x}'-\vec{p}-\vec{x}) \langle \vec{q}_f^o | t(z_o) | \vec{q}_i^o \rangle F_{nn}(\vec{Q}_f - \vec{Q}_i) \qquad (5)$$

where $\vec{q}^o_{i(f)}$ and z_o are functions of pion momenta $\vec{Q}_{i(f)}$ defined in π-nucleus c.m.s., F_{nn} is the nuclear form factor and $\vec{x}(\vec{x}')$ are nuclear momenta.

It follows from the conservation of the nuclear current, that the error connected with $\vec{k}_1 \sim \vec{k}_1^o$ and $\vec{k}_1' \sim \vec{k}_1'^o$ is of order $(m/M)^2 \sim 1/50$ only and in this sence, the choice made in eq.(4) is unique. Therefore, the nucleon motion can be taken into account very accurately retaining all the invariance properties as well as the factorization of the πN amplitude as indicated by eq.(5). An interesting feature of our effective energy z_o is its dependence on the scattering angle. Relativistic modification of our procedure will be published elsewhere.

PIONIC ATOMS

Energy shifts (ΔE_{nl}) and widths (Γ_{nl}) of the pionic atoms (caused by the strong interaction) are usually evaluated by means of the optical potential $W(\vec{Q}_f, \vec{Q}_i) \sim A \langle 0\vec{p}' | t(\tilde{E}) | \vec{p} 0 \rangle$. The standard choice $\tilde{E} = E \sim 0$ leads to real W and $\Gamma_{nl} = 0$. In order to explain the observed widths, one assumes the two-nucleon mechanism of the pion absorption and introduces the π-NN amplitude into optical potential (OP)[3]. The procedure is not very convincing and the calculated widths are systematically smaller than the experimental ones by a factor two.

OP becomes complex if constructed utilizing the matrix elements given by eq.(5). We evaluated

$$\Delta E_{1s} - \frac{i}{2}\Gamma_{1s} = \int \phi_{1s}(Q') W(\vec{Q}',\vec{Q}) \phi_{1s}(Q) \, d^3Q \, d^3Q'/(2\pi)^3 \qquad (6)$$

for ^{4}He atom in the Born approximation. The hydrogenic wave function was denoted as $\phi_{1s}(Q)$. Further, standard values of πN scattering lenghts $\alpha_1 = 0.182$ and $\alpha_3 = -0.109$ (in units $1/m$) were used. The result

$$\Delta E_{1s} = -64.5 \text{ eV} \qquad \Gamma_{1s} = 34.0 \text{ eV}$$

is very encouraging compared with the experimental one [4]

$$\Delta E_{1s}^{exp} = -(75.7 \pm 2) \text{ eV} \qquad \Gamma_{1s}^{exp} = (45 \pm 3) \text{ eV}.$$

REFERENCES
[1] A.K. Kerman, H. McManus and R.M. Thaler, Ann. of Phys. 8(1959)551.
[2] K.M. Watson, Phys. Rev. 89(1953)575.
[3] M. Krell and T.E.O. Ericson, Nucl. Phys. B11(1969)521.
[4] G. Backenstoss et al., Nucl. Phys. A232(1971)519.

IX Nucleon-Nucleon Interaction

SOME MESONIC ASPECTS OF THE NUCLEON-NUCLEON INTERACTION

R. VINH MAU

Université Pierre et Marie Curie, Paris 75230 Cedex 05

ABSTRACT

The dynamical calculations on the nucleon-nucleon interaction are presented and the latest results discussed.

Let me begin by recalling that, independently of any specific interaction models, the two nucleon interaction can be considered as mediated by the exchange of particles or systems of particles in the crossed (nucleon-antinucleon) channels. This is represented in figure 1.

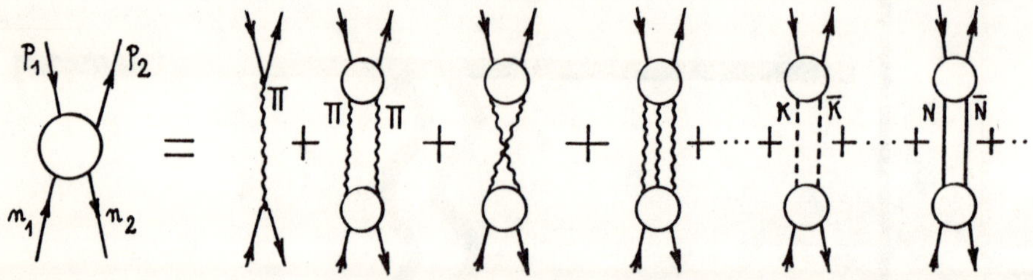

fig. 1. n_1, p_1 and n_2, p_2 are the momenta of the initial and final nucleons.

In this figure, the blobs of each term can contain anything allowed by conservation laws. For example, in the 2 pion exchange term each blob could contain one nucleon, one nucleon + any number of pions, a nucleonic resonance, etc...

Fixed energy dispersion relations can be used to write the following representation for the scattering amplitude :

$$M(s,t) = \frac{g^2}{t-\mu^2} + \frac{1}{\pi} \int_{4\mu^2}^{\infty} \frac{\rho_{2\pi}(s,t')}{t'-t} + \frac{1}{\pi} \int_{9\mu^2}^{\infty} \frac{\rho_{3\pi}(s,t')}{t'-t} + \text{etc.} \quad (1)$$

where $s = (n_1+p_1)^2$ and $t = (p_1-p_2)^2$.

Equivalently, in configuration space, the interaction* can be written as a series of terms :

$$V \simeq \frac{g^2 e^{-\mu r}}{r} + \int_{4\mu^2}^{\infty} \rho_{2\pi}(s,t') \frac{e^{-\sqrt{t'}r}}{r} dt' + \int_{9\mu^2}^{\infty} \rho_{3\pi}(s,t') \frac{e^{-\sqrt{t'}r}}{r} dt' + \text{etc..} \quad (2)$$

* I disregard spin and isospin complications throughout this talk for the sake of simplicity of presentation.

which correspond to the one pion exchange (OPE) contribution, the two pion exchange (TPE) contribution etc... The whole dynamics of the problem is contained in the spectral functions $\rho_{2\pi}$, $\rho_{3\pi}$ etc.. and their actual expressions depend, of course, on the contents of the blobs in fig. 1. For example, if one keeps only the one nucleon state in each blob of the TPE contribution one gets the socalled 4th order in perturbation theory.

Although at present no definite a priori statement can be made about the relative sizes of the various spectral functions, one can say, using phase space factor arguments, that the slope of the functions ρ'_is, at various thresholds, decreases with the mass of the exchanged systems. Consequently, if no unexpected strong enhancement occurs in the many particle exchanges, the spectral functions ρ'_is may have the shapes represented in figure 2. The series represented in figure 1 is then a series of contributions that have shorter and shorter ranges. (It is not a perturbation expansion!).

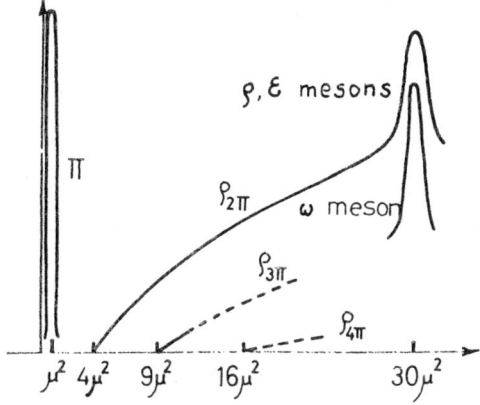

fig. 2. The spectral functions ρ'_is.

The long range forces are correctly described by the OPE and the next longest range forces are due to the two pion exchange (TPE). In the dynamical calculations that I am going to describe, the underlying belief is that most of the medium range forces should be correctly given by the TPE when the latter is properly depermined. Also the hope is that the long and medium range forces if accurately known would provide strong enough constraints to leave little freedom for an eventual phenomenological determination of the short range part of the interaction.

The main task is therefore to perform an accurate calculation of the TPE contribution i.e. the spectral function $\rho_{2\pi}$.

Avoiding heavy technical algebraic details[1,2], and taking liberties with mathematical rigor, a quick way of doing this is to proceed with the following steps :

i) In the t channel ($N\bar{N} \to N\bar{N}$), perform a partial wave expansion of $\rho_{2\pi}$:

$$\rho_{2\pi} = \sum (2\ell + 1) \, \text{Im} \, F_\ell^{N\bar{N} \to 2\pi \to N\bar{N}}(t') \, P_\ell(\cos\theta_s) \qquad (3)$$

where $F_\ell^{N\bar{N}\to 2\pi\to N\bar{N}}(t')$ are partial wave amplitudes for $N\bar{N}$ scattering with two pions as intermediate state.

ii) Express $\text{Im } F_\ell^{N\bar{N}\to 2\pi\to N\bar{N}}(t')$ in terms of $N\bar{N}\to 2\pi$ helicity amplitudes $f_\ell^{N\bar{N}\to 2\pi}$ through the unitarity condition

$$\text{Im } F_\ell^{N\bar{N}\to 2\pi\to N\bar{N}}(t') = \begin{pmatrix}\text{kinematic}\\\text{factor}\end{pmatrix} \left| f_\ell^{N\bar{N}\to 2\pi}(t'') \right|^2 \qquad (4)$$

which can be assumed to hold for $4\mu^2 \leq t'' \leq 50\mu^2$ since inelasticity is very weak up to the $K\bar{K}$ threshold.

iii) Use the analyticity properties of $f_\ell^{N\bar{N}\to 2\pi}$ to write the dispersion relation

$$\ln\left[f_\ell^{N\bar{N}\to 2\pi}(t')\right] = \frac{1}{\pi}\int_{4\mu^2}^{50\mu^2}\frac{\delta_\ell^{\pi\pi\to\pi\pi}(t'')}{t''-t'}dt'' + \frac{1}{\pi}\int\frac{\Phi_\ell^{N\bar{N}\to 2\pi}(t'')}{t''-t'}dt'' \qquad (5)$$

where $\Phi_\ell^{N\bar{N}\to 2\pi}$ is defined by $f_\ell^{N\bar{N}\to 2\pi} = \left|f_\ell^{N\bar{N}\to 2\pi}\right|e^{i\Phi_\ell^{N\bar{N}\to 2\pi}}$ and where the unitarity condition has been used again to identify $\Phi_\ell^{N\bar{N}\to 2\pi}$ with the $\pi\pi$ phase shifts $\delta_\ell^{\pi\pi\to\pi\pi}$, for $4\mu^2 \leq t'' \leq 50\mu^2$.

In the first integral of eqn.(5), the low angular momentum (S and P waves) $\delta_\ell^{\pi\pi}$ can be taken from theoretical models or from experimental studies ($\pi N \to \pi\pi N$, $K_{\ell 4}$ decay, $e^+e^- \to \pi\pi$, nucleon form factors etc... and the high angular momentum (D or higher waves) are known to be negligible.

In the second integral of eqn.(5), $f^{N\bar{N}\to 2\pi}$ can be obtained from the Froissart-Gribov formula since $t''<0$, namely

$$f_\ell^{N\bar{N}\to 2\pi}(t'') = \frac{1}{\pi}\int_{m^2}^{\infty} Q_\ell\left(\frac{2w+t''-2m^2-2\mu^2}{2(t''-4m^2)^{1/2}(t''-4\mu^2)^{1/2}}\right) \text{Im } M^{\pi N\to\pi N}(w,t'') dw$$

where $M(w,t'')$ is the physical πN scattering amplitude since $w \geq m^2$ and $t''<0$.

Schematically, one can summarize i) to iii) by :

$$\rho_{2\pi}(s,t') \sim \left|f_0^{N\bar{N}\to 2\pi}(t')\right|^2 + \left|f_1^{N\bar{N}\to 2\pi}(t')\right|^2 \cos\theta_s + \sum_{\ell\geq 2}(2\ell+1)\left|f_\ell^{N\bar{N}\to 2\pi}\right|^2 P_\ell(\cos\theta_s)$$

| $\pi\pi$ S wave phase shift (J=I=0) + πN phase shifts | $\pi\pi$ P wave phase shift (J=I=1) + πN phase shifts | πN phase shifts |

Note, that in the One Boson Exchange models, the last sum is set to zero and $\left|f_0^{N\bar{N}\to 2\Pi}\right|^2$ and $\left|f_1^{N\bar{N}\to 2\Pi}\right|^2$ are approximated respectively by $g_\sigma^2 \delta(t'-m_\sigma^2)$ and $g_\rho^2 \delta(t'-m_\rho^2)$.

The same procedure could, in principle, be repeated for $\rho_{3\Pi}$ and $\rho_{4\Pi}$ with, however, formidable complications. In the case of $\rho_{3\Pi}$, the ω meson exchange can be easily taken into account, other specific processes can also be included as we will see later.

The set of equations (3) to (5) forms the bulk of the relationships which via analyticity properties, unitarity and crossing connect the TPE contribution with the pion-nucleon and pion-pion interactions. The pion-nucleon scattering is very accurately known by various phase shift analyses[3]. The S and P wave pion-pion interaction has also been extensively studied these last few years[4].

By using these analyses as inputs, one automatically includes all the nucleonic resonances (through ΠN phase shifts) as well as the realistic S and P wave pion-pion interaction, and besides these the smooth background of both pion-nucleon and pion-pion scattering is also taken into account.

The ability of the OPE, of the above calculated TPE and of the ω exchange contributions (as part of the three pion exchange) to describe the long and intermediate range forces can be checked by calculating the low energy (up to 330 MeV) peripheral ($J \geq 2$) NN phase shifts[5] and by comparing them with the empirical ones[6]. The agreement is satisfactory.

For further progress two possibilities present themselves :

i) for the nucleon-nucleon scattering itself, one can perform a phase shift analysis similar to those done by the Livermore group[6], now with extra constraints on lower angular momentum phase shifts from the TPE and ω exchange. This program needs a long and exacting labour but is worth while in view of the coming experimental results from LAMPF, TRIUMPF, SIN etc... It is under investigation by the Paris group. It is also worth noting that the TPE contribution is imaginary and can provide inelasticity parameters.

ii) for further applications to nuclear many body systems, it is useful to define an equivalent potential which, once supplemented by a short range core, can be used in various nuclear calculations. Such a program has been carried out at Stony Brook[7] and in Paris[8] :

The TPE of the Stony Brook potential contains the 4th order contribution, the ρ exchange and a $f_0^{N\bar{N}\to 2\Pi}$ found by educated guess. The ω exchange is included with a coupling constant $g_\omega^2/4\Pi = 6.24$. This potential is regularised for high momentum transfer t (or small r) by a function obtained from certain multiple neutral vector meson exchange processes treated in the eikonal approximation. The calculated phase shifts are in good qualitative agreement with the experimental results. Some results are displayed for illustration in figure 3. The deuteron wave function is also computed and indicates a weaker short range repulsion than most phenomenological potentials.

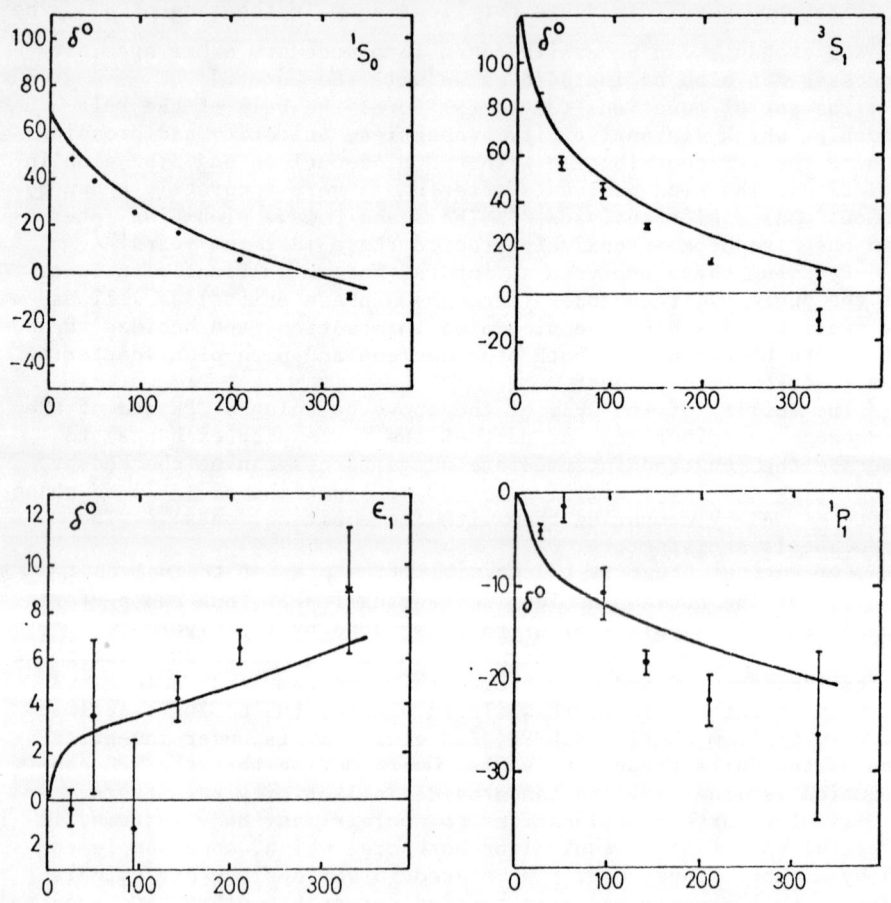

fig. 3. The Stony Brook NN phase shifts as compared with those of ref.(6)

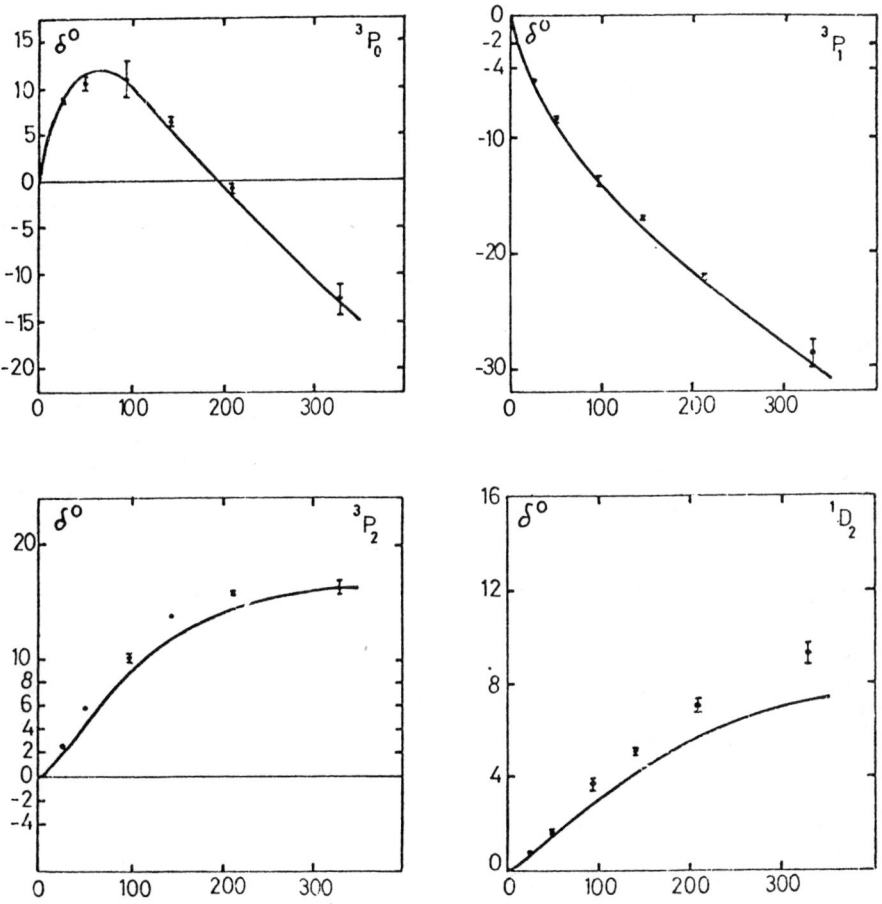

fig. 3. The Stony Brook NN phase shifts as compared with those of ref.(6)

The Paris group[8] calculated the TPE potential from the complete $\bar{\Pi}N$ amplitude as known in terms of phase shifts determined by the Glasgow group[3]. The helicity amplitudes $f_0^{N\bar{N}\to 2\Pi}$ and $f_1^{N\bar{N}\to 2\Pi}$ used here are consistent with observed $\Pi\Pi$ S and P waves. The ω coupling constant is taken to be $g_\omega^2/4\Pi = 9.52$. This somewhat larger value is due to the fact that the nucleonic resonance contributions are attractive and also that the $f_0^{N\bar{N}\to 2\Pi}$ used here induces more attraction than in the Stony Brook model. This theoretical long and medium range potential is cut off at $r \sim 0.8$ fm and replaced for $r < 0.8$ fm by a simple phenomenological soft core which contains in each isospin state six parameters (their values can be found in ref.[8] adjusted to fit the NN phases for all $J < 6$ as well as the deuteron parameters. In spite of the small number of adjustable parameters, the χ^2/data obtained are 2.5 for proton-proton scattering and 3.7 for neutron-proton scattering as good as are given by the well known purely phenomenological fits. The results of the fit are shown in figure 4 and in Table I.

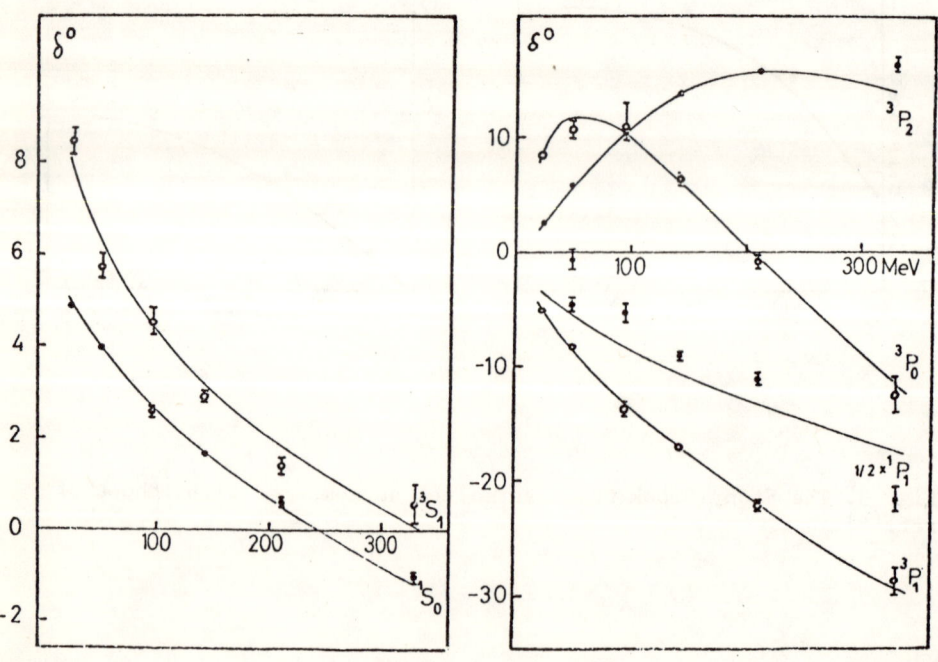

fig. 4. The Paris NN phase shifts as compared with those of ref.[6],[9]

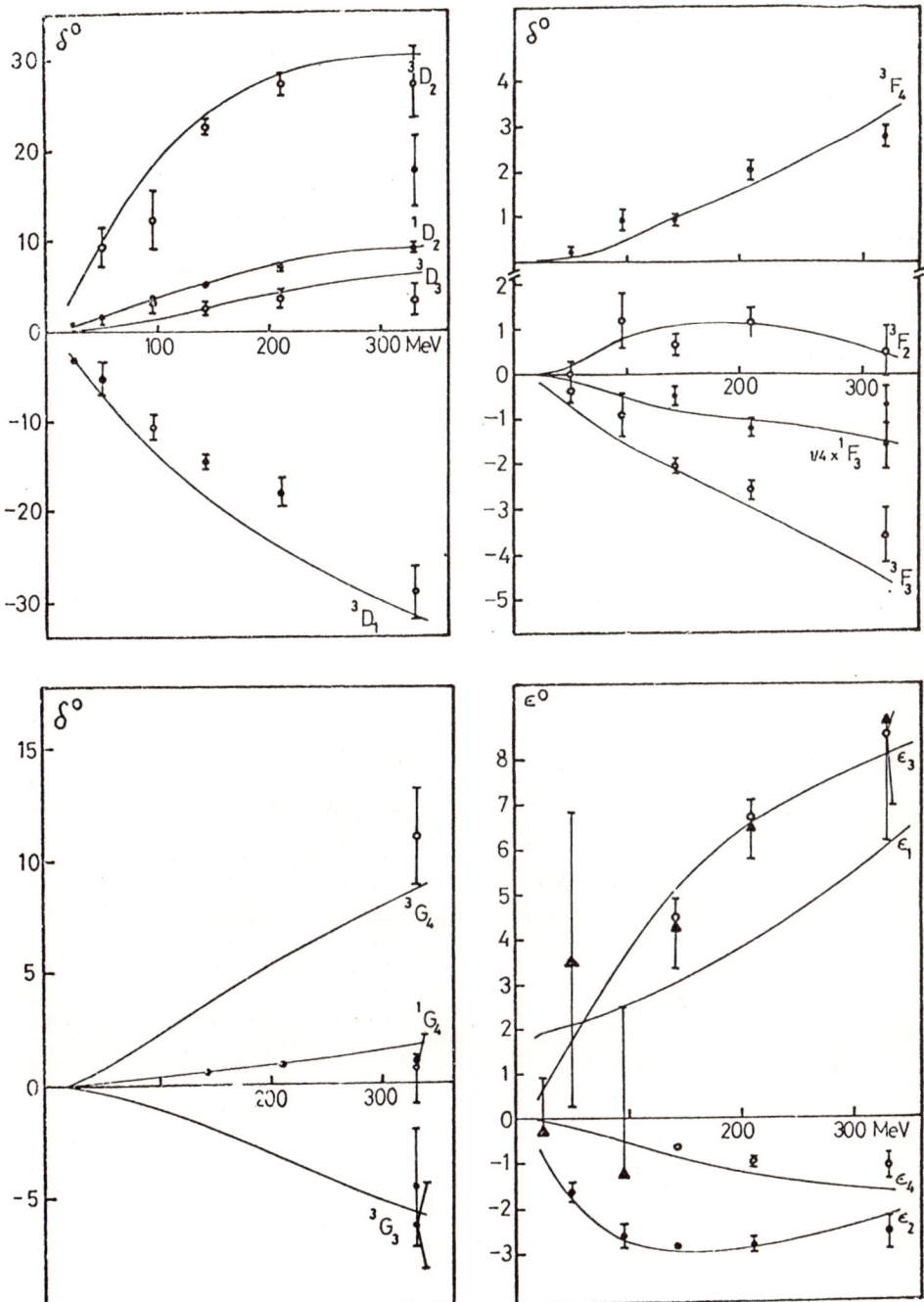

fig. 4. The Paris NN phase shifts as compared with those of ref.[6],[9]

E_D	Q_D	$P_D\%$	μ_D
-2.2246	.290	6.75	.8392
(-2.2246 ± .001)	(.2875 ± .002)		(.8574 ± .000006)

a_{np}	r_{np}	a_{pp}	r_{pp}
5.4179	1.753	-7.817	2.747
(5.413 ± .005)	(1.748 ± .005)	(-7.823 ± .01)	(2.794 ± .015)

Table 1. The deuteron and effective range parameters. Experimental results are given in brackets.

The potential is found to be significantly energy dependent in its central component which can be interpreted as some kind of non locality. However, this energy dependence is only linear and can be easily converted into simple p^2 or velocity dependence.

Both groups (Stony Brook and Paris) found some difficulties in fitting the $^3D_1$ phase at 142 MeV and 250 MeV. This can be attributed to the fact that the tensor potential in the triplet even subspace is too attractive. However, a long range part of the 3 pion exchange as represented in fig. 5 does provide a repulsion. The 3 pion contribution to the so called pion nucleon vertex correction (fig. 6a) has been calculated in the approximation shown by fig. 6b from the $N\bar{N} \to 2\Pi$ S and P waves[10].

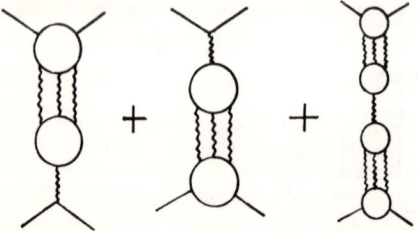

fig. 5. The longest range part of the 3Π exchange

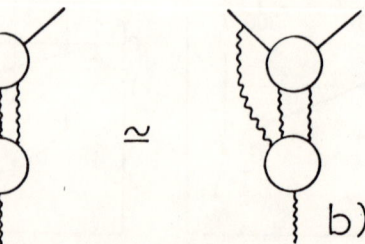

fig. 6. The 3Π contribution to the pion nucleon vertex correction

The calculation of other 3 pion exchange contributions to NN scattering represented by diagrams of fig. 7 is in progress[11]. The inputs here are the same as those used in the TPE calculations, namely, ΠN scattering, and ΠΠ S and P waves

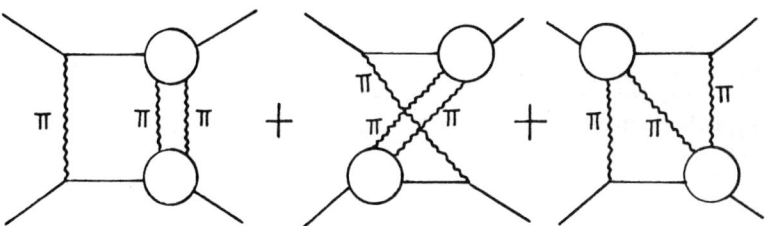

fig. 7. Some 3Π exchange contributions to NN scattering

Along similar lines, Πρ and Πω exchanges (fig. 8) are also being calculated[12]

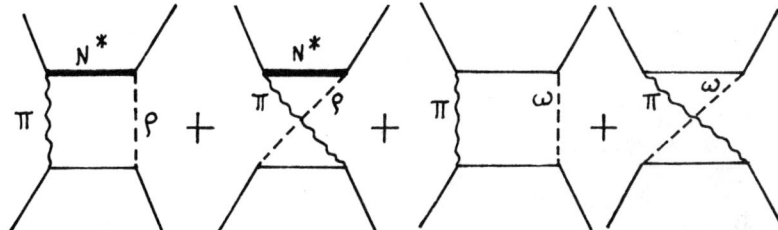

fig. 8. Πρ and Πω exchange contributions to NN scattering considered in ref.[12]

REFERENCES

(1) D. Amati, E. Leader and B. Vitale, Phys. Rev. 130, 750 (1963)
    W.N. Cottingham and R. Vinh Mau, Phys. Rev. 130, 735 (1963)

(2) G.E. Brown, J.W. Durso, Phys. Lett. 35B, 120 (1971)
    M. Chemtob, J.W. Durso and D.O. Riska, Nucl. Phys. B38,141(1972)
    W.N. Cottingham, M. Lacombe, B. Loiseau, J.M. Richard and
    R. Vinh Mau, Phys. Rev. D8, 800 (1973)
    G. Bohannon and P. Signell, Phys. Rev. D10, 815 (1974)
    G.B. Epstein and B.H.J. Mc Kellar, Phys. Rev. D10, 1005 (1974)

(3) See Particle Data Group, UCRL Report N° 20030

(4) See for example J.L. Basdevant, B. Bonnier, C.D. Froggatt, J.L. Petersen and C. Schomblond in the Proceedings of the 2nd Aix en Provence Internatioanl COnference on Elementary Particles 1973

(5) R. Vinh Mau, J.M. Richard, B. Loiseau, M. Lacombe and W.N. Cottingham, Phys. Lett. 44B, 1 (1972)

(6) M.H. Mac Gregor, R.A. Arndt and R.M. Wright, Phys. Rev. 182, 1714 (1969) and references cited herein

(7) A.D. Jackson, D.O. Riska and B. Verwest, Nucl. Phys. A249, 397 (1975)

(8) M. Lacombe, B. Loiseau, J.M. Richard, R. Vinh Mau, P. Pires and R. de Tourreil, Phys. Rev. D12, 1495 (1975)

(9) P. Signell and J. Holdeman, Phys. Rev. Lett. 27, 1393 (1971)
    R.A. Arndt, R.H. Hackman and L.D. Roper, Phys. Rev. C9, 555 (1974)

(10) B. Loiseau and W. Nutt, to be published in Nuclear Physics

(11) M. Lacombe, B. Loiseau, J.M. Richard and R. Vinh Mau, in preparation

(12) G.E. Brown, J.W. Durso, A.D. Jackson, D.O. Riska, private communication.

Carl Shakin (Brooklyn College): I assume your effective potential is to be inserted in the Schroedinger equation. Would some of the energy dependence be removed if you determined an effective potential for use in a relativistic two-body equation.
Vinh Mau: No. In fact if you take properly the kinematic factor in the potential in the Schroedinger equation it is equivalent to the Blankenbecler Sugar equation. In fact the Stony Brook group has used the B-S equation, but we have used rather the Schroedinger equation with proper kinematics.

W. Kloet (Los Alamos): Can this approach be used to get information about the pion production mechanism in nucleon-nucleon scattering and if so, what does it tell?
Vinh Mau: In principle it is possible but no one has tried.

Manoj K. Banerjee (Univ. of Maryland): The energy dependence of the potential reflects the role of intermediate Δ's etc. The resulting non-orthogonality of the wave functions in the 3 momentum space is also understandable since these are projections of the exact wave functions. Your prescription of equivalent hermitian non-local potential removes this non-orthogonality. Is this desirable?
Vinh Mau: In fact, you see this energy dependence reflects some of the non-locality of the potential, so if you solve the scattering states you can show (and I have no time to do so) if you have a linear energy dependence, then its equivalent to a p-squared dependent potential, and this potnetial is OK for the trouble you mentioned. (Further remark by Banerjee and somewhat lengthy response by Vinh Mau.)

M. Soyeur (Saclay): Have you got any results in nuclear matter using your nucleon-nucleon potential?
Vinh Mau: Yes, after having fit the two body problem, we went to the other extreme, namely a nuclear matter calculation, and we have some preliminary results for the static,   local part. But we have not yet got the proper result for the p-squared part because we want to make a G-matrix with a p-squared dependent potential and this involves some computational problems.

Herman Feshbach (MIT): The energy dependence is equivalent to the boundary condition model which Lomon and I introduced some time ago, if the coefficient of the linear energy dependence is unity. The phenomenological value you have obtained is close to one (.95 and .99) I might add that Lomon has applied the boundary condition model to the calculation of nucleon matter and finite nuclei.

G. N. Epstein (Michigan State Univ.): For years people have been trying to extract the deutron D-state probability (P_D) from experiment. It is well known that using just the one pion-exchange tail of the two nucleon potential one can get a rigorous theoretical constraint on P_D, i.e., that $P_D \gtrsim 3\%$. Now with all of this work on

the two-pion exchange potential it should be possible to improve this bound. Have you tried to do this, as distinct from simply quoting a value for P_D.
<u>Vinh Mau</u>: We have not done this.

AN OVERVIEW OF THE N-N SYSTEM[*]

G. E. Brown
State University of New York, Stony Brook, New York 11794

ABSTRACT

Components other than the one-pion exchange play a role in the interactions of nucleons which are important in pion-nucleon scattering. Especially the ρ-meson exchange interaction is crucial in a realistic description.

INTRODUCTION

Since Professor Vinh Mau has discussed the nucleon-nucleon interaction per se, I will direct my remarks to the relations between pion-nucleus scattering and the nucleon-nucleon interaction.

Traditionally, pion physicists have only crudely included effects of the nucleon-nucleon interaction; e.g., short-range correlations of often rudimentary type have been introduced to take into account effects from the repulsive short-range core in the nucleon-nucleon interaction. In multiple scattering models, the pion bounces merrily around between nucleons in the nucleus, Fig. 1, but nary a thought is given to the fact that the nucleons can interact through all of the other components of the nucleon-nucleon force.

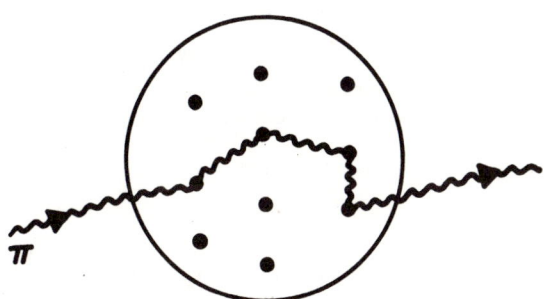

Fig. 1. Types of diagrams summed in the multiple scattering models. The dots represent nucleons; the wavy lines, the pion.

Once the pion is absorbed, however, two nucleons can communicate by exchanging any particle which carries isospin one. Examples are shown in Fig. 2.

---

[*]Supported by U.S.E.R.D.A. Contract No. AT(11-1)-3001.

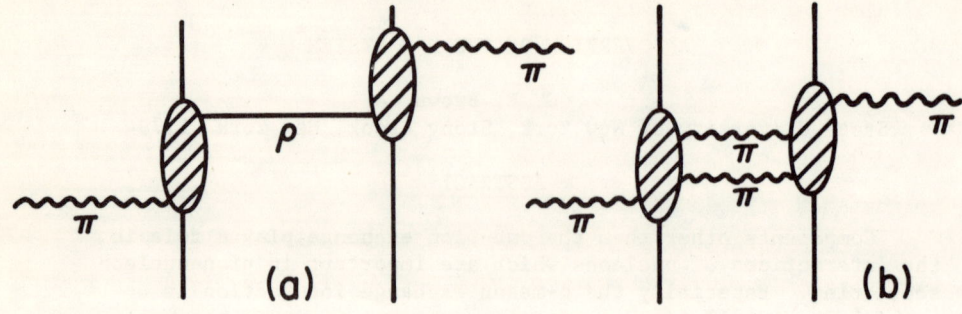

Fig. 2. Examples of possible double scattering processes, where the virtual particles are other than the pion.

For low-energy incident pions, the ρ-exchange potential in fig. 2a will not differ appreciably from the static potential in the nucleon-nucleon interaction, since $\omega_\pi \ll m_\rho$. [In momentum space, the denominator of the interaction, fig. 2a, will be $(k^2 + m_\rho^2 - \omega_\pi^2)$ whereas that of the static interaction will be $(k^2 + m_\rho^2)$.] The two pions in fig. 2b must carry isospin one, and, therefore, from Bose statistics, they are in a state of odd relative angular momentum. They will be chiefly in a relative P-state, and will, thus, carry the quantum numbers of the ρ-meson. Indeed, for the static potential between two nucleons, the two agencies, figs. 2a and 2b, can be combined to give a potential

$$V_\rho(r) = 2\{V_T^-(r) S_{12} + V_{SS}^-(r)(\sigma_1 \cdot \sigma_2)\} \tau_1 \cdot \tau_2 \tag{1}$$

where

$$V_\alpha^-(r) = \frac{1}{(2\pi)^2} \int_{4\mu^2}^\infty dt' \sqrt{t'} \, V_\alpha^-(t') R_\alpha(t') \frac{e^{-\sqrt{t'}r}}{\sqrt{t'}\, r}, \tag{1.1}$$

$$S_{12} = 3\sigma_1 \cdot \underline{r}\, \sigma_2 \cdot \underline{r}/r^2 - \sigma_1 \cdot \sigma_2$$

and

$$V_T^-(t') = \frac{1}{2} V_{SS}^-(t') = \frac{-\pi}{64M^2} \frac{(t'-4m_\rho^2)^{3/2}}{\sqrt{t'}} |f_-^1(t')|^2 \tag{1.2}$$

with M the nucleon mass. The functions R are:

$$R_T(t') = t' \left(1 + \frac{3}{\sqrt{t'}\, r} + \frac{3}{t' r^2}\right)$$

$$R_{SS}(t') = -t'. \tag{1.3}$$

The f_-^{1-} are the helicity amplitudes for the processes $N\bar{N} \to \pi\pi$. The variable $\sqrt{t'}$ plays the role of a (distributed) ρ mass.

In the above we have employed only that part of the ρ-exchange potential coming from the tensor coupling of the ρ. In the non-relativistic limit, the convection current coupling is independent of spin and would not contribute to the processes, fig. 2.

In the zero-width approximation, the ρ-exchange potential reduces to

$$(V_\rho)_{\text{tensor coupling}} = \frac{f_\rho^2}{4\pi}(m_\rho c^2)(\tau_1 \cdot \tau_2)$$

$$[\tfrac{2}{3}\sigma_1 \cdot \sigma_2 \left(\frac{e^{-m_\rho r}}{m_\rho r} - \frac{4\pi}{m_\rho^3}\delta(\underline{r})\right) - S_{12}\left(\frac{1}{(m_\rho r)^3} + \frac{1}{(m_\rho r)^2}\right.$$

$$\left. + \frac{1}{3(m_\rho r)}\right) e^{-m_\rho r}] \tag{1.4}$$

with

$$1.86 < \frac{f_\rho^2}{4\pi} < 4.86 \tag{1.5}$$

depending upon the authors, as we shall discuss. Suitable generalizations give the contribution to the transition potential between nucleons and $\Delta(1230)$ isobars.

We see, first of all, that the largest ambiguities in the nucleon-nucleon interaction, which arise from the exchange of systems of two pions in relative S-states (the distributed-mass σ term) do not arise directly here, because such isoscalar systems cannot transmit the isospin. Of course, these isoscalar, scalar degrees of freedom influence the correlation function between nucleons, but this effect is not very large.

Neglecting the processes, fig. 2, is like trying to calculate nucleon-nucleon scattering using only the one-pion exchange interaction.

The most important part of the potential (1) comes from exchanged masses $\sqrt{t'}$ of roughly the ρ-mass, m_ρ. In this case, the interaction is short range, and we can approximate the $\sigma_1 \cdot \sigma_2$ $\tau_1 \cdot \tau_2$ part by

$$\hat{V}_\rho = \frac{2}{3}\frac{f_\rho^2}{4\pi}(m_\rho c^2)\, \tau_1 \cdot \tau_2\, \sigma_1 \cdot \sigma_2 (A/m_\rho^3)\, \delta(\vec{r}_{12}) \tag{2}$$

where

$$\frac{4\pi A}{m_\rho^3} = \int \left(\frac{e^{-m_\rho r}}{m_\rho r} - \frac{4\pi}{m_\rho^3} \delta(\underline{r})\right) g(r) \, d^3r$$

$$= \int \frac{e^{-m_\rho r}}{m_\rho r} [g(r) - g(o)] \, d^3r. \tag{2.1}$$

Here $g(\underline{r})$ is the two-body correlation function from ω- and σ-meson exchanged. In deriving (2) and (2.1) it was assumed that the ρ-exchange potential was much weaker than the ω-exchange one. Although effects from the ρ-exchange are cut down greatly by the two-body correlation function $g(r)$, they are still very appreciable, as we shall see.

There is an analogy between removing the δ-function interaction in the one-pion-exchange potential, to take into account effects of short-range correlations[1] and adding the effects of ρ-exchange in the above way. The composite answer is summarized by modifying the Kisslinger potential.

$$4\pi a(\omega) \, \nabla \rho(\underline{r}) \nabla$$

$$\rightarrow 4\pi a(\omega) \, \nabla \frac{\rho(\underline{r})}{1 + (\xi_\rho \frac{2 f_\rho^2 m_\pi^2 A}{m_\rho^2 f_\pi^2} + \xi_\pi) \frac{4\pi a(\omega)}{3} \rho} \nabla$$

in a somewhat symbolic notation. Here ξ_ρ and ξ_π are factors which would be unity for zero-range ρ-nucleon and π-nucleon interactions, as we have assumed here. The S_{12} part of the ρ-exchange potential, eq.(1.4), does not contribute in the (pion) long-wavelength limit[2].

In Table I we give values for the ratio $2 f_\rho^2 m_\pi^2 A / m_\rho^2 f_\pi^2$ which determines the ratio of ρ- to π-contribution of the Lorentz-Lorenz correction, denominator of eq. (3).

Table I
Ratio $2 f_\rho^2 m_\pi^2 A / m_\rho^2 f_\pi^2$

Potential or Interaction	Ratio
Reid Soft-Core	.59
Hamada Johnston	.86
Vector Dominance	.60
Höhler-Pietarinen	1.56

The "effective" ρ-exchange parts of the Reid and Hamada-Johnston potentials is found by first subtracting off the one-pion exchange potential, and then taking the combination

$$g'(r) = \frac{1}{16} [V(^1P) + V(^3P) - V(^1S) - V(^3S)] \tag{4}$$

which would project out the piece

$$g'(r) \, \sigma_1 \cdot \sigma_2 \, \tau_1 \cdot \tau_2$$

in the interaction. In calculating all of the numbers in Table I, the two-body correlation function g(r) calculated[3] in Brueckner Theory from the Reid soft-core potential was used. This was, of course, inconsistent in the case of the Hamada-Johnston potentials. In the case of the vector-dominance model and the Höhler-Pietarinen[4] determination, no ω-exchange potential is given, and this seemed a reasonable procedure. The Höhler-Pietarinen determination is certainly the best one to date. The deviation between it and the vector-dominance model can be understood in terms of the contribution to the electromagnetic form factor from a direct coupling of photon to nucleon, not all of it going through the ρ-meson as intermediary. (This was found earlier by Iachello, Jackson and Lande[5].)

We see from Table I that we face a situation in which the ρ-contribution to the Lorentz-Lorenz correction is likely to be as large as the π-contribution.

Taking account of exchange between nucleons modifies quantitatively the contributions of π- and ρ-exchange[6]. This is part of the pion many-body problem.

Introducing the finite range of the π-nucleon, or π-nucleon-isobar interaction cuts down appreciably the pion Lorentz-Lorenz effect[7]. Introduction of finite-range effects into the ρ-nucleon, or ρ-nucleon-isobar interaction will "get the ρ-exchange potential more out of the way of the ω-exchange", giving an effectively longer-range ρ-interaction in eq. (2.1), and this will increase the ρ-contribution[8]. It seems clear, in summary, that one will have to live with a Lorentz-Lorenz correction somewhat larger than that proposed originally by Ericson and Ericson[1], and this has major consequences for π-nucleus scattering.

Of course, through the entire discussion, I've assumed that the ω-meson exchange provides the short-range repulsion which keeps nucleons apart, and starts the pionic Lorentz-Lorenz effect. There seem to be sufficient arguments in the theoretical derivations of the nucleon-nucleon force to motivate this. But note that if the repulsion is not as strong as in, say, the Reid potential, then the pionic contribution might go down, but it can be seen from Eq. (2.1) that the ρ-contribution would go up, unless the repulsion was so weak that $g(r) \simeq g(o)$ at short distances.

The pion-nucleus problem is then intimately related with the N-N interaction, chiefly with the ρ-like pieces of the N-N force. A suitable solution of the first requires knowledge of the latter.

REFERENCES

1. M. Ericson and T.E.O. Ericson, Ann. of Phys. $\underline{36}$, 323 (1966).

2. G. Baym and G.E. Brown, Nuclear Physics $\underline{A247}$, 395 (1975).

3. S.O. Bäckman, private communication.

4. G. Höhler and E. Pietarinen, Nucl. Phys. $\underline{B95}$, 210 (1975).

5. F. Iachello, A.D. Jackson and A. Lande, Phys. Letts. $\underline{43B}$, 191 (1973).

6. J. Delorme and M. Ericson, Phys. Letts., to be published.
   M. Thies, to be published.

7. J. Hüfner, J.M. Eisenberg and E.J. Moniz, Phys. Lett. $\underline{47B}$, 381 (1974).
   J. Hüfner, Proc. of Symposium on Interaction Studies in Nuclei, Mainz, February 1975, North-Holland Publ. Co. (1975) 709.

8. G. Baym, private communication.

9. M. Thies, Phys. Letts., to be published.

Mikkel Johnson (Los Alamos): How do the Lorentz-Lorentz and ρ-meson effects survive the very soft core potentials of Vinh Mau and Stony Brook?

G. Brown: I don't know that we have such soft potentials, but that's a very good point and that's why I like to have the antisymmetry to fall back upon. Because the antisymmetry keeps the relative wave function zero over quite a large distance, independent of the interaction. I'm sorry, I should say that with the softer core the rho meson contribution increases of course, because it's cut out by the core, so that for a long way if you soften the core there's again this stability; the pionic delta function is not cut out with the rho contribution increase.

L. Kisslinger (CMU): You have been making very strong statements on the basis of ρ exchange and the tensor type force. If one looks at the details of Vinh Mau's calculation of the NN force, the uncorrelated 2π exchanges often give larger effects and of opposite sign. Are you not referring to a program which will take many years to work out?

G. Brown: Well, this is why the formalism was cast in the form of helicity amplitudes, because that gives one the possibility of handling the two pion continuum, and we've actually put it into the calculations of a. You have to also put in the uncorrelated two-pion pieces because they are of much longer range than the rho, that's true.

Gerald A. Miller (Univ. of Washington): (1) I'd like to comment on the role of s-wave π-nucleon scattering, which you have ignored. At low energies the s-wave phase shifts are as large as the (3,3). It is only in the lowest order optical potential that these effects average out. In higher orders this does not happen and since there is no 0.7 there is no delta function. (2) Low energy π scattering is very forward peaked because of the Coulomb scattering. Coulomb interference effects destroy arguments about spherical symmetry. (3) I'd like to second F. Lenz's remarks. In particular, ρ-meson exchange plays a role in true meson absorption in which you have 2 nucleons emitted. If you have ρ exchange with no 2 nucleon cut, you must be incomplete.

G. Brown: As for the first part, I don't know about the p-wave two pion absorption, but the first parts were in Thies' talk; but, in my discussions with people I found out that they have not really understood the import of his talk, nor (and almost nobody has read) the Delorme and Ericson paper, and I believe the latter paper to be probably the most important work in this subject in the last year. So when people tell me that they do multiple scattering from some formalism starting with covariant equations and they put in scattering to all orders, and they calculate Fermi motion, they calculate this and that, and then they tell me that they've not yet put in the antisymmetry but they have a good fit. I'm not arguing with you now, this is a general comment, and I'm frankly very surprised because they've left out most of the problem.

Thomas (CERN): I'm not sure what to say. I'm afraid you sort of stunned me. There's a considerable amount of work on pi-nucleus scattering, much of which doesn't overlap with what you've had to say, but in which effects like Carl Shakin pointed out have been shown to be extremely important. And the theory in which you're working, in which the Lorentz-Lorenz effect has been derived is a theory in which the nucleons don't move, there are no recoil terms for the nucleon, the energy parameter in the t-matrix is fixed, and that has been shown in calculations to be an important parameter, and there are just so many effects which have been discussed by different groups and have been shown to be important, including the ones you've talked about, to dismiss out of hand the work that other people have done is just a little bit too much.

Brown: No, I didn't mean to dismiss it out of hand, and in fact, as I said in the beginning, the talk was going to be on the interrelation of the nucleon-nucleon interaction and the pion nucleon scattering, so I was discussing how the nucleon-nucleon interaction affected the multiple scattering. Now of course there are the other effects in nucleons scattering off of nuclei and so forth. We've also dealt with them for years, and especially in some problem like pions on deuterium where you don't have identical particles and the Pauli principle does not come into play, all these other effects are accentuated very much. But what I'm saying is that if you consider the pion moving from this point to this point, the two nucleons between this point and this point communicating by pion exchange, and only by pion exchange, that's like saying that the nucleon-nucleon interaction proceeds by one pion exchange. There's a richness in the nucleon-nucleon interaction which is not covered by the one pion exchange potential, in fact many very important features; and then I'm trying to summarize what else can be exchanged between these two points, and I point out that in a spin-isospin saturated system, that in the simple way that I've looked at what else that can be exchanged, changes the sign of the double scattering term. Now you can put in all of these other effects but I think that's the simple point which by and large has not been put in. Otherwise I agree with everything you said.

Eisenberg (Tel Aviv): In ^{4}He, where you advocate application of the Lorentz-Lorenz effect, especially as arising from antisymmetrization, the spatial wave functions are in fact symmetric in any pair and hence do not vanish at zero separation, whence, apparently, no Lorentz-Lorenz effect from antisymmetrization. Further detailed examination of the paper of Delorme and Ericson is in order.

Brown: Yes, I agree with a simple picture here. Here I'm quoting the paper of Delorme and Ericson and I cannot give you a simple picture offhand to answer this point, but it... (long remark by Banerjee).

THE EFFECT OF THE Δ(1236)-RESONANCE IN NN-SCATTERING, NUCLEAR AND NEUTRON MATTER INCLUDING ALL PARTIAL WAVES*

R. Machleidt**
State University of New York at Stony Brook, New York 11794

K. Holinde
Institut fuer Theoretische Kernphysik der Universitaet Bonn, D-5300 Bonn, West Germany.

ABSTRACT

Nuclear and neutron matter properties are calculated with one-boson-exchange-potentials which include the Δ-resonance in all partial waves. Many-body corrections in ($L \geq 1$)-partial wave states and those arising from the transition potential with two Δ are appreciable. In the nuclear matter calculations the prescription for the hole potential is varied leading to substantially different results. It is found that the many-body effects due to the Δ-resonance alone do not lead to the experimental saturation point.

The effect of the Δ(1236)-resonance in NN-scattering has been shown for all partial waves[1] and compared to the results of Jena and Kisslinger[2]. For this purpose a twice-iterated pion-exchange potential which couples the NN-channel with the NΔ- and ΔΔ-channel has been used. A strong cut-off for the NN - and NΔ -vertices is applied in these transition potentials. This cut-off stands at least in part for the rho-exchange. The transition potentials, describing the excitation of one or two Δ's are embedded into the one-boson-exchange (OBE) model for the NN-interaction[3,4]. A quantitative fit of the NN phase-shifts and the deuteron data is performed for different values of the cut-off mass Λ, namely 650, 550, and 450 MeV and different OBE models namely HM1[3] and HM2[4]. The fit of the NN-scattering phase shifts is mainly achieved by reducing the coupling constant of the sigma-meson, which describes the intermediate range attraction within the OBE model.

The properties of nuclear matter (Fig.1) and neutron matter (Fig.2) are calculated in the frame-work of lowest order Brueckner theory. Nuclear matter is computed using two different prescriptions for the hole potential: (I) $U(k < k_F) = BBP$[6]
(II) $U(k < k_F) = 0$.
($U(k > k_F) = 0$ in both cases). Prescription (II) is suggested by recent variational calculations[7]. It turns out that with prescription (I) the saturation points are lying on a line which does not pass through the experimental point (see Fig.1 big dots of dashed lines). With prescription (II) the experimental point can be reached by extrapolation (see Fig.1 big dots of solid lines).

*Supported in part by Deutsche Forschungsgemeinschaft.

**On leave of absence from Institut fuer Theoretische Kernphysik der Universitat Bonn, D-5300 Bonn, West Germany.

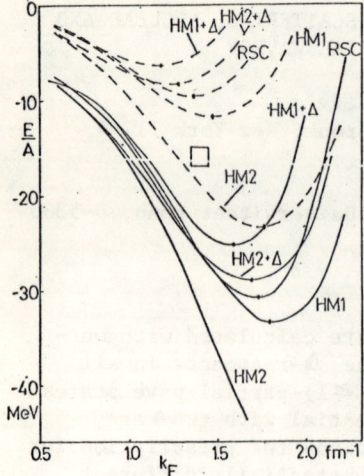

Fig. 1. Energy per particle in <u>nuclear</u> matter, E/A, as a function of the Fermi momentum, k_F, for several potential models: RSC[9], HM1[3], HM2[4]; cases HM1+Δ and HM2+Δ[8] use a cutoff mass of Λ = 650 MeV. The dashed lines give the results using prescription (I), the solid lines refer to prescription (II). (HM2 saturates at $-$ 55 MeV, 2.1 fm^{-1} for prescription (II).)

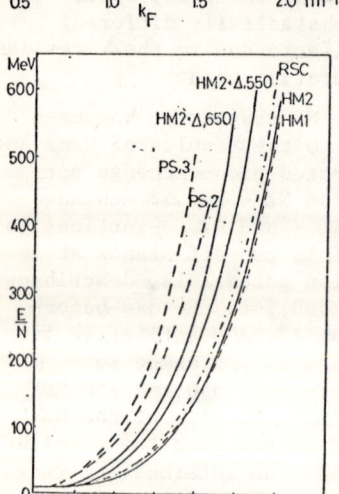

Fig. 2. Energy per particle in <u>neutron</u> matter, E/N, as a function of the Fermi momentum, k_F, for RSC HM1, HM2; HM2 + Δ with Λ = 550 MeV and Λ = 650 MeV; PS,2 and PS,3 are results from ref.[10].

REFERENCES

1. K. Holinde and R. Machleidt, to be published.
2. S. Jena and L.S. Kisslinger, Ann. of Phys. 85 (1974) 251.
3. K. Holinde and R. Machleidt, Nucl.Phys. A247 (1975) 495.
4. K. Holinde and R. Machleidt, Nucl.Phys. A256 (1976) 479, 497.
5. A.M. Green and J.A. Niskanen, Nucl.Phys. A249 (1975) 493.
6. H.A. Bethe, B.H. Brandow and A.G. Petschek, Phys.Rev. 129 (1963) 225.
7. V.R. Pandharipanda, R.B. Wiringa and B.D. Day, Phys.Lett. 57B (1975) 205.
8. K. Holinde and R. Machleidt, to be published.
9. R.V. Reid, Ann. of Phys. 50 (1968) 411.
10. V.R. Pandharipanda and R.A. Smith, Nucl. Phys. A237 (1975) 507.

T.-S. H. Lee (Argonne National Lab): Did your nuclear matter calculation include the Δ explicitly, or do you eliminate the Δ to get V_{eff} and then perform the nuclear matter calculation? If you use V_{eff}, how do you determine the starting energy in G?
R. Machleidt: We include the Δ explicitly in nuclear matter. The determination of the starting energy is straightforward and causes no problems.

Alan Goodman (CMU): You have considered the effect of Δ's in nuclear matter in lowest order Brueckner theory by calculating the diagrams below.

Lowest order
Brueckner diagrams

Do you think the effect of the Δ's in the Faddeev term, i.e. - the three body force term, might be the more important diagram?

Faddeev diagram

EVIDENCE FOR A Δ-NUCLEON VIRTUAL STATE

G. Alberi and F. Baldracchini
Istituto di Fisica Teorica, Miramare, Trieste

In a recent paper[1] the data for deuteron break-up induced by high energy hadrons have been collected and compared, in order to understand the known discrepancy between impulse approximation and experiment at high momenta of the spectator(Fig.1). The channel dependence of this effect, visible in Fig.1, can be reproduced qualitatively, assuming as main cause of the discrepancy a diagram of the type

that is a Δ-N virtual state, in the nucleon-nucleon final state interaction. It is now possible to test the above hypothesis using new exp. data for the charge exchange process $pD \to npp$ at 1.65 GeV/c[2]: the data are extracted from the high momentum tail and are displayed in terms of the invariant mass of the two protons (Fig.2). The above diagram may be calculated using the non relativistic approximation[3] for the propagator and the Hulthen wave function of the deuteron; for the $\Delta N \to NN$ amplitude, the K-matrix formalism

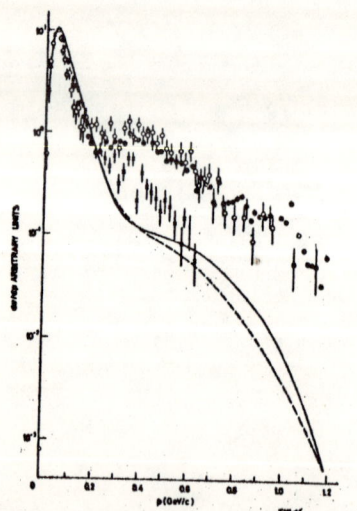

Fig.1. Spectator spectrum. Solid and dashed lines refer to hard and soft core Reid wave function; for exp. data see Ref.1.

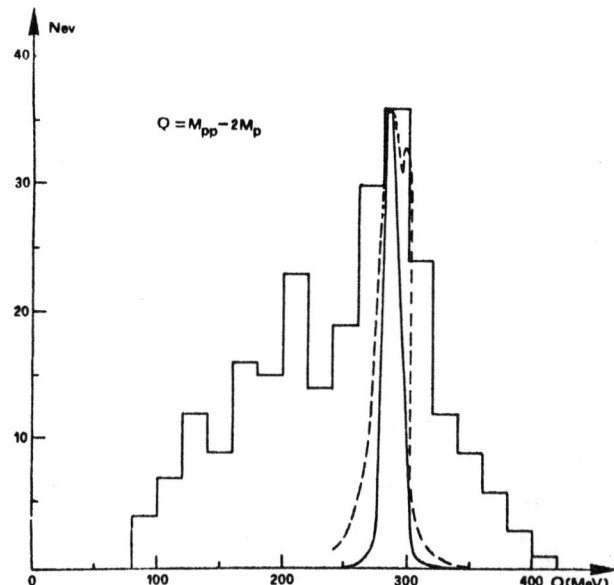

may be used. The results of Fig.2 are obtained, with this simple model using reasonable values for the Δ-N system in S-wave ($a=10$ f and $r_0=2.5$ f). The qualitative agreement between theory and experiment confirms the hypothesis of Ref.1.

Fig.2. Spectrum of the two proton mass, with $p_s > .3$ GeV/c and $|t| < .15$ (GeV/c)^2. The solide line is the result of the model with a Δ-N virtual state with $\Gamma = 0$ and the dashed line with $\Gamma = .14$ GeV.

REFERENCES

1. G.Alberi, V.Hepp, L.P.Rosa and Z.D.Thomè, Ref.TH 2113 CERN, to be published in Nuclear Physics B
2. B.S.Aladashvili et al, Dubna-Warsaw Collaboration, Preprint
3. I.S.Shapiro, Proc.Int.School of Phys."E.Fermi", Course 38, Academic Press, 1967

CHARGE ASYMMETRY OF NUCLEAR FORCES[†]

M. A. Alberg, E. M. Henley, G. A. Miller, and J. F. Walker[*]
Physics Department, University of Washington, Seattle, WA 98195

ABSTRACT

The charge asymmetry of nuclear forces due to real and virtual electromagnetic effects is considered for the reaction

$$n + p \to d + \pi^\circ . \tag{1}$$

An asymmetry about 90° in the angular distribution for reaction (1) can occur due to an isospin zero component of the initial state. An amplitude which does not conserve isospin in (1) is a measure of charge asymmetry since the system is a self-conjugate one. We consider the following sources for charge asymmetry: a) magnetic forces, b) virtual photopion production, c) π-η mixing, and d) radiative corrections to the π-nucleon (N) coupling constant. We compute the asymmetry in the angular distribution for the reaction through the above four mechanisms shown diagrammatically in Fig. 1.

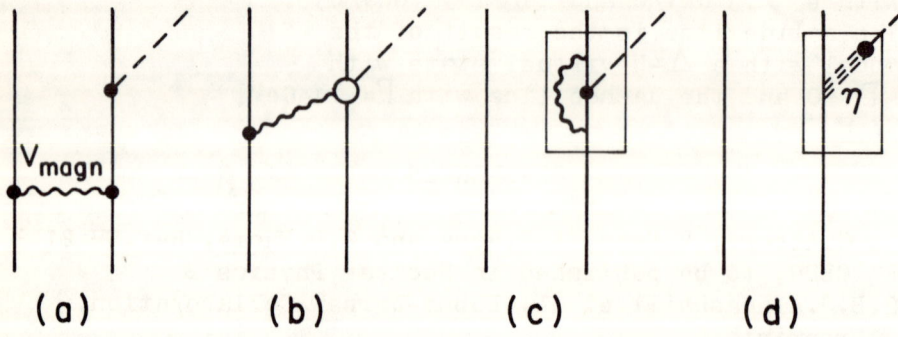

Mechanisms c) and d) can be computed simultaneously. In both cases the matrix is computed identically to that for normal pion production, e.g. $p + p \to \pi^+ + d$, except that the effective creation operator, which replaces the box at the pion vertex in Figs. 1c and d is that for an isoscalar particle with a coupling constant $\propto (g_{p\pi^\circ} - g_{n\pi^\circ})$

[†]Supported in part by Energy Research and Development Administration.
[*]Permanent address: Physics Department, University of Massachusetts, Amherst, MA 01002.

for Fig. 1c and to $\xi g_{N\pi O}$ in Fig. 1d where ξ is a measure of π-η mixing. The magnetic potential of Fig. 1a due to photon exchange can be added to the normal potential which distorts the initial state n-p wave function. This potential, which is taken to have the Hamada-Johnston form, is also used in computing the deuteron wave function. Final state interactions of the pion are taken into account by summing the multiple scattering expansion in the fixed scatterer approximation.

We intend to investigate the relative contributions of the various processes listed earlier and to compute their energy dependence. Preliminary estimates suggest that the asymmetry should grow with energy and might be of the order 10^{-4}-10^{-3}.

X PANEL: NUCLEAR STRUCTURE INFORMATION FROM MESON-NUCLEAR PHYSICS

J. Eisenberg, Chairman

T. Londergan, Sci. Secretary

G. Walker

G. Miller

A. Gal

J. Negele

P. Radvanyi

J. Schiffer

PANEL DISCUSSION: NUCLEAR STRUCTURE INFORMATION FROM MESON-NUCLEAR PHYSICS

The panel discussion was intended to focus on three pertinent questions in meson-nucleus interactions. For each of the three, a proponent set forth a claim as to the information which he thought would be forthcoming. A panel of "realists" then stated challenges to these theses, which the proponent tried to answer in a more extensive presentation, followed by queries from the panel and the floor. The three questions, on (π,π'), (π^+,π^-) and (K^-,π^-) reactions, were in no way intended to span the complete range of topics in medium-energy physics, but rather were meant to serve as case studies for the exploration of the issue at hand.

PION INELASTIC SCATTERING - George Walker

<u>Walker</u>:
<u>Premises</u>: One can obtain useful and special information from excitation energies and differential cross sections for final states reached in (π,π') that often cannot be obtained with other "elementary probes". Despite reaction mechanism uncertainties the information should be useful if the data is
(1) compared with other data (e,e'), (p,p'), (p,n), etc., and if
(2) theory and experiment are compared at several energies, including those above and below the 3,3 resonance. Although I shall discuss DWIA calculations, the qualitative feature of the predictions are likely to remain since they depend on these basic features: a) angular momentum selection rules; b) the dominance of the partial wave with $L=qr$, and c) the spin and isospin dependence of the π-nucleon interaction. Examples to be discussed are: (a) high-spin particle-hole $T=0$ states, (b) 2^+, $T=0$ states (giant E2?), and (c) $T_>$ giant dipole states.

<u>Radvanyi</u>: Which new properties can one obtain from pions which cannot be gotten with protons, alphas or heavy ions?

<u>Schiffer</u>: Shouldn't one first look at pion excitation of low-lying, well understood nuclear states before attempting to deal with highly collective levels? Further, (π^-,π^0) processes, which you will be advocating, are vastly more difficult than (n,p) reactions, which perhaps should be studied first.

<u>Negele</u>: Since I have been given a realist's license to ask outrageously iconoclastic questions, I would like to raise the following questions. At present, can we really use our knowledge of pion interactions to extract definitive information about nuclear structure, or rather, should we more properly use our considerable knowledge of nuclear structure to learn about pionic interactions? Specifically, you emphasize the utility of the $\sigma \cdot \nabla$ coupling in

measuring derivatives of collective form factors. At the very least, it seems to me we should insist upon detailed quantitative agreement with the proton form factors which can be extracted from high resolution elastic and inelastic electron scattering. Hence, I would like to hear concrete, specific proposals as to how you intend to calibrate your reaction mechanism and thereby use it as a reliable probe.

PION EXCITATION OF NUCLEAR COLLECTIVE MODES AND HIGH SPIN STATES[†]

G. E. Walker
Indiana University, Bloomington, IN. 47401

ABSTRACT

Despite the formidable challenge of understanding the pion-nucleus reaction mechanism, the pion may be a useful probe for studying particular nuclear collective modes and high spin states. This should be especially true if pion reaction data obtained in the energy region below 100 MeV as well as near the (3,3) resonance is compared with data obtained from, for example, electron and proton-nucleus reactions. We present results based on the distorted wave impulse approximation (DWIA) for pion-nucleus inelastic scattering from ^{16}O and simple charge exchange on ^{48}Ca. States not prominently excited by other probes are predicted to be observable. In particular $J^\pi = 3^-$ and $4^-, T = 0$ states appear prominently at large momentum transfer in inelastic scattering. The location and prominence of $2^+, T = 0$ states in the nuclear response function should be helpful in studying the degree of collectivity of quadrupole strength in nuclei. Deformation parameters can be deduced from macroscopic theories of inelastic scattering and compared with the results of other probes. The single charge exchange results should be useful in studying the location and degree of collectivity of the $T_>$ states arising from the splitting of the giant dipole resonance in $T \neq 0$ nuclei. The results presented depend primarily on angular momentum selection rules, the properties of bessel functions and the spin and isospin dependence of the microscopic pion-nucleon interaction and so should remain qualitatively valid.

INTRODUCTION

The pion-nucleus reaction mechanism is at present incompletely understood. As examples, uncertainties associated with the approximations adopted in obtaining the optical potential, multistep processes, a "correct" off-shell extrapolation of the pion-nucleon interaction, and indeed the adequacy of a non-field theoretic approach should cause one to have doubts regarding predictions based on the distorted wave impulse approximation (DWIA). Nevertheless it seems useful to apply existing phenomenological parametrizations of the π-nucleon interaction in many-body formalisms to study simple π-nucleus reactions. This enables one to contrast, given the information currently available, the expected pion results with data obtained (or anticipated) from other probes so that one can point more specifically to which pion experiments are most likely to lead to new or collaborative information about the nucleus. In this talk results of DWIA calculations of pion-nucleus inelastic scattering

[†]Work supported in part by the National Science Foundation.

and charge exchange will be presented and contrasted with the results predicted for other probes. This discussion is presented with the view that, historically, data on the location in energy of excited states and the magnitude and shape of the differential cross sections for reactions leading to a particular nuclear final state has been essential information for testing proposed models of the nucleus.

DETAILS AND RESULTS OF THE PION-NUCLEUS CALCULATIONS

The specific details and formulae for the calculations are discussed elsewhere.[1] We have used the DWIA to calculate differential cross sections for pion-nucleus inelastic scattering and charge exchange.

A fixed scatterer non-local pion-nucleon transition matrix[2] has been used both in obtaining the optical potential needed for the distorted waves and the inelastic transition operator. The optical potential obtained provides a good fit to elastic scattering.[2] In those cases where comparison is possible, the inelastic scattering predictions are in qualitative agreement with the results of inelastic scattering calculations of other authors.[3,4] We have used the harmonic oscillator shell model and the Tamn-Dancoff approximation to generate initial and final nuclear wavefunctions. Though an oversimplification these wave functions have been found adequate in the past for an interpretation of inelastic electron scattering experiments.[5,6,7] We do not make predictions for the very low lying collective $T = 0$ states of even-even nuclei (for which TDA p-h states are certainly inadequate).

The results of the calculations for (π,π') on ^{16}O are shown in Figs. 1(a,b). An oscillator parameter of $b = (\hbar/m\omega)^{\frac{1}{2}} = 1.77$ fm., a Serber-Yukawa residual interaction and 2s, 1d, 1f, 1p particle shells and 1p and 1s shell holes were used to obtain the ^{16}O particle-hole states.[7]

At forward angles (low momentum transfer) the $T = 1$, $J^\pi = 1^-$, giant dipole state here predicted at 23.26 MeV, dominates the spectrum. This state is seen experimentally in photoexcitation at 22.3 MeV. In general, where comparison with experimentally identified states is possible, we find that the states predicted by the present particle-hole model are correct in energy location to within 1-2 MeV.[5,6] Our experience has been that for $T = 1$ states the predicted levels have most often been $\frac{1}{2}$ to 1 MeV high. At larger angles high-spin states (3^- and 4^-) are found to dominate the spectrum. The dominance of high-spin states at large momentum transfers simply reflects the fact that larger angular momentum transfers, L, are dominant for large momentum transfer $q[L \approx qR_{nuc}]$.

While there are some similarities, there are important differences between the type of states previously seen or predicted to be prominent in inelastic electron scattering[5,6,7], photoexcitation[8], and inelastic proton scattering[9] and the states we predict will be strongly excited in inelastic pion scattering. In particular $T = 0$ states are much more important in the nuclear response to the pion

probe. This is particularly true for the high-spin states. The
reason that T = 1 high-spin states dominate the nuclear response
in inelastic electron scattering at high momentum transfers is that
for the energy range of electrons currently available high momentum
transfer implies moderately large angle where the transverse form
factor dominates over the longitudinal contribution. Since the
nucleon isovector magnetic moment is more than five times larger
than the isoscalar moment, $\Delta T = 1$ states dominate the contributions
of the transverse multipoles. For protons the spin and isospin
dependence of the effective interaction or pseudo potential adopted
suppresses the excitation of spin-flip T = 0 states compared to
spin-flip T = 1 states. Spin dependent forces (tensor, spin-orbit)
apparently play an important role in determining the nuclear
response to the proton probe at high momentum transfer. Of course,
the collective positive parity low-lying T = 0 states are relatively
strongly excited in inelastic pion, proton and inelastic electron
scattering (see later discussion).

Only the odd parity state differential cross sections have been
shown in fig. 1. The even parity $2\hbar\omega$ p-h states have not had spurious contaminations removed. Although one can easily identify the
states that are largely spurious, other $2\hbar\omega$ states may have their
cross sections inflated because of spurious state components. The
even parity state results for $T_\pi = 180$ MeV and states strongly
excited in the energy region below $E \leq 24$ MeV are 1) a 4^+ (~15 MeV),
T = 0 state should dominate the spectrum above q = 200 MeV/c and
2) in the region 17-23 MeV 2^+ T = 0 states are predicted to be
prominent in the range $q \approx 100$ MeV/c. That 2^+ and 4^+ T = 0 states
should be strongly excited can be traced back to the strong
$\Delta S = \Delta T = 0$ term present in the basic (P_{33}) pion-nucleon interaction. The strong relative excitation of 2^+ T = 0 states at low
momentum transfer implies the pion might be useful for studying a
proposed isoscalar quadrupole resonance.

Of course, these results depend on the dominance of one step
processes. If, for example, two step processes were also very
important the excitation spectrum could become quite muddied and
angular distributions associated with states reachable in single
step reactions could become significantly altered. Therefore one
should carry out the experiments in regions significantly below the
(3,3) resonance, for example, near $T_\pi = 70$ MeV. Looking at the
excitation spectrum at fixed momentum transfer as a function of the
incident energy of the pion from $T_\pi = 70 - 200$ MeV might help one
distinguish between single and multistep reactions, thereby clarifying the reaction mechanism and indirectly therefore clarifying
the underlying nuclear structure.

Since many of the states predicted to be strongly excited have
T = 0 it would be useful to also do single charge exchange - in
which case the T = 0 states should disappear thereby demonstrating
their isoscalar nature.

Other authors [3,10] have studied pion excitation of low lying
T = 0 collective states. Results of their calculations and subsequent comparison with experiment indicates that fits to the data

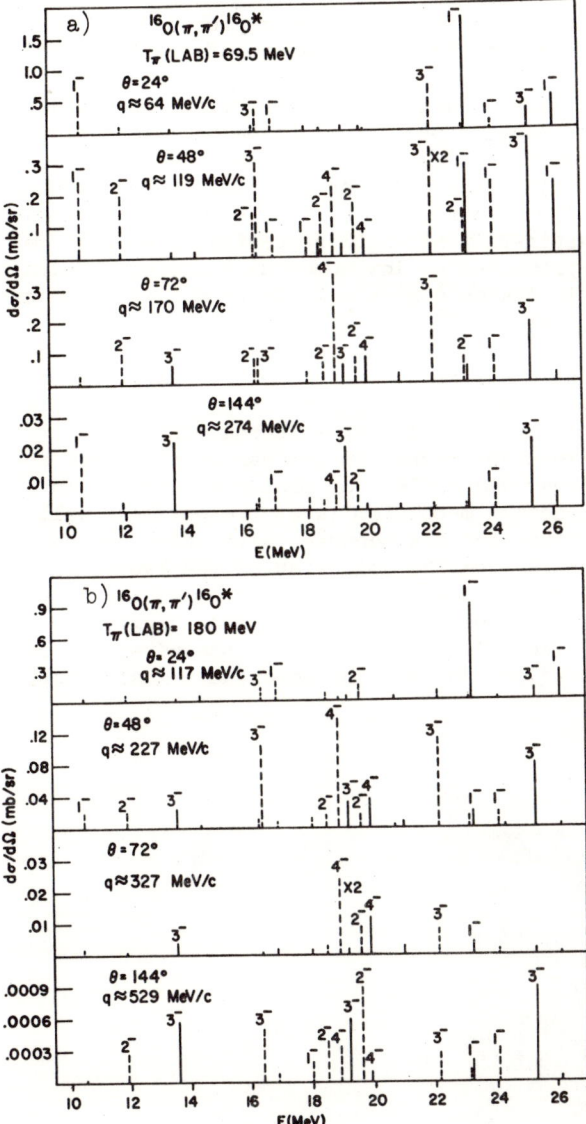

Fig. 1(a,b) Pion - ^{16}O inelastic scattering differential cross sections for odd parity particle - hole states as a function of the final nuclear excitation energy E shown for different scattering angles (momentum transfers). Solid lines correspond to T = 1 final nuclear excited states while T = 0 states are represented by dotted lines. Only odd parity states with appreciable cross sections are included.

are obtained by assuming energy independent β deformation parameters consistent with those obtained from other probes. It has been suggested[3] that because of the p wave nature of the interaction, derivatives of the nuclear collective form factor enter in the calculation and hence pion-nucleus inelastic scattering compared to other strongly absorbed nuclear probes, might be more sensitive to the form factor shape.

The results of the calculation for pion-charge exchange is shown in fig. 2 for $^{48}Ca(\pi^-,\pi^0)^{48}K^*$.[1] The $T_>$ dipole resonance states with non-spin-flip strength concentrated near 11 MeV are predicted to be prominent at low momentum transfer. Characteristically, high-spin states dominate at large angles and momentum transfers.

It is difficult to ascertain the analogue character of the $T_>$ states for inelastic scattering. Therefore, charge exchange reactions of the type (n,p), $(\mu^- + p_{nuc} \to \nu + n_{nuc})$, (π^-,π^0) are motivated. In the case of muon capture one has confidence in the understanding of the reaction mechanism but can only observe the final state from its decay via nucleon or photon emission. In the case of (π^-,π^0) the reaction mechanism is at present less well understood than for muon capture and the accuracy with which one can determine the energy of the π^0 from the detection of the two decay photons is still not definitively known. However, if possible, detection of the final decay particles would pin down the energies of the final states of $^{48}K^*$. The (π^-,π^0) reaction has the advantage over muon capture of allowing one to vary the momentum transfer delivered to the nucleus.

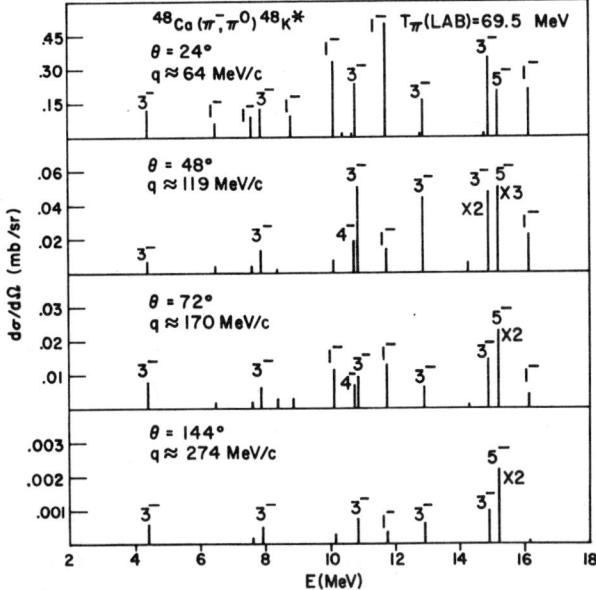

Fig. 2. The $^{48}Ca(\pi^-,\pi^0)^{48}K^*$ (pion single-charge exchange) differential cross sections as a function of the excitation of the final T = 5, $^{48}K^*$ nucleus (relative to the g.s. of ^{48}Ca) for $T_\pi(lab)$= 69.5 MeV. The differential cross sections are shown for four different scattering angles (momentum transfers). All states shown have T = 5. Only states with appreciable cross section are included in the diagram.

In conclusion, we argue that despite reaction mechanism complications, the pion may be useful in studying T = 0 high spin states as well as such collective excitations as a (possible) isoscalar quadrupole resonance and $T_>$ giant dipole states, and should provide useful information on deformation parameters associated with low lying T = 0 normal parity collective states. Comparison with results from other probes as well as a study of the dependence of the pion-induced nuclear excitation spectrum (over a range of momentum transfer ~50 - 450 MeV/c) as a function of the incident pion energy is desirable.

REFERENCES

1. M. K. Gupta and G. E. Walker, Nucl. Phys. A256, 444 (1976).
2. M. G. Piepho and G. E. Walker, Phys. Rev. C9, 1352 (1974).
3. T.-S. H. Lee and F. Tabakin, Nucl. Phys. A226, 253 (1974).
4. A. T. Hess and J. M. Eisenberg, Nucl. Phys. A241, 493 (1975).
5. I. Sick, E. B. Hughes, T. W. Donnelly, J. D. Walecka and G. E. Walker, Phys. Rev. Lett. 23, 1117 (1969).

6. T. W. Donnelly, J. D. Walecka, G. E. Walker and I. Sick, Phys. Lett. 32B, 545 (1970).
7. T. W. Donnelly and G. E. Walker, Ann. of Phys. 60, 209 (1970).
8. J. M. Wyckoff, B. Ziegler, Phys. Rev. 137, B576 (1965).
9. P. J. Moffa and G. E. Walker, Nucl. Phys. A222, 140 (1974); P. J. Moffa, Ph.D. thesis Indiana University (1974), unpublished
10. G. W. Edwards and E. Rost, Phys. Rev. Lett. 26, 785 (1971). See also E. Rost "Lectures from the LAMPF Summer School" July 23-26, 1973; LA-5443-C, 135.

Summary of Walker's answers: I have compared representative results for pions, protons and electrons in my talk. Alpha's are responsible for $\Delta S = \Delta T = 0$ transitions and therefore, for example, do not appreciably excite non-normal parity states.

The low lying collective states, while interesting to study, can be easily reached via other probes and do not require the $\sigma \cdot \nabla$ part of the pion operator even in pion inelastic scattering.

The $\sigma \cdot \nabla$ operator is worrisome to some because of the presence of the derivative term. In order to obtain collaborative information about the derivative matrix elements I have the following suggestions:
1) In medium energy inelastic proton scattering at high momentum transfer the short range two particle spin-orbit interaction plays a dominant role in the transition matrix element and
2) in (γ,p) interactions the convection current operator plays a dominant role.

Schiffer: The proton data with this resolution are just beginning to come in. Pion data of similar resolution are way in the future. At this stage, before analyzing the proton data, is it not premature to speculate on the (π,π') results? Also, the (n,p) data gives you the same information as (π^-,π^0), it is a factor of 10^6 easier to take than the pion data, yet it hasn't been done because it has been judged too difficult. Should we not save the harder pion experiments for a later data and concentrate on the nucleon experiments?

Walker: I certainly encourage the experimental community to perform the nucleon experiments, but I don't want to deny the pion its role in destiny. The pion experiments are presently being planned for Los Alamos.

Negele: Could you clarify your remarks about not being able to excite these states in electron scattering? Did that remark apply to the Stanford and MIT accelerators, or also to the Saclay machine? I assume there is sufficient energy at Saclay so that you could achieve high momentum transfer at forward angles.

Walker: I would agree with that point, in principle, and I would encourage people to do such experiments. It would be useful to have overlapping data with two different probes. But even if the longitudinal form factor dominates and T=0 and T=1 states were equally excited, one certainly doesn't excite T=0 spin flip states.

Radvanyi: Could someone from LAMPF give us the expected energy resolution of their π^0 spectrometer?

Heffner: About 2 MeV; later in this conference, I will present a talk about this subject.

Weber: I have a question for Dr. Negele. How would you look at a magnetic transition to a 1^+ state (e.g., in ^{12}C) in electron scattering, and separate the orbital from the spin-flip transition? In electron scattering you cannot do this, I believe, but Wilkin has proposed that pion scattering in the vicinity of the (3,3) resonance can be used to disentangle this.

Negele: In principle I support this, but first I suggest that one pick a case where one can use both probes and prove that you really can separate the two effects.

Kisslinger: I would like to ask a "realist's" question. I can understand 'collective' states which are defined in terms of low momentum transfer behavior. However, at the very large momentum transfers which you advocate, are these states still 'collective'?
Walker: A collective state is a state which exhausts the sum rule for that q.

Schiffer: Your nuclear structure approach presents a rather simplistic view of nuclear excited states. For example, the states of low spin lie in a region where the level density of excited states is enormous. If this is true, won't experimental results fragment so completely as to bear no resemblance to your simple particle-hole picture?
Walker: It is true that the shell-model picture is quite a simple one; however, for the states of higher spin the level density is very low, and for these states the shell model picture should be very accurate.

McVoy: One of the unique properties of pions as a probe is our ability to exploit the energy dependence of the pion interaction to learn structure information; by working at the (3,3) resonance energy we explore the nuclear surface, and by moving away from this energy we can probe the interior of the nucleus.
Walker: I completely agree.

Arvieux: I would like to announce that at SIN we have very recently taken experimental data for the (π,π') reaction for 150 and 190 MeV pions on ^{12}C. We see several excited states of ^{12}C in the region

19-22 MeV, where Walker has predicted excitation of high spin states. A contribution on this work will be presented later in this conference.

Schiffer: How long did it take to get that spectrum?

Arvieux: About one hour.

PION DOUBLE CHARGE EXCHANGE (DCE) - Gerald Miller

Miller:
Premises: (1) The (π^+,π^-) reaction must proceed on at least two nucleons.
(2) The elastic DCE final state is a member of the same isospin multiplet as the ground state (double analog). The final state has the same nuclear structure as the target, so that correlations in the initial and final states are the same.
(3) For heavy targets, calculations give an energy range (90-200 MeV) where the uncorrelated processes should be small; the pion excites many inelastic channels and elastic processes are suppressed.
(4) NN correlations allow the reaction to take place in many ways - (a) two π-N inelastic reactions followed by an NN interaction yielding the elastic DCE, or (b) virtual excitation consisting of two inelastic π-N reactions also leading to elastic DCE.
(5) The decisive feature is the overlap of the intermediate pion Green function with the correlation function - both have nodes which leads to a rapid energy dependence.
(6) Sample calculations by Miller and Spencer show the overlap to be largest at ~100 MeV. There the correlation effects dominate and give T^2 dependence.
(7) Distorted waves inhibit the uncorrelated process. However, the correlated process is insensitive to reasonable distorted waves.

Schiffer: How will you get something meaningful out of such a small cross section when there may be many second-order effects of comparable magnitude? Further, we know that pion absorption is also sensitive to correlations, and I wonder how that fits in with your analysis?

Radvanyi: In fact, the correlations between nucleons represent a variety of meson exchanges, the propagation of isobars, etc. Should an analysis of the DCE not include these processes in order to avoid contradictions? Also, what damping effects might smooth out the very fast variations of the cross section with energy?

Negele: I would like to ask you to focus on the crucial difference between sensitivity and accuracy. For years we have seen nucleon multiple scattering calculations which are sensitive to correlations in the sense that one gets different answers for different brands of correlation functions, but as yet there has been no convincing calculation which enables one to accurately determine the actual nucleon 2-body correlation function. Given our primitive ability to cope with the many body problem of a nucleus comprised of both mesons and nucleons, how are you going to disentangle correlation effects from other problems like intermediate ρ-exchange or time-ordered graphs in which many different mesons cross over one another in the intermediate states?

Secondly, in the much simpler case of ordinary isobaric analog states, the single most fundamental result, namely the Coulomb energy shift, is simply not understood from first principles. Although I still believe it is attributed to charge asymmetry in the nucleon force, this belief is not yet substantiated by a fundamental derivation of charge asymmetry in the nuclear force. Hence, is it not rather dangerous to prejudice your entire theory on the simplistic assumption that the structure of the double analog state is identical to that of the target, when the single analog state is so poorly understood?

NUCLEON-NUCLEON CORRELATIONS AND
ELASTIC PION DOUBLE CHARGE-EXCHANGE REACTIONS[*]

Gerald A. Miller
Physics Department, University of Washington, Seattle, WA 98195

INTRODUCTION

We would like to explain why the elastic pion double-charge exchange reaction (EDCE) offers us an unusual opportunity to learn about nucleon-nucleon interactions in the presence of the nuclear medium. The basic idea is very simple. In order to conserve charge this reaction must proceed on at least two nucleons. If these nucleons are correlated, that is, undergo interactions in the initial or final states (or both), then the elastic double charge-exchange reaction may proceed via two inelastic charge-exchange reactions. Whereas if there were no correlations, two inelastic reactions would leave the nucleus in an excited state. The presence of correlations vastly increases the number of allowed intermediate states and one is faced with the possibility that the elastic double charge exchange cross section is significantly enhanced.

Two other reasons for pursuing this in more detail are: (1) Accurate experiments will enable us to examine the reaction in which the final nuclear state is a member of the ground state isospin multiplet. This state, or double analog, is assumed to have the same structure as the target ground state and the question of whether the correlations occur in the initial or final state does not matter. In either case, one is observing ground state correlations. (2) It turns out that the energy dependence of the pi-nucleon cross-section allows us to determine a fairly large energy range in which that part of the EDCE reaction proceeding without correlations (which we call the uncorrelated process) is expected to be essentially negligible. Thus background processes are effectively turned off.

We now speculate on how one would hope to use these ideas. (I do not mention data as, to my knowledge, there is none at the interesting energies.) First notice that there have been several different calculations of the uncorrelated process for _heavy_ targets which show that, from energies of about 90 MeV to 200 (or higher) MeV, the uncorrelated process is very small. Furthermore, where one would naively expect the target dependence of the cross sections to be proportional to the square of the isospin (T), the calculations show that the cross sections generally decrease as the target size increases. This suppression, which occurs in many models, arises from the fact that in the vicinity of the pi-nucleon (3,3) resonance, the pion strongly excites many inelastic channels. Thus the uncorrelated process is small at energies where the pi-nucleon T-matrix is fairly large.

Now let us imagine that two-nucleon correlations occur. It is possible to have two successive inelastic charge-exchange reactions followed by a final state interaction between two nucleons which returns the nucleus to the ground state multiplet. One could also imagine a case in which the target nucleus is in a virtual excited state consisting of two particles and two holes which is then de-excited by two pi-nucleon scatterings. There are many other similar processes. The presence of NN interactions in the nucleus allows the elastic double charge exchange reaction to take place in many different ways.

There is another point that will help. A brief consideration of typical terms involving effects of two-nucleon correlation interactions in the medium shows that the results depend on the overlap of the intermediate pion's Green function (times a phase factor for the pion to go from one nucleon to the other) with the correlation function. Because of the presence of nodes in the pion's wave function and in the correlation function, this overlap is expected to have a strong dependence on the initial energy of the pion. Miller and Spencer[1] find, for example, that this overlap is largest in the region about 100 MeV. Because the uncorrelated process is very small at such an energy, the entire DCE cross section is the result of correlation effects. Furthermore at energies where the overlap is largest, one finds that the target dependence of the cross section is almost proportional to T^2. Thus the effects of correlations might have a clear signature. In a fairly narrow energy region the double charge exchange reaction is enhanced over its uncorrelated value and furthermore in that energy region it grows with target size (almost as T^2).

This dominance of the correlated process is made possible by the effect of distorted waves which inhibit the uncorrelated process. One must ask if the correlation effects depend sensitively on the particular models of distortion effects. We claim that reasonable models present very similar distorted waves and the value of the correlated effects are relatively insensitive to model choice. Thus we are optimistic.

QUALITATIVE DISCUSSION

It is particularly useful to focus on the observation of reactions in which the target is left in a state which is a member of the ground state multiplet. Such states (which for the (π^+,π^-) reaction is the double analog) are extremely interesting as these states provide a testing ground for ideas about isospin invariance and Coulomb energy differences. Here we examine the dynamical advantages of such elastic double charge exchange reactions. The double analogs couple to the continuum by charge dependent forces and are expected to have a narrow width. Hence these states, which have been observed in two-neutron transfer reactions on light targets ($A \leq 40$) should be readily observable when targets are irradiated with intense pion beams.

Reactions to the double analog are advantageous because the nucleon-nucleon (NN) interactions may be assumed to be the same whether occurring in the initial or final state. Then the second-order optical potential depends on integrals involving the ground-state correlation function and the interpretation of this reaction is simplified.

Now consider how one could use experimental data to learn something about two-nucleon interactions in the medium. First let us turn off the correlations and imagine that nuclear states are well described by single Slater determinants. The simplest process by which the double analog could be produced is

$$\pi^+ + \text{Target} \to \pi^0 + \text{Analog}$$
$$\pi^0 + \text{Analog} \to \pi^- + \text{Double Analog}$$
(1)

For this two-step process, the intermediate-state must be the analog. Thus the requirement that there be no initial or final NN interactions presents strong restrictions which severely inhibit the cross section.

The (theoretical) result of such restrictions is shown in Fig. 1

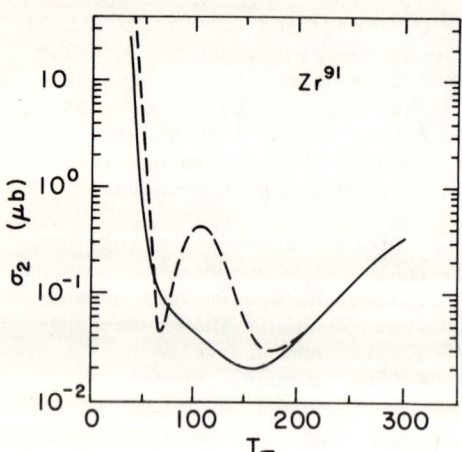

Fig. 1. Energy dependence:―――― Kisslinger,― ― ― local Laplacian.

which is taken from the calculations of Miller and Spencer.[1] The total double charge-exchange cross section (to the double analog) is plotted versus energy for both Kisslinger and Laplacian models of the optical potential. There are two features of these curves which are spurious. 1) The large cross section at low energies occurs because the difference between the S1 and S3 pi-nucleon phase shifts is larger than any other relevant combination at low energies. At low energies the impulse approximation to the optical potential has a very small imaginary part because the effects of true meson absorption are omitted and the isospin flip amplitude acts without inhibition. 2) The sharp dip which occurs for the local Laplacian calculation arises from a cancellation between S and P wave pi-nucleon amplitudes. This cancellation is suppressed in more sensible models in which the T-matrices for S and P wave scattering have different forms and do not interfere in so drastic a manner. Nevertheless, we will see below that the local Laplacian does have some validity.

The salient feature, for both models, is the decrease with energy (for E > 100 MeV) and small size of the cross section. This is the result of the broad pi-nucleon (3-3) resonance. As one approaches this energy region the pi-nucleon T-matrix and the resulting optical potential have large imaginary parts. Thus the resulting total reaction cross section is very large and displays a broad peak. However, it is likely that a pion may be removed from the elastic channels before or after the necessary charge-exchange reaction takes place. Absorption into other channels plays a dominant role in producing the low cross sections and overall structure in Fig. 1. The results of Fig. 2 show that the EDCE cross section decreases as the target size increases up to A = 88 and then levels off. This is in contrast with the naive expectations of the Born approximation in which the cross sections would go like T^2.

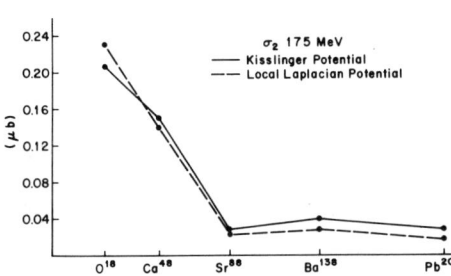

Fig. 2. Target dependence.

These results show that there is an energy region in which the uncorrelated process gives very small cross sections. It should be noted that this result does not seem to be model dependent as several other workers[2] find similar results.

Let us now turn on two-nucleon correlations. The number of allowed intermediate states is vastly increased. Two of the most important processes are shown schematically in Fig. 3 where the heavy line indicates the NN interaction in the medium and the wiggly line represents the pi-nucleon interaction. The charges of the various particles have not been specifically indicated because these processes contribute to elastic and single charge-exchange as well as the double charge-exchange reaction. Terms involving Pauli correlations and terms of second or higher order in the NN interaction are not shown because they generally give smaller effects than the terms shown in Fig. 3.

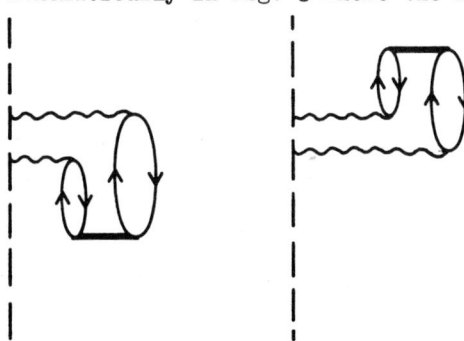

Fig. 3. Some effects of correlations.

The next step is to see whether there are <u>any</u> energies at which terms like those of Fig. 3 dominate. Any evaluation of such terms involves an integral over the distance, s, between the two-nucleons of the following form

$$\int d^3s\ C(s)\ g(s)\ e^{i\vec{k}\cdot\vec{s}} \propto \int ds\ sC(s)(e^{2iks}-1) \qquad (2)$$

where C(s) is the NN correlation function and the exponential arises from the phase difference of the pion wave function over the internucleon distance. The quantity g(s) is the Green's function of the intermediate pion, and includes the distortion caused by the first order pion-optical potential. The wave number, $\vec{k}$, is the complex momentum of the pion in the medium. Its s dependence is suppressed here. The effects of correlations depend sensitively on an overlap of the intermediate pion wave function with the correlation function. As both of these quantities have nodes, the integral is very sensitive to the incident pion energy. At different energies the integral depends on whether the node in the intermediate pion wave function matches the node in the correlation function resulting from the onset of the attractive forces. This effect not only gives a clear signature--the rapid energy variation--but also could be sensitive to different models of the correlation function. With the model[1] of the correlation function shown in Fig. 4, this integral is large in a narrow energy region about 100 MeV.

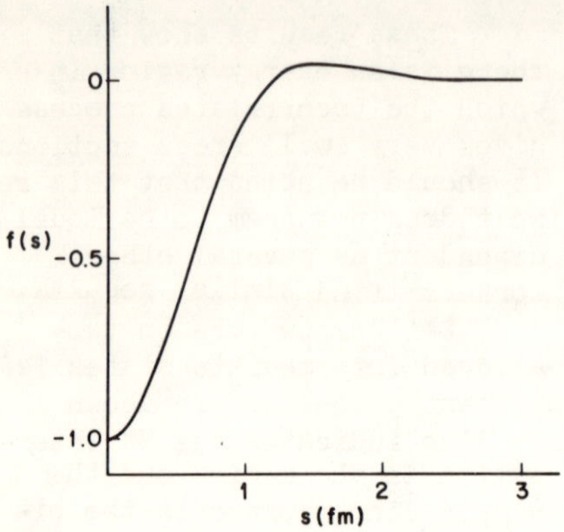

It is worthwhile to compare Eq. (2) to the expression that one would have in the uncorrelated process. Then the factor $C(s)$ is replaced by 1 and the required integral is

$$\int ds\, s\, (e^{2iks}-1) \qquad (3)$$

and there is no chance for a node matching of the e^{2iks} term.

Fig. 4. Defect function $f(s)$.

There are two other points to be made. The first involves sensitivity to the off-shell, pi-nucleon T-matrix. We want to learn about the short distance behavior of the relative two-nucleon relative wave function. If two nucleons are close together and the pion interacts with one of them, the pion is "close" to both nucleons. To make a reasonable calculation one must know the pi-nucleon wave function at small relative distances. It is precisely here where ignorance meets ignorance. However, the situation is not bleak.

The pi-nucleon wave function is related to the off-shell pi-nucleon T-matrix. A sum over intermediate pion momenta must be performed. If the off-shell nature of the T-matrix cuts off the high momentum contributions to the integral, the value of the calculated cross sections could be significantly changed. This point is illustrated in Fig. 5 for two extreme models, the no cutoff model and a Yukawa with a range of 300 MeV. The figure shows the contribution to the cross section as a function of the upper limit of the integral of Eq. (2). One sees the effect of short-range repulsion in suppressing short distance contributions and the dominance of contributions occurring between one and two Fermis. The 300 MeV model represents an extremely short range

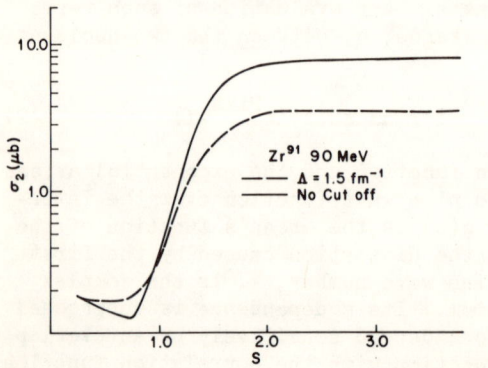

Fig. 5. Cross section vs. internucleon distance.

(in momentum space) and maximizes the suppression caused by the cutoff. Note that field theoretic estimates give much larger ranges which reduce the effect shown in Fig. 5. Even with the <u>extreme</u> short-range model one still finds a <u>very large</u> DCE cross section which is dominated by the effects of correlations.

The role of distortions must be discussed further. Recall that the use of distorted waves plays a dominant role in suppressing the uncorrelated process. One must verify that different models of distorted waves do not give very different values for the second-order terms. We claim that the values of the second-order terms are reasonably independent of the different models of distortion, so that with any reasonable model the effects of correlations are large at certain energies. Consider the thesis work of Phatak who compared the wave functions obtained from the Kisslinger, Laplacian and Londergan, McVoy and Moniz models of the off-shell T-matrix. As shown in Fig. 6, only the Kisslinger model is very different from the other two. The implication of this is that models which do not lead to infinite effective masses give similar distorted waves. Ernst and Miller[3] consider several very different form factors, $v(k)$, in the π-N separable off-shell model. A complete lack of sensitivity of pi-nucleus elastic angular distributions to $v(k)$

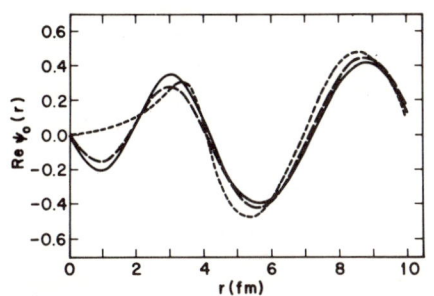

Fig. 6. Wave functions, 100 MeV, Ca^{40}. Kisslinger --- ; LMM ———— ; local Laplacian — — — .

is found. Furthermore, the distorted wave obtained from these different models are very similar. Figure 6 shows that wave functions obtained by two very different off-shell models, which avoid the high momentum pathologies of the Kisslinger potential, have similar behavior at the nuclear surface and, may be expected to give qualitatively similar correlation effects. Hence improvements in various first-order models, which do not result in drastic revision of the high momentum content of the pion wave function, may be expected to give small changes in correlation effects.

We present, as an example, the results of Miller and Spencer which were obtained with distortion crudely supplied by the local Laplacian potential. This is shown in Figs. 7 and 8. In the vicinity of 100 MeV the target dependence of the calculated cross section goes as something like T^2. Furthermore such behavior disappears at higher energies. Hence the behavior of these cross sections provides a clear signature. Because of the specific low energy deficiencies of the model mentioned above, no results are shown for energies less than 100 MeV. More reasonable calculations in which effects of true absorption are included are expected to suppress the low energy cross sections. (Estimates show that such annihilation effects are largest

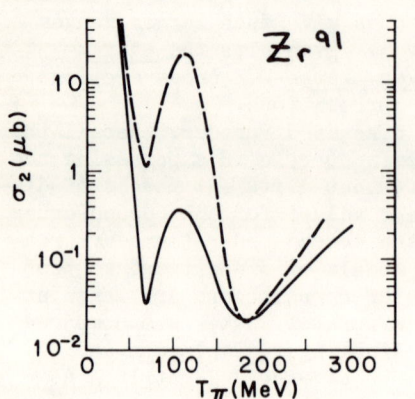

Fig. 7. DCE cross section. ——— Uncorrelated; — — — correlated.

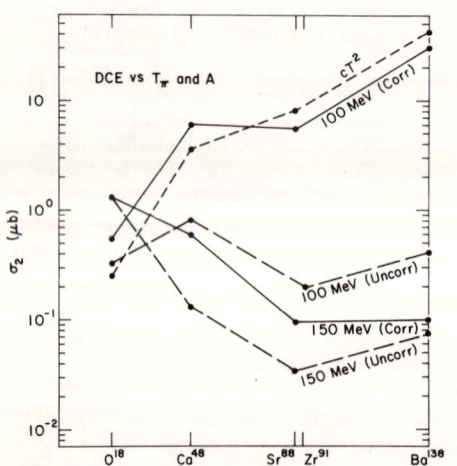

Fig. 8. Global properties ——— correlated; — — — uncorrelated. - - - cT^2

at 50 or 75 MeV but essentially vanish at energies greater than 100 MeV.) We expect to find a finite energy region in which the cross sections are large and increase with the isospin of the target. The finite range occurs from the behavior of the integral over the correlation function, because suppressive effects of the distortion increase as one approaches (3,3) resonance energy and because the low energy cross sections are assumed to be small. The exact position of the T^2 behavior depends on the particular model of the correlation functions as well as on the calculational details discussed above. However, an observation of such qualitative behavior could only result from the influence of NN interactions in the reaction.

We have argued about how extraction of information about NN correlations might proceed. We do not wish to imply that this extraction will be easy. Work must be done on considerations not discussed here. In particular, the role of production of intermediate rho mesons, the effects of the closure approximation and the sensitivity to the off-shell pinucleon T-matrix are presently under investigation by the author, E.M. Henley, and S.N. Yang.

The road to the correlation function is long. It may be that the best we hope for is to set limits on the gross features, such as the volume integral, of the correlation function. Even this would be extremely interesting.

We thank J.S. Blair and E.M. Henley for useful discussions.

REFERENCES

* Supported in part by E.R.D.A.
1. G.A. Miller and J.E. Spencer, Phys. Lett. <u>53B</u>, 329 (1974) and to published in Ann. Phys. (NY).
2. E. Rost and G. Edwards, Phys. Lett. <u>37B</u>, 247 (1971); D. Tow and J.M. Eisenberg, to be published in Nucl. Phys. A.
3. D.J. Ernst and G.A. Miller, Phys. Rev. <u>C12</u>, 1138 (1975).

<u>Negele</u>: I am worried by the very large number of possible reaction mechanisms which could be important in the double-charge-exchange process. For example, if I think in terms of diagrams, I can substitute a ρ meson for the π meson in intermediate scatterings, and recent work by the Stony Brook group suggests that this may be large. One can also permute the pion interaction with the nucleon-nucleon interactions which go into the g-matrix, and one can also think of graphs with different time orderings. How can you be sure you are evaluating the "correct" diagram?
<u>Miller</u>: First, the rho meson, although perhaps important at low energy, should not give large effects at pion energies of 100 MeV. Secondly, in my calculations I have examined terms with the NN interaction in both initial and final states. Finally, I emphasize that it is not necessary to calculate <u>all</u> possible diagrams in order to obtain this qualitative effect. Some of the diagrams you refer to have been included, and others we believe to be small in the energy region (100-200 MeV) where this effect is largest.

<u>Banerjee</u>: I would like to point out that one of the possible diagrams referred to by Negele is precisely the diagram responsible for the Ericson-Ericson Lorentz-Lorenz effect, and hence at least this diagram <u>must</u> be important to you.
<u>Miller</u>: Those diagrams have indeed been included in our calculation, which deals with the whole correlation function.

<u>Liu</u>: I believe that off-shell effects must be more important than you have indicated. For example, scattering of a 90 MeV pion from two correlated nucleons (each 100 MeV off shell) requires evaluation of the πN t-matrix below elastic threshold, requiring an enormous extrapolation of your form factors.
<u>Miller</u>: We find the results of our calculation to be very insensitive to even extreme choices of the πNN form factor.
<u>Liu</u>: Furthermore, you need to emply form factors which satisfy off-shell unitarity.
<u>Miller</u>: I agree; I believe our form factors certainly satisfy this constraint.

<u>Yavin</u>: In the light of the confusion which has developed over theoretical predictions of the (π^+,π^0) reaction, are you <u>sure</u> that you have calculated all small corrections which might change your results?

Miller: I am working on corrections, but I believe they will not affect the qualitative trends which I predict.

Kopaleishvili: Can we really obtain believable information from such reactions?
Miller: Eventually, by examining the sensitivity of our calculations to all parameters, we hope to set limits on some important features of the nucleon-nucleon correlation function.

Negele: I believe that following M. Goldhaber's example with the antiproton the real test of one's belief in a theory should be a willingness to make a public wager about the outcome of theoretical predictions.
Miller: I'm ready.

Miller: I'd like to summarize my responses to the realists:
To Dr. Schiffer: Although the cross section is small, it is not exotic, as it is dynamically the same as elastic scattering. The smallness is caused by the restriction on the final state. The "true" meson absorption decreases as a function of energy due to the rise of the suppressive effects of "regular" pion absorption (inelastic or quasielastic scattering). The same effect (increase of π-nucleon scattering) which suppresses the "true" meson absorption enhances the correlation term and there seems to be a clear separation as a function of energy.
To Dr. Radvanyi: The effects of meson degrees of freedom in the nucleus, which may come in at high momentum transfer, should not have the energy dependence I show here.
To Dr. Negele: The predicted cross section grows with target size ($\sim T^2$) in a narrow energy region. This is a definite signal of correlation effects. One does not have to calculate the cross section exactly. Discrepancies in Coulomb energy differences are about 5%. Such effects enter only when considering effects of charge dependent and asymmetric forces on the correlation function.

STRANGENESS EXCHANGE REACTIONS - Avraham Gal

Gal:

Premises: (1) It is suggested that the Heidelberg (K^-,π^-) hypernuclear spectrum, for pions observed in the forward direction, is dominated by quasi-free $K^-n \to \pi^- \Lambda$ interactions. We emphasize that $qR \ll 1$ is the coherence condition leading to analogs, but for the present experiment $qR \gtrsim 1$, and incoherent processes cannot be ignored.

(2) We wish to avoid the quasi-free region as much as possible, and concentrate on analog and supersymmetric states. The identification of the supersymmetric state gives a clue in some instances as to the exchange mixture of the ΛN force, and in other cases to its spin-orbit part.

Negele: The one aspect of your thesis which appears most relevant to the topic of this panel discussion is the hope of using hypernuclei to learn about Λ-nucleon interactions. Hence, I would like to remind you of two well-known problems in nuclear structure. From a wide variety of scattering data we have a tremendous amount of information about nuclear forces. At the very least, it is clear that they are attractive at long range and repulsive at short range. Now, when an experimental realist extracts the particle-particle, particle-hole, and hole-hole matrix elements of the effective interactions in nuclei from spectroscopic data, he tells me that the effective interaction is repulsive at long range and attractive at short range! In the face of this dilemma in the nuclear case, where the data is far more complete than you can ever expect for in hypernuclei, what sense does it make to seek information about Λ-N interactions from hypernuclei?

As a second example, consider spin-orbit splittings in nuclei. Although we know the two-body spin orbit force very well, it is impossible to calculate microscopicly the observed spin-orbit splittings for spin-unsaturated nuclei. Hence, if we were to work backwards from the experimental splittings with our present unsatisfactory theoretical methods, we would necessarily extract the wrong spin orbit force. Howe can you possibly expect to do better in hypernuclei than has been done in ordinary nuclei?

Schiffer: I am a little worried that, due to the vast effort and time required to obtain more data, the theoretical picture you and others are creating may tend to "mold" the data. I wonder if we could keep the mold somewhat flexible for a while?

QUASI-FREE FORMATION OF AND SUPERSYMMETRY
IN LIGHT HYPERNUCLEI*

A. Gal[†]
Department of Physics
University of Virginia, Charlottesville, Virginia 22901

ABSTRACT

A recent suggestion,[1] that the hypernuclear excitation spectrum observed[2] in (K^-,π^-) nuclear reactions in the forward direction is dominated by a quasi-free broad hump, is reviewed. The removal of this quasi-free peak is essential for identification of low energy analog[3] and non-analog special states. The future observation of supersymmetric[4] hypernuclear states will yield valuable information on the ΛN interaction. Results are presented for several p-shell hypernuclei.

QUASI FREE FORMATION

The $K^- \ ^A Z \to \pi^- \ ^A_\Lambda Z^*$ data for $A \leq 16$, presented a year ago at the Santa Fe meeting,[5] was considered by many as (i) proof that for small momentum transfer (~70 MeV/c for $K^-n \to \pi^-\Lambda$ at P_K = 900 MeV/c and π^- in the forward direction) the resulting hypernuclear spectrum is dominated by coherent excitations of strangeness analog states, and (ii) an indication that this situation will also occur for heavier nuclear targets. Both these expectations seem presently[2] to be negated by a careful elimination of the non-nuclear $K^- \to \pi^-\pi^0$ background and an extension of the previous measurements to ^{32}S and ^{40}Ca. As is clear from Fig. 1 the hypernuclear excitation spectrum is dominated by a broad hump. Whereas for $^{16}_\Lambda O^*$ one prominent sharp peak is clearly superposed on this broad hump, there is no compelling evidence in $^{40}_\Lambda Ca^*$ for any narrow peak.

In a very recent attempt to explain these features Dalitz and Gal suggested[1] that quasi-free (QF) $K^-n \to \pi^-\Lambda$ interactions dominate the observed hypernuclear (K^-,π^-) spectrum. The nuclear collision is viewed as occurring incoherently with kinematics essentially given by free particles. Quasi-free processes have been identified in (e,e') and (p,p') nuclear interactions for momentum transfer $q \gtrsim k_F$, where k_F is the Fermi momentum. Since the final baryon is a Λ hyperon, there is no Pauli restriction on its final states and such QF processes become possible even for $q \ll k_F$. These incoherent processes are expected to take over coherent processes for $qR \gtrsim 1$, a condition satisfied by the recent measurements.

For a first orientation, the shape of the QF spectrum is evaluated for a Fermi gas of target neutrons. The nuclear interactions

---
*Work supported, in part, by the U.S. National Science Foundation.
[†]On leave from the Hebrew University, Jerusalem, Israel.

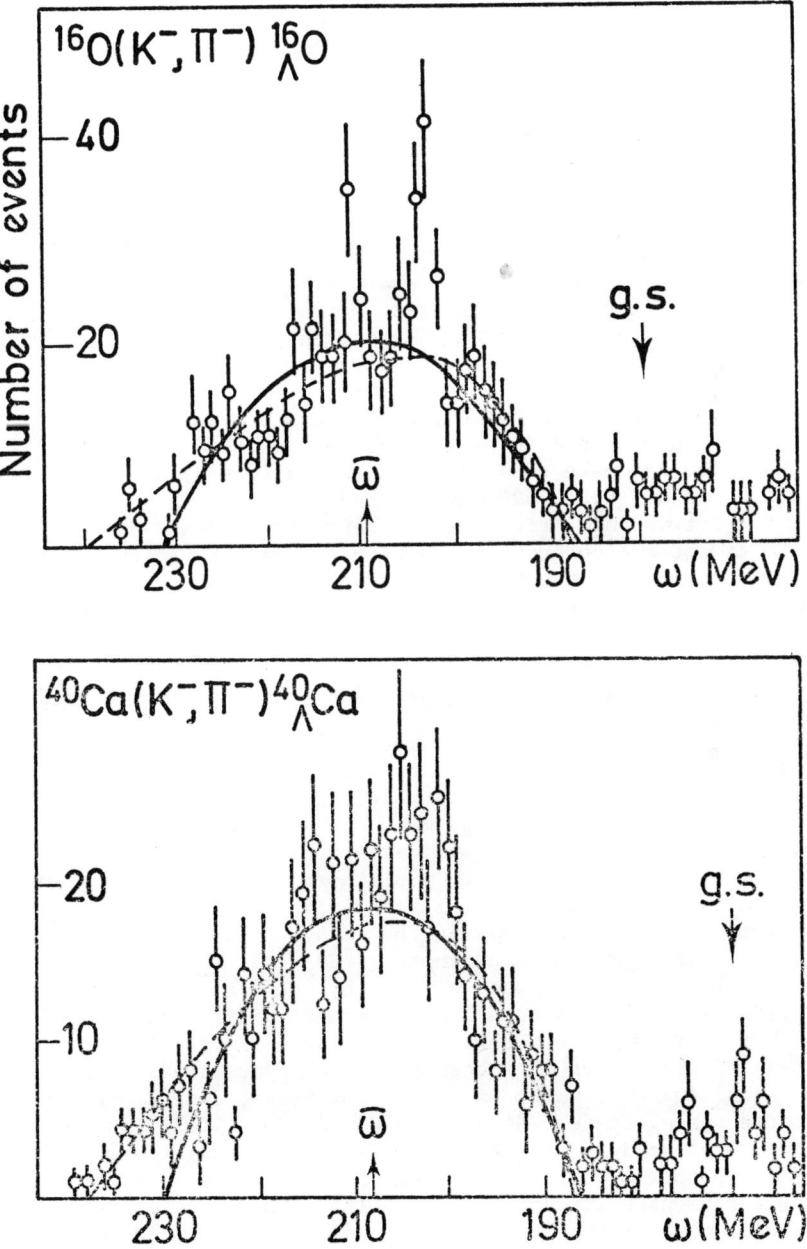

Fig.1. Excitation spectra for the (k^-,π^-) reaction on ^{16}O and ^{40}Ca are compared with a normalized calculated QF shape for a Fermi gas of neutrons. The solid line corresponds to the parabola mentioned in the text, whereas the dashed line gives the distortion of this parabola as described in the text.

are represented solely by a nuclear well of depth U_N for neutrons and depth U_Λ for the final Λ. Ignoring the variation of q with the momentum of the struck nucleon and the ΛN mass difference, the distribution predicted for the specific nuclear energy transfer Q (defined as the overall energy transfer $\omega = (E_{K^-} - E_{\pi^-})$ minus the mass difference $(M_\Lambda - M_N)$) is a parabola, centered on

$$\bar{Q} = (U_N - U_\Lambda) - \frac{1}{4}(\frac{1}{M_N} - \frac{1}{M_\Lambda})k_F^2 + q^2/2M_\Lambda \quad , \qquad (1)$$

vanishing beyond the interval $[\bar{Q} \pm qk_F/M_\Lambda]$. This distribution, normalized by eye to the observed height of the broad (K^-, π^-) hump is shown in Fig. 1 by a solid line. The dashed line in Fig. 1 shows the distortion of this parabola when the dependence of q on the momentum of the struck nucleon is considered. It gives a good account of the observed hump. The following comments are appropriate:

(i) The width of the QF peak is given, in the Fermi gas model, by qk_F/M_Λ. One may make this width quite small by going to the "magic" kaon momentum[6] $p_{K^-} \sim 420$ MeV/c, where the nuclear momentum transfer in the forward direction goes to zero (for assumed $U_N - U_\Lambda = 30$ MeV). We expect the coherent transition to dominate in this case, giving rise to most of the total strangeness exchange strength. We see that the usual coherence condition $qR \lesssim 1$ is insufficient when the final baryon is a Λ hyperon which is not obeyed by the Pauli principle; the more restrictive condition is $qk_F/M \lesssim \Delta E$ where ΔE is of the order of several MeV, signifying separation between single particle configurations.

(ii) The QF shape mentioned above includes also the coherent processes which may give rise to narrow strangeness analog states, but for which the simple Fermi gas model is inadequate. Since both the coherent strangeness exchange mechanism and the incoherent QF mechanism can have some final physical states in common, an interference between the two appropriate amplitudes is not ruled out and this may be the origin of the small subsidiary peak, or a shoulder, on the right of the narrow peak in the data.

(iii) For the present experiment $q^2/2M_\Lambda \sim 3.5$ MeV, quite a small value, and the QF peak is roughly given by the difference $U_N - U_\Lambda$. This implies a value of about 31 MeV for this difference, and with $U_\Lambda \sim 27$ MeV this is quite a reasonable value. Previously, Schiffer and Lipkin[7] implied that the parameter $U_N - U_\Lambda$ is, in first approximation, equal to the Q value of the narrow peaks, interpreted as strangeness analog states (SAS), in $^{12}_\Lambda C$ and $^{16}_\Lambda O$. Their interpretation would lead to $U_N - U_\Lambda \sim 20$ MeV. The most straightforward test of the QF hypothesis would be to observe the hypernuclear excitation spectrum for pions emitted with $\theta \neq 0°$. It is then predicted that the QF peak will move to higher values of $\bar{Q}$ with increasing θ, due primarily to the term $q^2/2M_\Lambda$. Thus, for the present experiment ($P_K = 900$ MeV/c) and pions emitted at $\theta = 20°$ in the laboratory frame, the QF peak is predicted to move about 40 MeV above the broad peak seen for $\theta = 0°$. The coherent excitations, however, are not expected then (q ~ 320 MeV/c) to dominate the low energy part of the spectrum and for this reason it becomes rather important to have a theoretical framework for the expected hypernuclear states. We note that in future experiments the QF structure will have to be separated out by theoretical means, since an experimental separation requires $q^2/2M \gtrsim qk_F/M$, i.e. $q \gtrsim 2k_F \simeq 540$ MeV/c, a very high value for q.

(iv) The reported constancy with A of the measured hypernuclear summed cross sections for $A \leq 40$ implies that the reactions $K^-n \to \pi\Lambda$ are limited to the periphery of the target nucleus. With the strong absorption of the ingoing K^- and outgoing π^- only a small fraction of the valent neutrons effectively participates in strangeness exchange. Absorption will obviously damp the formation of any particular hypernuclear final state. This holds, in particular, for the formation of coherent states derived from the SAS doorway state. In view of the strong absorption, the strongest coherent transition is expected for neutrons in the valent shell, irrespective of whether or not the energies of the various states $(n\ell)_n^{-1}(n\ell)_\Lambda$ will differ from that for the valent shell. Nevertheless, in going to heavier nuclei the degree of coherence for the valent transition is gradually destroyed, since the phase $e^{i\vec{q}\cdot\vec{r}_i}$ accompanying each possible production center $\vec{r}_i$ cannot be ignored; $qR \sim 1.7$ for ^{40}Ca with the present kinematics. It is then very likely that if Povh's experiment were carried out beyond ^{40}Ca, no sharp excitations would have been identified.

GENERAL EXPECTATIONS

We now turn to a discussion of the expected hypernuclear spectrum for larger values of q than to date, and when the QF hump is removed. In view of the strong absorption prevailing in strangeness exchange, the following discussion is limited to processes occuring on valent neutrons.

Without assuming any symmetry relationship between the ΛN and NN interactions, we follow Kerman and Lipkin[3] and introduce U-spin (+1/2 for a neutron, -1/2 for a Λ and 0 for a proton). Nuclei are trivially classified in U-spin multiplets with $U_3 = U_{max} \equiv N/2$. The reaction operator $\Sigma_j e^{i\vec{q}\cdot\vec{r}_j}\cdot U_-(j)$ transforms any nuclear state $|i\rangle$ into a mixture of $U_3 = U_{max} -1$ states with $U = U_{max}$ and $U = U_{max} -1$. For momentum transfer $qR \gg 1$ it is easy to show that, on the average, only a fraction (1/N) of any parent nuclear state is converted into $U = U_{max}$ hypernuclear states, whereas the rest, obviously the majority, is transformed into $U = U_{max} -1$ states. The $U = U_{max}$ states are "analog" states, though not necessarily of the g.s. target nucleus, since they can each be represented by $1/\sqrt{N}(U_-|i\rangle)$ for some nuclear state $|i\rangle$. The $U = U_{max} -1$ states are "non-analog" states; they cannot be obtained from any nuclear state by operating with U_-. The possibility exists, therefore, to form "non-analog" hypernuclear states, within a given configuration, whose space-spin symmetry is higher than that realized for the analog state. Such states, with higher symmetry are expected to lie below the SAS, for a wide range of reasonable exchange mixtures of the ΛN interaction.

For medium weight and heavy nuclei the situation may become similar to that observed in nuclear (p,n) reactions for large angles. For example, with 20 MeV incident protons and backward outgoing neutrons the nuclear momentum transfer is about 400 MeV/c, of the same order as will be achieved by observing π^- emitted in Povh's experiment at $\theta_L \sim 25°$. Thus, within the valent neutron configuration the doorway strangeness analog state will split into several components, mostly by mixing with neighboring non-analog states of the same configuration due to the residual ΛN interaction. Below, a group of non-analog states with higher symmetry is expected; these supersymmetric states[4] parallel the "isobaric configuration states" observed in CEX. Lowest, states of other configurations, where the Λ occupies a lower shell-model orbital, such as the 1s, may be found. Such a grouping into 3 essential structures is typical of (p,n) reaction, (a comment made to us earlier by Dr. A. Lane) but the main difference to recall is that instead of one analog state forming the upper structure in CEX, a distribution of analog strengths is expected to define the upper structure in SEX. We note that when the reaction operator concerns only excess neutrons, the resulting states for closed proton shells automatically have isospin value I-1/2; no SU(3) arguments are involved. Thus, there is no further I-spin splitting of this hypernuclear spectrum.

SUPERSYMMETRY IN THE P SHELL

Light hypernuclei, where both I-1/2 and I+1/2 states are expected to be formed in strangeness exchange on valent neutrons, deserve special treatment. Here we shall follow the recent analysis by Dalitz and Gal[4] for $^{16}_\Lambda O^*$ and $^{9}_\Lambda Be^*$ in the p-shell. We confine our attention to the first excited configuration $(1p)_\Lambda (1p)_n^{-1}$ in the parent nucleus, and in view of the expected good overlap of the 1p baryonic orbitals use Sakata SU(3) classification[3] in zeroth order; this need not be appropriate for heavier nuclei. The ground state configuration $(1s)_\Lambda (1p)_n^{-1}$ was discussed in ref.8. Consider $^{16}_\Lambda O^*$ 0+ states. One of the two appropriate basis states may be chosen to be SAS, defined by:

$$\psi_a = \frac{1}{\sqrt{3}} [\sqrt{2} \, (p_{3/2})_\Lambda (p_{3/2})_n^{-1} + (p_{1/2})_\Lambda (p_{1/2})_n^{-1}] \quad , \quad {}^1S_0 \, . \quad (2)$$

The orthogonal combination is given by:

$$\psi_s = \frac{1}{\sqrt{3}} [(p_{3/2})_\Lambda (p_{3/2})_n^{-1} - \sqrt{2} \, (p_{1/2})_\Lambda (p_{1/2})_n^{-1}] \quad , \quad {}^3P_0 \, . \quad (3)$$

The direct formation of the latter is forbidden in the forward direction because the transition $^{16}O_{g.s.}(^1S_0) \to {}^{16}_\Lambda O(^3P_0)$ requires spin flip. However, the two states (2), (3), may be mixed due to the residual ΛN interaction. The only ΛN two-body interaction capable of accomplishing this is the <u>antisymmetric</u> ΛN spin-orbit interaction. This will probably result in some non-zero value for the parameter $\delta\zeta \equiv \zeta_\Lambda - \zeta_N$, where ζ is the strength of the one-body spin-orbit interaction $\zeta s \cdot l$. By observing only <u>one</u> strong hypernuclear excitation in the appropriate energy range (Fig.1a), one concludes that $\delta\zeta$ cannot be significantly large. In fact, Dalitz and Gal show that, with a reasonable estimate for the ΛN exchange mixture of a spin-independent ΛN residual interaction, a bound of $-1.3 \leq \delta\zeta \leq 2$ MeV is obtained by requiring the relative intensity for production of ψ_s to be less than 5%. The energy separation varies then from 4.5 to 6.9 MeV, the state (3) lying lower than the SAS, eq.(2).

In this example, ψ_s is a non-analog state whose space symmetry is higher than that of the analog ψ_a. It is therefore called a supersymmetric state, although nuclear physicists usually call such a state "anti-analog." The term anti-analog is not uniquely defined, however, for the situation where many non-analog states appear. Supersymmetric states are generally expected to lie considerably below the appropriate analog state; the relatively weak spin dependence of the ΛN force is not going to affect this situation in a qualitative way.

Although other states with $J^\pi \neq 0^+$ are expected to be resolved in further experiments, all of them yielding some information on the ΛN interaction, the unique relationship between the two states (2), (3), offers one of the best direct estimates of the Λ central spin-orbit splitting. As stated above, the latter is believed to have the same sign and a comparable magnitude as the nucleon spin-orbit splitting.

The case of $^9_\Lambda Be^*$ is quite instructive and we shall outline the results of Dalitz and Gal in a simplified version. First, the spin-orbit interaction is probably not that important in the beginning of the p shell. This observation is confirmed by nuclear intermediate coupling calculations, leading to LS basis as a reasonable starting point for $A < 10$. ^9Be ground state is well described by a combination of $^{22}P_{3/2}$ and $^{22}D_{3/2}$ of the [4,1] Wigner supermultiplet, and for the sake of demonstration the D component will be suppressed. The fractional parentage of $^9Be(g.s.)$ is derived from existing nuclear calculations and agrees with the results of $^9Be(p,d)$ pick-up experiments. The situation is depicted qualitatively in Fig.(2a). The lowest group of states in ^8Be, serving as parents to $^9Be(g.s.)$, consists of $^8Be(g.s.)$ and $^8Be^*(2.9$ MeV$)$ both of which (I=0) are given, to a first good approximation, by a supermultiplet assignment [4]. The other group of parents is concentrated around 17 MeV excitation energy in ^8Be, with the supermultiplet [3,1] assignment. It includes both I=0 and I=1 levels. We now limit our attention to $^9_\Lambda Be^*(1p)_\Lambda \cdot (1p)_n^{-1}$ states with $J^\pi = 3/2^-$, since these include the coherent excitations observed to date.

Generalized supermultiplets in $^9_\Lambda Be^*$ are sketched in Fig.(2b). If the residual ΛN interaction were negligible with respect to the residual NN interaction in the corresponding states, one would have expected the two clusters of $^9_\Lambda Be^*$ states denoted by [4,1] to be separated by about 17 MeV from each other, as their corresponding ^8Be parent supermultiplets are. The $^9_\Lambda Be$ doorway $(1p)_\Lambda (1p)_n^{-1}$ analog state would then feed both of these clusters. As the ΛN residual interaction is increased, these two clusters approach each other and coincide in the limit of equal ΛN and NN residual interactions. In this limit Sakata SU(3) symmetry holds also for the Hamiltonian. In reality, the ΛN residual interaction is expected to be weaker than the NN residual interaction and the actual situation would be intermediate between the above limits. Experimentally, two prominent peaks are observed in $^9Be(K^-,\pi^-)$ (superimposed on what may be identified as QF hump), which are separated by about 11 MeV.[†] (footnote on next page)

This separation tells us something about the relative strength of the ΛN residual interaction. It does not reflect the I-spin splitting into I=0 and I=1 states in $^9_\Lambda Be^*$, since both upper and lower [4,1] clusters include I=0 states which are remnants of the doorway analog; these latter states, a_1 and a_2 in Fig.(2b), each involve large admixtures of non-analog states. In this picture, the upper observed $^9_\Lambda Be^*$

†(referring to previous page) The possibility that the QF structure itself is split into two components, due to the large symmetry energy of the 1p neutrons in ^9Be, is not completely ruled out by the present data.

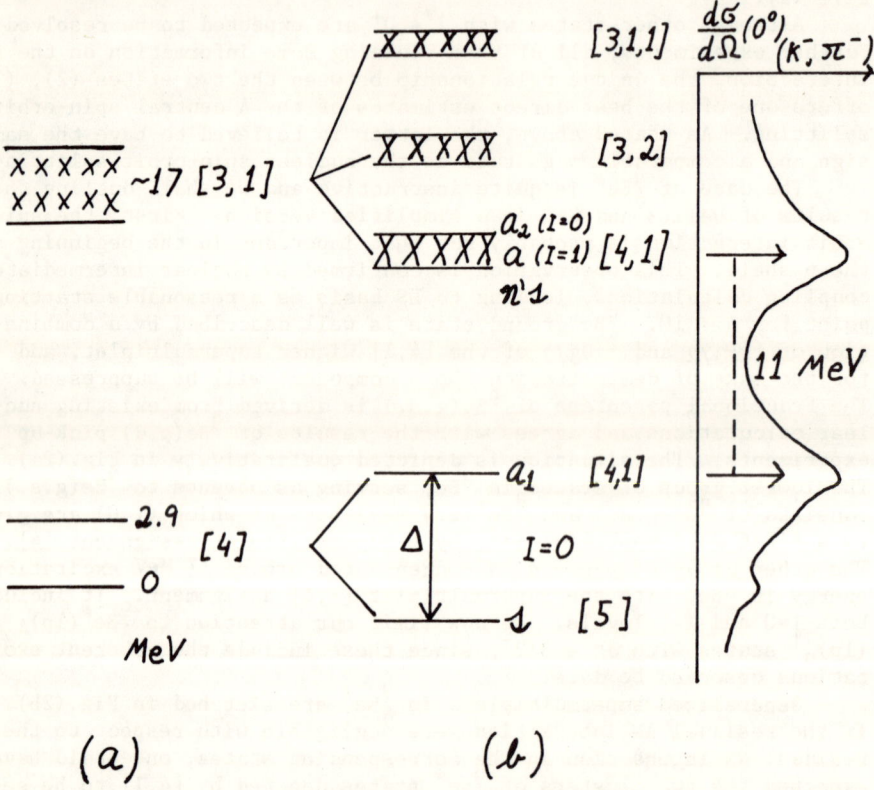

Fig.2. (a) ^8Be parents of ^9Be(g.s.). (b) Schematic representation of $^9_\Lambda Be^*$(3/2⁻) states. The two observed peaks, separated by about 11 MeV, are currently believed to originate from the analog state a and states a_1, a_2 which involve large admixtures of non-analog states. The upper [4,1] group contains other non-analog states, denoted by n's, which are not expected to be strongly coupled to a_2 and a, unless the spin dependence of the ΛN residual interaction is strong. The supersymmetric state s, with $(1p)^5$ space symmetry [5], has not yet been observed.

peak consists of both I=0 (a_2) and I=1 (a) states, whereas the lower peak consists of an I=0 (a_1) state. The spin dependence of the baryon-baryon interaction, ignored in the above qualitative discussion, is not expected to change the gross features of this interpretation, except for enforcing further mixings among the non-analogs (n's) and a_2, a. A sizeable spin-spin ΛN residual component would appreciably split the upper peak into its I=0 and I=1 components, but this does not seem to be the case as the experimental resolution is currently better than 2 MeV.

The relationship between the lower observed peak and the supersymmetric state s labelled by the maximal symmetry type [5] is less susceptible to uncertainties mentioned above. The latter state is uniquely expected to be the lowest $3/2^-$ in $^9_\Lambda Be^*$. The spacing Δ between the observed a_1 and the hitherto unobserved s is sensitive to the exchange-mixture of the ΛN interaction. For a fixed ΛN s-wave interaction strength, Dalitz and Gal found the following model dependence: $\Delta = 5.5$ MeV for no exchange mixture, whereas $\Delta = 8.6$ MeV for a Serber ΛN mixture.

A common feature in the above discussion of $^9_\Lambda Be^*$ and $^{16}_\Lambda O^*$ is the appearance of an hypernuclear doublet, the lower member of which is a supersymmetric state. The upper member of this doublet has been identified with the lowest narrow excitation observed. Although the lower member has not yet been observed, there can be no doubt about its existence. This belief derives from the rather high excitation energies observed for the relevant $^9_\Lambda Be^*$ (16 MeV) and $^{16}_\Lambda O^*$ (~21 MeV) states relative to the observed 0^+ $[(p_{3/2})_\Lambda (p_{3/2})_n^{-1}]^{12}_\Lambda C^*$ (11 MeV). For ^{12}C in the jj coupling limit[9] no 0^+ supersymmetric $^{12}_\Lambda C^*$ state emerges, so that no repulsion between two doublet members, which are degenerate in the absence of residual interactions, can occur. On the other hand, for both $^9_\Lambda Be^*$ and $^{16}_\Lambda O^*$ such a repulsion is expected, pushing one of the states to higher excitation energy, where it is observed. This mechanism roughly accounts for the higher excitation energy of $^9_\Lambda Be^*$ relative to that of $^{12}_\Lambda C^*$. The known 6 MeV spin-orbit splitting for A = 16 is then largely responsible to the even higher excitation energy of $^{16}_\Lambda O^*$.

A similar analysis may be carried out for other hypernuclei in the p-shell, and in higher nuclear shells such as the 2s - 1d shell. The simplified $^9_\Lambda Be^*$ discussion is rich enough to illustrate the main ingredients expected from such an analysis: (i) a strong mixing between analog and non-analog components, and (ii) a possible persistence of relatively pure low lying non-analog states, which are supersymmetric by the nature of their construction. The identification of such states would yield valuable information about the details of the ΛN residual interaction.

ACKNOWLEDGMENTS

Collaboration with R.H. Dalitz has led to crystalization of many of the concepts and techniques discussed above.

REFERENCES

1. R.H. Dalitz and A. Gal, "Quasi-Free Interactions in (K^-,π^-) Strangeness-Exchange Nuclear Reactions at 0°", submitted to Phys. Letters (May 1976).
2. W. Brueckner et al., "Strangeness Exchange Reactions on Nuclei", submitted to Phys. Letters (March 1976); see also K. Kilian, these proceedings.
3. A.K. Kerman and H.J. Lipkin, Ann. Phys. (N.Y.) 66, 738 (1971).
4. R.H. Dalitz and A. Gal, Phys. Rev. Letters 36, 362 (1976).
5. B. Povh, in "High Energy Physics and Nuclear Structure-1975", edited by D. Nagle et al. (AIP vol. 26, New York 1975) p.173; see also W. Brueckner et al., Phys. Letters 55B, 107 (1975).
6. H. Feshbach and A.K. Kerman, in "Preludes in Theoretical Physics", edited by A. de Shalit et al. (North Holland, Amsterdam 1966) p.260.
7. J.P. Schiffer and H.J. Lipkin, Phys. Rev. Letters 35, 708 (1975).
8. A. Gal, J.M. Soper and R.H. Dalitz, Ann. Phys. (N.Y.) 63, 53 (1971); ibid 72, 445 (1972).
9. J. Hüfner, S.Y. Lee and H.A. Weidenmüller, Phys. Letters 49B, 409 (1974).

Schiffer: Let me remind you of the situation with the analog states. For example, naively we expected the (p,n) cross section for isobaric analog states to change like (N-Z), and it does not. Why should we expect our naive models to hold for explaining hypernuclear systematics?

Gal: We expect to observe not only the analog state but also other levels, including the supersymmetric state which should have a very simple structure. With regard to the cross section, there is already a problem due to the very strong absorption, and we will probably be happy with our understanding of the cross sections for a given excitation, so long as they exhibit a smooth A-dependence.

Schiffer: I wish to point out that the experimental data are at present very sparse. With the exception of the Be data, the results are characterized by a single peak at Q = -(22-23) MeV. Is it not premature for detailed theoretical speculation?

Gal: It may be early, but if I don't do it, someone else will do it. There is presently evidence only for a single analog state for A = 12, 16, 32 and maybe 40. For ^{9}Be, the splitting can be explained as a symmetry-energy splitting. I further speculate that, for nuclei whose last neutron is less strongly bound (for all nuclei shown here, except ^{9}Be, the binding energy of the last neutron is roughly 15 MeV), there will be quite different systematics than that argued by you in the (K^-,π^-) reaction.

Gibson: To the extent that we cannot extract even the correct qualitative features of the basic Λ-N interaction from the study of s-shell hypernuclei why should we expect to extract believable features of that basic Λ-N interaction from p-shell hypernuclear spectroscopy?

Gal: It's a well-taken question. The problem in s-shell hypernuclei arose because we had only one measured state, whereas here we have an opportunity for an excitation spectrum, and will be less susceptible to such catastrophes. The problem you mentioned may have been partly resolved through consideration of the Σ component in hypernuclei, in analogy to N^*'s in ordinary nuclei.

Kisslinger: Regardless of whether or not we can determine the free ΛN force from hypernuclei, we can still make great progress towards a phenomenological understanding of hypernuclear behavior. In analogy, we obtain a great deal of nuclear structure information without a detailed, basic understanding of how this emerges from two-body considerations.

Gal: There is also some simplification due to the fact that there is only one hyperon in the nucleus, and the effective field which it feels can thus be derived from experiment. Indeed we may start with phenomenology and then progress.

GENERAL DISCUSSION

Radvanyi: I would like to inquire about N^*'s in nuclei, an important topic on which much experimental investigation has been undertaken. I have three questions: i) Is it really useful to assume existence of N^*'s in nuclear ground states? Have the problems of double-counting been resolved? ii) Can we see a difference between N^*'s which are formed in intermediate scatterings? iii) What 'Background' effects should be considered in these reactions?

Kisslinger: Meson degrees of freedom become important in medium energy physics. To the extent that we can go from mesons to the two-body force, we know about virtual N^*'s in nuclear ground states. An important part of the N-N strong interaction is associated with N^*'s. In experiments which produce real N^*'s from virtual N^*'s, apparently the form factors cut down the cross sections, and we are just beginning to understand the explanation for this suppression. This makes spectator experiments very difficult. Contributions from N^*'s come from high momentum transfer reactions where we are presently almost completely unable to understand anything.

Kamal: I would like to ask Dover what we can learn about nuclear structure from elastic π-nucleus processes, as contrasted with inelastic processes?

Dover: The same features which I discussed for reaction cross sections, regarding the geometric interpretation of pion cross sections, should show up in the elastic cross sections as well.

Johnson: In the vicinity of the resonance, the pion does not penetrate the nucleus, and you learn about the surface interaction. Beyond the resonance the pion interacts at higher density, and we could exploit this to examine the convergence of the multiple scattering series, for example.

Brown: I want to comment on the double charge exchange calculation. If you bring in a pion, one needs to evaluate the NN G-matrix at various q and finite ω; such a calculation has been performed by Bäckman and Weise, who simply took the Reid potential and calculated the G-matrix at finite ω, and found it to be very well approximated by a zero-range interaction. This is presumably the same operator needed by Jerry Miller, but none of the structure noted by Miller is observed.

Miller: We use a reasonable two-nucleon correlation function and πN information to compute this effect. If Brown wishes to provide us with a two-nucleon wave function, we would be happy to examine it further in the context of our calculation. Since we can derive our scattering equations from the LSZ formalism, we feel that we have a consistent theory for our calculation. The structure reported here deals with terms in which there is a pion in the intermediate state. Professor Brown's remarks are not addressed to these terms but to other or "true meson annihilation" effects, the importance of which decreases with increasing energy.

STATUS OF THE NEVIS SYNCHROCYCLOTRON FACILITY AND EXPERIMENTAL PROGRAM *

Derek W. Storm
Columbia University, New York, N.Y. 10027

INTRODUCTION

This paper will describe the status of the Nevis Synchrocyclotron Facility and its experimental program. The first item is to say that the machine works and has run for an experiment during the end of March and beginning of April of this year. For those of you who are unfamiliar with the accelerator, these are the basic specifications. More details will be given in the second section. The goals of the machine, as designed by J. Rainwater, are 560 MeV protons with beam intensity of 20 µA. The design repetition rate of the synchrocyclotron (S.C.) is 300 pulses/sec, and the extracted beam will have 50% duty factor. That is, the pulse period will be 3.3 msec, and the beam will be on for 1.7 msec, without micro-structure. Fast extraction is also possible. At present we have run 0.8 µA external beam at 75 pulses/sec with the full energy (which is not variable). In this running, we simply used every fourth pulse, so the acceleration time and extraction time were the same as for 300 pulses per second operation. Thus, the duty factor would be expected to be one-fourth of 50%, which it was. At present, we intend to run at 300 pulses per sec by the end of the year, and we expect to achieve 5 µA beams this July, by improving the ion source.

The experimental facilities are primarily oriented toward pion and muon experiments, although a low intensity scattered proton beam (which will be polarized) is also available. I will discuss these facilities below in more detail. We have a small but strong experimental program, which should start to grow as the machine becomes available during the summer. We welcome proposals from outside users; at present over half of our proposals are from groups composed entirely of non-Columbia people. The experimental program will be described in the Experimental Program section.

* The Nevis Facility and the research program at Columbia are supported by the National Science Foundation.

THE ACCELERATOR

The accelerator is a sector-focused S.C. It is converted from the old Nevis S.C. which was a 350 MeV machine with a 1 to 2 µA internal beam. Actually the only parts used from that machine are the magnet yoke and coils. The sector focusing enables one to increase the magnetic field with radius and thereby increase the final proton energy. At the same time, the increasing field permits a smaller frequency swing, which simplifies the design of the rf system. The sector focusing in the Nevis machine extends to very small radius, to give as strong as possible vertical focusing for the very low energy particles, since the intensity limitation in S.C.'s results from space charge blow-up of the beam a few turns after injection. In order to get the sector tips to such small radius, one iron sector is supported on alumina insulators inside the dee at rf potential. Fig. 1 is a diagram of the accelerator itself.

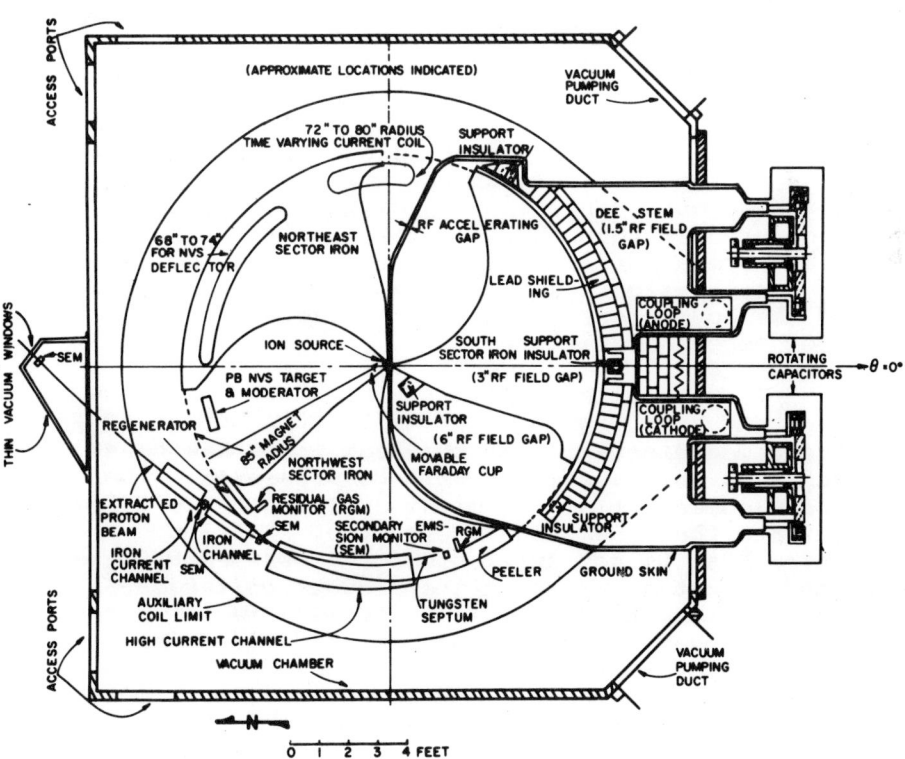

The extraction system is regenerative, with a magnetic septum. That means that when the beam reaches a certain position, the peeler and regenerator make it jump toward the magnetic channel, with amplitudes that increase on each turn (the extraction resonance). Finally most of the beam jumps the thin wall of the current channel (septum magnet) and is deflected out of the machine. The iron channel and current iron channel focus and guide the extracted beam into the external beam system. The beam can be brought into the extraction resonance by accelerating it, in which case a moderately fast (20 μsec) extraction takes place, or it can be brought there using the time varying bump coil in which case a slow (1 to 2 msec) extraction can occur. In the slow extraction mode, the beam is accelerated to an energy a little below that at which the fast extraction would occur, and the rf system is turned off, so the beam is left coasting. Then as the field in the time varying bump is ramped up, the coasting beam is slowly pushed into the peeler and begins to be extracted. The amount of beam entering the extraction resonance per unit time is determined by the time rate of change of the field in the bump and by the radial spread of beam in the accelerator. During the coasting period, the microstructure in the beam disappears, since it goes through many turns and has some energy spread. Thus the 50% duty factor is obtainable as long as there is time for extraction greater than the time for acceleration. These times are about 1.9 msec for extraction and 1.4 msec for acceleration. As mentioned above, we have obtained the equivalent of a 50% duty factor for 300 pulses per second operation.

BEAM LINES

A detailed discussion of the beam lines has been published,[1] so only a summary will be presented. The proton beam comes out of the machine, through two targets and then into the underground beam stop. Thus, the main proton beam is not available for experiments, unless one wanted to put a target in the production target location and look at it only from a few different directions. Three beam lines transport pions, muons, or scattered protons to the experimental area. The layout is shown in Fig. 2. Beams I and II are for low momentum pions or muons, and beam III is for high momentum pions or scattered protons. Unlike LAMPF or SIN, we do not have elaborate spectrometers or highly specialized beam lines. Some of the calculated properties of the beams delivered by these lines, for a 20 μA proton beam incident on a 3-cm thick (6 cm for muons) Be target, are

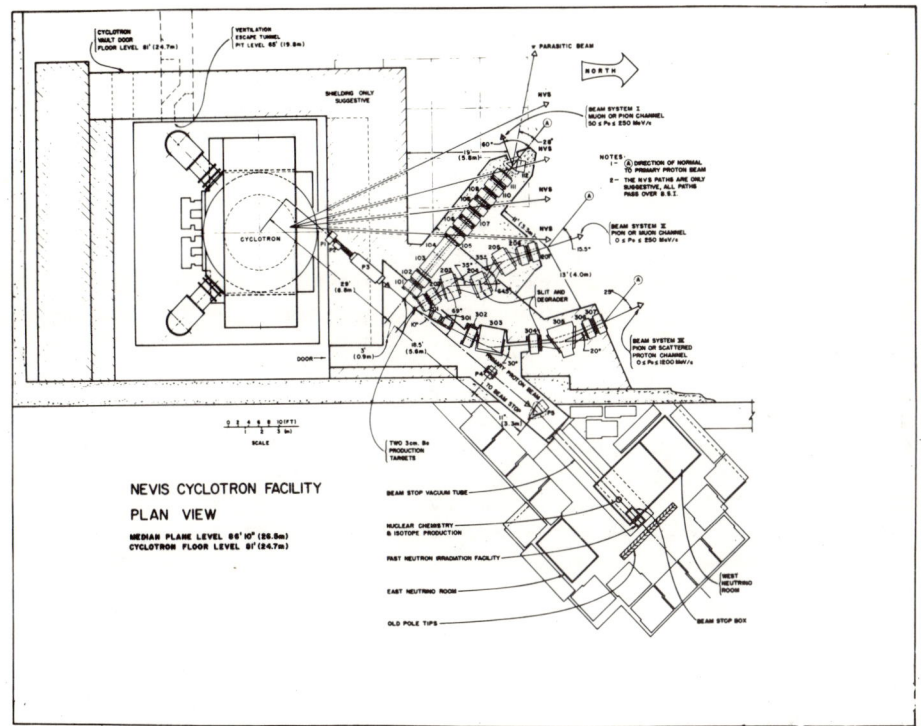

Fig. 2

given in Tables I and II. Since the predicted muon fluxes are based on cross sections with a fairly large uncertainty, it may be profitable to decrease the production angle for beam system I. We have several schemes for improving this beam system. The simplest of these can be implemented in about three months. At present, we have been compensating for the lower primary beam intensity by using somewhat thicker targets. Beam line II was used for the experiment that ran this spring. Because of a problem with the quadrupole triplet that focuses the primary proton beam on the production targets, we were unable to achieve the design intensity of beam line II. The other two beam lines are in the process of being tuned, and we expect to be using them in July. More details, including other momenta, purity, and emittance, are given in Ref. 1.

STATUS AND SCHEDULE

The machine has been run at 0.8 μA and we ran for a period of several weeks this spring at a 75 pulse per second rate and generally at or below 0.5 μA delivering about 1.5×10^4 negative pions/sec with beam line II.

After that run, our current channel broke because of a fault in its 300 ampere power supply. Rather than fix the channel, we have continued construction of a completely redesigned replacement which will be installed in June. At the same time, we are carrying out improvements in the rf system that will improve its reliability and also enable us to increase the repetition rate. The rate is presently limited by the power dissipation of some filters in the system that were provisional and are being replaced with higher power elements. During the run in the spring, the radiation background was higher than anticipated. Some of the excess was due to incomplete shielding, while the rest was due to some weaknesses that are being rectified. We expect to start the machine up in July and to run for about a month. About half of this time will be available for experiments. Then we will shut down for about a month to install the deflector for the neutron velocity spectrometer. After that, we intend to run steadily. Our goal is to be running 10 μA at 300 pulses/sec by the winter.

Table I

Muon Channel - Beam System I

Flux[1] x 10^5 particles/sec

	π^+	μ^+	π^-	μ^-	μ/π	<Pol>%
Forward Decay						
without final bend	380	73	95	19	0.2	-68
with final bend	24	26	6	7	1.1	-68
Backward Decay						
without final bend	400	21	100	6	0.05	+83
with final bend	0	15	0	4	pure	+83

[1] Fluxes are distributed over 8 in. x 8 in. surface. Cross sections used in flux calculations are derived from values in Refs. 2 and 3.

Table II

Pion Channels

Description	P_o (MeV/c)	Spot Size x y (in.)	Flux x 10^7 (part/sec) π^+ π^-		$\Delta P/P_o$ HWHM
Beam I					
small spot	200	1.0 x 1.5	7.6	2.0	5%
1% resolution	100	8.5 x 8.0	0.9	0.3	1%
Beam II					
	150	0.8 x 2.0	6.0	0.8	5%
	150	0.8 x 2.0	2.0	0.3	1%
Beam III					
good resolution	400	2.6 x 2.0	1.2	0.17	0.6%
poor resolution	400	6.0 x 2.0	12.0	1.7	5%

EXPERIMENTAL PROGRAM

As mentioned previously, the experimental program involves both Columbia people and outside users. Presently, there have been 21 proposals, some of which involve several experiments using the same setup. More than half of these are related to particle physics or muonium (mainly weak interactions): These will not be discussed at this time except to remark that they involve measurements of rare events or of high precision, and so need the intensity provided by the accelerator. Many of them also involve coincidence measurements, and will benefit from the 50% duty factor.

The selection of experiments is done through a program advisory committee, which consists mainly of non-Columbia physicists. Although the committee has a strong representation of particle physicists, there is by no means a bias against nuclear experiments. This committee reviews all proposals thoroughly. In nearly all cases proposals have had approval deferred until more information was made available to the committee, although most proposals were eventually approved. The program advisory committee will meet again this summer.

The proposals involving meson-nuclear physics come from as far east as New Haven and as far west as Tokyo with several from Columbia. Most of these nuclear experiments are coincidence experiments, in the sense that a beam particle trigger is involved in the measurement. Most do not, however, require coincidence measurements in the final state, although the duty factor of the accelerator will allow such experiments to be possible. Some of the proposed rare decay experiments do involve

such measurements, and we expect that in the future, there will be a growing interest in meson nuclear measurements involving final state coincidences.
The Kent State experiment on the spectra of neutrons produced by pion capture used a time of flight measurement, for which the duty factor was important. We hope that when they run again, the duty factor will be all the way up to 50%.

The meson nuclear experiments planned at Nevis basically fell into four categories. First, reactions induced by captured pions, second, reactions induced by pions in flight, third, muon capture reactions, and finally nuclear studies using mesic atoms. Rather than list experiments, I will mention some of the typical examples in each category. The experiment that was run this spring by the Kent State group involved a determination of neutron spectra produced in reactions where stopped π^- are captured, and falls into the first category. A Columbia group plans to survey $\pi^\pm$ elastic scattering by several nuclei from ^{28}Si to ^{208}Pb. In the third category are proposals by a William and Mary group to study radiative μ capture in ^{40}Ca and to study γ-ν correlations following μ capture. These are not nuclear physics experiments, but are intended to shed light on the weak pseudo-scalar coupling constant.
In the last category, a Columbia group plans to measure nuclear quadrupole moments via pionic atom studies this summer. Some later muonic atom studies will measure the magnetic hyperfine splitting in 2s-2p transitions, to learn about the distribution of the magnetic moment of the nucleus. We welcome additional interest in using the Nevis Facility and hope to hear from some of the participants in this conference.

REFERENCES

1. M.M. Holland, S.E. Metelits, Seventh International Conference on Cyclotrons, Zurich (1975), p. 299.
2. P.W. James et al, TRIUMF Report #VPN-75-1 (March, 1975).
3. W. Hirt et al, CERN Report #69-24 (September, 1969).

PRESENT STATUS OF BEAM LINES AT KEK

H. Hirabayashi, S. Kurokawa, A. Kusumegi, A. Maki*, S. Mikamo
T. Sato**, Y. Suzuki, M. Taino, K. Takamatsu, M. Takasaki,
K. Tsuchiya, and A. Yamamoto

KEK National Laboratory for High Energy Physics
Oho-machi, Tsukuba-Gun, Ibaraki, 300-32, Japan

ABSTRACT

The present status of the beam lines at KEK 12 GeV proton synchrotron is given. The accelerator recently has achieved successful operation to accelerate protons up to 10.4 GeV at the latest run. The time schedule of construction and the program of experiments are also given. A 4 GeV/c unseparated beam, π2, has been just installed at the internal target area. A beam, K1, up to 6 GeV/c from the fast extracted proton beam for the bubble chamber experiments is at the final stage of construction. Two experiments with the π2 beam and two experiments with the bubble chamber beam are approved for the initial phase of experiments. These experiments are expected to start in the middle of 1977. Two low momentum separated kaon beams, K2 and K3, up to 2 GeV/c as well as the splitting system of the slow extracted protons are now at the final stage of design and the construction has been partially started. The completion of these beam lines is scheduled in the spring of 1979. Three experiments utilizing the kaon beams are also approved.

INTRODUCTION

National Laboratory for High Energy Physics in Japan has been established in 1971 and KEK is the abbreviation of the Japanese name of the laboratory. The principal facility of KEK is a 12 GeV Proton Synchrotron. The machine has a cascade structure, i.e., 750 keV preinjector, 20 MeV linac, 500 MeV booster synchrotron and 12 GeV separated-function type main ring which has 4 superperiods[1]. The energy of accelerated protons in the machine has exceeded 10 GeV in the last March and it is anticipated to have the maximum energy with a sufficient intensity within one year.
Several beam lines are installed already around the main ring. Two beam lines are built in the area of the internal target for the counter experiments of initial phase. One is an unseparated secondary beam, π2, up to 4 GeV/c, and the other is a test beam line, T1, up to 2 GeV/c. Another beam line, K1, is installed from the exter-

---

* Present Address; Rutherford Laboratory
** Present Address; Stanford Linear Accelerator Center

nal target of the fast extracted proton beam, EP1, to the hydrogen bubble chamber. These beam lines, K1 and EP1, are near to operation.

Two more beam lines, K2 and K3, are about to start construction in the area of the slow extracted proton beam, EP2[2]. The completion of these lines are scheduled in the spring of 1979. The beam lines for counter experiments are shown in Fig. 1.

Two counter experiments with the π2 beam and two bubble chamber experiments with K1 beam have been approved to run from the middle of 1977 to 1978. Three counter experiments are approved to use the K2 and K3 beams from the slow extracted proton beam, EP2.

INTERNAL TARGET BEAM LINES

The counter experiments of initial phase will be performed with the internal target beam line, π2, of maximum momentum 4 GeV/c, unseparated. The beam line is not a special one ; however, it has been tried to get a reasonable luminosity in a circumstance of the reduction of a factor or even one order magnitude of circulating protons to avoid an unnecessary contamination of the accelerator.

The final design of the beam line, π2, is schematically shown in Fig. 2 and the parameters of the standard elements for secondaries are listed in Table I. The internal target is placed in the vacuum chamber and located 55 cm downstream of the III-5D quadrupole magnet of the main ring. The production angle of π2 beam has chosen at 10° with respect to the tangent of circulating protons. This is a compromise between minimizing the production angle and obtaining a resonable acceptance with our standard narrow quadrupole magnet, NQ420.

After the quadrupole doublet, NQ420 and Q420, a pair of standard bending magnets, 6D220×2, deflect the secondaries by 15° for the momentum selection at the intermediate focal plane, IF. A set of tungsten collimators is provided to define both the horizontal and vertical acceptances as well as the momentum bite. The calculated momentum dispersion is 1.08 cm/%. All those magnets are placed inside the main ring tunnel which is separated by shielding blocks from the counter experimental area.

Table I Prameters of the standard magnets

	Quadrupole magnet		Dipole magnet		
Type	Q220	NQ420, Q420	Type	6D220	8D320
Aperture	10 cm	20 cm	Gap	12 cm	15 cm
Length	100 cm	100 cm	Width	30 cm	40 cm
Pole shape	hyperbolic	hyperbolic	Length	100 cm	100 cm
Field gradient	1.7 kG/cm	1.0 kG/cm	Central field	20 kG	20 kG

The momentum recombination section of the π2 beam line consists of a pair of bending magnets, 8D320×2, and a quadrupole doublet, Q420×2. The beam is bent by 8° and focussed by the quadrupole doublet and bent further by 8° to obtain a complete momentum recombination at the final focus F0. The overall beam length is 31.3 m from the internal target to F0 and the horizontal and vertical magnification factors are 1.82 and 2.85, respectively. The results of first and second order calculations are shown in Fig. 3 together with the major parameters in Table II.

In practice, however, the approved experiments require having their experimental arrangements simultaneously on the floor and the beam switched over alternately. The solution has been looked for, by changing the parameters of the momentum recombination section, to have the one focus at F1, 4 m upstream of F0 and the other focus at F2, 4 m downstream of F0. Although the beam is not free from dispersion in these cases, it is still reasonably acceptable for the experimental groups.

With an internal target, Be of 1 mmφ and 10 mm long, a calculated target efficiency amounts to 32.5% with 8 GeV protons and to 37.5% with 12 GeV protons. The expected intensity is illustrated for π^- of 2 GeV/c to be $1.2 \times 10^5 \pi^-/10^{11}$ protons at 8GeV. The survey of the π2 beam will be started as soon as the insertion of the internal target into the main ring.

A test beam, T1, is provided for equipment tests from the same internal target at the direction of 23°. The beam has been furnished by assembling the elements at hand. The beam is almost parallel in both horizontal and vertical directions with the momentum upto 2 GeV/c and its intensity is approximately one order of magnitude lower than that of π2 beam.

FAST EXTRACTED PROTON BEAM AND BUBBLE CHAMBER BEAM LINE

Fast spill protons are extracted to the EP1 beam line from the long straight section II-2F of the 12 GeV proton synchrotron main ring with an angle of 6°. The extraction is performed by means of a multiturn beam shaving method using bump magnets, an electric septum and magnetic septa[4]. The EP1 beam line is provided for the secondary beams to the hydrogen bubble chamber and for the accelerator tuning. The extracted beam is convergent horizontally and divergent vertically. It has a horizontal waist at a point 3 m upstream of the first focussing element of the line and also has a vertical waist in the accelerator. Required properties of this line are at first to focus the fast spill beam to a small spot on the target and next to transport the full beam of large emittance to the dumper. The beam line is composed of a pair of doublets of quadrupole magnets and a couple of horizontal and vertical steering magnets along the beam length of 28.5 m. The focussing element is a Q220 type quadrupole magnet prepared for the primary proton beam. The magnet has an aperture of 10 cm in diameter with hyperbolic pole faces of 1 m long and a nominal field gradient of 1.7 kG/cm. The layout of the line is shown in Fig. 4. The estimated horizontal and vertical emittances of the fast spill beams are 5.4π mm·mr and 3.3π mm·mr, respectively[5]. With these values, a small spot of the beam such as 4 mmϕ could be obtained at the target position by tuning the focussing elements properly. For the full beam ejection having large emittance of circulating protons in the accelerator, the line is also capable of transporting it to the dumper without a significant loss. The EP1 beam line has been just installed in the T2 tunnel with three monitoring stations for fast extracted proton beam.

The K1 beam is a single stage electrostatically separated beam line up to 6 GeV/c for the bubble chamber experiments at the moment; however, it has been designed so that the line can be converted easily to a double stage beam line by adding a few elements. As shown in Fig. 4, it starts from the production target in the T2 tunnel with an angle of 3° to the proton beams. By means of the dipole magnets at the upstream end, the beam is bent to the direction of 10.8° with respect to the primary beam.

After passing a quadrupole doublet, the beam is conducted to the separator hall. In the hall, a pair of dipole magnets for 8.2°

Table II Characteristics of $\pi 2$ and K1 beams

Beam line	$\pi 2$	K1 (single stage)
Momentum range	1.0 - 4.3 GeV/c	1.0 - 6.0 GeV/c
Particles	π^+, π^-, p	π^-
Central production angle	10.0°	3.0°
Beam line length	31.3 m	84.9 m
Solid angle acceptance	0.594 msr	0.039 msr
Horizontal acceptance	± 9 mr	± 2.5 mr
Vertical acceptance	± 21 mr	± 5.0 mr
Momentum bite transmitted	± 1.0 %	± 0.5 %
Dispersion at momentum slit	-1.08 cm/%	-2.18 cm/%
Beam end characteristics		
horizontal magnification	1.816	
vertical magnification	2.851	
horizontal image size	22 mm	21 mm
vertical image size	14 mm	118 mm
dispersion	0.003 cm/%	
Yields expected	1.2×10^5 π^-	4.0×10^3 π^-
at momentum of	2.0 GeV/c	4.0 GeV/c
by 8 GeV protons of	10^{11}	10^{11}
on target of	1 cm long Be	10 cm long Be
Comments		With a 9 m dc separator, π^+ and p can be transported

bending, a 9 m dc separator and a couple of quadrupole magnets are installed in front of the momentum slit which is followed by another pair of dipole magnets for 8.4° bending and the final doublet of quadrupole magnets. All elements for bending and focussing are standard ones. The dc separator is used to separate the positive pions from protons for the experiments of initial phase. The total length of the line is 84.9 m. The horizontal focussing structure of this line is DFFD upstream of the momentum slit. After the slit, the beam is made slightly convergent in a narrow horizontal plane and parallel vertically with the final focussing elements. A typical acceptance of the line is 0.039 msr. The value is sufficient to perform an experiment with pions of a small momentum spread of ± 0.5 %. The major optical parameters are shown in Table II. The line is in the final stage of construction and is to be used for the first bubble chamber experiment with a 6 GeV/c π⁻beam. The layout of K1 line for the second phase of experiments is not fixed yet. A design has been proposed to built a double stage dc separated beam line as shown in Fig. 4 for the experiments with 2 to 4 GeV/c kaons. Another design has also been proposed to built a low momentum kaon and antiproton beam for the bubble chamber experiments.

SLOW EXTRACTED PROTON BEAM AND LOW MOMENTUM SEPARATED KAON BEAM LINES

From the long straight section III-2F of the accelerator, a slow spill proton beam will be extracted to the EP2 beam line by a half integer resonant extraction method[6]. Althouth the EP2 line is still under design, it should have a system to split the extracted protons into two independent production targets of the K2 and K3 beam lines. At present, a system utilizing an electric septum and several magnetic septa is considered as a horizontal beam splitting device. The focussing system of the EP2 line is composed of a doublet of Q220 quadrupole magnets in front of the splitting system which is followed by a triplet of the same Q220 quadrupole magnets in each branch of the beams. Assuming a possible emittance, a horizontal DFFDF focussing gives a beam spot of 3 mmϕ on each target. The length of each beam line from the extraction point to the production target is approximately 42 m. The line will be installed in the T3 tunnel and its extension to the counter experimental hall is shown in Fig. 1.

The K2 and K3 beams are the low momentum separated kaon beam lines. These two beam lines are at the final stage of design. The design of the K2 is almost finished. As to the K3, one feasible design of the line, which is only 13.56 m long, is presented here. Another design is now in progress by T. Kamae et al[7].

The K2 beam line is a 27.9 m long, variable acceptance beam line which can transport well separated kaons and antiprotons with momentum of 1 - 2 GeV/c. By adjusting the excitation of a quadrupole triplet placed next to the first diople magnet, the vertical acceptance of the line can be changed to the three values (±4.0 mr, ±6.7 mr, ±10.0 mr) without varying the horizontal acceptance (±40.0

mr). Therefore the three modes of operation with different acceptances are possible : K2-S of 0.5 msr, K2-M of 0.84 msr, and K2-L of 1.26 msr. The layout of the line is shown in Fig. 5. Secondary particles with a mean production angle of 0° are accepted. The momentum selected particles are separated by a 6 m dc separator with 60 kV/cm electric field and by a mass slit at the intermediate vertical focus. A sextupole magnet is used to correct the second order chromatic aberrations in order to reduce the vertical image size at the slit. The program DECAY TURTLE[8] is employed to estimate the π/K ratio and beam sizes at the final focus. The π^+/K^+ ratio of less than 1 and the beam size of 12 mm × 10 mm are obtained for K2-S at 2.0 GeV/c. The flux is calculated using the formula of Sanford and Wang[9], and the result gives the value of 1.83×10^5 K^+ per 10^{12} protons hitting on the 7.5 cm Cu target for the K2-S at 2.0 GeV/c. The parameters of the K2 are summarized in Table III.

The K3 beam is a beam of large acceptance and short length for 0.5 - 1.0 GeV/c separated kaons and antiprotons. The layout of the line is presented in Fig. 6. The mass separation is achieved by a 2.0 m dc separator of which the gap is 10 cm and the electric field is expected to be 60 kV/cm. The beam line transports secondary particles with a mean production angle of 0° similar to the K2 beam line. The horizontal and vertical acceptances are ±150 mr and ±8 mr, respectively. The latter can be increased to ±12 mr by widen-

Table III Characteristics of K2 and K3 beams

Beams	K2 (K2-S Mode)	K3
Momentum range	1.0-2.0 GeV/c	0.5-1.0 GeV/c
Particles	$K^+, K^-, \bar{p}$	$K^+, K^-, \bar{p}$
Central production angle	0°	0°
Beam line length	27.9 m	13.56 m
Solid angle	0.50 msr	3.8 msr (5.7 msr)
Horizontal acceptance	±40 mr	±150 mr
Vertical acceptance	±4.0 mr	±8.0 mr (12.0 mr)
Momentum bite transmitted	+2.0,-2.5 %	+2.5,-2.5 %
Separator	6.0 m, 60kV/cm	2.0 m, 60 kV/cm (45 kV/cm)
Separation/image size at mass slite(aberr.incl.)	2.78/1.12 mm at 2.0 GeV/c	10.76/5.83 mm at 0.8 GeV/c
Dispersion at momentum slit	2.186 cm/%	2.15 cm/%
Beam end characteristics		
horizontal magnification	2.359	2.616
vertical magnification	1.730	1.424
horizontal image size	12 mm	20 mm (22 mm)
vertical image size	8 mm	20 mm (30 mm)
dispersion	-0.092 cm/%	-0.407 cm/%
Yields expected with 10^{12} of 12 GeV protons	$1.83 \times 10^5 K^+$ at 2.0 GeV/C	$2 \times 10^5 K^+$ ($9 \times 10^4 K^+$) at 0.8 (0.6) GeV/C

ing the gap of the separator to 15 cm for the case of antiprotons and stopping kaons at the expense of the reduction of the separation due to the decrease of electric field strength. In this case, the solid angle acceptance can be increased to 5.7 msr from 3.7 msr. Since the line is designed to provide particles not only for scattering and reaction experiments but also for experiments with stopping kaons, it is necessary to make the length as short as possible to minimize the decay losses. The bending and focussing action of a rectangular dipole magnet with a positive n value (n = 1.724) is employed to save one quadrupole magnet and to shorten the length of the beam line. Two large aperture superconducting quadrupole magnets of Panofsky type will be placed in the momentum recombination section to shorten the length of the line further. The total length becomes 13.56 m. The intensity is estimated by extrapolating the values of ZGS Beam No. 42[10]. The quality of separation is calculated by the program DECAY TURTLE. The result shows the values of 8.8×10^4 K^+ with a π^+/K^+ ratio of 3.1 at 600 MeV/c and 2.0×10^5 K^+ with a π^+/K^+ ratio of 3.3 at 800 MeV/c. The parameters of the line are summarized in Table III.

TIME SCHEDULE AND EXPERIMENTS

As far as beam lines are concerned, it has been planned during the construction period of 1971-1976 to build one internal target beam line, π2, for the counter experiments, and the fast extracted proton beam line, EP1, with a secondary beam, K1, for the bubble chamber experiments of initial phase. For a maximum utilization of the accelerator, it is evidently necessary to add more beam lines, particulary, with the slow extracted protons. The budget was requested and has been approved since 1976.

It is anticipated that the accelerator runs for physics experiments of initial phase from the spring of 1977 till the summer of 1978. Two counter experiments and two bubble chamber experiments have been approved by the program advisory committee (PAC) as the initial phase experiments with π2 and K1 beams. These experiments are :
(1) Measurements of the differential cross section and polarization parameters in $\pi^- p \rightarrow \pi^0 n$ reaction from 1.8 to 2.0 GeV/c by the group of Kyoto.
(2) Measurements of the differential cross section and polarization parameters in $\pi^- p$ elastic scattering from 2 to 4 GeV/c by the joint group of Nagoya, Hiroshima, Osaka, Tokyo, Kyoto and KEK.
(3) Study of inelastic reactions in πN interactions at intermediate energy region with 300,000 pictures for $\pi^- p$ by the group of KEK.
(4) Study of three body reactions in diffraction dissociation with 200,000 pictures of $\pi^- p$ by the joint group of Tokyo Metro. Univ., Nagoya, Tokyo Univ. of Agri. and Tech. and Chuo Univ..

In addition, several minor experiments including a radiochemical study have been approved during the same period on a parasitic basis.

Although the first stage counter experiments will use the

internal target beam, π2, intensive studies have been done on the secondary beam lines from the slow extracted proton beam, EP2. The importance of low momentum kaon beams with high luminosity has been recognized above all and it has been decided to build two low momentum kaon beams, K2 and K3, from 1976 to 1978. Three experiments have been approved by PAC with those beam lines and are expected to run from 1978. These experiments are : (1) $K^+n \to K^+n$ scattering with the K2 beam, (2) Search for $K^+ \to \pi^+ \nu \bar{\nu}$ decay and (3) Investigation on hypernuclei, with the K3 beam.

REFERENCES

1. KEK Annual Report (1974).
2. H. Hirabayashi, S. Kurokawa and A. Kusumegi, to be reported in the Summer Study Meeting on Kaon Physics and Facilities at BNL (June 1-5, 1976).
3. J. Kishiro et al, to be reported.
4. J.D. Mccarthy and Y. Kimura, KEK 74-16.
5. L.N. Blumberg, private communication.
6. KEK Accelerator Dept. Internal Report.
7. T. Kamae et al, private communication.
8. K.L. Brown and Ch. Iselin, CERN 74-2 (1974).
9. J.R. Sanford and C.L. Wang, BNL Internal Report JRS-CLW1 (1967).
10. E. Colton, ZGS Internal Report EC-3 (1975).

721

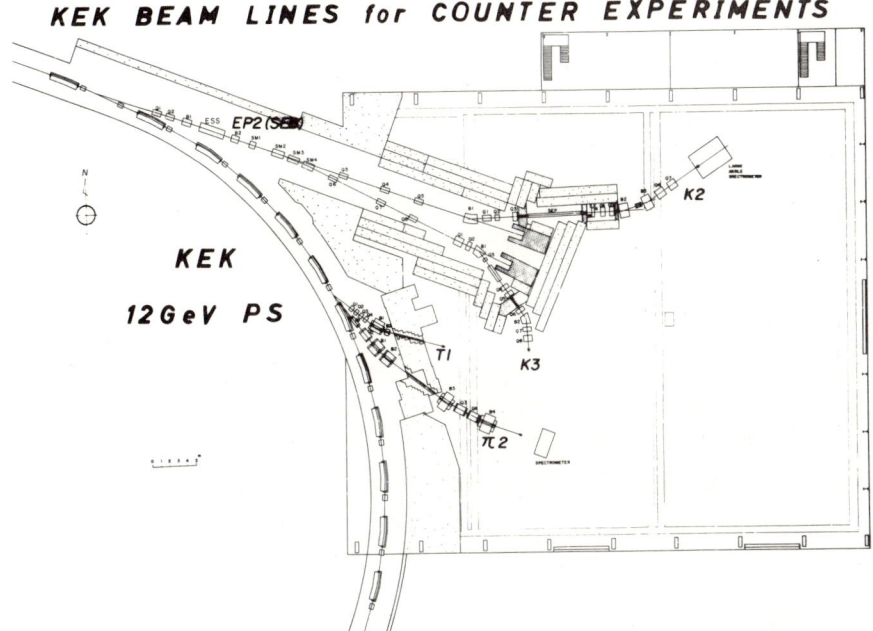

Figure 1 Floor plan of the KEK beam lines for counter experiments

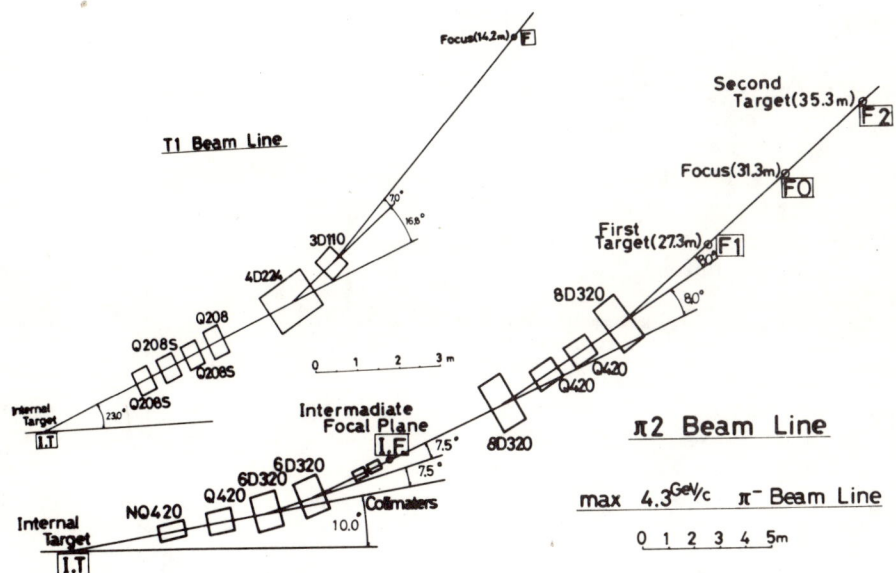

Figure 2 Schematic diagram of the π2 and the T1 beam lines

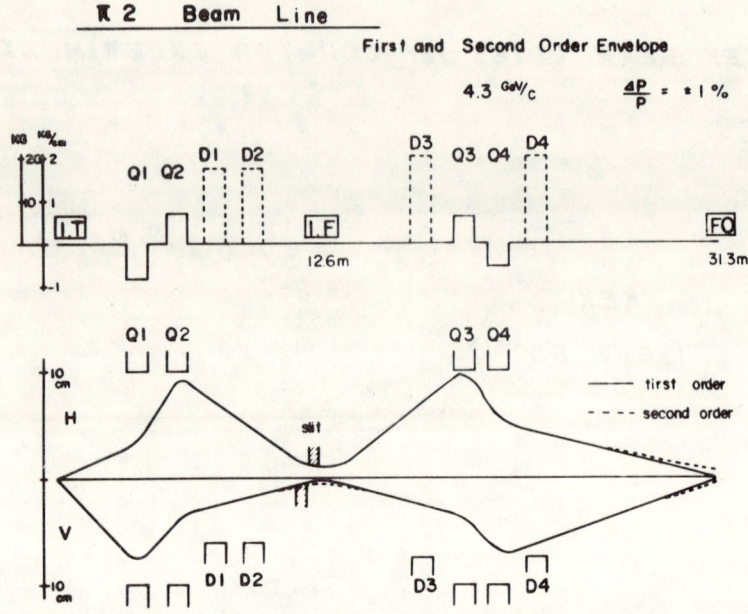

Figure 3 First and second oder envelopes of the π2 beam

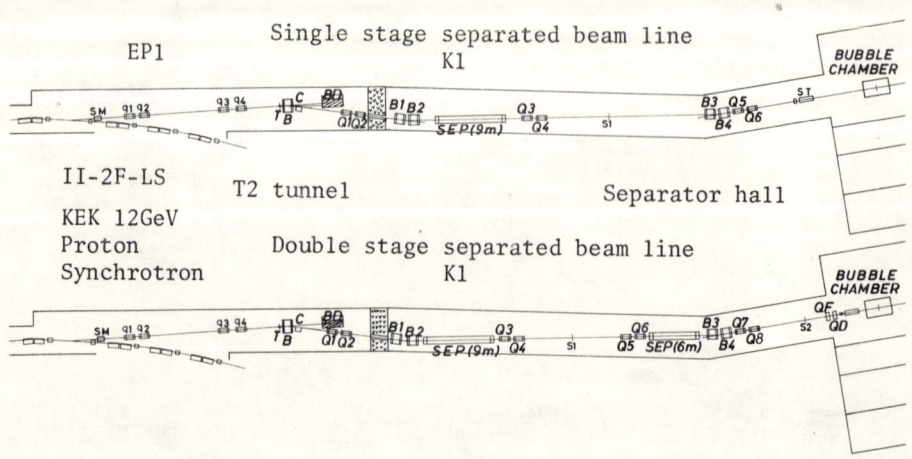

Figure 4 Layout of the EP1 and K1 beam lines

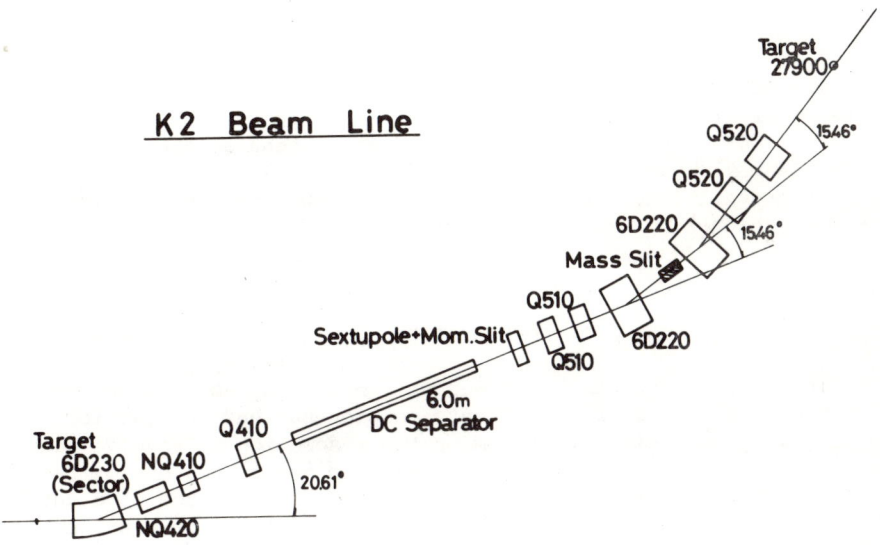

Figure 5 Schematic diagram of the K2 beam line

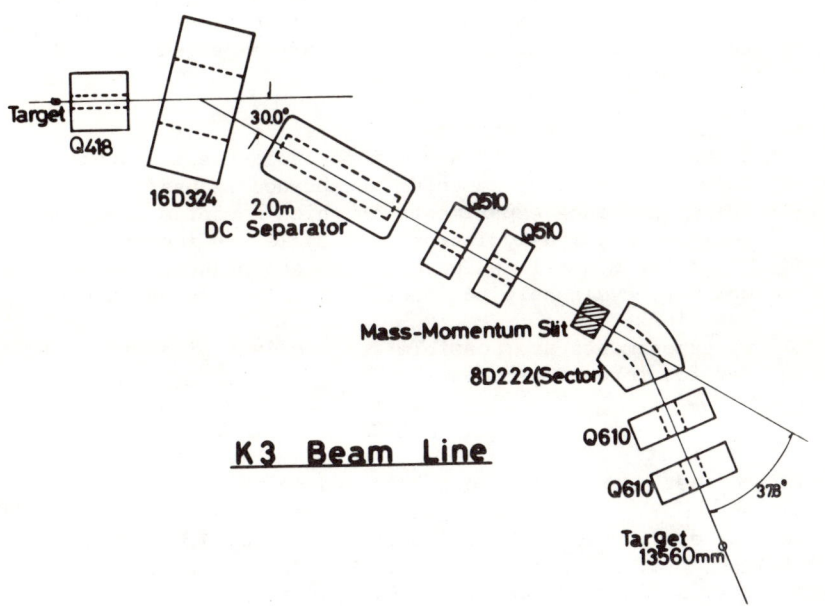

Figure 6 Schematic diagram of the K3 beam line

THE SIN PION SPECTROMETER

ETH/Zurich-Grenoble-Heidelberg-Karlsruhe-Neuchatel-SIN Collaboration[*]

ABSTRACT

The SIN pion spectrometer has been built to carry out a broad range of experiments on pion-nucleus interactions in the momentum range of 150-650 MeV/c. It is designed to subtend a large solid angle (16 msr), with a large momentum acceptance ($\pm$ 18% Δp/p), and with an overall momentum resolution, including the contribution from the pion channel, of 7×10^{-4} dp/p FWHM. At present an overall resolution of 1.3×10^{-3} dp/p has been achieved, over a momentum acceptance of $\pm 7.5\%$ Δp/p.

INTRODUCTION

The SIN pion spectrometer was designed to be able to be used in a number of types of experiments on pion-nucleus interactions. While having sufficiently good momentum resolution to resolve many nuclear states in inelastic scattering of pions in the energy range 50-350 MeV, it is also designed to take advantage of the good duty-cycle of the SIN cyclotron, by having the ability to be used in coincidence experiments of the type (π,π'N), in conjunction with a second spectrometer of some kind. Hence a design was chosen which had a large solid angle (16 msr), a large momentum acceptance ($\pm$18% Δp/p), a vertical assembly allowing a good angular range and floor space for a second arm, and with an overall resolution of less than 10^{-3} dp/p FWHM, including the contributions from the pion channel and the target. Experiments which are expected to be done may be grouped as follows:

π-nucleus elastic and inelastic scattering
π-nucleus "quasi-elastic" and summed inelastic measurements in which the large momentum acceptance is useful
(π,π'N) coincidence experiments in which momentum acceptance, solid angle, and cyclotron duty-cycle are important
(π^+,π^-) experiments in which momentum acceptance, solid angle and duty-cycle help to identify small cross-section reactions
(π^+,p) experiments are possible for incident pions of 50-100 MeV for forward protons.

PRESENT STATUS

In the past year the measurements and shimming of the dipoles

---

[*] J.P.Albanese, J.Arvieux, E.T.Boschitz, R.Corfu, J.P.Egger, P.Gretillat, C.H.Q.Ingram, C.Lunke, E.Pedroni, C.Perrin, J.Piffaretti, L.Pflug, E.Schwarz, C.Wiedner and J.Zichy

for the spectrometer and for the pion channel have been completed; the spectrometer and pion channel have been assembled and placed on the floor; the construction of two 1 metre multi-wire proportional chambers with cable delay-line read-out and 2mm resolution has been completed; a number of 10 and 20 cm chambers with cable delay-line or digital read-out and 1 mm resolution have been built; and in two beam periods we have successfully tuned and investigated the optics of the pion channel and the spectrometer and ascertained that there are no unexpected serious problems. The chambers in the beam-line with digital read-out operated with an overall efficiency of 80% for four planes, with a flux of 1×10^6 π^+/sec at the scattering target. Calibration measurements to empirically determine the solid angle of the spectrometer have been taken.

In the first beam period in autumn 1975, a momentum resolution of 1.6×10^{-3} dp/p FWHM was achieved, using a severely collimated beam (± 3 mr in the dispersive plane of the channel, ± 15 mr in the dispersive plane of the spectrometer), run directly into the spectrometer at 0°. After beam tuning in the second period in spring 1976, a direct beam resolution of 0.95×10^{-3} dp/p was achieved, using the full phase-space of the pion channel, and with the optical parameters of the spectrometer optimised for a small part of the focal plane. Under the same conditions, but with a single optical parametrisation of the spectrometer over the full focal plane presently available, a resolution of $1.2-1.5 \times 10^{-3}$ dp/p has been realised over $\Delta p/p = \pm 7.5\%$.

Spectra have been taken with 200 MeV pions scattered from a 2 gm.cm^{-2} ^{12}C target, where the observed resolution of 900 keV is largely accounted for by the Landau broadening in the target (700 keV). A computer break-down prevented the taking of any spectra with a thin target.

Background measurements have yielded negligible count-rates of less than 1% of target-in rates, even with the spectrometer at 30°. The simple, well-defined optics of the spectrometer permits a restriction of the permissible phase-space at the focal plane to eliminate most muons from pion decay in the spectrometer.

With the present 1 metre chambers at the focal region, the solid angle subtended by the spectrometer is uniform to $\pm$ 15% over a range of $\pm$ 5% $\Delta p/p$, which corresponds to excitation energies of the target nucleus of 0-30 MeV for 200 MeV pions.

LAYOUT AND SPECIFICATIONS OF CHANNEL AND SPECTROMETER

Fig 1 illustrates the layout of the pion channel and spectrometer and Table 1 summarises the important specifications. The optics of the pion channel is almost symmetric about the dispersed intermediate focal plane, F, where the horizontal magnification of the 2 mm wide production target is 1.1. A pion's position at this focal plane, measured by a chamber, C1, determines its momentum. The beam is then brought to a small achromatic focus at the scattering target, T, where its size is 9 mm (x) x 7 mm (y) FWHM. The chamber C2 determines the angle of incidence on the target. The beam contains an electrostatic separator, S, to remove the protons from the

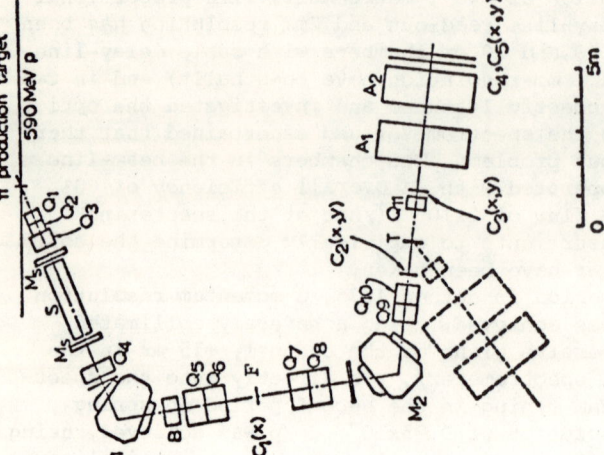

Fig. 1

Layout of high resolution π-channel

Q's are quadrupoles, M's bending magnets of the π-channel, A's bending magnets of the π-spectrometer, S electrostatic separator, C's multiwire proportional chambers. T is the scattering target.

Table 1. Specification of Pion Channel & Spectrometer.

	π Channel	Spectrometer
Momentum resolution	$5 \cdot 10^{-4}$ dp/p	$5 \cdot 10^{-4}$ dp/p
Solid angle	6 msr	16 msr
Momentum acceptance	±1.4%	±18%
Dispersion along focal plane	7.0 cm/%	5.5 cm/%
Focal plane inclination	90°	35°
Dispersion plane	Horizontal	Vertical
Momentum range	150 – 450 MeV/c	100 – 650 MeV/c
Total path length	22 metres	8 metres
Angular range (with auxiliary magnet)		$0° - 140°$ $(140° - 180°)$
Angular resolution	±35 mr y' ±25 mr x'	±0.5°
Angular divergence		±7 mr in elemental area
Spot size on target	9 mm FWHM x 7 mm FWHM y	
Pion flux on target for 100 μA protons	$1.5 \cdot 10^7 \pi^+$/sec $2 \cdot 10^6 \pi^-$/sec	at 300 MeV/c
	(x1/10 at 200 and 450 MeV/c)	
MWPC used (dimension × wire spacing)	200 × 1 mm (x)	200 × 1 mm (x,y)
	100 × 1 mm (x,y)	200 × 1 mm (x)
	100 × 1 mm	1000 × 2 mm (x)
		1000 × 2 mm (x)
	Both digital read-out.	All delay line read-out.

π^+ beam. The contamination at the intermediate focus is reduced from 400% to 5% at 300 MeV/c by the separator.

Surface coils at the entrance and exit of each of the two dipoles in the channel give curvatures to the effective field boundaries of these two magnets which are equivalent to sextupole fields. These are set to eliminate chromatic and spherical aberrations (in TRANSPORT notation $T_{1,26}$ and $T_{1,22}$) in the dispersive plane. This gives an intermediate focal plane perpendicular to the beam axis, and is necessary in order to produce a small spot at the final focus.

Figs 2 and 3 show a cross-section of the spectrometer and a schematic view of its optical properties in the dispersive plane. The spectrometer design is based on the SACLAY SPES II spectrometer but has been considerably modified. The 150 ton spectrometer is raised onto compressed air cushions for rotation over its $0°-140°$ angular range. The most critical mechanical alignment requirement for the spectrometer is that the two dipoles should be co-planar to within $\pm$ 0.1 mm and $\pm$ 1 mrad. After rotations the alignments have been checked and remain within tolerances.

The spectrometer is absolute in the sense that it measures the momentum of a particle independent of its momentum and phase space incident on the scattering target. Particles entering the spectrometer pass through a quadrupole and then a chamber, C3, before being dispersed and focussed by two dipole magnets. The chamber C3 acts as the source in the dispersed, x, plane optics, the particle being brought to a dispersed focus 2 metres long, inclined at 55° to the normal, some two metres beyond the second dipole. Each x-coordinate at C3 has its own focal plane (with a small y^2 dependence $\simeq 10^{-3}$ dp/p for y=10 cm at C3); these focal planes are slightly curved and almost parallel (fig 3). Because the beam spot at the target is small, the angular divergence at each coordinate at C3 is small ($\pm$ 7 mr) which removes the need to eliminate higher than second order optical aberrations in the spectrometer by complicated shaping of the dipole pole face edges. In addition, the coordinate at C3 determines the scattered particle's angle (to $\pm$ 7 mr), and specifies the expected angle at the focal plane to within $\pm$ 5 mr, independent of momentum. This last point provides a powerful means of eliminating muons created by pion decay in the spectrometer, nearly all of which have an inclination substantially different to that of the parent pion.

In the non-dispersive plane, the focussing is from target to focal plane, to maximise the solid angle.

CHAMBERS

The chambers C1 and C2 are 20 cm (x) and 10 cm (x,y) square with 1 mm wire spacing. C1 has two x-planes shifted by 0.5 mm that can be or'd. They were designed to handle up to 5×10^7 pions per second and to resolve events 20 ns apart (beam structure). In tests a resolution of 5.4 ns FWHM has been achieved. For high rates a special high-freon gas mixture is used. A special fast digital read-out[1] was developed as no existing system was fast enough (min. strobe width needed: 5 ns). Chamber C3 is 20 cm square

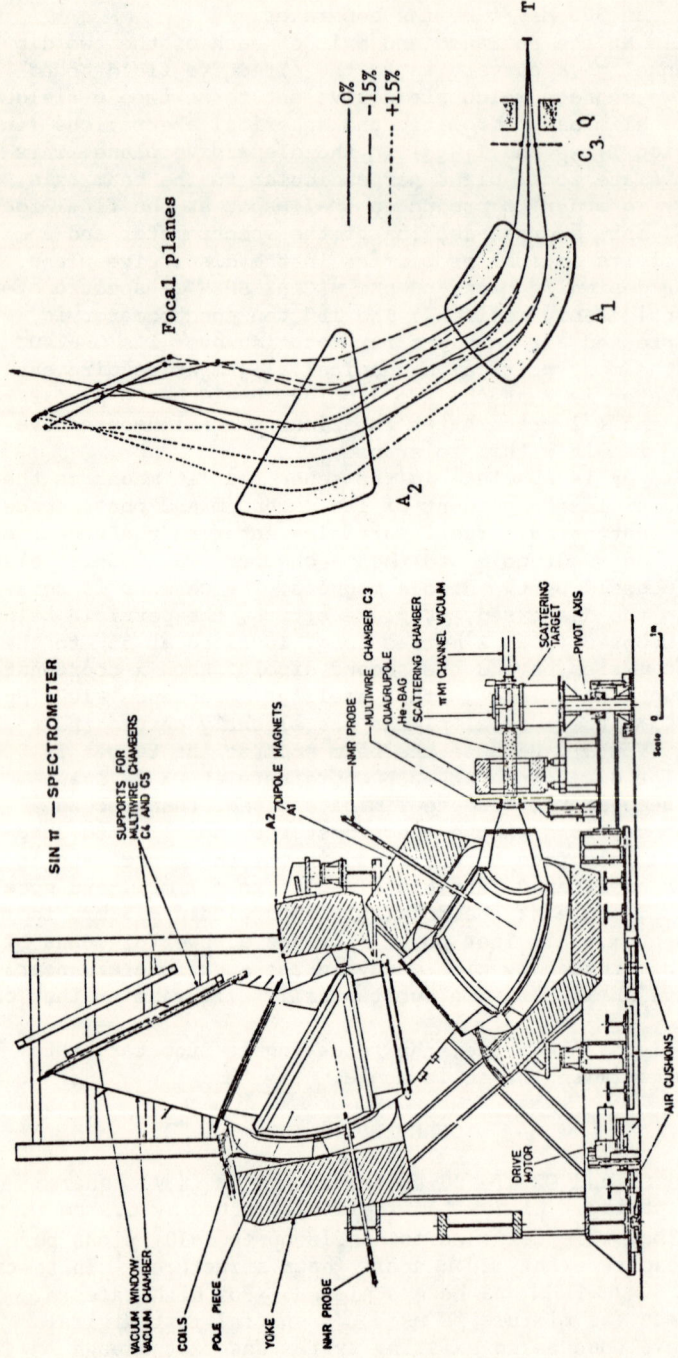

Fig 3 A schematic view of the optics of the spectrometer in the dispersive plane. The dispersed focal planes for three source points in the chamber C3 are shown, at an angle of 35° to the beam axis

Fig 2 Cross-section drawing of the spectrometer

with a printed-circuit delay-line read-out with 1 mm resolution. C4 and C5 are currently 100×25 cm^2 cable delay-line read-out chambers with 2 mm resolution. For all chambers a detection efficiency of 99% has been obtained.

BEAM TUNING AND SPECTROMETER OPTICS

The important optical aspects of the beam line are that the dispersed intermediate focal plane should lie along the chamber C1, and that the spot at the scattering target should be small. To this end, the final focus should be achromatic, and second order optical terms should be zero or small. The usual procedure of TRANSPORT calculations, followed by careful magnet measurement and alignment was followed. The beam tuning then consists of a scheme of empirical optimisation of the currents in the beam elements, to correct for small errors in field measurements and alignment. Pairs of beam parameters (first order TRANSPORT matrix elements) are adjusted by varying the currents in the appropriate pair of beam elements in an orthogonal sequence. The only second order beam parameters adjusted are those determining the inclination and curvature of the intermediate focal plane, and the corresponding parameters for the second half of the beam. These were adjusted using the surface coils in the dipoles.

In order to set the intermediate focus correctly, it is necessary to demand that the momentum is independent of angle in the dispersive plane (x'). The spectrometer was used to measure the momentum and the angle was measured using an extra chamber placed 60 cm downstream from C1. Although the vertical bend in the spectrometer conveniently decouples the first order optics from that of the beam-line, it was necessary to decouple the second order optics as the spectrometer was not sufficiently well understood that it could be used as a reliable measure of momentum indepedent of the angle of entry of a pion. This decoupling was effected by placing 3 mm of lead at the target position, and restricting the permissible acceptance into the spectrometer. Fig 4 illustrates how the first surface coils were adjusted to eliminate a x'^2 dependence of momentum, at the centre of C1 (i.e. $T_{1,22}$ is set to 0). At the end of the beam tuning, a spot at the scattering target of 9mm (x) x 7 mm(y) FWHM was obtained (fig 5), and the momentum resolution of the optics of the channel was estimated to be $\simeq \pm 5 \times 10^{-4}$ dp/p rms. Adjustments to the currents in the quadrupoles were in the range 0.5-2.5%, but the surface coils needed substantial corrections (50-100% changes in their currents).

The optics of the spectrometer is determined by parameterising the momentum in terms of the coordinates of the chambers C3, C4 and C5. The momentum-loss expression

$$P_{loss} = P_{channel} - P_{spectrometer}$$

should be zero in the case of the beam running directly through the spectrometer, with a peak width dtermined by the combined resolution of the channel and spectrometer. To optimise the parameterisation of the spectrometer optics, the measured width of this

730

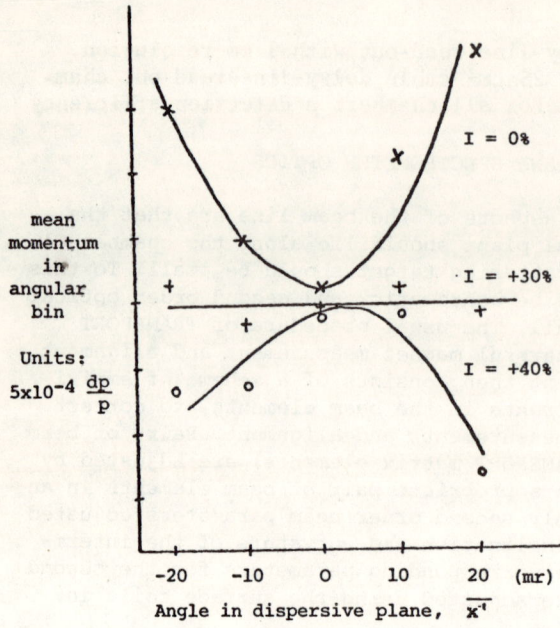

Fig 4 The dependence of the momentum of the beam on its angle in the dispersive plane for three settings of the current in the first surface coils on the dipole magnet M1. The value I = +30% sets the second order matrix element $T_{1,22} = 0$

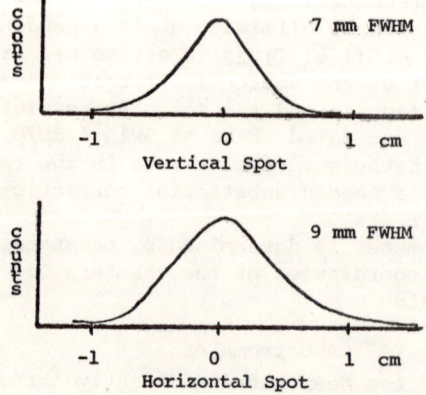

Fig 5 The measured beam spot size at the scattering target

expression is minimised using a linear least-squares-fit program, allowing the coefficients of the parametrisation to vary. Using the direct beam, for a single region of the focal plane thus illuminated an expression of the form

$$dp/p(spect) = \alpha x_3 + \beta x_3^2 + \gamma x_3^3 + \delta x_4 + \epsilon y_3^2 + \phi(x_4 - x_5)$$

where x_3 means the x coordinate at C3, etc, yields an overall resolution of 1×10^{-3} dp/p FWHM. However, the coefficients α, β, etc have to be parametrised further in terms of x_6 and x_6^2 to obtain a single expression for the full focal plane. The result of the first such full focal plane parametrisation, using the direct beam, is shown in Fig 6. the individual peaks have widths of $1.2-1.5 \times 10^{-3}$ dp/p FWHM. They correspond to six separate runs in which the spectrometer field has been varied in constant steps of 3% of the central value. The peak separation is constant to 1% over the full 15% $\Delta p/p$.

Using the direct beam to determine the spectrometer's optics has the disadvantage that the chamber C3 is not fully illuminated. Thus, in particular, the coefficients of the high-order terms (x_3^2, y_3^2) are not reliably determined. The spectrum shown in Fig 7 has peak widths (3.5×10^{-3} dp/p FWHM) substantially determined by the Landau broadening in the thick target. Apart from correcting for kinematic recoil of the target, no attempt has been made to optimise the optics using this spectrum. Although there is some evidence of poorly determined high order coefficients, there is nothing to suggest that an overall resolution of about 1.5×10^{-3} $\delta p/p$ FWHM would not have been reached with a thin target.

1. J. Pouxe, et al., Proc. of 2nd Ispra Nuclear Electronics Symposium, 1975.

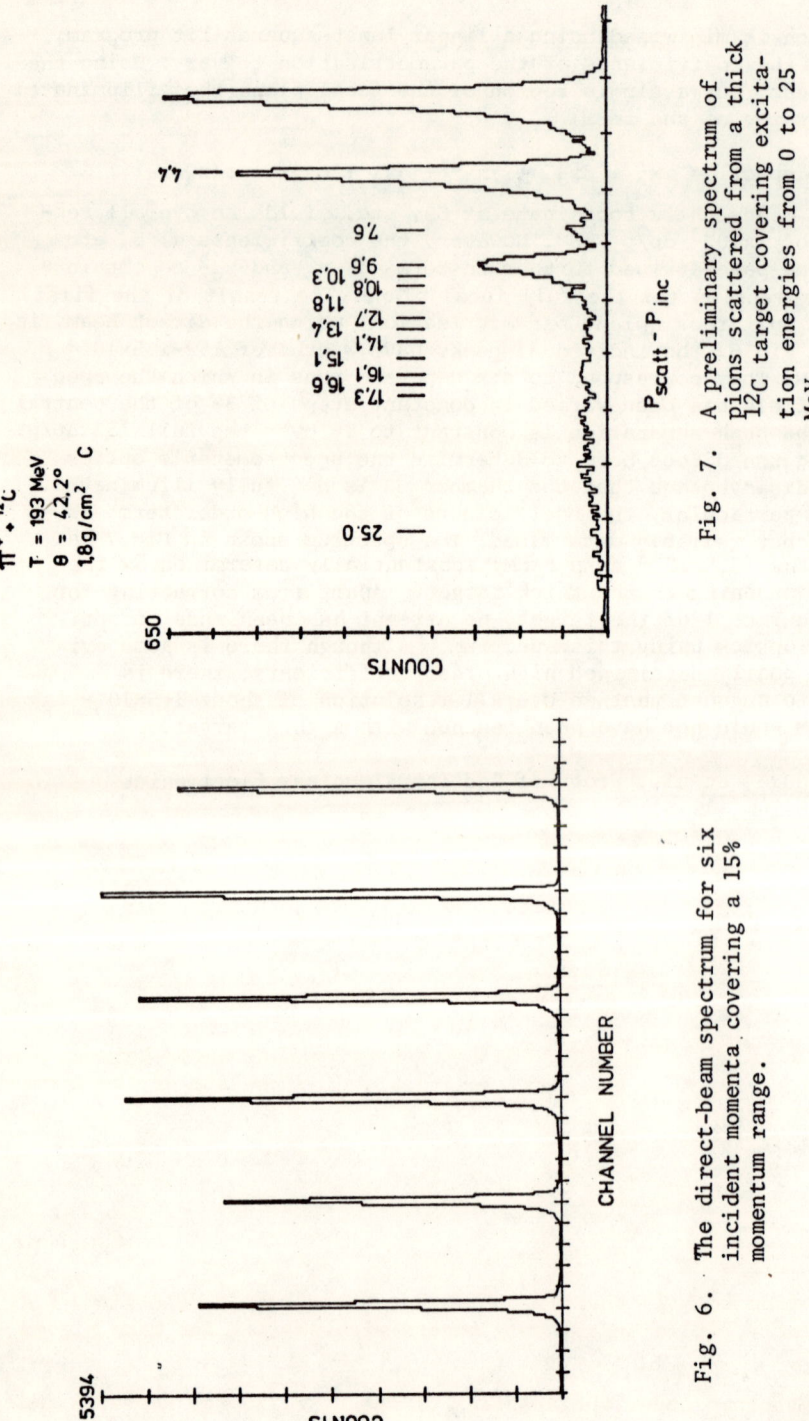

Fig. 6. The direct-beam spectrum for six incident momenta, covering a 15% momentum range.

Fig. 7. A preliminary spectrum of pions scattered from a thick ^{12}C target covering excitation energies from 0 to 25 MeV.

DESIGN OF A π^0 SPECTROMETER AT LAMPF*

R. Heffner
University of California, Los Alamos Scientific Laboratory
Los Alamos, New Mexico 87545

I. INTRODUCTION

This talk shall be concerned with a discussion of the salient features for the design of a π^0 spectrometer being constructed at LAMPF by groups from LASL, Tel Aviv University, and Case Western Reserve University. First, the basic concepts of π^0 detection will be explained, followed by a discussion of the expected performance and mechanical design of the instrument, and concluded with a list of intended experiments to be undertaken by our group.

The spectrometer is intended to be used for pion-charge-exchange and production reactions with nuclear targets, and so a π^0 energy resolution (ΔE_{π^0}) of at most a few MeV is desirable. The π^0 decays with about 99% probability into two gamma rays with a lifetime of 10^{-16} s, and thus the simultaneous detection of the decay gammas can be used to tag the π^0. Adding together the gamma-ray energies alone will not give the required energy resolution for the π^0 if one uses either sodium iodide (NaI) or lead-glass photon detectors, however. For lead glass the full width half maximum (FWHM) for the photon energy resolution ($\Delta E_\gamma/E_\gamma$) is given by $\Delta E_\gamma/E_\gamma \gtrsim 3/\sqrt{E_\gamma}$ and for NaI $\Delta E_\gamma/E_\gamma \gtrsim 0.04$, so that if one detects two 100-MeV gammas, the π^0 energy resolution is about 42 MeV for lead glass and 5-6 MeV for NaI.[1] A pair spectrometer could yield the required resolution, but the detection efficiency is too low for most applications. We shall therefore discuss an alternate solution.

An expression for the π^0 energy can be written as follows:

$$E_{\pi^0}^2 = \frac{2 m_{\pi^0}^2}{(1 - X^2)(1 - \cos \eta)}, \quad (1)$$

where m_{π^0} is the π^0 mass, η the laboratory opening angle between the gamma rays, and X is the difference divided by the sum of the gamma-ray energies:

$$X = \frac{E_{\gamma_1} - E_{\gamma_2}}{E_{\gamma_1} + E_{\gamma_2}}. \quad (2)$$

We note from Eq. (1) that if X = 0 then η completely determines E_{π^0}; consequently, if we accept only events with small X and make a good

---
*Work performed under the auspices of the U. S. Energy Research and Development Administration

measurement of η, we can expect an improved π^0 energy resolution even with large measured uncertainties in E_γ. Therefore, the gamma-ray directions in addition to their energies must be measured and the momentum vector of the π^0 reconstructed.

In the next section an overview of the detection scheme we have devised is presented. We have chosen to use lead-glass shower counters instead of NaI because (1) a reasonably large class of experiments is feasible with the resolution obtainable from a lead-glass system (Sec. III.A.), and (2) the lead glass is less expensive and far less susceptible to neutron and charged particle backgrounds than NaI.

II. MECHANICAL DESIGN

A schematic view of the spectrometer is shown in Figs. 1a and 1b where two photons of equal energy are shown to decay from a π^0 with 100-MeV kinetic energy. The photons convert to electrons in the lead-glass slabs (C), and their conversion points are measured as the electrons pass through multiwire-proportional chambers (MWPC) located behind each slab. The opening angle η can be reconstructed from the conversion points and an assumed decay point in the target. The shower is contained in the large total-energy lead-glass counter arrays (E)

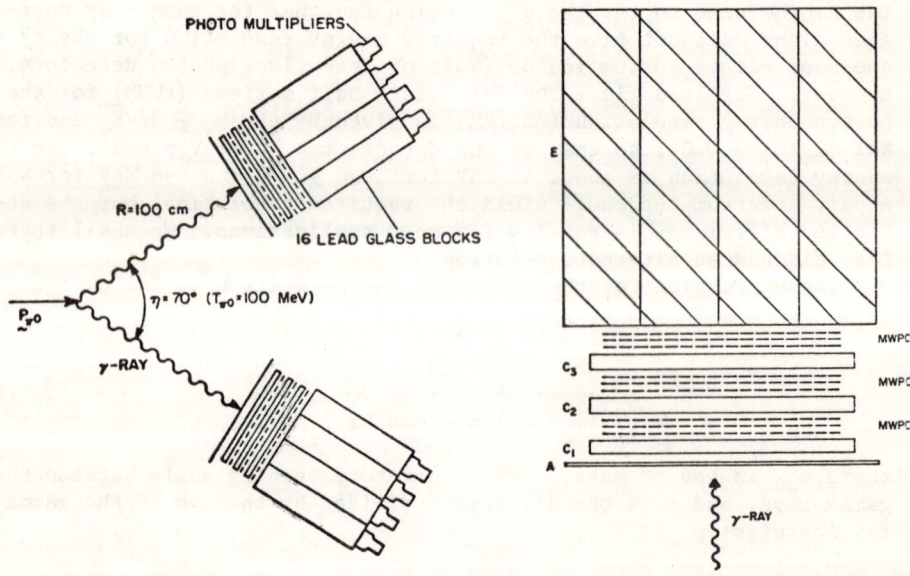

Fig. 1. (a--left, b--right) Figure 1a shows a schematic picture of the spectrometer set up to detect 100-MeV kinetic energy π^0. Figure 1b is a blowup of one of the spectrometer arms. The E detector is a 4 by 4 total-energy counter array of lead-glass blocks 14 radiation lengths deep (60 cm). Shown are three lead-glass converters (C_1, C_2, and C_3), each 0.6 radiation lengths thick backed by three planes of multiwire-proportional chambers (MWPC). A plastic scintillator (A) is used to reject charged particles from the target. The assemblies in Figs. 1a and 1b are drawn approximately to scale.

behind the converters. The signals from the Cherenkov light produced in each converter are added to the signals from the larger counter array, E, to form a total-energy signal. The array E consists of 16 lead-glass blocks, each of which is 15 cm by 15 cm by 60 cm (14 radiation lengths deep) so that the whole array forms a 60-cm cube. Each block is optically separated from its neighbors and is viewed by a 5-in. phototube coupled directly to the block end face. Each of the thin converters is 52 cm by 52 cm by 2.5 cm (0.6 radiation lengths thick) and is viewed by two photomultipliers.

We have written Monte Carlo programs to study the details of the shower development as well as other geometrical and physical factors which can influence the performance of the system. In the following a discussion of these studies and an explanation of our choice of design parameters and expectations for the instrumental resolution are presented.

III. DESIGN STUDIES

A. Energy Resolution

There are two general groups of factors which influence the π^0 energy resolution: (a) instrument-related factors and (b) target-beam size related factors. The latter occur for finite target and beam sizes because the decay point of the π^0 is not known exactly (occuring somewhere in the beam-target cross-sectional area), and thus the reconstruction of the gamma-ray directions is uncertain. In addition, the incident particle suffers energy loss in the target, affecting the opening angle of the decay gamma rays. These effects may dominate the π^0 energy resolution depending upon the choice of target density and thickness and are discussed in Ref. 2. We shall not detail them here.

The instrument-related factors are the photon energy resolution (ΔE_γ), the opening angle resolution ($\Delta \eta$), and the range of accepted X values (ΔX). The first two are independent measurements and so yield uncertainties in E_{π^0} which add in quadrature. For the choice of lead glass, one can safely assume $\Delta E_\gamma = 3\sqrt{E_\gamma}$ MeV (FWHM) as indicated above. Then one can show that if $\Delta \eta \leq 0.005$ radians FWHM, the contribution to the π^0 line shape for $|\Delta X| \leq 0.1$ (see below) from the uncertainty in η is negligible; that is, all the line broadening then comes from ΔE_γ.

The calculated π^0 line shape for different values of $|\Delta X|$ (and $\Delta \eta = 0.004$ radians FWHM) is shown in Fig. 2 for a 100-MeV π^0. For $|\Delta X| \leq 0.1$ the FWHM is $\simeq$ 1.0 MeV. No effects of finite beam-target sizes have been included. Note that for small $|\Delta X|$ the line shape has a small low-energy tail. Since $X = \beta \cos \theta^*$, where β is the π^0 velocity and θ^* the polar angle of one of the gamma rays in the π^0 rest frame, one can deduce that for $|\Delta X| \leq 0.1$, 12% of the events are accepted. We have decided that the range $0.1 \leq |\Delta X| \leq 0.2$ is a reasonable compromise between energy resolution and efficiency.

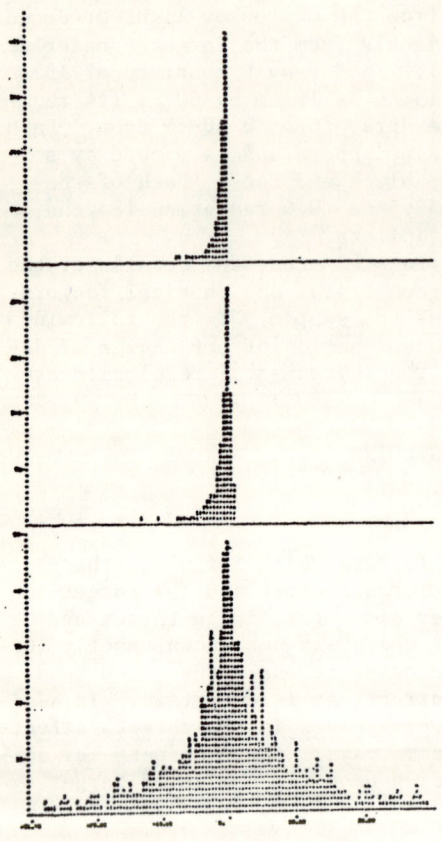

Fig. 2. (a--top, b--center, c--bottom) A Monte Carlo calculation showing the effect on the line shape of restricting the range of X values for a π^o kinetic energy of 100 MeV. The assumed gamma-ray resolution is $\Delta E_\gamma = 3\sqrt{E_\gamma}$ MeV FWHM, and the opening angle resolution $\Delta\eta = 0.004$ radians FWHM. No effects of finite target-beam size are included. The bin size is 0.5 MeV. In Fig. 2a $|\Delta X| < 0.01$ (1 in 86 events accepted), in Fig. 2b $|\Delta X| < 0.1$ (12 in 100 events accepted), and in Fig. 2c $|\Delta X| < 1$ (all events accepted).

B. Angular Resolution

The π^o angular resolution is dominated by the gamma-ray energy resolution and by the finite target-beam size effects; that is, the effect of the uncertainty in the opening angle (for $\Delta\eta \leq 0.005$ radians FWHM) is negligible. At low π^o momenta the angular resolution is worse than at high momenta and is estimated to range from about $9°$ FWHM at 30 MeV to about $4°$ FWHM at 300 MeV π^o kinetic energy. This is good enough to measure $(\pi^\pm,\pi^o)$ angular distributions for light- and medium-weight nuclei.

C. Shower Studies

1. Design of the Photon Converters. The choice of converter thickness involves a compromise between the probability of conversion and the spread of the shower which determines the spread in η. We have written a Monte Carlo code[3] to study these effects. Figure 3 displays some results for a 100-MeV photon incident on a converter of different thicknesses. Figure 3a shows the probability of having at

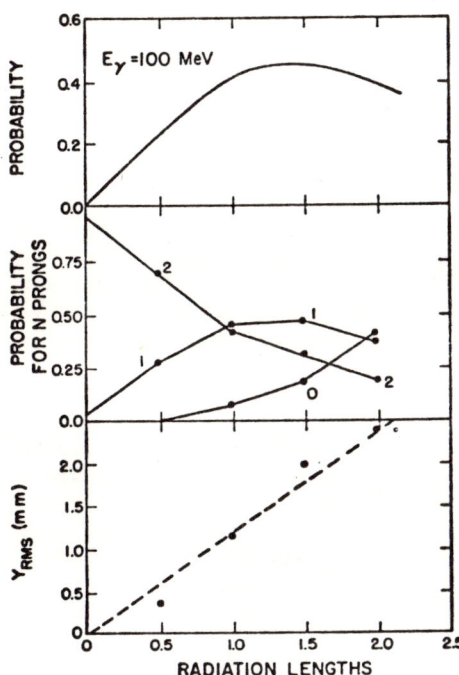

Fig. 3. (a--top, b--center, c--bottom) Monte Carlo results for a 100-MeV photon incident on a lead-glass converter of different thicknesses; the converter was chosen to have the same constituency as Schott F2 lead glass. Figure 3a gives the probability of at least one charged particle exiting the converter. Figure 3b shows the probability of having zero, one, or two charged particles in each event. Events with more than two charged particles constitute less than 5% probability at these thicknesses and are not plotted. Figure 3c shows the rms spread of the position of the centroid of charged particles which exit the converters. The dotted line is to guide the eye through the statistical fluctuations of the calculation.

least one charged particle exiting the converter. Figure 3b shows the composition of each charged particle event. Figure 3c displays the lateral spread of the charged particles as they leave the converter. Our constraints for a small uncertainty in the opening angle and a conversion efficiency of about 0.6 in each spectrometer arm lead to using three converters each 0.6 radiation lengths thick for each arm. We can then achieve the desired $\Delta\eta \leq 0.005$ radians FWHM for target-detector distances of 60 cm or more.

The converters in a single arm will be used in a logical "or" fashion. Each one will be backed by three planes of MWPC with wire spacing about 2 mm. Two of the MWPC will be oriented to measure the shower coordinates in the plane of the photon momenta (wires out of the paper in Fig. 1a), and the third MWPC will be orthogonal. The conversion point can then be obtained by an extrapolation of the shower tracks to an optimal plane within the converter.

2. Choice of Lead Glass and Phototubes. The results of our shower code studies in addition to tests[4] of lead-glass types at the Stanford Mark III electron accelerator have led to our choice of LF5-type[5] glass. This glass type was found to give better energy resolution than more dense glasses such as F2.[5] The choice of 14 radiation lengths depth for the E counters was made to contain a 500-MeV photon shower with negligible contribution (less than 5%) to the energy resolution from leakage out the back.

Performance tests[6] conducted at LAMPF for several kinds of 5-in. photomultiplier tubes have resulted in our decision to use the

EMI9618R tube for its superior gain, photoelectron collection efficiency, and pulse-height resolution. We plan to gain stabilize all of the tubes by using a system of temperature-stabilized light-emitting diodes, one for each glass block.

D. Overall Efficiency and Solid Angle

For a target-to-detector distance (R, in Fig. 1a) of 60 cm and an active area for each detector face of 30 cm^2, the spectrometer will subtend a solid angle of about 35×10^{-3} steradians. The combined conversion efficiency of 0.36 and the requirement for good resolution of $|\Delta X| \leq 0.2$ yield an overall efficiency of about 9%.

IV. SOME PROPOSED PHYSICS EXPERIMENTS

It is appropriate to conclude by listing some of the experiments in nuclear/particle physics to which the spectrometer is applicable and which have captured the interest of our group. The first experiment for which the spectrometer will be used is $\pi^- p \to \pi^0 n$ angular distributions for incident pion energies mostly below 100 MeV.[2] These measurements will nicely complement $\pi^\pm p$ elastic scattering measurements at low energies and thus help to determine the low-energy pion-nucleon phase shifts and S-wave scattering lengths. In addition, the reaction is a convenient one to calibrate the new spectrometer since a tagged π^0 can be obtained by detecting the coincident neutron.

Other experiments either proposed[7] or which will be proposed as the spectrometer development makes possible their execution include pion-charge-exchange reactions on ^2H, ^3H, ^3He, and heavier nuclei such as ^{13}C and ^{48}Ca. The experiments should be quite exciting since no charge-exchange angular distributions on nuclei have been measured before. For example, the study of neutron-matter radii should be particularly interesting.[8] π^0 production experiments are also planned.

REFERENCES

1. For a discussion of lead-glass resolution relevant to our counter design see C. A. Heusch, R. V. Kline, and S. J. Yellin, Nucl. Instrum. Methods 120, 237 (1974). The NaI resolution is from M. Hasinoff, private communication, regarding the University of British Columbia TINA detectors in operation at TRIUMF.
2. Addendum to LAMPF Research Proposal 181, "Measurement of the $\pi^- p \to \pi^0 n$ Angular Distribution at Low Energies and Calibration of the π^0 Spectrometer," J. D. Bowman and J. Alster, spokesmen.
3. J. D. Bowman, M. D. Cooper, R. Heffner, C. M. Hoffman, and M. Zaider, "Monte Carlo Simulation of Electromagnetic Shower Development," to be published.
4. T. Mast, private communication, Lawrence Berkeley Laboratory, Berkeley (1975).
5. LF5 glass has density 3.22 g/cm^3 and radiation length 4.11 cm. F2 glass has density 3.6 g/cm^3 and 3.16 cm radiation length. See Schott Optical Glass Catalog, Mainz (1975).
6. S. Gilad, M. D. Cooper, and M. A. Moinester, private communication, Los Alamos Scientific Laboratory, Los Alamos (1976).

7. LAMPF Research Proposal 181, "Angular Distribution Measurements for Pion-Nucleus Single Charge Exchange Reactions," J. Alster and J. D. Bowman, spokesmen.

8. G. A. Miller and J. Spencer, Phys. Lett. <u>53B</u>, 329 (1974).

PERFORMANCE OF EPICS PION CHANNEL

D. C. Slater and C. L. Morris
University of Virginia, Charlottesville, VA 22903

H. A. Thiessen and J. Källne
University of California, Los Alamos Sceintific Laboratory

C. Fred Moore and Joe E. Bolger
University of Texas at Austin, Austin, Texas 78712

S. Iversen and A. Obst
Northwestern University, Evanston, IL 60091

J. F. Amann
Carnegie-Mellon University, Pittsburgh, PA 15213

S. Verbeck and G. Burleson
New Mexico State University, Las Cruces, N. M. 88001

J. Peterson and S. Greene
University of Colorado, Boulder, Colorado 80302

ABSTRACT

A facility for pion scattering experiments called EPICS is under construction at LAMPF[1]. The channel part has been installed, and results from the first tests of this channel are reported.

TEXT

The EPICS pion channel[1] consists of four bending magnets, a crossed field separator, and three multipole trim magnets[2] located between the second and third dipole magnets. The multipole magnets are capable of producing independently adjustable quadrupole, sextupole, and octupole fields. Since the three magnets are series connected, the resolution of the channel is effectively tuned by a three parameter adjustment.

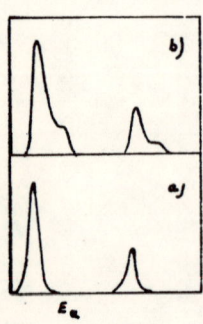

Fig. 1. ^{242}Cm spectrum with EPICS.

The first tests of the channel were made in October 1975. A ^{242}Cm alpha source located at the position of the pion production target; the source gives two α-lines at 6.1129 and 6.0696 MeV. The detector was a two dimensional read-out helical multiwire proportional counter[3] located at the focal plane of the channel. In these tests, a resolution of $\Delta p/p = 3.5 \times 10^{-4}$ (FWHM) was obtained after a straight forward adjustment of the multipoles; this figure includes source and detector contributions. The line shape with multipoles on and off is shown in Fig. 1a and b. During these

Table I Preliminary data on performance of
EPICS pion channel with separator on

Energy (MeV)	Polarity	Count Rate (Scaled to 1 ma and 3 cm target)	Fraction of Beam			
			π	μ	e	p
67.5	π^+	2.1×10^7 /sec	0.60	0.25	0.16	0.01
151.0	+	1.2×10^8	0.71	0.08	0.11	0.03
255.0	+	2.5×10^8	0.78	0.09	0.05	0.05
67.5	−	7.8×10^6	0.44	0.17	0.25	--
151.0	−	2.7×10^7	0.78	0.05	0.14	--
255.0	−	1.6×10^7				

tests, we also demonstrated that there was no noticeable change in the line width with the separator turned on. The alpha particle tests were repeated in January 1976 after the channel was covered with shielding. Refinements of our measuring technique allowed us to demonstrate a linewith of 1.9×10^{-4} at the center of the focal plane of the channel.

In March of 1976, the first pions were observed at the end of the channel. The pion beam was measured with regard to counting rate, phase space, and contamination. We observed, e.g., a rejection factor of 100:1 for protons with the separator on. The preliminary results on pion beam properties are summarized in Table I.

After a short period of further tests, the EPICS channel will be used for a program of experiments which do not require a spectrometer. Construction of the spectrometer is now proceeding at a reasonable pace, and completion of the spectrometer is expected to occur in February 1977.

REFERENCES

1. "A Proposal for EPICS: A High-Resolution Pion Beam and Spectrometer Facility for Nuclear Structure Research at LAMPF," LA-4534-MS, unpublished.
2. R. J. Peterson, J. J. Kraushaar, R. A. Ristinen, H. A. Thiessen, and M. E. Rickey, Nuclear Instruments and Methods 129 (1975) 47.
3. E. R. Flynn, S. Orbensen, N. Stein, H. A. Thiessen, D. M. Lee, and S. E. Sobottka, Nuclear Instruments and Methods 111 (1973) 67.

PERFORMANCE OF A LARGE STREAMER CHAMBER FOR THE OBSERVATION OF PARTICLE-NUCLEI INTERACTIONS WITH ALL CHARGED PARTICLES DETECTED*

A. S. Kanofsky[††,+], R. C. Allen[+], and A. Hasan[+]
Dept. of Physics, Lehigh University
Bethlehem, Penna. 18015
and
Brookhaven National Laboratory[†]
Upton, L. I., N. Y. 11973

We have performed an experiment to study the break-up of nuclei under pion and proton bombardment where all charged incoming and outgoing particles as well as nuclear fragments are observed. This was run in December, 1975 using the Argonne-Illinois streamer chamber facility at Argonne National Laboratory.

A schematic of the experiment is shown in fig. 1. The incoming beam particle and scattered particle signifying an event are defined by a logic signal, $B1 \cdot B2 \cdot \overline{S} \cdot B3 \cdot \overline{B4} \cdot H \cdot R$, where an interaction results in the incoming beam particle scattering off of a streamer chamber gas nucleus. There was, in addition, a beam Cerenkov counter to eliminate beam electrons. Here B1 and B2 are beam counters, S denotes slat counters which anti'd particles outside the beam region, B3 is a 1 1/2" x 1 1/2" hole anti counter mounted within the streamer chamber to define a small area for the beam, and B4 is a 2 1/2" x 2 1/2" counter that is used as an anti to determine that the beam particle was scattered along the beam path within the chamber. B4 is mounted just outside the chamber. Counter H is a circular hodoscope and R is an additional counter to reduce the spurious triggers due to two parallel particles passing through the chamber simultaneously, one of which follows close to the beam but just misses B4, the other of which passes through the outer circular counters of H, giving a false trigger. The H counter was usually connected with the inner circular counters (2", 6", 10" diameter) as anti's (to determine scattering from interactions) and the outer ones (14", 20" diameter) in coincidence.

The incoming beam particles used were π^+'s, π^-'s, and protons of 1.5 GeV/c to 2.5 GeV/c momentum. It was not possible to operate the beam below 1.5 GeV/c momentum since the long beam line gave rise to a large beam spread in the chamber due to coulomb scattering, interactions, and pion decays along the beam line. It was required to have a small beam diameter in order for the B4 anti to work effectively.

A gas to capture free ions, SF_6, was added to the chamber gases used, Ne and He, in small concentrations in order to obtain a memory time of about 2 μsec. A 6 KG. magnetic field was applied to the chamber. A 700 KV high voltage pulse was applied across the chamber between the central and side electrodes. It was applied for 20 nsec. and was generated by a Blumlein charged by a Marx generator.

Two lenses with different shutter openings were used in order to

*Supported in part by the U.S.E.R.D.A. under contract E(11-1)-2894.

be able to photograph both the lightly and heavily ionizing particles. The lightly ionized tracks are visible in the picture taken with the f2.0 lens, with the heavily ionized tracks greatly overexposed dark blobs, while the heavily ionized tracks are clearly visible and well defined in the f5.6 lens picture with the lightly ionized tracks not visible. The two pictures with the two lenses are adjacent on a single frame of film. Fig. 2 is a photograph of a typical event. A film speed of 6000 ASA was obtained using KODAK SO143 film developed with D19.

We show in figs. 3 to 5 the ranges of particles in Ne gas as a function of the momentum of the particle. The ranges of low energy α's are small enough for α particles to stop in the chamber, thus giving a measure of the energies. Particle discrimination is possible by measuring the momentum of the tracks. Then, as shown in fig. 6, the ionization, or equivalently, dE/dx, varies appreciably for the same momentum for α's, d's, and p's, and it is then possible to discriminate particles on the basis of their ionization.

In conclusion, using this new technique, it's possible to see the tracks due to all charged particles resulting from particle-nuclei collisions. The particle momenta and type can be determined using the track curvature and ionization. We thank Jerry Watson and his crew for ably operating the streamer chamber.

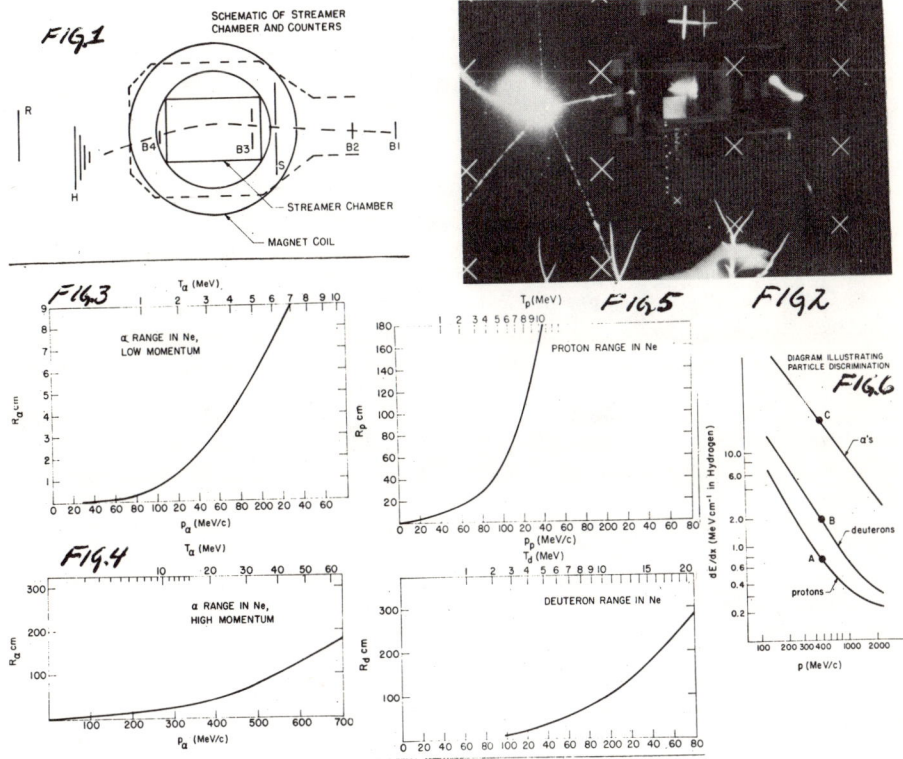

THE NUCLEAR MANY BODY PROBLEM AS A
RELATIVISTIC FIELD THEORY*

L. Wilets
Physics Department, University of Washington, Seattle WA 98195

1. INTRODUCTION

Interest in the nuclear many body problem as a relativistic field theory of elementary particles was stimulated, in large measure, by the discovery of high density neutron stars and by speculation over other forms of high density nuclear matter. Our own advent into the field was motivated by a desire to understand both high density nuclear matter, where such a theory is essential, and normal nuclear matter, where the effects are subtler but significant.

In contrast to many body theories which employ phenomenological two body <u>potentials</u>, the field theory approach (summarized in this paper) leads to an effective interaction which features:
  (1) "Proper" off-shell behavior.
  (2) Effects of the media; i.e., many-body forces. These arise not only from specific three-and-more-body diagrams, but also from the role of the media in modifying the nucleon and meson propagators.
  (3) Relativistic kinematics and dynamics, including retardation.

2. THE MODEL

The basic formulation of the model was presented in 1970 by Warren Brown, Robert Puff and myself.[1] The physics is completely described by the meson-nucleon vertex, Fig. 1, where any (and all) mesons can be included: π, ω, ρ, ϕ, ... We do not include, however, a fictitious σ meson. The vertices are assumed to be local, contact interactions. Meson-meson interactions are not included explicitly, although such interaction can occur within the model through, for example, the box diagram, Fig. 2.

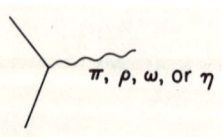

Fig. 1. Basic interactions.

We employ the Martin-Schwinger-Puff[2] Green's function formulation. Nucleon Green's functions are introduced by the expectation value of the Wick time ordered product of nucleon operators:

$$G_n(1...n;1'...n') = i^n <T\{\psi(1)...\psi(n)\bar\psi(n')...\bar\psi(1')\}>. \quad (1)$$

One can similarly introduce meson Green's functions and, for some purposes, it is convenient to introduce mixed meson-nucleon Green's functions. The index (n) refers to the set of coordinates $\vec r_n$, t_n, α_n and β_n (α_n and β_n are spin and isospin coordinates, respectively). The expectation value of any operator X is defined as the thermodynamic average,

$$\langle X \rangle = \frac{\text{tr } e^{-\beta(H-\mu N-\nu I)} X}{\text{tr } e^{-\beta(H-\mu N-\nu I)}} \qquad (2)$$

with $\beta = (kT)^{-1}$, μ the chemical potential and ν a Lagrange multiplier for isospin. We have, to date, restricted our considerations to ground state properties, $\beta \to \infty$.

A hierarchy of coupled equations is derived involving only nucleon Green's functions (and the bare meson Green's function). Thus the meson field functions do not appear explicitly. The first few equations are:

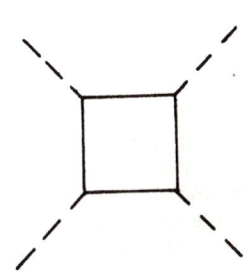

Fig. 2. Meson-meson scattering.

$$G_1(1;1') = G_1^0(1;1') + iG_1^0(1;2')\langle 23|v|45\rangle G_2(45;3^{\pm}1') \qquad (3a)$$

$$G_2(12;1'2') = G_1^0(1;1')G_1(2;2') - G_1^0(1;2')G_1(2;1')$$
$$+ iG_1^0(1;3)\langle 34|v|56\rangle G_3(256;4^{\pm}1'2') \qquad (3b)$$

$$G_3 = G_1^0 G_2 - G_1^0 G_2 + G_1^0 G_2 + iG_1^0 v G_4 \qquad (3c)$$

where

$$G_1^{0-1} = \gamma p + M_0 \quad -\gamma_0(\mu + \tfrac{1}{2}\tau_3 \nu) \qquad (4)$$

The form of these equations is identical with the non-relativistic, static potential problem, but here v is symmetric advanced-retarded, i.e. frequency-dependent; see Fig. 3.

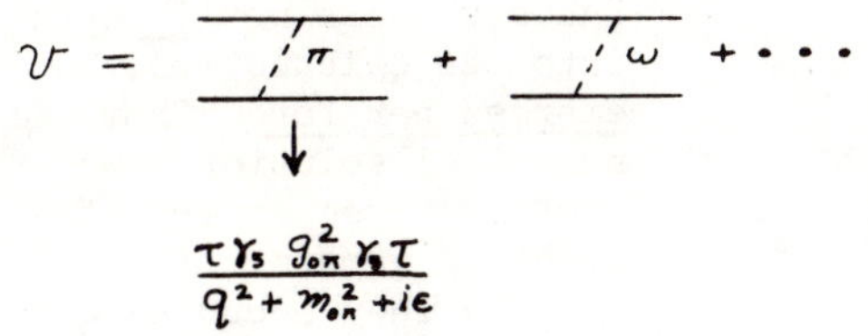

$$v = \underset{\substack{\downarrow \\ \frac{\tau \gamma_5 \, g_{\sigma\pi}^2 \, \gamma_5 \tau}{q^2 + m_{\sigma\pi}^2 + i\epsilon}}}{\underline{\pi}} + \underline{\omega} + \cdots$$

Fig. 3. Representation of the "bare" N-N interaction.

It is necessary to truncate the hierarchy of equations in order to make a calculation feasible. Factorization is the technique suggested by the structure and guided by the physics. Except in lowest order, however, one usually obtains approximations which are not expressible in terms of Feynman diagrams, but contain "anomalous diagrams."

3. HARTREE-FOCK IN VACUUM

The simplest factorization is obtained by replacing, in (3a),

$$G_2(45;31') \underset{\sim}{\sim} G_1(43)G_1(51') - G_1(41')G_1(53). \tag{5}$$

This is the Hartree-Fock approximation. This approximation <u>can</u> be identified with the summation of "uncrossed meson lines" diagrams of the form shown in Fig. 4. After mass and wave function renormalization the vacuum H-F equations were solved numerically including π and ω mesons.[1] Brown[3] has shown that the model is still renormalizable with the ρ-meson included; he then solved the H-F equations including the meson. The renormalized one-nucleon propagator is written in the usual Lehmann-Källen representation

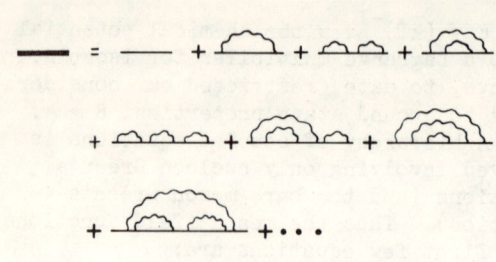

Fig. 4. The uncrossed meson line approximation for the nucleon propagator.

$$G_R(p) = \int \frac{A_R(K)dK}{\gamma \cdot p + K} + \frac{A_c}{\gamma \cdot p + K_c} + \frac{A_c^*}{\gamma \cdot p + K_c^*}. \tag{6}$$

The renormalized spectral function is shown in Fig. 5. Note the two spikes, which occur very close to the masses of the only known N* resonances with the same quantum numbers as the nucleon, namely the 1470 and 1780.

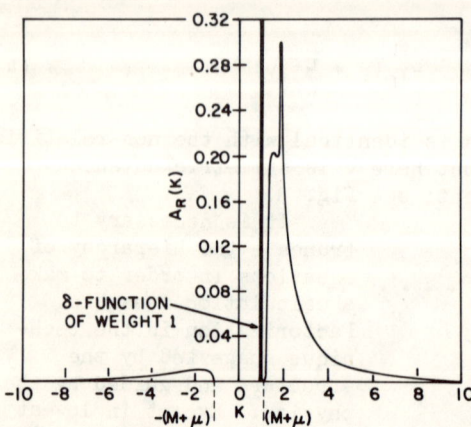

Fig. 5. The spectral function for the nucleon.

In addition to the real continuum, our numerical solution exhibited a conjugate pair of complex poles, as shown in Fig. 6. These "ghosts" were not <u>inserted</u> into the calculation, but <u>emerged per force</u> from the numerical solution. We proffer three interpretations of ghosts:

1) They occur as a consequence of the approximations, and will recede to infinity as the approximations improve.

2) As Lee and Wick have emphasized, ghosts are inevitable, and are essential for unitarity.

3) They always occur in a local relativistic field theory, and

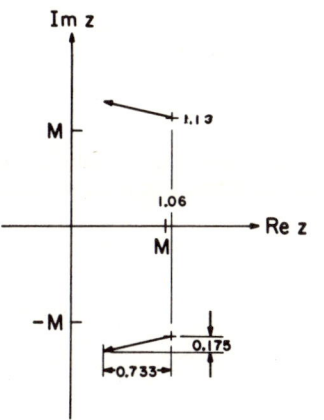

Fig. 6. Location of complex poles for the nucleon propagator.

indicate that such theories are unphysical.

We can live with either (1) or (2), but (3) would force us to seek other employment.

If we accept (1), we must for consistency retain the ghosts at every stage of approximation. If they eventually disappear, we expect or at least hope that the intermediate results converge rapidly.

If we accept the Lee-Wick arguments, we must <u>conclude</u> that we <u>began</u> with a non-Hermitian Hamiltonian. Our Hamiltonian is symmetric in the operators ψ and ψ^+. We used the usual equal time commutation relations for ψ and ψ^+, but nowhere did we assume that ψ^+ is the Hermitian conjugate of ψ. In retrospect, we must infer that they are not Hermitian conjugates. Certainly states with complex energies are physically unacceptable. Let me paraphrase Lee and Wick: The ghost states cannot be reached by physical processes. There must exist a representation where the Hamiltonian is reducible into the real world and the ghosts, with no interactions connecting the parts. Our representation is not that one, so our Hilbert space includes unphysical pieces.

Whether we accept interpretations (1) or (2), we must include the complex poles in our calculations. We have done so, and good things do emerge.

4. PI-NUCLEON SCATTERING

Bill Nutt and I[4] calculated pi-nucleon scattering in what we call the augmented Born approximation (ABA), Fig. 7, using the BPW nucleon propagator with ghosts. It is well known that the usual Born approximation, with the nucleon propagator corresponding to only the single pole at the nucleon mass, does very poorly. The inclusion of the ghosts accomplished the following:

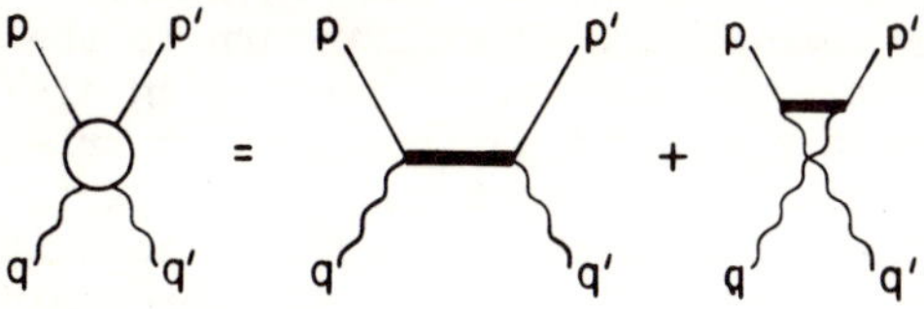

Fig. 7. Augmented Born approximation for π-N scattering.

1) There are two Adler-Weinberg conditions which give the values of certain invariant scattering amplitudes in the limit of zero pion mass (negative scattering energies).

One of these is zero, the other is $-g_\pi^2/M$ ignoring form factor effects of the order $(m_\pi/M)^2$. The usual Born approximation gives zero for both. The ABA gives zero for the first and $-0.80\, g_\pi^2/M$. The complex poles alone contribute $-0.96\, g_\pi^2/M$ and the continuum contributes $+0.16\, g_\pi^2/M$.

2) The cross sections in the I = 1/2 and I = 3/2 channels are shown in Figs. 8 and 9. Relative to the usual Born approximation,

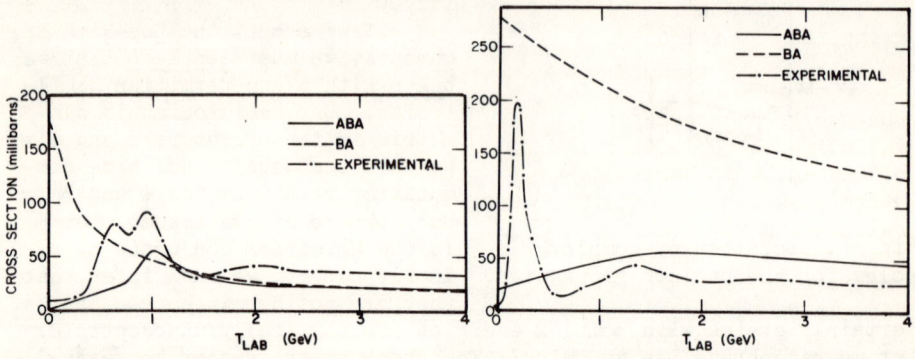

Fig. 8. π-N scattering cross sections, I = 1/2.

Fig. 9. π-N scattering cross sections, I = 3/2.

the threshold cross section is reduced by two orders of magnitude, in qualitative agreement with experiment. The standard argument is that one must invoke pair suppression, i.e. eliminate from the calculation the graphs shown in Fig. 10, since they dominate the threshold behavior. The calculation is then no longer relativistic. In our calculations, the complex poles effect the same result without any ad hoc assumptions. The overall agreement with experiment is only fair, and the 3-3 resonance is understandably missing completely. I will return to the 3-3 later. The pi-nucleon scattering diagram appears also in the two-pion contribution to the N-N potential. We suspect, since the pions there are virtual, that it is more important to satisfy the threshold and subthreshold behavior than the scattering resonances, even the 3-3 resonance.

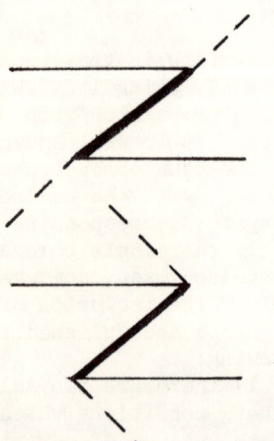

Fig. 10. Pair diagrams for π-N scattering.

5. N-N POTENTIALS

In his thesis, Nutt[5] calculated

nucleon-nucleon potentials following the reduction of Blankenbecler and Sugar.[6] The diagrams summed are shown in Fig. 11. The calculations include crossed pion lines and the first iteration of the ladder diagrams with the off-shell part of the two-nucleon propagator. The resultant potential is non-local and energy dependent.

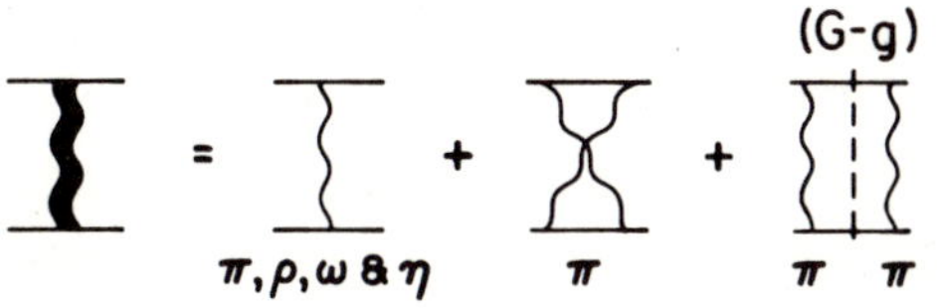

Fig. 11. Truncated Blankenbecler diagrams for N-N potential.

A similar calculation has been performed by Partovi and Lomon,[7] with the difference that they used the free nucleon propagator, $(\gamma p + M)^{-1}$, while Nutt used the BPW propagator including the complex poles. Both calculations used the same set of mesons, both excluding any σ. The fact that both gave strong attraction in the s-states raised skepticism. The Partovi-Lomon calculation was criticized on the grounds that the piece of the diagram corresponding to the π-N scattering is overestimated (no pair suppression). Nutt's calculation is not subject to this criticism and, in fact, shows even more attraction than Partovi and Lomon. See, for example, Figs. 12 and 13.

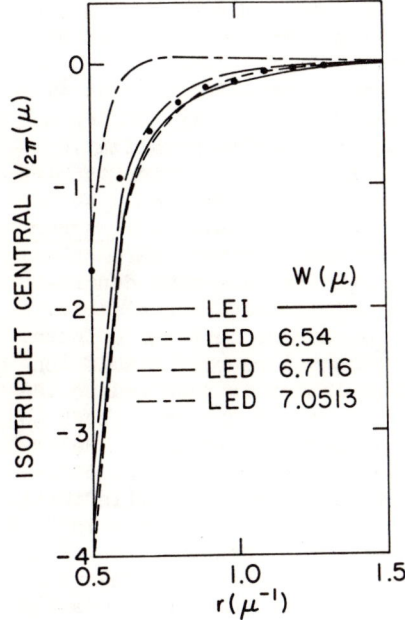

Fig. 12. Two pion contribution to N-N potential, $^1S_0$. Dots are Partovi-Lomon.[7]

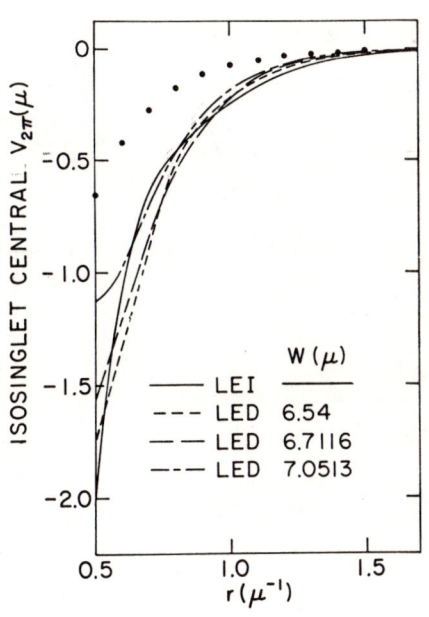

Fig. 13. Two pion contribution to N-N potential, $^3S_1$. Dots are Partovi-Lomon.[7]

The calculated N-N potential is in semi-quantitative agreement with the phenomenological potentials in the long and intermediate ranges, but an arbitrary core must be inserted to handle the short-range behavior and to reproduce scattering data. Nevertheless, the model prescribes the off-shell behavior outside the core. Nutt estimated that the effect of his off-shell behavior is to increase the binding energy of nuclear matter by the order of 1 MeV per nucleon compared with a static, local potential.

6. FINITE DENSITY HARTREE-FOCK

The ultimate goal of the program is finite density. There are subtle complications in using the formalism to calculate such important quantities as the density and energy, even in the low density limit where we know that the nucleons must behave like a free Fermi gas. The problem is that our nucleons are dressed. The Green's function is renormalized so that the pole at the nucleon mass has residue unity. The unrenormalized Green's function, however, has residue Z_2 (where you might think $0 < Z_2 < 1$, but in our theory $Z_2 = -\infty$). A simple minded calculation would fill a dilute Fermi sea with density Z_2. The resolution of the difficulty comes from what is known in nuclear physics as "blocking." The nucleon propagator is modified to first order in the density due to the occupation of the Fermi sea.

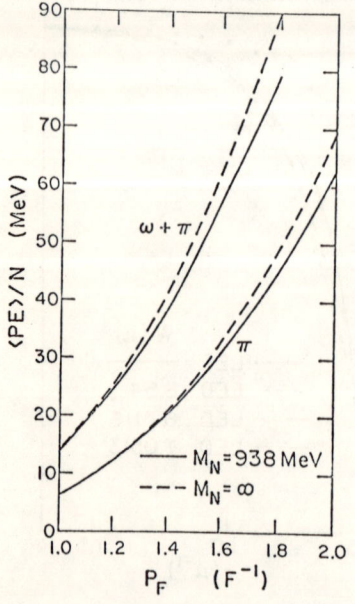

Fig. 14. Non-static Hartree-Fock energies.

With the help of a clue from the non-relativistic theory--supplied by Arthur Kerman--Puff, Chiang, Nutt, and I[8] derived in Hartree-Fock the usual expression for the density. After identifying the mass renormalization terms, a useful expression for the interaction energy was obtained. We evaluated the energy using the <u>vacuum</u> Green's function, in which case only the pole at the nucleon mass contributes. This is not self-consistent. While the results reflect the effects of retardation and recoil, a self-consistent calculation is required to include the effects of the media back onto nucleon and meson propagators. Nevertheless, we do find significant differences relative to the static (infinite nucleon mass) case, as is shown in Fig. 14. At normal density, the binding energy is increased by 1 MeV with pis and 2.9 MeV including both pis and omegas.

7. FUTURE DEVELOPMENTS

The immediate objectives of current and future calculations are the following:

1) Self-consistent calculation of the finite density Hartree-Fock problem.

2) Pi-nucleon scattering in the 3-3 channel. This is in progress in collaboration with Berndt Müller, Dan Koltun, and Ernest Henley. The Chew-Low diagrams are summed, but the BPW nucleon propagator is used in intermediate states.

3) The pi-nucleon vertex function is being studied in collaboration with Gerald Miller and William Nutt. An approximate integral equation for the vertex function is represented diagrammatically in Fig. 15. In a further approximation, we will sum the rainbow diagrams.

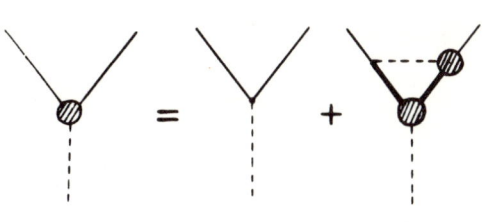

Fig. 15. Diagrammatic representation of integral equation for π-N vertex function.

4) The vacuum nucleon propagator should then be recalculated using the meson-nucleon vertex function and dressing of the meson propagators. What will happen to the complex poles? I suspect they will not recede, but we will keep an open mind.

5) The many-body problem should now be calculated in the approximation that preserves two-body correlations. That is, G_3 is factored into products of G_2 and G_1. As in the N-N problem, we will reduce the resulting Bethe-Salpeter equation to a Lipmann-Schwinger (or Bethe-Goldstone) equation by the truncated Blankenbecler-Sugar reduction and an adjustable core. Previous experience[9] with the non-relativistic, static potential problem lends encouragement that two-body correlations will be adequate.

In conclusion, I wish to reemphasize the importance of using dressed nucleon propagators in calculations involving nucleon intermediate states. In particular, the presence of complex poles very significantly affects physical processes.

REFERENCES

*Supported in part by E.R.D.A.

1. W.D. Brown, R.D. Puff, and L. Wilets, Phys. Rev. C2, 331 (1970).
2. P.C. Martin and J. Schwinger, Phys. Rev. 115, 1342 (1959); R.D. Puff, Ann. Phys. (N.Y.) 13, 317 (1961).
3. W.D. Brown, University of Maryland Technical Report No. 70-103, 1970 (unpublished).
4. W.T. Nutt and L. Wilets, Phys. Rev. D11, 110 (1975).

5. W.T. Nutt, Ph.D. Thesis, University of Washington (1974); and to be published in Ann. Phys. (N.Y.) (1976).
6. R. Blankenbecler and R. Sugar, Phys. Rev. $\underline{142}$, 1051 (1961).
7. M.H. Partovi and E.L. Lomon, Phys. Rev. $\underline{D2}$, 1999 (1970).
8. L. Wilets, R.D. Puff, W.T. Nutt, and D. Chiang, submitted to Phys. Rev. C.
9. R.D. Puff, A.S. Reiner, and L. Wilets, Phys. Rev. 149, 778 (1966); T.C. Foster, Phys. Rev. 149, 784 (1966).

BAG MODEL OF HADRON STRUCTURES

K. Johnson
Massachusetts Institute of Technology
Cambridge, Massachusetts   02139

(Manuscript not Received)

T. Mizutani (SIN): Since the MIT bag shows the Regge behavior, I wonder if it has some feature connected with the statistical bootstrap theory with maximal temperature (of Hagedorn type or Rizantschi type). (Incidentally, there seems to be some discussion related to this by Cabbibo).
K. Johnson: It is closely related to the dual resonance model. This is discussed in detail in a paper by C. Thorn and myself in Phys. Rev. D 13, (1934) 1976.

E. Ferreira (Pontificia Universidade Catolica, Rio De Janeiro): In you mass spectrum you obtain a too high value for the mass of the pion. Can you explain that?
K. Johnson: We expect that our approximations are the worst for the states where the Compton wave length of the particle is large in comparison to $\frac{R}{\pi}$ where R is the radius. This is the case for the pion which has a very low mass.

NEW DIRECTIONS FOR NUCLEAR PHYSICS - A PERSONAL VIEW

Maurice Goldhaber
Brookhaven National Laboratory
Upton, New York 11973

I have been asked to talk about the future, but this is easier if we first remind ourselves of the past. Let me do so in a largely qualitative way.

It is now widely believed that the cosmos in which we live originated some 15 billion years ago with a so-called "big bang". In this big bang relativistic energies were presumably available because the large red shifts found for some objects in the sky indicate that they recede with relativistic energies. All the particles we know and those we still hope to discover were probably created in the very beginning; nowadays we hardly find enough letters in the Latin and Greek alphabets, including some superscripts and subscripts, to describe them all. We, who are the metastable descendants of this super-alphabet-soup, clearly have a natural interest in the particles which form our prehistory! A selection of these particles is shown in Fig. 1, which some of you may have seen before. It is interesting that practically all these particles now play a role in the context of this conference, starting with the widest state represented here, the Δ, down to the sharpest decaying state we know, the neutron. Of course, the longer the particles live the more we can "manipulate" them.

You have probably all heard of the evidence for new particle families, J/Ψ etc.; "charmed" particles may also exist with lifetimes expected to be intermediate between those of the π^0 and the K_S^0, as I have indicated in Fig. 1. If I were to continue the lifetime axis indefinitely, I would finally come to lower limits for the lifetime of the proton. (See, e.g., M. Goldhaber, Proc. Am. Phil. Soc. 119, 24 (1975).) The proton lives more than 10^{23} years no matter how it decays, and for specific decay modes much larger partial lifetime limits have been obtained. For the decay mode $p \rightarrow \mu + X$, Reines and Crouch (Phys. Rev. Lett. 32, 493 (1974)) have found a partial lifetime $\gtrsim 10^{30}$ years. Now of course as soon as experimentalists find such limits, theoreticians come along to tease us and predict that the proton ought to decay with just about this lifetime or maybe ten times more slowly. Pati and Salam (Phys. Rev. Lett. 31, 661 (1973)) have published such an idea. With such lifetimes, observation of even one event per year requires a very large detector, and this is one of our problems for the future: how can we build huge detectors that will not be swamped by backgrounds?

Even if the proton were absolutely stable, the nuclei we are built of would not necessarily live forever. You know of course that radioactive nuclei will disappear after a sufficiently long time, but there are many nuclei you normally don't think of as disappearing which will also disappear, e.g., all those for which we haven't yet found double beta decay no matter how hard we have looked because our detectors are not sufficiently sensitive. All

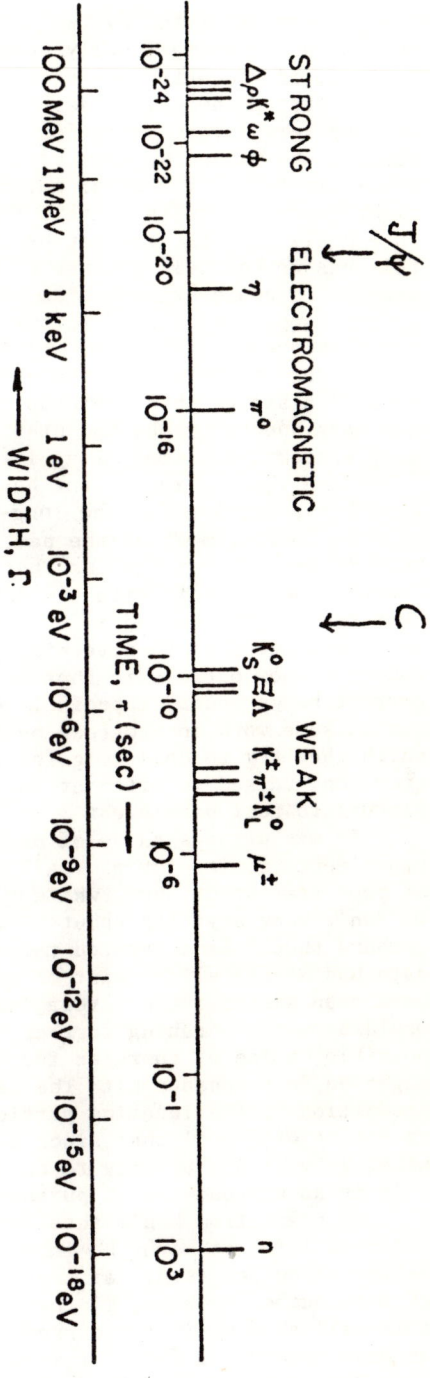

FIGURE 1

Lifetime (τ) for some of the "fundamental" particles and corresponding width (Γ). The interactions governing the decay of particles are strong, electromagnetic or weak, as indicated.

the isobars separated by two units of charge between which double beta decay is energetically possible, one way or the other, would be reduced, if you wait long enough, to a single one of that pair. We need better methods for measuring these very rare decays, searches involving very large quantities of material; maybe we should develop further the method we once used by building, e.g., huge calcium fluoride crystals. For other materials, too, attempts of this type might lead to results in this field. Where double beta decay results in a noble gas, the gas can be extracted from a very old mineral, and this has worked well because of negligible gas impurities. With more courage we could extract also non-gases and show that they have grown in the sample if we find either a) that the amounts of original impurities are negligible or b) that the chemical state of the decay end-product differs from that of the impurity; this depends on finding ways of distinguishing between decay product and impurity. Since modern mass spectrographs and other methods could, in principle, detect single atoms, extremely long lifetimes could, conceivably, be found. I believe that so far no one has had the courage to do this; it is probably a problem for the long-range future.

As I said, most of the particles represented in Fig. 1 play a role in the kind of physics that was discussed at this conference. Starting on the left with the delta, we heard in the talk by Weber that the original euphoria of three years ago concerning evidence for spectator Δ's did give rise to critical emphasis on backgrounds, and he is now optimistic that we know how to correct for some important backgrounds, especially in inclusive production. He also quoted some work in progress by A. S. Goldhaber and L. Kisslinger in which they try to come to grips with the important theoretical question, raised so often at this conference, of the transition from virtual to real particles.

It was perhaps naive of us to think that we could add up in one experiment results for $\pi^- + \Delta^{++}$ and $\pi^+ + \Delta^{++}$. We added these because of poor statistics, but, while $\pi^- + \Delta^{++}$ might form a reasonable vertex, we don't know anything about a triple-charge vertex and therefore perhaps should have treated these reactions separately. In fact the data had a very slight hint of a real effect for $\pi^- + \Delta^{++}$, which may have been washed out by averaging over other reactions, and this would be worth pursuing further. One might also try to find a possible choice of energies for the incoming particles at which they might be in resonance with the deltas. This might emphasize delta production if the incoming particles were not simultaneously in resonance with nucleons; since most of these resonances are rather wide, it will not be easy to choose useful resonances, but perhaps this is an approach worth pursuing.

We heard from Moniz that the mean free path of deltas is of the order of 1 fm; it might be interesting to study the production of deltas in nuclei (e.g., at Los Alamos proton energies) as a function of mass number because, if the delta cannot come out from very far, this will show up in the dependence of the production cross section on mass number; as far as I know such a study has not yet been made.

Some subjects we have heard about and will hear more about in the future are the many new kinds of spectroscopy that are emerging.

We heard about hyperfragments from Feshbach; there has also been a suggestion by A. A. Tyapkin (Dubna preprint) that, if charmed particles exist, then there might be nuclei corresponding to the hypernuclei but containing a charmed baryon instead of a hyperon, and these would have a much larger energy release. This would of course illustrate most explicitly that we have here a new quantum number which is conserved in strong interactions. Feshbach emphasized that we should look for hypernuclei in which a Σ^- or Σ^+ was formed instead of a Λ, but one problem is that in a nucleus with Z protons and N neutrons a Σ^- will very quickly be absorbed on a proton (making Λn), or correspondingly a Σ^+ on a neutron (making Λp). Therefore these states will be very wide, although still of considerable interest. The question arises whether there is a chance for the existence of sharp states containing a Σ^-. One possible state discussed in earlier conferences is $nn\Sigma^-$. There is still no good evidence for or against its existence. We might perhaps look at a reaction—this is a dream reaction—like $^3H + K^- \to nn\Sigma^- + \pi^+$, which would have a well-defined energy release if $nn\Sigma^-$ is a sharp state. Or, if a stable target is preferred, then $^4He + K^- \to nn\Sigma^- + p + \pi^+$ could be looked for, a more complicated reaction.

C. Dover, T. Londergan, and A. Gal are interested in trying to estimate the binding energy of the $nn\Sigma^-$ and other states, since something is known about these interactions if SU(3) arguments are used. But whether or not this state is bound, the chance that the more complicated $nn\Sigma^-\Sigma^-$ is bound is much larger. Here we have an α-particle-like structure which would clearly be bound more strongly. Still another super-α structure is $nn\Lambda\Lambda\Sigma^-\Sigma^-$, and you can go further. You can add two Ξ^-. By having chosen these charges correctly we have stabilized this super hypernucleus as far as strong interactions are concerned. Correspondingly you could have $pp\Sigma^+\Sigma^+\Lambda\Lambda\Xi^0\Xi^0$, and in this case you might go on adding charmed baryons, C^+C^+, and so on. So here are super alpha particles where you put everything into one pot, all with an S-wave function, and it is possible that these configurations would be strongly bound. Whether we can learn to calculate such things will be interesting to look into because for a while it will be very hard to produce anything of this kind. Maybe a fraction of the chain, e.g., $nn\Sigma^-\Sigma^-$, could be created by Ξ^- on neon in a bubble chamber, since Ξ^- beams have been produced. It could also be that in extremely high energy collisions, where we have already found that we can make antideuterons and anti-^3H, we might make such complex structures too. The mechanism for production of $\bar{D}$'s made in very high energy collisions is still incompletely understood. How are they formed dynamically? It could be that they are formed partly from $\bar{\Delta}$'s or other $\bar{N}^*$'s; if they are formed from such excited nucleons we can allow much higher relative momenta to form the deuterons. Maybe that would make it less astounding than the low momenta needed to form antideuterons from normal antiparticles. But that is still very much an open question and worth investigating further.

Another new spectroscopy still in its infancy is $\bar{B}B$ spectroscopy, in which an antibaryon and a baryon form states that are short-lived but still sufficiently well-defined for resonances to be seen. This idea was pioneered by I. S. Shapiro and his collaborators, and a number of experiments have confirmed the existence of sharp states near the nucleon-antinucleon threshold, the first being those by Kycia and collaborators and by Kalogeropoulos and collaborators, which give fairly good proof that near the $\overline{N}N$ threshold there are states of this sort which are fairly sharp, only a few MeV wide. Recently Kalogeropoulos and collaborators have developed a technique at Brookhaven which involves recording the time of flight of antineutrons, and they hope to measure the position of such states very accurately. Since Shapiro's calculations still made a number of simplifying assumptions, more detailed calculations have recently been carried out at Brookhaven (C. Dover and M. Goldhaber, submitted to Phys. Rev. D). These calculations must be considered as only illustrative of what we ultimately hope to achieve as experiments and theory improve.

The main idea of these calculations, which treat not only $\overline{N}N$ but also $\overline{\Lambda}\Lambda$, $\overline{\Sigma}\Sigma$, and $\overline{\Xi}\Xi$ (the whole SU(3) baryon octet), is to take the one-boson exchange potentials that have been fitted to nucleon-nucleon scattering and change the sign of those terms ascribed to mesons of odd G-parity (like the ω). In this way we get strong attractive forces where for nucleons there is a repulsive core. In the absence of a proper relativistic field theory, the potential approach still remains a useful tool of some predictive value. With the strong attractive forces you get binding, and of course for very strong binding you do not trust the calculation at all, but near thresholds they are perhaps a good guide to what might exist. You shouldn't take the predictions too literally. The results shown in Fig. 2 should be considered only illustrative; probably the energies are good only to 100 or 200 MeV. Close to the threshold for $\overline{N} + N$, states are predicted that have perhaps already been observed. In going away from the threshold, the knowledge is less definite, and the calculation becomes less reliable the further one goes. The width also increases, and this gives rise to less well-defined states. These are, incidentally, estimated widths; in the spirit of these illustrative calculations further assumptions must be made about the annihilation region, and so on. Wide resonances are known in this general region, but whether they have anything to do with the calculated states remains to be shown; they would be expected to have a large partial width into $\overline{B} + B$ breakup. This gives an impetus to more experiments. If some time in the far future we shall have good data on both position and width of these resonances, we would have a very interesting check of the whole idea.

Just to keep you from being too optimistic, Fig. 3 shows a comparison of some nucleon-antinucleon states calculated from the potential of Nagels, Rijken, and deSwart with some states calculated from the potential of Bryan and Phillips. You see that with two different potentials, both taken seriously in the literature, the states predicted move around considerably; thus, a few hundred MeV is about the best you can have as a guide. If one day we were to have experimental confirmation of these states, that would be of great importance to the ultimate understanding of the origins of the nucleon-nucleon force.

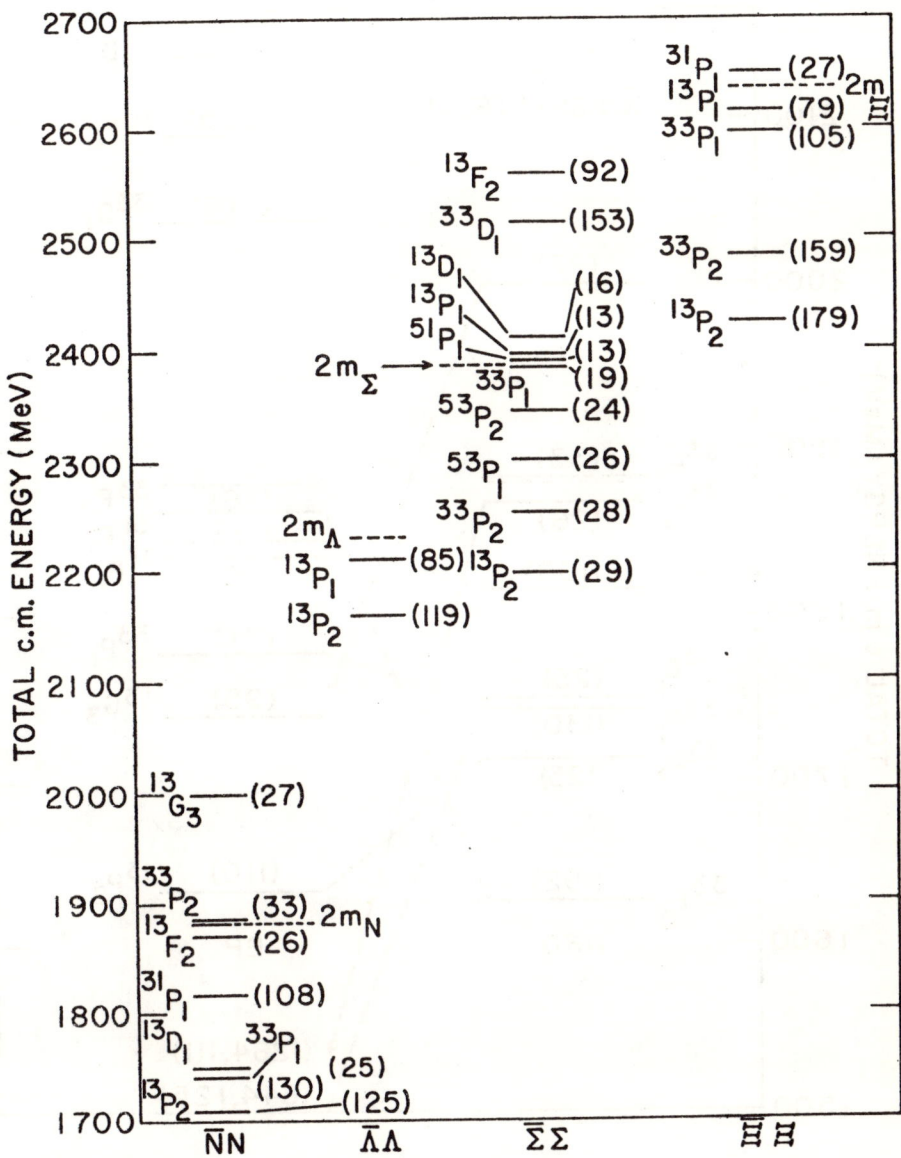

Fig. 2. Calculated energy spectra of $\bar{B}$-B states, using a potential obtained by the G-parity transformation from the BB potential of Nagels, Rijken and deSwart (Nijmegen preprint). The mesons corresponding to the states $^{2I+1,2S+1}L_J$ have parity $\pi = (-1)^{L+1}$, charge parity $C = (-1)^{L+S}$ and G-parity $G = C(-1)^I$. Bound L=0 states and the deeply bound $^{13}P_0$ states are omitted. The numbers in parentheses are very crude estimates of the total width of each state (in MeV).

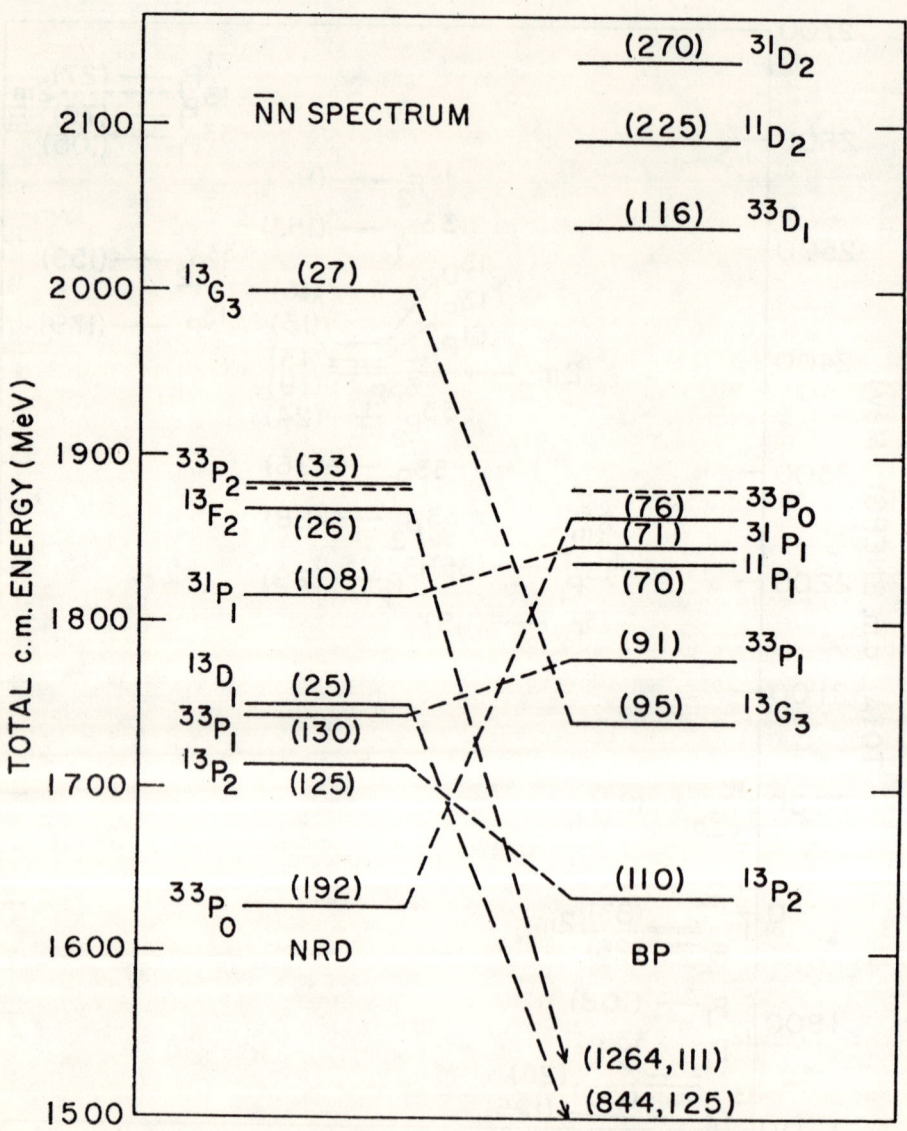

Fig. 3. Comparison of two different NN potential models. The right-hand spectrum [BP] follows from the Bryan-Phillips potential (see C. B. Dover, <u>Proceedings of the Fourth International Symposium on $\overline{N}N$ Interactions</u>, Syracuse, May 2-4, 1975, Vol. II Chap. VIII, pp. 37-91).

The dynamics of the formation of $\bar{B}B$ states in high energy collisions may have features in common with the dynamics of the formation of $\bar{D}$. The two "spectroscopies" can probably learn from each other. $\bar{B}B$ states would have played an important role in the big bang if it had been originally symmetrical in B and $\bar{B}$. G. Steigman has recently reviewed the evidence against such an initial symmetry between matter and antimatter in our cosmos (to appear in Am. Rev. of Astron. and Astrophys., Vol. 14).

One thing we have become aware of at this conference is a great increase in precision in reactions involving pions, either pion absorption giving very exact gamma energies or production of pions very close to threshold. We have heard from Tzara, and also G. T. Emery (Phys. Lett. 60, 351 (1976)) has published ideas, on how you might dream of doing precision work in the future. At the Indiana cyclotron, for example, with a resolution that Emery puts at 30 keV, one could study the production of negative pions in bound states, in S states, and sometimes in states not yet reached because the π^- mesons get absorbed from higher states. In this way we could also reach states around a nucleus that is normally not stable. Similarly, there are plans at the Bates Electron Linac to study e,e' reactions leading to bound states of π^-.

In past conferences you have often heard the cliché that not only can elementary particles be used to study nuclei but also that the nucleus is a laboratory for elementary particles. There are many examples of this. A thought that might be worth pursuing is the following: Let us expose a highly deformed odd-A nucleus aligned in some crystal (at low temperature) with the long axis alternatively parallel or at right angles to the direction of a beam of protons or pions, etc. This might be a very interesting way to study A-dependent effects because in considering such effects we are still often puzzled as to whether we are dealing with a complicated process, where simultaneous interactions with closely neighboring nucleons take place, or with multiple successive processes; in this way we might see a difference between such effects, but as far as I know such a search has not yet been started. Nuclear chemists might even be interested in studying fragments coming out of differently aligned nuclei. This might serve as a further check on their Monte Carlo calculations.

Let me finish by saying a few words about other far-out matters. Nuclear physicists have long accepted that neutron stars are in some way big nuclei and that these big nuclei may contain besides π-meson condensates Λ's and Σ's, because all the levels for neutrons are so highly filled that it pays to make new Fermi-particles, even considerably heavier ones. Now if charm is confirmed, we may also have to take into account a "charmed sea" of particles in neutron stars. This will deserve a closer look if and when charmed particles become a reality.

Another question concerns the existence of very heavy nuclei. By now many people have considered this question from a theoretical and experimental point of view. There is a possibility of a track of a very heavy nucleus: the "magnetic monopole" track reported by Price and his collaborators has an alternative interpretation.

In a recent talk at the New York meeting of the American Physical Society Price reported that a nucleus with $Z \approx 50\text{-}80$ and $A > 2000$ could be compatible with their data. If this interpretation remains viable, it will be up to us nuclear physicists to see whether we can in any way explain how a nucleus that heavy lives long enough to get to us from outer space, either directly or as a fragment produced in the atmosphere from a nucleus of still larger Z and A. I have no very good idea to offer on this, and I leave it with you as a little exercise for the future.

A. W. Thomas (CERN): (i) How many $N\bar{N}$ states are still in existence? I thought that only the state at 1932 MeV remained. (ii) Even a very small amount of absorption can completely destroy the resonances predicted by the OBEPs. Can the calculations of $N\bar{N} \to \pi\pi$ which have been used far from the physical region to give the 2π contribution to the NN force, be used to shed some light on the actual size of the annihilation cross section in different channels at the $N\bar{N}$ threshold?

M. Goldhaber: (i) The state at 1932 MeV may be a doublet (See C. B. Dover and S. H. Kahara, Phys. Lett., to be published) others are probably real too (Kalogeropoulos et al., to be published). (ii) the $\bar{N}N \to \pi\pi$ channel would only represent a small part of the annihilation cross section. An estimate of annihilation widths based on channels of the type $\bar{N}N \to M_i M_j$, where $M_i, M_j = \{\pi, \rho, \omega, \varepsilon, \text{etc.}\}$ has been made by Dover and Kahana.

AUTHOR INDEX

Afnan, I. R.   472
Alberg, M.   486, 668
Alberi, G.   666
Albu, M.   326
Alder, J. C.   624, 626, 628
Alexander, J. H.   468
Allen, R. C.   346, 742
Allred, J. C.   274, 446, 448
Amann, J. F.   740
Andrade, S. C. B.   532
Angelescu, T.   326
Arai, K.   548
Arthur, E. D.   342
Arvieux J.   25
Ashery, D.   307, 336
Aslanides, E.   204
Auerbach N.   322

Baba, K.   614
Bachelier, D.   262
Baldracchini, F.   666
Balea, O.   326
Balestra, F.   326
Ballagh, H. A.   65
Banerjee, M. K.   74, 119
Barnes, P. D.   281
Barrett, R. J.   342
Bassalleck, B.   272
Beer, G.   264
Berghofer, D.   632
Bernstein, A. M.   606, 608
Bertin, P. Y.   34, 332
Beurtey, R.   470
Bimbot, L.   260, 470
Binon, F. G.   348
Bolger, J. E.   740
Bonazzola, G. C.   536
Booth, E. C.   606, 608
Bosted, P.   608
Bouyssy, A.   538
Boyard, J. L.   262
Brack, M.   466
Brayshaw, D. D.   443
Bressani, T.   536

Brissaud, I.   260
Brown, G. E.   187, 655
Bruge, G.   470
Bryman, D.   264
Bunatian, G. G.   474
Burleson, G.   740

Cammarata, J. B.   74, 618
Chasan, B.   606, 608
Chen, T. W.   88
Cheon, Il-T.   188
Chiavassa, E.   536
Clement, J. M.   274, 446, 448
Cochavi, S.   307
Cooper, M. D.   237, 276
Corfu, P.   25
Coupat, B.   34, 332
Couvert, P.   310, 470

Dellacasa, G.   536
Deloff, A.   544, 546
Delorme, J.   190
Dennis, C. M.   330
DePommier, P.   632
Deutsch, J.   630
Dillig, M.   266
Domingo, J.   25
Donnelly, T. W.   618
Doss, K. G. R.   344
Dover, C. B.   82, 249
Dragoset, W. H.   274, 446, 448
Duclos, J.   34
Dumbrajs, O.   192
Dytman, S. A.   344

Eisenberg, J. M.   542
Eisenstein, R. A.   55, 84, 534
Endo, I.   548, 614
Engelhardt, D.   272
Ericson, M.   190
Ernst, D. J.   76

Falomkin, I. V.   326
Favart, D.   630

Fearing, H. W.   468
Felder, R. D.   274, 446, 448
Ferreira, E. M.   384, 452, 532
Feshbach, H.   521
Fetisov, V. N.   636
Figureau, A.   190, 616
Frascaria, R.   470
Friedman, W. A.   278
Friman, B. L.   182
Fujii, H.   614
Fujisaki, M.   614
Fujita, T.   540
Funsten, H. O.   330
Furic, M.   272
Furui, S.   320, 620

Gabathuler, K.   25
Gabioud, B.   624, 626, 628
Gal, A.   693
Gallio, M.   536
Garcilazo, H.   454
Garfagnini, R.   326
Garreta, D.   470
Gérard, A.   34, 332
Gibbs, W. R.   324, 328, 464, 622
Gibson, B. F.   328, 418, 464, 622
Gilad, S.   307
Ginocchio, J. N.   338
Glodis, P. F.   65
Goldhaber, M.   756
Goode, P.   463
Gordeev, V. A.   36
Greene, S.   740
Gretillat, P.   25
Guaraldo, C.   326

Hachenberg, F.   270
Haddock, R. P.   65
Hahn, Y.   456
Hasan, A.   742
Hasinoff, M. D.   316, 632
Heffner, R.   733
Heller, L.   93
Henley, E. M.   668
Hennino, T.   262
Herscovitz, V. E.   334
Hess, A. T.   324, 328
Hirabayashi, H.   713
Hirata, M.   180
Hirt, W.   25
Hivernat, A.   34

Hogstrom, K.   446, 448
Ho-Kim, Q.   68
Holinde, K.   663
Holmgren, H. D.   260
Hoock, D. W.   88
Hoop, F.   624
Hoshi, N.   540
Huber, M. G.   266
Hudomalj-Gabitzsch, J.   274, 446, 448
Hüfner, J.   270
Hungerford, E. V.   274, 446, 448

Ingram, Q.   25
Isabelle, D. B.   34, 332
Iversen, S.   740

Johnson, K.   755
Johnson, M.   76, 276
Joseph, C.   624, 626, 628
Jourdain, J. C.   262

Kadota, S.   614
Källne, J.   260, 740
Kanofsky, A. S.   346, 742
Karol, P. J.,   305
Kaufmann, W. B.   550
Kelly, R. L.   12
Kikugawa, M.   548
Kilian, K.   497
Kisslinger, L. S.   159, 184, 198
Klotz, W.   272
Koch, J. H.   591, 610
Koltun, D. S.   3
Kolybasov, V. M.   394
Koptev, V. P.   36
Kossler, W. J.   330
Krementzova, Y. N.   634
Kruglov, S. P.   36
Kubodera, K.   200
Kulbardis, A. A.   36
Kuo, T. T. S.   187
Kurokawa, S.   713
Kusumegi, A.   713
Kuzmin, L. A.   36

Landau, R. H.   86, 458
Lankford, W. F.   330
Lavine, J. P.   68
Law, J.   544, 546
Lazo, G.   346

Le Bornec, Y.   260, 470
Lee, J. K. P.   632
Lee, L. Y.   446
Lee, T.-S. H.   462
Legrand, D.   470
Lemmer, R. H.   82
Lenz, F.   180, 403
Leung, K. C.   65
Levedev, A. I.   634, 636
Lewis, C. W.   272
Li, K. K.   476
Lieb, B. J.   330
Lind, V. G.   330
Lipnik, P.   630
Liu, L. C.   186
Lo, J.   448
Locker, M. P.   340
Loude, J. F.   624, 626, 628
Lyashenko, V. I.   326

MacDonald, R.   632
Mach, R.   638
Machleidt, R.   663
Macq, P.   630
Maki, A.   713
Malov, Y. A.   36
Maris, T. A. J.   334
Mason, G. R.   264
Mathie, E.   264
Mayes, B. W.   446, 448
Measday, D. F.   632
Medicus, H.   624, 626, 628
McManus, H.   463
McVoy, K. W.   278
Mihul, A.   326
Mikamo, S.   713
Miller, G. A.   178, 668, 682
Miller, J.   34, 332
Miller, L. D.   268
Mizutani, T.   172
Moinester, M.   307
Moniz, E. J.   105
Moore, C. F.   740
Morel, N.   624, 626, 628
Morgenstern, J.   34, 332
Morlet, M.   470
Morris, C. L.   740
Mors, P. M.   334
Moss, G. A.   470
Mukhopadhyay, N. C.   172, 616
Murakami, A.   614

Murata, Y.   614
Musso, A.   536
Mutchler, G. S.   274, 446, 448
Myhrer, F.   340

Nagl, A.   612
Nakahara, K.   604
Nefkens, B. M. K.   65
Nguyen, V. G.   507
Nichitiu, F.   326, 460
Noble, J. V.   221
Noguchi, S.   614
Nutt, W. T.   78, 90, 186

Obst, A.   740
Ohashi, H.   604
Olin, A.   264
Olsson, M. G.   21
Oset, E.   318

Paras, N.   606
Pedroni, E.   25
Perrenoud, A.   624, 626, 628
Perroud, J. P.   624, 626, 628
Peterson, J.   740
Phillips, G. C.   274, 446, 448
Picard, J.   34, 332
Picker, H. S.   196
Piffarett, J. P.   25
Pilkuhn, H.   194, 550
Pinsky, L.   274, 448
Piragino, G.   326
Pirner, H. J.   270
Plendl, H. S.   330
Pol, Yu. S.   474
Pontecorvo, G. B.   326
Poutissou, J. M.   632
Poutissou, R.   632
Powers, R.   34, 552
Prieels, R.   630

Radomski, M.   175
Radvanyi, P.   262
Rao, K. S.   601
Reide, F.   260
Reitan, A.   316
Renker, D.   624, 626, 628
Rho, M.   146
Rinat, A. S.   450
Rinaudo, G.   536
Riska, D. O.   175, 466

Robertson, L. P.   264
Rockmore, R.   463
Rosa, L. P.   452
Roy-Stéphan, M.   262

Saghai, B.   332
Saharia, A.   184
Salomon, M.   316, 632
Sato, T.   713
Scherbakov, G. V.   36
Schlaile, H. G.   194
Schmitt, H.   624, 626, 628
Schneider, C.   334
Schwaller, P.   25
Scrimaglio, R.   326
Sedlak, J. E.   278
Seki, R.   80
Seraru, A.   326
Shakin, C. M.   186
Shamai, Y.   307
Shcherbakov, Y. A.   326, 365
Shoda, K.   604
Silbar, R. R.   297, 344
Slater, D. C.   740
Smith, D. E. A.   65
Sober, D. I.   65
Spuller, J.   632
Stephenson, Jr., G. J.   324, 328, 464, 622
Stith, J. H.   330
Storm, D. W.   706
Strakovsky, I. I.   36
Strassner, G.   624, 626, 628
Stronach, C. E.   330
Sumi, Y.   614
Suzuki, T.   632
Suzuki, Y.   713

Tabakin, F.   38, 84, 534
Taino, M.   713
Takamatsu, K.   713
Takasaki, M.   713
Takeutchi, F.   272
Tanner, N.   25
Tatischeff, B.   260, 470
Terrien, Y.   470
Thévenet, C.   190
Thies, M.   71
Thiessen, H. A.   740
Thomas, A. W.   86, 375, 450, 472
Thomé, Z. D.   452

Tran, M. T.   624, 626, 628
Trjasuchev, V. A.   636
Truöl, P.   581, 624, 626, 628
Tsuchiya, K.   713
Turchinetz, W.   606
Tzara, C.   566

Überall, H.   612
Ullrich, H.   272

Vaucher, B.   624
Verbeck, S.   740
Vernin, P.   34, 332
Vincent, J. S.   264
Vinh Mau, R.   642
Von Fellenberg, H.   624, 626, 628

Walker, G. E.   672
Walker, J. F.   668
Warneke, M.   274, 446, 448
Warszawski, J.   322
Weber, H. J.   130, 268
Weise, W.   466
Weng, W. T.   187
Wilets, L.   746
Williams, T. M.   274, 446, 448
Willis, N.   260, 470
Winkelmann, E.   624, 626, 628
Woloshyn, R. M.   610
Wu, C-S.   198

Yamamoto, A.   713
Yavin, A. I.   307
Yazaki, K.   180

Zaider, M.   307, 336
Zovko, N.   194
Zupancic, C.   624